PROGRESS IN BRAIN RESEARCH

VOLUME 120

NUCLEOTIDES AND THEIR RECEPTORS
IN THE NERVOUS SYSTEM

Other volumes in PROGRESS IN BRAIN RESEARCH

Volume 100: Neuroscience: From the Molecular to the Cognitive, by F.E. Bloom (Ed.) – 1994, ISBN 0-444-81678-X.

Volume 101: Biological Function of Gangliosides, by L. Svennerholm et al. (Eds.) – 1994, ISBN 0-444-81658-5.

Volume 102: The Self-Organizing Brain: From Growth Cones to Functional Networks, by J. van Pelt, M.A. Corner, H.B.M. Uylings and F.H. Lopes da Silva (Eds.) – 1994, ISBN 0-444-81819-7.

Volume 103: Neural Regeneration, by F.J. Seil (Ed.) – 1994, ISBN 0-444-81727-1.

Volume 104: Neuropeptides in the Spinal Cord, by F. Nyberg, H.S. Sharma and Z. Wiesenfeld-Hallin (Eds.) – 1995, ISBN 0-444-81719-0.

Volume 105: Gene Expression in the Central Nervous System, by A.C.H. Yu et al. (Eds.) – 1995, ISBN 0-444-81852-9.

Volume 106: Current Neurochemical and Pharmacological Aspects of Biogenic Amines, by P.M. Yu, K.F. Tipton and A.A. Boulton (Eds.) – 1995, ISBN 0-444-81938-X.

Volume 107: The Emotional Motor System, by G. Holstege, R. Bandler and C.B. Saper (Eds.) – 1996, ISBN 0-444-81962-2.

Volume 108: Neural Development and Plasticity, by R.R. Mize and R.S. Erzurumlu (Eds.) – 1996, ISBN 0-444-82433-2.

Volume 109: Cholinergic Mechanisms: From Molecular Biology to Clinical Significance, by J. Klein and K. Löffelholz (Eds.) – 1996, ISBN 0-444-82166-X.

Volume 110: Towards the Neurobiology of Chronic Pain, by G. Carli and M. Zimmermann (Eds.) – 1996, ISBN 0-444-82149-X.

Volume 111: Hypothalamic Integration of Circadian Rhythms, by R.M. Buijs, A. Kalsbeek, H.J. Romijn, C.M.A. Pennartz and M. Mirmiran (Eds.) – 1996, ISBN 0-444-82443-X.

Volume 112: Extrageniculostriate Mechanisms Underlying Visually-Guided Orientation Behavior, by M. Norita, T. Bando and B.E. Stein (Eds.) – 1996, ISBN 0-444-82347-6.

Volume 113: The Polymodal Receptor: A Gateway to Pathological Pain, by T. Kumazawa, L. Kruger and K. Mizumura (Eds.) – 1996, ISBN 0-444-82473-1.

Volume 114: The Cerebellum: From Structure to Control, by C.I. de Zeeuw, P. Strata and J. Voogd (Eds.) – 1997, ISBN 0-444-82313-1.

Volume 115: Brain Function in Hot Environment, by H.S. Sharma and J. Westman (Eds.) – 1998, ISBN 0-444-82377-8.

Volume 116: The Glutamate Synapse as a Therapeutical Target: Molecular Organization and Pathology of the Glutamate Synapse, by O.P. Ottersen, I.A. Langmoen and L. Gjerstad (Eds.) – 1998, ISBN 0-444-82754-4.

Volume 117: Neuronal Degeneration and Regeneration: From Basic Mechanisms to Prospects for Therapy, by F.W. van Leeuwen, A. Salehi, R.J. Giger, A.J.G.D. Holtmaat and J. Verhaagen (Eds.) – 1998, ISBN 0-444-82817-6.

Volume 118: Nitric Oxide in Brain Development, Plasticity, and Disease, by R.R. Mize, T.M. Dawson, V.L. Dawson and M.J. Friedlander (Eds.) – 1998, ISBN 0-444-82885-0.

Volume 119: Advances in Brain Vasopressin, by I.J.A. Urban, J.P.H. Burbach and D. De Wied (Eds.) – 1999, ISBN 0-444-50080-4.

PROGRESS IN BRAIN RESEARCH

VOLUME 120

NUCLEOTIDES AND THEIR RECEPTORS IN THE NERVOUS SYSTEM

EDITED BY

P. ILLES

Department of Pharmacology, University of Leipzig, D 04017 Leipzig, Germany

H. ZIMMERMANN

Biozentrum der J.W. Goethe-Universität, Zoologisches Institut, AK Neurochemie, D-60439 Frankfurt am Main, Germany

ELSEVIER
AMSTERDAM – LAUSANNE – NEW YORK – OXFORD – SHANNON – SINGAPORE – TOKYO
1999

ELSEVIER SCIENCE B.V.
Sara Burgerhartstraat 25
P.O. Box 211, 1000 AE Amsterdam, The Netherlands

First edition 1999

Library of Congress Cataloging in Publication Data
Nucleotides and their receptors in the nervous system / edited by P. Illes, H. Zimmermann. -- 1st ed.
     p.   cm. -- (Progress in brain research; v. 120)
   Includes bibliographical references and index.
   ISBN 0-444-50082-0
   1. Nucleotides--Physiological effect. 2. Nucleotides--Receptors.
  3. Neurochemistry. 4. Cellular signal transduction. I. Illes, P. (Peter) II. Zimmermann, Herbert, 1944–   .
  III. Series.
  QP376.P7 vol. 120
  [QP625.N89]
  612.8'2 s--dc21
  [573.8'485]

                                     99-23832
                                       CIP

0-444-50082-0(volume)
0-444-80104-9(series)

Printed in The Netherlands

List of Contributors

M.P. Abbracchio, Institute of Pharmacological Sciences, University of Milan, Via Balzaretti 9, 20133 Milan, Italy

G. Arslan, Department of Physiology and Pharmacology, Karolinska Institutet, 171 77 Stockholm, Sweden

E.A. Barnard, Molecular Neurobiology Unit, Royal Free Hospital School of Medicine, Rowland Hill Street, Hampstead, London NW3 2PF, UK and Department of Pharmacology, University of Cambridge, Tennis Court Road, Cambridge CB2 1QT, UK

A.R. Beaudoin, Département de Biologie, Faculté des Sciences, Université de Sherbrooke, Sherbrooke (Quebec) J1K 2R1, Canada

J.L. Boyer, University of North Carolina, Chapel Hill, NC, USA

R. Brambilla, Institute of Pharmacological Sciences, University of Milan, Via Balzaretti 9, 20133 Milan, Italy

A. Brand, Institut für Pharmazie der Universität Leipzig, Lehrstuhl Pharmakologie für Naturwissenschaftler, Brüderstr. 34, D-04103 Leipzig, Germany

N. Braun, AK Neurochemie, Biozentrum der J.W. Goethe-Universität, Marie-Curie-Str. 9, D-60439 Frankfurt am Main, Germany

M.J. Brodie, Department of Medicine, University of Glasgow, Western Infirmary, Glasgow G12 8QQ, UK

S.G. Brown, Autonomic Neuroscience Institute, Royal Free Hospital of Medicine, London NW3 2PF, UK

G. Burnstock, Autonomic Neuroscience Institute, Royal Free and University College Medical School, University College London, Royal Free Campus, Rowland Hill Street, London NW3 2PF, UK

C. Büttner, Pharmakologisches Institut für Naturwissenschaftler, Johann Wolfgang Goethe-Universität, Biozentrum N 260, Marie-Curie-Str. 9, D-60439 Frankfurt/Main, Germany

E. Camaioni, Molecular Recognition Section, LBC, NIDDK, NIH, Bethesda, MD 20892 USA

J.F. Cascalheira, Laboratory of Biochemistry, Department of Chemistry, University Beira Interior, 6200 Covilhã, Portugal

E. Castro, Departamento de Bioquimica, Facultad de Veterinaria, Universidad Complutense de Madrid, 28040 Madrid, Spain

F. Cattabeni, Institute of Pharmacological Sciences, University of Milan, Via Balzaretti 9, 20133 Milan, Italy

S. Ceruti, Institute of Pharmacological Sciences, University of Milan, Via Balzaretti 9, 20133 Milan, Italy

P. Chiozzi, Department of Experimental and Diagnostic Medicine, Section of General Pathology, University of Ferrara, Via Borsari 46, I-44100 Ferrara, Italy

A. Cornell-Bell, Cognetix/Viatech Imaging Laboratories, Ivoryton, CT 06442, USA

R.A. Cunha, Laboratory of Neurosciences, Faculty of Medicine, University of Lisbon, 1600 Lisbon, Portugal

S. Czeche, Department of Pharmacology, Biocentre Niederursel, University of Frankfurt, Marie-Curie-Strasse 9, D-60439 Frankfurt/Main, Germany

S. Damer, Department of Pharmacology, Biocentre Niederursel, University of Frankfurt, Marie-Curie-Strasse 9, D-60439 Frankfurt/Main, Germany

F. Di Virgilio, Department of Experimental and Diagnostic Medicine, Section of General Pathology, University of Ferrara, Via Borsari 46, I-44100 Ferrara, Italy

L. Diao, Department of Pharmacology, University of Colorado Health Sciences Center, 4200 E. 9th Avenue, Denver, CO 80262, USA

M. Díaz-Hernández, Departamento de Bioquímica, Facultad de Veterinaria, UCM, Ciudad Universitaria, 28040 Madrid, Spain

T.V. Dunwiddie, Neuroscience Program, Department of Pharmacology, C236, University of Colorado Health Sciences Center, Denver, CO 80262, USA

F.A. Edwards, Department of Physiology, University College London, Gower St., London WC1E 6BT, UK

Y.H. Ehrlich, The Program in Neuroscience, Department of Biology and the CSI/IBR Center for Developmental Neuroscience, The College of Staten Island, City University of New York, Staten Island, NY 10314, USA

A. Eichele, Pharmakologisches Institut für Naturwissenschaftler, Johann Wolfgang Goethe-Universität, Biozentrum N 260, Marie-Curie-Str. 9, D-60439 Frankfurt/Main, Germany

L.J. Erb, Department of Biochemistry, M121 Medical Sciences Building, University of Missouri-Columbia, Columbia, MO 65212, USA

D. Eschke, Institut für Pharmazie der Universität Leipzig, Lehrstuhl Pharmakologie für Naturwissenschaftler, Brüderstr. 34, D-04103 Leipzig, Germany

S. Falzoni, Department of Experimental and Diagnostic Medicine, Section of General Pathology, University of Ferrara, Via Borsari 46, I-44100 Ferrara, Italy

B.B. Fredholm, Department of Physiology and Pharmacology, Karolinska Institutet, 171 77 Stockholm, Sweden

C.J. Gallagher, Programme in Brain and Behaviour, Hospital for Sick Children, University of Toronto, Toronto, Ont. M5G 1X8, Canada

R.C. Garrad, Department of Biochemistry, M121 Medical Sciences Building, University of Missouri-Columbia, Columbia, MO 65212, USA

F.-P. Gendron, Département de Biologie, Faculté des Sciences, Université de Sherbrooke, Sherbrooke (Quebec) J1K 2R1, Canada

R. Gómez-Villafuertes, Departamento de Bioquímica, Facultad de Veterinaria, UCM, Ciudad Universitaria, 28040 Madrid, Spain

F.A. Gonzalez, Department of Chemistry, University of Puerto Rico, Rio Piedras, Puerto Rico

G. Grondin, Département de Biologie, Faculté des Sciences, Université de Sherbrooke, Sherbrooke (Québec) J1K 2R1, Canada

J. Gualix, Departamento de Bioquímica, Facultad de Veterinaria, UCM, Ciudad Universitaria, 28040 Madrid, Spain

D.-P. Guo, Molecular Recognition Section, LBC, NIDDK, NIH, Bethesda, MD 20892, USA

T.K. Harden, Department of Pharmacology, University of North Carolina School of Medicine, Chapel Hill, NC 27599-7365, USA

J.K. Hirsh, Department of Molecular Pharmacology and Biological Chemistry, Northwestern University Medical School, 303 East Chicago Avenue, Chicago, IL 60611, USA

C. Hoffmann, Molecular Recognition Section, LBC, NIDDK, NIH, Bethesda, MD 20892, USA

P. Illes, Department of Pharmacology, University of Leipzig, Härtelstrasse 16–18, D-04107 Leipzig, Germany

K. Inoue, Division of Pharmacology, National Institute of Health Sciences, 1-18-1 Kamiyoga, Setagaya, Tokyo 158, Japan

K.A. Jacobson, Molecular Recognition Section, Laboratory of Bioorganic Chemistry, National Institute of Diabetes, Digestive and Kidney Diseases, National Institutes of Health, Building 8A, Rm. B1A-19, Bethesda, MD 20892-0810, USA

S.Y. Jang, Molecular Recognition Section, LBC, NIDDK, NIH, Bethesda, MD 20892, USA

X.-D. Ji, Molecular Recognition Section, LBC, NIDDK, NIH, Bethesda, MD 20892, USA

Y. Kang, Research Service 151, VA Medical Center, 1201 NW 16th St., Miami, FL 33125, USA

C. Kennedy, Department of Physiology and Pharmacology, University of Strathclyde, 27 Taylor Street, Glasgow G4 0NR, UK

Y.-C. Kim, Molecular Recognition Section, LBC, NIDDK, NIH, Bethesda, MD 20892, USA

B.F. King, Autonomic Neuroscience Institute, Royal Free Hospital of Medicine, London NW3 2PF, UK

A. Kita, Division of Pharmacology, National Institute of Health Sciences, 1-18-1 Kamiyoga, Setagaya, Tokyo 158, Japan

H. Kittner, Department of Pharmacology and Toxicology, University of Leipzig, Härtelstrasse 16–18, D-04107 Leipzig, Germany

M. Klapperstück, Julius-Bernstein-Institute for Physiology, Martin-Luther-University Halle-Wittenberg, D-06097 Halle/S., Germany

H. Koch, Department of Pharmacology, University of Freiburg, Hermann-Herder-Strasse 5, D-79104 Freiburg, Germany

S. Koizumi, Division of Pharmacology, National Institute of Health Sciences, 1-18-1 Kamiyoga, Setagaya, Tokyo 158, Japan

E. Kornecki, Department of Anatomy and Cell Biology, The State University of New York, Downstate Medical Center, Brooklyn, NY 11203, USA

O. Krishtal, Department of Cellular Membranology, Bogomoletz Institute of Physiology, Bogomoletz St. 4, 252024 Kiev, Ukraine

U. Krügel, Department of Pharmacology and Toxicology, University of Leipzig, Härtelstrasse 16–18, D-04107 Leipzig, Germany

U. Lalo, Department of General Physiology of Nervous System, Bogomoletz Institute of Physiology, Bogomoletz St. 4, 252024 Kiev, Ukraine

G. Lambrecht, Department of Pharmacology, Biocentre Niederursel, University of Frankfurt, Marie-Curie-Strasse 9, D-60439 Frankfurt/Main, Germany

W. Laubinger, Institut für Neurobiochemie der Otto-von-Guericke-Universität Magdeburg, Leipziger Str. 44, 39120 Magdeburg, Germany

E.R. Lazarowski, Department of Pharmacology, University of North Carolina School of Medicine, Chapel Hill, NC 27599-7365, USA

B.T. Liang, University of Pennsylvania Medical Center, 3610 Hamilton Walk, Philadelphia, PA 19104, USA

M. Löhn, Institute for Medical Immunology, Martin-Luther-University Halle-Wittenberg, D-06097 Halle/S., Germany

F. Markwardt, Julius-Bernstein-Institute for Physiology, Martin-Luther-University Halle-Wittenberg, D-06097 Halle/S., Germany

J. Mateo, Departamento de Bioquímica, Facultad de Veterinaria, UCM, Ciudad Universitaria, 28040 Madrid, Spain

M. McCarthy, Research Service 151, VA Medical Center, Miami, FL 33125, USA

A. Meyer, Department of Pharmacology, University of Freiburg, Hermann-Herder-Strasse 5, D-79104 Freiburg, Germany

S. Mihaylova-Todorova, Department of Pharmacology, University of Nevada School of Medicine, Reno, NV 89557, USA

M.T. Miras-Portugal, Departamento de Bioquimica, Facultad de Veterinaria, Universidad Complutense de Madrid, 28040 Madrid, Spain

A. Mohanram, University of North Carolina, Chapel Hill, NC, USA

S. Moro, Molecular Recognition Section, LBC, NIDDK, NIH, Bethesda, MD 20892, USA

E. Mutschler, Department of Pharmacology, Biocentre Niederursel, University of Frankfurt, Marie-Curie-Strasse 9, D-60439 Frankfurt/Main, Germany

E. Nandanan, Molecular Recognition Section, LBC, NIDDK, NIH, Bethesda, MD 20892, USA

J.T. Neary, Research Service 151, VA Medical Center, Miami, FL 33125, USA

A. Nicke, Pharmakologisches Institut für Naturwissenschaftler, Johann Wolfgang Goethe-Universität, Biozentrum N 260, Marie-Curie-Str. 9, D-60439 Frankfurt/Main, Germany

P. Nickel, Department of Pharmaceutical Chemistry, University of Bonn, An der Immenburg 4, D-53121 Bonn, Germany

B. Niebel, Department of Pharmacology, Biocentre Niederursel, University of Frankfurt, Marie-Curie-Strasse 9, D-60439 Frankfurt/Main, Germany

K. Nieber, Institut für Pharmazie der Universität Leipzig, Lehrstuhl Pharmakologie für Naturwissenschaftler, Brüderstr. 34, D-04103 Leipzig, Germany

W. Nörenberg, Department of Pharmacology, University of Freiburg, Hermann-Herder-Strasse 5, D-79104 Freiburg, Germany

Y. Pankratov, Department of Cellular Membranology, Bogomoletz Institute of Physiology, Bogomoletz St. 4, 252024 Kiev, Ukraine

J. Pintor, Departamento de Bioquímica, Facultad de Veterinaria, UCM, Ciudad Universitaria, 28040 Madrid, Spain

W. Poelchen, Department of Pharmacology and Toxicology, University of Leipzig, Härtelstrasse 16–18, D-04107 Leipzig, Germany

R.S. Redman, Department of Molecular Pharmacology and Biological Chemistry, Northwestern University Medical School, 303 East Chicago Avenue, Chicago, IL 60611, USA

R. Reinhardt, Department of Pharmacology and Toxicology, University of Leipzig, Härtelstrasse 16–18, D-04107 Leipzig, Germany

G. Reiser, Institut für Neurobiochemie der Otto-von-Guericke-Universität Magdeburg, Leipziger Str. 44, 39120 Magdeburg, Germany

J. Rettinger, Pharmakologisches Institut für Naturwissenschaftler, Johann Wolfgang Goethe-Universität, Biozentrum N 260, Marie-Curie-Str. 9, D-60439 Frankfurt/Main, Germany

J.A. Ribeiro, Laboratory of Neurosciences, Faculty of Medicine, University of Lisbon, 1600 Lisbon, Portugal

D. Riemann, Institute for Medical Immunology, Martin-Luther-University Halle-Wittenberg, D-06097 Halle/S., Germany

S.J. Robertson, Department of Physiology, University College London, Gower St., London WC1E 6BT, UK

F.M. Ross, Institute of Biomedical and Life Sciences, Division of Neuroscience and Biomedical Systems, West Medical Building, University of Glasgow, Glasgow G12 8QQ, UK

A. Ruppelt, Max-Planck Institute for Experimental Medicine, Hermann-Rein Str. 3, D-37075, Göttingen, Germany

A. Rychkov, Kazan Medical Institute, Kazan, Russia

M.W. Salter, Programme in Brain and Behaviour, Hospital for Sick Children, University of Toronto, 555 University Avenue, Toronto, Ont. M5G 1X8, Canada

C. Santos-Berrios, Department of Chemistry, University of Puerto Rico, Rio Piedras, Puerto Rico

J.M. Sanz, Department of Experimental and Diagnostic Medicine, Section of General Pathology, University of Ferrara, Via Borsari 46, I-44100 Ferrara, Italy

R. Schäfer, Institut für Neurobiochemie der Otto-von-Guericke-Universität Magdeburg, Leipziger Str. 44, 39120 Magdeburg, Germany

G. Schmalzing, Pharmakologisches Institut für Naturwissenschaftler, Johann Wolfgang Goethe-Universität, Biozentrum N 260, Marie-Curie-Str. 9, D-60439 Frankfurt/Main, Germany

T.J. Searl, Department of Molecular Pharmacology and Biological Chemistry, Northwestern University Medical School, 303 East Chicago Avenue, Chicago, IL 60611, USA

A.M. Sebastião, Laboratory of Neurosciences, Faculty of Medicine, University of Lisbon, 1600 Lisbon, Portugal

D. Sieler, Department of Pharmacology and Toxicology, University of Leipzig, Härtelstrasse 16–18, D-04107 Leipzig, Germany

E.M. Silinsky, Department of Molecular Pharmacology and Biological Chemistry, Northwestern University Medical School, 303 East Chicago Avenue, Chicago, IL 60611, USA

P. Sneddon, Department of Physiology and Pharmacology, University of Strathclyde, 27 Taylor Street, Glasgow G4 0NR, UK

F. Soto, Max-Planck Institute for Experimental Medicine, Hermann-Rein Str. 3, D-37075, Göttingen, Germany

B. Sperlágh, Department of Pharmacology, Institute of Experimental Medicine, Hungarian Academy of Sciences, POB 67, H-1450 Budapest, Hungary

K. Starke, Department of Pharmacology, University of Freiburg, Hermann-Herder-Strasse 5, D-79104 Freiburg, Germany

T.W. Stone, Institute of Biomedical and Life Sciences, Division of Neuroscience and Biomedical Systems, West Medical Building, University of Glasgow, Glasgow G12 8QQ, UK

L.D. Todorov, Department of Pharmacology, University of Nevada School of Medicine, Reno, NV 89557, USA

M. Tsuda, Division of Pharmacology, National Institute of Health Sciences, 1-18-1 Kamiyoga, Setagaya, Tokyo 158, Japan

S. Ueno, Division of Pharmacology, National Institute of Health Sciences, 1-18-1 Kamiyoga, Setagaya, Tokyo 158, Japan

E.S. Vizi, Department of Pharmacology, Institute of Experimental Medicine, Hungarian Academy of Sciences, POB 67, H-1450 Budapest, Hungary

I. von Kügelgen, Department of Pharmacology, University of Freiburg, Hermann-Herder-Strasse 5, D-79104 Freiburg, Germany. Present address: Molecular Recognition Section, LBC, NIDDK, NIH, Bethesda, MD 20892, USA

M. Watanabe, Department of Molecular Pharmacology and Biological Chemistry, Northwestern University Medical School, 303 East Chicago Avenue, Chicago, IL 60611, USA

T.E. Webb, Molecular Neurobiology Unit, Royal Free Hospital School of Medicine, Rowland Hill Street, Hampstead, London NW3 2PF, UK and Department of Cell Physiology and Pharmacology, University of Leicester, University Road, P.O. Box 138, Leicester LE1 9HN, UK

G.A. Weisman, Department of Biochemistry, University of Missouri-Columbia, Columbia, MO 65212, USA

D.P. Westfall, Department of Pharmacology, University of Nevada School of Medicine, Reno, NV 89557, USA

T.D. Westfall, Department of Physiology and Pharmacology, University of Strathclyde, 27 Taylor Street, Glasgow G4 0NR, UK

S.S. Wildman, Autononic Neuroscience Institute, Royal Free Hospital School of Medicine, London NW3 2PF, UK

M. Williams, Neurological and Urological Diseases Research, D-464, Pharmaceutical Products Division, Abbott Laboratories, 100 Abbott Park Road, Abbott Park, IL 60064, USA

A.U. Ziganshin, Kazan Medical Institute, Kazan, Russia

H. Zimmermann, AK Neurochemie, Biozentrum der J.W. Goethe-Universität, Marie-Curie-Str. 9, D-60439 Frankfurt am Main, Germany

Preface

For more than 50 years ATP has been known as a universal source of chemical energy in cellular metabolism. Experiments performed during the past years demonstrate that extracellular nucleotides such as ATP, ADP, UTP, UDP and several diadenosine polyphosphates are extracellular signalling molecules. These compounds may leave the intracellular space via alternative routes including exocytotic release from nerve terminals, extrusion via transport systems from intact cells and efflux from injured or dying cells. Of the multiplicity of receptors for nucleotides so far identified two principal types can be differentiated. The P2X receptor is a ligand-activated cation channel, while the P2Y receptor is coupled to G proteins. Both types are widespread in the nervous system; this fact must have numerous physiological and pathological implications. As, however, the study of purinergic mechanisms for long only has been focused on the actions of the nucleoside adenosine, the contribution of nucleotides to the signalling systems was underestimated.

ATP is both a transmitter of its own right and a co-transmitter of acetylcholine and noradrenaline, for example. It mediates fast synaptic transmission via P2X receptors and slow synaptic signals via P2Y receptors. A negative or positive modulation of transmitter release may occur subsequent to the stimulation of P2 purinoceptors located at the nerve terminals or the respective cell somata. All these effects have consequences for the regulation of vegetative and behavioural functions. Specifically, in the peripheral and central nervous system purinoceptors are involved in nociception. In addition, ATP affects cell proliferation, acts in glial scar formation and can inhibit tumour growth. Enzymes located to the cell surface (ecto-nucleotidases) terminate the extracellular nucleotide signal. They catalyse the hydrolysis of the nucleotides with the nucleoside appearing as the final hydrolysis product. Evidence is also accumulating that the function of surface-located proteins can be modified via the action of ecto-protein kinases. At present a pharmacology of the nucleotide signalling system is being developed. Of particular interest is the production of receptor subtype-specific antagonists and of drugs that selectively affect the extracellular lifetime of the nucleotide.

This volume offers a comprehensive update and overview of nucleotide release, the structure and function of nucleotide receptors, nucleotide-metabolizing ecto-enzymes as well as the physiological functions of nucleotides in the nervous system. It is based on the proceedings of a IUPHAR Satellite Conference held in Leipzig, Germany, August 1–2, 1998.

The Conference generated a very positive response from its more than 150 participants from all over the world. Many thanks are due to the Scientific Committee, composed of Maria P. Abbracchio, Eric A. Barnard, Geoffrey Burnstock, T. Kendall Harden, Kazuhide Inoue, Kenneth A. Jacobson, Günter Lambrecht, Ernst Mutschler, R. Alan North and Klaus Starke for the care and input they put forth for this project. The logistics of the conference were the responsibility of the Organization Committee at the University of Leipzig. We would like to acknowledge the effective work of the secretariat supervised by Eleonore Schönfeld and of all coworkers and graduate students of the Department of Pharmacology whose efforts granted a smooth and effective meeting. Our particular thanks go to Karen Nieber and Robert Reinhardt for effectively solving uncounted minor and major practical problems that emerged during the planning of the conference.

The success of the conference greatly depended on financial contributors and sponsors. The conference was held under the auspices of the International Union of Pharmacological Sciences (IUPHAR) which held its XIIIth International Congress in Munich, Germany, July 26–31, 1998. The main sponsors of the conference in Leipzig were the Deutsche Forschungsgemeinschaft and the Bundesministerium für Bildung und Forschung. Additional financial support was obtained from Abbott, Arzneimittelwerk Dresden, Bayer, Byk Gulden Lomberg, Carl Zeiss, Dresdner Bank, Eppendorf-Netheler-Hinz, Grünenthal, Merck, Novartis and Science Products.

Finally we would like to thank the staff of Elsevier Science for their support and professional efforts in putting together this volume.

Peter Illes and Herbert Zimmermann
Leipzig and Frankfurt am Main, Germany, 1998

Contents

List of Contributors   v

Preface   xi

I. Signalling via nucleotides in the nervous system

1. Current status of purinergic signalling in the nervous system
   G. Burnstock (London, UK)   3

2. Modulation of purinergic neurotransmission
   P. Sneddon, T.D. Westfall, L.D. Todorov, S. Mihaylova-Todorova,
   D.P. Westfall and C. Kennedy (Glasgow, UK)   11

II. Molecular biology of P2Y receptors

3. Molecular biology of P2Y receptors expressed in the nervous system
   T.E. Webb and E.A. Barnard (London, UK)   23

4. P2Y receptors in the nervous system: Molecular studies of a $P2Y_2$ receptor
   subtype from NG108-15 neuroblastoma x glioma hybrid cells
   G.A. Weisman, R.C. Garrad, L.J. Erb, C. Santos-Berrios and F.A.
   Gonzalez (Columbia, MO, USA and Rio Piedras, Puerto Rico) . .   33

5. Nucleotide radiolabels as tools for studying P2Y receptors in membranes
   from brain and lung tissue
   G. Reiser, W. Laubinger and R. Schäfer (Magdeburg, Germany) . .   45

III. Molecular biology of P2X receptors

6. Evolving view of quarternary structures of ligand-gated ion channels
   A. Nicke, J. Rettinger, C. Büttner, A. Eichele, G. Lambrecht and
   G. Schmalzing (Frankfurt am Main, Germany)   61

7. Cloning, functional characterization and developmental expression of a P2X
   receptor from chick embryo
   A. Ruppelt, B.T. Liang and F. Soto (Göttingen, Germany and
   Philadelphia, PA, USA) 81

IV. Development of nucleotide analogues

8. Developments in P2 receptor targeted therapeutics
   M. Williams (Abbott Park, IL, USA) 93

9. Novel ligands for P2 receptor subtypes in innervated tissues
   G. Lambrecht, S. Damer, B. Niebel, S. Czeche, P. Nickel, J.
   Rettinger, G. Schmalzing and E. Mutschler (Frankfurt am Main,
   Germany). 107

10. Molecular recognition in P2 receptors: Ligand development aided by
    molecular modeling and mutagenesis
    K.A. Jacobson, C. Hoffmann, Y.-C. Kim, E. Camaioni,
    E. Nandanan, S.Y. Jang, D.-P. Guo, X.-D. Ji, I. von Kügelgen,
    S. Moro, A.U. Ziganshin, A. Rychkov, B.F. King, S.G. Brown,
    S.S. Wildman, G. Burnstock, J.L. Boyer, A. Mohanram and
    T.K. Harden (Bethesda, MD, USA, Kazan, Russia, London, UK
    and Chapel Hill, NC, USA) 119

V. Release of nucleotides

11. Release of ATP and UTP from astrocytoma cells
    T.K. Harden and E.R. Lazarowski (Chapel Hill, NC, USA) 135

12. Quantal ATP release from motor nerve endings and its role in neurally
    mediated depression
    E.M. Silinsky, J.K. Hirsh, T.J. Searl, R.S. Redman and M.
    Watanabe (Chicago, IL, USA). 145

13. Receptor- and carrier-mediated release of ATP of postsynaptic origin:
    Cascade transmission
    E.S. Vizi and B. Sperlágh (Budapest, Hungary). 159

VI. Presynaptic modulation by nucleotides of neurotransmitter release

14. P2-receptors controlling neurotransmitter release from postganglionic
    sympathetic neurones
    I. von Kügelgen, W. Nörenberg, H. Koch, A. Meyer, P. Illes and
    K. Starke (Freiburg and Leipzig, Germany). 173

15. Adenine nucleotides as inhibitors of synaptic transmission: Role of localised
    ectonucleotidases
    A.M. Sebastião, R.A. Cunha, J.F. Cascalheira and J.A. Ribeiro
    (Lisbon and Covilhã, Portugal) 183

16. The functions of ATP receptors in the synaptic transmission in the
    hippocampus
    K. Inoue, S. Koizumi, S. Ueno, A. Kita and M. Tsuda (Tokyo,
    Japan) 193

VII. Physiology of nucleotide function

17. Electrophysiological analysis of P2-receptor mechanisms in rat sympathetic
    neurones
    W. Nörenberg, I. von Kügelgen, A. Meyer and P. Illes (Freiburg
    and Leipzig, Germany). 209

18. P2 receptor-mediated activation of noradrenergic and dopaminergic neurons
    in the rat brain
    H. Kittner, U. Krügel, W. Poelchen, D. Sieler, R. Reinhardt,
    I. von Kügelgen and P. Illes (Leipzig and Freiburg, Germany) . . . 223

19. ATP receptor-mediated component of the excitatory synaptic transmission in
    the hippocampus
    Y. Pankratov, U. Lalo, E. Castro, M.T. Miras-Portugal and O.
    Krishtal (Kiev, Ukraine and Madrid, Spain). 237

20. Nucleotide and dinucleotide effects on rates of paroxysmal depolarising
    bursts in rat hippocampus
    F.M. Ross, M.J. Brodie and T.W. Stone (Glasgow, UK) 251

VIII. Physiology of nucleoside function

21. The function of A_2 adenosine receptors in the mammalian brain: Evidence for
    inhibition vs. enhancement of voltage gated calcium channels and
    neurotransmitter release
    F.A. Edwards and S.J. Robertson (London, UK) 265

22. An adenosine A_3 receptor-selective agonist does not modulate calcium-
    activated potassium currents in hippocampal CA1 pyramidal
    neurons
    T.V. Dunwiddie, K.A. Jacobson and L. Diao (Denver, CO and
    Bethesda, MD, USA) 275

23. Brain hypoxia: Effects of ATP and adenosine
    K. Nieber, D. Eschke and A. Brand (Leipzig, Germany) 287

IX. Nucleotide effects on neuronal differentiation and glial proliferation

24. Adenosine and P2 receptors in PC12 cells. Genotypic, phenotypic and
individual differences
G. Arslan and B.B. Fredholm (Stockholm, Sweden) 301

25. Nucleotide receptor signalling in spinal cord astrocytes: Findings and
functional implications
C.J. Gallagher and M.W. Salter (Toronto, Canada) 311

26. Trophic signaling pathways activated by purinergic receptors in rat and
human astroglia
J.T. Neary, M. McCarthy, A. Cornell-Bell and Y. Kang (Miami,
FL and Ivoryton, CT, USA) 323

27. Signalling mechanisms involved in P2Y receptor-mediated reactive
astrogliosis
M.P. Abbracchio, R. Brambilla, S. Ceruti and F. Cattabeni (Milan,
Italy) 333

X. Immunomodulatory effects of ATP

28. Purinoceptors in human B-lymphocytes
F. Markwardt, M. Klapperstück, M. Löhn, D. Riemann, C. Büttner
and G. Schmalzing (Halle/S. and Frankfurt am Main, Germany). . 345

29. The P2Z/P2X$_7$ receptor of microglial cells: A novel immunomodulatory
receptor
F. Di Virgilio, J.M. Sanz, P. Chiozzi and S. Falzoni (Ferrara, Italy) 355

XI. Ecto-nucleotidases and ecto-protein kinases

30. Ecto-nucleotidases – molecular structures, catalytic properties, and functional
roles in the nervous system
H. Zimmermann and N. Braun (Frankfurt am Main, Germany) . . 371

31. Immunolocalization of ATP diphosphohydrolase in pig and mouse brains, and
sensory organs of the mouse
A.R. Beaudoin, G. Grondin and F.-P. Gendron (Sherbrooke,
Canada). 387

32. Diadenosine polyphosphates, extracellular function and catabolism
M.T. Miras-Portugal, J. Gualix, J. Mateo, M. Díaz-Hernández, R.
Gómez-Villafuertes, E. Castro and J. Pintor (Madrid, Spain) 397

33. Ecto-protein kinases as mediators for the action of secreted ATP in the brain
Y.H. Ehrlich and E. Kornecki (Staten Island and Brooklyn, NY,
USA) 411

Subject Index 427

Signalling via nucleotides in the nervous system

P. Illes and H. Zimmermann (Eds.)
Progress in Brain Research, Vol 120

CHAPTER 1

Current status of purinergic signalling in the nervous system

Geoffrey Burnstock

Autonomic Neuroscience Institute, Royal Free and University College Medical School, University College London, Royal Free Campus, Rowland Hill Street, London NW3 2PF, UK. Tel: + 44 171 830 2948; Fax: + 44 171 830 2949; e-mail: g.burnstock@ucl.ac.uk

Introduction

Drury and Szent-Györgyi demonstrated way back in 1929 the potent extracellular actions of ATP and, in the 1950s, Pamela Holton presented the first hint of a transmitter role for ATP in the nervous system by demonstrating release of ATP during antidromic stimulation of sensory nerves (Holton, 1959). Then, in my laboratory in Melbourne in 1970, we proposed that non-adrenergic, non-cholinergic (NANC) nerves supplying smooth muscle of the gut and bladder utilised ATP as a neurotransmitter (Burnstock et al., 1970, 1972). The experimental evidence included: mimickry of the NANC nerve mediated response by ATP; measurement of release of ATP during stimulation of NANC nerves with luciferin–luciferase luminometry; histochemical labelling of subpopulations of neurones in the gut and the bladder with quinacrine, a fluorescent dye known to selectively label high levels of ATP bound to peptides; the demonstration that the slowly-degradable analogue of ATP, α,β-methylene ATP, which produces selective desensitisation of the ATP receptor, blocks the responses to NANC nerve stimulation. The term 'purinergic' and the evidence for purinergic transmission in a wide variety of systems was presented in an early pharmacological review (Burnstock, 1972).

Purinergic receptors

Implicit in the concept of purinergic neurotransmission is the existence of postjunctional purinergic receptors. A brief history of the development of the purinoceptor nomenclature follows. Purinergic receptors were first defined in 1976 (Burnstock, 1976) and two years later a basis for distinguishing two types of purinoceptor, identified as P1 and P2 (for adenosine and ATP/ADP, respectively), was proposed (Burnstock, 1978). At about the same time, the two subtypes of the P1 (adenosine) receptor were recognised (Van Calker et al., 1979; Londos et al., 1980), but it was not until 1985 that a proposal suggesting a basis for distinguishing two types of P2 receptor (P2X and P2Y) was proposed (Burnstock and Kennedy, 1985). A year later, Gordon (1986) tentatively named two further P2 purinoceptor subtypes, namely a P2T receptor selective for ADP on platelets and a P2Z receptor on macrophages. Further subtypes followed, perhaps the most important of which being the P2U receptor which could recognise pyrimidines such as UTP as well as ATP (O'Connor et al., 1991). In 1994 Mike Williams made the point that a classification of P2 purinoceptors based on a 'random walk through the alphabet' was not satisfactory, and Abbracchio and Burnstock (1994), on the basis of studies of transduction mechanisms (Dubyak,

1991) and the cloning of P2Y (Lustig et al., 1993; Webb et al., 1993) and later P2X purinoceptors (Brake et al., 1994; Valera et al., 1994), proposed that purinoceptors should be considered to belong to two major families: a P2X family of ligand-gated ion channel receptors and a P2Y family of G protein-coupled purinoceptors. This nomenclature has been widely adopted and currently seven P2X subtypes and about eight P2Y receptor subtypes are recognised. Since a number of these receptors are sensitive to pyrimidines as well as purines, it was decided in 1996 to adopt the term 'P2 receptor' rather than 'purinoceptor' (Fredholm et al., 1996).

P2X receptors

Although no new P2X receptors have been recognised for over two years, now there is general recognition that important issues still need to be resolved in this field, including:

(1) whether three (Nicke et al., 1998) or four (Rassendren et al., 1997) subunits make up the ion pores;

(2) whether both homomultimers and heteromultimers form these pores, and in what combinations and proportions;

(3) the roles of the many spliced variants of different P2X subtypes which are beginning to be described (see Burnstock et al., 1998), many of which are inactive;

(4) interesting issues are emerging about the nature of the $P2X_7$ receptor (formerly P2Z) in terms of its involvement in apoptosis when expressed on external plasma membranes, but not when internalised (unpublished observations). There is also the question of why $P2X_7$ receptors become dominant in pathological tissues.

In addition, there is much still to be learned about the distribution of P2X receptor subtypes in different tissues. The knowledge of the distribution based on northern blots and in situ hybridisation methodology (Collo et al., 1996) is valuable, but now that selective antibodies are becoming available, immunolabelling at both light and electron microscope levels is giving significant new information in this field (Vulchanova et al., 1996; Lê et al., 1998; Llewellyn-Smith and Burnstock, 1998;

Loesch and Burnstock, 1998). For example, $P2X_1$ receptors were previously not considered to be present in the brain; however, recent cytochemical studies of the cerebellum show clear evidence of synaptic transmission involving this receptor (see Fig. 1). It has also become clear that ATP receptors in vascular smooth muscle are not only of subtype $P2X_1$, but $P2X_2$ and $P2X_4$ receptors have also been demonstrated (Nori et al., 1998).

P2Y receptors

There are important recent developments concerning P2Y receptor subtypes:

(1) The $P2Y_1$ subtype was originally cloned from the chick (Webb et al., 1993) where 2-methylthio ATP was the most potent agonist, but the mammalian $P2Y_1$ subtype appears to be ADP-selective (Leon et al., 1997).

(2) ATP and UTP are equipotent at the $P2Y_2$ receptor (Lustig et al., 1993) and this receptor has, therefore, been equated with the old P2U receptor. However, the rat $P2Y_4$ receptor was recently cloned and at this receptor, too, ATP and UTP are equipotent; nevertheless, they can be distinguished because the $P2Y_2$ receptor is suramin-sensitive, while the $P2Y_4$ receptor is suramin-insensitive, but is blocked by Reactive Blue 2 (Bogdanov et al., 1998).

(3) It has been suggested that the $P2Y_3$ receptor is an orthologue of $P2Y_6$, which is UDP selective, while $P2Y_{5,9 \text{ and } 10}$ receptors have not yet been shown to be functional.

(4) The $P2Y_7$ receptor has been identified as a leukotriene B_4 receptor (Yokomizo et al., 1996).

(5) No mammalian homologue has yet been identified for the $P2Y_8$ receptor cloned from the frog embryo (Bogdanov et al., 1997).

(6) Most recently, a $P2Y_{11}$ receptor has been identified which appears to accommodate two alternative second messenger transduction mechanisms, one coupled to phospholipase, the other to adenylate cyclase (Communi et al., 1997).

(7) The current assessment of ATP receptors on platelets is that $P2X_1$ and $P2Y_1$ are both present and also a $P2Y_{A-C}$ receptor negatively coupled

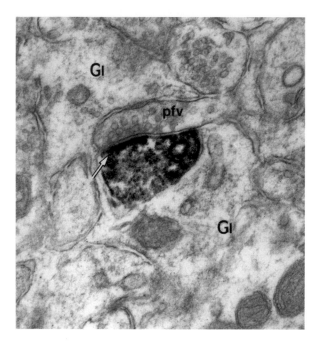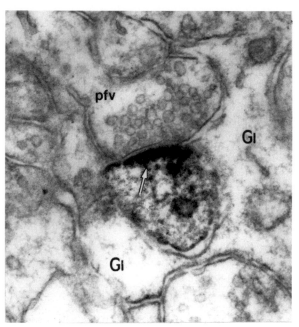

Fig. 1. Imunohistochemistry of synapses in the molecular layer of rat cerebellum showing $P2X_1$-positive labelling of postsynaptic membranes (arrows) of the dendritic spines of Purkinje cells opposed to prejunctional varicosities (containing vesicles) of parallel fibres (*pfv*) of the ascending granule cell axons (*Gl*, neuroglia process). Right micrograph, $\times 53,000$. Left micrograph,. $\times 74,000$. (From Loesch and Burnstock, 1998).

to adenylate cyclase, which has yet to be cloned (Fagura et al., 1998).

(8) A separate receptor has been claimed for diadenosine polyphosphates, at least in the hippocampus, named P2D (Miras-Portugal ct al., 1996) but this receptor, too, has still to be cloned.

Physiology

Short-term signalling

Most studies of fast signalling in the nervous system have been concerned with the role of ATP as a cotransmitter (see Burnstock, 1996a).

Sympathetic neurotransmission

It is now well-established, largely from studies of the vas deferens (Sneddon and Westfall, 1984) and later from a variety of blood vessels (see Burnstock, 1990), that ATP and noradrenaline (NA) are cotransmitters in sympathetic nerves. ATP, when released, acts on P2X ion channel receptors to produce excitatory junctional potentials and Ca^{2+} influx leading to rapid contraction, while NA acts usually via α_1 adrenoceptors which are G protein-coupled, involving IP_3 second messenger systems, Ca^{2+} influx and slow contractions. Neuropeptide Y (NPY) is a major component of the 'chemical coding' of most sympathetic nerves but in many tissues does not act as a cotransmitter, but rather as a neuromodulator of the release and/or of activity of the principal transmitters ATP and NA (see Burnstock, 1995). The proportions of ATP and NA released from sympathetic nerves vary enormously between different tissues and different species and under different pathophysiological conditions (see later). For example, in intestinal arterioles only ATP acts as a neurotransmitter, while the NA released acts prejunctionally to modulate the release of the principal transmitter (Evans and Surprenant, 1992). In other vessels, for example in renal and ear arteries, NA is the dominant transmitter.

Parasympathetic neurotransmission

In parasympathetic nervous control of the urinary bladder, ATP and acetylcholine (ACh) act as cotransmitters in a comparable way to sympathetic cotransmission. Thus, ATP produces the fast, initial response via P2X ion channel receptors and the slow component of the response is mediated by muscarinic receptors acting through the G protein-coupled system (Hoyle and Burnstock, 1985).

Motor neurotransmission to skeletal muscle

It has been known for many years that ATP is stored and released together with ACh from motor nerve terminals. In the adult skeletal neuromuscular junction, ACh appears to be the sole neuro-transmitter acting through nicotinic receptors, while the released ATP acts both postjunctionally to potentiate the action of ACh and prejunctionally after breakdown to adenosine via P1 (A_1) receptors to modulate the release of ACh (Silinsky, 1984; Lu and Smith, 1998). However, in the developing myotube, ATP as well as ACh acts through ion channel receptors (Kolb and Wakelam, 1983); ATP responsiveness also reappears after denervation (Wells et al., 1995).

Sensory-motor nerves

It has been known since the early work of Lewis that, during axon reflex activity, substances are released during antidromic impulses down sensory collaterals to dilate vessels in the skin. It is now clear that sensory motor nerves are extremely widespread, supplying tissues such as lung, heart, gut and blood vessels and it appears that ATP is a cotransmitter together with calcitonin gene-related peptide and substance P in subpopulations of these nerves (see Rubino and Burnstock, 1996).

Non-adrenergic, non-cholinergic enteric nerves

There is good evidence from a number of laborato-ries now that NANC inhibitory nerves in the gut utilise three transmitters, ATP, nitric oxide (NO) and vasoactive intestinal polypeptide (VIP) in variable proportions in different sphincteric and non-sphincteric regions of the gut (e.g. Belai and Burnstock, 1994). In general, ATP evokes the fastest response, NO the next fastest, while VIP usually produces slow tonic responses.

Nerve/nerve synapses in ganglia and CNS

While there were several early reports of the potent actions of ATP on autonomic ganglion cells (e.g. Akasu et al., 1983; Theobald and de Groat, 1989; Allen and Burnstock, 1990), it was not until 1992 that excitatory postsynaptic potentials (EPSPs) were recorded in the coeliac ganglion which were reversibly blocked by suramin (Evans et al., 1992; Silinsky et al., 1992). These findings were rapidly followed by a paper showing purinergic EPSPs in the medial habenula (Edwards et al., 1992) and neuroscientists began to recognise the widespread presence of purinergic synaptic transmission. Other examples have followed: in enteric ganglia (LePard et al., 1997), in sympathetic and sensory (see Evans and Surprenant, 1996; Vizi et al., 1997) and pelvic ganglia (Zhong et al., 1998), and in the CNS (see Gibb and Halliday, 1996). In addition, there have been numerous reports of potent effects of ATP and its analogues on neurones of the central nervous system in culture or in slices (see Abbracchio, 1997) and in situ hybridization and immunohis-tochemical localisation of various P2 receptor subtypes in different regions of the brain (Bo and Burnstock, 1994; Collo et al., 1996; Vulchanova et al., 1996; Lê et al., 1998; Loesch and Burnstock, 1998).

Long-term (trophic) actions of purines and pyrimidines

There are many examples of the potent long-term effects of ATP, UTP and related compounds on neurones and glial cells (see Neary et al., 1996). There are also many examples of trophic effects of nucleosides and nucleotides on peripheral nerve, smooth muscle and epithelial cell proliferation, growth and differentiation (see Abbracchio and Burnstock, 1998).

In this report I want to stress the growing evidence for trophic roles of nucleotides in cell signalling during embryonic neural development (see Burnstock, 1996b). For example, in our own laboratory we have cloned, sequenced and characterised a new receptor (P2Y$_8$) which appears to be related to the development of the nervous system (Bogdanov et al., 1997). From northern blots and in situ hybridisation, this receptor was shown to be expressed at Stage 13 at the beginning of neurulation and to show little expression by the end of neurulation at Stage 21. It reappeared during secondary neurulation in the tail bud at Stage 28. The receptor is unusual in the length of the C-terminal (216 amino acids compared with 16–67 in all the other known P2Y receptor subtypes) and with the recombinant receptor expressed in transfected *Xenopus* oocytes it showed equipotency with ATP, UTP, CTP, ITP and GTP. In more recent studies in our laboratory (Meyer et al., 1999), the P2Y$_1$ receptor has been localised in the chick forebrain as well as in limb buds, mesonephros of the kidney and brachial plexus. We also have recent evidence for extensive expression of P2X subtypes in different nervous tissues at different stages of chick embryonic development.

Plasticity–pathophysiology

It is well known that the autonomic nervous system shows a high degree of plasticity during development and ageing in nerves that remain following trauma and surgery, and in disease situations (Burnstock, 1991). For example, while parasympathetic purinergic transmission of the urinary bladder has been demonstrated in a variety of laboratory animals, purinergic transmission in the human bladder is minimal, even though purinergic receptors have been shown to be present in the smooth muscle (Bo and Burnstock, 1995). However, there are several reports that show an increase of up to 40% of the purinergic component of parasympathetic cotransmission in pathological human bladder in interstitial cystitis (Palea et al., 1993), outflow obstruction (Smith and Chapple, 1994) and neurogenic bladder (Wammack et al., 1995). Similarly, in spontaneously hypertensive rats (SHRs), a significantly greater cotransmitter role for ATP compared to NA has been demonstrated in blood vessels (Vidal et al., 1986; Bulloch and McGrath, 1992; Brock and Van Helden, 1995).

A major emerging role for purines in the nervous system appears to be in nociception. A receptor was cloned in 1995 by our laboratory (Chen et al., 1995) and at about the same time by Lewis et al. (1995) which has been shown immunohistochemically to be located selectively on small nociceptive sensory neurones in dorsal root, nodose and trigeminal ganglia (Vulchanova et al., 1996; Bradbury et al., 1998). It has been shown that the small nociceptive neurones are largely of the non-peptide-containing subpopulation which are labelled by isolectin B4. Central projections of these neurones are immunostained for P2X$_3$ receptors in inner lamina II of the dorsal horn of the spinal cord, and peripheral projections in the skin and tongue have also been shown to be immunopositive for P2X$_3$ (Bo et al., 1999). It has been recently proposed (Burnstock, 1999) that in tubes such as ureter, gut, salivary and bile ducts, and sacs such as bladder and lung, the pain caused by distension works through a purinergic mechanosensory transduction mechanism, i.e. that the epithelial cells lining this plexus release ATP to act on P2X$_{2/3}$ nociceptive receptors on subepithelial sensory nerve terminals which relay impulses to the CNS to be registered as pain. There is already supportive evidence for this concept in the bladder (see Ferguson et al., 1997; Morrison et al., 1998a).

This hypothesis and the evidence that mechano-distortion of membranes leads to release of ATP from other cell types, such as vascular endothelial cells (Bodin et al., 1991; Milner et al., 1992), epithelial cells (Watt et al., 1998), osteoblasts (Bowler et al., 1998; Morrison et al., 1998b) and marginal cells of the stria vascularis in the inner ear (White et al., 1995), raises the key question, namely what is the mechanism of ATP transport?

There are several hints that ATP binding cassette (ABC) proteins, sulphonylurea receptors and the cystic fibrosis transmembrane conductance regulator (CFTR) channels may be implicated, but the precise mechanism remains to be resolved.

Future directions

The field of nucleotides and their receptors in the nervous system is still in its infancy and I predict a rapid expansion of interest in this field in the coming years. Some of the directions these studies are likely to take are the development of transgenic mice with absent or enhanced P2 receptor subtypes, behavioural studies of the effects of purines and pyrimidines and related compounds applied to the brain, trophic roles of nucleotides in cell development and death in various pathophysiological systems, resolution of ATP membrane transport mechanisms and the development of selective agonists and antagonists that work in vivo for therapeutic application for a variety of diseases.

References

Abbracchio, M.P. (1997) ATP in brain function. In: K.A. Jacobson and M.F. Jarvis (Eds.), *Purinergic Approaches in Experimental Therapeutics*. Wiley-Liss, New York, pp. 383–404.

Abbracchio, M. and Burnstock, G. (1994) Purinoceptors: Are there families of P_{2X} and P_{2Y} purinoceptors? *Pharmacol. Ther.*, 64: 445–475.

Abbracchio, M.P. and Burnstock, G. (1998) Purinergic signalling: Pathophysiological roles. *Jpn. J. Pharmacol.*, 78: 113–145.

Akasu, T., Hirai, K. and Koketsu, K. (1983) Modulatory actions of ATP on membrane potentials on bullfrog sympathetic ganglion cells. *Brain Res.*, 258: 313–317.

Allen, T.G.J. and Burnstock, G. (1990) The actions of adenosine 5′-triphosphate on guinea-pig intracardiac neurones in culture. *Br. J. Pharmacol.*, 100: 269–276.

Belai, A. and Burnstock, G. (1994) Evidence for coexistence of ATP and nitric oxide in non-adrenergic, non-cholinergic (NANC) inhibitory neurones in the rat ileum, colon and anococcygeus muscle. *Cell Tissue Res.*, 278: 197–200.

Bo, X. and Burnstock, G. (1994) Distribution of [^{3}H]α,β-methylene ATP binding sites in rat brain and spinal cord. *NeuroReport*, 5: 1601–1604.

Bo, X. and Burnstock, G. (1995) Characterization and autoradiographic localization of [^{3}H]α,β-methylene adenosine 5′-triphosphate binding sites in human urinary bladder. *Br. J. Urol.*, 76: 297–302.

Bo, X., Alavi, A., Xiang, Z., Oglesby, I.B., Ford, A.P.D.W. and Burnstock, G. (1999). Localization of $P2X_2$ and $P2X_3$ receptor immunoreactive nerves in rat taste buds. *NeuroReport* (In press).

Bodin, P., Bailey, D.J. and Burnstock, G. (1991) Increased flow-induced ATP release from isolated vascular endothelial but not smooth muscle cells. *Br. J. Pharmacol.*, 103: 1203–1205.

Bogdanov, Y.D., Dale, L., King, B.F., Whittock, N. and Burnstock, G. (1997) Early expression of a novel nucleotide receptor in the neural plate of *Xenopus* embryos. *J. Biol. Chem.*, 272: 12583–12590.

Bogdanov, Y., Rubino, A. and Burnstock, G. (1998) Characterisation of subtypes of the P2X and P2Y families of receptors in the foetal human heart. *Life Sci.*, 62: 697–703.

Bowler, W.B., Tattersall, J.A., Hussein, R., Dixon, C.J., Cobbold, P.H. and Gallagher, J.A. (1998) Release of ATP by osteoblasts: modulation by fluid shear forces. *Bone*, 22: 3S (abstract).

Bradbury, E., McMahon, S.B. and Burnstock, G. (1998) The expression of $P2X_3$ purinoceptors in sensory neurons: Effects of axotomy and glial-derived neurotrophic factor. *Mol. Cell. Neurosci.*, 12: 256–268.

Brake, A.J., Wagenbach, M.J. and Julius, D. (1994) New structural motif for ligand-gated ion channels defined by an inotropic ATP receptor. *Nature*, 371: 519–523.

Brock, J.A. and Van Helden, D.F. (1995) Enhanced excitatory junction potentials in mesenteric arteries from spontaneously hypertensive rats. *Pflügers Arch.*, 430: 901–908.

Bulloch, J.M. and McGrath, J.C. (1992) Evidence for increased purinergic contribution in hypertensive blood vessels exhibiting co-transmission. *Br. J. Pharmacol.(Suppl.)*, 107: 145 pp.

Burnstock, G. (1972) Purinergic nerves. *Pharmacol. Rev.*, 24: 509–581.

Burnstock, G. (1976) Purinergic receptors. *J. Theor. Biol.*, 62: 491–503.

Burnstock, G. (1978) A basis for distinguishing two types of purinergic receptor. In: R.W. Straub and L. Bolis (Eds.), *Cell Membrane Receptors for Drugs and Hormones: A Multidisciplinary Approach*. Raven Press, New York, pp. 107–118.

Burnstock, G. (1990) Noradrenaline and ATP as cotransmitters in sympathetic nerves. *Neurochem. Int.*, 17: 357–368.

Burnstock, G. (1991) Plasticity in expression of cotransmitters and autonomic nerves in aging and disease. In: P.S. Timiras and A. Privat (Eds.), *Plasticity and Regeneration of the Nervous System. Advances in Experimental Medicine and Biology*. Plenum Press, New York, pp. 291–301.

Burnstock, G. (1995) Noradrenaline and ATP: cotransmitters and neuromodulators. *J. Physiol. Pharmacol.*, 46: 365–384.

Burnstock, G. (1996a) (Guest Ed.). Purinergic Neurotransmission. *Semin. Neurosci.*, 8: 171–257.

Burnstock, G. (1996b) Purinoceptors: Ontogeny and Phylogeny. *Drug Dev. Res.*, 39: 204–242.

Burnstock, G. (1999) Release of vasoactive substances from endothelial cells by shear stress and purinergic mechanosensory transduction. *J. Anat.*, 194: 335–342.

Burnstock, G. and Kennedy, C. (1985) Is there a basis for distinguishing two types of P_2-purinoceptor? *Gen. Pharmacol.*, 16: 433–440.

Burnstock, G., Campbell, G., Satchell, D. and Smythe, A. (1970) Evidence that adenosine triphosphate or a related nucleotide is the transmitter substance released by non-adrenergic inhibitory nerves in the gut. *Br. J. Pharmacol.*, 40: 668–688.

Burnstock, G., Dumsday, B. and Smythe, A. (1972) Atropine resistant excitation of the urinary bladder: The possibility of transmission via nerves releasing a purine nucleotide. *Br. J. Pharmacol.*, 44: 451–461.

Burnstock, G., King, B.F. and Townsend-Nicholson, A. (1998) *P1 and P2 receptor update: News-sheet 8*. Autonomic

Neuroscience Institute, Royal Free Hospital School of Medicine, London.

Chen, C.-C., Akopian, A.N., Sivilotti, L., Colquhoun, D., Burnstock, G. and Wood, J.N. (1995) A P2X purinoceptor expressed by a subset of sensory neurons. *Nature*, 377: 428–431.

Collo, G., North, R.A., Kawashima, E., Merlo-Pich, E., Neidhart, S., Surprenant, A. and Buell, G. (1996) Cloning of $P2X_5$ and $P2X_6$ receptors and the distribution and properties of an extended family of ATP-gated ion channels. *J. Neurosci.*, 16: 2495–2507.

Communi, D., Govaerts, C., Parmentier, M. and Boeynaems, J.M. (1997) Cloning of a human purinergic P2Y receptor coupled to phospholipase C and adenylyl cyclase. *J. Biol. Chem.*, 272: 31969–31973.

Drury, A.N. and Szent-Györgyi, A. (1929) The physiological activity of adenine compounds with special reference to their action upon the mammalian heart. *J. Physiol.*, 68: 213–237.

Dubyak, G.R. (1991) Signal transduction by P_2-purinergic receptors for extracellular ATP. *Am. J. Respir. Cell Mol. Biol.*, 4: 295–300.

Edwards, F.A., Gibb, A.J. and Colquhoun, D. (1992) ATP receptor-mediated synaptic currents in the central nervous system. *Nature*, 359: 144–147.

Evans, R.J. and Surprenant, A. (1992) Vasoconstriction of guinea-pig submucosal arterioles following sympathetic nerve stimulation is mediated by the release of ATP. *Br. J. Pharmacol.*, 106: 242–249.

Evans, R.J. and Surprenant, A. (1996) P2X receptors in autonomic and sensory neurons. *Semin. Neurosci.*, 8: 217–225.

Evans, R.J., Derkach, V. and Surprenant, A. (1992) ATP mediates fast synaptic transmission in mammalian neurons. *Nature*, 357: 503–505.

Fagura, M.S., Dainty, I.A., McKay, G.D., Kirk, I.P., Humphries, R.G., Robertson, M.J., Dougall, I.G. and Leff, P. (1998) $P2Y_1$-receptors in human platelets which are pharmacologically distinct from $P2Y_{ADP}$-receptors. *Br. J. Pharmacol.*, 124: 157–164.

Ferguson, D.R., Kennedy, I. and Burton, T.J. (1997) ATP is released from rabbit urinary bladder epithelial cells by hydrostatic pressure changes—a possible sensory mechanism? *J. Physiol. (Lond.)*, 505: 503–511.

Fredholm, B.B., Burnstock, G., Harden, T.K. and Spedding, M. (1996) Receptor Nomenclature. *Drug Dev. Res.*, 39: 461–466.

Gibb, A.J. and Halliday, F.C. (1996) Fast purinergic transmission in the central nervous system. *Semin. Neurosci.*, 8: 225–232.

Gordon, J.L. (1986) Extracellular ATP: effects, sources and fate. *Biochem. J.*, 233: 309–319.

Holton, P. (1959) The liberation of adenosine triphosphate on antidromic stimulation of sensory nerves. *J. Physiol.*, 145: 494–504.

Hoyle, C.H.V. and Burnstock, G. (1985) Atropine-resistant excitatory junction potentials in rabbit bladder are blocked by α,β-methylene ATP. *Eur. J. Pharmacol.*, 114: 239–240.

Kolb, H.-A. and Wakelam, M.J.O. (1983) Transmitter-like action of ATP on patched membranes of cultured myoblasts and myotubes. *Nature*, 303: 621–623.

Lê, K.T., Villeneuve, P., Ramjaun, A.R., McPherson, P.S., Beaudet, A. and Séguela, P. (1998) Sensory presynaptic and widespread somatodendritic immunolocalization of central ionotropic P2X ATP receptors. *Neuroscience*, 83: 177–190.

Leon, C., Hechler, B., Vial, C., Leray, C., Cazenave, J.P. and Gachet, C. (1997) The $P2Y_1$ receptor is an ADP receptor antagonized by ATP and expressed in platelets and megakaryoblastic cells. *FEBS Lett.*, 403: 26–30.

LePard, K.J., Messori, E. and Galligan, J.J. (1997) Purinergic fast excitatory postsynaptic potentials in myenteric neurons of guinea pig: distribution and pharmacology. *Gastroenterology*, 113: 1522–1534.

Lewis, C., Neldhart, S., Holy, C., North, R.A., Buell, G. and Surprenant, A. (1995) Coexpression of $P2X_2$ and $P2X_3$ receptor subunits can account for ATP-gated currents in sensory neurons. *Nature*, 377: 432–435.

Llewellyn-Smith, I.J. and Burnstock, G. (1998) Ultrastructural localization of $P2X_3$ receptors in rat sensory neurons. *NeuroReport*, 9: 2245–2250.

Loesch, A. and Burnstock, G. (1998) Electron-immunocytochemical localization of the $P2X_1$ receptors in the rat cerebellum. *Cell Tissue Res.*, 294: 253–260.

Londos, C., Cooper, D.M. and Wolff, J. (1980) Subclasses of external adenosine receptors. *Proc. Natl. Acad. Sci. USA*, 77: 2551–2554.

Lu, Z. and Smith, D.O. (1998) Adenosine 5'-triphosphatase increases acetylcholine channel opening frequency in rat skeletal muscle. *J. Physiol.*, 436: 45–56.

Lustig, K.D., Shiau, A.K., Brake, A.J. and Julius, D. (1993) Expression cloning of an ATP receptor from mouse neuroblastoma cells. *Proc. Natl. Acad. Sci. USA*, 90: 5113–5117.

Meyer, M.P., Clarke, J.D.W., Patel, K., Townsend-Nicholson, A. and Burnstock, G. (1999). Selective expression of purinoceptor $cP2Y_1$ suggests a role for nucleotide signalling in development of the chick embryo. *Dev. Dynam.*, 214: 152–158.

Milner, P., Bodin, P., Loesch, A. and Burnstock, G. (1992) Increased shear stress leads to differential release of endothelia and ATP from isolated endothelial cells from 4- and 12-month-old male rabbit aorta. *J. Vacs. Res.*, 29: 420–425.

Miras-Portugal, M.T., Castro, E., Mateo, J. and Pintor, J. (1996) The diadenosine polyphosphate receptors: P2D purinoceptors. In: D.J. Chadwick and J.S. Goode (Eds.), *P2 Purinoceptors: Localization, Function and Transduction Mechanisms, Ciba Foundation Symposium, Vol. 198*. John Wiley and Sons, Chichester, pp. 35–47.

Morrison, J.F.B., Namasivayam, S. and Eardley I. (1998a) ATP may be a natural modulator of the sensitivity of bladder mechanoreceptors during slow distention. In: *1st International Consultation on Incontinence, Monaco, 28 June–1 July 1998*. World Health Organisation, p. 84 (Abstract).

Morrison, M.S., Turin, L., King, B.F., Burnstock, G. and Arnett, T.R. (1998b) ATP is a potent stimulator of the activation and formation of rodent osteoclasts. *J. Physiol.*, 511: 495–500.

Neary, J., Rathbone, M., Cattabeni, F., Abbracchio, M. and Burnstock, G. (1996) Trophic actions of extracellular nucleotides and nucleosides on glial and neuronal cells. *Trends Neurosci.*, 19: 13–18.

Nicke, A., Baumert, H.G., Rettinger, J., Eichele, A., Lambrecht, G., Mutschler, E. and Schmalzing, G. (1998) $P2X_1$ and $P2X_3$ receptors form stable trimers: A novel structural motif of ligand-gated ion channels. *EMBO J.*, 17: 3016–3028.

Nori, S., Fumagalli, L., Bo, X., Bogdanov, Y. and Burnstock, G. (1998) Coexpression of mRNAs for $P2X_1$, $P2X_2$ and $P2X_4$ receptors in rat vascular smooth muscle: An in situ hybridization and RT–PCR study. *J. Vacs. Res.*, 35: 179–185.

O'Connor, S.E., Dainty, I.A. and Leff, P. (1991) Further subclassification of ATP receptors based on agonist studies. *Trends Pharmacol. Sci.*, 12: 137–141.

Palea, S., Artibani, W., Ostardo, E., Trist, D.G. and Pietra, C. (1993) Evidence for purinergic neurotransmission in human urinary bladder affected by interstitial cystitis. *J. Urol.*, 150: 2007–2012.

Rassendren, F., Buell, G., Newbolt, A., North, R.A. and Surprenant, A. (1997) Identification of amino acid residues contributing to the pore of a P2X receptor. *EMBO J.*, 16: 3446–3454.

Rubino, A. and Burnstock, G. (1996) Capsaicin-sensitive sensory-motor neurotransmission in the peripheral control of cardiovascular function. *Cardiovasc. Res.*, 31: 467–479.

Silinsky, E.M. (1984) On the mechanism by which adenosine receptor activation inhibits the release of acetylcholine from motor nerve endings. *J. Physiol.*, 346: 243–256.

Silinsky, E.M., Gerzanich, V. and Vanner, S.M. (1992) ATP mediates excitatory synaptic transmission in mammalian neurones. *Br. J. Pharmacol.*, 106: 762–763.

Smith, D.J. and Chapple, C.R. (1994) In vitro response of human bladder smooth muscle in unstable obstructed male bladders: a study of pathophysiological causes. *Neurourol. Urodyn.*, 13: 414–415.

Sneddon, P. and Westfall, D.P. (1984) Pharmacological evidence that adenosine triphosphate and noradrenaline are co-transmitters in the guinea-pig vas deferens. *J. Physiol. (Lond.)*, 347: 561–580.

Theobald, R.J., Jr. and de Groat, W.D. (1989) The effects of purine nucleotides on transmission in vesical parasympathetic ganglia of the cat. *J. Autonom. Pharmacol.*, 9: 167–182.

Valera, S., Hussy, N., Evans, R.J., Adami, N., North, R.A., Surprenant, A. and Buell, G. (1994) A new class of ligand-gated ion channel defined by P_{2X} receptor for extracellular ATP. *Nature*, 371: 516–519.

Van Calker, D., Müller, M. and Hamprecht, B. (1979) Adenosine regulates via two different types of receptors, the accumulation of cyclic AMP in cultured brain cells. *J. Neurochem.*, 33: 999–1005.

Vidal, M., Hicks, P.E. and Langer, S.Z. (1986) Differential effects of α,β-methylene ATP on responses to nerve stimulation in SHR and WKY tail arteries. *Naunyn-Schmiedeberg's Arch. Pharmacol.*, 322: 384–390.

Vizi, E.S., Liang, S.-D., Sperlágh, B., Kittel, A. and Juranyi, Z. (1997) Studies on the release and extracellular metabolism of endogenous ATP in rat superior cervical ganglion: Support for neurotransmitter role of ATP. *Neuroscience*, 79: 893–903.

Vulchanova, L., Arvidsson, U., Riedl, M., Wang, J., Buell, G., Surprenant, A., North, R.A. and Elde, R. (1996) Differential distribution of two ATP-gated channels (P2X receptors) determined by imunocytochemistry. *Proc. Natl. Acad. Sci. USA*, 93: 8063–8067.

Wammack, R., Weihe, E., Dienes, H.-P. and Hohenfeller, R. (1995) Die neurogene Blase in vitro. *Akt. Urol.*, 26: 16–18.

Watt, W.C., Lazarowski, E.R. and Boucher, R.C. (1998) Cystic fibrosis transmembrane regulator-independent release of ATP. Its implications for the regulation of $P2Y_2$ receptors in airway epithelia. *J. Biol. Chem.*, 273: 14053–14058.

Webb, T.E., Simon, J., Krishek, B.J., Bateson, A.N., Smart, T.G., King, B.F., Burnstock, G. and Barnard, E.A. (1993) Cloning and functional expression of a brain G-protein coupled ATP receptor. *FEBS Lett.*, 324: 219–225.

Wells, D.G., Zawisa, M.J. and Hume, R.I. (1995) Changes in responsiveness to extracellular ATP in chick skeletal muscle during development and upon denervation. *Dev. Biol.*, 172: 585–590.

White, P.N., Thorne, P.R., Housley, G.D., Mockett, B.E., Billett, T.E. and Burnstock, G. (1995) Quinacrine staining of marginal cells in the stria vascularis of the guinea-pig cochlea: a possible source of extracellular ATP? *Hear. Res.*, 90: 97–105.

Yokomizo, T., Izumi, T., Chang, K., Takuwa, Y. and Shimizu, T. (1996) a G-protein-coupled receptor for leukotriene B_4 that mediates chemotaxis. *Nature*, 387: 337–352.

Zhong, Y., Dunn, P.M., Xiang, Z., Bo, X. and Burnstock, G. (1998) Pharmacological and molecular characterisation of P2X purinoceptors in rat pelvic ganglion neurons. *Br. J. Pharmacol.*, 125: 771–781.

P. Illes and H. Zimmermann (Eds.)
Progress in Brain Research, Vol 120

CHAPTER 2

Modulation of purinergic neurotransmission

P. Sneddon[1,*], T.D. Westfall[1], L.D. Todorov[2], S. Mihaylova-Todorova[2], D.P. Westfall[2] and C. Kennedy[1]

[1]*Department of Physiology and Pharmacology, University of Strathclyde, 27 Taylor Street, Glasgow G4 DNR, UK*
[2]*Department of Pharmacology, University of Nevada School of Medicine, Reno, NV 89557, USA*

Introduction

In the past 25 years a wide variety of experimental evidence has been put forward to support the purinergic nerve hypothesis. However, the major conceptual advance has been the realization that in most instances adenosine 5'-triphosphate (ATP) acts as a cotransmitter with substances such as noradrenaline (NA), acetylcholine (ACh), various peptides and probably nitric oxide. Indeed, this concept of multiple neurotransmitters should probably now be regarded as a general rule. ATP acts as a cotransmitter with NA and neuropeptide Y in sympathetic nerves innervating smooth muscle in tissues such as the vas deferens and arteries. In the parasympathetic nervous system ATP acts with ACh, notably in the urinary bladder. In the enteric nervous system ATP acts with various neurotransmitters, including VIP, neuropeptide Y and nitric oxide, to influence gut motility and exocrine secretions. More recently, evidence for purinergic neurotransmission in the central nervous system (Edwards et al., 1992) has attracted considerable interest, although the physiological role for ATP in this instance has yet to be established. Purinergic signaling has also been identified in various components of the sensory nervous system where its involvement in autonomic sensory-motor reflexes, sensory afferent transmission, and its

actions in dorsal root and dorsal horn, indicate potentially important roles for ATP in the perception of pain, touch, temperature etc. (see Kennedy and Leff (1995) and Thorne and Housley (1996) for details).

There are several recent, excellent and comprehensive reviews of the actions of ATP as a neurotransmitter, in particular see Burnstock (1996). The purpose of this article is to highlight some recent developments in the understanding of ATP as a cotransmitter and which point to new areas for future research. Four main topics of purinergic research are emphasized; the storage and release of ATP and its regulation, the structure and classification of P2-receptor subtypes, the postjunctional effector mechanisms by which ATP mediates its neurotransmitter actions and the mechanism of inactivation of the neurotransmitter actions of ATP by ATPascs.

Prejunctional aspects of purinergic neurotransmission

Storage of ATP

ATP is present in various amounts in all types of synaptic vesicles, even where it is not thought to act as a neurotransmitter, such as in somatic motor nerves in skeletal muscle. This may indicate that ATP in vesicles has an alternative physiological function. ATP is continuously produced in mitochondria, mainly via oxidative phosphorylation

*Corresponding author. Tel.: +44 0141 548 2674; fax: +44 0141 552 2562; e-mail: p.sneddon@strath.ac.uk

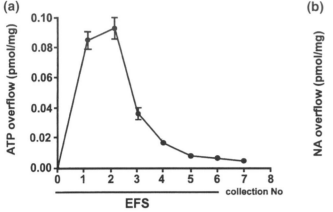 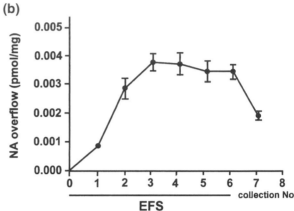

Fig. 1. The timecourse of overflow of ATP (a) and NA (b) during electrical field stimulation (EFS) for 1 min at 8 Hz in the guinea-pig isolated vas deferens. The numbers on the abscissa indicate samples of superfusate collected at 10 s intervals before (0) during (1–6) and after the stimulation period. Note the different scales for the ordinates of the two graphs. (Data from Todorov et al., 1996.)

utilizing glucose. In the cytoplasm of nerve terminals, the concentration of ATP is estimated to be about 10 mM, but a specific transporter in the vesicle membrane is thought to be needed to facilitate its accumulation into the neurotransmitter vesicles (see Sperlagh and Vizi (1996) for further details).

Biochemical studies in the 1960s and 1970s estimated the concentration of ATP in NA-containing vesicles to be 1–200 μM. Therefore, ATP is present at much lower concentrations than NA, which outnumbers it by a molecular ratio of between 2:1 and 50:1. However, the recent release studies described below support the view that synaptic vesicles in sympathetic nerves are not homogeneous and imply that another population of vesicles may exist which are rich in ATP.

Release of ATP

Many studies have measured the release of ATP as a neurotransmitter from a wide variety of preparations, such as the sympathetic nerves of vas deferens and arteries, cholinergic nerves of guinea-pig ileum, gallbladder, myenteric plexus and taenia coli, cultured sympathetic neurons and mammalian brain slices of hippocampus or habenula (see Sperlagh and Vizi (1996) for review). Endogenous ATP overflow is now usually measured by HPLC combined with UV detection or by the luciferin-luciferase assay and the amount of stimula-

tion-evoked ATP overflow varies greatly between tissues, from less than 1, to over 10,000 pmol/gram tissue/min, depending on the parameters of stimulation. However, these estimates of ATP overflow should be treated with caution, since they do not take into account other possible sources of ATP release, such as endothelial cells, glial cells or smooth muscle cells. Nor do they take into account the degradation of ATP by ATPases. Obviously, in tissues where there is high ecto-ATPase activity, the amount of ATP release will be considerably underestimated. This would appear to be the case in guinea-pig vas deferens, as will be described later.

Westfall and co-workers have refined a high performance liquid chromatography based method of endogenous purine detection which involves converting purines in the superfusate to their etheno (ε)-derivatives, producing a more sensitive measurement of ε-ATP, ε-adenosine 5′-diphosphate (ε-ADP), ε-adenosine 5′-monophosphate (ε-AMP) and ε-adenosine using fluorescence detection. Using this method they have provided strong evidence that all the purine overflow they detect comes from nerves and not from other sources such as smooth muscle (see Todorov et al., 1996). They have combined this with simultaneous electrochemical detection of NA. Figure 1 shows the time-course of overflow of ATP and NA during stimulation at 8 Hz for 1 min in the guinea-pig isolated vas deferens. ATP overflow peaks at about 20 s then subsides to low levels. NA overflow is initially low, climbing gradually to a steady level after about 30 s, and is well maintained throughout the 60 s stimulation period. At its peak level, the amount of ATP overflow is much greater than that of NA. Thus, the time-courses of overflow of ATP and NA are completely different and the overflow ratio of the two transmitters changes during the stimulation period. This is consistent with the suggestion that ATP and NA are released from separate populations of synaptic vesicles. It is interesting to note that whilst ATP overflow appears to be transient, the purinergic electrical response of the tissue to nerve stimulation, the excitatory junction potentials (EJPs) are well maintained during a train of stimulation.

Prejunctional modulation of cotransmitter release

The release of ATP from sympathetic nerves has been the most extensively studied model of pre-junctional receptor-mediated modulation of purinergic neurotransmission. Prejunctional α_2-adrenoceptors are the earliest and clearest example of negative feedback by a neurotransmitter. In the guinea-pig isolated vas deferens, α_2-adrenoceptor antagonists produced a substantial increase in the neurogenic overflow of NA, but only a relatively small increase in ATP overflow. Conversely, stimulation of prejunctional A_1-receptors by exogenously applied agonists, increases neurogenic overflow of ATP much more than that of NA (Todorov et al., 1996; Von Kugelgen, 1996 and references therein). This differential modulation of the overflow of ATP and NA is consistent with the suggestion that they may be stored in distinct populations of vesicles. Whilst it is clear that exogenously-applied adenosine inhibits transmitter release via prejunctional A_1-receptors, it is not clear whether endogenous adenosine influences neuro-transmission. There is also the possibility that ATP *per se* can also modulate transmitter release via prejunctional P2-receptors (see Von Kugelgen, 1996) or perhaps via a putative novel P3- receptor (Shinozuka et al., 1988).

Inhibition of ATP release from sympathetic nerves is also produced by agonists acting via β_2-adrenoceptors, prostaglandin E_2 receptors, neuropeptide Y receptors and arterial natriuretic peptide receptors. Conversely, ATP release is enhanced by agonists acting at angiotensin receptors and endothelin ET-3 receptors. Functional studies on neurogenic contractions also indicate that dopamine D2, muscarinic M1 and μ, δ and κ-opioid receptors inhibit release of ATP. The physiological relevance, if any, of these receptors is not known.

Postjunctional ATP receptors

Postjunctional receptors for ATP are designated P2-receptors, to distinguish them from adenosine receptors, classified as P1-receptors. The 1994 provisional subclassification by IUPHAR into P_{2x}, P_{2y}, P_{2u}, P_{2t}, P_{2z} and P_{2d} has now been superseded, and P2-receptor classification is now essentially based on results of molecular biology studies determining the structure of the receptors. The current classification (Abbracchio and Burnstock, 1994; summarized in Fredholm et al., 1997) subdivides P2-receptors into P2X-receptors, which form intrinsic cationic channels, and P2Y-receptors, which are G-protein coupled. The P2X-receptors are in turn divided into seven subclasses, $P2X_1$ to $P2X_7$. (The $P2X_7$ receptor was formerly known as the P2Z-receptor.) There are also numerous subclasses of cloned P2Y-receptors, up to eleven subtypes (designated $P2Y_1$ to $P2Y_{11}$) have been proposed, depending on the criteria used.

The $P2X_1$ to $P2X_6$ receptors have 379–472 amino acids, with a predicted tertiary structure of two transmembrane segments, a large extracellular loop and intracellular C- and N-termini (see Fig. 2). The $P2X_7$ receptor (595 amino acids) has a similar structure, but with a much larger intracellular C-terminus. This is strikingly different from that of any of the other known ligand-gated ionotropic receptors, such as the nicotinic, glutamate, glycine, $GABA_A$ or $5HT_3$ receptors. Each P2X-receptor subtype can form functional homomultimeric ion channels when expressed in mammalian cell lines or *Xenopus* oocytes. The subunit composition of most native receptors is still to be determined, but initial observations suggest that heteromultimers are needed to mimic the functional characteristics of some native P2X-receptors (see Kennedy and Leff (1995) and North (1996) for details).

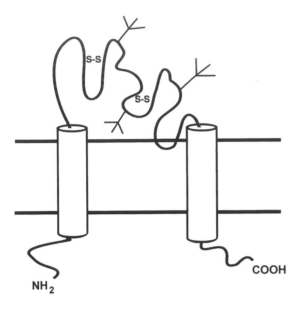

Fig. 2. General schematic representation of the predicted tertiary structure of P2 receptors of the subtype P2X$_1$ to P2X$_6$. The proposed receptor has two transmembrane spanning regions, a large extracellular loop with at least three *N*-glycosylation sites, three disulphide bridges and intracellular C and N termini. (Modified from Brake et al., 1994).

Northern blotting and in-situ hybridization techniques have been used to map the distribution of each P2X mRNA. In relation to purinergic neurotransmission, the most significant findings are that the P2X$_1$ mRNA is present predominately in smooth muscle, P2X$_2$ mRNA is widespread in the CNS, P2X$_3$ mRNA is located in sensory neurons of the trigeminal, nodose and dorsal root ganglia, and P2X$_5$ mRNA has been detected in ganglia, brain and heart. The P2X$_4$ and P2X$_6$ mRNAs have been colocalized in the brain.

In contrast to the enormous success of molecular biology in classifying P2-receptors, the ability of pharmacological agents to identify the functional aspects of P2-receptors has been relatively poor. For example, the agents available to block P2X-receptors, such as PPADS, suramin and pyridoxyl-5′-phosphate, generally have poor selectivity and low or moderate potency. The situation for P2Y-receptors is even less satisfactory, although recently a group of selective P2Y1-antagonists has been identified (Boyer et al., 1996). A new generation of potent, selective P2-receptor antagonists is urgently required in order that the physiological relevance of the various P2-receptor subclasses can be determined, and the potential therapeutic benefits explored.

Postjunctional purinergic effector mechanisms

When ATP acts as a cotransmitter, its actions are generally synergistic with the other major cotransmitter(s) utilized by the nerve. The physiological benefits of having two or more substances, each exciting or inhibiting the effector tissue, are probably related to utilizing different effector mechanisms with different time-courses of action. In general, ATP produces the fast, transient response to nerve stimulation, whereas the other cotransmitter, NA, ACh or nitric oxide produces a slower, but better maintained change in mechanical activity. This is well illustrated by considering the actions of ATP on vascular smooth muscle, as shown in Fig. 3a,b and c. The rapid, transient time-course of the functional response to ATP is best understood using electrophysiological analysis. Figure 3a shows the effect of exogenous ATP application to a single, dissociated smooth muscle cells of the rat tail artery (Evans and Kennedy, 1994). ATP (1 μM) initiated a large inward current in less than 3 ms, which reached a peak and began to decline within 100 ms, even in the continued presence of ATP. This suggests that in the intact tissue, rapid desensitization of the P2X-receptor could be an important mechanism in terminating the actions of ATP during a burst of nerve activity. The inward current is abolished by the P2-receptor antagonist suramin (Evans and Kennedy, 1994; McLaren et al., 1995a).

Intracellular microelectrode recording from the intact rat tail artery (Fig. 3b) shows that each sympathetic nerve stimulus produced a rapidly rising and decaying, transient depolarization, or EJP. The EJPs are also abolished by suramin (McLaren et al, 1995b). A train of stimuli to the sympathetic nerves also produces a slow depolarization, which is mediated by NA and abolished by a-adrenoceptor antagonists such as phentolamine.

The vasoconstrictor response of arteries to sympathetic nerve stimulation varies considerably (see Sneddon (1995) and references therein). For example, in rat tail artery ATP contributes less than

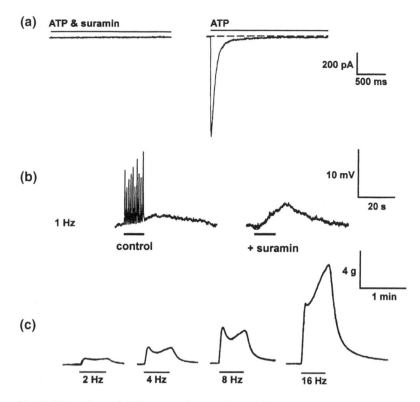

Fig. 3. The actions of ATP on vascular smooth muscle. The upper panel (a) shows a whole-cell patch-clamp recording from a single smooth muscle cell dissociated from the rat tail artery. Application of ATP (1 μM) for 2 s induced a transient inward current (right) which was abolished in the presence of suramin, 100 μM (left) (Evans and Kennedy, 1994). The centre panel (b) shows an intracellular microelectrode recording of the electrical response of the rat isolated tail artery to stimulation of the sympathetic nerve at 1 Hz for 10 s. Each stimulus produced a rapid EJP (left) which was abolished by suramin, 100 μM (right). The slow depolarization was abolished by α-adrenoceptor antagonists (not shown) (McLaren et al., 1995b). The lower panel (c) shows the mechanical response of the dog isolated mesenteric artery to sympathetic nerve stimulation at 2–16 Hz. Subsequent pharmacological investigation (not shown) indicated that ATP contributed primarily to the initial phasic component of the contraction, whilst the sary, tonic component was mediated by NA (Machlay et al., 1988).

10% to the sympathetic vasoconstriction, whereas in rabbit saphenous artery the response is almost entirely purinergic. Most arteries lie somewhere between these to extremes. Figure 3c shows that in dog mesenteric artery, sympathetic nerve stimulation at 2–16 Hz produced biphasic responses. The initial, transient contraction is predominantly purinergic, whilst the secondary component is predominantly mediated by NA (Machlay et al., 1988). A similar pattern of cotransmission is repeated in many other sympathetically innervated smooth muscles (for further examples see Sneddon, 1995). The parasympathetic innervation of the urinary bladder of various species provides a good example of the cotransmitter actions of ATP and

ACh. Again, ATP activates P2X-receptors, producing rapid EJPs and action potentials, which underlie the phasic contraction. ACh acts on muscarinic receptors to produce a slow depolarization and more maintained contraction, which are abolished by atropine (Brading and Mostwin, 1989).

In the enteric nervous system ATP acts on P2Y-receptors on the smooth muscle to produce inhibitory junction potentials (IJPs). These potentials have a longer latency of onset (about 100 ms) and slower rise time (about 500 ms) than the EJPs mediated by P2X-receptors. This is consistent with the view that the P2Y-receptors are G-protein coupled. It is often assumed that some enteric nerves co-release ATP with VIP and/or nitric oxide, however, there is no direct evidence that ATP and the other agents come from the same nerve. An interesting, recent example was found in the guinea-pig internal anal sphincter. In this tissue the IJP is biphasic, with the initial component inhibited by suramin (indicating involvement of P2Y-receptors), whereas the slower, secondary component is inhibited by the nitric oxide synthase inhibitors such as L-NAME, indicating involvement of nitric oxide (Rae and Muir, 1996). Different regions of the gastrointestinal tract utilize these and other cotransmitters to differing degrees (for details see Hoyle and Burnstock, 1989; Keef et al., 1994).

Inactivation of ATP as a neurotransmitter

The postjunctional actions of ATP are curtailed by its sequential dephosphorylation to ADP, AMP and adenosine. The first step, the removal of the terminal phosphate of ATP, to produce ADP, is dependent upon the action of an ATPase or apyrase (see Ziganshin et al., 1994; Plesner, 1995; Zimmerman, 1996), but it is not clear if this is mediated only by ecto-enzymes attached to the outer membrane surface of the effector cell (i.e. an ecto-ATPase or ecto-apyrase) or if there are multiple forms and sources of the ATPase and apyrase, such as that proposed to be released from sympathetic nerves (see Todorov et al., 1997). Recently several cDNAs for ecto-ATPase and ecto-apyrase were cloned. (Maliszewski et al., 1994; Wang and Guidotti, 1996; Kegel et al., 1997; Kirley, 1997). The deduced amino acid sequence indicates a 66-kDa monomer with a predicted structure of two transmembrane-spanning regions, and a single ATPase site on a large extracellular region of a protein which has four putative N-glycosylation sites (Fig. 4).

An important advance in understanding the functional significance of ATPase activity was the development of ARL 67156 (6-N,N-diethyl-D-β,γ-dibromomethyleneATP) as an inhibitor of ecto-ATPase activity in human blood, with a pIC_{50} of 4.62 (Crack et al., 1995). A similar value, 5.1, was found in smooth muscle membranes of the rat vas deferens (Khakh et al., 1995). In functional studies, μM concentrations of ARL 67156 poten-

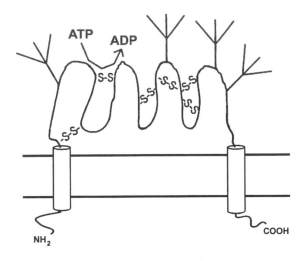

Fig. 4. The deduced tertiary structure of smooth muscle ecto-ATPase (based on Kirley, 1997). Like the P2X receptor shown in Fig. 2, the ecto-ATPase has intracellular C and N-termini, two transmembrane-spanning regions and a large extracellular loop. The putative positions of six disulphide bridges, four N-glycosylation sites and a single site for ATP hydrolysis are also indicated on the extracellular loop.

tiated contractions of the rabbit central ear artery to exogenous ATP, but not those evoked by the stable analogue α,β-methyleneATP (Crack et al., 1995).

There is now convincing evidence that ATPases modulate the purinergic component of neurogenic responses in some tissues. We have recently published the first study on the effects of ARL 67156 on ATP released as a neurotransmitter (Westfall et al., 1996). In the guinea-pig isolated vas deferens, sympathetic nerve stimulation evokes an initial phasic contraction which is predominantly purinergic. ARL 67156 rapidly and reversibly almost doubled this response, both in the absence or presence of the α_1-adrenoceptor antagonist prazosin. In the presence of ARL 67156 (100 μM) the peak response to exogenous ATP was increased by over 60%, whilst that to α,β-methyleneATP was unchanged.

We have also measured the overflow of endogenous ATP, ADP, AMP and adenosine in samples of superfusate collected every 10 s during a 60 s train of nerve stimulation at 8 Hz in the guinea-pig isolated vas deferens using combined HPLC/fluorescence detection (see Todorov et al. (1996) for methods). Figure 5 shows that under control conditions only small amounts of ATP were

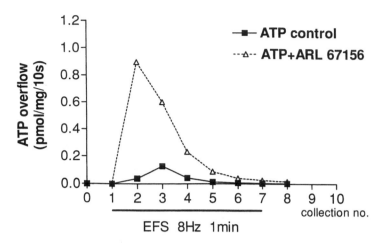

Fig. 5. The overflow of ATP from guinea-pig isolated vas deferens during electrical field stimulation of the sympathetic nerves (EFS) at 8 Hz for 1 min. Under control conditions (■) only small amounts of ATP were recovered in the superfusate, since it is rapidly broken down to adenosine. ARL 67156 (100 μM) enhanced ATP overflow by up to 450% (Δ). The numbers on the abscissa indicate the samples of superfusate collected at 10 s intervals before (0) during (1–6) and after (7–8) nerve stimulation.

detected in the superfusate during electrical field stimulation of the sympathetic nerves, but large amounts of adenosine were collected. However, in the presence of 100 μM ARL 67156, the amount of ATP increased dramatically, by about 450%. This indicates that under control conditions, most of the neuronally released ATP is very rapidly converted to adenosine and that ARL 67156 inhibits this breakdown.

An exciting new development in this area is the proposal that in addition to the established ecto-ATPase on the smooth muscle membrane, stimulation of the sympathetic nerves releases a soluble ATPase into the extracellular space (Todorov et al., 1997). When the guinea-pig vas isolated deferens was superfused with a solution containing exogenous ε-ATP (100 nM), almost no degradation was detected (see Fig. 6). However, if the sympathetic nerves were stimulated concomitantly (8 Hz for 60 s), then almost all the exogenous ε-ATP was broken down to ε-ADP, ε-AMP and ε-adenosine, suggesting that some aspect of neurotransmission enhances ε-ATP hydrolysis. This was inhibited by adding Cd^{++} to, or omitting Ca^{++} from, the superfusate during nerve stimulation, consistent with the vesicular release of a factor that promotes ε-ATP breakdown.

Subsequent experiments examined whether nerve stimulation released stable enzymes into the superfusate which could be collected and subsequently assayed for nucleotidase activity in the absence of the tissue. Superfusate collected from vasa deferentia before nerve stimulation did not promote significant degradation of exogenous ε-ATP (100 nM). However, samples of superfusate taken during nerve stimulation had substantial nucleotidase activity, and promoted the breakdown of about 90% of the exogenous ε-ATP to ε-ADP, ε-AMP and ε-adenosine. Exogenous noradrenaline, phenylephrine, methoxamine, ATP or α,β-methyleneATP, each caused large contractions of the vas deferens, but did not release nucleotidase activity into the superfusate.

The nucleotidase activity in superfusate samples was abolished by incubation at 80°C or pH 4.0 and greatly reduced by incubation at 4°C. The ATPase activity in superfusate samples was not reduced by known inhibitors of intracellular ATPases such as ouabain, oligomycin, or vanadate, suggesting that "leakage" of these enzymes into the superfusate during stimulation does not contribute to the ATPase activity in the superfusate. ARL 67156 (100 μM) reduced by about 50% the ATPase activity in the superfusate samples from stimulated tissues.

The nucleotidases released by nerve stimulation promoted hydrolysis of ATP to ADP, AMP and adenosine and it is yet to be determined whether

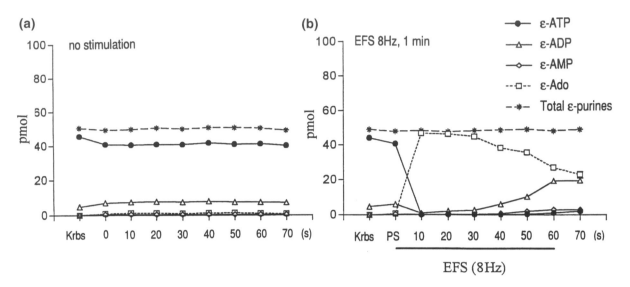

Fig. 6. The breakdown of exogenously added etheno-ATP (ε-ATP) to ε-ADP, ε-AMP and ε-adenosine in the guinea-pig isolated vas deferens. Under control conditions (a) there is little breakdown of ε-ATP. However, during electrical field stimulation of the sympathetic nerves (EFS) at 8 Hz for 1min (b) ε-ATP is rapidly converted to ε-adenosine. The samples indicated on the abscissa are as follows: Krebs, levels of purines in superfusing Krebs solution. PS, pre-stimulation levels. 10–60, samples taken 10–60 s after the beginning of EFS. 70, a sample taken 10 s after cessation of stimulation (Todorov et al., 1997).

this is due to the action of a single enzyme or by several enzymes with different specificities. The physiological significance of the releasable nucleotidase enzyme(s) also remains to be established and there are a number of questions which such a mechanism raises. How is the ATPase enzyme stored and released? Does it come only from the nerve, or also from other sites? If it is stored in neurotransmitter vesicles in the nerve terminal, how could ATP be stored in the same site? Are ATPases released from other nerve types, or just from sympathetic nerves where ATP is a cotransmitter? What is the relative importance of the releasable nucleotidase and the membrane bound ecto-ATPases located on the smooth muscle surface in inactivation of ATP during purinergic neurotransmission? If potent, selective inhibitors of the releasable ATPase can be identified then the physiological significance of these enzymes in modulating purinergic neurotransmission could be determined.

Summary and conclusions

During the past 25 years ATP has become accepted as an important neurotransmitter at a wide variety of neuroeffector junctions, usually acting as a cotransmitter with NA, ACh, nitric oxide or a neuropeptide such as NPY or VIP. The details of the storage and release of ATP with its cotransmitters has yet to be resolved. However, recent studies indicate that there is more than one population of storage vesicles in the nerves, since the release of the various cotransmitters varies over time and can be differentially modulated by drugs. The subclassification of P2 receptors has advanced dramatically in the past few years due to the use of molecular biology methods allowing the cloning and expression of 14 different subclasses of P2 receptors, seven P2X and seven P2Y. Determination of the functional significance of the various receptor subtypes would be helped by the development of selective agonists and antagonists. The neurotransmitter action of ATP at visceral and vascular smooth muscle P2X receptors has been elucidated in considerable detail. ATP induces a transient inward current via ligand-gated channels, which produces EJPs, action potentials and a phasic contraction of the effector tissue. ATP's neurotransmitter actions appear to be curtailed by the action of ATPases. It has been assumed that this ATPase activity is due to membrane bound ecto-

ATPases on the surface of the effector tissue, however, the recently identified soluble ATPase released during nerve stimulation could also be involved in inactivation of ATP. The relative importance of ecto-ATPase and the releasable ATPase is yet to be determined.

Acknowledgments

We thank Astra plc and the Wellcome Trust (CK, PS, TDW), the Caledonian Research Foundation (CK) and the Carnegie Trust (PS), the NIH and American Heart Foundation (DPW, LDT), for financial support. We are also grateful to the Scottish Hospital Endowments Research Trust for refurbishing the laboratories in which some of these experiments were performed.

References

Abbracchio, M.P. and Burnstock, G. (1994) Purinoceptors: Are there families of P2X and P2Y purinoceptors? *Pharmac. Ther.*, 64: 445–475.

Brading, A.F. and Mostwin, J.C. (1989) Electrical and mechanical responses of guinea-pig bladder muscle to nerve stimulation. *Br. J. Pharmacol.*, 98: 1083–1090.

Boyer, J.L., Romero-Avila, T., Schachter, J.B. and Harden, T.K. (1996) Identification of competitive antagonists of the P2Y1 receptor. *Mol. Pharmacol.*, 50: 1323–1329.

Brake, A.J., Wagenbach, M.J. and Julien, D. (1994) New structural motif for ligand-gated ion channels defined by an ionotropic ATP receptor. *Nature*, 371: 519–523.

Burnstock, G. (Ed.) (1996) *Seminars in the Neurosciences. Purinergic Neurotransmission*, Vol. 8.

Crack, B.E., Pollard, C.E., Beukers, M.W., Roberts, S.M., Hunt, S.F., Ingall, A.H., McKenchnie, K.C.W., Ijzerman, A.P. and Leff, P. (1995) Pharmacological and biochemical analysis of FPL 67156, a novel, selective inhibitor of ecto-ATPase. *Br. J. Pharmacol.*, 114: 475–481.

Edwards, F.A., Gibb, A.J. and Colquhoun, D. (1992) ATP-mediated synaptic currents in the central nervous system. *Nature*, 359: 144–147.

Evans, R.J. and Kennedy, C. (1994) Characterization of P_2-purinoceptors in the smooth muscle of the rat tail artery: A comparison between contractile and electrophysiological responses. *Br. J. Pharmacol.*, 113: 853–860.

Fredholm, B.B., Abbrachio, M.P., Burnstock G., Dubyak, G.R., Harden, T.K., Jacobson, K.A., Schwabe, U. and Williams, M. (1997) Towards a revised nomenclature for P1 and P2 receptors. *Trend Pharm. Sci.*, 18: 79–82.

Hoyle, C.H.V. and Burnstock, G. (1989) Neuromuscular transmission in the gastrointestinal tract. In: J.D. Wood (Ed.), *Handbook of Physiology*, Section 6: *The Gastrointestinal System*, vol. 1. American Physiological Society, Bethesda. pp. 435–464.

Keef, K.D., Shuttleworth, C.W.R., Xue, C., Bayguinov, O., Publicover, N.G. and Sanders, K.M. (1994) Relationship between nitric oxide and vasoactive intestinal polypeptide in enteric inhibitory neurotransmission. *Neuropharmacology*, 33: 1303–1314.

Kegel, B., Braun, N., Heine, P., Maliszewski, C.R. and Zimmerman, H. (1997) An ecto-ATPase and an ecto-ATP diphosphohydrolase are expressed in rat brain. *Neuropharmacology*, 36: 1189–1200.

Kennedy, C. and Leff, P. (1995) Painful connection for ATP. *Nature*, 377: 385–386.

Khakh, B.S., Michel, A.D. and Humphrey, P.P.A. (1995) Inhibition of ecto-ATPase and Ca-ATPase in rat vas deferens by P_2-purinoceptor antagonists. *Br. J. Pharmacol.*, 115: 2 pp.

Kirley, T.L. (1997) Complementary DNA cloning and sequencing of the chicken muscle ecto-ATPase. *J. Biol. Chem.*, 272: 1076–1081.

Machalay, M., Dalziel, H.H. and Sneddon, P. (1988) Evidence for ATP as a cotransmitter in dog mesenteric artery. *Eur. J. Pharmacol.*, 147: 83–92.

Maliszewski, C.R. et al., (1994) The CD39 lymphoid cell activation antigen. Molecular cloning and structural characterisation. *J. Immunol.*, 153: 3574–3583.

McLaren, G.J., Sneddon, P. and Kennedy, C. (1995a) Comparison of the electrophysiological actions of ATP and UTP in single dissociated cells of the rat tail artery. *Br. J. Pharmacol.*, 116: 55 pp.

McLaren, G.J., Sneddon, P. and Kennedy, C. (1995b) The effects of suramin on purinergic and noradrenergic neurotransmission in the isolated rat tail artery. *Eur. J. Pharmacol* 227: 56–62.

North, R.A. (1996) P2X purinoceptor plethora. *Seminars in the Neurosciences. Purinergic Neurotransmission*, 8: 187–194.

Plesner, L. (1995) Ecto-ATPases: Identities and functions. *Int. Rev. Cytol.*, 158: 141–214.

Rae, M.G. and Muir, J.C. (1996) Neuronal mediators of inhibitory junction potentials and relaxation in the guinea-pig internal and sphincter. *J. Physiol.*, 493: 517–527.

Shinozuka, K., Bjur, R.A. and Westfall, D.P. (1988) Characterization of prejunctional purinoceptor on adrenergic nerves of the rat caudal artery. *Naunyn-Schmiedeberg's Arch. Pharmacol.*, 388: 221–227.

Sneddon, P. (1995) Cotransmission. In: D.A. Powis, S.J. Bunn (Eds.), *Neurotransmitter Release and its Modulation*. Cambridge University Press, Cambridge, pp. 22–35.

Sperlagh, B. and Vizi, E.S. (1996) Neuronal synthesis, storage and release of ATP. *Seminars in the Neurosciences. Purinergic Neurotransmission*, 8: 175–186.

Thorne, P.R. and Housley, G.D. (1996) Purinergic signaling in sensory systems. *Seminars in the Neurosciences. Purinergic Neurotransmission*, 8: 233–246.

Todorov, L.D., Mihaylova-Todorova, S., Raviso, G.L., Bjur, R.A. and Westfall, D.P. (1996) Evidence for the differential release of the cotransmitters ATP and noradrenaline from sympathetic nerves of the guinea-pig vas deferens. *J. Physiol.*, 496.3: 731–748.

Todorov, L.D., Mihaylova-Todorova, S., Westfall, T.D., Sneddon, P., Kennedy, C., Bjur, R.A. and Westfall, D.P. (1997) Neuronal release of soluble nucleotidases and their role in neurotransmitter inactivation. *Nature* 387: 76–79.

Von Kugelgen, I. (1996) Modulation of ATP release through presynaptic receptors. *Seminars in the Neurosciences. Purinergic Neurotransmission*, 8: 247–257.

Wang, T.-F. and Guidotti, G. (1996) CD39 is an ecto-(Ca^{2+}, Mg^{2+})-apyrase. *J. Biol. Chem.*, 271, 9898–9901.

Westfall T.D., Kennedy, C. and Sneddon, P. (1996) Enhancement of sympathetic purinergic neurotransmission in the guinea-pig isolated vas deferens by the novel ecto-ATPase inhibitor ARL 67156. *Br. J. Pharmacol.*, 117: 867–872.

Ziganshin, A.U., Hoyle, C.H.V. and Burnstock, G. (1994) Ecto-enzymes and metabolism of extracellular ATP. *Drug Dev. Res.*, 32: 134–146.

Zimmerman, H. (1996) Biochemistry, localisation and functional roles of ecto-nucleotidases in the nervous system. *Progress in Neurobiol.*, 49: 589–618.

Molecular biology of
P2Y receptors

P. Illes and H. Zimmermann (Eds.)
Progress in Brain Research, Vol 120
© 1999 Elsevier Science BV. All rights reserved

CHAPTER 3

Molecular biology of P2Y receptors expressed in the nervous system

T.E. Webb*[#] and E.A. Barnard[‡]

Molecular Neurobiology Unit, Royal Free Hospital School of Medicine, Rowland Hill Street, Hampstead, London NW3 2PF, UK
Present address: [#] Department of Cell Physiology and Pharmacology, University of Leicester, University Road, P.O. Box 138,
Leicester LE1 9HN, UK. [‡] Department of Pharmacology, University of Cambridge, Tennis Court Road, Cambridge CB2 1QT, UK

Introduction

Although the action of purine compounds on the cardiovascular system was observed in the late 1920s (Drury and Szent-Györgyi, 1929) evidence for the role of such compounds as neurotransmitters was not presented for a further three decades (Holton and Holton, 1953, 1954; Holton, 1959). In 1963, synapses were identified in the autonomic nervous system which were neither adrenergic nor cholinergic in origin (Burnstock et al., 1963) and Burnstock later proposed the concept of purinergic

---

*Corresponding author. Tel.: 44(0)116 252 2929; fax: 44(0)116 252 5045; email: tew2@le.ac.uk

[1]The IUPHAR nomenclature committee has recommended that the notation P2Y be used to denote recombinant sequences that encode G protein coupled receptors for extracellular nucleotides which have been shown in some way to be functional receptors (Vanhoutte et al., 1996). Where this demonstration has not yet been made, lowercase notation is therefore used here (e.g. p2y$_5$). Non-mammalian sequences which can be shown to express functional receptors can, in circumstances where a mammalian homologue has not yet been demonstrated, be used in the series in uppercase, where this does not compromise an existing mammalian nomenclature, and we follow this ruling here. However, in some such cases where the original authors have not assigned a subtype number and have used lower case we continue their notation.

nerves that utilise ATP or its derivatives as transmitters or cotransmitters at specific cell surface receptors activated by either adenosine or nucleotides: the P$_1$ and P$_2$ receptors respectively (Burnstock, 1972, 1976, 1978). The receptors for extracellular nucleotides can be divided into two families on the basis of the mechanism of signal transduction after nucleotide binding to the receptor: the P2X receptor family in which the nucleotide gates an intrinsic ion channel, and the P2Y receptor family where the effect of nucleotide binding is mediated through a heterotrimeric G protein and is linked to the modulation of intracellular second messengers. In this chapter we will review the current status of the P2Y receptor family in terms of the recombinant receptor sequences that have been isolated, their distribution on cell types in the nervous system and provide examples of some of the responses that have been characterised due to their activation in these locations.

Molecularly defined P2Y receptor subtypes

Since the cloning of the first two members of the P2Y receptor family—the P2Y$_1$ and P2Y$_2$ receptors (Lustig et al., 1993; Webb et al., 1993)—a number of related sequences have been isolated from species ranging from *Xenopus laevis* to *Homo sapiens*. On the basis of sequence analysis alone 12 subtypes of P2Y receptor have been proposed across these species boundaries.[1] However, as noted below some may yet prove to be species orthologues while others have yet to be unambiguously characterised as functional P2Y receptors. Those predicted members that await functional confirmation of their assignment are p2y$_5$ (Webb et al., 1996c; Li et al., 1997), p2y$_5$-like (Janssens et al., 1997) and the sequences deposited in the database described as p2y$_9$ (identical to p2y$_5$-like) and p2y$_{10}$. The receptor sequence originally known as P2Y$_7$ (Akbar et al., 1996) has since been shown to encode a leukotriene B$_4$ receptor (Yokomizo et al., 1997). However, the P2Y$_7$ polypeptide sequence contains a nucleotide binding motif and binds nucleotides (Akbar et al., 1996); when

expressed by transfection it gave a response to ATP which could have been due to an ATP-mediated release of leukotriene.

Thus, at the time of writing the P2Y receptor family is composed of a maximum of eight functionally defined receptor subtypes. The sequence relatedness that these receptor polypeptides share is illustrated in Fig. 1.

A general description of the pharmacological properties of the P2Y receptor family members can be found in North and Barnard (1997). Based upon that information, and up-dated by recent reports, this family of receptors can be divided into four subtypes: the adenine nucleotide specific receptors,

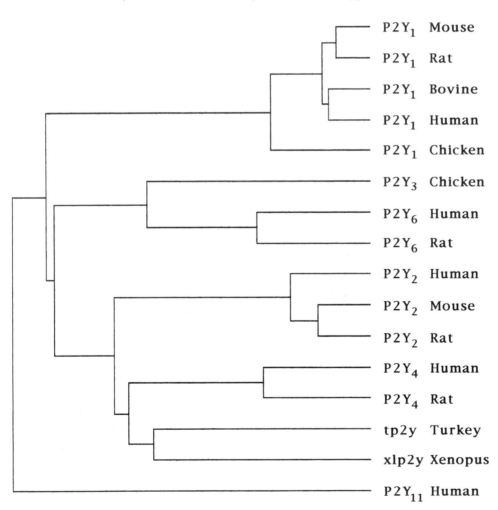

Fig. 1. Dendogram representing the primary sequence relatedness within the P2Y receptor family. The plots were obtained from the multiple sequence alignment program CLUSTAL (Higgins and Sharp, 1988).

P2Y$_1$ and P2Y$_{11}$ (Webb et al., 1993; Communi et al., 1997; the uridine nucleotide preferring receptors, P2Y$_3$ and P2Y$_6$ (Chang et al., 1995; Webb et al., 1996b), those at which both UTP and ATP are highly active, P2Y$_2$ and P2Y$_4$ (Lustig et al., 1993; Communi et al., 1995) and finally those which are activated by all triphosphate nucleotides, xlp2y and tp2y (Bogdanov et al., 1997; Boyer et al., 1997, 1998). However, some questions or controversies or unusual features arise in certain cases within these divisions. We consider here each of these issues in turn.

P2Y$_1$: is this an ADP or ATP receptor?

The P2Y$_1$ receptor was originally observed to be most strongly activated by, among common nucleotides, 2-Methylthio ATP (2MeS-ATP) and ATP (Webb et al., 1993; Simon et al., 1995). However, the commercial samples of ATP then available, although described as of high purity, were found to contain considerable amounts of ADP and the use of freshly purified nucleotides has led to strong evidence that ADP is the most potent natural agonist for the P2Y$_1$ receptor and that ATP is in fact a competitive antagonist (Leon et al., 1997; Hechler et al., 1998b). The previously observed stronger activation by 2MeS-ATP was likewise found to be due to its content of, or breakdown to, 2MeS-ADP. An intrinsic agonist activity of ATP at the P2Y$_1$ receptor has, on the other hand, been maintained by T.K. Harden and co-workers (Schachter et al., 1996; Palmer et al., 1998).

P2Y$_3$ and P2Y$_6$: could these be species orthologues?

The P2Y$_3$ receptor cDNA was cloned from the chick brain (Webb et al., 1996b) and has not as yet been investigated in mammals. It is clearly a functional receptor: its activation produced in oocytes or transfected cell lines channel coupling and Ca^{2+} mobilisation from stores respectively. It is nearest in sequence to the mammalian P2Y$_6$ receptor and since both have the highest agonist activity with UDP, the question arises whether P2Y$_3$ and P2Y$_6$ are species equivalents of the same gene, i.e. orthologue proteins. On present informa-

tion there is no basis for assigning them thus. The percentage amino acid identity between these two is ~ 60%, but this compares poorly with the 85% identity between another chicken/human pair, that of the P2Y$_1$ receptors. As noted above both the P2Y$_3$ and the P2Y$_6$ receptors prefer UDP as an agonist, but other activities do not match well between them. The cloning of a true P2Y$_6$ type from chicken or a true P2Y$_3$ type from a mammal would dispose of this possibility.

Species differences in agonist activity at the P2Y$_4$ receptor

Both the mammalian P2Y$_4$ and P2Y$_2$ receptors can, in fact, act about equally well with both UTP and ATP, when these two receptors from the rat are compared (Webb et al., 1998a). However, the P2Y$_4$ receptor appeared to be in a different sub-set from the P2Y$_2$ receptor when only the human P2Y$_4$ form was known (Nicholas et al., 1996) since there ATP was ~ 50-fold lower in potency than UTP and behaved differently, in that the ATP response (both for Ca^{2+} mobilisation and IP$_3$ increase) occurred only after a delay of ~ 10 min and not immediately as with UTP (Lazarowski et al., 1997). At the rat P2Y$_4$ receptor ATP is essentially equipotent with UTP and the response to its application is not delayed. Small changes in the P2Y$_4$ sequence between rat and human could account for the differences seen (Webb et al., 1998a).

Dual signal transduction mechanisms

Where a second messenger coupling has been identified for recombinant P2Y receptors, this has been the formation of IP$_3$. However, in two recent cases two pathways for one receptor have been reported. For the P2Y$_{11}$ receptor, transfection into astrocytoma cells produced IP$_3$ increase and, when expressed in CHO-K1 cells, dose dependent stimulation of adenylate cyclase was observed (Communi et al., 1997). In the case of the tp2y receptor, activation leads to increase in IP$_3$ and also inhibition of adenylate cyclase (Boyer et al., 1997, 1998). More information on the mechanism of activation of these dual systems is required, in order to relate them to the G proteins proposed to be coupled to other P2Y receptors.

How many more P2Y receptor subtypes remain to be cloned?

There are other subtypes of G protein coupled receptor for extracellular nucleotides which have been deduced to exist in native tissues or cells which are not covered by the above categories. At least two further subtypes of P2Y receptor have yet to be identified at the molecular level. The first of these has long been known as the P2T receptor, most extensively characterised on platelets (Hourani and Hall, 1996). This receptor has recently been further defined in functional terms by the laboratories of C. Gachet and S. Kunapuli (Jin and Kunapuli, 1998; Hechler et al., 1998b). The second is the "P2D" receptor activated by diadenosine polyphosphates (Miras Portugal et al., 1996).

Detection of P2Y receptor subtypes in the nervous system

There has been a distinct lack of progress in the investigation of the location of P2Y receptor proteins in the nervous system due to a lack of receptor specific antibodies. Radiolabelled dATPαS has been used to map the distribution of 2MeS-ATP sensitive, UTP insensitive binding sites in both the chick and rat brain and the use of this radioligand for this purpose has been validated (Simon et al., 1997; Webb et al., 1998b). Due to high nonspecific binding other radiolabelled nucleotides have not proved useful in clarifying the expression and distribution of P2Y receptor proteins in the nervous system.

However, a number of these receptor subtypes have been detected in cells of the nervous system by the application of molecular biological techniques. Indeed, we originally cloned three P2Y receptors subtypes from brain RNA—the avian $P2Y_1$ and $P2Y_3$ receptors and the rat $P2Y_4$ receptor (Webb et al., 1993, 1996b, 1998a) and the $P2Y_2$ receptor was first cloned from the mouse neuroblastoma/rat glioma cell line NG108-15 (Lustig et al., 1993). Northern hybridisation analysis revealed that, of the two avian receptors, the $P2Y_1$ mRNA was extremely abundant in the brain compared to the $P2Y_3$ transcript. An in situ hybridisation study performed on the one-day-old chick brain demon-

strated that the avian $P2Y_1$ receptor transcript has a highly abundant and widespread pattern of expression (Webb et al., 1998b). Furthermore, at cellular resolution a large proportion of the hybridisation signal could be detected over the perikaya, indicating a neuronal localisation for this transcript (Webb et al., 1998b). In comparison, and in agreement with the Northern hybridisation analysis, no significant region-specific hybridisation signal has been detected for the $P2Y_3$ receptor transcript by this technique, suggesting a glial location (T.E. Webb, unpublished observation).

The mammalian $P2Y_1$ receptor mRNA has likewise been detected in brain by Northern hybridisation analysis (Tokuyama et al., 1995; Ayyanathan et al., 1996) and again has a region-specific distribution as determined by in situ hybridisation (T.E. Webb, unpublished). The $P2Y_4$ transcript could not be detected by Northern hybridisation analysis of rat brain RNA, but was detectable by RT–PCR in whole brain RNA and in the RNA of cultured cortical astrocytes (Webb et al., 1998a). The low abundance of this transcript in the adult rat brain was confirmed by in situ hybridisation, with the transcript being detected mainly in non-neuronal cell types. For example, the highest level of expression was detected in the pineal gland, with the ventricular system (including the choroid plexus) moderately labelled. The $P2Y_2$ and $P2Y_6$ receptor transcripts are barely detectable by Northern hybridisation analysis of whole brain RNA, although both mRNAs are expressed at higher levels outside the CNS (Lustig et al., 1993; Parr et al., 1994; Chang et al., 1995). The mammalian P2Y receptor subtype $P2Y_{11}$ does not appear to be expressed in the brain (Communi et al., 1997). The amphibian xlp2y receptor mRNA is expressed during embryonic development and specifically during neuralisation (Bogdanov et al., 1997). The tissue distribution of the tp2y receptor transcript awaits investigation.

Functional evidence for P2Y receptors in the nervous system

There is a wealth of functional evidence for the presence of P2Y receptors on neurons, glial cell types and endothelial cells of the nervous system;

some illustrative examples are noted below and others can be found in Barnard et al. (1997).

There are well-characterised metabotropic responses to nucleotides in established cell lines such as the neuroblastoma–glioma cell hybrid NG108-15, the rat glioma C6 and its sub-clone C6-2B and the pheochromocytoma PC12. Indeed, as mentioned above, the $P2Y_2$ receptor was originally cloned from the NG108-15 cell line (Lustig et al., 1993) and we have detected the expression of a further two P2Y receptor subtypes ($P2Y_1$ and $P2Y_6$) in this cell line by RT-PCR (Fig. 2). A number of metabotropic signal transduction pathways are evoked by the application of nucleotides to these cell lines. For example, for NG108-15 cells, as might be expected, activation of phospholipase C and subsequent mobilisation of intracellular calcium have been well documented (Lin et al., 1993; Lin 1994; Reetz and Reiser, 1994). However, ATP, ATPγS and ADPβS are also able to stimulate adenylate cyclase in these cells (Matsuoka et al., 1995) while UTP application has also been reported to inhibit forskolin-stimulated cAMP accumulation (Song and Chueh, 1996). Both ATP and UTP, presumably acting via the $P2Y_2$ receptor in these cells, are also able to inhibit the M-type potassium currents and high threshold calcium currents (Filippov et al., 1994; Filippov and Brown,

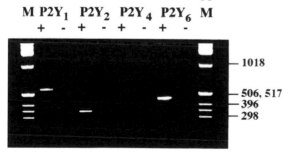

Fig. 2. Expression of P2Y receptor transcripts in the NG108-15 cell line. Agarose gel electrophoresis of RT-PCR products from RNA prepared from differentiated NG108-15 cells. (M), size markers: 1 Kb ladder (Gibco BRL), appropriate sizes are indicated. (+), PCR reaction using a cDNA reaction that included reverse transcriptase. (−), PCR reaction using a cDNA reaction from which reverse transcriptase was excluded. The receptor-specific primer pairs used are indicated above each pair of lanes. Amplification products were cloned and sequenced to confirm their identity. For methological details refer to Harper et al. (1998). The cells were cultured and differentiated as described previously (Filippov et al., 1994).

1996). An adenine nucleotide-selective receptor has been extensively pharmacologically characterised (although its molecular identity remains to be determined) in the C6 line and couples to the inhibition of adenylate cyclase (Boyer et al., 1993; Lin and Chuang, 1993; Boyer et al., 1995). $P2Y_2$ and $P2Y_6$ receptor mediated activation of phospholipase C have also been noted in this cell line (Munshi et al., 1993; Lazarowski and Harden, 1994). In the PC12 cell line the application of nucleotides increases phospholipase C activity and raises the intracellular calcium concentration by acting at $P2Y_1$ and $P2Y_2$ receptors (Murrin and Boarder, 1992; Nikodijevic et al., 1994; Desouza et al., 1995). Furthermore, ATP, at concentrations lower than 100 μM, stimulates adenylate cyclase activity, while higher concentrations are inhibitory (Yakushi et al., 1996; Murayama et al., 1998).

Responses mediated by P2Y receptors have also been detected in neurons from brain slice preparations. For example, in neocortical neurons the application of ATP leads to an increase of intracellular calcium which is not wholly dependent on the presence of extracellular calcium (Lalo et al., 1998).

Intracellular calcium increases have also been noted in cultured astrocytes from the dorsal horn, hippocampus and cerebellum in response to the application of ATP (Bernstein et al., 1998; Idestrup and Salter, 1998; Jimenez et al., 1998). However, it appears that astrocytes from different locations express distinct receptor types; while ATP and UTP were active at astrocytes from the dorsal horn of the spinal cord, UTP was inactive at those isolated from the hippocampus. Extracellular nucleotides also stimulate DNA synthesis via a protein kinase C/tyrosine kinase activation dependent pathway (Neary et al., 1998) and increased immediate early gene expression via P2Y receptors has also been noted (Priller et al., 1998). In activated microglia the application of nucleotides leads to P2Y receptor mediated opening of potassium channels (Nörenberg et al., 1997).

Within the peripheral nervous system both myelinating and non-myelinating Schwann cells express P2Y receptors coupled to intracellular calcium increases. Again, which receptors are expressed depends on the cell type: a suramininsenstive UTP-activated receptor is expressed by the myelinating cells, whereas non-myelinating cells express a $P2Y_1$ receptor (Mayer et al., 1998). Cultured neurons from the superior cervical ganglion express a uridine nucleotide receptor (probably $P2Y_6$) after one week in culture, the activation of which leads to inhibition of the M-type potassium current but not voltage dependent calcium channels (Boehm, 1998). The $P2Y_1$ receptor has also been detected in the dorsal root and trigeminal ganglia of the embryonic chick (Nakamura and Strittmatter, 1996).

Blood–brain barrier endothelial cells can also express a number of P2Y receptor subtypes in vitro. For example, three P2Y receptors are expressed in clonal lines derived from rat brain endothelial cells: a $P2Y_2$ receptor coupled to phospholipase C and calcium mobilisation, a $P2Y_1$-like receptor that increases intracellular calcium but in a mechanism independent of phospholipase C, plus a third receptor that is activated solely by adenine nucleotides and couples negatively to adenylate cyclase (Frelin et al., 1993; Feolde et al., 1995; Webb et al., 1996a; Hechler et al., 1998a, 1998b). In unpassaged primary cell cultures of rat brain capillary endothelial cells a further response has been noted, namely that ATP can augment the stimulatory effect of forskolin on adenylate cyclase (Albert et al., 1997; Anwar et al., 1999). Furthermore, in these cells UDP and UTP were equipotent in their ability activate phospholipase C, indicating the presence of a $P2Y_6$ receptor response in these cells, which has since been confirmed by RT-PCR (Albert et al., 1997; Anwar et al., 1999).

Physiological roles for P2Y receptors in the nervous system

While P2Y receptor neuromodulation has not been demonstrated directly at the synapse there is a large body of evidence to suggest that it does occur, as other papers in this volume indicate. Nucleotides are able to modulate channel activity via P2Y receptor activation. This does not appear to be limited to a particular channel or receptor type. The recombinant $P2Y_2$ receptor is able to inhibit both the N type calcium channel and the M-type potassium channel in a heterologous expression system and the $P2Y_6$ receptor, expressed in the same system also shares these dual coupling properties (Filippov et al., 1997, 1998a, 1999). Nucleotides have also been implicated in the activation of an outwardly rectifying K^+ channel in a number of neuronal cell types (Ikeuchi and Nishizaki, 1996a, 1996b). Thus, it can be envisaged that activation of a P2Y receptor could have either an inhibitory or excitatory neuromodulatory role depending on whether the receptor was pre- or postsynaptic.

Nucleotides are also well placed to be involved in other forms of cell to cell communication in the nervous system. Astrocytes have a close anatomic and functional relationship with both neurons and the endothelial cells of the blood brain barrier. The indicated presence of P2Y receptors on all of these cell types would allow each to respond to and possibly control the function of the others by this common signalling molecule—i.e. ATP and other nucleotides released under physiological or pathological conditions. The ubiquitous occurrence in the brain of P2Y receptors on astrocytes may, therefore, serve in a mechanism for the detection of an early stage of cell injury, when leakage of ATP and other nucleotides readily occurs: evidence for such an autocrine/paracrine signalling role of nucleotides has been reported (Nakamura and Strittmatter, 1996; Schlosser et al., 1996).

Acknowledgments

We thank Brenda Browning for providing the differentiated NG108-15 cells. This work was supported by the Wellcome Trust, the Leverhulme Trust and the European Union.

References

Akbar, G.K.M., Dasari, V.R., Webb, T.E., Ayyanathan, K., Pillarisetti, K., Sandhu, A.K., Athwal, R.S., Daniel, J.L., Ashby, B., Barnard, E.A. and Kunapuli, S.P. (1996) Molecular-cloning of a novel P2 purinoceptor from human erythroleukemia cells. *J. Biol. Chem.*, 271: 18363–18367.

Albert, J.L., Boyle, J.P., Roberts, J.A., Challiss, R.A.J., Gubby, S.E. and Boarder, M.R. (1997) Regulation of brain capillary endothelial cells by P2Y receptors coupled to Ca^{2+}, phospholipase C and mitogen-activated protein kinase. *Br. J. Pharmacol.*, 122: 935–941.

Anwar, Z., Albert, J.L., Gubby, S.E., Boyle, J.B., Roberts, J.A., Roalfe, G.J., Webb, T.E. and Boarder, M.R. (1999) Regulation of cyclic AMP by P_2 receptors: Studies on P2Y transfected 1321N1 cells and brain capillary endothelial cells. (Submitted for publication.)

Ayyanathan, K., Webb, T.E., Sandhu, A.K., Athwal, R.S., Barnard, E.A. and Kunapuli, S.P. (1996) Cloning and chromosomal localisation of the human $P2Y_1$ purinoceptor. *Biochem. Biophys. Res. Commun.*, 218: 783–788.

Barnard, E.A., Simon, J. and Webb, T.E. (1997) Nucleotide receptors in the nervous system—An abundant component using diverse transduction mechanisms. *Molec. Neurobiol.*, 15: 103–129.

Bernstein, M., Behnisch, T., Balschun, D., Reymann, K.G. and Reiser, G. (1998) Pharmacological characterisation of metabotropic glutamatergic and purinergic receptors linked to Ca^{2+} signalling in hippocampal astrocytes. *Neuropharmacology*, 37: 169–178.

Boehm, S. (1998) Selective inhibition of M-type potassium channels in rat sympathetic neurons by uridine nucleotide preferring receptors. *Br. J. Pharmacol.*, 124: 1261–1269.

Bogdanov, Y.D., Dale, L., King, B.F., Whittock, N. and Burnstock, G. (1997) Early expression of a novel nucleotide receptor in the neural plate of Xenopus embryos. *J. Biol. Chem.*, 272: 12583–12590.

Boyer, J.L., Lazarowski, E.R., Chen, X.H. and Harden, T.K. (1993) Identification of a P2Y purinergic receptor that inhibits adenylyl-cyclase. *J. Pharmacol. and Exp. Ther.*, 267: 1140–1146.

Boyer, J.L., Mohanram, A., Deleney, S., Waldo, G.L. and Harden, T.K. (1998) Signalling mechanism and pharmacological selectivity of an avian P2Y receptor. *Archives of Pharmacol.*, 358: p. 10.19.

Boyer, J.L., Otuel, J.W., Fischer, B., Jacobson, K.A. and Harden, T.K. (1995) Potent agonist action of 2-thioether derivatives of adenine-nucleotides at adenylyl cyclase-linked P2Y purinoceptors. *Br. J. Pharmacol.*, 116: 2611–2616.

Boyer, J.L., Waldo, G.L. and Harden, T.K. (1997) Molecular cloning and expression of an avian G protein-coupled P2Y receptor. *Mol. Pharmacol.*, 52: 928–934.

Burnstock, G. (1972) Purinergic nerves. *Pharmacol. Rev.*, 24: 509–581.

Burnstock, G. (1976) Do some nerve cells release more than one transmitter? *Neuroscience*, 1: 239–248.

Burnstock, G. (1978) A basis for distinguishing two types of purinergic receptor. In: R.W. Straub and L. Bolis (Eds.), *Cell Membrane Receptors for Drugs and Hormones: A multidisciplinary approach.* Raven Press, New York pp. 107–118.

Burnstock, G., Campbell, G., Bennett, M. and Holman, M.E. (1963) Inhibition of the smooth muscle of the taenia coli. *Nature*, 200: 581–582.

Chang, K.G., Hanaoka, K., Kumada, M. and Takuwa, Y. (1995) Molecular-cloning and functional-analysis of a novel P_2 nucleotide receptor. *J. Biol. Chem.*, 270: 26152–26158.

Communi, D., Govaerts, C., Parmentier, M. and Boeynaems, J.M. (1997) Cloning of a human purinergic P2Y receptor coupled to phospholipase C and adenylyl cyclase. *J. Biol. Chem.*, 272: 31969–31973.

Communi, D., Pirotton, S., Parmentier, M. and Boeynaems, J.M. (1995) Cloning and functional expression of a human uridine nucleotide receptor. *J. Biol Chem.*, 270: 30849–30852.

Desouza, L.R., Moore, H., Raha, S. and Reed, J.K. (1995) Purine and pyrimidine nucleotides activate distinct signaling pathways in PC12 cells. *J. Neurosci. Res.*, 41: 753–763.

Drury, A.N. and Szent-Györgyi, A. (1929) The physiological activity of adenine coupounds with special reference to their action upon the mammalian heart. *J. Physiol.*, 68: 213–237.

Feolde, E., Vigne, P., Breittmayer, J.P. and Frelin, C. (1995) ATP, a partial agonist of atypical P2Y purinoceptors in rat-brain microvascular endothelial-cells. *Br. J. Pharmacol.*, 115: 1199–1203.

Filippov, A.K. and Brown, D.A. (1996) Activation of nucleotide receptors inhibits high-threshold calcium currents in NG108-15 neuronal hybrid-cells. *Eur. J. Neurosci.*, 8: 1149–1155.

Filippov, A.K., Selyanko, A.A., Robbins, J. and Brown, D.A. (1994) Activation of nucleotide receptors inhibits M-type K-current [I-K(M)] in neuroblastoma x glioma hybrid-cells. *Eur. J. Physiol.*, 429: 223–230.

Filippov, A.K., Webb, T.E., Barnard, E.A. and Brown, D.A. (1997) Inhibition by heterologously-expressed $P2Y_2$ nucleotide receptors of N-type calcium currents in rat sympathetic neurones. *Br. J. Pharmacol.*, 121: 849–851.

Filippov, A.K., Webb, T.E., Barnard, E.A. and Brown, D.A. (1998a) $P2Y_2$ nucleotide receptors expressed heterologously in sympathetic neurons inhibit both N-type Ca^{2+} and M-type K^+ currents. *J. Neurosci.*, 18: 5170–5179.

Filippov, A.K., Webb, T.E., Barnard, E.A. and Brown, D.A. (1999) Dual coupling of heterologously-expressed rat $P2Y_6$ nucleotide receptors to N-type Ca^{2+} and M-type K^+ currents in rat sympathetic neurones. *Br. J. Pharmacol.*, 126: 1009–1017.

Frelin, C., Breittmayer, J.P. and Vigne, P. (1993) ADP induces inositol phosphate-independent intracellular Ca^{2+} mobilization in brain capillary endothelial-cells. *J. Biol. Chem.*, 268: 8787–8792.

Harper, S., Webb, T.E., Charlton, S.J., Ng, L.L. and Boarder, M.R. (1998) Evidence that $P2Y_4$ nucleotide receptors are involved in the regulation of rat aortic smooth muscle cells by UTP and ATP. *Br. J. Pharmacol.*, 124: 703–710.

Hechler, B., Leon, C., Vial, C., Vigne, P., Frelin, C., Cazenave, J.P. and Gachet, C. (1998a) The $P2Y_1$ receptor is necessary for adenosine 5'-diphosphate-induced platelet aggregation. *Blood*, 92: 152–159.

Hechler, B., Vigne, P., Leon, C., Breittmayer, J.P., Gachet, C. and Frelin, C. (1998b) ATP derivatives are antagonists of the

P2Y$_1$ receptor: Similarities to the platelet ADP receptor. *Mol. Pharmacol.*, 53: 727–733.

Holton, P. (1959) The liberation of adenosine triphosphate on antidromic stimulation of the sensory nerves. *J. Physiol.*, 145: 491–504.

Holton, F.A. and Holton, P. (1953) The possibility that ATP is a transmitter at sensory nerve ending. *J. Physiol.*, 119: 50–51P.

Holton, F.A. and Holton, P. (1954) The capillary dilator substances in dry powders of spinal roots; a possible role for adenosine triphosphate in chemical transmission from nerve endings. *J. Physiol.*, 126: 124–140.

Hourani, S.M.O. and Hall, D.A. (1996) P2T purinoceptors—ADP receptors on platelets. *Ciba Foundation Symposia*, 198: 53–70.

Higgins D.G. and Sharp P.M. (1988) CLUSTAL: A package for performing multiple sequence alignments on a microcomputer. *Gene*, 73: 237–244.

Idestrup, C.P. and Salter, M.W. (1998) P2Y and P2U receptors differentially release intracellular Ca^{2+} via the phospholipase C inositol 1,4,5-triphosphate pathway in astrocytes from the dorsal spinal cord. *Neuroscience*, 86: 913–923.

Ikeuchi, Y. and Nishizaki, T. (1996a) ATP-regulated K$^+$ channel and cytosolic Ca^{2+} mobilisation in cultured rat spinal neurons. *Eur. J. Pharmacol.*, 302: 163–169.

Ikeuchi, Y. and Nishizaki, T. (1996b) P2 purinoceptor-operated potassium channel in rat cerebellar neurons. *Biochem. Biophys. Res. Commun.*, 218: 67–71.

Janssens, R., Boeynaems, J.M., Godart, M. and Communi, D. (1997) Cloning of a human heptahelical receptor closely related to the P2Y$_5$ receptor. *Biochem. Biophys. Res. Commun.*, 236: 106–112.

Jimenez, A.I., Castro, E., Delicado, E.G. and Miras Portugal, M.T. (1998) Potentiation of adenosine 5'-triphosphate calcium responses by diadenosine pentaphosphate in individual rat cerebellar astrocytes. *Neurosci. Letts.*, 246: 109–111.

Jin, J.G. and Kunapuli, S.P. (1998) Coactivation of two different G protein-coupled receptors is essential for ADP-induced platelet aggregation. *Proc. Natl. Acad. Sci. USA*, 95: 8070–8074.

Lalo, U., Voitenko, N. and Kostyuk, P. (1998) Iono- and metabotropically induced purinergic calcium signalling in rat neocortical neurons. *Brain Res.*, 799: 285–291.

Lazarowski, E.R. and Harden, T.K. (1994) Identification of a uridine nucleotide-selective G-protein linked receptor that activates phospholipase-C. *J. Biol. Chem.*, 269: 11830–11836.

Lazarowski, E.R., Homolya, L., Boucher, R.C. and Harden, T.K. (1997) Direct demonstration of mechanically induced release of cellular UTP and its implication for uridine nucleotide receptor activation. *J. Biol. Chem.*, 272: 24348–24354.

Leon, C., Hechler, B., Vial, C., Leray, C., Cazenave, J.P. and Gachet, C. (1997) The P2Y$_1$ receptor is an ADP receptor antagonized by ATP and expressed in platelets and megakaryoblastic cells. *Febs Letts.*, 403: 26–30.

Li, Q., Schachter, J.B., Harden, T.K. and Nicholas, R.A. (1997) The 6H1 orphan receptor, claimed to be the p2y5 receptor, does not mediate nucleotide-promoted second messenger responses receptor. *Biochem. Biophys. Res. Commun.*, 236: 455–460.

Lin, W.W. (1994) Heterogeneity of nucleotide receptors in NG108-15 neuroblastoma and C6 glioma-cells for mediating phosphoinositide turnover *J. Neurochem.*, 62: 536–542.

Lin, W.W. and Chuang, D.M. (1993) Endothelin-induced and ATP-induced inhibition of adenylyl-cyclase activity in C6 glioma cells—role of Gi and calcium. *Mol. Pharmacol.*, 44: 158–165.

Lin, T.A., Lustig, K.D., Sportiello, M.G., Weisman, G.A. and Sun, G.Y. (1993) Signal transduction pathways coupled to a P2U receptor in neuroblastoma X glioma (NG108-15) cells. *J. Neurochem.*, 60: 1115–1125.

Lustig, K.D., Shiau, A.K., Brake, A.J. and Julius, D. (1993) Expression cloning of an ATP receptor from mouse neuroblastoma-cells. *Proc. Natl. Acad. Sci. USA*, 90: 5113–5117.

Matsuoka, I., Zhou, Q., Ishimoto, H. and Nakanishi, H. (1995) Extracellular ATP stimulates adenylyl-cyclase and phospholipase C through distinct purinoceptors in NG108-15 cells. *Mol. Pharmacol.*, 47: 855–862.

Mayer, C., Quasthoff, S. and Grafe, P. (1998) Differences in the sensitivity to purinergic stimulation of myelinating and non-myelinating Schwann cells in peripheral human and rat nerve. *Glia* 23: 374–382.

Miras Portugal, M.T., Castro, E., Mateo, J. and Pintor, J. (1996) The diadenosine polyphosphate receptors—P2D purinoceptors. *Ciba Foundation Symposia*, 198: 35–52.

Munshi, R., Debernardi, M.A. and Brooker, G. (1993) P2U purinergic receptors on C6-2B rat glioma-cells—modulation of cytosolic Ca^{2+} and cAMP levels by protein-kinase-C. *Mol. Pharmacol.*, 44: 1185–1191.

Murayama, T., Yakushi, Y., Watanabe, A. and Nomura, Y. (1998) P2 receptor-mediated inhibition of adenylyl cyclase in PC12 cells. *Eur. J. Pharmacol.* 348: 71–76.

Murrin, R.J.A. and Boarder, M.R. (1992) Neuronal nucleotide receptor linked to phospholipase-C and phospholipase-D—stimulation of PC12 cells by ATP analogs and UTP. *Mol. Pharmacol.*, 41: 561–568.

Nakamura, F. and Strittmatter, S.M. (1996) P2Y$_1$ purinergic receptors in sensory neurons—contribution to touch-induced impulse generation. *Prod. Natl. Acad. Sci. USA*, 93: 10465–10470.

Neary, J.T., McCarthy, M., Kang, Y. and Zuniga, S. (1998) Mitogenic signaling from P$_1$ and P$_2$ purinergic receptors to mitogen-activated protein kinase in human fetal astrocyte cultures. *Neurosci. Letts.*, 242: 159–162.

Nicholas, R.A., Watt, W.C., Lazarowski, E.R., Li, Q. and Harden K. (1996) Uridine nucleotide sensitivity of three phospholipase C-activating P2 receptors: Identification of a

UDP-selective, a UTP-selective, and an ATP- and UTP-specific receptor. *Mol. Pharmacol.*, 50: 224–229.

Nikodijevic, B., Sei, Y., Shin, Y.M. and Daly, J.W. (1994) Effects of ATP and UTP in pheochromocytoma PC12 cells – evidence for the presence of three P2 receptors, only one of which subserves stimulation of norepinephrine release. *Cellular and Mol. Neurobiol*, 14: 27–47.

Nörenberg, W., Cordes, A., Blohbaum, G., Frohlich, R. and Illes, P. (1997) Coexistence of purino- and pyrimidinoceptors on activated rat microglial cells. *Br. J. Pharmacol.*, 121: 1087–1098.

North, R.A. and Barnard, E.A. (1997) Nucleotide receptors. *Curr. Opin. Neurobiol.*, 7: 346–357.

Palmer, R.K., Boyer, J.L., Schachter, J.B., Nicholas, R.A. and Harden, T.K. (1998) Agonist action of adenosine triphosphates at the human $P2Y_1$ receptor. *Mol. Pharmacol.*, 54: 1118–1123.

Parr, C.E., Sullivan, D.M., Paradiso, A.M., Lazarowski, E.R., Burch, L.H., Olsen, J.C., Erb, L., Weisman, G.A., Boucher, R.C. and Turner, J.T. (1994) Cloning and expression of a human-P2U nucleotide receptor, a target for cystic-fibrosis pharmacotherapy. *Proc. Nat. Acad. Sci. USA*, 91: 3275–3279.

Priller, J., Reddington, M., Haas, C.A. and Kreutzberg, G.W. (1998) Stimulation of P2Y-purinoceptors on astrocytes results in immediate early gene expression and potentiation of neuropeptide action. *Neuroscience*, 85: 521–525.

Reetz, G. and Reiser, G. (1994) Cross-talk of the receptors for bradykinin, serotonin, and ATP shown by single-cell Ca^{2+} responses indicating different modes of Ca^{2+} activation in a neuroblastoma x glioma hybrid cell-line. *J. Neurochem.*, 62: 890–897.

Schachter, J.B., Li, Q., Boyer, J.L., Nicholas, R.A. and Harden, T.K. (1996) Second messenger cascade specificity and pharmacological selectivity of the human $P2Y_1$ purinoceptor. *Br J. Pharmacol.*, 118: 167–173.

Schlosser, S.F.Burgstahler, A.D. and Nathanson, M.H. (1996) Isolated rat hepatocytes can signal to other hepatocytes and bile duct cells by release of nucleotides. *Proc. Natl. Acad. Sci. USA*, 93: 9948–9953.

Simon J., Webb T.E. and Barnard E.A. (1995) Characterization of a P2Y purinoceptor in the brain. *Pharmacol. and Toxicol.*, 76: 302–307.

Simon J., Webb T.E. and Barnard, E.A. (1997) Distribution of [^{35}S]dATPαS binding sites in the adult rat neuraxis. *Neuropharmacology*, 36: 1243–1251.

Song, S.L. and Chueh, S.H. (1996) P2 purinoceptor-mediated inhibition of cyclic-amp accumulation in NG108-15 cells. *Brain Res.*, 734: 243–251.

Tokuyama, Y., Hara, M., Jones, E.M.C., Fan, Z., Bell, G.I. (1995) Cloning of rat and mouse P-2Y purinoceptors. *Biochem. Biophys. Res. Commun.*, 211: 211–218.

Vanhoutte, P.M., Humphrey, P.P.A. and Spedding, M. (1996) International Union of Pharmacology recommendations for nomenclature of new receptor subtypes. *Pharmacol. Rev.*, 48: 1–2.

Webb, T.E., Feolde, E., Vigne, P., Neary, J.T., Runberg, A., Frelin, C. and Barnard, E.A. (1996a) The P2Y purinoceptor in rat brain microvascular endothelial cells couples to inhibition of adenylate cyclase. *Br. J. Pharmacol.*, 119: 1385–1392.

Webb, T.E., Henderson, D., King, B.F., Wang, S., Simon, J., Bateson, A.N., Burnstock, G. and Barnard, E.A. (1996b) A novel G protein-coupled P2 purinoceptor ($P2Y_3$) activated preferentially by nucleoside diphosphates. *Mol. Pharmacol.*, 50: 258–265.

Webb, T.E., Kaplan, M.G., Barnard, E.A. (1996c) Identification of 6H1 as a P2Y purinoceptor – $P2Y_5$ *Biochem. Biophys. Res. Commun.*, 219: 105–110.

Webb, T.E., Henderson, D., Roberts, J. and Barnard E.A. (1998a) Molecular cloning and characterisation of the rat $P2Y_4$ receptor. *J. Neurochem.*, 71: 1424–1434.

Webb, T.E., Simon, J. and Barnard, E.A. (1998b). Regional distribution of [^{35}S]2′-deoxy 5′-o-(1-thio) ATP binding sites and the $P2Y_1$ mRNA within the chick brain. *Neuroscience*, 84: 825–837.

Webb, T.E., Simon, J., Krishek, B.J., Bateson, A.N., Smart, T.G., King, B.F., Burnstock,G. and Barnard, E.A. (1993) Cloning and functional expression of a brain G-protein-coupled ATP receptor. *Febs Letts.*, 324: 219–225.

Yakushi, Y., Watanabe, A., Murayama, T. and Nomura, Y. (1996) P_2 purinoceptor-mediated stimulation of adenylyl cyclase in PC12 cells. *Eur. J. Pharmacol.*, 314: 243–248.

Yokomizo, T., Izumi, T., Chang, K., Takuwa, Y., Shimizu, T. (1997) A G-protein-coupled receptor for leukotriene B4 that mediates chemotaxis. *Nature*, 387: 620–624.

P. Illes and H. Zimmermann (Eds.)
Progress in Brain Research, Vol 120
© 1999 Elsevier Science BV. All rights reserved

P2Y receptors in the nervous system: Molecular studies of a P2Y$_2$ receptor subtype from NG108–15 neuroblastoma x glioma hybrid cells

Gary A. Weisman[1],*, Richard C. Garrad[1], Laurie J. Erb[1], Cynthia Santos-Berrios[2] and Fernando A. Gonzalez[2]

[1]*Department of Biochemistry, University of Missouri-Columbia, Columbia, MO 65212, USA*
[2]*Department of Chemistry, University of Puerto Rico, Rio Piedras, Puerto Rico*

P2Y receptors: signaling and function

Extracellular nucleotides activate responses in virtually all of the cell types that comprise the central and peripheral nervous system. These cellular responses are mediated by P2 nucleotide receptors belonging to two separate receptor superfamilies. In the nervous system, nucleotides can activate both P2X receptors, which are ligand-gated ion channels, and G protein-coupled P2Y receptors. Within the past six years, multiple P2X and P2Y receptor subtypes have been cloned and the diversity of P2 nucleotide receptors in the nervous system is becoming more apparent. The P2X receptors by virtue of their ability to affect ion gradients apparently play an important role in neurotransmission (for reviews see Neary et al., 1996; Barnard et al., 1997). In contrast, the functional relevance of P2Y receptor expression in the nervous system is less well understood.

Recent research indicates that several distinct G protein-coupled P2Y receptor subtypes are expressed in cells of the nervous system and other tissues (for reviews see Dubyak and El-Moatassim, 1993; Barnard et al., 1997; Heilbronn et al., 1997;

Weisman et al., 1998). The specific P2Y receptor subtypes that are present in each cell type are not well defined, since many of the studies were performed prior to the availability of P2Y receptor clones when the structural diversity of these receptors was not yet recognized. Nonetheless, the identity of some P2Y receptor subtypes has been satisfactorily defined for some cell types in the central and peripheral nervous system. A P2Y$_1$ receptor that is selective for adenine nucleotides is expressed in primary astrocytes (Neary et al., 1988, 1991; Pearce et al., 1989; Bruner and Murphy, 1990; Kastritsis et al., 1992; Shao and McCarthy, 1993; Ciccarelli et al., 1994). Other studies have indicated the presence of P2Y$_1$ receptors in brain (Illes et al., 1996) and sensory (Nakamura and Strittmatter, 1996) neurons. P2Y$_2$ receptors (formerly termed P$_{2U}$), at which adenosine 5′-triphosphate (ATP) and uridine 5′-triphosphate (UTP) are equipotent and equiefficacious agonists, are expressed in immortalized astrocytes (Wu and Sun, 1997), NG108–15 neuroblastoma x glioma cells (Ehrlich et al., 1988; Lin et al., 1993; Filippov et al., 1994; Filippov and Brown, 1996) and N1E-115 neuroblastoma cells (Iredale et al., 1992).

Other studies have documented the presence of G protein-coupled P2Y receptors without a clear assignment of receptor subtype. P2 receptors that

---

*Corresponding author. Tel.: +1-573-882-5005; fax:
+1-573-884-4597; e-mail: weismang@missouri.edu

are selective for pyrimidine nucleotides have been described in rat superior cervical ganglia that also express P2X receptors (Connolly et al., 1993; Connolly, 1994). Immortalized Schwann cells have been shown to express multiple P2Y receptor subtypes, including the $P2Y_2$ receptor (Berti-Mattera et al., 1996). Similarly, dorsal horn astrocytes express P2Y receptors whose pharmacological profiles resemble $P2Y_1$ and $P2Y_2$ receptors (Ho et al., 1995). G protein-coupled P2Y receptors also are expressed in Bergmann glial cells, and oligodendrocytes from cortex and retina have P2Y receptors with characteristics of the $P2Y_1$ and $P2Y_2$ subtypes, respectively (Kirischuk et al, 1995a,b). Receptors for diadenosine polyphosphate (Ap_4A) have been described in adrenal chromaffin cells, components of the peripheral nervous system (Castro et al., 1994; Pintor and Miras-Portugal, 1995). From these reports, it appears evident that G protein-coupled P2Y receptors are widely expressed in cells that comprise the central and peripheral nervous system. However, more research is required to conclusively define all of the P2Y receptor subtypes found in these tissues.

Activation of G protein-coupled P2Y receptors in most cells increases the activity of phospholipase C leading to the intracellular generation of inositol 1,4,5-trisphosphate (IP_3) from the plasma membrane phospholipid phosphatidylinositol 4,5-bisphosphate (PIP_2). IP_3 formation induces the release of calcium from intracellular storage sites causing a rise in the cytoplasmic free calcium ion concentration, thereby activating calcium-dependent processes in the cell. In some tissues, $P2Y_2$ receptors can couple to phospholipase C_β via the α-subunit of G_q protein by a pertussis toxin-sensitive mechanism, whereas coupling of the $P2Y_1$ receptor is apparently pertussis toxin-insensitive (Boarder et al., 1995). The diversity of G proteins that can couple to each P2Y receptor subtype in the nervous system remains to be elucidated and may reflect the divergence of responses to nucleotides that are seen between individual cell types. The breakdown of PIP_2 linked to P2Y receptors also leads to the stimulation of protein kinase C (PKC) by virtue of the formation of diacylglycerol, a PKC activator. Phosphorylation of intracellular proteins by PKC in response to P2Y receptor activation has a variety of

effects on cellular functions (Boarder et al., 1995). Other signal transduction pathways can be linked to P2Y receptor activation. In C6 glioma cells, activation of a P2Y receptor can lead to inhibition of isoproteronol- and forskolin-stimulated adenylyl cyclase activity via a pertussis toxin-sensitive G protein (Boyer et al. 1993; Lin and Chuang, 1994). A similar signal transduction pathway has been described for P2Y receptors in immortalized Schwann cells (Berti-Mattera et al., 1996).

Physiological endpoints of P2Y receptor activation have been defined in some cell types in the nervous system. $P2Y_1$ receptors regulate K^+ channels in neurons from rat colliculus, medulla oblongata, cerebellum and hippocampus that likely involves the $\beta\gamma$ subunits of a PTX-insensitive G protein (Ikeuchi and Nishizaki, 1995, 1996; Ikeuchi et al., 1995, 1996). Activation of G protein-coupled $P2Y_2$ receptors in NG108–15 neuroblastoma x glioma cells induces IP_3-dependent calcium mobilization (Lin et al., 1993) leading to inhibition of M-type K^+ currents by a pertussis toxin-insensitive mechanism (Filippov et al., 1994). $P2Y_2$ receptor activation in NG108–15 cells also inhibits PTX-sensitive N-type Ca^{2+} currents (Filippov and Brown, 1996; Filippov et al., 1997) and PTX-insensitive L-type Ca^{2+} currents (Filippov and Brown, 1996, Barnard et al., 1997). These results suggest that $P2Y_2$ receptors can activate multiple G proteins to mediate different responses in a single cell type. $P2Y_1$ and $P2Y_2$ receptor activation is mitogenic to cells via stimulation of mitogen-activated protein (MAP) kinases (Gonzalez et al., 1990; Boarder et al., 1995; Soltoff et al., 1998) and similar pathways have been described in astrocytes (Rathbone et al., 1992; Neary and Zhu, 1994). ATP also increases prostaglandin synthesis in primary rat astrocytes (Gebicke-Haerter et al., 1988; Pearce et al., 1989), which is consistent with the ability of P2Y receptor-mediated calcium mobilization to activate phospholipase A2 and arachidonic acid release from plasma membrane phospholipid. Recently, P2Y receptor activation has been linked to the turning response of embryonic *Xenopus* spinal cord growth cones that is dependent upon activation of protein kinase C (Fu et al., 1997). A more complete description of the functional consequences of P2Y receptor activation in the nervous system has been provided recently (Barnard et al., 1997).

Molecular studies with the NG108–15 cell P2Y$_2$ receptor

Among the variety of potential P2Y receptor subtypes that may be expressed in the nervous system, P2Y$_1$ and P2Y$_2$ receptors were the first to be cloned. A P2Y$_1$ receptor cDNA was isolated from a chick embryonic whole brain cDNA library by homology screening with oligonucleotide probes to sequences conserved among G protein-coupled receptor genes (Webb et al., 1993). A P2Y$_2$ receptor was cloned from a NG108–15 neuro-blastoma x glioma cell cDNA library by expression cloning in *Xenopus laevis* oocytes (Lustig et al., 1993), and functionally expressed in mammalian cells (Erb et al., 1993). Recently, P2Y$_4$ (Nguyen et al., 1995; Communi et al., 1995) and P2Y$_6$ (Chang et al. 1995; Communi et al., 1996) receptors have been cloned and characterized by functional expression. Similar to the P2Y$_2$ receptor, P2Y$_4$ and P2Y$_6$ receptors can be activated by uridine nucleotides.

P2Y receptors have been detected in the brain. Utilizing Northern or ribonuclease protection analysis, mRNA for the P2Y$_1$ receptor was detected in avian and mammalian brain (Webb et al., 1993; Filtz et al., 1994; Tokuyama et al., 1995). Low levels of P2Y$_2$ receptor mRNA were detected in mouse (Lustig et al., 1993), and human brain (Parr et al., 1994). The P2Y$_2$ receptor has been shown to be upregulated in response to ligation of salivary gland (Turner et al., 1997), upregulated as an immediate early gene response in activated thymo-cytes (Koshiba et al., 1997) and downregulated during in vitro differentiation of human myeloid leukocytes (Martin et al., 1997). Downregulation of P2Y$_2$ receptor expression in differentiated cells could explain the low level of P2Y$_2$ receptor mRNA found in mature brain. Clearly more work is needed to provide definitive information about the distribution of P2Y receptor subtypes in the central and peripheral nervous system in mature and develop-ing tissue under normal and pathological conditions. Little information is available about the distribution of other P2Y receptor subtypes, although mRNA for a cloned P2Y$_3$/P2Y$_6$ receptor

subtype has been found in chick brain and spinal cord (Barnard et al., 1997). Since other novel clones for potential P2Y receptor subtypes have been isolated in the past months, the complement of P2Y receptors in the nervous system may prove to be diverse.

The availability of cloned P2Y receptor cDNAs has facilitated studies to define structure and function relationships for these G protein-coupled receptors. Towards this end, we have stably expressed the cloned NG108–15 cell P2Y$_2$ receptor in human K562 erythroleukemia cells that are devoid of endogenous P2Y receptor activity (Erb et al., 1993, 1995). As in NG108–15 cells, the expressed P2Y$_2$ receptor in K562 cells was acti-vated with equal potency and efficacy by ATP and UTP determined by measuring changes in the intracellular calcium free ion concentration (Erb et al., 1993). In an attempt to conclusively identify the P2Y$_2$ receptor protein, cDNA for the receptor was modified to incorporate a hexahistidine epitope tag at the C-terminus which enabled purification of the expressed receptor from membrane extracts of K562 cell transfectants by chromatography on Ni^{2+}-charged Sepharose columns (Erb et al., 1993). Incorporation of amino acids at the C-terminus did not affect P2Y$_2$ receptor signaling as compared to the wild-type receptor. The hexahistidine-tagged P2Y$_2$ receptor protein was followed in the purifica-tion scheme through its photoaffinity labeling with [^{32}P]-3′-O-(4-benzoyl)benzoyl ATP (BzATP). Our earlier studies had identified BzATP as a potent agonist of the P2X$_7$ (P2Z) receptor (Gonzalez et al., 1989; Erb et al., 1990), but BzATP is relatively ineffective at P2Y$_2$ receptors. However, the ability of BzATP to be crosslinked to ATP-binding pro-teins in the presence of long-wavelength ultraviolet light makes it a suitable tool for labeling P2Y receptors (Boyer and Harden, 1989; Erb et al., 1993). Using this approach, a 53 kDa [^{32}P]-BzATP-labeled, hexahistidine-tagged P2Y$_2$ receptor protein was isolated from K562 cells expressing the recombinant P2Y$_2$ receptor (Erb et al., 1993). Furthermore, a 53 kDa protein was intensely lab-eled by [^{32}P]-BzATP in K562 cells transfected with the P2Y$_2$ receptor cDNA, but not in untransfected K562 cells (Erb et al., 1993), evidence that this protein is the P2Y$_2$ receptor. [^{32}P]-BzATP labeling studies with turkey erythrocyte plasma membranes have suggested that the P2Y$_1$ receptor, like P2Y$_2$, is a 53 kDa protein (Boyer et al., 1990). The apparent molecular weight of 53 kDa for the P2Y$_2$ receptor suggests that ~20% of its mass is due to N-linked glycosylation at sites in its extracellular domain. A two-dimensional structure of the P2Y$_2$ receptor, derived from the nucleotide sequence of the P2Y$_2$ receptor cDNA and hydropathicity analysis of the predicted amino acid sequence of the encoded protein is shown in Fig. 1.

Pharmacological studies with P2Y receptor subtypes have been hampered by a lack of high affinity, selective P2Y receptor antagonists. Pyridoxal-phosphate-6-azophenyl-2′,4′ disulphonic acid (PPADS) and suramin have been described as

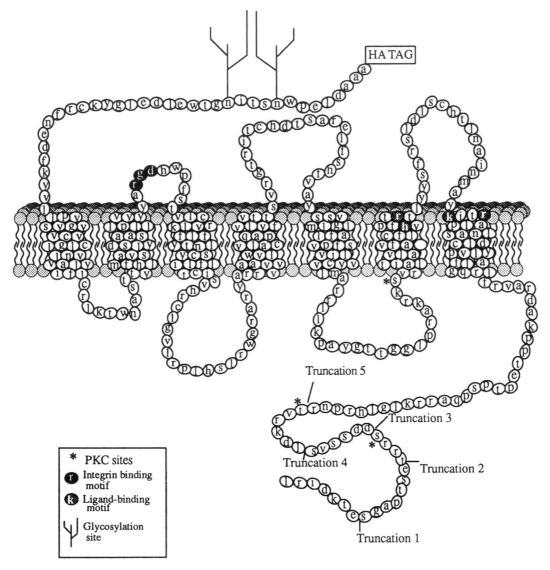

Fig. 1. Predicted secondary structure of the wild-type P2Y$_2$ receptor. Truncations (arrows) indicate positions of deletion sites for mutant receptors described in this manuscript (see Tables 1 and 2). Receptors were N-terminally tagged with the HA epitope from influenza virus, and consensus phosphorylation sites for protein kinase C are shown (*). Amino acids in the 6th and 7th transmembrane domains that affect agonist potency and specificity are shown in gray, and an RGD, integrin-binding motif in the first extracellular loop is shown in black.

antagonists of cloned P2Y receptors (Charlton et al., 1996a,b), although they are clearly not subtype-selective and their mode of action is unclear. In an attempt to learn more about the ligand binding site of the cloned NG108–15 cell $P2Y_2$ receptor, a series of mutant receptors were produced by site-directed mutagenesis of $P2Y_2$ receptor cDNA and expressed in a clonal line of human 1321N1 astrocytoma cells that are devoid of endogenous P2Y receptor activity (Erb et al., 1995). Expression of receptor mutants in 1321N1 cells was chosen over K562 cells since 1321N1 cells exhibited a 4–5-fold higher level of intracellular calcium mobilization in response to $P2Y_2$ receptor activation as compared to K562 cells (Erb et al., 1993, 1995). This may be due to higher expression levels in 1321N1 cells, which express the $P2Y_2$ receptor at ~ 20,000 mol/cell (Garrad et al., 1998). Amino acids targeted for substitution were chosen by comparison of sequence homologies and differences between the adenine nucleotide-selective chick $P2Y_1$ receptor (Webb et al., 1993) and the UTP-binding murine and human $P2Y_2$ receptors (Lustig et al., 1993; Parr et al., 1994). It was postulated that positively charged amino acids in the $P2Y_2$ receptor could serve as counter ions for binding the negatively charged receptor agonists. Expression of mutant $P2Y_2$ receptors in 1321N1 cells identified receptors with altered agonist potency profiles as compared to the wild-type receptor (Erb et al., 1995). Substitution of the amino acids, [262]His, [265]Arg and [292]Arg, with the neutral amino acids leucine or isoleucine diminished ATP and UTP potencies. Substitution of [289]Lys with Arg conserved the positive charge but increased the potencies of ADP and UDP and decreased potencies of ATP and UTP. The location of these four basic amino acids is depicted in Fig. 1 (see gray residues in transmembrane domains 6 and 7). These results indicate that differences in agonist potency profiles among the P2Y receptors may reside within minor sequence differences between the subtypes, since a conservative substitution could turn a nucleoside 5′-triphosphate-preferring receptor into one that exhibits highest potency for nucleoside 5′-diphosphates. To date, positively charged amino acids in the 6th and 7th transmembrane domains appear to be conserved

among all of the established P2Y receptor subtypes (Weisman et al., 1997). Molecular modeling of the $P2Y_1$ and $P2Y_2$ receptors utilizing information provided by several structure models available for the seven transmembrane-spanning proteins rhodopsin and bacteriorhodopsin (Erb et al., 1995; van Rhee et al., 1995, 1998) has helped delineate the ligand binding site of P2Y receptors. Recently, additional amino acids in the 3rd and 6th transmembrane domains have been shown to participate in ligand binding by the $P2Y_1$ receptor (Jiang et al., 1997).

P2Y receptors undergo agonist-induced desensitization (Wilkinson et al., 1994), similar to other members of the G protein-coupled receptor (GPCR) superfamily. P2Y receptor desensitization can be caused by treatment of cells with phorbol myristate acetate (PMA), an activator of protein kinase C (PKC) (Boarder et al., 1995), suggesting that phosphorylation of the P2Y receptor can lead to desensitization. The mechanisms of GPCR desensitization have been studied extensively for the β_2-adrenergic receptor (Freedman and Lefkowitz, 1996), where it has been shown that agonist-induced desensitization is due to phosphorylation of serine and threonine residues in the carboxy-terminus catalyzed by G protein-coupled receptor kinase (GRK). Furthermore, β_2-adrenergic receptor coupling to adenylyl cyclase leads to agonist-induced increases in cAMP and activation of protein kinase A (PKA). In turn, PKA-dependent phosphorylation of the receptor potentiates its desensitization and sequestration (Fredericks et al., 1996). Recovery from desensitization of G protein-coupled β_2-adrenergic receptors requires receptor internalization (sequestration) and dephosphorylation by intracellular phosphatases (Yu et al., 1993; Pippig et al., 1995). Receptors that are not re-sensitized are degraded, and this downregulation necessitates de novo protein synthesis to re-establish receptor levels. It seems likely that similar pathways underlie P2Y receptor desensitization, sequestration, re-sensitization and downregulation. However, little is known about the molecular mechanisms that regulate these pathways for P2Y receptors.

The C-terminus of the $P2Y_2$ receptor contains two consensus phosphorylation sites for PKC (another site is present in the third intracellular loop) (see Fig. 1) and potential phosphorylation sites for GRK. We investigated the role of the serine- and threonine-rich C-terminus of the $P2Y_2$ receptor in agonist-induced receptor desensitization and sequestration. Towards this end, a series of five truncation mutants (see Fig. 1 for truncation sites) of the NG108–15 cell $P2Y_2$ receptor cDNA was constructed using the polymerase chain reaction (PCR). In addition, nucleotides encoding an N-terminal, hemagglutinin (HA) epitope-tag were incorporated into the cDNA. The truncated receptor cDNAs were expressed in human 1321N1 astrocytoma cells and the kinetics and dose response of UTP-induced desensitization of receptor-mediated calcium mobilization were determined in cells expressing the epitope-tagged mutant or wild-type $P2Y_2$ receptors (Garrad et al., 1998). The presence of the HA-epitope tag in the extracellular domain of the $P2Y_2$ receptor enabled determinations of cell surface receptor density and agonist-induced sequestration by monitoring the level of epitope tag on intact cells using an immunofluorescence detection assay. Results with the wild-type $P2Y_2$ receptor and truncated receptor mutants indicated that the potency and efficacy for UTP-induced calcium mobilization were relatively unaffected even for the most truncated receptor. In contrast, deletion of 18 or more amino acids from the C-terminus increased by approximately 30-fold the concentration of UTP necessary to desensitize the receptor (Table 1). The rate and extent of UTP-induced receptor sequestration also were decreased as a function of increasing length of C-terminal truncations (Table 1). In addition, the time for recovery from sequestration was markedly increased with the longest truncation mutant as compared to the wild-type receptor (Garrad et al., 1998). Results clearly indicated that receptor desensitization and sequestration were independent processes. For example, maximal receptor desensitization was obtained even though more than 50% of the original receptor complement remained on the cell surface (Garrad et al., 1998). Furthermore, PKC activation with PMA caused desensitization of the $P2Y_2$ receptor and to a lesser extent the truncation mutants, but had no effect on receptor sequestration (Table 2). These results support the hypothesis that $P2Y_2$ receptor desensitization is caused by phosphorylation of consensus sites for PKC, whereas sequestration requires additional determinants within the C-terminus or other regions of the receptor. It remains to be shown whether phosphorylation of individual or multiple serine and threonine residues in the $P2Y_2$ receptor is sufficient for desensitization. The mechanism that regulates $P2Y_2$ receptor sequestration, and the role of GRKs must still be determined.

A property unique for the $P2Y_2$ receptor among GPCRs has been proposed; that is its potential to interact directly with integrins. This hypothesis is originally based on sequence analysis of recombinant $P2Y_2$ receptors (Lustig et al., 1993; Parr et al.,

TABLE 1

EC_{50}/IC_{50} values for activation and desensitization of UTP-induced increases in $[Ca^{2+}]_i$, and $t_{1/2}$ values and maximum levels of receptor sequestration

Recombinant $P2Y_2$ receptors	Concentration response EC_{50} (UTP, μM)	Concentration response IC_{50} (UTP, μM)	Estimated $t_{1/2}$ values of sequestration (min)	Maximal receptor sequestration after 180 min (%)
	Activation	Desensitization		
Wild type	0.25 ± 0.03	0.43 ± 0.1	<5	90 ± 1
Truncation 1	0.47 ± 0.07	1.23 ± 0.8	$5–10$	90 ± 1
Truncation 2	0.61 ± 0.23	3.37 ± 1.8	45	80 ± 1
Truncation 3	0.05 ± 0.01	7.78 ± 4.8	60	80 ± 2
Truncation 4	0.10 ± 0.02	12.3 ± 3.3	150	50 ± 2
Truncation 5	0.18 ± 0.04	9.75 ± 5.1	120	67 ± 14

Values used were determined as described in Garrad et al. (1998). Estimated $t_{1/2}$ values for receptor sequestration represent the time taken to sequester half of the receptors from the cell surface.

TABLE 2

Effects of protein kinase C activation by PMA on $P2Y_2$ receptor desensitization and sequestration

Recombinant $P2Y_2$ receptors	PMA (1 μM) effects	
	Desensitization 0 = no desensitization 100 = maximum desensitization	Sequestration
Wild type	100	None
Truncation 1	88	None
Truncation 2	n/a	None
Truncation 3	n/a	None
Truncation 4	40	None
Truncation 5	40	None

Data were obtained as described in Garrad et al. (1998). n/a = not available.

1994) that indicated the presence of the integrin-binding motif arginine–glycine–aspartic acid (RGD) in the first extracellular loop (see Fig. 1), raising the possibility that the $P2Y_2$ receptor may be involved in cell-to-cell adhesion and communication through binding to integrins. We assessed this possibility in studies with an 18-mer peptide encompassing the RGD domain of the $P2Y_2$ receptor ($P2Y_2^{93-110}$) or with a similar peptide ($P2Y_2^{93-110}E^{97}$) in which the RGD domain was replaced with RGE (arginine–glycine–glutamic acid), a motif that does not have high affinity for integrins (Pierschbacher and Ruoslahti, 1984). $P2Y_2^{93-110}$ and $P2Y_2^{93-110}E^{97}$ were coupled to alde-hyde-modified fluorescent beads and incubated with human K562 erythroleukemia cells that we found to express α_5, α_v, β_1, β_3, and β_5 integrin subunits. The results indicated that $P2Y_2^{93-110}$-coated beads, but not $P2Y_2^{93-110}E^{97}$-coated beads, bound to K562 cells. As expected, beads coated with fibronectin or vitronectin, natural ligands for these integrins, also bound to K562 cells. Furthermore, soluble $P2Y_2^{93-110}$ inhibited K562 cell binding of $P2Y_2^{93-110}$-coated and vitronectin-coated beads but not fibronectin-coated beads, suggesting that the $P2Y_2^{93-110}$ peptide was interacting with an integrin receptor that binds vitronectin. After testing a variety of antibodies, $P2Y_2^{93-110}$-coated bead binding to K562 cells was found to be inhibited by monoclonal antibodies to the $\alpha_v\beta_3$ integrin, but not by antibodies to the $\alpha_5\beta_1$ integrin. This study suggests that the RGD domain of the $P2Y_2$ receptor

binds to the $\alpha_v\beta_3$ integrin (vitronectin receptor) but not the $\alpha_5\beta_1$ integrin (fibronectin receptor).

To assess whether the intact $P2Y_2$ receptor could interact with the $\alpha_v\beta_3$ integrin, we constructed an RGE mutant of the $P2Y_2$ receptor. A dose response curve using human 1321N1 astrocytoma cells transfected with the RGE mutant $P2Y_2$ receptor indicated that this receptor requires approximately 1000 times higher concentrations of UTP or ATP to stimulate the mobilization of intracellular calcium, as compared to the wild-type $P2Y_2$ receptor. One possible explanation for this result is that the RGD domain binds to the $\alpha_v\beta_3$ integrin in 1321N1 cells and this binding alters the conformation of the $P2Y_2$ receptor to increase ligand-binding affinity. Another possibility is that integrin interaction with the RGD domain of the $P2Y_2$ receptor assists in localizing it to focal adhesions to more efficiently activate intracellular signaling pathways. To determine whether the $P2Y_2$ receptor localizes to focal adhesions, cells expressing either the N-terminal HA-epitope-tagged wild type or RGE mutant $P2Y_2$ receptors were probed for co-localization of the receptors with vinculin, a cytoskeletal protein that is thought to associate exclusively with focal adhesion complexes. Immunofluorescence experiments with antibodies to the HA epitope and vinculin indicated that co-localization of the $P2Y_2$ receptor and vinculin occurred in cells expressing the wild-type $P2Y_2$ receptor, but not in cells expressing the RGE mutant receptor. Taken together, these results indicate that the RGD domain of the $P2Y_2$ receptor is necessary for localizing the receptor to focal adhesions and suggests that this localization occurs via a direct $\alpha_v\beta_3/P2Y_2$ receptor interaction. This association may provide an efficient means to transduce intracellular signals from extracellular nucleotides thereby activating protein kinases or other enzymes to mediate cellular responses such as proliferation or differentiation under conditions where $P2Y_2$ receptors are expressed.

The functional relevance of the putative $\alpha_v\beta_3$ integrin/$P2Y_2$ receptor interactions is unknown. Recent studies have indicated that the $\alpha_v\beta_3$ integrin receptor is expressed at high levels in angiogenic endothelial cells and that antagonists of $\alpha_v\beta_3$ (e.g. soluble anti-$\alpha_v\beta_3$ antibodies and certain RGD-containing peptides) cause apoptosis and inhibition of angiogenesis in these cells (Brooks et al., 1994a, 1994b). Angiogenesis is not only essential for embryonic development and wound repair, but also for the growth of solid tumors, and its inhibition has become an important approach in the treatment of cancer. In the nervous system, cells of the oligodendrocyte lineage also express $\alpha_v\beta_3$ (Milner and Ffrench-Constant, 1994) and a variety of integrins including $\alpha_v\beta_3$ are variably expressed in nonneoplastic neural cells and tumors of the peripheral and autonomic nervous system (Mechtersheimer et al., 1994). One hypothesis is that the $P2Y_2$ receptor may be expressed predominantly during differentiation or tissue injury and down-regulated in mature tissues. Thus, interactions between $\alpha_v\beta_3$ and the $P2Y_2$ receptor may be confined to specific stages during development or the wound healing process.

Conclusions

The cloning and expression of P2Y receptors has provided the means to address the relevance of these receptors in neurological function. A clearer definition of the diversity and distribution of each P2Y receptor subtype expressed among the cell types that comprise the central and peripheral nervous systems is required, but it has already been established that the $P2Y_1$, $P2Y_2$ and several other subtypes are represented. There also is a need to determine P2Y receptor subtype expression patterns in vivo under a variety of physiological and pathological conditions, including cell growth and differentiation, tissue damage and neurological disorders. Ultimately, P2Y receptors may prove to be ideal targets for drug therapies in the nervous system once a fuller understanding of the structural basis for receptor function has evolved. Already, we are beginning to define P2Y receptor determinants for ligand binding, which may help in the design of potent receptor agonists and subtype-selective antagonists. Similarly, the delineation of P2Y receptor interactions leading to receptor desensitization, sequestration and downregulation may help control these processes to increase the effectiveness of receptor agonists. The potential that $P2Y_2$ receptors, in particular, may be localized to focal adhesions to participate in cell-to-cell communication suggests that these receptors may have functions yet to be recognized.

References

Barnard, E.A., Simon, J. and Webb, T.E. (1997) Nucleotide receptors in the nervous system: An abundant component using diverse transduction mechanisms. *Mol. Neurobiol.*, 15: 103–129.

Berti-Mattera, L.N., Wilkins, P.L., Madhun, Z. and Suchovsky, D. (1996) P_2-purinergic receptors regulate phospholipase C and adenylate cyclase activities in immortalized Schwann cells. *Biochem. J.*, 314: 555–561.

Boarder, M.R., Weisman, G.A., Turner, J.T. and Wilkinson, G.F. (1995) G protein-coupled P_2 purinoceptors: From molecular biology to functional responses. *TiPS*, 16: 133–139.

Boyer, J.L. and Harden, T.K. (1989) Irreversible activation of phospholipase C-coupled P_{2Y}-purinergic receptors by 3'-0-(4-benzoyl)benzoyl adenosine 5'-triphosphate. *Mol. Pharmacol.*, 36: 831–835.

Boyer, J.L., Cooper, C.L. and Harden, T.K. (1990) [^{32}P]3'-0-(4-benzoyl)benzoyl ATP as a photoaffinity label for a phospholipase C-coupled P_{2Y}-purinergic receptor. *J. Biol. Chem.*, 265: 13515–13520.

Boyer, J.L., Lazarowski, E.R., Chen, X-H. and Harden, T.K. (1993) Identification of a P_{2Y}-purinergic receptor that inhibits adenylyl cyclase. *J. Pharmacol. Exp. Ther.*, 267: 1140–1146.

Brooks, P.C., Clark, R.A.F. and Cheresh, D.A. (1994a) Integrin $\alpha_v\beta_3$ antagonists promote tumor regression by inducing apoptosis of angiogenic blood vessels. *Cell*, 79: 1157–1165.

Brooks, P.C., Clark, R.A.F. and Cheresh, D.A. (1994b) Requirement of vascular integrin $\alpha_v\beta_3$ for angiogenesis. *Science*, 264: 569–571.

Bruner, G. and Murphy, S. (1990) ATP-evoked arachidonic acid mobilization in astrocytes is via a P_{2Y}-purinergic receptor. *J. Neurochem.*, 55: 1569–1575.

Castro, E., Tomé, A.R., Miras-Portugal, M.T. and Rosário, L.M. (1994) Single-cell fura-2 microfluorometry reveals different purinoceptor subtypes coupled to Ca^{2+} influx and intracellular Ca^{2+} release in bovine adrenal chromaffin and endothelial cells. *Pflugers Arch.* 426: 524–533.

Chang, K., Hanaoka, K., Kumada, M. and Takuwa, Y. (1995) Molecular cloning and functional analysis of a novel P_2 nucleotide receptor. *J. Biol. Chem.*, 270: 26152–26158.

Charlton, S.J., Brown, C.A., Weisman, G.A., Turner, J.T., Erb, L. and Boarder, M.R. (1996a) PPADS and suramin as antagonists at cloned P_{2Y}- and P_{2U}-purinoceptors. *Br. J. Pharmacol.*, 118: 704–710.

Charlton, S.J., Brown, C.A., Weisman, G.A., Turner, J.T., Erb, L. and Boarder, M.R. (1996b) Cloned and transfected $P2Y_4$ receptors: characterization of a suramin and PPADS insensitive response to UTP. *Br. J. Pharmacol.*, 119: 1301–1303.

Ciccarelli, R., Di Iorio, P., Ballerini, P., Ambrosini, G., Giuliani, P., Tiboni, G.M. and Caciagli, F. (1994) Effects of exogenous ATP and related analogues on the proliferation rate of dissociated primary cultures of rat astrocytes *J. Neurosci. Res.*, 39: 556–566.

Communi, D., Pirotton, S., Parmentier, M. and Boeynaems, J.-M. (1995) Cloning and functional expression of a human uridine nucleotide receptor. *J. Biol. Chem.*, 270: 30849–30852.

Communi, D., Parmentier, M. and Boeynaems, J.-M. (1996) Cloning, functional expression and tissue distribution of the human P2Y$_6$ receptor. *Biochem. Biophys. Res. Comm.*, 222: 303–308.

Connolly, G.P., Harrison, P.J. and Stone, T.W. (1993) Action of purine and pyrimidine nucleotides on the rat superior cervical ganglion. *Br. J. Pharmacol.*, 110: 1297–1304.

Connolly, G.P. (1994) Evidence from desensitization studies for distinct receptors for ATP and UTP on the rat superior cervical ganglion. *Br. J. Pharmacol.*, 112: 357–359.

Dubyak, G.R. and El-Moatassim, C. (1993) Signal transduction via P$_2$-purinergic receptors for extracellular ATP and other nucleotides. *J. Physiol.*, 265: C577–C606.

Ehrlich, Y.H., Snider, R.M., Kornecki, E., Garfield, M.G. and Lenox, R.H. (1988) Modulation of neuronal signal transduction systems by extracellular ATP. *J. Neurochem.*, 50: 295–301.

Erb, L., Lustig, K.D., Ahmed, A.H., Gonzalez, F.A. and Weisman, G.A. (1990) Covalent incorporation of 3'-O-(4-benzoyl)benzoyl ATP into a P$_2$ purinoceptor in transformed mouse fibroblasts. *J. Biol. Chem.*, 265: 7424–7431.

Erb, L., Lustig, K.D., Sullivan, D.M., Turner, J.T. and Weisman, G.A. (1993) Functional expression and photoaffinity labeling of a cloned P$_{2U}$ purinergic receptor. *Proc. Natl. Acad. Sci., USA*, 90: 10449–10453.

Erb, L., Garrad, R., Wang, Y., Quinn, T., Turner, J.T. and Weisman G.A. (1995) Site-directed mutagenesis of P$_{2U}$ purinoceptors: Positively charged amino acids in transmembrane helices 6 and 7 affect agonist potency and specificity. *J. Biol. Chem.*, 270: 4185–4188.

Filippov, A.K., Selyanko, A.A., Robbins, J. and Brown, D.A. (1994) Activation of nucleotide receptors inhibits M-type K current IKM in neuroblastoma x glioma hybrid cells. *Eur. J. Phys.*, 429: 223–230.

Filippov, A.K. and Brown, D.A. (1996) Activation of nucleotide receptors inhibits high-threshold calcium currents in NG108–15 neuroblastoma x glioma hybrid cells. *Eur. J. Neurosci.*, 8: 1149–1155.

Filippov, A.K., Webb, T.E., Barnard, E.A. and Brown, D.A. (1997) Inhibition by heterogenously-expressed P2Y$_2$ nucleotide receptors of N-type calcium currents in rat sympathetic neurones. *Br. J. Pharmacol.*, 121: 849–851.

Filtz, T.M., Li, Q., Boyer, J.L., Nicholas, R.A. and Harden, T.K. (1994) Expression of a cloned P$_{2Y}$ purinergic receptor that couples to phospholipase C. *Mol. Pharmacol.*, 46: 8–14.

Fredericks, Z.L., Pitcher, J.A. and Lefkowitz, R.J. (1996) Identification of the G protein-coupled receptor kinase phosphorylation sites in the human β$_2$-adrenergic receptor. *J. Biol. Chem.*, 271: 13796–13803.

Freedman, N.J. and Lefkowitz, R.J. (1996) Desensitization of G protein-coupled receptors. *Recent Prog. in Hormone Res.*, 51: 319–351.

Fu, W.M., Tang, Y.B. and Lee, K.F. (1997) Turning of nerve growth cones induced by the activation of protein kinase C. *NeuroRep.*, 8: 2005–2009.

Garrad, R.C., Otero, M.A., Erb, L., Theiss, P.M., Clarke, L.L., Gonzalez, F.A., Turner, J.T. and Weisman, G.A. (1998) Structural basis of agonist-induced desensitization and sequestration of the P2Y$_2$ nucleotide receptor: consequences of truncation of the C-terminus. *J. Biol. Chem.* 273: 2943–2944.

Gebicke-Haerter, P.J., Wurster, S., Schobert, A. and Hertting, G. (1988) P$_2$-purinoceptor induced prostaglandin synthesis in primary rat astrocyte cultures. *Naunyn-Schmiedeberg's Arch. Pharmacol.*, 338: 704–707.

Gonzalez, F.A., Ahmed, A.H., Lustig, K.D., Erb, L. and Weisman, G.A. (1989) Permeabilization of transformed mouse fibroblasts by 3'-O-(4-benzoyl)benzoyladenosine 5'-triphosphate and the desensitization of the process. *J. Cell. Physiol.*, 139: 109–115.

Gonzalez, F.A., Wang, D.-J., Huang, N.-N. and Heppel, L.A. (1990) Activation of early events of the mitogenic response by a P$_{2Y}$ purinoceptor with covalently bound 3'-O-(4-benzoyl)-benzoyladenosine 5'-triphosphate. *Proc. Natl. Acad. Sci., USA*, 87: 9717–9721.

Heilbronn, E., Knoblauch, B.H.A. and Müller, C.E. (1997) Uridine nucleotide receptors and their ligands: Structural, physiological, and pathophysiological aspects, with special emphasis on the nervous system. *Neurochem. Res.*, 22: 1041–1050.

Ho, C., Hicks, J. and Salter, M.W. (1995) A novel P$_2$-purinoceptor expressed by a subpopulation of astrocytes from the dorsal spinal cord of the rat. *Br. J. Pharmacol.*, 116: 2909–2918.

Ikeuchi, Y., Nishizaki, T. and Okada, Y. (1995) A P$_2$ purinoceptor activated by ADP in rat medullary neurons. *Neurosci. Lett.*, 198: 71–74.

Ikeuchi, Y. and Nishizaki, T. (1995) The P$_{2Y}$ purinoceptor-operated potassium channel is possibly regulated by the βγ subunits of a pertussis toxin-insensitive G-protein in cultured rat inferior colliculus neurons. *Biochem. Biophys. Res. Commun.*, 214: 589–596.

Ikeuchi, Y. and Nishizaki, T. (1996) P$_2$ purinoceptor-operated potassium channel in rat cerebellar neurons. *Biochem. Biophys. Res. Commun.*, 218: 67–71.

Ikeuchi, Y., Nishizaki, T., Mori, M. and Okada, Y. (1996) Adenosine activates the K$^+$ channel and enhances cytosolic Ca^{2+} release via a P$_{2Y}$ purinoceptor in hippocampal neurons. *Eur. J. Pharmacol.*, 304: 191–199.

Illes, P., Nieber, K. and Nörenberg, W. (1996) Electrophysiological effects of ATP on brain neurones. *Autonom. Pharmacol.*, 16: 407–411.

Iredale, P.A., Martin, K.F., Alexander, S.P., Hill, S.J. and Kendall, D.A. (1992) Inositol 1,4,5-trisphosphate generation and calcium mobilisation via activation of an atypical P2 receptor in the neuronal cell line, N1E-115. *Br. J. Pharmacol.*, 107: 1083–1087.

Jiang, Q., Guo, D., Lee, B.X., van Rhee, A.M., Kim, Y.C., Nicholas, R.A., Schachter, J.B., Harden, T.K. and Jacobson, K.A. (1997) A mutational analysis of residues essential for ligand recognition at the human $P2Y_1$ receptor. *Mol. Pharmacol.*, 52: 499–507.

Kastritsis, C.H.C., Salm, A.K. and McCarthy, K. (1992) Stimulation of the P_{2Y} purinergic receptor on type 1 astroglia results in inositol phosphate formation and calcium mobilization. *J. Neurochem.*, 58: 1277–1284.

Kirischuk, S., Möller, T., Voitenko, N., Kettenmann, H. and Verkhratsky, A. (1995a) ATP-induced cytoplasmic calcium mobilization in Bergmann glial cells. *J. Neurosci.*, 15: 7861–7871.

Kirischuk, S., Scherer, J., Kettenmann, H. and Verkhratsky, A. (1995b) Activation of P_2-purinoreceptors triggered Ca^{2+} release from $InsP_3$-sensitive internal stores in mammalian oligodendrocytes. *J. Physiol.*, 483: 41–57.

Koshiba, M., Apasov, S., Sverdlov, V., Chen, P., Erb, L., Turner, J.T., Weisman, G.A. and Sitkovsky, M.V. (1997) Transient up-regulation of $P2Y_2$ nucleotide receptor mRNA expression is an immediate early gene response in activated thymocytes. *Proc. Natl. Acad. Sci. USA*, 94: 831–836.

Lin, T.A., Lustig, K.D., Sportiello, M.G., Weisman, G.A. and Sun, G.Y. (1993) Signal transduction pathways coupled to a P_{2U} receptor in neuroblastoma x glioma (NG108-15) cells. *J. Neurochem.*, 60: 1115–1125.

Lin, W.-W. and Chuang, D.M. (1994) Different signal transduction pathways are coupled to the nucleotide receptor and the P_{2Y} receptor in C_6 glioma cells. *J. Pharmacol. Exp. Ther.*, 269: 926–931.

Lustig, K.D., Shiau, A.K., Brake, A.J. and Julius, D. (1993) Expression cloning of an ATP receptor from mouse neuroblastoma cells. *Proc. Natl. Acad. Sci., USA*, 90: 5113–5117.

Martin, K.A., Kertesy, S.B. and Dubyak, G.R. (1997) Down-regulation of P_{2U}-purinergic nucleotide receptor messenger RNA expression during *in vitro* differentiation of human myeloid leukocytes by phorbol esters or inflammatory activators. *Mol. Pharmacol.*, 51: 97–108.

Mechtersheimer, G., Barth, T., Quentmeier, A. and Möller, P. (1994) Differential expression of β_1, β_3 and β_4 integrin subunits in nonneoplastic neural cells of the peripheral and autonomic nervous system and in tumors derived from these cells. *Lab. Invest.*, 70: 740–752.

Milner, R. and Ffrench-Constant, C. (1994) A developmental analysis of oligodendroglial integrins in primary cells: changes in α_v-associated β subunits during differentiation. *Development*, 120: 3497–3506.

Nakamura, F. and Strittmatter, S.M. (1996) $P2Y_1$ purinergic receptors in sensory neurons: contribution to touch-induced impulse generation. *Proc. Natl. Acad. Sci., USA*, 93: 10,465–10,470.

Neary, J.T., Laskey, R., van Breemen, C., Blicharska, J., Norenberg, L.O.B. and Norenberg, M.D. (1991) ATP-evoked calcium signal stimulates protein phosphorylation/dephosphorylation in astrocytes. *Brain Res.*, 566: 89–94.

Neary, J.T., van Breemen, C., Forster, E., Norenberg, L.O.B. and Norenberg, M.D. (1988) ATP stimulates calcium influx in primary astrocyte cultures. *Biochem. Biophys. Res. Comm.*, 157: 1410–1416.

Neary, J.T. and Zhu, Q. (1994) Signaling by ATP receptors in astrocytes. *NeuroRep.*, 5: 1617–1620.

Neary, J.T., Rathbone, M.P., Cattabeni, F., Abbracchio, M.P. and Burnstock, G. (1996) Trophic actions of extracellular nucleotides and nucleosides on glial and neuronal cells. *Trends Neurosci.*, 19: 13–18.

Nguyen, T., Erb, L., Weisman, G.A., Marchese, A., Heng, H.H.Q., Garrad, R.C., George, S.R., Turner, J.T. and O'Dowd, B.F. (1995) Cloning, expression and chromosomal localization of the human uridine nucleotide receptor gene. *J. Biol. Chem.*, 270: 30845–30848.

Parr, C.E., Sullivan, D.M., Paradiso, A.M., Lazarowski, E.R., Burch, L.H., Olsen, J.C., Erb, L., Weisman, G.A., Boucher, R.C. and Turner, J.T. (1994) Cloning and expression of a human $P2_U$ nucleotide receptor, a target for cystic fibrosis pharmacotherapy. *Proc. Natl. Acad. Sci., USA*, 91: 3275–3279.

Pearce, B., Murphy, S., Jeremy, J., Morrow, C. and Dandona, P. (1989) ATP-evoked Ca^{2+} mobilisation and prostanoid release from astrocytes: P_2-purinergic receptors linked to phosphoinositide hydrolysis. *J. Neurochem.*, 52: 971–977.

Pierschbacher, M.D. and Rouslahti, C. (1984) Cell attachment activity of fibronectin can be duplicated by small fragments of the molecule. *Nature*, 309: 30–33.

Pintor, J. and Miras-Portugal, M.T. (1995) P_2 purinergic receptors for diadenosine polyphosphates in the nervous system. *Gen. Pharmacol.*, 26: 229–235.

Pippig, S., Andexinger, S. and Lohse, M.J. (1995) Sequestration and recycling of β_2-adrenergic receptors permit receptor resensitization. *Mol. Pharmacol.*, 47: 666–676.

Rathbone, M.P., Middlemiss, P.J., Kim, J.K., Gysbers, J.W., DeForge, S.P., Smith, R.W. and Hughes, D.W. (1992) Adenosine and its nucleotides stimulate proliferation of chick astrocytes and human astrocytoma cells. *Neurosci. Res.*, 13: 1–17.

Shao, Y. and McCarthy, K.D. (1993) Regulation of astroglial responsiveness to neuroligands in primary culture. *Neurosci.*, 55: 991–1001.

Soltoff, S.P., Avraham, H., Avraham, S. and Cantley, L. (1998) Activation of $P2Y_2$ receptors by UTP and ATP stimulates mitogen-activated kinase activity through a pathway that involves related adhesion focal tyrosine kinase and protein kinase C. *J. Biol. Chem.*, 273: 2653–2660.

Tokuyama, Y., Hara, M., Jones, E.M.C., Fan, Z. and Bell, G.I. (1995) Cloning of rat and mouse $P2_Y$ purinoceptors. *Biochem. Biophys. Res. Comm.*, 211: 211–218.

Turner, J.T., Weisman, G.A. and Camden, J.M. (1997) Upregulation of $P2Y_2$ nucleotide receptors in rat salivary gland cells during short-term culture. *Am. J. Physiol.*, 273: C1100–C1107.

van Rhee, A.M., Fischer, B., van Galen, P.J.M. and Jacobson, K.A. (1995) Modelling the P_{2Y} purinoceptor using rhodopsin as template. *Drug Des. and Disc.*, 13: 133–154.

van Rhee, A.M., Jacobson, K.A., Garrad, R., Weisman, G.A. and Erb, L. (1998) P_2 receptor modeling and identification of ligand binding sites. In: J.T. Turner, G.A. Weisman and J.S. Fedan (Eds.), *P₂ Nucleotide Receptors*, Humana Press, Totowa, NJ, USA, pp. 135–166.

Webb, T.E., Simon, J., Krishek, B.J., Bateson, A.N., Smart, T.G., King, B.F., Burnstock, G. and Barnard, E.A. (1993) Cloning and functional expression of a brain G-protein-coupled ATP receptor. *FEBS Lett.*, 324: 219–225.

Weisman, G.A., Turner, J.T., Clarke, L.L., Gonzalez, F.A., Otero, M., Garrad, R.C. and Erb, L. (1997) P2 nucleotide receptor structure and function. In: L. Plesner, T.L. Kirley, and A.F. Knowles (Eds.), *Ecto-ATPases*. Plenum Press, New York, pp. 231–237.

Weisman, G.A., Gonzalez, F.A., Erb, L., Garrad, R.C. and Turner, J.T. (1998) The cloning and expression of G protein-coupled P2Y nucleotide receptors. In: J.T. Turner, G.A. Weisman and J.S. Fedan (Eds.), *P₂ Nucleotide Receptors*, Humana Press, Totowa, NJ, USA, pp. 63–79.

Wilkinson, G.F., Purkiss, J.R. and Boarder, M.R. (1994) Differential heterologous and homologous desensitization of two receptors for ATP (P_{2Y} purinoceptors and nucleotide receptors) coexisting on endothelial cells. *Mol. Pharmacol.*, 45: 731–736.

Wu, J.M. and Sun G.Y. (1997) Effects of IL-1β on receptor-mediated poly-phosphoinositide signaling pathway in immortalized astrocytes (DITNC). *Neurochem. Res.*, 22: 1309–1315.

Yu, S.S., Lefkowitz, R.J. and Hausdorff, W.P. (1993) β-Adrenergic receptor sequestration: A potential mechanism of receptor resensitization. *J. Biol. Chem.*, 268: 337–341.

P. Illes and H. Zimmermann (Eds.)
Progress in Brain Research, Vol 120
© 1999 Elsevier Science BV. All rights reserved

Nucleotide radiolabels as tools for studying P2Y receptors in membranes from brain and lung tissue

Georg Reiser*, Werner Laubinger and Rainer Schäfer

Institut für Neurobiochemie der Otto-von-Guericke-Universität Magdeburg, Leipziger Str. 44, 39120 Magdeburg, Germany

Introduction

Extracellular nucleotides act as transmitters or modulators through P2 receptors, which can be distinguished into two families, P2X and P2Y, on the basis of their signalling mechanism (Abbracchio and Burnstock, 1994). The ionotropic P2X receptors, ligand-gated, non-selective cation channels (reviewed in Abbracchio and Burnstock, 1994; Burnstock, 1997) recognize purine nucleotides (North, 1996) and mediate fast excitatory transmission at synapses (Edwards et al., 1992; Sun and Stanley, 1996). Rapid neurotransmission by P2X receptors was also determined in cell cultures of sympathetic neurones (Evans et al., 1992) and DRG neurones (Robertson et al., 1996). Several P2X receptor subtypes have been cloned (reviewed in Burnstock, 1997).

The metabotropic, G-protein-coupled P2Y receptor family (Burnstock, 1997) shows varied sensitivity to purine or pyrimidine nucleotides. The $P2Y_2$ (formerly P2U) subtype is activated similarly by uridine and adenine nucleotides, whereas the $P2Y_1$ and $P2Y_3$ receptor subtypes have a strong preference for purine, and the $P2Y_4$ and $P2Y_6$ subtypes for pyrimidine nucleotides (Burnstock, 1997).

*Corresponding author. Tel.: +49-391-6713088; fax: +49-391-6713097; e-mail: georg.reiser@medizin.uni-magdeburg.de

Nucleotides are also of great importance in lung physiology. In airways Cl^- secretion and subsequent passive water secretion determine the composition of mucus and the mucociliary clearence in lung. Normally, Cl^- secretion is mediated by cAMP-regulated Cl^- channels of cystic fibrosis transmembrane conductance regulator (CFTR). In patients suffering from the hereditary disease cystic fibrosis this pathway of Cl^- secretion is impaired, but can be circumvented by an alternative Cl^- transport coupled to intracellular Ca^{2+} mobilization after P2 receptor activation (Hwang et al., 1996). These P2 receptors ($P2Y_2$) which are localized in the apical membrane of lung epithelial cells are activated equipotently by the nucleotides ATP and UTP.

UTP is already successfully applied as a therapeutic agent in the treatment of cystic fibrosis to stimulate this way of Cl^- secretion (Olivier et al., 1996). Therefore, identification, localization, and quantification of nucleotide-activated P2 receptors in lung is very important for understanding the treatment of lung diseases like cystic fibrosis. Hydrolyzing enzymes in lung lead to rapid degradation of UTP. Thus, there is a great need to find more stable nucleotides that can replace UTP as $P2Y_2$ receptor agonist or that can activate different P2 receptors in lung. In pharmacological studies of the heterologously expressed human $P2Y_2$ receptor it was shown that the long-lived diadenosine polyphosphate Ap_4A is an agonist equally potent as UTP or ATP (Lazarowski et al., 1995). Therefore,

Ap$_4$A binding sites in lung are of considerable interest in the approach to find new strategies to improve treatment of cystic fibrosis. However, so far only sparse information exists about P2 receptors in lung (Laubinger and Reiser, 1998). In order to develop therapeutically active nucleotides it is necessary to further identify and quantify P2 receptor binding sites in lung and to fractionate distinct P2 receptors after membrane solubilization for further characterization.

Many physiological processes are mediated by P2Y receptors. In contrast to the P2X receptors which were localized in the different brain areas or in other tissues by subtype specific antibodies (Vulchanova et al., 1996, 1997), the localization of P2Y receptors is not yet possible. The existence of P2Y receptors in brain was demonstrated by binding studies (Simon et al., 1995; Schäfer and Reiser, 1997), by autoradiography with [^{35}S]dATPαS (Simon et al., 1997; Webb et al., 1998), by molecular cloning studies (reviewed in Abbracchio and Burnstock, 1994; Burnstock, 1997) and also indicated by physiological experiments. P2Y receptors (i) reduce noradrenaline release on postganglionic sympathetic neurones (Von Kügelgen et al., 1994), (ii) induce nitric oxide-dependent cGMP production in the neuronal cell line NG 108-15 (Reiser, 1995), (iii) activate microglia (Illes et al., 1996) and rat sensory neurones (Krishtal et al., 1988) and (iv) elicit Ca^{2+}-oscillations associated with Ca^{2+}-influx in rat glioma cells (Czubayko and Reiser, 1996a, 1996b). Moreover, in the CNS in neurones and glial cells, P2Y receptors induce cell differentiation, apoptosis, mitogenesis and release or synthesis of cytokines and neurotrophic factors (reviewed in Neary et al., 1996).

The quantification and localization of the P2Y receptors in the different tissues is a prerequisite for their use as potential therapeutic targets for the treatment of many diseases. Unfortunately, up to date no subtype-specific or selective agonists for P2Y receptors are available. Only one antagonist, N^6-methyl 2'-deoxyadenosine 3', 5'-bisphosphate, selective for the recombinant P2Y$_1$ receptor has been reported most recently (Boyer et al., 1998).

We therefore examined different radiolabelled nucleotides ([^{35}S]ATPαS, [^{3}H]UTP, [^{3}H]Ap$_4$A, [^{3}H]α,β-MeATP) as possible tools for the characterization of P2Y receptors from rat lung and cortical synaptosomes determining (i) the specificity of nucleotide binding, and (ii) the density of receptor binding sites. Furthermore, we used stimulation of [^{35}S]GTPγS binding to G-protein(s) by the different non-radiolabelled nucleotides to demonstrate the existence of functional metabotropic P2Y receptors in the membrane fractions of lung and brain.

Materials and methods

Materials

ATP, ADP, AMP, ATPγS, deoxyATP, ADPβS, α,β-MeATP, ADPβS, Ap$_2$A, Ap$_3$A, Ap$_4$A, Ap$_5$A, Ap$_6$A, GTP, GTPγS, GDP and UTP were from Sigma (Deisenhofen, Germany), ATPαS from Calbiochem (Bad Soden, Germany) and 2-MeSATP from RBI (Biotrend Chemikalien GmbH, Cologne, Germany). Radiolabelled [^{35}S]ATPαS (> 1000 Ci/ mmol; 37 TBq/mmol), [^{3}H]UTP (36.37 Ci/mmol; 1.3 TBq/mmol) and [^{35}S]GTPγS (1250 Ci/mmol; 46.3 TBq/mmol) were from NEN (Brussels, Belgium), [^{3}H]α,β-MeATP (26 Ci/mmol; 962 GBq/mmol) and [^{3}H]Ap$_4$A (11.4 Ci/mmol; 422 GBq/mmol) were from Amersham (Braunschweig, Germany).

Preparation of rat cortex synaptosomal and plasmalemma membranes and lung membranes

Preparation of synaptosomal and plasmalemma membranes was started by dissecting cortices from whole rat brain. They were homogenized in 15 volumes of 0.32 M sucrose, 10 mM Hepes-NaOH pH 7.4 with 10 strokes with a motor-driven pestle (800 rpm). Nuclei and cellular debris were pelleted by centrifuging at $1000 \times g_{max}$ for 5 min. The synaptosomal membranes were enriched by centrifugation of the supernatant at $17,000 \times g_{max}$ (SS34-rotor, Sorvall) for 20 min. Plasma membranes were then prepared from the resulting supernatant by further centrifugation at $44,000 \times g_{max}$ for 20 min. The synaptosomal and plasma membrane pellets were further purified as described recently (Schäfer and Reiser, 1997) by sucrose density gradient centrifugation and stored at a protein concentration of 10 to 15 mg/ml at $-80°C$.

Lungs dissected from rats were washed with sufficient amounts of ice-cold PBS (137 mM NaCl, 2.6 mM KCl, 8.1 mM Na_2HPO_4, and 1.4 mM KH_2PO_4, pH 7.4). Receptor-containing plasma membranes were prepared as described earlier by Laubinger and Reiser (1998). Protein content was determined according to Lowry et al. (1951) using bovine serum albumin as a standard.

Binding experiments

In saturation binding experiments of [^{35}S]ATPαS or [^{3}H]α,β-MeATP total binding was measured by incubating synaptosomal membranes or plasma membranes (25 μg of protein) for 35 min at 4°C in an incubation medium containing 25 mM HEPES, pH 7.4, 50 mM NaCl, 5 mM KCl (binding buffer) with increasing concentrations of [^{35}S]ATPαS (0.1–100 nM) or [^{3}H] α,β-MeATP (0.3–200 nM), respectively in a final assay volume of 0.1 ml. In competition binding experiments binding of 1 nM [^{35}S]ATPαS, 1 nM [^{35}S]UTPαS or 3 nM [^{3}H]α,β-MeATP was determined in binding buffer under the same conditions used for saturation studies in the presence of various non-labelled agonists and antagonists in the concentration range indicated. Similarly, binding of labelled nucleotides to lung membranes (25–50 μg of protein) was measured by incubation in binding buffer (final volume 0.1 ml; 0.2 ml for the [^{3}H]UTP and [^{3}H]Ap$_4$A experiments) with 10 nM [^{3}H]UTP; 10 nM [^{3}H]Ap$_4$A, 1 nM [^{35}S]ATPαS, 3 nM [^{3}H]α,β-MeATP for competition binding experiments. Equilibrium binding was achieved for all radioligands used after about 30 min and binding remained constant for at least 2 h. The incubation was terminated by pelleting the membranes by centrifugation at $25,000 \times g_{max}$ for 12 min. To remove unbound radioligand, pelleted membranes were quickly washed twice (within 10–15 s) with 150 μl of ice-cold binding buffer (300 μl for the [^{3}H]UTP and [^{3}H]Ap$_4$A experiments). Further washes did not affect the amount of ligand bound to membranes. Membrane-bound radioligand was measured in the pellet after solubilization with 100 μl 1.5% (w/w) SDS. In

saturation and competition studies nonspecific binding was determined in the presence of 100 μM of the respective non-radioactively labelled ligands and was 5–15% of total binding. Total binding never exceeded 15% of the ligand concentration used.

The [^{35}S]GTPγS binding assay was performed as described above for the other nucleotides with the following modifications. For lung membranes the binding buffer (10 mM HEPES pH 7.4, 100 mM NaCl and 10 mM $MgCl_2$) contained 140 pM [^{35}S]GTPγS, 0.1 μM GDP and the appropriate amount of Ap$_n$A ($n = 2$, 4 or 5) in a total volume of 0.1 ml. For brain membranes the binding buffer was modified to 500 pM [^{35}S]GTPγS, 2 μM GDP, and ATP or ATPαS. Incubation was started by addition of lung membranes (10 μg of protein) or brain membranes (5 μg of protein) and continued for 45 min at 30°C. Assays were terminated by centrifugation, pellets were washed twice with 150 μl of binding buffer, and analysed as described above. Non-specific binding was determined in the presence of 10 μM GTPγS.

Data analysis

The binding data of saturation and displacement of binding were analysed by nonlinear regression using the RADLIG program (RADLIG program, Biosoft Corp.) with models using one or two binding sites. Experiments were performed with duplicate or triplicate determinations.

Results

Characterization of the [^{35}S]ATPαS binding sites in brain synaptosomes and in lung membranes

The radiolabel [^{35}S]ATPαS was found to be a useful nucleotide to detect ATP binding sites of some P2Y receptors. The [^{35}S]ATPαS binding sites in brain synaptosomes were characterized by using agonists active at P2 receptors. The efficacy in the displacement of high affinity [^{35}S]ATPαS binding of the different nucleotides was highest for ATPαS ($K_i = 21 ± 5$ nM) and ATP ($K_i = 29 ± 5$ nM). They displace binding of 1 nM [^{35}S]ATPαS to synaptosomal membranes with equal potency, followed by ADPβS($K_i = 52 ± 7$nM), 2-MeSATP($K_i = 64 ± 7$nM) and 2′-deoxyATP ($K_i = 107 ± 14$ nM), whereas α,β-

MeATP ($K_i = 518 \pm 119$ nM) was much less potent (Fig. 1A). GTP did not displace [^{35}S]ATPαS binding at concentrations below 100 μM and, like α,β-MeATP and UTP, only partially displaced binding of [^{35}S]ATPαS even at concentrations of up to 1 mM.

The different P2 receptor-active compounds used to compete for high affinity [^{35}S]ATPαS binding in rat synaptosomal membranes demonstrated a rank order of potency of (ATPαS, ATP > ADPβS, 2-MeSATP > 2′-deoxyATP ≫ GTP, α,β-MeATP), which corresponds to the classification of P2Y$_1$ receptors.

Analysis of [^{35}S]ATPαS binding sites in lung membranes was less clear-cut than in brain synaptosomes. In lung [^{35}S]ATPαS displacement by unlabelled ATPαS revealed a single high affinity binding site for [^{35}S]ATPαS with a K_i value of 244 ± 88 nM and a B_{max} value of 165 ± 28 pmol/mg protein (Laubinger and Reiser, 1998). The specificity of the [^{35}S]ATPαS binding site was investigated by application of several other nucleotides known to be potent activators of P2 receptors (Fig. 1B). ATP and 2-MeSATP were one order of magnitude less effective displacing agents than unlabelled ATPαS with K_i values of 1.96 ± 0.30 μM and 2.5 ± 0.6 μM, respectively. Interestingly both α,β-MeATP, a specific ligand of some P2X receptors, and ADPβS were very potent displacers of [^{35}S]ATPαS. As ADPβS is claimed to be a potent agonist of P2Y$_1$ receptors rather than of P2X receptors our heterologous displacement data are in contrast to our data analysing the homologous ligand ATPαS. The assumption that ATPαS binds to more than only one binding site was further substantiated by the incomplete displacement of ATPαS by α,β-MeATP (70% to 80%) and the extraordinary high density of ATPαS binding sites (about 165 pmol per mg protein). Thus, in lung tissue apart from binding experiments additional physiological investigations are necessary to elucidate the proper identity of [^{35}S]ATPαS binding sites.

Characterization of the [^{3}H]UTP binding site in lung membranes

In lung membranes the specific binding activity of [^{3}H]UTP at a concentration of 10 nM was about 0.16 pmol/mg protein (Fig. 2). Data analysis revealed a single high affinity binding site for [^{3}H]UTP with a K_d of 43 ± 9 nM (Laubinger and Reiser, 1998). This binding site was very specific for UTP. Among the nucleotides tested only the pyrimidine nucleotide UDP could displace [^{3}H]UTP from its binding site, although much weaker than UTP. At a concentration of 100 μM only 65% of the receptor-bound [^{3}H]UTP were displaced by UDP.

In physiological experiments some UTP receptors, as P2Y$_2$, are almost equally stimulated by UTP and ATP (Burnstock, 1997). In lung membranes, however, neither ATP nor ATP analogues, like ATPαS, 2-MeSATP, and α,β-MeATP were able to displace UTP from its binding site. On the contrary, ATP or ATP analogues had even a significantly stimulatory effect on [^{3}H]UTP binding at concentrations above 1 μM. Thus, in lung membranes in the presence of 10 μM ATPαS more than twice as much [^{3}H]UTP was bound than in the absence of other nucleotides (Fig. 2).

Evidence for a specific [^{3}H]Ap$_4$A binding site in rat lung

As Ap$_4$A specifically activates the recombinant human P2Y$_2$ receptor (Lazarowski et al., 1995), [^{3}H]Ap$_4$A was applied to lung membranes in order to test whether there are specific Ap$_4$A binding sites which might be identical to or distinct from known P2Y$_2$ receptors. [^{3}H]Ap$_4$A bound to membranes from rat lung in a time and concentration dependent manner. At a concentration of 10 nM maximal binding of [^{3}H]Ap$_4$A was reached after an incubation time of 35 min at 4°C and remained constant for at least 60 min. Saturation binding studies with increasing concentrations of [^{3}H]Ap$_4$A from 0.1 to 200 nM revealed a single high-affinity binding site for Ap$_4$A with a K_d of 91 ± 9 nM (Laubinger and Reiser, 1998). At a concentration of 10 nM, [^{3}H]Ap$_4$A bound to lung membranes with a specific binding density of about 0.6 pmol/mg protein.

The specificity of the [^{3}H]Ap$_4$A binding site was investigated by applying nucleotides known to be potent agonists of different P2 receptors (Fig. 3). The mononucleotide ATP, a potent agonist of P2X and P2Y receptors, had a 10-times lower affinity for this binding site than Ap$_4$A, giving a K_i value of

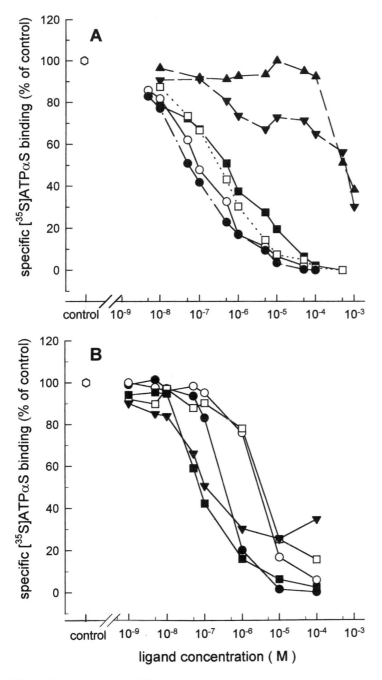

Fig. 1. Specific binding of [^{35}S]ATPαS. Homologous (●) and heterologous displacement of [^{35}S]ATPαS (1 nM) binding determined with (A) rat brain cortical synaptosomal membranes (25 μg of protein) and (B) lung membranes (25 μg of protein). Competing ligands were ATP (○), α,β-MeATP (▼), ADPβS (■), 2-MeSATP (□) and GTP (▲). 100% binding represents 0.32 (A) and 0.49 (B) pmol/mg protein, respectively.

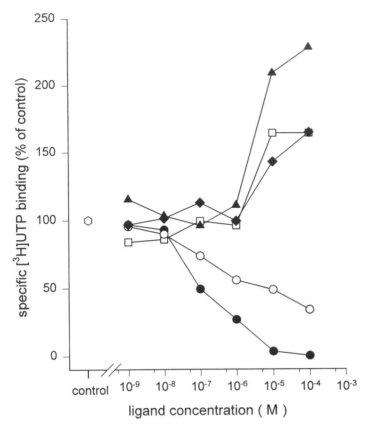

Fig. 2. Binding of 10 nM [³H]UTP to rat lung membranes (50 μg of protein) in the simultaneous presence of various nucleotides. 100% binding (corresponds to 0.16 pmol/mg protein) was competed by UTP (●) and UDP (○) and was enhanced by ATP (□), ATPαS (▲) and 2-MeSATP (◆).

0.9 μM ($n = 3$). α,β-MeATP, a very specific agonist of some P2X receptors was only a poor displacer at relatively high concentrations (above 100 μM) and UTP was unable to displace [³H]Ap₄A, even at high concentrations. Thus, these data reveal that [³H]Ap₄A binding sites which are not identical with P2Y₂ receptors can be labelled in lung membranes.

The influence of GTP on the binding of [³H]Ap₄A to lung membranes was investigated to explore whether signal transduction of G-protein coupled P2Y receptors is involved. Preincubation of lung membranes with 10 μM GTP or the hydrolysis-resistant GTP analogue GTPγS in the presence of 5 mM MgCl₂ showed additional low affinity binding sites for Ap₄A with a K_i value of 5.1 μM in addition to remaining high affinity binding sites.

Effect of diadenosine polyphosphates on [³⁵S]ATPαS binding in brain synaptosomes

Diadenosine polyphosphates had been reported to exert physiological effects on native and recombinantly expressed P2X- and P2Y-receptors (Pintor et al., 1993, 1996; Lazarowski et al., 1995). Therefore we examined the effect of diadenosine polyphosphates on high affinity [³⁵S]ATPαS binding of synaptosomal membranes (Fig. 4). None of the diadenosine polyphosphates tested (Ap₃A, Ap₄A, Ap₅A, and Ap₆A) showed high affinity for the [³⁵S]ATPαS binding site in synaptosomal membranes. Only Ap₃A with a K_i value of 7.2 ± 1.3 μM ($n = 3$) showed a reasonable affinity for the [³⁵S]ATPαS binding site. Interestingly, the diadenosine polyphosphates Ap₄A, Ap₅A, and Ap₆A enhanced [³⁵S]ATPαS binding gradually within the concen-

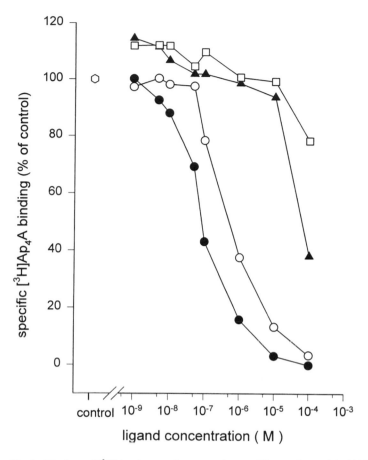

Fig. 3. Binding of [³H]Ap₄A to rat lung membranes (50 μg of protein). 100% binding (0.63 pmol/mg protein) was inhibited at the concentrations indicated by Ap₄A (●), ATP (○), α,β-MeATP (▲) and UTP (□).

tration range from 1 to 50 μM. Binding of 1 nM [³⁵S]ATPαS was increased in a concentration-dependent manner by Ap₄A, Ap₅A, Ap₆A with a maximal stimulation above control binding of 15%, 60%, and 60%, respectively. Thus, the diadenosine polyphosphates having a longer phosphate chain seem to interact with [³⁵S]ATPαS binding proteins, which are likely to be P2Y receptors at an allosteric site.

Ap₄A receptor-stimulated binding of [³⁵S]GTPγS in lung membranes

Stimulation of [³⁵S]GTPγS binding by application of receptor ligands is a useful tool in studying G-protein-coupled receptors (Akam et al., 1997). Indirect evidence for the interaction of Ap₄A

binding sites with G-proteins was already obtained from the preincubation experiments using GTPγS/MgCl₂. Therefore, further evidence for the activation of G-proteins by P2 receptors binding diadenosine polyphosphates was sought from experiments testing the effect of increasing concentrations of Ap$_n$A ($n = 2,4,5$) on [³⁵S]GTPγS binding to lung membranes. In initial experiments the conditions for [³⁵S]GTPγS binding were optimized; applying different concentrations of [³⁵S]GTPγS, GDP, and NaCl, stimulatory effects on [³⁵S]GTPγS binding to lung membranes caused by the P2 receptor ligand Ap₄A (100 μM) were investigated. With 140 pM [³⁵S]GTPγS highest Ap₄A-stimulated [³⁵S]GTPγS binding was found at a concentration of 0.1 μM GDP, independent of a low (10 mM) or a high (100 mM) concentration of NaCl in the incubation assay. Maximal stimulation was reached

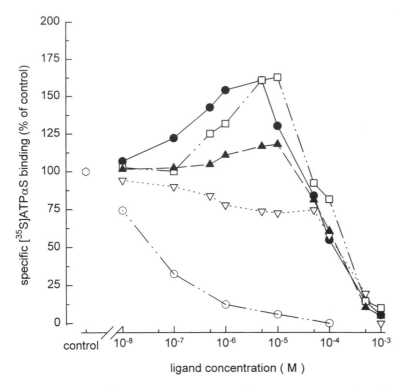

Fig. 4. Binding of [^{35}S]ATPαS (0.5 nM) to rat brain cortex synaptosomal membranes (25 μg of protein) in the presence of various concentrations of diadenosine polyphosphates. 100% binding (0.23 pmol/mg protein) was affected by ATPαS (○), Ap$_3$A (▽), Ap$_4$A (▲), Ap$_5$A (□) and Ap$_6$A (●).

after an incubation time of 45 min at 30°C (data not shown).

Ap$_4$A stimulated [^{35}S]GTPγS binding to lung membranes (Fig. 5A) which indicates the activation of G-proteins by P2 receptors specifically binding diadenosine polyphosphates. Stimulation was concentration dependent with a maximal stimulation of [^{35}S]GTPγS binding of $63.5 \pm 8.9\%$ ($n = 5$) over basal binding (binding in absence of Ap$_n$A) at a concentration of 1 μM Ap$_4$A. Similar stimulation rates were found with the diadenosine polyphosphates Ap$_2$A and Ap$_5$A.

Stimulation of [^{35}S]GTPγS binding by nucleotides to brain plasma membranes and synaptosomal membranes.

The stimulation of [^{35}S]GTPγS binding to heterotrimeric G-proteins by agonists has been used also in brain to confirm the functional existence of different G-protein-coupled receptors (Traynor and Nahorski, 1995; Sim et al., 1998). We employed agonist-stimulated [^{35}S]GTPγS binding to demonstrate the presence of P2Y receptors in plasma membranes and synaptosomal membranes from rat brain and to test whether ATPαS and ATP exhibit the same agonist efficacy for G-protein activation by P2Y receptors. In the presence of 2 μM GDP, ATPαS enhanced binding of 0.5 nM [^{35}S]GTPγS in brain plasma membranes in a concentration dependent manner maximally by $42 \pm 4.0\%$ over basal. Similarly ATP stimulated [^{35}S]GTPγS binding up to $34 \pm 3.2\%$ over basal (Fig. 5B). Stimulation of [^{35}S]GTPγS binding was also found in synaptosomal membranes with a somewhat lower efficacy for both agonists. At a concentration of 0.1 μM ATPαS and ATP, [^{35}S]GTPγS binding was stimulated up to 21% over basal (data not shown). These results demonstrate that ATPαS has the same potency in stimulating [^{35}S]GTPγS binding as ATP in both membrane fractions.

Density and affinity of nucleotide binding sites in rat brain synaptosomal membranes and lung membranes

The synaptosomal membranes and plasma membranes obtained from rat brain cortex were examined whether they contained high affinity binding sites for [^{35}S]ATPαS or [^{3}H]α,β-MeATP. The analysis of the binding data of [^{35}S]ATPαS to synaptosomal membranes at a concentration range from 0.1 to 100 nM clearly revealed the existence of a high affinity binding site ($K_d = 22.2 \pm 9.1$ nM; $n = 5$) with a maximal binding capacity (B_{max}) of 14.8 ± 2.3 pmol/mg protein. When specific binding of [^{35}S]ATPαS was examined by displacement of binding by non-labelled ATPαS, the analysis of the data also revealed a K_i-value of 21 ± 5 nM ($n = 7$). Binding of [^{35}S]ATPαS to the plasma membranes also exhibited a high affinity binding site with K_d of 29.1 ± 5.8 nM ($n = 3$) and a maximal binding capacity (B_{max}) of 9.51 ± 2.17 pmol/mg protein. Thus, similar densities of [^{35}S]ATPαS binding sites

with almost identical affinity are found for synaptosomal and plasma membranes.

Analysis of the binding of the P2X receptor-selective ligand [^{3}H]α,β-MeATP to both membrane fractions revealed a high affinity binding site with $K_d = 13.7 \pm 1.8$ nM ($n = 5$) for synaptosomal membranes whereas plasma membranes contained a high affinity binding site with a slightly lower affinity ($K_d = 38.4 \pm 10.2$ nM; $n = 3$) with maximal binding capacities of $B_{max} = 6.34 \pm 0.28$ pmol/mg protein for synaptosomal membranes and of $B_{max} = 4.81 \pm 1.2$ for plasma membranes.

Similarly densities of nucleotide receptors in membranes of rat lung were determined by incubating lung membranes with radioactively labelled purine and pyrimidine nucleotide agonists at concentrations ranging from 0.1 to 200 nM. The binding data of saturation studies were analysed by non-linear regression models with one or two binding sites. This analysis yielded a single binding site for each of the nucleotides under investigation. The K_d values were 43.7 ± 9.1 nM for [^{3}H]UTP,

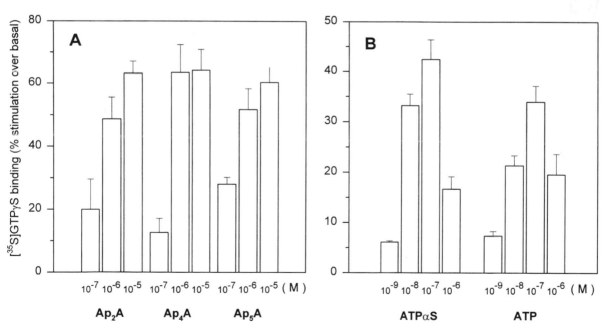

Fig. 5. G-protein coupling of nucleotide ligand binding sites in rat lung membranes (A) and in rat brain cortex plasma membranes (B). (A) Lung membranes (10 μg of protein) were incubated with 0.14 nM [^{35}S]GTPγS and indicated concentrations of agonists Ap$_2$A, Ap$_4$A, Ap$_5$A. Data are presented as % stimulation over basal of specific binding in the absence of agonist. Unspecific binding was determined in the presence of 10 μM GTPγS. (B) Brain cortex plasma membranes (5 μg of protein) were incubated with 0.5 nM [^{35}S]GTPγS and the indicated concentrations of ATP and ATPαS. Unspecific binding was measured in the presence of 100 μM GTPγS.

243.7 ± 87.9 nM for [^{35}S]ATPαS, 56.5 ± 10.0 nM for [^{3}H]α,β-MeATP, and 91.2 ± 8.8 nM for [^{3}H]Ap$_4$A. Maximal binding capacities (B_{max}) were 1.6 ± 0.9 pmol/mg protein for [^{3}H]UTP, 164.7 ± 27.6 pmol/mg protein for [^{35}S]ATPαS, 5.7 ± 1.0 pmol/mg protein for [^{3}H]α,β-MeATP, and 6.9 ± 1.6 pmol/mg protein for [^{3}H]Ap$_4$A, respectively.

Discussion

So far the lack of receptor-specific ligands for nucleotide P2Y receptors enables localization and quantification of these receptors only by means of labelled nucleotides. Such attempts were successful in some cases whereas in others the results turned out to be erroneous as dicussed below. Moreover, the method of radioligand binding analysis of the receptors may also yield some new insights of possible physiological significance. Thus we propose that the enhancement of (i) [^{3}H]UTP binding in lung membranes by adenine nucleotides (Fig. 1) and (ii) [^{35}S]ATPαS binding in brain synaptosomal membranes by diadenosine polyphosphate nucleotides (Ap$_4$A, Ap$_5$A, Ap$_6$A) suggests a potentiation of the respective nucleotide effects. In line with this interpretation, recently Jiménez et al. (1998) reported a potentiation of the ATP-induced Ca^{2+} response in rat cerebellar astrocytes by Ap$_5$A.

Nucleotide radiolabels have been used successfully to localize and to quantify P2Y receptors in brain tissue (Simon et al., 1997). Saturable and specific binding sites for [^{35}S]dATPαS were detected in adult rat brain membranes and were found to have characteristics expected for those of P2Y$_1$ receptors. These receptors are abundant and widely distributed within the neuraxis (Simon et al., 1997). Using the ligand [^{35}S]ATPαS for rat brain synaptosomal and plasma membranes we obtained similar values for K_d and B_{max}, also with properties indicative for P2Y receptors, as discussed below. Moreover, a detailed analysis of the distribution of P2Y$_1$ receptors was carried out in chick brain where Webb et al. (1998) compared [^{35}S]dATPαS binding by in-vitro autoradiography with in-situ hybridization localizing P2Y$_1$ messenger RNA.

High affinity [^{35}S]ATPαS binding sites were detected in synaptosomal and plasma membranes from rat brain cortex. The rank order in the displacement potency of high affinity [^{35}S]ATPαS binding to rat brain synaptosomes shows an affinity profile characteristic for P2Y$_1$ receptors. The K_i-values of the most potent agonists (ATPαS, ATP, ADPβS) are below or within the range normally reported for the EC$_{50}$ values found for these compounds in physiological responses. However, the rank order of the displacement potency of the different P2 receptor agonists by itself does not prove that the [^{35}S]ATPαS binding proteins indeed represent P2Y receptors as membranes contain numerous proteins which can bind [^{35}S]ATPαS including ecto-ATPases, nicotinic acetylcholine receptors (β-subunit), the family of ABC proteins (e.g. CFTR, P-glycoprotein) and certainly also P2X receptors.

Recently, specific labelling of binding sites for [^{3}H]α,β-MeATP on recombinant P2X receptors expressed in CHO-K1 cells with [^{35}S]ATPγS or [^{33}P]ATP was shown (Michel et al., 1996). We therefore used [^{3}H]α,β-MeATP to examine if this ligand labels the same P2Y-like binding sites detected with [^{35}S]ATPαS in synaptosomal membranes (Schäfer and Reiser, 1997). The [^{3}H]α,β-MeATP binding sites were found to be completely different from the P2Y receptor-like sites detected by [^{35}S]ATPαS shown by (i) the different affinity profile of the P2 receptor agonists at the [^{3}H]α,β-MeATP binding site, and (ii) the opposite regulation of binding by the divalent cations Ca^{2+} and Mg^{2+} (Schäfer and Reiser, 1997). Furthermore, the binding affinity of [^{35}S]ATPαS was decreased in the presence of Mg^{2+}/GTPγS, as expected for the binding of a radioligand to a G-protein-coupled receptor. This suggests that the high affinity [^{35}S]ATPαS binding proteins in brain comprise indeed P2Y receptors.

The presence of functional G-protein coupled P2Y receptors in synaptosomal membranes and plasma membranes could be demonstrated by the stimulation of [^{35}S]GTPγS binding with ATPαS or ATP, as shown here, and the selective regulation of [^{35}S]ATPαS binding by GTPγS/Mg^{2+} reported recently (Schäfer and Reiser, 1997). This proves the usefulness of [^{35}S]ATPαS to label and identify

P2Y receptors in membrane preparations from rat brain. Nevertheless, the identity of the binding sites in brain membranes with certain subtypes of the P2Y receptors has to be confirmed by molecular cloning studies, immunochemistry with subtype-specific antibodies, which are not yet available or by the use of new selective antagonists, such as N^6-methyl 2′deoxyadenosine 3′,5′-bisphosphate for P2Y$_1$ receptor (Boyer et al., 1998).

However, [^{35}S]ATPαS cannot be used as a general ligand to specifically characterize P2Y receptors in every tissue, since the complex binding characteristics of [^{35}S]ATPαS to rat lung plasma membranes do not allow to discriminate between specific binding to either P2Y or P2X receptors (Laubinger and Reiser, 1998). This is also true for the binding characteristics of [^{3}H]α,β-MeATP (see Michel et al., 1996) and [^{35}S]ATPγS or [α-^{32}P]ATP (Motte et al., 1996). Binding to rat brain and rat vas deferens possibly reflected different P2X receptors, while binding to a 5′-nucleotidase was demonstrated for an endothelial-derived cell line (see Michel et al. 1996). Similarly, Motte et al. (1996) concluded from the displacement of [^{35}S]ATPγS or [α-^{32}P]ATP binding by 2-MeSATP that two thirds of the binding sites on bovine aortic endothelial membranes did not correspond to P2Y receptors.

In lung Ap$_4$A receptor sites might be of great physiological significance. By radioligand analysis we have detected a specific G-protein coupled Ap$_4$A receptor binding site in rat lung membranes. Diadenosine polyphosphates had been reported to interact either with P2Y receptors (Castro et al., 1994; Ralevic et al., 1995) and P2X receptors (Pintor et al., 1996) or with a separate receptor which is specific for diadenosine polyphosphates (Pintor and Miras-Portugal, 1995). Nucleotide binding receptors play a crucial role in the regulation of Cl$^-$ and water transport processes in lung, because in lung epithelial cells Cl$^-$ channels can be regulated by the nucleotides ATP and UTP via activation of P2 receptors (P2Y$_2$) which are located in the apical membrane of the epithelial cells (Hwang et al., 1996). UTP is already successfully applied as therapeutic agent in treatment of cystic fibrosis (Olivier et al., 1996). However, rapid degradation of UTP makes it necessary to find more stable nucleotides that can be used as P2Y receptor agonists. One possible candidate is the relatively stable diadenosine polyphosphate Ap$_4$A (Lazarowski et al., 1995).

So far only little is known about the densities and characteristics of different P2 receptors in lung tissue (Laubinger and Reiser, 1998). Binding studies using the radiolabelled nucleotide [^{3}H]Ap$_4$A may be a helpful tool for the characterization of Ap$_4$A receptors. Until now, it is still completely unclear whether there are distinct P2 receptors for Ap$_4$A in lung or whether Ap$_4$A is merely a potent agonist of P2Y receptors already identified. As [^{3}H]Ap$_4$A was not displaced by α,β-MeATP or by UTP, which are specific ligands for P2X and P2Y$_2$ receptors, respectively, this is the first evidence for the existence of a possibly new receptor specific for Ap$_4$A in lung membranes. Nevertheless, only purification of this receptor binding site and separation from other nucleotide binding sites can finally prove the existence of a distinct receptor for Ap$_4$A.

In synaptosomal membranes all diadenosine polyphosphates tested possessed low affinity for the P2Y receptor-like [^{35}S]ATPαS binding sites, indicating the existence of separate receptors specific for diadenosine polyphosphates. High affinity binding sites specific for Ap$_4$A in rat brain had been reported (Rodríguez-Pascual et al., 1997). However, the enhancement of [^{35}S]ATPαS binding to synaptosomal membranes by Ap$_5$A and Ap$_6$A in the concentration range of 1 to 100 μM may be consistent with the finding reported very recently where ATP-induced Ca^{2+} responses of rat cerebellar astrocytes could be potentiated by costimulation with 0.1 μM Ap$_5$A (Jiménez et al., 1998), indicating a new mechanism of action for diadenosine polyphosphates possibly due to a cooperative effect on the P2Y receptor-like [^{35}S]ATPαS binding proteins.

In the past in some cases analysis based solely on radioligand nucleotide binding turned out to be erroneous. P2Y$_5$ and P2Y$_7$ which were by sequence homology and by [^{35}S]dATPαS ligand binding claimed to be genuine nucleotide receptors (Webb et al., 1996; Akbar et al., 1996) were found to be activated physiologically by a completely different agonist (Yokomizo et al., 1997) or questioned recently on the basis of a physiological analysis of dATPαS being a partial P2Y$_1$ receptor agonist (Schachter and Harden, 1997). The latter authors moreover doubt whether a large quantity of high-affinity binding sites can be attributed to P2Y receptor subtypes. Here we see that in lung membranes the maximal binding capacity for [^{35}S]ATPαS is very high. Therefore, these binding sites certainly comprise non-P2 receptor binding sites. However, for the enriched fraction of synaptosomal membranes from brain the number of binding sites probably largely corresponds to P2Y receptor sites.

Radiolabelled nucleotides are a valuable tool for analysing the density of certain P2Y receptors in membranes or after solubilization and purification of P2Y receptors. However, we have to be cautious about interpreting the binding data obtained with radiolabelled nucleotides. A definitive characterization of P2Y receptors will only be possible by combining these results with those obtained from physiological experiments.

Conclusions

Extracellular nucleotides have a wide range of physiological effects in many different tissues through activation of specific P2 receptors, which comprise transmitter gated ion channels (P2X receptors) and G-protein linked metabotropic P2Y receptors. So far there is hardly any knowledge about the distribution and density of P2Y receptors. We prepared membranes from rat lung, and plasma membranes or synaptosomal membranes from rat brain cortex for radioligand binding analysis. [^{3}H]UTP was used to label uridine nucleotide selective P2Y receptor types in lung membranes. Stimulation of [^{3}H]UTP binding ($K_d = 43$ nM) by ATP at concentrations above 1 μM indicates possible potentiation of UTP effects by other nucleotides. Similarly, in rat synaptosomal membranes, binding of [^{35}S]ATPαS ($K_d = 22$ nM) which was used to label preferentially P2Y$_1$ receptors was enhanced by the additional presence of diadenosine polyphosphates. In both tissues, lung and brain, binding density for UTP was considerably lower than for [^{3}H]α,β-MeATP, a P2X specific ligand, for [^{35}S]ATPαS and for [^{3}H]Ap$_4$A. In brain membranes ligand specificity of ATPαS binding was consistent with P2Y$_1$ receptor pharmacology, whereas in lung membranes ATPαS seemed to label various types of receptors. [^{3}H]Ap$_4$A could identify a new receptor type in lung membranes ($K_d = 91$ nM) which was not apparent in brain under the same conditions. G-protein coupling of this receptor binding site was confirmed by stimulation of [^{35}S]GTPγS binding by Ap$_4$A equipotently with ligand binding affinity. Thus, nucleotide labels can be used for tissue-specific analysis of P2Y receptor types.

Acknowledgement

We gratefully acknowledge the expert technical assistance of P. Hennig. The work was supported by a project grant from Deutsche Forschungsgemeinschaft (Re 563/9-1), a grant from the Mukoviszidose e.V. and Fonds der Chemischen Industrie.

References

Abbracchio, M.P. and Burnstock, G. (1994) Purinoceptors: are there families of P2X and P2Y purinoceptors? *Pharmacol. Ther.,* 64: 445–475.

Akam, E.C., Carruthers, A.M., Nahorski, S.R. and Challiss, R.A.J. (1997) Pharmacological characterization of type 1α metabotropic glutamate receptor-stimulated [^{35}S]-GTPγS binding. *Br. J. Pharmacol.,* 121: 1203–1209.

Akbar, G.K.M., Dasari, V.R., Webb, T.E., Ayyanathan, K., Pillarisetti, K., Sandhu, A.K., Athwal, R.S., Daniel, J.L., Ashby, B., Barnard, E.A. and Kunapuli, S.P. (1996) Molecular cloning of a novel P2 purinoceptor from human erythroleukemia cells. *J. Biol. Chem.,* 271: 18363–18367.

Boyer, J.L., Mohanram, A., Camaioni, E., Jacobson, K.A. and Harden, T.K. (1998) Competitive and selective antagonism of P2Y$_1$ receptors by N^6-methyl 2′-deoxyadenosine 3′,5′-bisphosphate. *Br. J. Pharmacol.,* 124: 1–3.

Burnstock, G. (1997) The past, present and future of purine nucleotides as signalling molecules. *Neuropharmacology,* 36: 1127–1139.

Castro, E., Tomé, A.R., Miras-Portugal, M.T. and Rosario, L.M. (1994) Single cell fura-2 microfluorometry reveals different purinoceptor subtypes coupled to Ca^{2+} influx and intracellular Ca^{2+} release in bovine adrenal chromaffin and endothelial cells. *Pflügers Arch.,* 426: 524–533.

Czubayko, U. and Reiser, G. (1996a) P$_{2U}$ nucleotide receptor activation in rat glial cell line induces [Ca^{2+}]$_i$ oscillations which depend on cytosolic pH. *Glia,* 16: 108–116.

Czubayko, U. and Reiser, G. (1996b) Desensitization of P$_{2U}$ receptor in neuronal cell line. Differential control by the agonists ATP and UTP, as demonstrated by single-cell Ca^{2+} responses. *Biochem. J.,* 320: 215–219.

Edwards, F.A., Gibb, A.J. and Colquhoun, D. (1992) ATP receptor-mediated synaptic currents in the central nervous system. *Nature,* 359: 144–147.

Evans, R.J., Derkach, V. and Surprenant, A. (1992) ATP mediates fast synaptic transmission in mammalian neurons. *Nature,* 357: 503–505.

Hwang, T. H., Schwiebert, E. M. and Guggino, W. B. (1996) Apical and basolateral ATP stimulates tracheal epithelial chloride secretion via multiple purinergic receptors. *Am. J. Physiol.,* 270: C1611–C1623.

Illes, P., Nörenberg, W. and Gebicke-Haerter, P.J. (1996) Molecular mechanisms of microglial activation. B. Voltage- and Purinoceptor-operated channels in microglia. *Neurochem. Int.,* 29: 13–24.

Jiménez, A.I., Castro, E., Delicado, E.G. and Miras-Portugal, M.T. (1998) Potentiation of adenosine 5'-triphosphate calcium responses by diadenosine pentaphosphate in individual rat cerebellar astrocytes. *Neurosci. Lett.,* 246: 109–111.

Krishtal, O.A., Marchenko, S.M., Obukhov, A.G. and Volkova, T.M. (1988) Receptors for ATP in rat sensory neurones: The structure–function relationship for ligands. *Br. J. Pharmacol.,* 95: 1057–1062.

Laubinger, W. and Reiser, G. (1998) Differential characterization of binding sites for adenine and uridine nucleotides in membranes from rat lung as possible tools for studying P2 receptors in lung. *Biochem. Pharmacol.,* 55: 687–695.

Lazarowski, E. R., Watt, W. C., Stutts, M. J., Boucher, R. C. and Harden, T. K. (1995) Pharmacological selectivity of the cloned human P_{2U}-purinoceptor: potent activation by diadenosine tetraphosphate. *Br. J. Pharmacol.,* 116: 1619–1627.

Lowry, O.H., Rosebrough, M.H., Farr, A.L. and Randall, R.J. (1951) Protein measurement with the folin reagent. *J. Biol. Chem.,* 193: 265–275.

Michel, A.D., Lundström, K., Buell, G.N., Surprenant, A., Valera, S. and Humphrey, P.P.A. (1996) The binding characteristics of a human bladder recombinant P_{2X} purinoceptor, labelled with [^{3}H]-αβmeATP, [^{35}S]-ATPγS or [^{33}P]-ATP. *Br. J. Pharmacol.,* 117: 1254–1260.

Motte, S., Swillens, S. and Boeynaems, J.M. (1996) Evidence that most high-affinity ATP binding sites on aortic endothelial cells and membranes do not correspond to P2 receptors. *Eur. J. Pharmacol.,* 307: 201–209.

Neary, J.T., Rathbone, M.P., Cattabeni, F., Abbracchio, M.P. and Burnstock, G. (1996) Trophic actions of extracellular nucleotides and nucleosides on glial and neuronal cells, *Trends Neurosci.,* 19: 13–18.

North, R.A. (1996) P2X purinoceptor plethora. *Semin. Neurosci.,* 8: 187–194.

Olivier, K. N., Bennett, W. D., Hohnecker, K. W., Zeman, K. L., Edwards, L. J., Boucher, R. C. and Knowles, M. R. (1996) Acute safety and effects on mucociliary clearance of aerosolized uridine 5'-triphosphate +/ − amiloride in normal human adults. *Am. J. Respir. Crit. Care Med.,* 154: 217–223.

Pintor, J., Diaz-Rey, M.A. and Miras-Portugal, M.T. (1993) Ap₄A and ADP-β-S binding to P2-purinergic receptors present on rat brain synaptic terminals. *Br. J. Pharmacol.,* 108: 1094–1099.

Pintor, J. and Miras-Portugal, M.T. (1995) A novel receptor for diadenosine polyphosphates coupled to calcium increase in rat midbrain synaptosomes. *Br. J. Pharmacol.,* 115: 895–902.

Pintor, J., King, B.F., Miras-Portugal, M.T. and Burnstock, G. (1996) Selectivity and activity of adenine dinucleotides at recombinant $P2_{X2}$ and $P2_{Y1}$ purinoceptors. *Br. J. Pharmacol.,* 119: 1006–1012.

Ralevic, V., Hoyle, C.H.V. and Burnstock, G. (1995) Pivotal role of phosphate chain length in vasoconstrictor versus vasodilator action of adenine dinucleotides in rat mesenteric arteries. *J. Physiol.,* 483: 703–713.

Reiser, G. (1995) Ca^{2+}- and nitric oxide-dependent stimulation of cyclic GMP synthesis in neuronal cell line induced by P2-purinergic/pyrimidinergic receptor. *J. Neurochem.,* 64: 61–68.

Robertson, S.J., Rae, M.G., Rowan, E.G. and Kennedy, C. (1996) Characterization of a P2X-purinoceptor in cultured neurones of the rat dorsal root ganglion. *Br. J. Pharmacol.,* 118: 951–956.

Rodríguez-Pascual, F., Cortes, R., Torres, M., Palacios, J.M. and Miras-Portugal, M.T. (1997) Distribution of [^{3}H]diadenosine tetraphosphate binding sites in rat brain. *Neuroscience,* 77: 247–255.

Schachter, J.B. and Harden, T.K. (1997) An examination of deoxyadenosine 5'(α-thio)triphosphate as a ligand to define P2Y receptors and its selectivity as a low potency partial agonist of the P2Y1 receptor. *Br. J. Pharmacol.,* 121: 338–344.

Schäfer, R. and Reiser, G. (1997) Characterization of [^{35}S]-ATPαS and [^{3}H]-a,β-MeATP binding in rat brain cortical synaptosomes: Regulation of ligand binding by divalent cations. *Br. J. Pharmacol.,* 121: 913–922.

Sim, L.J., Liu, Q., Childers, S.R. and Selley, D.E. (1998) Endomorphin-stimulated [^{35}S]GTPγS binding in rat brain: Evidence for partial agonist activity at μ-opioid receptors. *J. Neurochem.,* 70: 1567–1576.

Simon, J., Webb, T.E. and Barnard, E.A. (1995) Characterization of a P_{2Y} purinoceptor in the brain. *Pharmacol. Toxicol.,* 76: 302–307.

Simon, J., Webb, T.E. and Barnard, E.A. (1997) Distribution of [^{35}S]dATPαS binding sites in the adult rat neuraxis. *Neuropharmacol.,* 36: 1243–1251.

Sun, X.P. and Stanley, E.F. (1996) An ATP-activated, ligand-gated ion channel on a cholinergic presynaptic nerve terminal. *Proc. Natl. Acad. Sci. USA,* 93: 1859–1863.

Traynor, J.R. and Nahorski, S.R. (1995) Modulation by μ-opioid agonists of guanosine-5'-O-(3-[^{35}S]thio)triphosphate binding to membranes from human neuroblastoma SH-SY5Y cells. *Mol. Pharmacol.,* 47: 848–854.

Von Kügelgen, I., Späth, L. and Starke, K. (1994) Evidence for P_2-purinoceptor-mediated inhibition of noradrenaline release in rat brain cortex. *Br. J. Pharmacol.,* 113: 815–822.

Vulchanova, L., Arvidsson, U., Riedl, M., Wang, J., Buell, G., Surprenant, A., North, R.A. and Elde, R. (1996) Differential distribution of two ATP-gated ion channels (P_{2X} receptors) determined by immunocytochemistry. *Proc. Natl. Acad. Sci. USA,* 93: 8063–8067.

Vulchanova, L., Riedl, M.S., Shuster, S.J., Buell, G., Surprenant, A., North, R.A. and Elde, R. (1997) Immunohistochemical study of the $P2X_2$ and $P2X_3$ receptor subunits in rat and monkey sensory neurons and their central terminals. *Neuropharmacol.,* 36: 1229–1242.

Webb, T.E., Kaplan, M.G. and Barnard, E.A. (1996) Identification of 6H1 as a P_{2Y} purinoceptor: $P2Y_5$. *Biochem. Biophys. Res. Commun.,* 219: 105–110.

Webb, T.E., Simon, J. and Barnard, E.A. (1998) Regional distribution of [^{35}S]2′-deoxy 5′-O-(1-thio)ATP binding sites and the P2Y1 messenger RNA within the chick brain. *Neuroscience,* 84: 825–837.

Yokomizo, T., Izumi, T., Chang, K. Takuwa, Y. and Shimizu, T. (1997) A G-protein-coupled receptor for leukotriene B4 that mediates chemotaxis. *Nature,* 387: 620–624.

Molecular biology of
P2X receptors

P. Illes and H. Zimmermann (Eds.)
Progress in Brain Research, Vol 120
© 1999 Elsevier Science BV. All rights reserved

CHAPTER 6

Evolving view of quaternary structures of ligand-gated ion channels

Annette Nicke, Jürgen Rettinger, Cora Büttner, Annette Eichele, Günter Lambrecht, and Günther Schmalzing*

Pharmakologisches Institut für Naturwissenschaftler, Johann Wolfgang Goethe-Universität, Biozentrum N 260, Marie-Curie-Str. 9, D-60439 Frankfurt/Main, Germany

Introduction

Last year has brought significant progress in our insight into the structure of ion channels. First, with the high resolution crystallography of inward rectifier K^+ channels, we have now for the first time the benefits of a structure that provides details of the ion pore at 3.4 Å resolution, answering such fundamental biophysical questions of how channel gating and ion selectivity work. Second, congruent with the K^+ channel-like re-entrant loop of ionotropic glutamate receptors, there is recent evidence from both biochemical and electrophysiological data that glutamate receptors share a tetrameric architecture with K^+ channels. Hence, also in regard of their quaternary structure, glutamate receptors are dissimilar to the members of the nicotinic acetylcholine receptor superfamily (encompassing acetylcholine, serotonin, glycine and GABA receptors), which all exhibit a pentameric architecture. Likewise, a tetrameric architecture has been described this year for epithelial Na^+ channels including the related FRMamide-gated Na^+ channel of *Helix aspersa*, which also resemble K^+ channels in having a re-entrant loop that contributes to the ion pore.

Finally, recent biochemical data suggest that the third major class of ligand-gated ion channels, the ATP-gated P2X receptors, neither follow the architectural pattern of nicotinic acetylcholine receptors nor of glutamate receptors, but rather exhibit a novel trimeric architecture. We are therefore confronted with an unexpected repertoire of at least three distinct pore forming arrangements of subunits of ligand-gated ion channels that exhibit similar channel properties such as non-selective permeability for cations, not considering the ionotropic receptors for glycine and GABA, which are selective for anions.

The scope of this chapter is to review the current view of the quaternary structure and subunit composition of the three major classes of ligand-gated ion channels, with special focus on P2X receptors. The topics covered also include K^+ channels because of the emerging structural similarity of ion conduction pathways in K^+ channels and glutamate receptors and the initial assumption that P2X receptors too possess a K^+ channel-like H5-domain.

Distinct classes of voltage-gated and ligand-gated ion channels

According to their mechanism of activation, ion channels can be divided into two major classes: (i) voltage-gated channels and (ii) ligand-gated chan-

*Corresponding author. Tel.: +49-69-798 29365; fax: +49-69-798 29374; e-mail: Schmalzing@em.uni-frankfurt.de

nels. Voltage-gated channels are activated by a change in membrane potential, whereas ligand-gated channels are activated by binding of a neurotransmitter to a receptor domain. In addition, several examples of channel modulation by protein–protein interaction have emerged recently. In these cases, the binding of a ligand to a receptor protein is somehow communicated to a tightly associated but entirely different pore-forming subunit to modulate channel opening.

Nearly all known ligand- and voltage-gated channels are organised by symmetric or pseudo-symmetric arrangements of subunits or domains. Voltage-gated Na^+ and Ca^{2+} channels, for instance, are formed by single large polypeptides (>2000 amino acids) with several homologous repeats, whereas K^+ channels and ligand-gated ion channels are formed by the assembly of several, typically four to five homologous subunits. An exception is ClC–O, a voltage-gated chloride channel, which is a two-pore homo-dimer, as originally proposed on the basis of sedimentation analysis (Middleton et al., 1994), and later confirmed by single channel analysis (Ludewig et al., 1996; Middleton et al., 1996). The two chloride-conducting pores are identical but physically distinct and largely independent and must be formed off-axis. In all other known cases, polar residues of each subunit or domain contribute to form an ion conducting hydrophilic pathway along the central symmetry axes. Channel properties such as conductance and ion selectivity depend on the profile created by the encircling domains or subunits. Therefore, if the pore is formed within the subunit assembly rather than being intrinsic to an individual polypeptide chain, knowledge of the subunit composition and subunit stoichiometry is fundamental for any structural modelling of the ion pore. Likewise, kinetic models attempting to describe the interactions of agonists and antagonists with ligand-gated ion channels critically depend on the number of subunits needed to form the functional receptor channel, since each subunit may carry independent or co-operatively interacting ligand-binding sites. If a receptor channel consists of multiple subunits from the same subclass, considerable receptor diversity may be generated through the combinatorial assembly of subunits encoded by multiple homologous genes giving rise to an array of homo- and hetero-oligomeric structures. Additional structural requirements that are essential for ion channel function are: (i) one or more recognition sites or sensors that recognise the respective stimuli; (ii) a gate that opens or closes in response to a stimulus; (iii) an allosteric transition mechanism that regulates the gate by conformational changes upon the reception of a stimulus.

In general, neurotransmitter-gated ion channels are less ion selective than voltage-gated channels. Moreover, consistent with their capabilities to bind extracellular ligands, a large fraction of the mass of neurotransmitter-gated ion channels lies on the extracellular side of the plasma membrane as compared with voltage-gated ion channels, the masses of which are predominantly exposed to the cytoplasm. On the basis of their amino acid sequences, ligand-gated ion channels have been grouped into three major classes (North, 1996b): (i) the nicotinic acetylcholine receptor superfamily comprising the ionotropic receptors for acetylcholine, serotonin, glycine and GABA (γ-aminobutyric acid); (ii) the glutamate receptor family; (iii) the P2X receptor family. The paradigmatic nicotinic acetylcholine receptor, for which a wealth of information exists, exhibits a pentameric stoichiometry. Extensive amino acid sequence similarity among members of the superfamily of neurotransmitter-gated channels suggests conservation of structural design including membrane topology and pentameric architecture. Despite lack of sequence similarity, glutamate receptors have been thought to share the pentameric architecture of nicotinic acetylcholine receptors. Recent results show, however, that glutamate receptors share structural similarity with K^+ channels and that only four subunits are involved in the formation of functional glutamate receptor channels. Analysis of P2X receptors by complementary approaches unexpectedly revealed a trimeric subunit organisation not hitherto observed for any other ion channel (Nicke et al., 1998).

In addition, the cDNAs of two ionotropic receptors have been cloned recently, which may be considered to constitute additional classes of ligand-gated ion channels. One is an ionotropic receptor from *Helix aspersa* nervous tissue opened by a peptide (FMRFamide) for Na^+ (Lingueglia et al., 1995). The other, the most recent addition of a ligand-gated ion channel, is the capsaicin receptor designated VR1 (for vanilloid receptor subtype 1), which is activated not only by vanilloid compounds such as capsaicin, the main pungent ingredient in hot chilli peppers, but also by thermal stimuli (Caterina et al., 1997). VR1 shares distant sequence similarity with the TRP-family of ion channels (Caterina et al., 1997), which have been proposed to be activated in response to Ca^{2+} store depletion to mediate Ca^{2+} entry into cells by a process named capacitative Ca^{2+} entry. Also the tetrameric cyclic nucleotide-gated K^+ channels (reviewed by Zagotta and Siegelbaum, 1996) and the two major types of ligand-gated Ca^{2+} channels, the ryanodine receptor and the inositol 1,4,5-trisphosphate ($InsP_3$) receptor, which are also tetramers (reviewed by Mikoshiba, 1997) represent ligand-gated ion channels. They differ from the classical ligand-gated ion channels in that their receptor domains are orientated towards the cytoplasm and that at least the ryanodine receptor and the $InsP_3$ receptor are located on intracellular membranes rather than on the plasma membrane.

Voltage-gated and inward rectifier K^+ channels are tetramers

The voltage-activated K^+, Na^+ and Ca^{2+} channels are responsible for the generation and propagation of electrical signals in cell membranes. K^+ channels are at least 10,000 times more permeant for K^+ than for Na^+. They are assembled from pore-forming α subunits, which may be associated with auxiliary β subunits. Most of the K^+ channel α subunits can be assigned to two distantly related superfamilies encompassing voltage-activated channels, Kv, and inwardly rectifying channels, Kir (reviewed by Roeper and Pongs, 1996). α subunits of Kv channels posses six membrane-spanning segments (S1–S6) and between segments S5 and S6 an approximately 20 amino acid long lipophilic domain that extends from the extracellular surface part through the membrane without spanning it, thus forming a deep invagination. This sequence, called P loop or H5 domain (for fifth hydrophobic region), forms the narrow part of the ion pore and includes the K^+ channel signature sequence of

eight amino acids (TMTTVGYG) highly conserved among all K^+ channel subunits. Mutation of H5 amino acids disrupts the channel's ability to discriminate between K^+ and Na^+, but has little effect on the gating properties, indicating that H5 confers K^+ selectivity (Kavanaugh et al., 1991; Yool and Schwarz, 1991).

Sequence comparison shows that a single K^+ channel polypeptide resembles each of the four internally homologous repeats of Na^+ and Ca^{2+} channels (see Marban et al., 1998 for a recent review of Na^+ channels). The emergence of a single K^+ channel subunit as the equivalent of one quarter of an Na^+ channel polypeptide has led to the suggestion that functional K^+ channels may be multimers. Whereas Na^+ channels apparently fold from the four domains of one contiguous polypeptide to form a pseudotetramer, K^+ channels can be imagined to attain a similar structure by a symmetrical assembly of four individual subunits around a central pore. First experimental evidence for a tetrameric nature of K^+ channels (now confirmed crystallographically) was obtained by studying the interaction of the scorpion toxin charybdotoxin with coexpressed wild type and toxin-insensitive mutant Shaker K^+ channel subunits (MacKinnon, 1991). A tetrameric structure was supported by electrophysiological analysis of concatenated multimers consisting of four Kv channel subunits (Hurst et al., 1992; Liman et al., 1992). Low resolution electron microscopy of recombinant Shaker K^+ channels purified from insect Sf9 cells directly demonstrated a four-fold symmetric tetramer, with a large central vestibule supposed to constitute part of the ion conducting pathway (Li et al., 1994).

Inwardly rectifying channels are only distantly related to voltage-gated K^+ channels and can be considered to be shortened versions of voltage-gated K^+ channels that lack the first four membrane-spanning segments of Kv channels. Their simpler structural motif of only two transmembrane segments flanking an H5 domain (S5–H5–S6, Fig. 1E) appears to represent the core design of a K^+-selective pore. Like voltage-gated K^+ channels, inward rectifiers have been found by functional analysis to have a tetrameric nature based on their differential sensitivity to voltage-

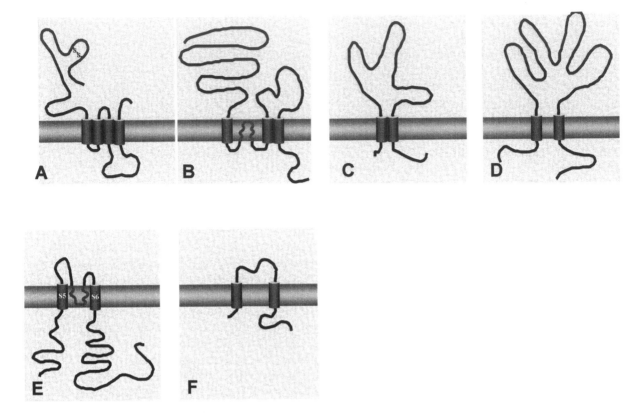

Fig. 1. Models of transmembrane organisation of ion channel pore-forming subunits. (A) Nicotinic acetylcholine receptor; in the acetylcholine receptor superfamily, each subunit spans the membrane four times. The hydrophilic domain separating M3 from M4 faces the cytoplasm. (B) Glutamate receptor; the M2 region does not cross the membrane, but it contributes to the lining of the pore. (C) P2X receptor. (D) Epithelial Na^+ channel. (E) Inward rectifying K^+ channel Kir. (F) Mechanosensitive ion channel of *E. coli*.

dependent pore block by spermine (Glowatzki et al., 1995) and on single-channel properties of concatenated multimers consisting of three or four inward rectifier subunits (Yang et al., 1995). A biochemical study based upon chemical cross-linking, immunoprecipitation and sedimentation analysis provided physical evidence for a tetra-meric organisation of the native brain inwardly rectifying K^+ channel Kir 2.2 (Raab-Graham and Vandenberg, 1998). Within heteromeric channels, the relative positions of subunits have profound effects on channel properties (Pessia et al., 1996).

Additional K^+ channels have been identified that can be considered to be composed of combinations of Kv and Kir. TwiK (for tandem of P domains in a weak inward rectifying K^+ channel) channels contain the equivalent of two Kir subunits in tandem in one polypeptide and are thought to form the tetramer through the assembly of two dimer subunits (Ketchum et al., 1995). ToK (for two P region containing outwardly rectifying K^+ channel) channels correspond to a Kv channel subunit followed by a Kir subunit. Also Kv channels with an additional (seventh) transmembrane region at their N terminus have been described.

Inward rectifier channels exist also in an ATP-sensitive form named K_{ATP} channels that play a critical role in regulating insulin release (Babenko et al., 1998). Although initially believed to be ATP-sensitive channels in their own right (Inagaki et al., 1995), these members of the inward rectifier family are now known to exist in a complex with the sulfonylurea receptor which confers ATP-sensitivity (Clement et al., 1997). Sulfonylurea receptors turned out to belong to the ATP-binding cassette (ABC) superfamily with multiple transmembrane-spanning domains and two nucleotide-binding folds. Due to the physical association of the two quite different types of subunits, changes in cytoplasmic ATP sensed by the sulfonylurea receptor are transmitted to the inward rectifier channels to alter K^+ conductance and hence membrane excitability. A molecular mass of 950 kDa was estimated by sedimentation analysis for the sulfonylurea receptor/inward rectifier complex, consistent with an octameric stoichiometry arranged of four inward rectifier subunits and four sulfonylurea receptors (Clement et al., 1997).

Direct protein–protein interactions between the G protein $\beta\gamma$ subunit and G protein-gated Kir channels (GIRKs 1–4) are thought to be responsible for the generation of I_{KACh}, a K^+ mediated current elicited by acetylcholine. Also GIRKs possess a tetrameric subunit organisation, as shown by chemical cross-linking of purified cardiac channels composed of the pore-forming GIRK1 and GIRK4 subunits (Corey et al., 1998).

Nicotinic acetylcholine superfamily members (Cys-loop receptors) share a pentameric architecture

The nicotinic acetylcholine receptor (nAChR) is considered the prototype of a gene superfamily that also includes ionotropic receptors for glycine, GABA and 5 HT (Betz, 1990). Members of the nAChR superfamily share sequence (25–60% sequence identity) and structural similarities (Fig. 1A). Obviously, they have evolved from a common ancestor, probably a homo-oligomer (Ortells and Lunt, 1995). The subunits are made up of 450–700 amino acids, each having a processed N terminal signal peptide and a large glycosylated N terminal ectodomain which carries the neurotransmitter binding site (Hucho et al., 1996). Moreover, all superfamily members contain in the N terminal extracellular region a signature pattern termed the Cys-loop (Karlin and Akabas, 1995) consisting of a conserved disulphide bond separated by a stretch of 13 amino acids (Cockcroft et al., 1990). Four 25–30 amino acid stretches (M1 to M4) span the membrane and place the C terminus on the extracellular side of the plasma membrane (for review see Dani and Mayer, 1995).

The muscle type nicotinic acetylcholine receptor (nAChR). nAChRs from *Torpedo* electric organ and muscle, collectively designated muscle-type

nAChRs, combine from a total of five distinct, yet homologous subunits ($\alpha 1$, $\beta 1$, γ, δ and ε) with a fixed stoichiometry of $(\alpha 1)_2\beta 1\gamma\delta$ in fetal muscle and $(\alpha 1)_2\beta 1\varepsilon\delta$ in adult muscle. Replacement of γ for ε occurs during muscle development, conferring a change in nAchR properties (Mishina et al., 1986). The acetylcholine binding sites have been localised at the α/δ and α/γ interfaces (for discussion and references see Galzi and Changeux, 1994; Hucho et al., 1996).

Muscle type nAChR subunits share 80% sequence identity in *Torpedo* and man, indicating that they are remarkably conserved during evolution. The extraordinarily high density of nAChRs in the electric organ of the electric ray *Torpedo californica* (and related species) has greatly facilitated the characterisation of muscle-type nAChR allowing early determination of its five subunit stoichiometry of 2α, 1β, 1γ and 1δ chain by preparative gel electrophoresis (Lindstrom et al., 1979) and microsequencing (Raftery et al., 1980). Part of the nAChRs of *Torpedo* exist as dimers of two nAChR pentamers, formed by cross-linking of two nAChRs by a disulfide bridge between the δ subunits within each pentamer. The heterologous expression of nAChRs was accomplished already in 1981 by injection of mRNA extracted from *Torpedo marmorata* into *Xenopus* oocytes (Sumikawa et al., 1981). Moreover, using sedimentation analysis, it was shown that the four *Torpedo* nAChR subunits assembled in *Xenopus* oocytes to a complex of the same size (9S) as native *Torpedo* nAChR (Sumikawa et al., 1981).

Electron microscopy combined with image analysis of tubular crystals of *Torpedo* postsynaptic membranes having receptors arranged in helical arrays across their surfaces provided direct evidence that the nAChR consists of five subunits arranged around a five-fold pseudosymmetric axis in a barrel-like structure (Brisson and Unwin, 1985). Upon improvement of resolution to 9 Å, the contour of the pore-forming structure emerged to be a helical bundle, in which each of the five subunits contributes one transmembrane helix, presumably M2, surrounding a central water-filled pore that conducts cations when its opening is triggered by ligand binding (Unwin, 1995). The functional importance of M2 for ion conductivity is documented by a wealth of biochemical and electrophysiological data (Hucho et al., 1996). Though it is generally accepted that M2 is the main contributor to the pore lining, M1 can also contribute to the pore-forming region, at least when the channel is open (Akabas and Karlin, 1995).

Neuronal nAChRs. Neuronal nAChRs of mammals form from a currently identified repertoire of seven neuron-specific α (α2–α7, α9) and three β isoforms (β2–β4, reviewed in Galzi and Changeux, 1995; Karlin and Akabas, 1995; McGehee and Role, 1995). The neuronal subunits share sequence similarity with α1 and β1 of muscle, but unlike muscle-type nAChR which incorporate four types of subunits, neuronal nAChRs operate as hetero-oligomeric complexes including two types of subunits only, α and β, or even as homo-oligomeric complexes of α subunits (α7, α8 or α9) alone. Sedimentation analysis combined with determination of the relative amounts of each subunit showed that neuronal hetero-oligomeric nAChRs assemble according to the general stoichiometry of $(\alpha4)_2(\beta2)_3$ (Anand et al., 1991). The same stoichiometry was deduced from the number of subconductance levels generated by co-injection of wild type and mutant subunits for neuronal α4 and β1 subunits (Cooper et al., 1991). Also the size of the homomeric α7 nAChR formed in *Xenopus* oocytes was found by sedimentation analysis to be compatible with a pentamer (Anand et al., 1993). It is largely unknown, however, which particular subunits are incorporated into native neuronal nAChRs. Transcripts for neuronal nAChR subunits are widespread in the nervous system and often show an overlapping distribution. Electrophysiological recordings in native tissues have shown a high diversity of physiological and pharmacological properties of neuronal nAChRs which might be due to different subunit combinations, post-ranslational modifications or alternative stoichiometries and positional arrangements of subunits.

The ionotropic serotonin receptor (5-HT₃R). Like neuronal nAChRs, 5-HT₃R mediates fast excitatory responses in mammalian neurons (for a review, see Jackson and Yakel, 1995). 5-HT₃R cDNA was originally cloned from the mouse neuroblastoma x chinese hamster embryonic brain cell hybrid NCB-20 and found to encode a 487 amino acid protein including an N terminal signal peptide (Maricq et al., 1991). 5-HT₃R-A exhibits highest similarity with nAChR subunits, particularly α7 with 30% amino acid similarity (Jackson and Yakel, 1995). The close relationship with the nAChR superfamily is also apparent from the finding that functional properties of nAChR α7 and 5-HT₃R-A can be combined into a chimeric α7/5-HT₃ receptor by the exchange of homologous elements (Eiselé et al., 1993) and that 5-HT₃R-A and nAChR α4 subunits can coassemble into a novel type of heteromeric ion channel (van Hooft et al., 1998). Heterologously expressed 5-HT₃R-A subunits form homo-oligomeric receptors with high efficiency, suggesting that homo-oligomeric 5-HT₃R may exist in native tissues. However, electrophysiological properties of homo-oligomeric 5-HT₃Rs differ from that of 5-HT₃Rs of various native tissues, implying that additional 5-HT₃ subunit isoforms may exist (van Hooft et al., 1997). Indeed, very recently a new isoform termed 5-HT₃R-B was identified that when co-expressed with 5-HT₃R-A exhibited electrophysiological characteristics similar to those of native neuronal 5-HT₃ channels (Davies et al., 1999).

Masses of 250–280 kDa have been found by size exclusion chromatography or non-reducing SDS-PAGE for purified 5-HT₃Rs from native sources such as NCB20 cells (McKernan et al., 1990) and pig cerebral cortex (Fletcher et al., 1998), respectively, and interpreted to be indicative of a pentameric subunit organisation. Also size exclusion chromatography of a hexahistidyl-tagged mouse 5-HT₃R purified from mammalian cells yielded a mass of 280 kDα, compatible with a pentameric homo-oligomer (Hovius et al., 1998). The native 5-HT₃R of NG108-15 cells, which exhibited a mass of 370 kDa when analysed by size exclusion chromatography (Boess et al., 1992), was visualised by electron microscopic image analysis to consist of five subunits arranged symmetrically around a central cavity (Boess et al., 1995). Likewise, electron microscopic visualisation of purified recombinant 5-HT₃R showed pentameric doughnut-shaped particles (Green et al., 1995).

The ionotropic glycine receptor (GlyR). The glycine receptor (GlyR) is a ligand-gated chloride channel, which mediates synaptic inhibition in spinal cord and certain brain regions (for reviews, see Betz, 1990; Kuhse et al., 1995). Native GlyRs purified by strychnine affinity chromatography from rat spinal cord consist of only two types of subunits, a ligand-binding α subunit and a structural β subunit. A third protein of 93 kDa named gephyrin (for Greek bridge) copurifies with the receptor channel (Pfeiffer et al., 1982) without forming part of it (Langosch et al., 1988) and anchors GlyRs at the synapse (Kirsch et al., 1993). By chemical cross-linking and sedimentation analysis, the native GlyR from spinal cord has been shown to be a pentamer with a defined subunit stoichiometry of 3α and 2β subunits (Langosch et al., 1988). To date, three different α subunits (α1–α3) have been identified in rats and humans (Kuhse et al., 1995). Heterologous expression of α1, α2 or α3 in *Xenopus* oocytes or mammalian cell lines generates functional homo-oligomeric channels. In contrast to the invariant 3 : 2 stoichiometry of αβ hetero–oligomeric GlyRs, assembly of adult α1 subunits and fetal α2 subunits yields GlyRs with variable subunit ratios, depending on the relative abundance of each of both subunits (Kuhse et al., 1993). This indicates that the glycine receptor α subunits must have similar surfaces for oligomer formation in contrast to the muscle type nAChR α1 subunit, which cannot combine to homo-oligomers. The different assembly behaviour of GlyR α and β subunits could be attributed to several short amino acid sequences ("assembly boxes") preceding the conserved Cys-loop (Kuhse et al., 1993).

The GABA_A receptor (GABA_AR). γ-Aminobutyric acid type A receptors are the major sites of fast synaptic inhibition in the mammalian brain (for reviews, see Sigel and Buhr, 1997; Costa, 1998). Like glycine receptors, GABA_ARs include an intrinsic Cl⁻ channel that when opened, mostly hyperpolarises and occasionally depolarises neuronal membranes, depending on the existing Cl⁻ gradient. To date, at least 16 GABA_A receptor subunits have been identified by molecular cloning in mammalian tissues and categorised in five subunit classes including α1–α6, β1–β4, γ1–γ3, δ, ε, and π (Costa, 1998). In addition, splice variants are known for β4 and γ2 subunits. Most of the GABA_A subunits can form with low efficiency functional GABA-gated receptors when expressed individually in *Xenopus* oocytes, but the pharmacological properties of these homomeric GABA_A receptors differ from that of native GABA_A receptors. In general, coexpression of α and β plus one or more of the γ, δ or ε subunit types is needed to reconstitute the principal features of GABA_A receptors found in native tissues such as differing responses to a variety of ligands, particularly benzodiazepines and barbiturates (Costa, 1998). Hence, there is a considerable potential for structural diversity. While the stoichiometry and subunit arrangement of native GABA_ARs have not been clearly determined, it is likely that most are composed of combinations of α, β and either γ, δ or ε subunits (Costa, 1998).

From the ratio of α and β chains in SDS polyacrylamide gels, the GABA_A receptor purified from bovine cerebral cortex by benzodiazepine affinity chromatography has been concluded to have a tetrameric architecture (Mamalaki et al., 1987). In contrast, native GABA_A receptors purified from porcine brain have been shown by electron microscopic image analysis to have a pentameric structure (Nayeem et al., 1994). Electrophysiological analysis of functionally tagged α3, β2 and γ2 subunits suggested the most likely stoichiometry to be two α3, one β2 and two γ2 (Backus et al., 1993), whereas an analysis of tandem constructs of α6–β2 GABA subunits lead to the conclusion that the α6β2γ2 GABA_AR consists of two α, two β and one γ (Im et al., 1995). An electrophysiological analysis exploiting an increase in GABA sensitivity in relation to the number of incorporated mutant α1, β2 or γ2 subunits, was also consistent with a pentamer composed of two α, two β and one γ (Chang et al., 1996)

The GABA_C receptor (GABA_CR). A further GABA_AR-related subunit class named ρ (Cutting et al., 1991) with three known isoforms (ρ1–ρ3) is currently considered to constitute a separate subfamily of GABA receptors, GABA_CRs, which may be phylogenetically older than the GABA_ARs (for reviews, see Bormann and Feigenspan, 1995; Johnston, 1996; Enz and Cutting, 1998). In contrast to GABA_AR subunits, which only express efficiently as hetero–oligomers (see above), ρ subunits have the propensity to form efficiently homo-oligomeric GABA-gated Cl⁻ channels in *Xenopus* oocytes. Interestingly, the electrophysiological and pharmacological properties of receptors consisting solely of ρ subunits are very similar to those of the native GABA$_c$Rs characterised in retinal cells, suggesting that ρ subunits are the sole components of GABA$_c$R (Enz and Cutting, 1998). GABA$_c$ receptors do not assemble with GABA$_A$ α, β or γ subunits (Enz and Cutting, 1998).

Ionotropic glutamate receptors share a tetrameric architecture with K⁺ channels

Ionotropic glutamate receptors are the principal excitatory neurotransmitter receptors in the vertebrate brain assembled from large polypeptide subunits of 900-1500 residues (Seeburg, 1993; Hollmann and Heinemann, 1994; Barnard, 1997). Based on electrophysiological and pharmacological criteria and amino acid sequences, glutamate receptors are subdivided into *N*-methyl-D-aspartic acid (NMDA) receptors and non-NMDA receptors for α-amino-3-hydroxy-5-methyl-4-isoxazole propionic acid (AMPA, GluR 1–4) and kainate (GluR 5–7 and KA-1 and KA-2). Each subclass includes several disparate subunits identified by molecular cloning such that physiological diversity may be generated by appropriate co-expression of the subunits. Native NMDA receptors, for instance, are composed of NR1 and NR2 subunits, although NR1 subunits are capable of forming functional homo-oligomers in *Xenopus* oocytes (Moriyoshi et al., 1991).

Like the nAChR superfamily members, glutamate receptor subunits contain a signal peptide at the N terminus, assigning the N terminus to the extracellular side; the N-terminal ectodomain is more than twice as long as that of nAChRs. Ionotropic glutamate receptors lack the characteristic N-terminal Cys-loop (Cockcroft et al., 1993) and exhibit no obvious sequence similarity with nAChRs. Nevertheless, the existence of at least some general structural similarities between both receptor classes was suggested by hydropathy analysis indicating that glutamate receptor subunits possess four hydrophobic domains, M1–M4, at similar relative spacing like the nAChR super-family members. M1–M4 were initially assigned as membrane-spanning domains. Direct experimental examination, however, revealed a fundamental difference in transmembrane topology, since unlike the situation for nAChRs, only M1, M3, and M4 of glutamate receptors cross the membrane. M2 of glutamate receptors is not a transmembrane domain, but rather forms a loop structure with both ends facing the cytoplasm (Fig. 1B), thus resembling the pore-forming re-entrant loop of the voltage-activated K⁺ channels, rotated 180° with respect to the plasma membrane (Wood et al., 1995). In fact, the second hydrophobic domain, formerly M2, of glutamate receptors has significant sequence similarity extending over a region of 34 amino acids to the pore-forming region of K⁺ channels (Wood et al., 1995) and has been shown to play an important role in channel lining (Kuner et al., 1996).

Initial biochemical studies including chemical cross-linking and size-exclusion chromatography revealed non-NMDA glutamate receptors to exist as large oligomeric proteins with masses of 590–730 kDa (Blackstone et al., 1992; Wenthold et al., 1992; Brose et al., 1993). Using earlier structural studies on nAChRs as guides, these data have generally been interpreted as an indication of a pentameric organisation. A pentameric structure was also deduced from the analysis of the inhibition profile of non-NMDA glutamate receptors assembled from two mutant subunits with differential sensitivities to ion channel blockers (Ferrer-Montiel and Montal, 1996). Likewise, single channel recordings from *Xenopus* oocytes expressing wild type and mutant subunits were interpreted to indicate that NMDA receptors are pentamers composed of three NR1 and two NR2 subunits (Premkumar and Auerbach, 1997).

Because of the voltage-gated channel-like pore-forming motif, other investigators proposed the widely assumed relationship between glutamate receptors and nAChRs to be incorrect (Wo and Oswald, 1995). Indeed, a recent biochemical study exploiting chemical cross-linking of native AMPA receptors yielded products of approximately 200, 300 and 400 kDa as compared to 100 kDa of the monomer, indicating that AMPA receptors have a tetrameric structure (Wu et al., 1996). Moreover, several recent functional studies are consistent with a tetrameric rather than a pentameric architecture. These include an analysis of hetero-oligomeric NMDA receptors formed from different ratios of co-expressed wild type with low affinity or negative dominant mutants of the glycine binding NR1 subunit and the glutamate binding NR2 subunit (Laube et al., 1998) and of AMPA receptors formed from mutant and wild-type GluR1 subunit (Mano and Teichberg, 1998). A tetrameric structure was also deduced by counting the distinct electrophysiological states that a homo-oligomeric AMPA receptor passes through as successively more agonist binding sites become occupied by an agonist (Rosenmund et al., 1998). The shared tetrameric structure of glutamate receptors and K^+ channels may reflect a similar organisation of the ion pore, which is indicated by the similarity in primary sequence and secondary structure between the re-entrant loops of glutamate receptors and K^+ channels (Wood et al., 1995).

Architecture of P2X receptors

P2X receptors are ligand-gated ion channels that open an intrinsic ion channel pore when challenged with extracellular ATP. The channel forming ATP receptors in rat and human are encoded by seven

identified genes named $P2X_1$–$P2X_7$ (Table 1). The cloning of the different P2X channel genes, which give rise to ATP-gated ion channels in heterologous systems, have been the subject of several recent reviews (Buell et al., 1996b; Barnard et al., 1997; North and Barnard, 1997; Soto et al., 1997). For a discussion of the electrophysiological and pharmacological properties of cloned P2X receptors, the reader is referred to these reviews. Here, we introduce only the information that is required to provide the nomenclature and the necessary background for the structural studies.

P2X receptors possess two transmembrane domains connected by a large ectodomain. Because of the absence of an N terminal signalling sequence, the N terminus of P2X receptor subunits has been assigned to the intracellular side of the membrane (Valera et al., 1994). Hydrophobicity plots predict a conserved overall structure for $P2X_1$–$P2X_7$ with two membrane-spanning domains (M1 and M2) that flank a large ectodomain (Fig. 1C) supposed to comprise the ATP binding site (North and Barnard, 1997). The two membrane spanning domains place the C terminal tail of P2X receptors on the cytoplasmic side of the membrane. The cytoplasmic orientation of the N and C termini has been demonstrated by immunofluorescence

TABLE 1

Properties of rat P2X isoforms

Isoform	Originally cloned from	Accession number	Amino acid deviations within one isoform	Length (aa)	Mass (Da)	N-X-S/T	N terminus (aa)	Ecto-domain (aa)	C terminus (aa)	References
$P2X_1$	Vas deferens	X80477		399	44,954	5	30	280	45	(Valera et al., 1994)
$P2X_2$	PC12 cells	U14414		472	52,613	3	30	279	119	(Brake et al., 1994)
		Y09910								(Brändle et al., 1997)
$P2X_3$	Dorsal root ganglia	X90651	I96	397	44,371	4	24	276	53	(Chen et al., 1995)
		X91167	M96		44,389					(Lewis et al., 1995)
$P2X_4$	Hippocampus	X91200	V137,S136,E237,305A	388	43,512	7	29	285	30	(Bo et al., 1995)
	Superior cervical ganglia	X87763	V137,S136,G237,305G		43,426					(Buell et al., 1996)
		U32497	R137,L136,E237,305G		43,581					(Séguéla et al., 1996)
		X93565	V137,S136,E237,305G		43,498					(Soto et al., 1996a)
		U47031	V137,S136,E237,305G		43,498					(Wang et al., 1996)
$P2X_5$	Celiac ganglia	X97328	191F, 396Q	455	51,508	3	30	285	96	(Garcia-Guzman et al., 1996)
	Heart	X92069	191S, 396R		51,476	3				(Collo et al., 1996)
$P2X_6$	Superior cervical ganglia	X92070		379	42,447	3	31	277	27	(Collo et al., 1996)
	Brain	X97376			42,447					(Soto et al., 1996b)
$P2X_7$	Autonomic ganglia	X95882		595	68,386	6	27	286	239	(Surprenant et al., 1996)

Length of N terminus:, length of the cytoplasmic N terminal tail from the initiating methionine up to the predicted beginning of M1.
Length of C terminus:, length of the cytoplasmic C-terminus from the end of M2.

studies (Torres et al., 1998). The identification of residues that contribute to the channel pore (Rassendren et al., 1997) and the usage of four out of five naturally occurring N-glycosylation sites of $P2X_1$ (Nicke et al., 1998) and mutational N-linked glycosylation analysis on $P2X_2$ (Newbolt et al., 1998; Torres et al., 1998) all indicate that the predicted membrane topology is essentially correct. For rat His-$P2X_1$, it is very likely that N285 is the natural site that is not used, since the corresponding sequon NLS is followed by a P (Nicke et al., 1998), which is known to impair N-glycosylation at the X and Y position of the consensus N-X-T/S-Y (Gavel and Von Heijne, 1990).

P2X polypeptides are between 379 and 595 amino acids in length. A survey about the length and putative number of N-glycosylation sites of different P2X receptor subunits is given in Table 1. The cytoplasmic C terminal tails show no apparent sequence similarity, indicating that they might provide specific properties. Overall, the first 400 amino acids of $P2X_1$–$P2X_7$ share >30% sequence identity. The ectodomain contains 67 amino acids that are completely conserved among all seven receptors including 14 G residues, 10 C residues (six of which are found in an approximately 50 amino acids long C-rich region comprising amino acids 115–165), six K residues and two R residues. The regularly spaced C residues have been suggested to form disulphide-bonded loop structures (Brake et al., 1994; Hansen et al., 1997).

P2X receptors exist as homo- and hetero-oligomers. When heterologously expressed, all cloned P2X isoforms give rise to functional channels, which have distinctive kinetic and pharmacological properties. Recombinant homo-oligomeric $P2X_1$ and $P2X_2$ receptors have properties that closely resemble those observed natively in smooth muscle and phaeochromocytoma cells, respectively, the original source of $P2X_1$ and $P2X_2$ cDNAs (Brake et al., 1994; Evans et al., 1995). Also $P2X_3$ efficiently forms homo-oligomeric channels in heterologous cells (Chen et al., 1995; Lewis et al., 1995), but their pharmacological profile does not fit that of the native receptor phenotype of sensory neurons, from which the $P2X_3$ cDNA was derived and which contains mRNA for at least three additional P2X isoforms,

$P2X_1$ $P2X_2$ and $P2X_4$ (Khakh et al., 1995). Co-expression of $P2X_3$ with $P2X_2$, but not with $P2X_1$ or $P2X_4$, turned out to constitute a novel P2X phenotype with properties similar to that seen in sensory neurons, providing indirect evidence for the capability of P2X receptors to hetero-polymerise and the existence of hetero-oligomeric P2X receptors in native tissues (Lewis et al., 1995). Physical association was found by coprecipitation of $P2X_2$ with $P2X_3$ after transfection in Sf9 cells (Radford et al., 1997).

$P2X_5$ and $P2X_6$ are the sole P2X isoforms that are poorly expressed as homo-oligomers, which is suggestive of hetero-polymerisation (Collo et al., 1996). Indeed, co-expression of $P2X_4$ and $P2X_6$ subunits in *Xenopus* oocytes has been shown to result in the formation of a novel pharmacological phenotype of P2X receptor (Lê et al., 1998). Physical interaction of the two subunits was demonstrated by coprecipitation of epitope-tagged $P2X_4$ and $P2X_6$. Since $P2X_4$ and $P2X_6$ are major central subunits in mammalian CNS with highly overlapping mRNA distribution at both regional and cellular levels, $P2X_4$/$P2X_6$ hetero-oligomers are likely to occur in native tissues.

Binding domains for ATP, PPADS and suramin. The parts of the molecule forming the ATP binding site are currently unknown, and there are no obvious sequence homologies with other known ATP-binding proteins. A recent sequence comparison suggests distant structural similarity with aminoacyl tRNA synthetases, which catalyse the esterification of specific tRNAs with particular amino acids in the expense of ATP (Freist et al., 1998). For all ATP binding motifs of the synthetases, corresponding sequences have been found in the P2X ectodomains including an R residue that may bind to the α phosphate moiety and R, K and H residues that may interact with the γ phosphate group (Freist et al., 1998).

Electrostatic interaction of the negatively charged phosphate groups of ATP with positively charged amino acid side chains is not unexpected, since negatively charged phosphate groups are essential elements in all known P2X agonists. The six conserved K and two R residues are clearly candidates critical for ATP binding. K residues in general are known to play an important role in nucleotide binding of nucleotide binding proteins (Duda et al., 1993; Koonin, 1993; Traut, 1994). In addition, the best known P2X receptor antagonists pyridoxalphosphate-6-azophenyl-2′,4′-disulfonic acid (PPADS) and suramin (Lambrecht, 1996; Soto et al., 1997) carry by themselves highly acidic groups, suggesting that the sulphonic acid groups substitute for the phosphate groups in binding. Hence, an alternative approach for the identification of the ATP binding site may be the mapping of residues that contribute to the binding of PPADS or suramin. Sites and domains identified so far to be involved in the interaction of PPADS and suramin with P2X receptors are summarised in Fig. 2. A first clue came from the observation that PPADS and suramin were rather weak blockers of heterologously expressed P2X$_4$ and P2X$_6$ receptors (Buell et al., 1996a; Collo et al., 1996). Sequence comparison showed that P2X$_4$ and P2X$_6$ carry E and L, respectively, at position 249 rather than K, which is found at the equivalent position of the

PPADS sensitive isoforms P2X$_1$ and P2X$_2$. Indeed, when a K residue was introduced at position 249 in place of E or L, both P2X$_4$ and P2X$_6$ became PPADS sensitive, presumably by allowing Schiff base formation with the aldehyde of PPADS (Buell et al., 1996a; Collo et al., 1996), as observed for the parent compound, pyridoxal phosphate (Cake et al., 1978).

In view of the above results it is surprising that PPADS is an effective antagonist at the human P2X$_4$, which like rat P2X$_4$ has E rather than K at position 249 (Garcia-Guzman et al., 1997). Using chimeras between rat and human P2X$_4$, PPADS sensitivity could be assigned to a 102 amino acids long domain of the extracellular loop of human P2X$_4$ (residues 81–183), which contained one K residue (K127) exclusively present in the human but not in the rat P2X$_4$ sequence (Garcia-Guzman et al., 1997). However, replacement of the equivalent N127 of rat P2X$_4$ by K did not confer PPADS sensitivity. These results indicate that a variety of

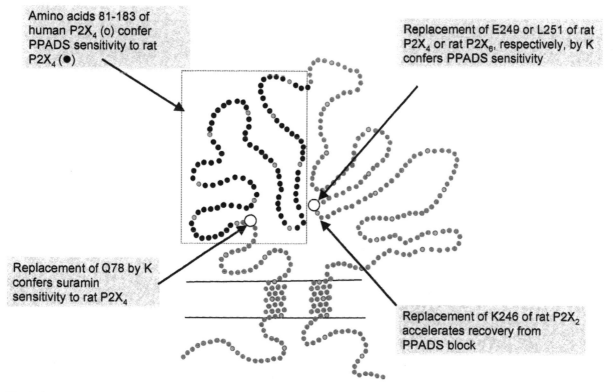

Amino acids 81-183 of human P2X$_4$ (o) confer PPADS sensitivity to rat P2X$_4$ (●)

Replacement of E249 or L251 of rat P2X$_4$ or rat P2X$_6$, respectively, by K confers PPADS sensitivity

Replacement of Q78 by K confers suramin sensitivity to rat P2X$_4$

Replacement of K246 of rat P2X$_2$ accelerates recovery from PPADS block

Fig. 2. Map of domains and residues of P2X receptors identified to contribute to the binding of PPADS and suramin. References are given in the text.

distant sites contribute to PPADS sensitivity. A single residue (K78 of human $P2X_4$) has been shown to confer sensitivity to inhibition by suramin. Rat $P2X_4$ which has E at position 78 rather than K is 50-fold less sensitive to suramin than the human one (Garcia-Guzman et al., 1997).

Trimers or tetramers? The quaternary structure of the P2X receptor. First evidence to suggest that ATP-gated ion channels are multimers came from electrophysiological experiments. In sensory neurons, the onset kinetics in response to low concentrations of ATP fit the assumption that ATP must bind to each of three identical, non-interacting binding sites to open the channel, whereas data from vas deferens and cardiac muscle are fitted to a two site model with positive cooperativity (Bean, 1992). First biochemical efforts to determine the quaternary structure of P2X receptors were undertaken before P2X receptors were cloned. A 62 kD protein of digitonin-solubilised membranes of rat vas deferens was identified by photoaffinity labelling with $[^3H]$-α,β-mATP and shown by sedimentation analysis to sediment at 12.1 *S* (Bo et al., 1992). Using nAChRs as guides this finding was interpreted as indicative for a pentameric complex (Bo et al., 1992). Once cloned and sequenced, the striking resemblance of the overall transmembrane topology of P2X with inwardly rectifying K^+ channels became apparent and led to the general expectation that P2X receptors too possess a tetrameric architecture. Such a view gained some support by the observation that P2X receptors possess a highly conserved lipophilic domain immediately before M2 (Fig. 3), which has been postulated to constitute a re-entry loop like the H5 domain of K^+ channels and hence to contribute to the pore (Brake et al., 1994). Even more intriguing, rat $P2X_1$ was found to possess an eight amino acid sequence (TMTTIGSG) closely resembling the H5 signature sequence (TMTTVGYG) of K^+ channels (Valera et al., 1994). It should be mentioned, however, that the H5-related sequence of $P2X_1$ and the lipophilic domain of $P2X_2$ preceding M2 are not related with each other and are located at adjacent, but distinct parts of the polypeptide chains (Fig. 3).

In a first study that addressed the quaternary structure of a recombinant P2X receptor, the $P2X_2$ extracellular domain was overexpressed in bacteria with an N terminal hexahistidyl tag, solubilised in urea, purified by Ni^{2+} affinity chromatography, and refolded in vitro by sulfitolysis and dialysis (Kim et al., 1997). The resulting product was found by equilibrium centrifugation, dynamic light scattering and gel filtration chromatography to have a mass of 132 kDa, 144 kDa and 160 kDa, respectively. Taking a predicted mass of 29 kDa for the monomer into account, it was concluded that the refolded $P2X_2$ ectodomain forms stable tetramers in solution (Kim et al., 1997), though masses of 144 kDa or 160 kDa could also be reconciled with a pentamer.

We have assessed the quaternary structure of full-length P2X receptors synthesised in *Xenopus*

```
                                                            M2
rP2X1        G-TNRRHLFKVFGIHFDILVDGKAGKFDIIP|TMTTIGSGIGIFGVATVLCDLLLL|HIL
H5 signature                            ----|TMTTVGYG----------------|----
                                            |****.* *                |
rP2X2        GTTTTRTLIKAYGIRIDVIVHGQAGKFSLIP|TIINLATALTSIGVGSFLCDWILL|TFM
rP2X3        NGSEYRTLLKAFGIRFDVLVYGNAGKFNIIP|TIISSVAAFTSVGVGTVLCDIILL|NFL
rP2X4        AGKEQRTLTKAYGIRFDIIVFGKAGKFDIIP|TMINVGSGLALLGVATVLCDVIVL|YCM
rP2X5        NGVEFRDLMKAYGIRFDVIVNGKAGKFSIIP|TVINIGSGLALMGAGAFFCDLVLI|YLI
rP2X6        SGVESRSLLKLYGIRFDILVTGQAGKFALIP|TAITVGTGAAWLGMVTFLCDLLLL|YVD
rP2X7          MEKRTLIKAFGVRFDILVFGTGGKFDIIQ|LVVYIGSTLSYFGLATVCIDLIIN|TYA
```

Fig. 3. Sequence comparison of P2X subunits. Amino acid alignment of the putative second transmembrane domain (M2) of rat $P2X_1$ to $P2X_7$. The complete M2 sequence is boxed. Conserved amino acids are shown in bold type, and conserved positively charged amino acids are in bold and underlined. The sequence AYGIRIDVIVHGQA (highlighted in black) of rat $P2X_2$ has been suggested to form a re-entry loop (Brake et al., 1994).

laevis oocytes by two complementary approaches, chemical cross-linking and blue native polyacrylamide gel electrophoresis (Nicke et al., 1998). Hexahistidyl-tagged $P2X_1$ and $P2X_3$ were purified by Ni^{2+} affinity chromatography from cRNA-injected oocytes. The His tag did not change the electrophysiological properties of the $P2X_1$ receptor. As cross-linkers we used bifunctional derivatives of PPADS, which is unique among the various compounds identified as P2X antagonists in forming by virtue of its aldehyde group a Schiff base with K residues. Of the cross-linkers used, DIPS and CL II can be considered to be composed of two PPADS-like molecules linked by spacers of distinct length and conformational flexibilities. The design of DIPS and CL II was based on the assumption that the PPADS moiety may favour cross-linking by the bivalent-ligand-mechanism

(Portoghese, 1989). Like PPADS itself, we expected that DIPS or CL II become non-covalently bound to one binding site on the P2X receptor and then form a Schiff base with a critical K residue. Once immobilised by one $C = N$ bond, the free aldehyde group of the second PPADS-like moiety of DIPS or CL II can efficiently react with any K residue of a neighbouring P2X chain. CL II produced covalently-linked $P2X_1$ trimers at concentrations as low as 1 μM. At 100 μM, DIPS (Fig. 4) like CL II cross-linked digitonin-solubilised His-$P2X_1$ and His-$P2X_3$ quantitatively to homo-trimers.

As an additional approach, we exploited blue native PAGE (Schägger and von Jagow, 1991) for the determination of the quaternary structure of P2X receptors. When analysed by blue native PAGE, P2X receptors purified in digitonin or

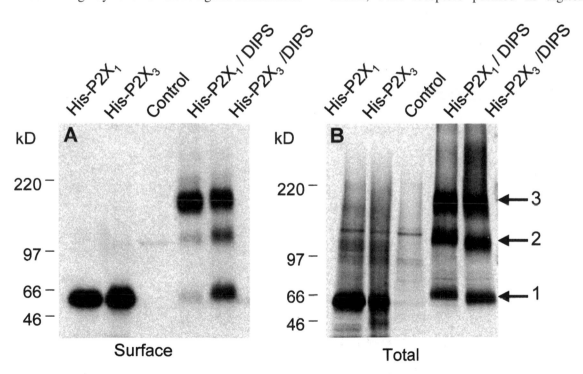

Fig. 4. Cross-linking of detergent-solubilised and purified His-$P2X_1$ and His-$P2X_3$ yields dimers and trimers. Oocytes injected with the indicated cRNAs were kept for 3 days at 19°C and then labelled with K-reactive membrane impermeant ^{125}I-sulpho-SHPP (**A**). Subsets of the same oocytes were labelled overnight with $[^{35}S]$methionine and chased for 48 h (**B**). Digitonin extracts were then prepared and P2X receptors were purified by Ni^{2+} agarose chromatography. While still bound to the beads, P2X receptors were cross-linked at 100 μM DIPS. The labile Schiff base was subsequently reduced with $NaBH_4$ to a stable amine bond. After elution with non-denaturing elution buffer, samples were supplemented with SDS sample buffer and 20 mM DTT and resolved by SDS–PAGE (4–10% acrylamide gradient gel) followed by autoradiography. Numbered arrows indicate positions of P2X monomers, dimers, and trimers.

dodecyl-β-D-maltoside migrated entirely as non-covalently linked homo-trimers (Fig. 5). P2X monomers remained undetected soon after synthesis, indicating that trimerisation occurred in the endoplasmic reticulum. The plasma membrane form of His-P2X$_1$ was also identified as a homo-trimer (Fig. 5, surface). If n-octylglucoside was used for P2X receptor solubilisation, homo-hexamers were observed, suggesting that trimers can aggregate to form larger complexes. Taken together, these findings lead to the conclusion that trimers represent an essential element of P2X

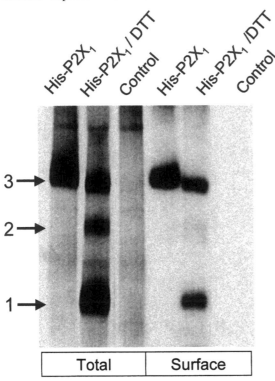

Fig. 5. His-P2X1 migrates as trimer when analysed by blue native PAGE. Oocytes injected with His-P2X$_1$ cRNA were kept for 3 days at 19°C and then labelled with membrane impermeant ^{125}I-sulpho-SHPP (surface). Subsets of the same oocytes were labelled with [^{35}S]methionine (total). His-P2X$_1$ was purified by Ni^{2+} NTA chromatography from 1% digitonin extracts of each oocyte group, eluted with non-denaturing elution buffer, and resolved by blue native PAGE (5–13% acrylamide gradient gel) followed by autoradiography. Left lane, native His-P2X1 incubated without DTT migrates as a trimer. Central lane, DTT induces an almost complete dissociation of native His-P2X$_1$ into dimers and monomers. Numbered arrows indicate positions of P2X monomers, dimers, and trimers.

receptor structure. Blue native PAGE as a method for the determination of the oligomeric state of ion channels was validated using the α$_2$βγδ nicotinic acetylcholine receptor, which migrated as the expected pentamer (Nicke et al., 1998). The usefulness of blue native PAGE was further substantiated in a study about assembly intermediates of nAChR subunits in *Xenopus* oocytes (Nicke et al., 1999).

How can a trimeric arrangement of P2X subunits form a functional ion channel? Helical wheel plots illustrated the potential of M2 of P2X$_2$ to form an amphipathic α helix (Brake et al., 1994; North, 1996c). Like the ionotropic receptors for acetylcholine and glutamate, P2X receptors are non-selectively permeable to cations including Ca^{2+} and possess an effective pore diameter slightly larger than that of the nAChR (Evans et al., 1996). Given that the pore motif of P2X receptors consists also of a bundle of α helices contributed by M2, the proposed pore-forming segment (Rassendren et al., 1997), a tetrameric or pentameric subunit organisation would be the most plausible one. Recent data, however, obtained from extensive cysteine accessibility scanning imply that at least part of M2 (T336–G243) of rat P2X$_2$ exists as a β strand (Rassendren et al., 1997) or as non-α- and non-β-structure (Egan et al., 1998). In the latter study, the data were reconciled with an outwardly facing loop with G342 at the apex of the loop (Egan et al., 1998). Since P2X receptors apparently do not possess a classical α helical bundle motif, there seems to be sufficient latitude for an architecture distinct from the canonical tetrameric and pentameric design.

Architecture of ion channels that share with P2X receptors the same overall structural organisation

Because of lack of sequence similarity of P2X receptors with other currently known proteins, P2X receptors are often classified as members of a superfamily of proteins that share the same overall membrane topology. Besides the inward rectifier K$^+$ channels, these are the epithelial Na$^+$ channel/degenerin gene superfamily (Fig. 1D) and the mechanosensitive channel of E. coli (mscL, Fig. 1F) (North, 1996a).

Tetrameric or nonameric architecture of epithelial Na$^+$ channels (ENaC). ENaC is composed of three homologous subunits, termed α, β, and γ, all consisting of two large transmembrane domains linked by an approximately 460 amino acids long extracellular loop (Canessa et al., 1994). All three subunits participate in channel formation as the absence of any one subunit results in a significant reduction of Na$^+$ current expression in *Xenopus* oocytes. ENaC has recently been reported to exist as hetero-tetramers of two α, one β and one γ subunit (Firsov et al., 1998). With respect to the pore structure there is evidence that ENaCs contain a hairpin structure similar to the lipophilic H5 segment of K$^+$ channels (Renard et al., 1994). An electrophysiological approach to determine the subunit stoichiometry utilising mutant subunits that display significant differences in sensitivity to channel blockers from the wild-type channel also indicated that ENaC is a tetrameric channel with an α2βγ stoichiometry (Kosari et al., 1998). However, there is also electrophysiological and biochemical evidence for a hetero-nonameric architecture of this channel formed by three α, three β and three γ subunits (Snyder et al., 1998).

Architecture of the FMRFamide peptide-gated Na$^+$ channel. One branch of the epithelial Na$^+$ channel/degenerin gene superfamily, the FMRFamide peptide-gated Na$^+$ channel from *Helix aspersa* nervous tissue (Lingueglia et al., 1995), represents the first example of a peptide-gated ion channel. When expressed in *Xenopus* oocytes, it encodes a homo-oligomeric FMRFamide-activated Na$^+$ channel that can be blocked by amiloride. The corresponding protein shares a very low but unambiguous sequence identity with the previously cloned epithelial Na$^+$ channel subunits and *Caenorhabditis elegans* degenerins, and it displays the same overall structural organisation. Cross-linking by intersubunit disulphide bonds following cell homogenisation and solubilisation with Triton X-100 or after use of bifunctional reagents resulted in the formation of covalent multimers that contained up to four subunits. A hydrodynamic analysis also indicated that the solubilised channel complex is a tetramer (Coscoy et al., 1998).

Homo-pentameric architecture of the mechanosensitive ion channel of E. coli. A third channel protein with only two transmembrane domains, the mechanosensitive ion channel of E. coli has one of the largest channel pores yet described with an open pore diameter of about 40 Å (Cruickshank et al., 1997). A homo-hexameric subunit organisation has been found by chemical cross-linking (Blount et al., 1996) and electron microscopic examination of two-dimensional crystals of detergent-solubilised recombinant mechanosensitive ion channels combined with image analysis at 15 Å resolution (Saint et al., 1998). However, the recent determination of the structure of a mechanosensitive ion channel homolog from *Mycobacterium tuberculosis* by X-ray crystallography to 3.5 Å resolution leaves little doubt that mechanosensitive ion channels exhibit a pentameric subunit organisation like nAChRs and their relatives (Chang et al., 1998).

Conclusions

Taken together, recent data suggest that each of the three major classes of ligand-gated ion channels is characterised by its own architecture, i.e. trimers (P2X receptors), tetramers (glutamate receptors) and pentamers (nicotinic acetylcholine receptor superfamily). This lack of a shared overall structural design is remarkable, since most of these ligand-gated ion channels exhibit a similar non-selective permeability for cations, which seems not inevitably to necessitate structural diversity of the array of pore-forming subunits. Functional diversity of K$^+$ channels, for instance, is brought about by a modular combination of the pore forming S5–H5–S6 module with other elements such as a voltage sensor and/or a regulatory binding domain.

Whereas previous views of the quaternary structure of ligand-gated ion channels were guided by cation selectivity and transmembrane topology folding, the pore structure now emerges as the structural determinant of the quaternary structure of a channel protein. The pore of nicotinic acetylcholine receptors is formed by the amino acids contributed by a pentameric α helical bundle, which constitutes one of the few (known) folds in channel structure. K$^+$ channel-like pore loops, on the other hand, appear to constitute the structural signature for a superfamily of tetrameric ion channels including K$^+$ channels, ionotropic glutamate receptors and cyclic nucleotide-gated channels. The trimeric structure of P2X receptors may imply the existence of an additional pore-forming motif distinct from the α helical bundle of nicotinic acetylcholine receptors and the pore loop motif of glutamate receptors.

Acknowledgements

Research in the authors' laboratories is supported by grants of the Deutsche Forschungsgemeinschaft (La 350/7-2, Schm 536/2-1, and Graduiertenkolleg GRK 137: "Arzneimittelentwicklung und -analytik") and the Fonds der Chemischen Industrie.

References

Akabas, M.H. and Karlin, A. (1995) Identification of acetylcholine receptor channel-lining residues in the M1 segment of the α-subunit. *Biochemistry,* 34: 12496–12500.

Anand, R., Conroy, W.G., Schoepfer, R., Whiting, P. and Lindstrom, J. (1991) Neuronal nicotinic acetylcholine receptors expressed in *Xenopus* oocytes have a pentameric quaternary structure. *J. Biol. Chem.,* 266: 11192–11198.

Anand, R., Peng, X. and Lindstrom, J. (1993) Homomeric and native α7 acetylcholine receptors exhibit remarkably similar but non-identical pharmacological properties, suggesting that the native receptor is a heteromeric protein complex. *FEBS Lett.,* 327: 241–246.

Babenko, A.P., Aguilar-Bryan, L. and Bryan, J. (1998) A view of sur/KIR6.X, KATP channels. *Annu. Rev. Physiol.,* 60: 667–687.

Backus, K.H., Arigoni, M., Drescher, U., Scheurer, L., Malherbe, P., Möhler, H. and Benson, J.A. (1993) Stoichiometry of a recombinant GABA$_A$ receptor deduced from mutation-induced rectification. *Neuroreport,* 5: 285–288.

Barnard, E.A. (1997) Ionotropic glutamate receptors: New types and new concepts. *Trends Pharmacol. Sci.,* 18: 141–148.

Barnard, E.A., Simon, J. and Webb, T.E. (1997) Nucleotide receptors in the nervous system. An abundant component using diverse transduction mechanisms. *Mol. Neurobiol.,* 15: 103–129.

Bean, B.P. (1992) Pharmacology and electrophysiology of ATP-activated ion channels. *Trends Pharmacol. Sci.,* 13: 87–90.

Betz, H. (1990) Ligand-gated ion channels in the brain: The amino acid receptor superfamily. *Neuron,* 5: 383–392.

Blackstone, C.D., Moss, S.J., Martin, L.J., Levey, A.I., Price, D.L. and Huganir, R.L. (1992) Biochemical characterization and localization of a non-*N*-methyl-D-aspartate glutamate receptor in rat brain. *J. Neurochem.,* 58: 1118–1126.

Blount, P., Sukharev, S.I., Moe, P.C., Schroeder, M.J., Guy, H.R. and Kung, C. (1996) Membrane topology and multimeric structure of a mechanosensitive channel protein of *Escherichia coli. EMBO J.,* 15: 4798–4805.

Bo, X., Simon, J., Burnstock, G. and Barnard, E.A. (1992) Solubilization and molecular size determination of the P2X purinoceptor from rat vas deferens. *J. Biol. Chem.,* 267: 17581–17587.

Bo, X.N., Zhang, Y., Nassar, M., Burnstock, G. and Schoepfer, R. (1995) A P2X purinoceptor cDNA conferring a novel pharmacological profile. *FEBS Lett.,* 375: 129–133.

Boess, F.G., Beroukhim, R. and Martin, I.L. (1995) Ultrastructure of the 5-hydroxytryptamine$_3$ receptor. *J. Neurochem.,* 64: 1401–1405.

Boess, F.G., Lummis, S.C. and Martin, I.L. (1992) Molecular properties of 5-hydroxytryptamine$_3$ receptor-type binding sites purified from NG108-15 cells. *J. Neurochem.,* 59: 1692–1701.

Bormann, J. and Feigenspan, A. (1995) GABA$_C$ receptors. *Trends Neurosci.,* 18: 515–519.

Brake, A.J., Wagenbach, M.J. and Julius, D. (1994) New structural motif for ligand-gated ion channels defined by an ionotropic ATP receptor. *Nature,* 371: 519–523.

Brändle, U., Spielmanns, P., Osteroth, R., Sim, J., Surprenant, A., Buell, G., Ruppersberg, J.P., Plinkert, P.K., Zenner, H.P. and Glowatzki, E. (1997) Desensitization of the P2X$_2$ receptor controlled by alternative splicing. *FEBS Lett.,* 404: 294–298.

Brisson, A. and Unwin, P.N. (1985) Quaternary structure of the acetylcholine receptor. *Nature,* 315: 474–477.

Brose, N., Gasic, G.P., Vetter, D.E., Sullivan, J.M. and Heinemann, S.F. (1993) Protein chemical characterization and immunocytochemical localization of the NMDA receptor subunit NMDA R1. *J. Biol. Chem.,* 268: 22663–22671.

Buell, G., Collo, G. and Rassendren, F. (1996b) P2X receptors: An emerging channel family. *Eur. J. Neurosci.,* 8: 2221–2228.

Buell, G., Lewis, C., Collo, G., North, R.A. and Surprenant, A. (1996a) An antagonist-insensitive P2X receptor expressed in epithelia and brain. *EMBO J.,* 15: 55–62.

Cake, M.H., DiSorbo, D.M. and Litwack, G. (1978) Effect of pyridoxal phosphate on the DNA binding site of activated hepatic glucocorticoid receptor. *J. Biol. Chem.,* 253: 4886–4891.

Canessa, C.M., Schild, L., Buell, G., Thorens, B., Gautschi, I., Horisberger, J.D. and Rossier, B.C. (1994) Amiloride-sensitive epithelial Na$^+$ channel is made of three homologous subunits. *Nature,* 367: 463–467.

Caterina, M.J., Schumacher, M.A., Tominaga, M., Rosen, T.A., Levine, J.D. and Julius, D. (1997) The capsaicin receptor: A heat-activated ion channel in the pain pathway. *Nature,* 389: 816–824.

Chang, G., Spencer, R.H., Lee, A.T., Barclay, M.T. and Rees, D.C. (1998) Structure of the MscL homolog from *Mycobacterium tuberculosis*: A gated mechanosensitive ion channel. *Science,* 282: 2220–2226.

Chang, Y., Wang, R., Barot, S. and Weiss, D.S. (1996) Stoichiometry of a recombinant GABA$_A$ receptor. *J. Neurosci.,* 16: 5415–5424.

Chen, C.C., Akopian, A.N., Sivilotti, L., Colquhoun, D., Burnstock, G. and Wood, J.N. (1995) A P2X purinoceptor expressed by a subset of sensory neurons. *Nature,* 377: 428–431.

Clement, J.P., Kunjilwar, K., Gonzalez, G., Schwanstecher, M., Panten, U., Aguilar-Bryan, L. and Bryan, J. (1997) Association and stoichiometry of K_{ATP} channel subunits. *Neuron,* 18: 827–838.

Cockcroft, V.B., Lunt, G.G. and Osguthorpe, D.J. (1990) Modelling of binding sites of the nicotinic acetylcholine receptor and their relation to models of the whole receptor. *Biochem. Soc. Symp.,* 57: 65–79.

Cockcroft, V.B., Ortells, M.O., Thomas, P. and Lunt, G.G. (1993) Homologies and disparities of glutamate receptors: A critical analysis. *Neurochem. Int.,* 23: 583–594.

Collo, G., North, R.A., Kawashima, E., Merlo-Pich, E., Neidhart, S., Surprenant, A. and Buell, G. (1996) Cloning of $P2X_5$ and $P2X_6$ receptors and the distribution and properties of an extended family of ATP-gated ion channels. *J. Neurosci.,* 16: 2495–2507.

Cooper, E., Couturier, S. and Ballivet, M. (1991) Pentameric structure and subunit stoichiometry of a neuronal nicotinic acetylcholine receptor. *Nature,* 350: 235–238.

Corey, S., Krapivinsky, G., Krapivinsky, L. and Clapham, D.E. (1998) Number and stoichiometry of subunits in the native atrial G-protein-gated K^+ channel, IKACh. *J. Biol. Chem.,* 273: 5271–5278.

Coscoy, S., Linguelia, E., Lazdunski, M. and Barbry, P. (1998) The Phe-Met-Arg-Phe-amide-activated sodium channel is a tetramer. *J. Biol. Chem.,* 273: 8317–8322.

Costa, E. (1998) From $GABA_A$ receptor diversity emerges a unified vision of GABAergic inhibition. *Annu. Rev. Pharmacol. Toxicol.,* 38: 321–350.

Cruickshank, C.C., Minchin, R.F., Le Dain, A.C. and Martinac, B. (1997) Estimation of the pore size of the large-conductance mechanosensitive ion channel of *Escherichia coli. Biophys. J.,* 73: 1925–1931.

Cutting, G.R., Lu, L., O'Hara, B.F., Kasch, L.M., Montrose-Rafizadeh, C., Donovan, D.M., Shimada, S., Antonarakis, S.E., Guggino, W.B., Uhl, G.R. and Kazazian, H.H. (1991) Cloning of the γ-aminobutyric acid (GABA) ρ1 cDNA: A GABA receptor subunit highly expressed in the retina. *Proc. Natl. Acad. Sci. USA,* 88: 2673–2677.

Dani, J.A. and Mayer, M.L. (1995) Structure and function of glutamate and nicotinic acetylcholine receptors. *Curr. Opin. Neurobiol.,* 5: 310–317.

Davies, P.A., Pistis, M., Hanna, M.C., Peters, J.A., Lambert, J.J., Hales, T.G. and Kirkness, E.F. (1999) The $5-HT_{3B}$ subunit is a major determinant of serotonin-receptor function. *Nature,* 397: 359–363.

Duda, T., Goraczniak, R.M. and Sharma, R.K. (1993) Core sequence of ATP regulatory module in receptor guanylate cyclases. *FEBS Lett.,* 315: 143–148.

Egan, T.M., Haines, W.R. and Voigt, M.M. (1998) A domain contributing to the ion channel of ATP-gated $P2X_2$ receptors identified by the substituted cysteine accessibility method. *J. Neurosci.,* 18: 2350–2359.

Eiselé, J.L., Bertrand, S., Galzi, J.L., Devillers-Thiéry, A., Changeux, J.P. and Bertrand, D. (1993) Chimaeric nicotinic-serotonergic receptor combines distinct ligand binding and channel specificities. *Nature,* 366: 479–483.

Enz, R. and Cutting, G.R. (1998) Molecular composition of $GABA_C$ receptors. *Vision Res.,* 38: 1431–1441.

Evans, R.J., Lewis, C., Buell, G., Valera, S., North, R.A. and Surprenant, A. (1995) Pharmacological characterization of heterologously expressed ATP-gated cation channels (P_{2x} purinoceptors). *Mol. Pharmacol.,* 48: 178–183.

Evans, R.J., Lewis, C., Virginio, C., Lundstrom, K., Buell, G., Surprenant, A. and North, R.A. (1996) Ionic permeability of, and divalent cation effects on, two ATP-gated cation channels (P2X receptors) expressed in mammalian cells. *J. Physiol. (Lond.),* 497: 413–422.

Ferrer-Montiel, A.V. and Montal, M. (1996) Pentameric subunit stoichiometry of a neuronal glutamate receptor. *Proc. Natl. Acad. Sci. USA,* 93: 2741–2744.

Firsov, D., Gautschi, I., Merillat, A.M., Rossier, B.C. and Schild, L. (1998) The heterotetrameric architecture of the epithelial sodium channel (ENaC). *EMBO J.,* 17: 344–352.

Fletcher, S., Lindstrom, J.M., McKernan, R.M. and Barnes, N.M. (1998) Evidence that porcine native $5-HT_3$ receptors do not contain nicotinic acetylcholine receptor subunits. *Neuropharmacology,* 37: 397–399.

Freist, W., Verhey, J.F., Stühmer, W. and Gauss, D.H. (1998) ATP binding site of P2X channel proteins: Structural similarities with class II aminoacyl-tRNA synthetases. *FEBS Lett.,* 434: 61–65.

Galzi, J.L. and Changeux, J.P. (1994) Neurotransmitter-gated ion channels as unconventional allosteric proteins. *Curr. Opin. Struct. Biol.,* 4: 554–565.

Galzi, J.L. and Changeux, J.P. (1995) Neuronal nicotinic receptors: molecular organization and regulations. *Neuropharmacology,* 34: 563–582.

Garcia-Guzman, M., Soto, F., Gomez-Hernandez, J.M., Lund, P.E. and Stühmer, W. (1997) Characterization of recombinant human $P2X_4$ receptor reveals pharmacological differences to the rat homologue. *Mol. Pharmacol.,* 51: 109–118.

Garcia-Guzman, M., Soto, F., Laube, B. and Stühmer, W. (1996) Molecular cloning and functional expression of a novel rat heart P2X purinoceptor. *FEBS Lett.,* 388: 123–127.

Gavel, Y. and Von Heijne, G. (1990) Sequence differences between glycosylated and non-glycosylated Asn-X-Thr/Ser acceptor sites: implications for protein engineering. *Protein Eng.,* 3: 433–442.

Glowatzki, E., Fakler, G., Brändle, U., Rexhausen, U., Zenner, H.P., Ruppersberg, J.P. and Fakler, B. (1995) Subunit-dependent assembly of inward-rectifier K^+ channels. *Proc. Roy. Soc. Lond. Ser. B,* 261: 251–261.

Green, T., Stauffer, K.A. and Lummis, S.C. (1995) Expression of recombinant homo-oligomeric 5-hydroxytryptamine$_3$ receptors provides new insights into their maturation and structure. *J. Biol. Chem.,* 270: 6056–6061.

Hansen, M.A., Barden, J.A., Balcar, V.J., Keay, K.A. and Bennett, M.R. (1997) Structural motif and characteristics of the extracellular domain of P2X receptors. *Biochem. Biophys. Res. Commun.,* 236: 670–675.

Hollmann, M. and Heinemann, S. (1994) Cloned glutamate receptors. *Annu. Rev. Neurosci.,* 17: 31–108.

Hovius, R., Tairi, A.P., Blasey, H., Bernard, A., Lundström, K. and Vogel, H. (1998) Characterization of a mouse serotonin 5-HT$_3$ receptor purified from mammalian cells. *J. Neurochem.,* 70: 824–834.

Hucho, F., Tsetlin, V.I. and Machold, J. (1996) The emerging three-dimensional structure of a receptor. The nicotinic acetylcholine receptor. *Eur. J. Biochem.,* 239: 539–557.

Hurst, R.S., Kavanaugh, M.P., Yakel, J., Adelman, J.P. and North, R.A. (1992) Cooperative interactions among subunits of a voltage-dependent potassium channel. Evidence from expression of concatenated cDNAs. *J. Biol. Chem.,* 267: 23742–23745.

Im, W.B., Pregenzer, J.F., Binder, J.A., Dillon, G.H. and Alberts, G.L. (1995) Chloride channel expression with the tandem construct of α6-β2 GABA$_A$ receptor subunit requires a monomeric subunit of α6 or γ2. *J. Biol. Chem.,* 270: 26063–26066.

Inagaki, N., Tsuura, Y., Namba, N., Masuda, K., Gonoi, T., Horie, M., Seino, Y., Mizuta, M. and Seino, S. (1995) Cloning and functional characterization of a novel ATP-sensitive potassium channel ubiquitously expressed in rat tissues, including pancreatic islets, pituitary, skeletal muscle, and heart. *J. Biol. Chem.,* 270: 5691–5694.

Jackson, M.B. and Yakel, J.L. (1995) The 5-HT$_3$ receptor channel. *Annu. Rev. Physiol.,* 57: 447–468.

Johnston, G.A.R. (1996) GABAC receptors: Relatively simple transmitter-gated ion channels? *Trends Pharmacol. Sci.,* 17: 319–323.

Karlin, A. and Akabas, M.H. (1995) Toward a structural basis for the function of nicotinic acetylcholine receptors and their cousins. *Neuron,* 15: 1231–1244.

Kavanaugh, M.P., Varnum, M.D., Osborne, P.B., Christie, M.J., Busch, A.E., Adelman, J.P. and North, R.A. (1991) Interaction between tetraethylammonium and amino acid residues in the pore of cloned voltage-dependent potassium channels. *J. Biol. Chem.,* 266: 7583–7587.

Ketchum, K.A., Joiner, W.J., Sellers, A.J., Kaczmarek, L.K. and Goldstein, S.A.N. (1995) A new family of outwardly rectifying potassium channel proteins with two pore domains in tandem. *Nature,* 376: 690–695.

Khakh, B.S., Humphrey, P.P.A. and Surprenant, A. (1995) Electrophysiological properties of P2X purinoceptors in rat superior cervical, nodose and guinea-pig coeliac neurones. *J. Physiol. (Lond.),* 484: 385–395.

Kim, M., Yoo, O.J. and Choe, S. (1997) Molecular assembly of the extracellular domain of P2X$_2$, an ATP-gated ion channel. *Biochem. Biophys. Res. Commun.,* 240: 618–622.

Kirsch, J., Wolters, I., Triller, A. and Betz, H. (1993) Gephyrin antisense oligonucleotides prevent glycine receptor clustering in spinal neurons. *Nature,* 366: 745–748.

Koonin, E.V. (1993) A superfamily of ATPases with diverse functions containing either classical or deviant ATP-binding motif. *J. Mol. Biol.,* 229: 1165–1174.

Kosari, F., Sheng, S., Li, J., Mak, D.O., Foskett, J.K. and Kleyman, T.R. (1998) Subunit stoichiometry of the epithelial sodium channel. *J. Biol. Chem.,* 273: 13469–13474.

Kuhse, J., Betz, H. and Kirsch, J. (1995) The inhibitory glycine receptor: architecture, synaptic localization and molecular pathology of a postsynaptic ion-channel complex. *Curr. Opin. Neurobiol.,* 5: 318–323.

Kuhse, J., Laube, B., Magalei, D. and Betz, H. (1993) Assembly of the inhibitory glycine receptor: Identification of amino acid sequence motifs governing subunit stoichiometry. *Neuron,* 11: 1049–1056.

Kuner, T., Wollmuth, l.P., Karlin, A., Seeburg, P.H. and Sakmann, B. (1996) Structure of the NMDA receptor channel M2 segment inferred from the accessibility of substituted cysteines. *Neuron,* 17: 343–352.

Lambrecht, G. (1996) Design and pharmacology of selective P2-purinoceptor antagonists. *J. Auton. Pharmacol.,* 16: 341–344.

Langosch, D., Thomas, L. and Betz, H. (1988) Conserved quarternary structure of ligand-gated ion channels: The postsynaptic glycine receptor is a pentamer. *Proc. Natl. Acad. Sci. USA,* 85: 7394–7398.

Laube, B., Kuhse, J. and Betz, H. (1998) Evidence for a tetrameric structure of recombinant NMDA receptors. *J. Neurosci.,* 18: 2954–2961.

Lê, K.T., Babinski, K. and Séguéla, P. (1998) Central P2X$_4$ and P2X$_6$ channel subunits coassemble into a novel heteromeric ATP receptor. *J. Neurosci.,* 18: 7152–7159.

Lewis, C., Neidhart, S., Holy, C., North, R.A., Buell, G. and Surprenant, A. (1995) Coexpression of P2X$_2$ and P2X$_3$ receptor subunits can account for ATP-gated currents in sensory neurons. *Nature,* 377: 432–435.

Li, M., Unwin, N., Stauffer, K.A., Jan, Y.N. and Jan, L.Y. (1994) Images of purified Shaker potassium channels. *Curr. Biol.,* 4: 110–115.

Liman, E.R., Tytgat, J. and Hess, P. (1992) Subunit stoichiometry of a mammalian K$^+$ channel determined by construction of multimeric cDNAs. *Neuron,* 9: 861–871.

Lindstrom, J., Merlie, J. and Yogeeswaran, G. (1979) Biochemical properties of acetylcholine receptor subunits from *Torpedo californica. Biochemistry,* 18: 4465–4470.

Lingueglia, E., Champigny, G., Lazdunski, M. and Barbry, P. (1995) Cloning of the amiloride-sensitive FMRFamide peptide-gated sodium channel. *Nature,* 378: 730–733.

Ludewig, U., Pusch, M. and Jentsch, T.J. (1996) Two physically distinct pores in the dimeric ClC-0 chloride channel. *Nature,* 383: 340–343.

MacKinnon, R. (1991) Determination of the subunit stoichiometry of a voltage-activated potassium channel. *Nature,* 350: 232–235.

Mamalaki, C., Stephenson, F.A. and Barnard, E.A. (1987) The GABA$_A$/benzodiazepine receptor is a heterotetramer of homologous α and β subunits. *EMBO J.,* 6: 561–565.

Mano, I. and Teichberg, V.I. (1998) A tetrameric subunit stoichiometry for a glutamate receptor-channel complex. *Neuroreport,* 9: 327–331.

Marban, E., Yamagishi, T. and Tomaselli, G.F. (1998) Structure and function of voltage-gated sodium channels. *J. Physiol. (Lond.)*, 508: 647–657.

Maricq, A.V., Peterson, A.S., Brake, A.J., Myers, R.M. and Julius, D. (1991) Primary structure and functional expression of the 5HT$_3$ receptor, a serotonin-gated ion channel. *Science*, 254: 432–437.

McGehee, D.S. and Role, L.W. (1995) Physiological diversity of nicotinic acetylcholine receptors expressed by vertebrate neurons. *Ann. Rev. Physiol.*, 57: 521–546.

McKernan, R.M., Gillard, N.P., Quirk, K., Kneen, C.O., Stevenson, G.I., Swain, C.J. and Ragan, C.I. (1990) Purification of the hydroxytryptamine 5-HT$_3$ receptor from NCB20 cells. *J. Biol. Chem.*, 265: 13572–13577.

Middleton, R.E., Pheasant, D.J. and Miller, C. (1994) Purification, reconstitution, and subunit composition of a voltage-gated chloride channel from *Torpedo* electroplax. *Biochemistry*, 33: 13189–13198.

Middleton, R.E., Pheasant, D.J. and Miller, C. (1996) Homo-dimeric architecture of a CIC-type chloride ion channel. *Nature*, 383: 337–340.

Mikoshiba, K. (1997) The InsP$_3$ receptor and intracellular Ca^{2+} signaling. *Curr. Opin. Neurobiol.*, 7: 339–345.

Mishina, M., Takai, T., Imoto, K., Noda, M., Takahashi, T., Numa, S., Methfessel, C. and Sakmann, B. (1986) Molecular distinction between fetal and adult forms of muscle acetyl-choline receptor. *Nature*, 321: 406–411.

Moriyoshi, K., Masu, M., Ishii, T., Shigemoto, R., Mizuno, N. and Nakanishi, S. (1991) Molecular cloning and characterization of the rat NMDA receptor. *Nature*, 354: 31–37.

Nayeem, N., Green, T.P., Martin, I.L. and Barnard, E.A. (1994) Quaternary structure of the native GABA$_A$ receptor determined by electron microscopic image analysis. *J. Neurochem.*, 62: 815–818.

Newbolt, A., Stoop, R., Virginio, C., Surprenant, A., North, R.A., Buell, G. and Rassendren, F. (1998) Membrane topology of an ATP-gated ion channel (P2X receptor). *J. Biol. Chem.*, 273: 15177–15182.

Nicke, A., Bäumert, H.G., Rettinger, J., Eichele, A., Lambrecht, G., Mutschler, E. and Schmalzing, G. (1998) P2X$_1$ and P2X$_3$ receptors form stable trimers: A novel structural motif of ligand-gated ion channels. *EMBO J.*, 17: 3016–3028.

Nicke, A., Rettinger, J., Mutschler, E. and Schmalzing, G. (1999) Blue native PAGE as a useful method for the analysis of the assembly of distinct combinations of nicotinic acetylcholine receptor subunits. *J. Recept. Signal. Transduct. Res.*, 19: 493–507.

North, R.A. (1996a) Families of ion channels with two hydrophobic segments. *Curr. Opin. Cell Biol.*, 8: 474–483.

North, R.A. (1996b) P2X receptors: A third major class of ligand-gated ion channels. *Ciba Found. Symp.*, 198: 91–105.

North, R.A. (1996c) P2X purinoceptor plethora. *Semin. Neurosci.*, 8: 187–194.

North, R.A. and Barnard, E.A. (1997) Nucleotide receptors. *Curr. Opin. Neurobiol.*, 7: 346–357.

Ortells, M.O. and Lunt, G.G. (1995) Evolutionary history of the ligand-gated ion-channel superfamily of receptors. *Trends Neurosci.*, 18: 121–127.

Pessia, M., Tucker, S.J., Lee, K., Bond, C.T. and Adelman, J.P. (1996) Subunit positional effects revealed by novel hetero-meric inwardly rectifying K$^+$ channels. *EMBO J.*, 15: 2980–2987.

Pfeiffer, F., Graham, D. and Betz, H. (1982) Purification by affinity chromatography of the glycine receptor of rat spinal cord. *J. Biol. Chem.*, 257: 9389–9393.

Portoghese, P.S. (1989) Bivalent ligands and the message-address concept in the design of selective opioid receptor antagonists. *Trends Pharmacol. Sci*, 10: 230–235.

Premkumar, L.S. and Auerbach, A. (1997) Stoichiometry of recombinant N-methyl-D-aspartate receptor channels inferred from single-channel current patterns. *J. Gen. Physiol.*, 110: 485–502.

Raab-Graham, K.F. and Vandenberg, C.A. (1998) Tetrameric subunit structure of the native brain inwardly rectifying potassium channel Kir 2.2. *J. Biol. Chem.*, 273: 19699–19707.

Radford, K.M., Virginio, C., Surprenant, A., North, R.A. and Kawashima, E. (1997) Baculovirus expression provides direct evidence for heteromeric assembly of P2X$_2$ and P2X$_3$ receptors. *J. Neurosci.*, 17: 6529–6533.

Raftery, M.A., Hunkapiller, M.W., Strader, C.D. and Hood, L.E. (1980) Acetylcholine receptor: Complex of homologous subunits. *Science*, 208: 1454–1457.

Rassendren, F., Buell, G., Newbolt, A., North, R.A. and Surprenant, A. (1997) Identification of amino acid residues contributing to the pore of a P2X receptor. *EMBO J.*, 16: 3446–3454.

Renard, S., Lingueglia, E., Voilley, N., Lazdunski, M. and Barbry, P. (1994) Biochemical analysis of the membrane topology of the amiloride-sensitive Na$^+$ channel. *J. Biol. Chem.*, 269: 12981–12986.

Roeper, J. and Pongs, O. (1996) Presynaptic potassium channels. *Curr. Opin. Neurobiol.*, 6: 338–341.

Rosenmund, C., Stern-Bach, Y. and Stevens, C.F. (1998) The tetrameric structure of a glutamate receptor channel. *Science*, 280: 1596–1599.

Saint, N., Lacapere, J.J., Gu, L.Q., Ghazi, A., Martinac, B. and Rigaud, J.L. (1998) A hexameric transmembrane pore revealed by two-dimensional crystallization of the large mechanosensitive ion channel (MscL) of Escherichia coli. *J. Biol. Chem.*, 273: 14667–14670.

Schägger, H. and von Jagow, G. (1991) Blue native electro-phoresis for isolation of membrane protein complexes in enzymatically active form. *Anal. Biochem.*, 199: 223–231.

Seeburg, P.H. (1993) The TINS/TiPS Lecture. The molecular biology of mammalian glutamate receptor channels. *Trends Neurosci.*, 16: 359–365.

Séguéla, P., Haghighi, A., Soghomonian, J.J. and Cooper, E. (1996) A novel neuronal P$_{2X}$ ATP receptor ion channel with widespread distribution in the brain. *J. Neurosci.*, 16: 448–455.

Sigel, E. and Buhr, A. (1997) The benzodiazepine binding site of GABA$_A$ receptors. *Trends Pharmacol. Sci.,* 18: 425–429.

Snyder, P.M., Cheng, C., Prince, L.S., Rogers, J.C. and Welsh, M.J. (1998) Electrophysiological and biochemical evidence that DEG/ENaC cation channels are composed of nine subunits. *J. Biol. Chem.,* 273: 681–684.

Soto, F., Garcia-Guzman, M., Gomez-Hernandez, J.M., Hollmann, M., Karschin, C. and Stühmer, W. (1996a) P2X$_4$: An ATP-activated ionotropic receptor cloned from rat brain. *Proc. Natl. Acad. Sci. USA,* 93: 3684–3688.

Soto, M., Garcia-Guzman, M., Karschin, C. and Stühmer, W. (1996b) Cloning and tissue distribution of a novel P2X receptor from rat brain. *Biochem. Biophys. Res. Commun.,* 223: 456–460.

Soto, F., Garcia-Guzman, M. and Stühmer, W. (1997) Cloned ligand-gated channels activated by extracellular ATP (P2X receptors). *J. Membr. Biol.,* 160: 91–100.

Sumikawa, K., Houghton, M., Emtage, J.S., Richards, B.M. and Barnard, E.A. (1981) Active multi-subunit ACh receptor assembled by translation of heterologous mRNA in *Xenopus* oocytes. *Nature,* 292: 862–864.

Surprenant, A., Rassendren, F., Kawashima, E., North, R.A. and Buell, G. (1996) The cytolytic P$_{2Z}$ receptor for extracellular ATP identified as a P$_{2X}$ receptor (P2X$_7$). *Science,* 272: 735–738.

Torres, G.E., Egan, T.M. and Voigt, M.M. (1998) Topological analysis of the ATP-gated ionotrophic P2X$_2$ receptor subunit. *FEBS Lett.,* 425: 19–23.

Traut, T.W. (1994) The functions and consensus motifs of nine types of peptide segments that form different types of nucleotide-binding sites. *Eur. J. Biochem.,* 222: 9–19.

Unwin, N. (1995) Acetylcholine receptor channel imaged in the open state. *Nature,* 373: 37–43.

Valera, S., Hussy, N., Evans, R.J., Adami, N., North, R.A., Surprenant, A. and Buell, G. (1994) A new class of ligand-gated ion channel defined by P$_{2x}$ receptor for extracellular ATP. *Nature,* 371: 516–519.

van Hooft, J.A., Kreikamp, A.P. and Vijverberg, H.P. (1997) Native serotonin 5-HT$_3$ receptors expressed in *Xenopus* oocytes differ from homopentameric 5-HT$_3$ receptors. *J. Neurochem.,* 69: 1318–1321.

van Hooft, J.A., Spier, A.D., Yakel, J.L., Lummis, S.R. and Vijverberg, H.M. (1998) Promiscuous coassembly of serotonin 5-HT$_3$ and nicotinic α4 receptor subunits into Ca^{2+}-permeable ion channels. *Proc. Natl. Acad. Sci. USA,* 95: 11456–11461.

Wang, C.Z., Namba, N., Gonoi, T., Inagaki, N. and Seino, S. (1996) Cloning and pharmacological characterization of a fourth P2X receptor subtype widely expressed in brain and peripheral tissues including various endocrine tissues. *Biochem. Biophys. Res. Commun.,* 220: 196–202.

Wenthold, R.J., Yokotani, N., Doi, K. and Wada, K. (1992) Immunochemical characterization of the non-NMDA glutamate receptor using subunit-specific antibodies. Evidence for a hetero-oligomeric structure in rat brain. *J. Biol. Chem.,* 267: 501–507.

Wo, Z.G. and Oswald, R.E. (1995) Unraveling the modular design of glutamate-gated ion channels. *Trends Neurosci.,* 18: 161–168.

Wood, M.W., VanDongen, H.M. and VanDongen, A.M. (1995) Structural conservation of ion conduction pathways in K channels and glutamate receptors. *Proc. Natl. Acad. Sci. USA,* 92: 4882–4886.

Wu, T.Y., Liu, C.I. and Chang, Y.C. (1996) A study of the oligomeric state of the α-amino-3-hydroxy-5-methyl-4-isoxazolepropionic acid-preferring glutamate receptors in the synaptic junctions of porcine brain. *Biochem. J.,* 319: 731–739.

Yang, J., Jan, Y.N. and Jan, L.Y. (1995) Determination of the subunit stoichiometry of an inwardly rectifying potassium channel. *Neuron,* 15: 1441–1447.

Yool, A.J. and Schwarz, T.L. (1991) Alteration of ionic selectivity of a K$^+$ channel by mutation of the H5 region. *Nature,* 349: 700-704.

Zagotta, W.N. and Siegelbaum, S.A. (1996) Structure and function of cyclic nucleotide-gated channels. *Annu. Rev. Neurosci.,* 19: 235–263.

P. Illes and H. Zimmermann (Eds.)
Progress in Brain Research, Vol 120
© 1999 Elsevier Science BV. All rights reserved

CHAPTER 7

Cloning, functional characterization and developmental expression of a P2X receptor from chick embryo

Anja Ruppelt[1], Bruce T. Liang[2] and Florentina Soto[1,*]

[1]*Max-Planck Institute for Experimental Medicine, Hermann-Rein Str. 3, D-37075, Göttingen, Germany*
[2]*University of Pennsylvania Medical Center, 3610 Hamilton Walk, Philadelphia, PA 19104, USA*

Introduction

Fast excitatory neuronal responses elicited by extracellular ATP have been observed in many regions of the nervous system, including medial habenula (Edwards et al., 1992), dorsal horn of the spinal cord (Bardoni et al., 1997), locus coeruleus (Harms et al., 1992) and in peripheral ganglia (Evans et al., 1992; Silinsky and Gerzanich, 1994) raising the possibility that ionotropic ATP receptors might play an important role in specific neuron-to-neuron synapses. In the periphery, ATP is co-stored in secretory granules together with catecholamines (Sneddon and Westfall, 1984; Burnstock, 1996) and acetylcholine (Zimmermann, 1982) and is secreted upon nerve stimulation (Silinsky, 1975). Additionally, ATP is released in the heart under stressful condition such as hypoxia, supplying an additional source of the extracellular transmitter in pathological situations (Clemens and Forrester, 1981). Responses to extracellular ATP have been found in a number of isolated cells and tissues including smooth muscle (Benham and Tsien, 1987; Friel, 1988; Kennedy et al., 1996), adrenal gland (Castro et al., 1994), pituitary gland (Tomic et al., 1996) and heart and muscle myocytes (Kolb and Wakelam, 1983; Hume and Honig, 1986; Friel and Bean, 1988).

*Extracellular ATP evokes responses through two subclasses of P2-receptors—P2X and P2Y. The P2X subclass consists of ligand-gated ion channels, of which seven subunits have been cloned ($P2X_1$– $P2X_7$) whereas the P2Y subclass is coupled to GTP-binding proteins, with 11 putative members ($P2Y_1$–$P2Y_{11}$) cloned so far (North and Barnard, 1997; Burnstock et al., 1998). The binding of agonist to P2X receptors opens an intrinsic ion channel which is non-selective for monovalent cations and shows an appreciable permeability to Ca^{2+} (Rogers and Dani, 1995; Garcia-Guzman et al., 1997). Cloned P2X receptor subtypes have been functionally characterized by heterologous expression in *Xenopus laevis* oocytes and in mammalian cell lines (Buell et al., 1996; North and Barnard, 1997).

P2X-like responses have been found in a wide variety of tissues. Accordingly, expression of at least one member of the P2X family has been detected in every tissue analyzed (Soto et al., 1997). The main P2X receptors found in adult rat CNS by in situ hybridization are $P2X_4$ and $P2X_6$ whose expression pattern overlapps in many areas of the brain (Collo et al., 1996). There is evidence that the expression of P2X receptors might be developmentally regulated. For instance, some studies performed on rat brain have shown that the $P2X_2$ mRNA level decreases during the development of neonates to adulthood (Kidd et al., 1995; Collo et al., 1996).

---

*Corresponding author. Tel.: +49-551-3899-624; fax: +49-551-3899-644; e-mail: soto@mail.mpiem.gwdg.de

Chick embryos have long been among the most widely used experimental animals in developmental biology. Studies have been performed showing the presence of ATP responses in several chick tissues including ciliary ganglia neurons (Abe et al., 1995), isolated skeletal muscle myocytes and cardiac myocytes (Kolb and Wakelam, 1983; Podrasky et al., 1997). Here, we describe the cloning and functional characterization of a new P2X subunit isolated from embryonic chick brain. The corresponding protein is highly homologous to the mammalian P2X$_4$ receptor subtypes (human and rat) with approximately 75% sequence identity to both subunits. Chick P2X$_4$ (chP2X$_4$) transcripts are present in both brain and heart embryonic tissue during development. The functionality of chP2X$_4$ receptor was assayed by heterologous expression in *Xenopus laevis* oocytes, a system that has been successfully used to describe the mammalian P2X receptors. The homomeric receptor shows functional properties that are very similar to rat and human P2X$_4$, with slight differences in sensitivity to ATP and suramin.

Methods

cDNA cloning

A partial cDNA sequence for the chP2X$_4$ receptor (~400 bp) was initially obtained from RNA isolated from cardiac ventricular myocytes cultured from chick embryos 14 days in ovo (Podrasky et al., 1997) using degenerate PCR following the conditions previously described (Soto et al., 1996). The resulting PCR fragment was labelled by random priming using [α-^{32}P]dCTP (Rediprime kit, Amersham) and used to screen 1×10^6 recombinants of a custom made 17-day embryo brain cDNA library (Stratagene), following standard protocols (Sambrook et al., 1989). We isolated one 1.7-kb-long phage containing the complete open reading frame of 1155 bp. The sequence from both strands was determined by fluorescent DNA sequencing (Applied Biosystems). The nucleotide sequence of chP2X$_4$ cDNA has been submitted to Genbank with accession number Y18008.

Expression in Xenopus oocytes and electrophysiology

For functional expression the full-length chP2X$_4$ cDNA trimed of untranslated regions was cloned into the pSGEM vector (Soto et al., 1996). Plasmid DNA was purified using the Wizard DNA system (Promega). Capped cRNA was transcribed in vitro with T7 RNA polymerase (Promega) in the presence of the cap analog m^{7}G(5')ppp(5')G (Boehringer Mannheim), using 5 µg of NheI-linearized DNA. The cRNA was examined on ethidium bromide-stained denaturing agarose gels to assure the presence of a single, undegraded band of the expected size. The final cRNA concentration was in the range of 0.25 mg/ml—as estimated visually by comparing it to a known amount of molecular weight standards—and was used directly for injection into defolliculated *Xenopus* oocytes. Oocyte isolation and handling were performed using standard techniques (Stühmer, 1992).

Two electrode voltage-clamp recordings were performed 2–5 days after cRNA injection. The standard Mg^{2+} solution (Mg^{2+}-Ringer) superfusing the oocytes contained 115 mM NaCl, 2.8 mM KCl, 1.8 mM MgCl$_2$, 100 µM BaCl$_2$, 10 mM HEPES, pH = 7.2. The drugs were prepared as concentrated stocks in 100 mM HEPES pH = 7.2, and stored at $-20°$C until use. ATP (disodium salt), CTP (sodium salt), UTP (sodium salt), α,β-methylene ATP ($\alpha\beta$-meATP, lithium salt) and β,γ-methylene ATP ($\beta\gamma$-meATP, sodium salt) were obtained from Sigma; GTP (disodium salt) was purchased from Fluka; ADP (free acid) was obtained from Boehringer Mannheim; 2-methylthio ATP (2MeSATP, tetrasodium salt) was obtained from Research Biochemicals Inc.; suramin was purchased from Calbiochem.

Agonists and antagonists were applied using a gravitational flow system. The small volume of the bath in the recording chamber (< 100 µl) and the high rate of perfusion (7 ml/min) allowed a rapid exchange of solutions. The recovery of the current was complete after < 6 min of wash-out, even at the highest concentrations of ATP used. Consequently, we allowed a 6 min wash-out period between two successive recordings. Suramin was co-applied with ATP while perfusing with the standard Mg^{2+}

solution. Current values obtained in the presence of suramin were normalized to the current elicited with the agonist in the absence of the antagonist.

Voltage and current electrodes were filled with 2M KCl solution and had resistances of 0.5–1.5 MΩ. All experiments were performed at room temperature (18–22°C). Currents were filtered at 50 Hz and recorded using a Turbo TEC-10CD amplifier (npi electronics) and Pulse software (HEKA). For all the data error bars represent the standard error of the mean (SEM). The ramp I–V relations were performed by applying voltage-ramps (-70 to $+70$ mV, 200 ms) when the current has reached the maximal value. The ramps were corrected for leak currents by substracting them from the ramps that were obtained in the absence of agonist application. When fast ramps were performed, the current was filtered at 2 kHz. Steady-state I–V was obtained by steeping from -70 mV to the desired potential (10 s voltage jumps) and then applying 100 μM ATP. The leak current was obtained by performing the same pulse in the absence of agonist and subsequently substracting it from the current obtained in the presence of ATP.

Tissue expression analysis by RT–PCR

Selected tissues were dissected from chick embryos at various developmental stages and immediately frozen in liquid nitrogen. Total RNAs were extracted from the tissues by the one-step method (Chomczynski and Sacchi, 1987). After isolation, 100 μg total RNA were treated with 50 U RNASe free DNAse I (Promega) to remove any contaminant genomic DNA. cDNA was prepared from 5 μg of total RNA as previously described, using Superscript Plus Reverse Transcriptase (Gibco). A control reaction in the absence of reverse transcriptase was also performed for each tissue to be used as a negative control for the RT–PCR. The PCR thermal profile was 5 min at 94°C and 30 cycles of 40 s at 94°C, 40 s at 53°C and 45 s at 72°C. The PCR primers were:

5′-TCGGCTGGGTCTTTCTGT-3′ (forward) and 5′-TCTGTCTGGGCATCATAG-3′ (reverse), flanking a region 537 bp long. The PCR products were separated in an agarose gel and visualized with ethidium bromide staining. The specificity of the amplification bands was confirmed by high stringency hybridization. The DNA used to generate a $[\alpha^{32}P]dCTP$ random-labelled probe (Rediprime, Amersham) corresponded to nucleotides 237 to 934 of the chP2X$_4$ open reading frame.

Results and discussion

Reverse transcription and PCR amplification of isolated cardiomyocyte total RNA from a 14-day-old chick embryo with degenerate oligonucleotides for P2X receptor sequences resulted in the amplification of one band of approximately 400 bp. Sequencing revealed that this band contained a DNA fragment encoding a partial protein sequence that showed the highest sequence identity to the fourth members of the P2X receptor family. Oligonucleotides designed to match exactly the new isolated sequence were used to PCR cDNA prepared from RNA isolated from 14-day-old embryo total brain. The cloning and sequencing of the obtained PCR fragment indicated the expression of the same P2X subtype in chick brain. The full-length cDNA was obtained by screening a 17-day-old embryo brain cDNA library. One of the isolated phages contained 1.7-kb-long cDNA with a 384 amino acid open reading frame (ORF) flanked by 65 bp of 5′-untranslated region and 480 bp of 3′-untranslated region (Fig. 1). The position of the initiator methionine was assigned by comparison to the mammalian P2X$_4$ subtypes since an upstream stop codon in frame with the protein sequence was not found. The amino acid sequence identity was the highest (aproximately 75%) between the isolated cDNA and the human and rat P2X$_4$ subtypes so we called the new receptor chick P2X$_4$ (chP2X$_4$, Fig. 2). When compared to the rest of cloned P2X subtypes, the lowest amino acid sequence identity was toward the P2X$_6$ subtype (43%) while the highest was to P2X$_5$ with 52% amino acid sequence identity. The amino acid changes are scattered all along the sequence. Most of the non-identical amino acids are conservative substitutions or appear in other P2X receptors. However, there is a notable exception. In position 129 of P2X$_4$ receptors there is an aspartic residue

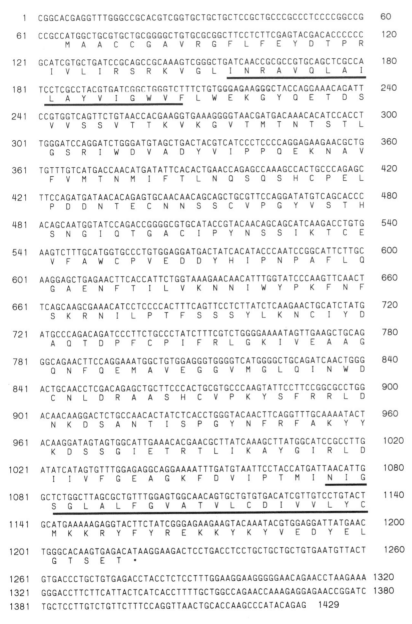

Fig. 1. Nucleotide and predicted amino acid sequence (single letter code) of chick P2X$_4$ receptor. The underlined sequences are the putative transmembrane domains.

that is conserved in all the P2X receptor sequences. However, in the case of chP2X$_4$, the aspartic acid is substituted for an asparagine. A 4.5-kb-long phage (183) that contains an additional asparagine at position 128 of human and rat P2X$_4$ was also isolated from the brain library. However, this phage did not contain an ORF but coding sequences

intercalated with non-coding sequences without any detected exon-intron border. A 385-amino acids long P2X subunit isolated from chick embryo has been reported in a recently published abstract (Liu and Hume, 1997). We have no information on the sequence, but the additional amino acid could be due to the presence of this additional asparagine.

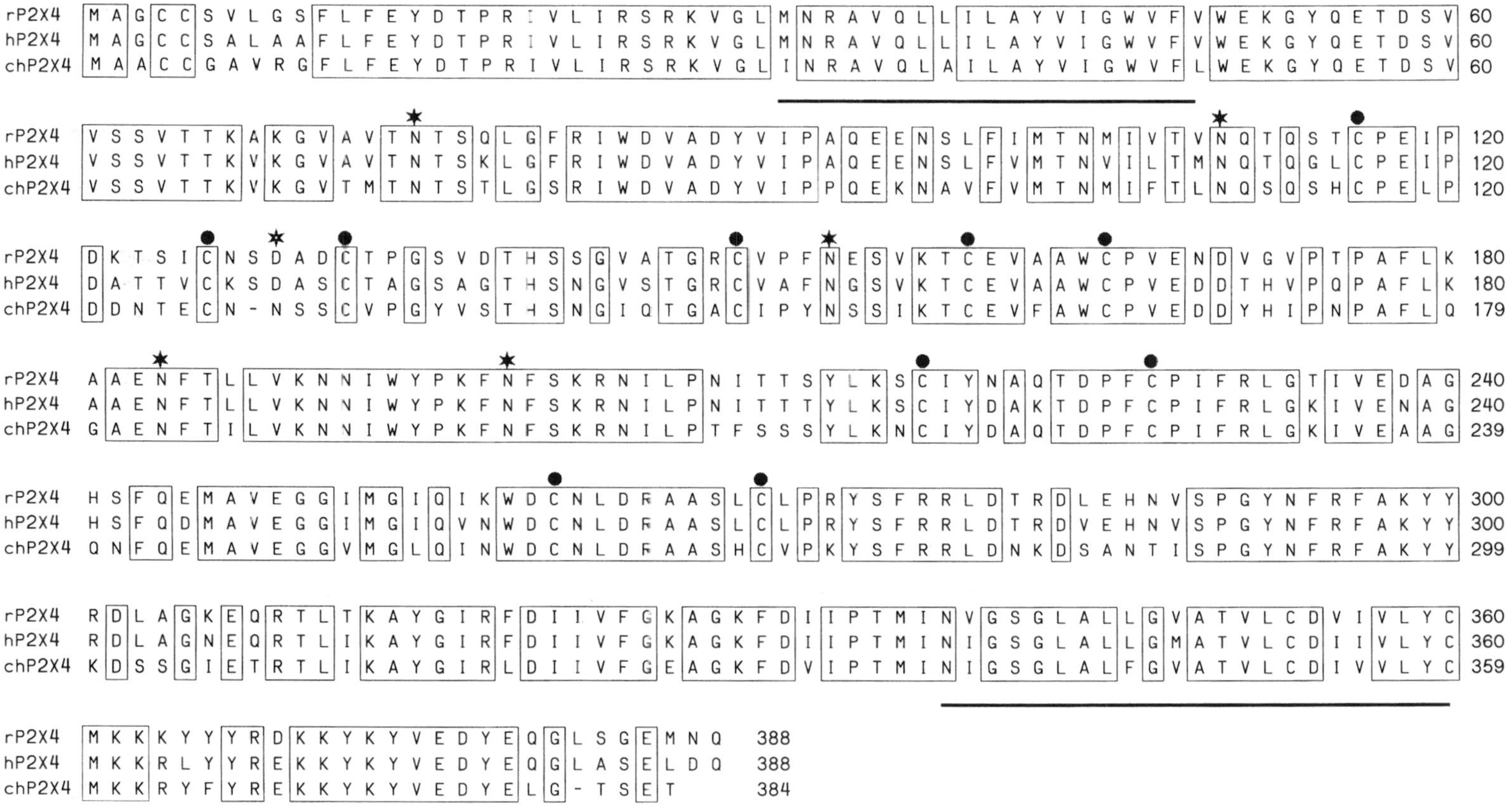

```
rP2X4   M A G C C S V L G S F L F E Y D T P R I V L I R S R K V G L M N R A V Q L L I L A Y V I G W V F V W E K G Y Q E T D S V   60
hP2X4   M A G C C S A L A A F L F E Y D T P R I V L I R S R K V G L M N R A V Q L L I L A Y V I G W V F V W E K G Y Q E T D S V   60
chP2X4  M A A C C G A V R G F L F E Y D T P R I V L I R S R K V G L I N R A V Q L A I L A Y V I G W V F L W E K G Y Q E T D S V   60

rP2X4   V S S V T T K A K G V A V T N T S Q L G F R I W D V A D Y V I P A Q E E N S L F I M T N M I V T V N Q T Q S T C P E I P   120
hP2X4   V S S V T T K V K G V A V T N T S K L G F R I W D V A D Y V I P A Q E E N S L F V M T N V I L T M N Q T Q G L C P E I P   120
chP2X4  V S S V T K V K G V T M T N T S R I W D V A D Y V F V M T N M I F T L N Q S H C P E L P   120

rP2X4   D K T S I C N S D A D C T P G S V D T H S S G V A T G R C V P F N E S V K T C E V A A W C P V E N D V G V P T P A F L K   180
hP2X4   D A T T V C K S D A S C T A G S A G T H S N G V S T G R C V A F N G S V K T C E V A A W C P V E D D T H V P Q P A F L K   180
chP2X4  D D N T E C N - N S S C V P G Y S V S T S N G I Q T G A C I P Y N S S I K T C E V F A W C P V E D D Y H I P N P A F L Q   179

rP2X4   A A E N F T L L V K N N I W Y P K F N F S K R N I L P N I T T S Y L K S C I Y N A Q T D P F C P I F R L G T I V E D A G   240
hP2X4   A A E N F T L L V K N N I W Y P K F N F S K R N I L P N I T T T Y L K S C I Y D A K T D P F C P I F R L G K I V E D A G   240
chP2X4  G A E N F T I L V K N N I W Y P K F N F S K R N I L P T F S S S Y L K N C I Y D A Q T D P F C P I F R L G K I V E A A G   239

rP2X4   H S F Q E M A V E G G I M G I Q I K W D C N L D R A A S L C L P R Y S F R R L D T R D L E H N V S P G Y N F R F A K Y Y   300
hP2X4   H S F Q D M A V E G G I M G I Q V N W D C N L D R A A S L C L P R Y S F R R L D V E H N V S P G Y N F R F A K Y Y   300
chP2X4  Q N F Q E M A V E G G V M G L Q I N W D C N L D R A A S H C V P K Y S F R R L D N K D S A N T I S P G Y N F R F A K Y Y   299

rP2X4   R D L A G K E Q R T L T K A Y G I R F D I I V F G K A G K F D I I P T M I N V G S G L A L L G V A T V L C D V I V L Y C   360
hP2X4   R D L A G N E Q R T L T K A Y G I R F D I I V F G K A G K F D I I P T M I N I G S G L A L L G M A T V L C D I I V L Y C   360
chP2X4  K D S S G I E T R T L I K A Y G I R L D I I V F G E A G K F D V I P T M I N I G S G L A L F G V A T V L C D I I V L Y C   359

rP2X4   M K K Y Y Y R D K K Y K Y V E D Y E Q G L S G E M N Q   388
hP2X4   M K K R L Y Y R E K K Y K Y V E D Y E Q G L A S E L D Q   388
chP2X4  M K K R Y F Y R E K K Y K Y V E D Y E L G - T S E T   384
```

Fig. 2. Amino acid sequence alignment of rat, human and chick P2X$_4$ receptors. The putative transmembrane domains are underlined. The boxes enclose the amino acids that are identical in all three receptor sequences. Filled circles mark the position of the 10 conserved cysteines. Stars indicate the position of the putative N-glycosylation sites of chP2X$_4$.

This small disagreement in sequence might reflect inter-allelic differences. We are currently performing analysis on the expression using PCR on cDNA, in order to determine which clone is actually expressed.

Protein sequence analysis demonstrated a topology corresponding to that of all the cloned P2X receptors, with two hydrophobic domains flanking a large extracellular loop that contains 10 conserved cysteine residues (Brake et al., 1994; Newbolt et al., 1998). In addition there are six consensus sites for N-glycosylation, N-X-[S/T],

five of which are conserved between all P2X$_4$ receptors (Fig. 2).

The functional properties of chP2X$_4$ were determined by expression in *Xenopus laevis* oocytes. The oocytes were recorded 2–5 days after cRNA injection using the two-electrode voltage-clamp configuration. Only in cRNA injected oocytes, application of ATP evoked a fast-activating inward current that slowly desensitized in the presence of the agonist (Fig. 3A). Dose-response analysis gave half maximal effective concentration of 13.7 ± 2.1 μM ATP with a Hill coefficient of

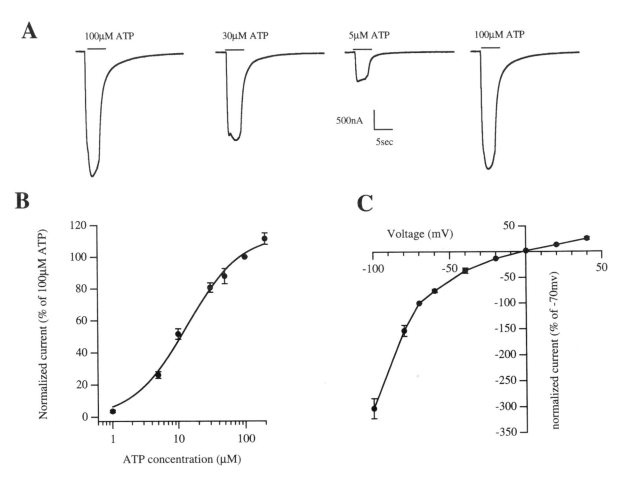

Fig. 3. Functional expression of recombinant chick P2X$_4$ receptors in *Xenopus* oocytes. (A) Representative inward currents elicited by application (5 s) of various ATP concentrations. Holding potential is -70 mV. (B) Dose-response curve of chP2X$_4$ for ATP ($n = 4$–13). The continuous line for ATP is the fit to the data using the equation, $I = I_{max}/(1 + (EC_{50}/L)^{nH})$, where "$I$" is the actual current for a ligand concentration ("L"), "nH" is the Hill coefficient and "I_{max}" the maximal current: $EC_{50} = 13.7 \pm 2.1$ μM, nH $= 1.1 \pm 0.1$. The amount of current obtained at each ATP concentration was normalized to the current obtained with 100 μM ATP. (C) Maximal currents elicited by 100 μM ATP at different holding voltages (normalized to the response at -70 mV, $n = 3$–6).

1.1 ± 0.1 ($n = 4$–13, Fig. 3B). The chP2X$_4$ receptor exhibits a slightly lower affinity for ATP compared to that of the human and rat P2X$_4$ (7.4 and 6.9 μM, respectively) (Soto et al., 1996; Garcia-Guzman et al., 1997). This could be due to the amino acid sequence differences in the extracellular domain. The rank order of potency for different nucleotide analogs (100 μM) follows a pattern that is identical to the rat and human homologue clones: ATP $\geqslant$ 2MeSATP = CTP > $\alpha\beta$-meATP (% current values: 100, 14.2 ± 1.4 ($n = 11$), 15.2 ± 1.3 ($n = 10$), $8.6 \pm (n = 9)$). However, we found 2MeSATP and CTP to be much less effective for chP2X$_4$ than for the mammalian counterparts (Soto et al., 1996; Garcia-Guzman et al., 1997). The nucleotides GTP and ADP (at 100 μM) were almost ineffective, being able to elicit only 3% of the current obtained with 100 μM ATP ($n = 3$). Not detectable responses were obtained with $\beta\gamma$-meATP and UTP.

Ramp I–V relations recorded in Mg^{2+}-Ringer buffer (see Methods) showed a reversal potential of -2.4 ± 0.7 mV ($n = 5$). When half of the Na$^+$ concentration was substituted by K$^+$ the current reversed at -4.5 ± 0.6 mV ($n = 5$). Both reversal potentials are not significantly different (Student's test, $p < 0.05$) so we conclude that chP2X$_4$ is equally permeable to Na$^+$ and K$^+$ ions. The ramp I–V relation differs from the steady-state I–V curve shown in Fig. 3C. In the case of the steady state measurements, a strong inward rectification at positive potentials is observed while the ramp I–V relation presented a linear behavior over the range of voltages tested (not shown). These data indicate that the rectification of the current is probably due to a voltage-dependence of the gating as has been shown for P2X$_2$ receptors (Zhou and Hume, 1998). However, the linear response of the ramp I–V shows that in contrast to P2X$_2$ there is no apparent voltage-dependence of the single channel conductance.

The currents elicited through chP2X$_4$ by 5 μM ATP were blocked by $25 \pm 4\%$ ($n = 8$) in a reversible manner when suramin (200 μM) was co-applied with the agonist. The low affinity of suramin for chP2X$_4$ is a shared feature between all the members of the P2X$_4$ subgroup. However, there are some small differences between the different P2X$_4$ receptors. Thus, hP2X$_4$ is more sensitive to suramin than

the chick homologue (45% of the control current block) while the rat receptor shows less sensitivity to suramin (10% control current block) when exposed to the same concentration of suramin and ATP (Garcia-Guzman et al., 1997).

ATP has a potent depolarizing action on cultured myotubes derived from 11-day-old embryo pectoral muscle, a time in which the pharmacological and physiological characteristics of those receptors have been described by Hume and collaborators (Hume and Honig, 1986; Thomas and Hume, 1990; Thomas et al., 1991). Some of the pharmacological characteristics of the current are coincident with the ones we obtained for chP2X$_4$. For instance, the ATP-evoked current showed low sensitivity to $\alpha\beta$-meATP and $\beta\gamma$-meATP (Hume and Honig, 1986). However, the long lasting desensitization of the native myocyte response (Thomas and Hume, 1990) argues against homomeric chP2X$_4$ subunits forming the native skeletal muscle receptor.

To investigate if there is any regulation in the expression of chP2X$_4$ during development, we isolated brain and heart tissues from chick embryos that were 4, 10 and 14 days in ovo. These embryos ages correspond to stage 23–24, 36 and 40 of Hamburger and Hamilton, respectively (Hamburger and Hamilton, 1951). For both tissues at every age analyzed, the expression of chP2X$_4$ was detected by Southern blot as a band of 537 bp that corresponded to the size of the expected PCR fragment (Fig. 4). Additionally, a smaller band of 395 bp was also hybridizing with the chP2X$_4$ probe under high stringency conditions. Cloning and sequence analysis reveals the presence of a shorter variant of chP2X$_4$ lacking 142 bp from the extracellular domain. So far, the only splice variant described for P2X$_4$ has the 5′-untranslated region and the first 90 amino acids of hP2X$_4$ replaced by 35 amino acids of a protein of the hsp-90 family (Dhulipala et al., 1998). Genomic DNA analysis are necessary to determine if the shorter cDNA detected is a real splice variant of chP2X$_4$.

The presence of P2X$_4$ receptors in mammalian brain and heart has been reported by Northern Blot and RT–PCR (Soto et al., 1997). Additionally, in situ hybridization and inmunocytochemical analysis revealed that the rP2X$_4$ subunit is widely distributed in neurons in many areas of the CNS

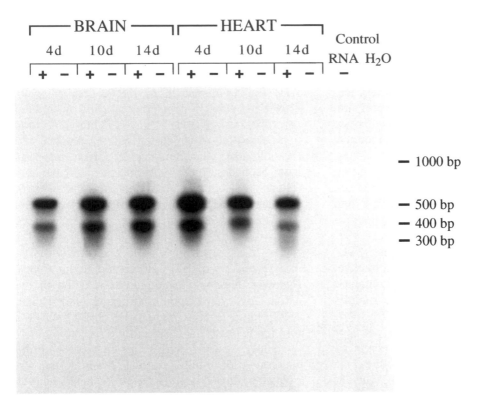

Fig. 4. Analysis of chick P2X$_4$ mRNA expression by RT–PCR and Southern blot autoradiography using a chP2X$_4$-specific probe. (+) and (−) denote presence or absence of superscript reverse transcriptase, respectively. RNA- correspond to the PCR performed with 1 μl of the reverse transcription mix in the absence of added RNA and H$_2$O is the negative control of the PCR reaction.

(Collo et al., 1996; Lê et al., 1998). In contrast, the presence of mammalian P2X$_4$ receptors transcripts in heart cannot be assigned to a specific cell type. P2X$_1$ subtype is the only P2X receptor so far that is known to be localized in the membrane of cardiac myocytes (Vulchanova et al., 1996).

P2X$_2$ receptor transcript levels change from neonatal to adult in rat brain (Kidd et al., 1996; Collo et al., 1996). Additionally, responses of rat nucleus coeruleus neurons to 2MeSATP increase with age, reaching a maximum at postnatal day 18 (Wirkner et al., 1998). Moreover, changes in the contractile response elicited by ATP in embryo skeletal muscle with age have also been reported. Thus, the sensitivity to ATP varies from a strong contraction on day 6 to no response on day 17. The functional change seems to be dependent upon innervation, since unilateral denervation leads to the maintenance of the response to ATP (Wells et al., 1995). Comparable changes in the number and

distribution of nicotinic acetylcholine receptors in the myocyte membrane during synaptogenesis have been described (Berg et al., 1972).

We attempted to check if there were changes in chP2X$_4$ RNA levels at days 4, 10 and 14 in ovo. In case of the heart, day 4–5 is the onset of cholinergic innervation (Kirby et al., 1988a) while days 9–10 harbour the onset of sympathetic innervation (Kirby et al., 1980b). Although we could not find any extreme effect in the expression levels, we can not discard different functional effects of ATP on different embryonic ages. For instance, in chick cardiomyocytes, muscarinic receptor number measured by ligand binding stays constant during development. However, the receptors are not functional until the vagus innervation has reached the heart (Galper et al., 1977). Determination of possible differences in the mRNA levels and protein localization in brain and heart of chick embryos at different developmental stages are now necessary. Moreover, it should now be interesting to see if any functional regulation of P2X-mediated effects occurs during embryonal development.

Acknowledgments

We thank Kerstin Borchardt for expert technical assistance and Hendrik Knötgen for kind advice on the preparation of chick embryos. We are indebted to Prof. Walter Stühmer for continous support and advice and for critically reading the manuscript.

References

Abe, Y., Sorimachi, M., Itoyama, Y., Furukawa, K. and Akaike, N. (1995) ATP responses in the embryo chick ciliary ganglion cells. *Neuroscience*, 64: 547–551.

Bardoni, R., Goldstein, P.A., Lee, C.J., Gu, J.G. and McDermott, A.B. (1997) ATP P2X receptors mediate fast synaptic transmission in the dorsal horn of the rat spinal cord. *J. Neurosci.*, 17: 5297–5304.

Benham, C. D. and Tsien, R. W. (1987) A novel receptor-operated Ca^{2+} permeable channel activated by ATP in smooth muscle. *Nature*, 328: 275–278.

Berg, D. K., Kelly, R. B., Sargent, P. B., Williamson, P. and Hall, Z. W. (1972) Binding of α-bungarotoxin to acetylcholine receptor in mammalian muscle. *Proc. Natl. Acad. Sci. USA*, 69: 147–151.

Brake, A. J., Wagenbach, M. J. and Julius, D. (1994) New structural motif for ligand-gated ion channels defined by an ionotropic ATP receptor. *Nature*, 371: 519–523.

Buell, G., Collo, G. and Rassendren, F. (1996) P2X receptors: An emerging channel family. *Eur. J. Neurosci.*, 8: 2221–2228.

Burnstock, G. (1996) P2 purinoceptors: Historical perspective and classification. In: *P2 Purinoceptors: Localization, Function and Transduction Mechanisms*. Ciba Foundation Symposium Vol. 198, Wiley, Chichester, pp. 1–34.

Burnstock, G., King, B. F. and Townsend-Nicholson, A. (1998) P1 and P2 receptor update news-sheet.

Castro, E., Tome, A. R., Miras-Portugal, M. T. and Rosario, L. M. (1994) Single-cell fura-2 microfluorometry reveals different purinoceptor subtypes coupled to Ca^{2+} influx and intracellular Ca^{2+} release in bovine adrenal chromaffin and endothelial cells. *Pflügers Arch.*, 426: 524–533.

Chomczynski, P. and Sacchi, N. (1987) Single-step method of RNA isolation by acid guanidinium thiocyanate-phenol chloroform extraction. *Anal. Biochem.*, 162: 156–159.

Clemens, M. G. and Forrester, T. (1981) Appearance of adenosine triphosphate in the coronary sinus effluent from isolated working rat heart in response to hypoxia. *J. Physiol.* 312: 143–158.

Collo, G., North, R.A., Kawashima, E., Merlo-Pich, E., Neidhart, S., Surprenant, A. and Buell, G. (1996) Cloning of P2X5 and P2X6 receptors and the distribution and properties of an extended family of ATP-gated ion channels. *J. Neurosci.*, 16: 2495–2507.

Dhulipala, P. D. K., Wang, Y. X. and Kotlikoff, M. I. (1998) The human $P2X_4$ receptor gene is alternatively spliced. *Gene*, 207: 259–266.

Edwards, F. A., Gibb, A. J. and Colquhoun, D. (1992) ATP receptor mediated synaptic currents in the central nervous system. *Nature*, 359: 144–147.

Evans, R. J., Derkach, V. and Surprenant, A. (1992) ATP mediates fast synaptic transmission in mammalian neurons. *Nature*, 357: 503–505.

Friel, D. D. (1988) An ATP-sensitive conductance in single smooth muscle cells from the rat vas deferens. *J. Physiol.*, 401: 361–380.

Friel, D. D. and Bean, B. P. (1988) Two ATP-activated conductances in bullfrog atrial cells. *J. Gen. Physiol.*, 91: 1–27.

Galper, J. B., Klein, W. and Catterall, W. A. (1977) Muscarinic acetylcholine receptors in developing chick heart. *J. Biol. Chem.*, 252: 8692–8699.

Garcia-Guzman, M., Soto, F., Gomez-Hernandez, J.M., Lund, P.E. and Stühmer, W. (1997) Characterization of recombinant human $P2X_4$ receptor reveals pharmacological differences to the rat homologue. *Mol. Pharmacol.*, 51: 109–118.

Hamburger, V. and Hamilton, H. L. (1951) A series of normal stages in the development of the chick embryo. *J. Morphol.*, 88: 49–92.

Harms, L., Finta, E. P., Tschopl, M. and Illes, P. (1992) Depolarization of rat locus coeruleus neurons by adenosine 5'-triphosphate. *Neuroscience*, 48: 941–952.

Hume, R. I. and Honig, M. G. (1986) Excitatory action of ATP on embryonic chick muscle. *J. Neurosci.*, 6: 681–690.

Kennedy, C., McLaren, G. J., Westfall, T. D. and Sneddon, P. (1996) ATP as a co-transmitter with noradrenaline in sympathetic nerves: function and fate. In: *P2 Purinoceptors: Localization, Function and Transduction Mechanisms*. Ciba Foundation Symposium, Vol. 8, Wiley, Chichester, pp. 223–235.

Kidd, E. J., Grahames, C. B. A., Simon, J., Michel, A. D., Barnard, E. A. and Humphrey, P. P. A. (1995) Localization of P2X purinoceptor transcripts in the rat nervous system. *Mol. Pharmacol.*, 48: 569–573.

Kirby, M. L., Weidman, T. A. and McKenzie, J. W. (1980a) An ultrastructural study of the cardiac ganglia in the bulbar plexus of the developing chick heart. *Dev. Neurosci.*, 3: 174–184.

Kirby, M. L., McKenzie, J. W. and Weidman, T. A. (1980b) Developing innervation of the chick heart: a histofluorescence and light microscopy study of sympathetic innervation. *Anatom. Rec.*, 196: 333–340.

Kolb, H. A. and Wakelam, M. J. O. (1983) Transmitter-like action of ATP on patched membranes of cultured myoblasts and myotubes. *Nature*, 303: 621–623.

Lê, K.T., Villeneuve, P., Ramjaun, A.R., McPherson, P.S., Beaudet, A. and Séguéla, P. (1998) Sensory presynaptic and widespread somatodendritic immunolocalization of central ionotropic P2X ATP receptors. *Neuroscience*, 83: 177–190.

Liu, H. and Hume, R. I. (1997) *Cloning and expression of a P2X Receptor from the Chick Embryo.* 27th Annual Meeting of the Society for Neuroscience. New Orleans, 151.3.

Newbolt, A., Stoop, R., Virginio, C., Surprenant, A., North, R. A., Buell, G. and Rassendren, F. (1998) Membrane topology of an ATP-gated ion channel (P2X receptor). *J. Biol. Chem.,* 273: 15177–15182.

North, R. A. and Barnard, E. A. (1997) Nucleotide receptors. *Current Neurobiol.* 7: 346–357.

Podrasky, E., Xu, D. and Liang, B. T. (1997) A novel phospholipase C- and cyclic AMP-independent positive ionotropic mechanism via the P2 purinergic receptor in cardiac ventricular myocyte. *Am. J. Physiol.,* 273: H2380–H2387.

Rogers, M. and Dani, J.A. (1995) Comparison of quantitative calcium flux through NMDA, ATP and ACh receptor channels. *Biophys. J.,* 68: 501–506.

Sambrook, J., Fritsch, E. F. and Maniatis, T. (1989) *Molecular Cloning. A Laboratory Manual,* 2nd ed. Cold Spring Harbor Laboratory Press, New York.

Sneddon, P. and Westfall, D. P. (1984) Pharmacological evidence that adenosine triphosphate and noradrenaline are co-transmitters in the ginea pig vas deferens. *J. Physiol.,* 347: 561–580.

Silinsky, E. M. (1975) On the association between transmitter secretion and the release of adenine nucleotided from mammalian motor nerve terminals. *J. Physiol.,* 247: 145–162.

Silinsky, E. M. and Gerzanich, V. (1994) On the excitatory effects of ATP and its role as neurotransmitter in coelic neurons of the guinea pig. *J. Physiol.,* 477: 117–127.

Soto, F., Garcia-Guzman, M., Gomez-Hernandez, J. M., Hollmann, M., Karschin, C. and Stühmer, W. (1996) P2X$_4$: An ATP-activated ionotropic receptor cloned from rat brain. *Proc. Natl. Acad. Sci. USA,* 93: 3684–3688.

Soto, F., Garcia-Guzman, M. and Stühmer, W. (1997) Cloned ligand-gated channels activated by extracellular ATP (P2X receptors). *J. Mem. Biol.,* 160: 91–100.

Stühmer, W. (1992) Electrophysiological recording from *Xenopus* oocytes. In: B. Rudy and L.E. Iverson (Eds.), *Methods in Enzymology.* Academic Press, 6277 Sea Harbor Drive, Orlando, Vol. 207, pp. 319–345.

Thomas, S. A. and Hume, R. I. (1990) Irreversible desensitization of ATP responses in developing chick skeletal muscle. *J. Physiol.,* 430: 373–388.

Thomas, S. A., Zawisa, M. J., Lin, X. and Hume, R. I. (1991) A receptor that is highly specific for extracellular ATP in developing chick skeletal muscle *in vitro. Br. J. Pharmacol.,* 103: 1963–1969.

Tomic, M., Jobin, R. M., Vergara, L. A. and Stojilkovic, S. S. (1996) Expression of purinergic receptor channels and their role in calcium signaling and hormone-release in pituitary gonadotrophs—integration of P2 channels in plasma membrane-derived and endoplasmic reticulum-derived calcium oscillations. *J. Biol. Chem.,* 271: 21200–21208.

Vulchanova, L., Arvidsson, U., Riedl, M., Wang, J., Buell, G., Surprenant, A., North, R. A. and Elde, R. (1996) Differential distribution of two ATP-gated ion channels (P2X receptors) determined by immunocytochemistry. *Proc. Natl. Acad Sci. USA,* 93: 8063–8067.

Wells, D. G., Zawisa, M. J. and Hume, R. I. (1995) Changes in responsiveness to extracellular ATP in chick skeletal muscle during development and upon denervation. *Dev. Biol.,* 172: 585–590.

Wirkner, K., Franke, H., Inoue, K. and Illes, P. (1998) Differential age-dependent expression of a2 adrenoceptors and P2X purinoceptor functions in rat locus coeruleus neurons. *Naunyn-Schmiedeberg's Arch. Pharmacol.,* 357: 186–189.

Zimmermann, H. (1982) Biochemistry of the isolated cholinergic vesicle. In: R.L. Klein, H. Lagercrantz and H. Zimmermann (Eds.), *Neurotransmitter vesicles.* Academic Press, New York, pp. 271–304.

Zhou, Z. and Hume, R. I. (1998) Two mechanisms for inward rectification of current flow through the purinoceptor P2X$_2$ class of ATP-gated channels. *J. Physiol.,* 507: 353–364.

Development of nucleotide analogues

P. Illes and H. Zimmermann (Eds.)
Progress in Brain Research, Vol 120
© 1999 Elsevier Science BV. All rights reserved

CHAPTER 8

Developments in P2 receptor targeted therapeutics

Michael Williams

Neurological and Urological Diseases Research, D-464, Pharmaceutical Products Division, Abbott Laboratories, 100 Abbott Park Road, Abbott Park, IL 60064, USA. Tel.: +847- 937-8186; fax: +847- 937- 9195; e-mail: mike. williams@abbott.com

Introduction

It is now clearly established that endogenous ATP and its breakdown products, ADP and adenosine, the latter acting via the P1 receptor G-protein receptor coupled family (Williams and Burnstock, 1997), act as extracellular signaling molecules. ATP as well as the uracil congener, UTP, produce their effects via interactions with various members of the P2 receptor family, the P2X, a ligand gated ion channel (LGIC) family and the P2Y, a G-protein coupled receptor (GPCR) family (Buell et al., 1996; North, 1996; North and Barnard, 1997). While eight possible members of the P2X receptor family have been described to date with a number of splice variants, and more than 11 P2Y receptors have also been reported (Burnstock et al., 1998), only seven P2X receptors ($P2X_{1-7}$) and four P2Y receptors ($P2Y_1$, $P2Y_2$, $P2Y_4$ and $P2Y_6$) are currently accepted as molecularly distinct entities that elicit functional responses.

The ligands available to characterize P2 receptors are highly limited in terms of their receptor affinity, selectively, biostability and pharmacokinetics. Thus the pharmacology of P2 receptors is currently based on the rank order potency of a series of ATP based agonists, since antagonists for either P2X or P2Y receptors have questionable efficacy and selectivity (Humphrey et al., 1995; Bhagwat and Williams, 1997). The latter situation is, however, gradually changing with an increased focus on compound identification using high throughput screening and traditional medicinal chemistry approaches (Boyer et al., 1996; Bultmann et al., 1996a, 1996b; Wittenberg et al., 1996; Gargett and Wiley, 1997; Camaioni et al., 1998; Humphreys et al., 1998; Jacobson et al., 1998; Pintor et al., 1998; Virginio et al., 1998).

ATP plays a critical role in tissue homeostasis, fast excitatory neurotransmission, tissue development, pain transmission, macrophage apoptosis, platelet aggregation, astroglial cell function and processes involving metastasis formation (Chadwick and Goode, 1996) while UTP, acting via the $P2Y_2$ receptor is a potent and selective modulator of mucocilliary transport (Knowles et al., 1991; Bennett et al., 1996; Shaffer et al., 1998). The $P2Y_4$ and $P2Y_6$ receptors are also uracil nucleotide (pyrimidine) sensitive (Communi and Boeynaems, 1997). Together with the $P2Y_2$ receptor, the $P2Y_4$ and $P2Y_6$ receptors mediate the actions of extracellular UTP and UDP (Anderson and Parkinson, 1997).

In defining the role of adenine and uracil nucleotides in CNS function and the potential therapeutic utility of agonists and antagonists for the various P2 receptor subtypes, there are some important lessons that can be learnt from the research activities devoted to identifying novel therapeutics acting via P1 receptors. Over the past 25 years, many pharmaceutical companies have targeted P1 receptor agonists and antagonists as potential therapeutic agents. The primary diseases identified that were thought to have a P1 receptor component included hypertension and a variety of CNS disorders that included epilepsy, schizophrenia, pain, depression and anxiety (Williams, 1995). The many compounds that were advanced to clinical trial status suffered from a poor understanding of disease pathophysiology, target selectivity, unmet medical need and clinical endpoints such that compound after compound was dismissed as a viable entity in early trials due to unexpected side effect liabilities.

Thus the CNS sedative actions of adenosine limited the utility of P1 agonists as antihypertensive agents despite their profound effects on blood pressure, while conversely, agonists with hypotensive activity had limited utility in treating CNS disorders. With the approach of the 21st century few compounds acting via P1 receptors have reached the market place. In fact, adenosine remains the most successful P1 receptor ligand with utility in the treatment of supraventricular tachycardia and for cardiac imaging where its short plasma half life produces the desired effects on cardiac function while circumventing the various side effects mentioned above.

In advancing the concept of using P2 ligands for the treatment of central and peripheral nervous system disorders, it is imperative that compounds are evaluated on a hierarchical scale, recognizing the inherent complexity of the brain and how much there is still to be learnt about the molecular substrates of psychiatric and neurological disorders (Williams, 1994). This complexity contrasts with tissues like kidney, heart and lung where disease targets, therapeutic utility and clinical endpoints are somewhat more distinct and thus more easily testable. Studies of P2 receptor ligands in these relatively simpler systems will aid in understanding efficacy and side effect liabilities of new agents that may, in time, be targeted for CNS disorders, a point recently made in regard to UTP receptor function by Connolly et al. (1997).

The development of antagonists for P2 receptors, in essence antagonists of ATP, will be confounded if such entities antagonize ATP actions across molecular targets other than P2 receptors especially if these involve basic cellular energy production. There is thus considerable work to be done in understanding the structural requirements for compounds interacting with the various classes of P2 receptors and how these differ from the require-ments for recognition by members of the ATP binding cassette protein family (Al-Awqati, 1995; Delombe and Escande, 1996), ectonucleotidases (Zimmermann, 1996; Kegel et al., 1997), ATP-modulated potassium channels (Edwards and Weston, 1993; Freedman and Lin, 1996) and the plethora of enzymes that utilize ATP for their function.

ATP and UTP availability

In assessing therapeutic prospects of ligands interacting with ATP (and UTP) receptors, a major issue has been delineating the source of extracellular ATP, whether its availability is uniquely increased in relation to discrete disease processes and what the dynamics of its presence in the extracellular space are in regard to its breakdown. The latter reflects the inactivation of ATP (and UTP) by ectonucleotidase activity (Ziganshin et al., 1994; Plesner, 1995) and the subsequent production of ADP, AMP and adenosine.

In the case of ATP, the term inactivation is very much a misnomer inasmuch as the products of ATP hydrolysis formed via a purinergic cascade (Williams, 1995) all have their own functional activities. In the case of adenosine, its CNS sedating activity is in direct contrast to the excitatory actions of ATP (Silinsky et al., 1992; Edwards et al., 1992; Evans et al., 1992). Furthermore, in the broader framework of ATP-modulated proteins, ATP-sensitive potassium channels are activated by a reduction in intracellular ATP levels (Edwards and Weston, 1993). Thus as P2 receptor mediated responses decrease as ATP is hydrolyzed to adenosine, P1 mediated responses and ATP-sensitive potassium channel-mediated responses are enhanced.

For UTP, while uracil has been reported to modulate dopaminergic systems in animal models, there is currently no evidence for a "P1"-like receptor for uracil (Williams and Burnstock, 1997).

Ecto-ATPases comprise a group of approximately 11 enzymes that metabolize ATP, diadenosine polyphosphates like Ap5A, and NAD (Zimmermann, 1996). Ecto-ATPases preferentially hydrolyze ATP to ADP and ecto-apyrases convert both ATP and ADP to AMP. Ecto-5′-nucleotidase converts AMP to adenosine. In myeloid leuckocytes, P2Y receptors, ecto-apyrases and ecto-5′-nucleotidase show stage specific transient expression (Dubyak et al., 1996; Clifford et al., 1997). CD39 is an antigen/ecto-apyrase that is only present on activated lymphocytes. (Wang and Guidotti, 1996). Ecto-ATPases, ecto-apyrases and ecto-5′-nucleotidases function to limit the extracellular actions of ATP (and presumably UTP) by enhancing its removal, as well as by producing adenosine, which can functionally antagonize some of the actions of ATP.

In the guinea-pig vas deferens, soluble nucleotidases are released from neurons together with ATP and norepinephrine (Todorov et al., 1997) suggesting that ATP inactivation is increased by nerve activity. Study of ectoATPase/ecto-apyrase function is limited by the lack of selective inhibitors of their activity. ARL 67156 is the only selective inhibitor of ecto-ATPase (Crack et al.,1995) although its potential for interaction with P2 receptors has been questioned.

Compound identification

Without exception, all agonists for the P2 receptor are analogs of ATP and UTP (Jacobson and van Rhee, 1997). The antagonist ligands identified to date that include PPADS, DIDS, various blue dyes that include Evans, trypan and reactive blue-2, suramin and XAMR 0271 have been identified empirically and, as noted, they are neither very selective, even for the P2 receptor family (Bhagwat and Williams, 1997), nor very potent. Furthermore, there is a major concern that the use of these agents can confound receptor characterization due to their ability to inhibit ectonucleotidase activity (Kennedy and Leff, 1995). The search for newer agents is therefore dependent on a more systematic evaluation of the current agents, or to the serendipity associated with a high throughput screening approach. The latter is confounded however by a lack of reliable binding assays for any of the P2 receptor subtypes and species nuances in receptor pharmacology (Chessell et al., 1997, 1998; Humphreys et al., 1998).

Using the first approach, novel analogs of DIDS (Bultmann et al., 1996a), Evans blue and trypan

blue (Wittenburg et al., 1996) and suramin (Bultmann et al., 1996b) have been identified and characterized for their relative effects on receptor function and ectonucleotidase activity. Thus β-INS is a DIDS derivative selective for rat P2X receptors, NH01 is the desmethyl form of Evans blue dye and is also selective for rat P2X receptors and NF023 is a truncated analog of suramin again selective for rat P2X receptors. All three compounds had reduced ectonucleotidase activity versus their actions on the rat vas deferens P2X receptor.

A series of trinitrophenyl (TNP) substituted nucleotides have been identified as non-competitive, reversible antagonists of $P2X_1$ and $P2X_3$ receptors (Virginio et al., 1998). TNP–ATP which has antagonist activity in the 1 nM range has been characterized as an allosteric modulator and shows weak activity at $P2X_4$ and $P2X_7$ receptor constructs. In terms of the structure activity relationship for antagonist activity, guanine can substitute for adenine, e.g. TNP–GTP and there is no difference in activity between TNP–ATP and its ADP and AMP analogs. TNP-adenosine is, however, inactive.

The ATP analogs, A3P5PS and A3P5P were identified as partial agonists/competitive antagonists at the turkey erythrocyte $P2Y_1$ receptor by empirical evaluation (Boyer et al., 1996). From these interesting leads, MRS 2179 has been identified as a full $P2Y_1$ receptor antagonist with an IC_{50} value of 330 nM (Camaioni et al., 1998). Modification of the P2 antagonist, PPADS resulted in MRS 2220 a moderately selective $P2X_1$ receptor antagonist that is 10-fold more potent than PPADS (Jacobson et al., 1998). MRS 2559 and MRS 2560 are additional PPADS derived ligands that are selective for P2X receptors. ARL 67085 (Leff et al., 1997) is 2-alkylthio substituted bioisostere of ATP that is a selective antagonist at the ADP-sensitive $P_{2T}/P2Y_{AC}$ receptor involved in platelet aggregation.

KN-62 is an isoquinoline inhibitor of calcium–calmodulin dependent protein kinase-II (CamK-II) with an IC_{50} value of 900 nM. The compound is also a potent antagonist of $P2X_7$ receptor function with an IC_{50} value of 9–13 nM (Gargett and Wiley, 1997). KN-04, an analog of KN-62 that is inactive as a CamK-II inhibitor can also block $P2X_7$

receptor mediated responses (IC_{50} = 13 nM) indicating a specific effect of this pharmacophore on the $P2X_7$ receptor that is distinct from its ability to inhibit kinase activity. Further examination of these isoquinolines in HEK-293 cells transfected with human or rat recombinant $P2X_7$ receptors showed that while KN-62 had an IC_{50} value of approximately 100 nM in an ethidium flux functional assay at the human $P2X_7$ receptor, it was without effect on the transfected rat receptor (Humphreys et al., 1998b) highlighting potential species dependent differences in pharmacology at the $P2X_7$ receptor. Ip_5I is a diinosine polyphosphate that has potent (IC_{50} ~ 10 nM) P2 receptor antagonist activity (G. Burnstock, personal communication; Pinto et al., 1998).

The ability to evaluate the activity of compounds at transfected human P2 receptors in functional ion flux assays using techniques like the fluorescent imaging plate reader (FLIPR) that are compatible with high throughput screening approaches to compound evaluation may be anticipated to build on these early leads to develop novel and potent P2 receptor pharmacophores. However, great care must be taken to avoid a circular approach to compound identification where the receptor defines the ligand and the ligand is then used, a priori, to define the receptor (Black, 1987). Problems can arise when this is done in an absolute sense, assigning the activity of one ligand to a single receptor in the absence of information to support the assignment. For instance, while BzATP is widely reported as a selective agonist at the $P2X_7$ receptor (EC_{50} = 18 μM), it is far more potent at the $P2X_1$ (EC_{50} = 1.9 nM) and $P2X_3$ (EC_{50} = 98 nM) receptors (van Biesen et al., 1998).

LGICs typically have associated ligand binding sites, activation of which can markedly alter receptor gating characteristics, The most notable modulator of an LGIC is the benzodiazepine, diazepam, that activates the $GABA_A$ receptor, resulting in anxiolytic, anticonvulsant, muscle relaxant and hypnotic activity. In addition to the putative allosteric actions of TNP–ATP, other compounds have allosteric effects on P2X receptors. The amiloride analog, HMA (5-(N-N hexamethylene)amiloride), which has been reported to block pore formation at the human $P2X_7$ receptor in the micromolar concentration range (Nuttle and Dubyak, 1994; Chessell et al., 1998) can potentiate the effects of ATP on mouse microglial $P2X_7$ receptors inducing a maximal effect on current flow (Chessell et al., 1997), a phenomenon thought to reflect large pore formation (Di Virgilio, 1995).

Excitatory neurotransmission

The potential role of ATP as a neurotransmitter was implicit in the purinergic hypothesis proposed by Burnstock (1972) and the nucleotide was shown to have excitatory effects in toad spinal cord (Phillis and Kirkpatrick, 1978), rat trigeminal nucleus (Salt and Hill, 1983), dorsal horn (Jahr and Jessell, 1983), rat locus ceruleus (Tschopl et al.,1992) and rat nucleus tractus solitarius (Ueno et al., 1992). However, more direct evidence for the role of ATP as fast excitatory transmitter was provided in studies in guinea pig celiac ganglion (Evans et al.,1992; Silinsky et al., 1992) and rat medial habenula (Edwards et al., 1992) where ATP-induced currents were shown to be blocked with suramin. Autoradiographic and in situ hybridization studies have shown widespread distribution of both P2X and P2Y receptors in the central and peripheral nervous systems (Abbracchio, 1997). In the dorsal horn, the exitatory effects of ATP are mediated via P2X receptors (Bardoni et al., 1997).

The functional role of ATP in nervous tissue remains at a very early stage of investigation, again as a result of the paucity of selective antagonists that could be used in vivo to characterize ATP-mediated responses. P2 receptors are present on neurons, astroglia, microglia and oligodendroglia. The potential role of ATP in the development and maturation of nervous tissue has been well defined (Burnstock, 1996a). However, the involvement of P2 receptor mediated processes in disease-associated pathophysiology has yet to be elucidated. In addition to the complexity of effects of ATP due to the purinergic cascade (Williams, 1995), a number of other neurotransmitters are released together with ATP. These include acetylcholine, norepinephrine, glutamate, GABA and neuropeptide Y (Sperlagh and Vizi, 1996) depending on the transmitter repertoire of the neuron.

Epilepsy

The possibility that the ability of ATP to mediate fast excitatory neurotransmission might play a role in seizure generation is suggested by the fact that ATP, given i.c.v., at high doses can evoke tonic–clonic convulsions (Buday et al.,1961). Similarly, microinjection of α,β-methylene ATP and β,γ-methylene ATP into the prepiriform cortex induced generalized motor seizures that were similar to those produced by either NMDA or bicuculline when administered in the same manner (Knutsen and Murray, 1997). ATP had no effect in this paradigm due to its degradation by ecto-ATPase activity to adenosine. When ATP was microinjected into the prepiriform cortex in combination with the xanthine adenosine antagonist, 8-SPT, to block P1 receptor-mediated effects, the mean seizure score went from 6% to 43% of the maximum response. $P2X_2$, $P2X_4$ and $P2X_6$ receptors are expressed in the prepiriform cortex (Kidd et al.,1995; Seguela et al., 1996; Collo et al., 1996) suggesting that a P2X receptor antagonist may represent a potentially novel approach to the treatment of epilepsy. However, it is not yet known whether P2 receptor mediated events are involved in the pathophysiological events related to epilepsy per se and to what extent the potent and specific effects of endogenous adenosine on seizure generation (During and Spencer, 1992) rely on endogenous ATP as a source. Certainly any P2 receptor ligands advanced to obtain proof of principle for the role of ATP in seizure generation would have to be devoid of effects on ecto-ATPase activity since any inhibition of this class of enzyme would prevent the formation of adenosine.

Pain

The hyperalgesic actions of ATP have been known for over 30 years (Burnstock and Wood, 1996; Sawynok, 1997). Peripheral administration of ATP produces pain (Collier et al., 1966; Bleehan and Keele, 1977; Bleehan, 1978, 1981).These effects appear to involve P2X receptors although P2Y receptors in dorsal horn astrocytes respond to ATP by increasing Ca^{2+} levels (Salter and Hicks, 1994). ATP can also enhance the production of prosta-

glandins (Needleman et al., 1974) which also produce pain. Burnstock (1996b) has suggested that locally produced ATP may contribute to the pain associated with causalgia, reflex sympathetic dystrophy, migraine, angina, lumbar, pelvic and cancer pain.

The discovery of the $P2X_3$ receptor (Chen et al., 1995; Lewis et al., 1995) which is selectively localized to sensory pathways in trigeminal, nodose and dorsal root ganglia provides a unique target for the development of novel analgesic agents which in this instance would be $P2X_3$ receptor antagonists. The non-specific P2 receptor antagonists, suramin and various blue dyes have antinociceptive activity in the mouse hot plate and writhing tests (Driessen et al., 1994), although, as noted, the P2 receptor selectivity of these compounds is questionable.

$P2X_3$ receptors show a modest increase on the side ipsilateral to the injured nerve in rats that developed neuropathic pain following the Bennett procedure suggesting an increase in receptor expression or a redistribution of $P2X_3$ receptors (Kassotakis et al., 1997), while neonatal capcasicin treatment reduces $P2X_3$ mRNA in the DRG (Chen et al., 1995) and can abolish ATP-mediated acute nociception (Bland-Ward and Humphrey, 1997). $P2X_2$ receptor immunoreactivity is also present in the DRG (Vulchanova et al., 1996) and there is considerable evidence for the existence of functional heteromers of $P2X_2$ and $P2X_3$ receptor subunits (Lewis et al., 1995; Radford et al., 1997). The $P2X_3$ receptor has an overlapping localization with the recently cloned vanilloid VR-1 receptor in lamina II of the dorsal horn of the spinal cord (Elde, 1998).

The effects of ATP may be facilitatory with those of glutamate (Li and Perl, 1995) and in the dorsal horn, activation of presynaptic P2X receptors can evoke glutamate release (Gu and MacDermott, 1997).

In addition to the algesic actions of ATP, the nucleotide has also been characterized as a messenger for innocuous mechanical stimuli (Fyffe and Perl, 1984). In a tissue culture system of tooth-pulp afferent (nociceptive) and muscle stretch receptor (non-nociceptive) rat sensory neurons, Cook et al. (1997) showed that ATP, acting via distinct P2X receptors, had both nociceptive and non-nociceptive roles in sensory neurons. Furthermore, it has been reported (Nakamura and Strittmatter, 1996) that $P2Y_1$ receptors may be involved in tactile responses in sensory neurons.

ATP may be a prime mediator of neurogenic inflammation via its actions on P2 receptors on neutrophils, macrophages and monocytes leading to cytokine (Dubyak and El-Moatassim, 1993) and prostaglandin (Needleman et al., 1974) release. In this context, it is of interest that the putative $P2Y_7$ receptor cloned from human erythroleukemia cells (Akbar et al., 1996) was subsequently found (Yokomizo et al., 1997) to be the leukotriene B_4 receptor, suggesting that the receptors for prostanoids and ATP may be functionally related.

When administered together with nitrous oxide, ATP can mimic the effects of the inhalation anesthetic, enflurane (Fukunaga, 1997) and can also reduce the amount of inhalation anesthetic required for anesthesia. In addition, ATP can provide analgesic relief in the postoperative setting via formation of adenosine.

Auditory and ocular function

Perilymphatic ATP, probably acting via P2Y receptors, can depress the sound-evoked gross compound action potential of the auditory nerve and the distortion product otoacoustic emission, the latter a measure of the active process of the outer hair cells (Kujawa et al., 1994). Several splice variants of the $P2X_2$ receptor have been localized to the cochlea of the ear. The $P2X_{2-1}$ (Housley et al., 1995) and $P2X_{2-3R}$ receptors (Salih et al., 1998) are found in rat and the $P2X_{2-1}$, $P2X_{2-2}$ and $P2X_{2-3}$ receptors have been reported in guinea-pig (Parker et al., 1998). The rat splice variants of the $P2X_2$ receptor are associated with the endolymphatic surface of the cochlear endothelium, an area associated with sound transduction (Housley, 1998). P2Y receptor expression is also observed in the marginal cells of the stria vascularis, a tissue involved in providing the ionic and electrical gradients of the cochlea. Little is currently known about the pharmacology and receptor associated physiology of hearing and vestibular function, however, it has been suggested that ATP may regulate fluid homeostasis, hearing sensitivity and

development. With the increase in the level of ambient noise in the environment, the use of headphones with personal stereos, video devices and computers and an evolving understanding of the epidemology of auditory disorders like tinnitus and Meniere's disease, this is a fertile area for further exploration especially when the molecular targets, like $P2X_3$ receptors in sensory pathways, are uniquely associated with a tissue.

Apoptosis

The role of the $P2X_7$ receptor in macrophage function has been well documented by the work of Dubyak and El Moatassim (1993) and Di Virgilio (1995). Originally known as the P_{2Z} receptor, the $P2X_7$ receptor is a unique member of the P2X receptor family functioning as a non-selective ion pore in mast cells, platelets, macrophages and lymphocytes. Monocytes can fuse with one another to form multinucleated giant cells (MGCs) a type of macophage seen in granulomatous inflammation. IFN-γ and LPS can stimulate MCG formation via induction of $P2X_7$ receptor expression. The $P2X_7$ receptor is also present in superior cervical ganglion and spinal cord (Surprenant et al., 1996) suggesting a role in CNS function and indeed cerebral artery occlusion has been found to result in an increase in $P2X_7$ immunoreactive cells in the penumbral region around the stroke (Collo et al., 1997).

Macrophages can be activated by IFNγ and LPS (Humphreys and Dubyak, 1996) leading to a loss in ecto-ATPase activity and an upregulation of $P2X_7$ receptor expression. They thus become more susceptible to the cytolytic actions of extracellular ATP. ATP-induced $P2X_7$ receptor-mediated apoptotic events are involved in macrophage killing of mycobacterium (Lammas et al., 1997). This antimicrobial activity of ATP may have potential utility in the treatment of tuberculosis and may also provide a basic understanding of $P2X_7$ receptor mediated apoptotic events that can be applied to more complex cell systems like the brain.

The release and maturation of IL-1β from macrophages is stimulated by ATP acting via a $P2X_7$ receptor (Perregaux and Gabel, 1994; Di Virgilio, 1995; Di Virgilio et al., 1996). This involves ATP activation of the cysteine protease/ caspase, interleukin-1β convertase (ICE) that may be involved in the initiation of apoptosis, and provides additional evidence for a key role of ATP in modulating immune system function. Apoptotic cell death is a process that is important in both embryogenesis and in removing cancerous or infected cells from tissues (Chow et al., 1997). There are several reports indicating that ATP may have potential utility in the treatment of cancer, and there have been Phase II trials with ATP in cancer patients (Rapaport, 1997).

In cystic fibrosis (CF) , there is a reduction in the incidence of breast cancer and melanoma and an increase in GI malignancies (Neglia et al., 1995). CF is the most common lethal recessive genetic disease found in Caucasians and is due to a defect in the expression of the ABC protein, CFTR (cystic fibrosis transmembrane conductance regulator), an ATP channel. Nude mice homozygous for CFTR ($nu^{-/-}CFTR^{-/-}$) had red blood cell ATP levels that were nearly two-fold higher than controls (Abraham et al., 1996). These mice showed a reduction in breast cancer implantability with a slower tumor growth rate. From these studies, Abraham et al., (1996) suggest that a "non permissive host micro-environment" exists that inhibits or slows tumor growth and this environment is the result of elevated endogenous ATP levels. To explain the paradoxical finding of a decrease in breast cancer and melanoma and an increase in GI malignancies, in CF patients, the authors speculate that when there is a defect in functional CFTR as in CF, there is a reduced excretion of ATP into the gut, which reduces gut ATP levels and allows tumor proliferation. Conversely, blood ATP is elevated in CF leading to an ATP-induced tumor suppression.

ATP has been in clinical trials for the treatment of cancer and cancer cachexia (Rapaport and Fontaine, 1989; Rappaport, 1997). Administration of ATP by a 96 h infusion at 50 μg/kg/min every 28 days, resulted in an expansion of red blood cell ATP pools that correlated with cytostatic and cytotoxic effects on tumor growth, inhibition of cachexia, improvements in organ function, analgesia, enhancement of superoxide production and modulation of blood flow. While Phase II trials reported positive effects on quality of life outcomes, these

trials were put on hold for unknown reasons. In vitro, ATP, acting via a receptor mediated cAMP signalling pathway, can transform promyelocytic leukemia cells into normal white blood cells (Lawson, 1996).

Again, while these studies are not directly applicable to nervous tissue function apart from the preliminary findings of an upregulation of the $P2X_7$ receptor following stroke (Collo et al., 1997), they may provide clues as to a potential role of ATP-mediated events in tumor formation in nervous tissue.

Neuro-urology

The urinary bladder is under the control of both the sympathetic and parasympathetic nervous systems and ATP has been shown (Burnstock et al., 1978; Dean and Downie, 1978) to mimic the effects of parasympathetic nerve stimulation resulting in bladder contraction. ATP appears to act via P2X receptors that are present in the smooth muscle of the urinary bladder detrusor that is involved in bladder emptying (Chancellor et al., 1992). Malfunctions of the detrusor can lead to urge urinary incontinence. Micturition involves relaxation of the urethra where ATP functions as a cotransmitter with nitric oxide (NO). NO mediates the first stage of relaxation (Pinna et al., 1998) with ATP acting via P2 receptors to mediate the second phase of the voiding response. P2X receptors are present in the bladder urothelium and there is serosal release of ATP in rabbit bladder resulting from hydrostatic pressure changes associated with bladder filling (Yoshimura and de Groat, 1997). Hashimoto and Kokubun (1995) have shown that 15% of the neurogenic contraction in rat urinary bladder is mediated by muscarinic receptors while 50% is mediated by P2X receptor mechanisms. Since muscarinic antagonists are currently the mainstay in the treatment of urge urinary incontinence (UUI), and have typical muscarinic side effects limitations, this finding together with an increased understanding of the role of P2 receptors in bladder function suggests that P2X receptor antagonists may be potentially useful as novel treatment modalities for UUI.

Trophic actions

Trophic factors are required to ensure neuronal viability and regeneration. Neural injury leads to increases in growth factors, especially those for polypeptides that act via receptor tyrosine kinases, e.g. fibroblast growth factors, epidermal growth factor and platelet-derived growth factor (Neary et al., 1996). ATP can act in combination with growth factors to stimulate astrocyte proliferation and potentially contribute to the process of reactive astrogliosis, a hypertrophic/hyperplastic response that is frequently seen in brain trauma, stroke/ischemia, seizures and various neurodegenerative disorders. It is presently unknown whether this process is beneficial in neuronal repair.

In reactive astrogliosis, astrocytes undergo process elongation and express GFAP (glial fibrillary acidic protein), an astrocyte specific intermediate filament protein. There is also an increase in astroglial cellular proliferation (Norenberg, 1994). ATP and adenosine can induce DNA synthesis in human astrocytes. ATP can also increase GFAP and AP-1 complex formation in astrocytes (Neary, 1996; Neary et al., 1996). These effects of ATP were quantitatively similar to those seen with bFGF. In addition, ATP and GTP can induce trophic factor (NGF, NT-3, FGF) synthesis in astrocytes and neurons. The observed activity of GTP is not consistent with any known P2 receptor profile. Nonetheless, these studies have led to the identification of a synthetic purine, AIT-082 that also upregulates neurotrophin production and can enhance working memory and restore age-induced memory deficits in mice (Rathbone et al., 1998). At the morphological level, AIT-082 can increase neurite complexity in hippocampal neuron cultures. Preliminary studies (Di Iorio et al.,1998) have shown that co-infusion of AIT-082 with NMDA into rat caudate nucleus reduced neuronal loss. AIT-082 is currently in Phase II clinical trials for Alzheimer's disease (AD) under the auspices of the NIA Alzheimer's Consortium initiative. Propentofylline is a compound that can potentiate the actions of adenosine, inhibit phosphodiesterase activity and enhance NGF production (Rudolphi et al., 1997). This compound has been approved for human use and has reportedly shown efficacy in AD. It is also rumored to be showing a trend to arresting AD progression. Microglia are the resident immunocytes of the brain and are usually present in a resting state. When activated, by trauma and infection they transform to a macrophage-like state and produce cytokines (IL-1, IL-6) and superoxide anions (Thomas, 1992).

Diabetes

ATP can stimulate pancreatic insulin release via a glucose dependent mechanism by activating a P2Y receptor (Loubatrieres-Mariani et al., 1997). From these studies, it has been suggested that a P2Y receptor agonist may improve glucose tolerance and have potential as an antiidiabetic drug. A patent for the use of ATP in the treatment of diabetes has been issued (Rapaport, 1996). ATP can also affect insulin secretion by interactions with ATP-sensitive potassium channels in islet β-cells. ADP can antagonize the ATP inhibition of these channels by binding to the second nucleotide binding site on the associated sulfonylurea receptor (SUR; Nichols et al., 1996) thus activating K_{ATP} channels and inhibiting insulin secretion. From a nervous system viewpoint, elevated blood glucose levels can impact nerve cell viability, a situation that can lead to diabetic neuro- and retinopathies.

Lung function

ATP and UTP are potent stimulants of chloride secretion in airway epithelium (Mason et al., 1993) and mucin glycoprotein release from epithelial goblet cells (Lethem et al., 1993) acting via the $P2Y_2$ receptor. While both ATP and UTP are equipotent at this P2 receptor, UTP has the advantage that its breakdown product, unlike adenosine has no ancillary pharmacology. UTP is being developed as INS 316 in an inhalation formulation to enhance mucociliary clearance for the potential treatment of cystic fibrosis and chronic bronchitis. Controlled clinical studies (Shaffer et al., 1998) have shown that INS 316 stimulates mucociliary clearance and sputum expectoration in smokers, non-smokers and patients with chronic bronchitis in a dose-dependent manner. Molecular modeling of the $P2Y_2$ receptor has shown that UTP occupies a binding cleft formed by helices 2, 3, 5, 6 and 7 in this GPCR and that 4-thio-UTP was five- fold more potent as an agonist than UTP with an EC_{50} value of 30 nM in stimulating inositol phosphate formation in 1321N astrocytoma cells transfected with the human $P2Y_2$ receptor (Siddiqi et al., 1998).

ATP also appears to have a potential role in asthma via its actions on bronchial innervation. ATP can trigger a reflex bronchconstriction via activation of a P2X receptor on vagal C fibers (Katchanov et al., 1998). ATP and UTP can potentate IgE-mediated mast cell histamine release, effects involving P2Y receptors (Schulman et al., 1998).

Mitochondrial function

The role of ATP in mitochondrial function and its production by the electron transport chain is a well established event. What is not known is whether P2 receptors are present in mitochondria which may have far reaching implications as mitochondrial dysfunction is found to underlie CNS disease states like Alzheimer's and Parkinson's diseases (Williams and Davis, 1997). A preliminary report (Katsuragi et al., 1998) has suggested that P2 receptor-mediated ATP release may involve mitochondrial stores.

Other targets for ATP action

As already discussed, when ATP is released into the extracellular milieu it has the potential to interact with a variety of molecular targets in addition to the P2 receptor superfamily. These targets include the various members of the ATP binding cassette (ABC) family of proteins as well as a plethora of ATP-sensitive receptors that are also targets for the actions of ATP. Thus in studying the effects of ATP and UTP ligands in both in vitro and in vivo systems, the potential effects of the nucleotide triphosphates as well as their breakdown products on ATP and adenosine receptor targets must be viewed in the context of the activation/modulation of other ATP-sensitive macromolecules. With the evolving molecular diversity of K_{ATP} channels, P2

receptors, ectoATPases, etc., the potential list of target proteins through which ATP (and UTP) may produce their physiological effects is approaching three digits. Given the diversity of structures involved (North, 1996) it should not be impossible to identify selective antagonists that will potentially lead to novel therapeutics once such tools are used to define and characterize the role of P2 receptors in disease pathophysiology.

Future challenges

While the pioneering work of Boucher's group in developing aerosol formulations of UTP for the treatment of cystic fibrosis (Bennett et al., 1996) and, potentially, chronic bronchitis, is an elegant and simple approach to rectifying defects in mucocilliary function, it is also an important proof of principle for the field of P2 receptor therapeutics.

In reading this overview, the neuroscientist may be puzzled as to the inclusion of a variety of non-CNS disease states that involve P2 receptor-mediated events. However, as pointed out in the introduction, there is much to be learnt from the effects of ATP (and UTP) in these relatively more simple systems that may aid in the choice of appropriate and 'doable' targets for the development of novel CNS therapeutics.

At the present time, developing novel $P2X_3$ receptor antagonists for use in the treatment of pain appears to be the most logical opportunity given the unique localization of this receptor in sensory ganglia. However, further work is required to understand the significance of $P2X_7$ receptor upregulation in stroke and indeed, whether the plasticity of immune system P2 receptors (Dubyak et al., 1996; Clifford et al., 1997) may be applicable to the CNS and, accordingly, be disease state related.

The pioneering work of Burnstock coupled with the cloning of the P2 receptor family has provided a plethora of interesting new drug discovery targets. With the identification, characterization and optimization of potent and selective antagonists for these receptors, the challenge of defining their role in CNS and peripheral nervous tissue function and disease will make P2 receptors an area of intense basic and applied biomedical research interest into the next century.

102

References

Abbraccio, M.P. (1997) ATP in the brain. In: K.A. Jacobson and M.F. Jarvis (Eds.), *Purinergic Approaches in Experimental Therapeutics*. Wiley-Liss, New York, pp. 383–404.

Abraham, E.H., Vos, P., Kahn, J., Grubman, S.A., Jefferson, D. M., Ding, I. and Okunieff, P. (1996) Cystic fibrosis hetero- and homozygosity is associated with inhibition of breast cancer growth. *Nature Med.*, 2: 593–596.

Akbar, G.K.M., Dasari, V.R., Webb, T.E., Ayyanathan, K., Pillarisetti, K., Sandhu, A.K., Athwal, R.S., Daniel, J.L., Ashby, B., Barnard, E.A. and Kunapuli, S.P. (1996) Molecular cloning of a novel P_2 purinoceptor from human erythroleukemia cells. *J. Biol. Chem.*, 270: 26152–26158.

Al-Awqati, Q. (1995) Regulation of ion channels by ABC transporters that secrete ATP. *Science*, 269: 805–806.

Anderson, C.M. and Parkinson, F.E. (1997) Potential signalling roles for UTP and UDP: sources, regulation and release of uracil nucleotides. *Trends Pharmacol. Sci.*, 18: 387–392.

Bardoni, R., Goldstein, P., Lee, J., Gu, J.G. and MacDermott, A.B. (1997) ATP P2X receptors mediate fast synaptic neurotransmission in the dorsal horn of the spinal cord. *J. Neurosci.*, 17: 5297–5304.

Bennett, W.D., Olivier, K.N., Zeman, K.L., Hohneker, K.W., Boucher, R.C. and Knowles M.R. (1996) Effect of uridine 5'-triphosphate (UTP)+amiloride on mucociliary clearance in adult cystic fibrosis. *Am. J. Respir. Crit. Care Med.*, 153: 1796–1801.

Bhagwat, S.S. and Williams, M. (1997) P2 purine and pyrimidine receptors: Emerging superfamilies of G protein coupled and ligand-gated ion channel receptors. *Eur. J. Med. Chem.*, 32: 183–913.

Black, J.W. (1987) Should we be concerned about the state of hormone receptor classification? In: J.W. Black, D.H. Jenkinson, V.P. Gerskowitch and A.R. Liss (Eds.), *Perspectives on Receptor Classification*. New York, pp. 11–15.

Bland-Ward, P.A. and Humphrey, P.P.A. (1997) Acute nociception mediated by hindpaw P2X receptor activation in the rat. *Br. J. Pharmacol.*, 122: 365–371.

Bleehan, T. (1978) The effects of adenine nucleotides on cutaneous afferent nerve activity. *Br. J. Pharmacol.*, 62: 573–577.

Bleehan, T. and Keele, C.A. (1977) Observations on the alogenic actions of adenosine compounds on human blister base preparation. *Pain,* 3: 367–377.

Boyer, J.L., Romero-Avila, T., Schachter, J.B. and Harden, T.K. (1996) Identification of competitive antagonists of the $P2Y_1$ receptor. *Mol. Pharmacol.*, 50: 1323–1329.

Buday, P.V., Carr, C.J. and Myga, T.S. (1961) A pharmacologic study of some nucleosides and nucleotides. *J. Pharm. Pharmacol.*, 13: 290–299.

Buell, G., Collo, G. and Rassendren, F. (1996) P2X receptors: An emerging channel family. *Eur. J. Neurosci.*, 8: 2221–2228.

Bultmann, R., Pause, B., Wittenberg, H., Kurz, G. and Starke, K. (1996a) P_2-purinoceptor antagonists: I. Blockade of P_2-purinoceptor subtypes and ecto-nucleotidases by small

isothocyanato sulphonates. *Naunyn Schmiedeberg's Arch. Pharmacol.*, 354: 481–490.

Bultmann, R., Wittenberg, H., Pause, B., Kurz, G., Nickel, P. and Starke, K. (1996b) P_2-purinoceptor antagonists: III. Blockade of P_2-purinoceptor subtypes and ecto-nucleotidases by compounds related to suramin. *Naunyn Schmiedeberg's Arch. Pharmacol.*, 354: 498–504..

Burnstock, G (1972) Purinergic nerves. *Pharmacol. Rev.*, 24: 509–581.

Burnstock, G. (1996a) Purinoceptors: Ontogeny and phylogeny. *Drug Dev. Res.*, 39: 204–242.

Burnstock, G. (1996b) A unifying purinergic hypothesis for the initiation of pain. *Lancet* 347: 1604–1605.

Burnstock, G. and Wood, J.N. (1996) Purinergic receptors: Their role in nociception and primary afferent neurotransmission. *Curr. Opin. Neurobiol.*, 6: 526–532.

Burnstock, G., Cocks, T., Crowe, R. and Kasakov, I. (1978) Purinergic innervation of the guinea-pig urinary bladder. *Br. J. Pharmacol.*, 63: 125–138.

Burnstock, G., King, B.F. and Townsend-Nicholson, A. (1998) P1 and P2 Receptor update News sheet 8 Autonomic Neuroscience Institute, Royal Free Hospital School of Medicine, London, June, 1998.

Camaioni, E., Boyer, J.L., Mohanram, A., Harden, T.K. and Jacobson, K.A. (1998) Deoxyadenosine bisphosphate derivatives as potent antagonists at $P2Y_1$ receptors. *J. Med. Chem.*, 41: 183–190.

Chadwick, D.J. and Goode, J.A. (1996) *P2 Purinoceptors: Localization, Function and Transduction Mechanisms*, CIBA Foundation Symposium 198, John Wiley, Chichester, UK.

Chancellor, M.B., Kaplan S.A. and Blaivas, J.G. (1992). The cholinergic and purinergic components of detrussor contracility in a whole rabbit bladder model. *J. Urol.*, 148: 906–909.

Chen, C.C., Akopian, A.N, Sivilotti, L., Colquhoun, D., Burnstock, G. and Wood, J.N. (1995) A P_{2X} purinoceptor expressed by a subset of sensory neurons. *Nature*, 377: 428–431.

Chessell, I.P., Michel, A.D. and Humphrey, P.P.A. (1997) Properties of the pore-forming $P2X_7$ purinoceptor in mouse NTW8 microglial cells. *Br. J. Pharmacol.*, 121: 1429–1437.

Chessell, I.P., Michel, A.D. and Humphrey, P.P.A. (1998) Effects of antagonists at the human recombinant $P2X_7$ receptor. *Br. J. Pharmacol.*, 124: 1314–1320.

Chow, S.C. Kass, G.E.N. and Orrenius, S. (1997) Purines and their role in apoptosis. *Neuropharmacol.*, 36: 1149–1156.

Clifford, E.E., Martin, K.A., Dalal, P., Thomas, R. and Dubyak, G.R. (1997) Stage specific expression of P2Y receptors, ecto-apyrase and ecto-5'-nucleotidase in myeloid leukocytes. *Am. J. Physiol.*, 273: (*Cell Physiol.*, 42:), C973–C987.

Collier, H.O.J., James, G.W.L. and Schnieder, C. (1966) Antagonism by aspirin and femanates of bronchoconstriction and nociception by adenosine-5'-triphosphate. *Nature* 212: 411–412.

Collo, G., North, R.A., Kawashima, E., Merlo-Pich, E., Neidhart, S., Surprenant, A. and Buell, G. (1996) Cloning of $P2X_5$ and $P2X_6$ receptors and their distribution and properties for an extended family of ATP-gated ion channels. *J. Neurosci.*, 16: 2495–2507.

Collo, G., Neidhart, S., Kawashima, E., Kosco-Vilbois, M., North, R.A. and Buell, G. (1997) Tissue distribution of P2X$_7$ receptors. *Neuropharmacol.*, 36: 1277–1283.

Communi, O. and Boeynaems, J.M. (1997) Receptors responsive to extracellular pyrimidine nucleotides *Trends Pharmacol. Sci.*, 18: 83–86.

Connolly, G.P., Abbott, N.A., Deamine, C. and Duly, J.A. (1997) Investigation of receptors responsive to pyrimidines. *Trends Pharmacol. Sci.*, 18: 413.

Cook, S.P., Vulchanova, L., Hargreaves, K.M., Elde, R. and McClesky, E.W. (1997) Distinct ATP receptors on pain-sensing and stretch-sensing neurons. *Nature* 387: 505–508.

Crack, B.E., Pollard, C.E., Beukers, M.W., Roberts, S.M., Hunt, S.F., Ingall, A.H., McKechnie, K.C.W., IJzerman, A.P. and Leff, P. (1995) Pharmacological and biochemical analysis of FPL 67156, a novel, selective inhibitor of ecto-ATPase. *Br. J. Pharmacol.*, 114: 475–481.

Dean, D.M. and Downie, J.W. (1978) Contribution of adrenergic and 'purinergic' neurotransmission to contraction in rabbit detrusor. *J. Pharmacol. Exp. Ther.*, 207: 43–437.

Delombe S. and Escande, D. (1996) ATP-binding cassette proteins as targets for drug discovery. *Trends Pharmacol. Sci.*, 17: 273–275.

Di Iorio, P., Ciccarelli, R., Ballerini, P., Battaglia, G., Bruno, V., Middlemiss, P.J., Rathbone, M.P. and Caciagi, F. (1998) Effect of guanosine and guanosine-like drugs on NMDA-induced unilateral lesion of rat striatum in vivo. *Drug Dev. Res.*, 43: 53.

Di Virgilio F. (1995) The purinergic P$_{2Z}$ receptor: an intriguing role in immunity, inflammation and cell death. *Immunol. Today*, 16: 524–528.

Di Virgilio, F., Ferrari, D., Chiozzi, P., Falzoni, S., Sanz, J.M., dal Susion, M., Mutini. C., Hanau, S. and Baricordi, O.R. (1996) Purinoceptor function in the immune system. *Drug Dev. Res.*, 39: 319–329.

Driessen, B., Reimann, W., Selve, N., Friderichs, E. and Bultmann, R. (1994) Antinociceptive activity of intrathecally administered P2-purinoceptor antagonists in arts. *Brain Res.*, 666: 182–188.

Dubyak, G.R. and El Moatassim, C. (1993) Signal transduction via P$_2$-purinergic receptors for extracellular ATP and other nucleotides. *Am. J. Physiol.*, 265: C577–C606.

Dubyak, G.R., Clifford, E.E., Humphreys, B.D., Kertsey, S.B. and Martin, K.A. (1996) Expression of multiple ATP subtypes during the differentiation and inflammatory activation of myeloid leukocytes. *Drug Dev. Res.*, 39: 269–278.

During, M.J. and Spencer, D.D. (1992) Adenosine: a potential mediator of seizure arrest and postictal refractoriness. Anal. Neurol. 32: 618–624.

Edwards, F.A., Gibba, A.J. and Colquhoun, D. (1992) ATP-receptor-mediated synaptic currents in the central nervous system. *Nature* 359: 144–147.

Edwards, G. and Weston, A.H. (1993) The pharmacology of ATP-sensitive potassium channels. *Ann. Rev. Pharmacol. Toxicol.*, 33: 597–637.

Elde, R. (1998) Presentation at Spring Pain Research Conference, Grand Cayman, BWI, April, 1998.

Evans, R.J., Derkach, V. and Surprenant, A. (1992) ATP mediates fast synaptic transmission in mammalian neurons. *Nature* 357: 503–505.

Freedman, J.E. and Lin, T-J. (1996) ATP-sensitive potassium channels: diverse functions in the central nervous system. *Neuroscientist*, 2: 145–152.

Fukunaga, A.F. (1997) Purines in anesthesia. In: K.A. Jacobson and M.F. Jarvis (Eds.), *Purinergic Approaches in Experimental Therapeutics*. Wiley-Liss, New York, pp. 471–493.

Fyffe, R.E.W. and Perl, E.R. (1984) Is ATP a central synaptic mediator for certain primary afferent fiber from mammalian skin? *Proc. Natl. Acad. Sci. USA*, 81: 6890–6893.

Gargett, C.E. and Wiley, J.S. (1997) The isoquinoline derivative KN-62 a potent antagonist of the P2Z receptor of human lymphocytes. *Br. J. Pharmacol.*, 120: 1483–1490.

Gu, J.G. and MacDermott, A.B. (1997) Activation of ATP P2X receptors elicits glutamate release from sensory neuron synapses. *Nature* 389: 749–753.

Hashimoto, M. and Kokubun, S. (1995) Contribution of P$_2$-purinoceptors to neurogenic contraction of rat urinary bladder smooth muscle. *Br. J. Pharmacol.*, 115: 636–640.

Housley, G.D. (1998) Extracellular nucleotide signalling in the inner ear. *Mol. Neurobiol.*, 16: 21–48.

Housley, G.D., Greenwood, D.J. Bennett, T. and Ryan, A.F. (1995) Identification of a short form of the P2xR1-purinoceptor subunit produced by alternative splicing in the pituitary and cochlea. *Biochem. Biophys. Res. Comm.*, 212: 501–508.

Humphrey, P.P.A., Buell, G., Kennedy, I., Khakh, B.S., Michel, A D , Surprenant, A. and Triese, D.J. (1995) New insights on P$_{2X}$ purinoceptors. *Naunyn-Schimdebergs Arch. Pharmacol.*, 352: 585–596.

Humphreys, B.D. and Dubyak, G.R. (1996) Induction of the P2Z/P2X$_7$ nucleotide receptor and associated phospholipase D activity by lipopolysaccharide and IFN-γ in the human THP-1 monocytic cell line. *J. Immunol.*, 157: 5627–5637.

Humphreys, B.D., Virginio, C., Surprenant, A., Rice, J. and Dubyak, G.R. (1998) Isoquinolines as antagonists of the P2X$_7$ nucleotide receptor: high selectivity for the human versus rat receptor homologues. *Mol. Pharmacol.*, 54: 22–32.

Jacobson, K.A. and van Rhee, A.M. (1997) Development of selective purionceptor agonists and antagonists. In: K.A. Jacobson and M.F. Jarvis (Eds.), *Purinergic Approaches in Experimental Therapeutics*. Wiley-Liss, New York, pp. 101–128.

Jacobson, K.A., Kim, Y.-C., Wildman, S.S., Monhanram, A., Harden, T.K., Boyer, J.L., King, B.F. and Burnstock, G. (1998) A pyridoxine cyclic phosphate and its 6-azoaryl derivative selectively potentiate and antagonize activation of P2X$_1$ receptors. *J. Med. Chem.*, 41: 2201–2206.

Jahr, C.E. and Jessel, T.M. (1983) ATP excites a subpopulation of rat dorsal horn neurones. *Nature*, 304: 730–732.

Kassotakis, L.C., Navakovic, S.D., Oglesby, I.B., Ford, A.P.D.W. and Hunter, J.C. (1997) Immunocytochemical localization of P2X$_3$ receptors in normal and neuropathic rats. ASPET/BPS Meeting, San Diego, CA, March, *Pharmacologist*, Abstract 191.

Katchanov, G., Xu, J., Schulman, E.S., Pelleg, A. (1998) ATP-triggered reflex bronchoconstriction. *Drug Dev. Res.*, 43: 45.

Katsuragi, T., Fujiki, S., Sato, C., Usune, S., Segawa, M. and Honda, K. (1998) Signalling pathway via mitochondria in ATP-evoked ATP release system. *Naunyn-Schimebergs Arch. Pharmacol.*, 358: Suppl. R 127.

Kegel, B., Braun, N., Heine, P. Maliszewski, C.R. and Zimmmmermann, H. (1997) An ecto-ATPase and an ecto-ADP diphosphohydrolase are expressed in rat brain. *Neuropharmacol.*, 36: 1189–1200.

Kennedy, C. and Leff, P. (1995) How should P2X receptors be classified pharmacologically*?* *Trends Pharmacol. Sci.*, 16: 168–174.

Kidd, E.J., Grahames, B.A., Simon, J., Michel, A.D., Barnard, E.A. and Humphrey, P.P.A. (1995) Localization of P2X purinoceptor transcripts in the rat nervous system. *Mol. Pharmacol.*, 45: 569–573.

Knowles, M.R., Clarke, L.L. and Boucher, R.C. (1991) Activation by extracellular nucleotides of chloride secretion in the airway epithelia of patients with cystic fibrosis. *New Eng. J. Med.*, 325: 533–538.

Knutsen, L.J.S. and Murray, T.F. (1997) Adenosine and ATP in epilepsy. In: K.A. Jacobson and M.F. Jarvis (Eds.), *Purinergic Approaches in Experimental Therapeutics*. Wiley-Liss, New York, pp. 423– 447.

Kujawa, S.G., Erostegui, C., Fallon, M., Christ, J. and Bobin, R.P. (1994) Effects of adenosine 5′ triphosphate and related agonists on cochlear function. *Hear Res.*, 76: 87–100.

Lammas, D.A., Stober, C., Harvey, C.J., Kendrick, N., Panchalingam, S. and Kumararatne, D.S. (1997) ATP-induced killing of mycobacteria bu human macrophages is mediated by purinergic P2Z (P2X$_7$) receptors. *Immunity*, 7: 433–444.

Lawson, M. (1996) Australians show potential of common cell compound as ant-cancer treatment. *Bioworld Inter.*, December 4th, 1996, p. 8.

Lethem, M.I., Dowell, M.L., Van Scott, M., Yankaskas, J.R., Egan, T., Boucher, R.C. and Davis, C.W. (1993) Nucleotide regulation of goblet cells in human airway epithelial cells. Normal exocytosis in cystic fibrosis. *Am. J. Respir. Cell. Mol. Biol.*, 9: 315–322.

Lewis, C., Neidhart, S., Holy, C., North, R.A., Buell, G. and Surprenant, A. (1995) Coexpression of P2X2 and P2X3 receptor subunits can account for ATP- gated currents in sensory neurons. *Nature*, 377: 432–435.

Leff, P., Robertson, M.J. and Humphries, R.G. (1997) The role of ADP in thrombosis and the therapeutic potential of P_{2T} receptor antagonists as novel antithrombotic agents. In: K.A. Jacobson and M.F. Jarvis (Eds.), *Purinergic Approaches in Experimental Therapeutics*. Wiley-Liss, New York, pp. 203–216.

Li, J. and Perl, E.R. (1995) ATP modulation of synaptic transmission in the spinal substantia gelatinosa. *J. Neurosci.*, 15: 3357–3365.

Loubatrieres-Mariani, M.-M., Hillaire-Buys, D., Chapal, J., Betrand, G. and Petit, P. (1997) P2 purinoceptor agonists: new insulin secretogogues potentially useful in the treatment of non-insulin-dependent diabetes mellitus. In: K.A. Jacobson and M.F. Jarvis (Eds.), *Purinergic Approaches in Experimental Therapeutics*. Wiley-Liss, New York, pp. 253–260.

Mason, S.J., Olivier, K.N., Bellinger, D., Meuten, D.J., Pare, P.D., Knowles, M.R. and Boucher, R.C. (1993) Studies of absorption and acute and chronic effects of aerosolized and parenteral uridine 5′ triphosphate (UTP) in animals. *Am. Rev Respir. Dis.*, 147: A27.

Nakamura, F. and Strittmatter, S.M. (1996) P2Y$_1$ purinergic receptors in sensory neurons: contribution to touch induced impulse generation. *Proc. Natl. Acad. Sci. USA*, 93: 10465–10470.

Neary, J.T. (1996) Trophic actions of extracellular ATP on astrocytes, synergistic interactions with fibroblast growth growth factors and underlying signal transduction mechanisms. In: D.J. Chadwick and J.A. Goode (Eds.), *P2 Purinoceptors: Localization, Function and Transduction Mechanisms*, CIBA Foundation Symposium 198. John Wiley, Chichester, UK.

Neary, J.T., Rathbone, M.P., Cattabeni, F., Abbracchio, M.P. and Burnstock, G. (1996) Trophic actions of extracellular nucleotides and nucleosides on glial and neuronal cells. *Trends Neurosci.*, 19: 13–18.

Neglia, J.P., FitzSimmons, S.C., Maisonneuve, P., Schoni, M.H., Schoni-Affolter, F., Corey, M., Lowenfels, A.B. and The Cystic Fibrosis and Cancer Study Group (1995) The risk of cancer among patients with cystic fibrosis. *New Eng. J. Med.*, 332: 494–499.

Needleman, P., Minkes, M.S. and Douglas, J.R. (1974) Stimulation of prostaglandin biosynthesis by adenine nucleotides. *Circ. Res.*, 34: 455–460.

Nichols, C.G., Shyng, S.-L, Nestorowicz, A., Glaser, B., Clement IV, J.P., Gonzalez, G., Aguilar-Bryan, L., Permutt, M.A. and Bryan, J. (1996) Adenosine diphosphate as an intracellular regulator of insulin secretion. *Science*, 272: 1785–1787.

Norenberg, M.D. (1994) Astrocytic responses to CNS injury. *J. Neuropathol. Exp. Neurol.*, 53: 213–220.

North, R.A. (1996) P2X purinoceptor plethora. *Seminar Neurosci.*, 8: 187–194.

North, R.A. and Barnard, E.A. (1997) Nucleotide receptors. *Curr. Opin. Neurobiol.*, 7: 346–357.

Nuttle, L.C. and Dubyak, G.R. (1994) Differential activation of action channels and non-selective pores by macrophage P2Z purinergic receptors expressed in Xenopus oocytes. *J. Biol. Chem.*, 269: 13988–13996.

Parker, M.S., Larroque, M.L., Campbell, J.M., Bobbin, R.P. and Deininger, P. (1998) Novel variant of the P2X$_2$ receptor from the guinea pig organ of Corti. *Hear. Res.*, 121: 62–70.

Perregaux, D. and Gabel, C.A. (1994). Interleukin 1-β maturation and release in response to ATP and nigericin. *J. Biol. Chem.*, 2669: 15195–15203.

Phillis, J.W. and Kirkpatrick, J.R. (1978) The actions of adenosine and various nucleosides and nucleotides on the isolated toad spinal cord. *Gen. Pharmacol.*, 9: 239–247.

Pinna, C., Puglis, C. and Burnstock, G. (1998) ATP and vasoactive intestinal polypeptide relaxant responses in hamster isolated proximal urethra. *Br. J. Pharmacol.*, 124: 1069–1074.

Pintor, J. Gomez-Villafurtes, R., Gualix, J. and Miras-Portugal, M.T. (1998) Antagonistic properties of diinosine polyphosphates on dinucleotide and ATP receptors. *Drug Dev. Res.*, 43: 57.

Plesner, L. (1995) Ecto-ATPases: identities and function. *Inter. Rev. Cytol.*, 158: 141–214.

Radford, K.M., Virginio, C., Surprenant, A., North, R.A. and Kawasgima, E. (1997) Baculovirus expression provides direct evidence for heteromeric assembly of P2X$_2$ and P2X$_3$ receptors. *J. Neurosci.*, 17: 6529–6533.

Rapaport, E. (1996) Method of treatment of diabetes mellitus by administration of adenosine-5′-triphosphate and other adenine nucleotides. US Patent 5547942, August 20th, 1996.

Rapaport, E. (1997) ATP in the treatment of cancer. In: K.A. Jacobson and M.F. Jarvis (Eds.), *Purinergic Approaches in Experimental Therapeutics*. Wiley-Liss, New York, pp. 545–553.

Rapaport, E. and Fontaine, J. (1989) Anticancer activities of adenine nucleotides in mice are mediated through expansion of erythrocyte ATP pools. *Proc. Natl. Acad. Sci. USA*, 86: 1662–1666.

Rathbone, M.P. and Middlemass, P.J. (1998) Physiology and pharmacology of natural and synthetic nonadenine-based purines in the nervous system. *Drug Dev. Res.*, 45: 356–372.

Rudolphi, K., Park, C.K. and Rother, M. (1997) Propentofylline (HW 285), a neuroprotective glial cell modulator: Pharmacologic profile. *CNJ Drug Rev.*, 260–277.

Salih, S.G., Houslay, G.D., Burton, L.D. and Greenwood, D. (1998) P2X$_2$ receptor subunit expression in a subpopulation of cochlear 1 spiral ganglion cells. *NeuroReport* 9: 279–282.

Salt, T.E. and Hill, R.G. (1983) Excitation of single sensory neurons in the rat caudate trigeminal nucleus by iontophoretically applied adenosine-5′-triphosphate. *Neurosci. Letts.*, 35: 53–57.

Salter, M.W. and Hicks, J.L. (1994) ATP-evoked increases in in intracellular calcium in neurons and glia from the dorsal spinal cord, *J. Neurosci.*, 14: 1563–1575.

Sawynok, J. (1997) Purines and nociception. In: K.A. Jacobson and M.F. Jarvis (Eds.), *Purinergic Approaches in Experimental Therapeutics*. Wiley-Liss, New York, pp. 495–513.

Schulman, E., Glaum, M., Post, T. Wang, Y., Mohnaty, J. and Pelleg, A. (1998) P2-purinoceptor–mediated enhancement of Anti-IgE–stimulated human lung mast cell histamine release. *Drug Dev. Res.*, 43: 40.

Seguela, P., Haghighi, A., Soghomonian, J.J. and Cooper, E. (1996) A novel neuronal P2X ATP receptor ion channel with widespread distribution in the brain. *J. Neurosci.*, 16: 448–455.

Shaffer, S., Jacobus, K., Foy, C., Pue, C., Donohue, J., Bennett, W., Ye, H., Graham, C., Noone, P. and Drutz, D. (1998) Controlled clinical studies indicate that INS 316 (uridine 5′-triphopshate) a P2Y$_2$ receptor agonist stimulates mucociliary clearance and enhances sputum expectoration.

Siddiqi, S.M., Shaver, S.R., Pendergast, W., Yerxa, B.R., Croom, D.K., Dougherty, R.W., James, M.K., Jones, A.C. and Rideout, J.L. (1998) Molecular modeling of the P2Y$_2$ purinoceptor and synthesis of selected ligands. *Drug Dev. Res.*, 43: 33.

Silinsky, E.M., Gerzanich, V. and Vanner, S.M. (1992). ATP mediates excitatory synaptic transmission in mammalian neurones. *Br. J. Pharmacol.*, 106: 762–763.

Sperlagh, B. and Vizi, E.S. (1996) Neuronal synthesis, storage and release of ATP. *Seminar Neurosci.*, 8: 175–186.

Sperlagh, B., Hasko, G., Nemeth, Z. and Vizi, E.S. (1998) Effect of P2X$_7$ receptor activation on LPS induced nitric oxide production from RAW 264.7 macrophage cell line. *Drug Dev. Res.*, 43: 41.

Surprenant, A., Rassendren, F., Kawashima, E., North, R.A. and Buell, G. (1996) The cytolytic P2Z receptor for extracellular ATP identified as a P2X receptor (P2X$_7$). *Science*, 272: 735–738.

Thomas, W.E. (1992) Brain macrophages: evaluation of microglia and their functions. *Brain Res. Rev.*, 17: 61–74.

Tordorov, L.D., Mihaylova-Todorova, S. Westfall, T.D., Sneddon, P., Kennedy, C., Bjur, R.A. and Westfall, D.P. (1997) Neuronal release of soluble nucleotidases and their role in neurotransmitter inactivation. *Nature* 387: 76–79.

Tschopl, M., Harms, L., Norenberg, W. and Illes, P. (1992) Excitatory effects of adenosine 5′-triphosphate on rat locus coerleus neurons. *Eur. J. Pharmacol.*, 213: 71–77.

Ueno, S., Harata, N., Inoue, K. and Akaike, N. (1992) ATP-gated current in dissociated rat nucleus solitarii neurons. *J. Neurophysiol.*, 68: 778–785.

Van Biesen, T., Lynch, K.J., Burgard, E., Touma, E., Metzger, R., Kage, K., Yu, H., Bianchi, B., Alexander, K., Park, H.S., Niforatos, W., Jarvis, M.F. and Kowaluk, E.A. (1998) Pharmacological characterization of recombinant ionotropic P2X receptors. *Naunyn-Schimdebergs Arch. Pharmacol.*, 358: Suppl. R 129.

Virginio, C., Roberston, G., Suprenant, A. and North, R.A. (1998) Trinitrophenyl-substituted nucleotides are potent antagonists selective for P2X$_1$, P2X$_3$ and heteromeric P2X$_{2/3}$ receptors. *Mol. Pharmacol.*, 53: 969–973.

Vulchanova, L., Arvidsson, U., Riedl, M., Wang, J., Buell, G., Surprenant, A., North, R.A. and Elde, R. (1996) Differential distribution of two ATP-gated ion channels (P2X receptors) determined by immunocytochemistry. *Proc. Natl. Acad. Sci. USA*, 93: 8063–8067.

Wang, T.F. and Guidotti, G. (1996) CD39 is an ecto-(Ca^{2+}Mg^{2+})-apyrase. *J. Biol. Chem.*, 271: 9898–9901.

Williams, M. (1995) Purines in Central Nervous System Function: Targets for therapeutic intervention. In: F.E. Bloom and D.J. Kupfer (Eds.), *Psychopharmacology. The Fourth Generation of Progress*. Raven, New York, pp. 643–655.

Williams, M. (1994). The decade of the brain initiative and health care reform: Irreconcilable differences? *Exp. Opin. Invest. Drugs*, 3: 947–954.

Williams, M. and Burnstock, G. (1997) Purinergic neurotransmission and neuromodulation: a historical perspective. In: K.A. Jacobson and M.F. Jarvis (Eds.), *Purinergic Approaches in Experimental Therapeutics*. Wiley-Liss, New York, pp. 1–27.

Williams, M. and Davis, R.E. (1997) Alzheimer's disease and related dementias: prospects for treatment. *Exp. Opinion Invest. Drugs*, 6: 735–757.

Wittenberg, H., Bultmann, R., Pause, B., Ganter, C., Kurz, G. and Starke, K. (1996) P_2–purinoceptor antagonists: II. Blockade of P_2–purinoceptor subtypes and ecto-nucleotidases by compounds related to Evans blue and trypan blue. *Naunyn Schmiedeberg's Arch. Pharmacol.*, 354: 491–497.

Yokomizo, T., Izumi, T., Chang, K., Takuwa, Y. and Shimizu, T. (1997) A G-protein-coupled receptor for leukotriene B_4 that mediates chemotaxis. *Nature* 387: 620–624.

Yoshimura, N. and de Groat, W.C. (1997) Neural control of the lower urinary tract. *Inter. J. Urol.*, 4: 111–125.

Zimmermann, H. (1996) Extracellular purine metabolism. *Drug Dev. Res.*, 39: 337–352.

Ziganshin, A.U., Hoyle, C.H.V. and Burnstock, G. (1994) Ectoenzymes and metabolism of extracellular ATP *Drug Dev. Res.*, 32: 134–146.

P. Illes and H. Zimmermann (Eds.)
Progress in Brain Research, Vol 120

CHAPTER 9

Novel ligands for P2 receptor subtypes in innervated tissues

Günter Lambrecht[1,*], Susanne Damer[1], Beate Niebel[1], Sittah Czeche[1], Peter Nickel[2], Jürgen Rettinger[1], Günther Schmalzing[1] and Ernst Mutschler[1]

[1]*Department of Pharmacology, Biocentre Niederursel, University of Frankfurt, Marie-Curie-Strasse 9, D-60439 Frankfurt/Main, Germany*
[2]*Department of Pharmaceutical Chemistry, University of Bonn, An der Immenburg 4, D-53121 Bonn, Germany*

Introduction

There is now wide acceptance that ATP (and other nucleotides) may be released into the extracellular fluid by exocytosis from nerve terminals or secretory cells. Thus, extracellular ATP can act as a neurotransmitter or modulator in a variety of peripheral tissues and cells, in autonomic ganglia and in the central nervous system (Zimmermann, 1994; Brake and Julius, 1996; Burnstock, 1997). The cell surface receptors which mediate the effects of extracellular ATP, the P2 receptors, have been divided by pharmacological techniques into five main subtypes, namely P_{2X}, P_{2Y}, P_{2U}, P_{2T} and P_{2Z} (Burnstock and Kennedy, 1985; Abbracchio and Burnstock, 1994; Fredholm et al., 1994; Chen et al., 1995; Brake and Julius, 1996). In a new subclassification scheme, P2 receptor subtypes fall, based on transduction pathways, into two classes: these are the G protein-coupled receptors (P2Y; these include P_{2Y}, P_{2U} and possibly P_{2T} subtypes) and the receptors with integral ion channels (P2X; these include P_{2X} and P_{2Z} subtypes) (Abbracchio and Burnstock, 1994; Burnstock and King, 1996; Barnard et al., 1997; Fredholm et al., 1997). The

recent molecular cloning of P2 receptors removed the last doubts about nucleotides being true extracellular mediators. At present, there are seven known subtypes of P2X and at least four subtypes of P2Y receptors (Burnstock and King, 1996; Barnard et al., 1997; Fredholm et al., 1997; North and Barnard, 1997; Soto et al., 1997).

One of the current challenges in the P2 receptor field is to relate the cloned receptors to the diverse native responses using the relative potencies of various agonists and antagonists. However, fuller understanding of the functions subserved by discrete receptor subtypes has been slowed by several constraints, e.g. the impurity and contamination of commercial nucleotides (Humphrey et al., 1995; Buell et al., 1996; North and Barnard, 1997). One of the major difficulties in this field is the lack of potent and subtype-selective, competitive antagonists. Although a number of compounds have been tested for this purpose, including suramin, pyridoxalphosphate-6-azophenyl-2',4'-disulfonic acid (PPADS), cyclic pyridoxine-$\alpha^{4,5}$-monophosphate-6-azophenyl-2',5'-disulfonic acid (MRS 2220), reactive blue 2, evans blue, trypan blue, 4,4'-diisothiocyanatostilbene-2,2'-disulfonate (DIDS), 2-propylthio-D-β,γ-difluoromethylene ATP (ARL 66096), reactive red 2 and adenosine 3'-phosphate 5'-phosphosulfate (PAPS) and its analogue MRS 2179, none is ideal (Bültmann and Starke, 1995;

---
*Corresponding author. Tel.: +49-69-79829366; fax: +49-69-79829374; e-mail: lambrecht@em.uni-frankfurt.de

Chen et al., 1995; Humphrey et al., 1995; Boyer et al., 1996; Lambrecht, 1996; Bhagwat and Williams, 1997; Hansmann et al., 1997; Bültmann et al., 1998; Camaioni et al., 1998; Jacobson et al., 1998; Moro et al., 1998; Tuluc et al., 1998). All compounds are limited in their usefulness by their irreversibility of the antagonism or by their lack of potency, subtype-selectivity and P2 receptor specificity. Their suspected ability to inhibit some ecto-nucleotidases and thereby protect ATP and other nucleotides from degradation also complicates their use (Humphrey et al., 1995; Ziganshin et al., 1996). The $2',3'-O-(2,4,6$-trinitrophenyl)-substituted analogues of ATP (TNP-ATP), ADP and AMP have recently been reported to be highly potent (IC_{50} values = 1–36 nM) and selective antagonists at recombinant $P2X_1$, $P2X_3$ and $P2X_2/P2X_3$ (heteromeric) receptors, but at least 1000-fold less effective at $P2X_2$, $P2X_4$ and $P2X_7$ receptors (Virginio et al., 1998). However, the antagonism is non-competitive and nothing is known about the P2 receptor specificity of these fluorescent ATP analogues and of their activity at P2Y receptor subtypes and at ecto-nucleotidases.

In the face of this situation, the development of novel P2 receptor antagonists will still be an ambitious undertaking in receptor pharmacology and medicinal chemistry. To this end, structure–activity studies using a series of suramin analogues (Fig. 1) were conducted in our laboratory. These studies culminated in the symmetrical $3'$-urea of 8-(benzamido)naphthalene-1,3,5-trisulfonic acid (NF023, Fig. 1). NF023 was found to be a highly specific P2 receptor antagonist showing selectivity

for native P2X receptors ($pA_2 = 5.5$–5.7) in smooth muscle preparations (Lambrecht, 1996; Bültmann et al., 1996; Ziyal et al., 1997). NF023 was also a selective and highly potent $P2X_1$ antagonist in studies with the cloned P2X receptor subtypes expressed in *Xenopus laevis* oocytes (pIC_{50} values: $rP2X_1 = 6.62$, $rP2X_2 = < 4.3$, $rP2X_3 = 5.1$, $rP2X_4 = < 4.0$; Soto et al., 1999). In order to increase the potency of NF023 at the $P2X_1$ receptor and, hence, to alter its functional selectivity in favour of the $P2X_1$ subtype, we have synthesized and pharmacologically characterized a series of second generation suramin analogues with $4'$-aminobenzoyl-linkages of the two phenyl rings "1" and "2". Two of these compounds, NF279 and NF031 (Fig. 2), were the subject of a more detailed pharmacological investigation. Suramin and NF023 were used as reference drugs.

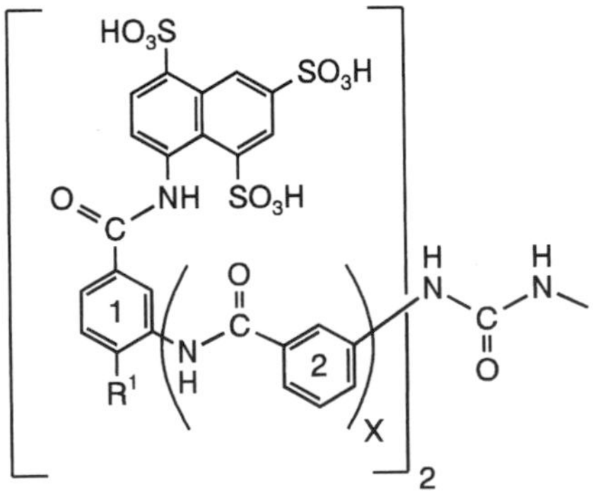

Material and methods

Experimental methods will not be described since these are detailed in the literature: rat vas deferens ($P2X_1$ receptors: Khakh et al., 1995; Bültmann et al., 1996; Westfall et al., 1997; Tuluc et al., 1998); guinea-pig taenia coli (P2Y receptors: Windscheif et al., 1995; Bültmann et al., 1996); guinea-pig ileum longitudinal smooth muscle (neuronal P2X receptors and postjunctional $P2Y_1$ receptors: Kennedy and Humphrey, 1994; Czeche et al., 1998b); *Xenopus laevis* oocytes (recombinant $P2X_1$ receptors and ecto-nucleotidase: Ziganshin et al., 1996; Nicke et al., 1998). To test the P2 receptor specificity of NF279 and NF031, we investigated their effects on responses to various agonists in rat vas deferens (noradrenaline-induced contractions: Bungardt et al., 1992), guinea-pig ileum (inhibition

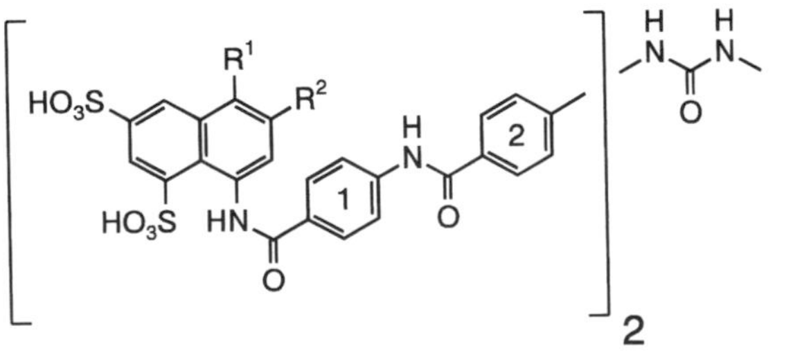

Fig. 1. Chemical structure of suramin ($R^1 = CH_3$, $X = 1$) and NF023 ($R^1 = H$, $X = 0$).

Fig. 2. Chemical structure of NF279 ($R^1 = SO_3H$, $R^2 = H$) and NF031 ($R^1 = H$, $R^2 = SO_3H$).

of neurogenic contractions by 2-chloro-N[6]-cyclo-pentyladenosine as well as contractions induced by histamine and arecaidine propargyl ester: Kennedy and Humphrey, 1994; Pfaff et al., 1995) and guinea-pig taenia coli (nicotine- or 2-chloro-adenosine-induced relaxations in carbachol-precontracted preparations: Iselin et al., 1988; Bültmann and Starke, 1995; Prentice and Hourani, 1997). Data are presented as means from at least three independent experiments.

P2X receptors

Rat vas deferens

In prostatic segments of rat vas deferens, α,β-methylene ATP (α,β-mATP; 10 μM) elicited rapid and transient contractions amounting to 1500 mg tension generated in control experiments (all experiments pooled). NF279, NF031, NF023 and suramin inhibited the contractions evoked by α,β-mATP in a concentration-dependent manner and nearly abolished them at the highest concentration used (Fig. 3). pIC_{50} values are summarized in Table 1.

NF279 was significantly more potent (six-fold) than NF031 and NF023, which in turn were more

potent than suramin. As a result, the rank order of antagonist potency in rat vas deferens (including PPADS; $pIC_{50} = 5.36$, data not shown) was: NF279 > PPADS > NF031 = NF023 > suramin. These data confirm the view that the P2 receptors mediating contraction in prostatic segments of rat vas deferens share some properties with the cloned $P2X_1$ subtype (Colo et al., 1996; Soto et al., 1997): sensitivity and fast desensitization to α,β-mATP and high potency (selectivity) of PPADS, NF279 and NF023.

The time course of action of NF279 and the reversibility of antagonism on successive responses of rat vas deferens to α,β-mATP was also examined (Fig. 4). Following reproducible control contractions to α,β-mATP (10 μM), the addition of NF279 (3 μM) induced a reduction in the magnitude of responses to α,β-mATP (added at 30- or 60-min intervals) of 89%. This effect was reversed on repeated washout during 90–120 min when responses to α,β-mATP recovered to their respective pre-NF279 controls. Over the same time period responses of control tissues to 10 μM α,β-mATP remained uniform (Fig. 4).

In order to elucidate the mechanism of antagonism by NF279 in rat vas deferens,

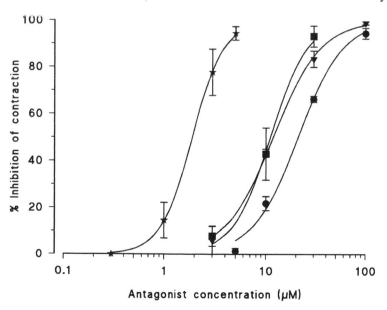

Fig. 3. Concentration-response curves for inhibition of α,β-methylene ATP (10 μM)-induced isometric contractions in prostatic segments of rat vas deferens by suramin (●), NF023 (▼), NF279 (★) and NF031 (■). Antagonists were incubated for at least 60 min.

TABLE 1

Potency estimates of P2 receptor antagonists obtained in tissues or cells that are endowed with P2X or P2Y receptors or ecto-nucleotidases

Compound	P2X$_1$ (ratVD)[a] pIC$_{50}$		P2X (GPI)[b] pA$_2$		P2Y (GPTC)[c] pA$_2$		P2Y$_1$ (GPI)[d] pA$_2$		ectoN[e] % inhibition	
Suramin	4.68	(10–100)[f]	4.89	(30–300)	4.46	(300)[f]	5.68	(10–100)	41	(300)[f]
NF023	4.93	(3–100)[f]	4.45	(30–100)	4.13	(300)[f]	4.74	(100–1000)	31	(300)[f]
NF279	5.72	(0.3–5)[f]	5.95	(3–30)	4.25	(100)[f]	4.97	(30)	66	(300)[f]
NF031	4.96	(3–50)[g]	5.85	(10–100)	3.46[g]	(300)	4.56	(100)	53[g]	(300)

IC$_{50}$ values (pIC$_{50}$ = − log IC$_{50}$) are the molar concentrations necessary to inhibit the response to a single dose of agonist by 50%. pA$_2$ values were determined according to Arunlakshana and Schild (1959). Concentrations (μM) of antagonists are given in parentheses.
[a] Inhibition of contractions to single doses of α,β-mATP (10 μM; tension generated in control experiments = 1500 mg) in prostatic segments of rat vas deferens (ratVD).
[b] Inhibition of indirect (acetylcholine-mediated) contractions to α,β-mATP (0.1–30 μM; EC$_{50}$ = 1.1 μM, 797 mg tension generated in control experiments) in guinea-pig ileum (GPI).
[c] Inhibition of relaxant responses to ADPβS (EC$_{50}$ = 0.63 μM) in carbachol (0.1–0.3 μM)-precontracted guinea-pig taenia coli (GPTC).
[d] Inhibition of contractions to ADPβS (0.3–300 μM; EC$_{50}$ = 4 μM, 1200 mg tension generated in control experiments) in the presence of atropine (0.3 μM) in guinea-pig ileum (GPI).[e] Inhibition of ATP (100 μM) breakdown by ecto-nucleotidases (ectoN) in Xenopus laevis oocytes. EctoN activity was studied by measuring the production of inorganic phosphate (production of P$_i$ in control incubations = 4.5 nmol/30 min/cell).
[f] Data from Damer et al. (1998a).
[g] Data from Damer et al. (1998b).

concentration–response curves of α,β-mATP were determined in the absence and presence of NF279. As shown in Fig. 5, the concentration–response curves did not reach a well defined maximum even at 100 μM α,β-mATP in the control experiments. Incubation of the rat vas deferens with NF279

(1–10 μM; 60- to 120-min exposure) produced a concentration–dependent antagonism of contractile responses to α,β-mATP. Because of the incomplete nature of the agonist concentration-effect curves, equi-effective concentrations of α,β-mATP in the absence and presence of NF279 were measured at

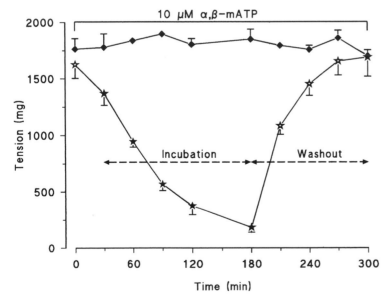

Fig. 4. Kinetics of onset (incubation) and offset (washout) of antagonism by NF279 ($\star$, 3 μM) in prostatic segments of rat vas deferens; $\blacklozenge$ indicates data from time-matched control experiments. The diagram shows inhibition of isometric contractile responses to single doses of α,β-methylene ATP (α,β-mATP; 10 μM).

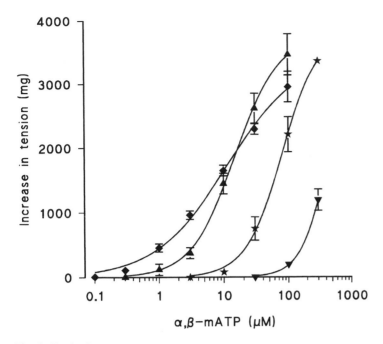

Fig. 5. Single dose concentration-response curves for isometric contractile responses to α,β-methylene ATP (α,β-mATP) in the absence (◆) and in the presence of NF279 1 μM (▲), 3 μM (★) and 10 μM (▼) in prostatic segments of rat vas deferens (60- to 120-min exposure of NF279).

the 1000 mg tension response level, and individual dose-ratios were calculated.

Mean pA_2 estimates derived from each concentration of NF279 were significantly different from each other and increased with the NF279 concentration (6.10 at 1 μM, 6.38 at 3 μM and 7.02 at 10 μM). The mechanism responsible for this noncompetitive behavior of NF279 is unknown. It may result when the lower NF279 concentrations have not reached equilibrium (Leff et al., 1990). However, the pA_2 values calculated at 1 μM NF279 were similar after 60- or 120-min incubation time. Further experiments are needed to clarify this issue.

Guinea-pig ileum

It has been reported that P2X-like receptors are located on cholinergic nerves in the guinea-pig ileal longitudinal smooth muscle (Kennedy and Humphrey, 1994), the stimulation of which causes release of acetylcholine and, subsequently, a tetrodotoxin- and atropine-sensitive contraction via postjunctional muscarinic M_3 receptors (Czeche et al., 1998a,b) (Fig. 6). In the present study, α,β-

mATP (0.1–30 μM; $EC_{50} = 1.1$ μM, 797 mg tension generated in control experiments) elicited concentration-dependent contractile responses. NF279, NF031, NF023 and suramin shifted the concentration-response curve of α,β-mATP to the right in a parallel fashion. The antagonism was reversible and surmountable (illustrated for NF279 in Fig. 7, together with the corresponding Schild plot, the slope of which (1.15) was not significantly different from unity). pA_2 values are summarized in Table 1.

NF279 and NF031 were significantly more potent (10-fold) than suramin, which in turn was more potent than NF023. As a result, the rank order of antagonist potency at the α,β-mATP-sensitive P2X receptors in guinea-pig ileum (including PPADS; $pK_B = 6.26$; Czeche et al., 1998b), PPADS > NF279 = NF031 > Suramin > NF023, is different from that found in rat vas deferens. The pharmacological properties of the neuronal P2X receptors in guinea-pig ileum (sensitivity to α,β-mATP and effectiveness of antagonists) do not correspond to those observed at recombinant homomultimeric P2X receptors characterized so far

(Soto et al., 1997, and this study). These native channels in guinea-pig ileum might be heteropolymers of the cloned P2X receptor subunits, such as P2X$_2$/P2X$_3$ receptors (Lewis et al., 1995), or may contain additional subunits not yet cloned.

Recombinant P2X$_1$ receptors

To elucidate the P2X selectivity of NF279, *Xenopus laevis* oocytes were injected with the cRNA for rat P2X$_1$ receptors. Two to four days after injection, the influence of NF279 on current responses to ATP (1 μM) was studied. ATP-induced inward currents were measured with the two-electrode voltage clamp technique (Nicke et al., 1998). NF279 concentrations were applied 10 s prior to ATP

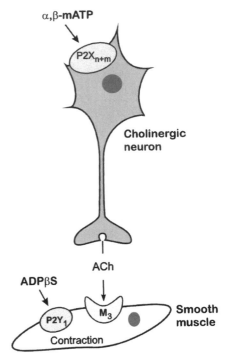

Fig. 6. Functional heterogeneity of P2 receptors mediating contraction in guinea-pig ileal longitudinal smooth muscle. P2 receptor agonists contract ileal preparations by acting at two different receptors. One is selective for α,β-mATP and appears to be a neuronal P2X receptor, the stimulation of which causes indirect contraction via postjunctional muscarinic M$_3$ receptors. These native channels might be heteropolymers of the cloned P2X receptor subunits or may contain additional subunits not yet cloned. The other P2 receptor selectively activated by ADPβS has the properties of native and recombinant P2Y$_1$ subtypes and appears to be situated on the smooth muscle.

application. All experiments were performed at a holding potential of -60 mV and at room temperature (22–24°C).

Increasing concentrations of NF279 (0.01–3 μM) reduced and finally abolished the current evoked by ATP in oocytes expressing the P2X$_1$ subtypes (pIC$_{50}$ = 7.26) (Damer et al., 1998a). As a result, the following P2X$_1$ receptor selectivity profile was obtained: TNP–ATP > NF279 > NF023 > PPADS > suramin (Collo et al., 1996; Soto et al., 1997, 1999; Boarder and Hourani, 1998; Virginio et al., 1998).

P2Y receptors

Guinea-pig taenia coli

The guinea-pig taenia coli is a prototype tissue for the study of P2Y receptors (Burnstock and Kennedy, 1985; Windscheif et al., 1995; Bültmann et al., 1996). Relaxations caused by 5'-O-(2-thiodiphosphate) (ADPβS) are mediated exclusively by this subtype (Dudek et al., 1995) which has not so far been cloned. However, studies with novel P2 receptor agonists and antagonists (Burnstock et al., 1994; Bültmann et al., 1996; Lambrecht, 1996) bear out the view that the P2Y receptor in the taenia coli is different from the P2Y$_1$ receptor in turkey erythrocytes (Nicholas et al., 1996; Camaioni et al., 1998).

In the guinea-pig taenia coli precontracted by carbachol (0.1–0.3 μM), suramin, NF023, NF279 and NF031 (60-min exposure) antagonized relaxant responses to ADPβS (EC$_{50}$ = 0.63 μM) in a parallel and surmountable manner. pA$_2$ values are summarized in Table 1.

The resulting pA$_2$ estimates for NF279 (4.25) and NF023 (4.13) were similar, these two compounds being about six-fold more potent than NF031 and about two-fold less potent than suramin. As a result, the ADPβS-sensitive P2Y receptor in the taenia coli is characterized by the following rank order of antagonist potency (including PPADS; pA$_2$ = 4.59, Lambrecht, 1996): PPADS = suramin > NF279 = NF023 > NF031. The potency estimates for the antagonists employed in this study at the P2Y receptors in the taenia coli were clearly lower (up to 50-fold) compared to that

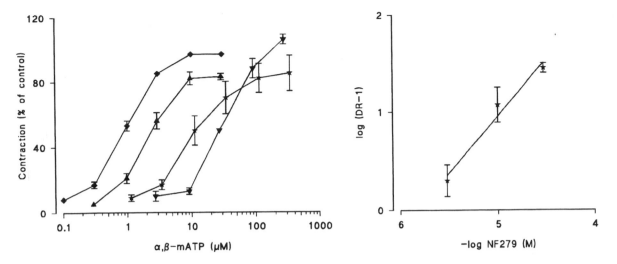

Fig. 7. Concentration-response curves for α,β-methylene ATP (α,β-mATP) in guinea-pig ileum (left panel) in the absence (◆) and presence of NF279 3 μM (▲), 10 μM (★) and 30 μM (▼). The right panel shows the corresponding Schild plot.

found at P2X receptors in rat vas deferens and guinea-pig ileum (Table 1).

Guinea-pig ileum

In guinea-pig ileum, ADPβS (10 μM) caused contractions, which were partially reduced by TTX (1 μM; 32%) and atropine (300 nM; 45%). These results demonstrate that ADPβS acts at P2 receptors located at two sites in the ileal longitudinal muscle: a P2 receptor located on cholinergic neurons and a postjunctional P2 receptor on ileal smooth muscle (Fig. 6). Thus, to isolate the stimulatory action of ADPβS at the postjunctional receptor pharmacologically, atropine (300 nM) was present throughout the antagonist experiments. Under these experimental conditions, α,β-mATP did not cause contraction of the ileal preparation in concentrations up to 10 μM (Kennedy and Humphrey, 1994; Czeche et al., 1998b).

NF279, NF031, NF023 and suramin shifted the concentration-response curve of ADPβS (EC$_{50}$ = 4 μM) to the right in a parallel fashion. The antagonism was reversible and surmountable (illustrated for NF032 in Fig. 8, left panel). Where three concentrations were tested (NF023 and suramin), Schild plots were linear (Fig. 8, right panel) and the slopes (1.01 and 0.94, respectively) did not differ significantly from unity. pA$_2$ values are summarized in Table 1.

The resulting pA$_2$ estimates for NF279, NF023 and NF031 were similar, these three compounds being less potent (up to 13-fold) than suramin. The potency of all antagonists examined at the ADPβS-sensitive P2 receptors in guinea-pig ileum was higher (up to 17-fold) than that found in the taenia coli (Table 1). As a result, the rank order of antagonist potency in the ileal preparation (including PPADS; pK$_B$ = 6.20, Czeche et al., 1998b), PPADS > suramin > NF279 ≥ NF023 ≥ NF031, is different from that obtained in the taenia coli. The pharmacological properties of the postjunctional P2 receptors which subserve contractions in guinea-pig ileum (insensitiveness to α,β-mATP, selectivity for ADPβS and high potency of PPADS and suramin) indicate that these receptors are very similar to the native and recombinant P2Y$_1$ receptors (Fig. 6) (Lambrecht, 1996; Nicholas et al., 1996; Barnard et al., 1997).

Ecto-nucleotidases

Folliculated *Xenopus laevis* oocytes readily metabolize naturally occuring nucleotides (Ziganshin et al., 1996). In the present study, where ATP (100 μM) was used as the substrate for oocyte ecto-nucleotidases, the generation of inorganic phosphate by the ecto-enzymes was 4.5 nmol/30 min/cell. Suramin, NF023, NF279 and NF031 (300 μM) inhibited significantly and to approx-

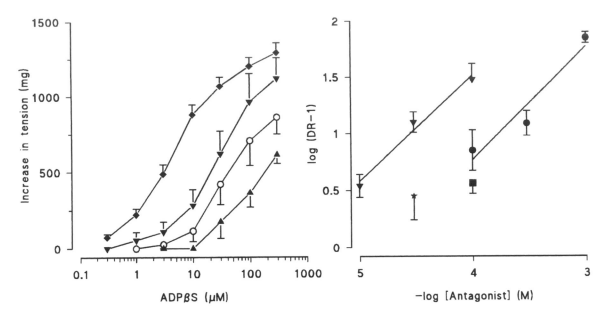

Fig. 8. Concentration-response curves of ADPβS in guinea-pig ileal longitudinal smooth muscle preparations (left panel) in the absence (◆) and in the presence of NF023 (▼ 100 μM , ◯ 300 μM, ▲ 1000 μM). The right panel shows the corresponding Schild plots for the antagonism by NF023 (●) and suramin (▼), the slopes of which were not significantly different from unity. Only one concentration of NF279 (★ 30 μM) and NF031 (■ 100 μM) was tested (right panel). Antagonists were incubated for at least 60 min, and isometric contractile responses were obtained to single doses of ADPβS in the presence of atropine (300 nM).

imately a similar extent the breakdown of ATP by oocytes (Table 1). In principle, these results indicate that caution should be adopted when using high concentrations of these compounds (>30 μM) in P2 receptor classification studies. The propensity for their receptor antagonistic and ecto-nucleotidase inhibitory properties to self-cancel when metabolically unstable agonists are used can be the basis for misinterpretation. However, if used at concentrations lower than 30 μM, effective blockade of $P2X_1$ receptors by NF279 and NF031 may be achieved, with little or no inhibition of ecto-nucleotidases. The challenges for drug discovery will be the design of highly potent and selective P2 receptor antagonists, lacking the ability to inhibit ecto-nucleotidases (Humphrey et al., 1995).

P2 receptor specificity of NF279 and NF031

At 100 μM, NF279 and NF031 did not interact with α_{1A}-adrenoceptors in rat vas deferens, adenosine A_1, histamine H_1 and muscarinic M_3 receptors in guinea-pig ileum and adenosine A_{2B} as well as

neuronal nicotinic receptors in guinea-pig taenia coli (Table 2). Thus, the antagonism of NF279 and NF031 appears to be specific for P2 receptors.

Summary and conclusions

Among suramin analogues, the properties of P2 receptor subtype blockade and ecto-nucleotidase inhibition appear to be controlled by different structural parameters (Fig. 1 and 2, Table 1; Van Rhee et al., 1994; Beukers et al., 1995; Bültmann et al., 1996; Damer et al., 1998a, 1998b; and this study): the molecular size of the compounds, the position of the sulfonic acid residues in the naphthalene rings, the substitution pattern of the benzoyl moieties and the 3'- or 4'-aminobenzoyl-linkages of the phenyl rings "1" and "2". As a result, compounds with different receptor selectivity profiles were obtained. A maximum in potency at and selectivity for $P2X_1$ receptors is reached in NF279, which is a specific P2 receptor antagonist and the compound with the highest $P2X_1$ vs. P2Y receptor and ecto-nucleotidase selectivity presently available.

TABLE 2

P2 receptor specificity of NF279 and NF031. Potencies (pEC$_{50}$ values) of various agonists at different receptors in rat vas deferens (ratVD; α_{1A}-adrenoceptors), guinea-pig ileum (GPI; A$_1$, H$_1$ and M$_3$ receptors) and guinea-pig taenia coli (GPTC; A$_{2B}$ and neuronal nicotinic receptors) in the absence (controls) and in the presence of NF279 and NF031 (100 μM)

Receptor/Agonist/Tissue	pEC$_{50}$			
	Control	NF279	Control	NF031
α_{1A}/noradrenaline/ratVD	5.97	5.82	5.85	5.79
A$_1$/CCPA[a]/GPI	8.46	8.58	8.45	8.48
A$_{2B}$/2-chloroadenosine/GPTC[b]	4.56	4.46	4.63	4.43
H$_1$/histamine/GPI	6.04	6.04	6.11	6.17
M$_3$/APE[c]/GPI	7.53	7.60	7.44	7.49
N/nicotine/GPTC	4.45	4.26	4.57	4.44

EC$_{50}$ values (pEC$_{50}$ = $-$ log EC$_{50}$) were calculated from semi-logarithmic concentration-response curves by fitting a logistic function to the data.

[a] 2-Chloro-N^6-cyclopentyladenosine.
[b] In the presence of the A$_1$-selective antagonist 8-cyclopentyl-1,3-dipropylxanthine (100 nM).
[c] Arecaidine propargyl ester.

Acknowledgements

The authors thank the Fonds der Chemischen Industrie (Germany) and the Deutsche Forschungsgemeinschaft (grant numbers La 350/7-1 and Schm 536/2-1) for financial support. The skillful technical assistance of Mrs. Caren Hildebrandt is gratefully acknowledged.

References

Abbracchio, M.P. and Burnstock, G. (1994) Purinoceptors: Are there families of P2X and P2Y purinoceptors? *Pharmacol. Ther.*, 64: 445–475.

Arunlakshana, O. and Schild, H.O. (1959) Some quantitative uses of drug antagonists. *Br. J. Pharmacol.*, 14: 48–58.

Barnard, E.A., Simon, J. and Webb, T.E. (1997) Nucleotide receptors in the nervous system: An abundant component using diverse transduction mechanisms. *Molec. Neurobiol.*, 15: 103–130.

Beukers, M.W., Kerkhof, C.J.M., Van Rhee, M.A., Ardanuy, U., Gurgel, C., Widjaja, H., Nickel, P., IJzerman, A.P. and Soudijn, W. (1995) Suramin analogs, divalent cations and ATPγS as inhibitors of ecto-ATPase. *Naunyn-Schmiedeberg's Arch. Pharmacol.*, 351: 523–528.

Bhagwat, S.S. and Williams, M. (1997) P2 purine and pyrimidine receptors: Emerging superfamilies of G-protein-coupled and ligand-gated ion channel receptors. *Eur. J. Med. Chem.*, 32: 183–193.

Boarder, M.R. and Hourani, S.M.O. (1998) The regulation of vascular function by P2 receptors: Multiple sites and multiple receptors. *Trends Pharmacol. Sci.*, 19: 99–107.

Boyer, J.L., Romero-Avila, T., Schachter, J.B. and Harden, T.K. (1996) Identification of competitive antagonists of the P2Y$_1$ receptor. *Mol. Pharmacol.*, 50: 1323–1329.

Brake, A.J. and Julius, D. (1996) Signaling by extracellular nucleotides. *Annu. Rev. Cell Dev. Biol.*, 12: 519–541.

Buell, G., Collo, G. and Rassendren, F. (1996) P2X receptors: An emerging channel family. *Eur. J. Neurosci.*, 8: 2221–2228.

Bültmann, R. and Starke, K. (1995) Reactive red 2: a P$_{2Y}$-selective purinoceptor antagonist and an inhibitor of ecto-nucleotidase. *Naunyn-Schmiedeberg's Arch. Pharmacol.*, 352: 477–482.

Bültmann, R., Wittenburg, H., Pause, B., Kurz, G., Nickel, P. and Starke, K. (1996) P$_2$-purinoceptor antagonists: III. blockade of P$_2$-purinoceptor subtypes and ecto-nucleotidases by compounds related to suramin. *Naunyn-Schmiedeberg's Arch. Pharmacol.*, 354: 498–504.

Bültmann, R., Tuluc, F. and Starke, K. (1998) On the suitability of adenosine 3'-phosphate 5'-phosphosulphate as a selective P2Y receptor antagonist in intact tissues. *Eur. J. Pharmacol.*, 351: 209–215.

Bungardt, E., Buschauer, A., Moser, U., Schunack, W., Lambrecht, G. and Mutschler, E. (1992) Histamine H$_1$ receptors mediate vasodilation in guinea-pig ileum resistance vessels: characterization with computer-assisted videomicroscopy and new selective agonists. *Eur. J. Pharmacol.*, 221: 91–98.

Burnstock, G. (1997) The past, present and future of purine nucleotides as signalling molecules. *Neuropharmacology*, 36: 1127–1139.

Burnstock, G. and Kennedy, C. (1985) Is there a basis for distinguishing two types of P$_2$-purinoceptor? *Gen. Pharmacol.*, 16: 433–440.

Burnstock, G. and King, B.F. (1996) Numbering of cloned P2 purinoceptors. *Drug Dev. Res.*, 38: 67–71.

Burnstock, G., Fischer, B., Hoyle, C.H.V., Maillard, M., Ziganshin, A.U., Brizzolara, A.L., von Isakovics, A., Boyer, J.L., Harden, T.K. and Jacobson, K.A. (1994) Structure activity relationships for derivatives of adenosine-5'-triphosphate as agonists at P$_2$ purinoceptors: Heterogeneity within P$_{2X}$ and P$_{2Y}$ subtypes. *Drug Dev. Res.*, 31: 206–219.

Camaioni, E., Boyer, J.L., Mohanram, A., Harden, T.K. and Jacobson, K.A. (1998) Deoxyadenosine bisphosphate derivatives as potent antagonists at P2Y$_1$ receptors. *J. Med. Chem.*, 41: 183–190.

Chen, Z.-P., Levy, A. and Lightman, S.L. (1995) Nucleotides as extracellular signalling molecules. *J. Neuroendocrinol.*, 7: 83–96.

116

Collo, G., North, R.A., Kawashima, E., Merlo-Pich, E., Neidhart, S., Surprenant, A. and Buell, G. (1996) Cloning of P2X$_5$ and P2X$_6$ receptors and the distribution and properties of an extended family of ATP-gated ion channels. *J. Neurosci.*, 16: 2495–2507.

Czeche, S., Niebel, B., Mutschler, E. and Lambrecht, G. (1998a) Neuronal P2X-like receptors mediate cholinergic contraction via postjunctional muscarinic M$_3$-receptors in guinea-pig ileal longitudinal smooth muscle. *Naunyn-Schmiedeberg's Arch. Pharmacol.*, 357 (Suppl.): R23.

Czeche, S., Niebel, B., Mutschler, E., Nickel, P. and Lambrecht, G. (1998b) P2-receptor heterogeneity in the guinea-pig ileal longitudinal smooth muscle preparation. *Drug Dev. Res.*, 43: 52.

Damer, S., Niebel, B., Czeche, S., Nickel, P., Ardanuy, U., Schmalzing, G., Rettinger, J., Mutschler, E. and Lambrecht, G. (1998a) NF279: A novel potent and selective antagonist of P2X receptor-mediated responses. *Eur. J. Pharmacol.*, 350: R5–R6.

Damer, S., Niebel, B., Nickel, P., Ardanuy, U., Mutschler, E. and Lambrecht, G. (1998b) NF279 and NF031, new selective P2X-receptor antagonists. *Naunyn-Schmiedeberg's Arch. Pharmacol.*, 357 (Suppl.): R34.

Dudek, O., Bültmann, R. and Starke, K. (1995) Two relaxation-mediating P$_2$-purinoceptors in guinea-pig taenia caeci. *Naunyn-Schmiedeberg's Arch. Pharmacol.*, 351: 107–110.

Fredholm, B.B., Abbracchio, M.P., Burnstock, G., Daly, J.W., Harden, T.K., Jacobson, K.A., Leff, P. and Williams, M. (1994) VI. Nomenclature and classification of purinoceptors. *Pharmacol. Rev.*, 46: 143–156.

Fredholm, B.B., Abbracchio, M.P., Burnstock, G., Dubyak, G.R., Harden, T.K., Jacobson, K.A., Schwabe, U. and Williams, M. (1997) Towards a revised nomenclature for P1 and P2 receptors. *Trends Pharmacol. Sci.*, 18: 79–82.

Hansmann, G., Bültmann, R., Tuluc, F. and Starke, K. (1997) Characterization by antagonists of P2-receptors mediating endothelium-dependent relaxation in the rat aorta. *Naunyn-Schmiedeberg's Arch. Pharmacol.*, 356: 641–652.

Humphrey, P.P.A., Buell, G., Kennedy, I., Khakh, B.S., Michel, A.D., Surprenant, A. and Trezise, D.J. (1995) New insights on P$_{2X}$ purinoceptors. *Naunyn-Schmiedeberg's Arch. Pharmacol.*, 352: 585–596.

Iselin, C.E., Martin, J.-L., Magistretti, P.J. and Ferrero, J.D. (1988) Stimulation by nicotine of enteric inhibitory nerves and release of vasoactive intestinal peptide in the taenia of the guinea-pig caecum. *Eur. J. Pharmacol.*, 148: 179–186.

Jacobson, K.A., Kim, Y.-C., Wildman, S.S., Mohanram, A., Harden, T.K., Boyer, J.L., King, B.F. and Burnstock, G. (1998) A pyridoxine cyclic phosphate and its 6-azoaryl derivative selectively potentiate and antagonize activation of P2X$_1$ receptors. *J. Med. Chem.*, 41: 2201–2206.

Kennedy, I. and Humphrey, P.P.A. (1994) Evidence for the presence of two types of P$_2$ purinoceptor in the guinea-pig ileal longitudinal smooth muscle preparation. *Eur. J. Pharmacol.* 261: 273–280.

Khakh, B.S., Surprenant, A. and Humphrey, P.P.A. (1995) A study on P$_{2X}$ purinoceptors mediating the electrophysiological and contractile effects of purine nucleotides in rat vas deferens. *Br. J. Pharmacol.*, 115: 177–185.

Lambrecht, G. (1996) Design and pharmacology of selective P$_2$-purinoceptor antagonists. *J. Auton. Pharmacol.*, 16: 341–344.

Leff, P., Wood, B.E. and O'Connor, S.E. (1990) Suramin is a slowly-equilibrating but competitive antagonist at P$_{2X}$-receptors in the rabbit isolated ear artery. *Br. J. Pharmacol.*, 101: 645–649.

Lewis, C., Neidhart, S., Holy, C., North, R.A., Buell, G. and Surprenant, A. (1995) Coexpression of P2X$_2$ and P2X$_3$ receptor subunits can account for ATP-gated currents in sensory neurons. *Nature*, 377: 432–435.

Moro, S., Guo, D., Camaioni, E., Boyer, J.L., Harden, T.K. and Jacobson, K.A. (1998) Human P2Y$_1$ receptor: Molecular modeling and site-directed mutagenesis as tools to identify agonist and antagonist recognition sites. *J. Med. Chem.*, 41: 1456–1466.

Nicholas, R.A., Lazarowski, E.R., Watt, W.C., Li, Q., Boyer, J. and Harden, T.K. (1996) Pharmacological and second messenger signalling selectivities of cloned P2Y receptors. *J. Auton. Pharmacol.*, 16: 319–323.

Nicke, A., Bäumert, H.G., Rettinger, J., Eichele, A., Lambrecht, G., Mutschler, E. and Schmalzing, G. (1998) P2X$_1$ and P2X$_3$ receptors form stable trimers: A novel structural motif of ligand-gated ion channels. *EMBO J.*, 17: 3016–3028.

North, R.A. and Barnard, E.A. (1997) Nucleotide receptors. *Current Opinion Neurobiol.*, 7: 346–357.

Pfaff, O., Hildebrandt, C., Waelbroeck, M., Hou, X., Moser, U., Mutschler, E. and Lambrecht, G. (1995) The (S)-(+)-enantiomer of dimethindene: A novel M$_2$-selective muscarinic receptor antagonist. *Eur. J. Pharmacol.*, 286: 229–240.

Prentice, D.J. and Hourani, S.M.O. (1997) Adenosine analogues relax guinea-pig taenia caeci via an adenosine A$_{2B}$ receptor and a xanthine-resistant site. *Eur. J. Pharmacol.*, 323: 103–106.

Soto, F., Garcia-Guzman, M. and Stühmer, W. (1997) Cloned ligand-gated channels activated by extracellular ATP (P2X receptors). *J. Membr. Biol.*, 160: 91–100.

Soto, F., Lambrecht, G., Nickel, P., Stühmer, W. and Busch, A.E. (1999) Antagonistic properties of the suramin analogue NF023 at heterologously expressed P2X receptors. *Neuropharmacology*, 38: 141–149.

Tuluc, F., Bültmann, R., Glänzel, M., Frahm, A.W. and Starke, K. (1998) P2-receptor antagonists: IV. blockade of P2-receptor subtypes and ecto-nucleotidases by compounds related to reactive blue 2. *Naunyn-Schmiedeberg's Arch. Pharmacol.*, 357: 111–120.

Van Rhee, A.M., Van der Heijden, M.P.A., Beukers, M.W., IJzerman, A.P., Soudijn, W. and Nickel, P. (1994) Novel competitive antagonists for P$_2$-purinoceptors. *Eur. J. Pharmacol.*, 268: 1–7.

Virginio, C., Robertson, G., Surprenant, A. and North, R.A. (1998) Trinitrophenyl-substituted nucleotides are potent antagonists selective for P2X$_1$, P2X$_3$, and heteromeric P2X$_2$/$_3$ receptors. *Mol. Pharmacol.*, 53: 969–973.

Westfall, T.D., McIntyre, C.A., Obeid, S., Bowes, J., Kennedy, C. and Sneddon, P. (1997) The interaction of diadenosine polyphosphates with P$_{2X}$-receptors in the guinea-pig isolated vas deferens. *Br. J. Pharmacol.*, 121: 57–62.

Windscheif, U., Pfaff, O., Ziganshin, A.U., Hoyle, C.H.V., Bäumert, H.G., Mutschler, E., Burnstock, G. and Lambrecht, G. (1995) Inhibitory action of PPADS on relaxant responses to adenine nucleotides or electrical field stimulation in guinea-pig taenia coli and rat duodenum. *Br. J. Pharmacol.*, 115: 1509–1517.

Ziganshin, A.U., Ziganshina, L.E., King, B.F., Pintor, J. and Burnstock, G. (1996) Effects of P2-purinoceptor antagonists on degradation of adenine nucleotides by ecto-nucleotidases in folliculated oocytes of xenopus laevis. *Biochem. Pharmacol.*, 51: 897–901.

Zimmermann, H. (1994) Signalling via ATP in the nervous system. *Trends Neurosci.*, 17: 420–426.

Ziyal, R., Ziganshin, A.U., Nickel, P., Ardanuy, U., Mutschler, E., Lambrecht, G. and Burnstock, G. (1997) Vasoconstrictor responses via P2X-receptors are selectively antagonized by NF023 in rabbit isolated aorta and saphenous artery. *Br. J. Pharmacol.*, 120: 954–960.

P. Illes and H. Zimmermann (Eds.)
Progress in Brain Research, Vol 120
© 1999 Elsevier Science BV. All rights reserved

CHAPTER 10

Molecular recognition in P2 receptors: Ligand development aided by molecular modeling and mutagenesis

Kenneth A. Jacobson[1],*, Carsten Hoffmann[1], Yong-Chul Kim[1], Emidio Camaioni[1], Erathodiyil Nandanan[1], Soo Yeon Jang[1], Dan-Ping Guo[1], Xiao-duo Ji[1], Ivar von Kügelgen[1], Stefano Moro[1], Airat U. Ziganshin[2], Alexei Rychkov[2], Brian F. King[3], Sean G. Brown[3], Scott S. Wildman[3], Geoffrey Burnstock[3], Jose L. Boyer[4], Arvind Mohanram[4] and T. Kendall Harden[4]

[1]*Molecular Recognition Section, LBC, NIDDK, NIH, Bethesda, MD 20892 USA.*
[2]*Kazan Medical Institute, Kazan, Russia.*
[3]*Autononic Neuroscience Institute, Royal Free Hospital of Medicine, London, England*
[4]*University of North Carolina, Chapel Hill, NC, USA*

Introduction

The contemporary development of drugs that modulate G protein-coupled receptors (GPCRs) is based not only on organic synthesis, but also on structural studies of the protein targets. GPCRs contain seven membrane-spanning helical domains (TMs). The structural characterization of ligand interactions with specific regions of GPCRs presently requires mutagenesis (van Rhee and Jacobson, 1996) in conjunction with molecular modeling (Moro et al., 1998), since direct analysis has not yet been achieved. As molecular modeling of cloned GPCR sequences using a rhodopsin template has been refined, it has become possible to generate hypotheses for location of the binding sites that are consistent with mutagenesis results and ligand specificities. To obtain an energetically refined 3D structure of the ligand–receptor complex, we have introduced a new computational

approach, a *"cross docking"* procedure, which simulates the reorganization of the native receptor induced by the ligand (Moro et al., 1998).

Mutagenesis and model for binding of agonists at human P2Y₁ receptors

Extracellular nucleotides may act in cellular signaling through two families of membrane-bound P2 receptors: P2Y subtypes, G protein-coupled receptors (GPCRs) activated by both adenine and uracil nucleotides; and P2X subtypes, ligand-gated ion channels activated principally by adenine nucleotides (Fredholm et al, 1994). As many as seven subtypes have been cloned within each family. We have selected the human P2Y₁ receptor as a model system for development of ligands with the aid of molecular modeling and mutagenesis (Jiang et al., 1997; Moro et al., 1998). The P2Y₁ receptor is a membrane bound G protein-coupled receptor of the rhodopsin family and is stimulated by extracellular adenine nucleotides.

The P2Y₁ receptor subtype is a phospholipase C-activating receptor (Schachter et al, 1996) present

*Corresponding author. Tel.: + 301-496-9024; fax: + 301-480-8422; e-mail: kajacobs@helix.nih.gov

in heart, skeletal muscle, and various smooth muscles. At this receptor, the potency order for activation is 2-methylthioadenosine 5′-diphosphate (2-MeSADP) > 2-(hexylthio)adenosine 5′-mono-phosphate (HT-AMP) > ADP > ATP, while AMP and UTP are inactive (Boyer et al., 1996a).

Within the TM domains

In order to ascertain which residues of the human P2Y$_1$ receptor are involved in ligand recognition (Fig. 1), we have mutated (Jiang et al., 1997; Hoffmann et al., 1999) the transmembrane helical domains (TM 3, 5, 6, and 7) and the extracellular loops (EL). The mutant receptors expressed in COS-7 cells were measured for stimulation of phospholipase C (PLC) in the presence of 2-MeSADP, an agonist which activates the wild-type receptor with an EC$_{50}$ value of 2 nM, and other adenine nucleotides. A cluster of positively charged amino acids, Lys and Arg residues near the exofacial side of TMs 3 and 7 and to a lesser extent TM6, predicted by molecular modeling to coordinate the phosphate moieties of nucleotide ligands in the human P2Y$_1$ receptor, were replaced with alanine and, in some cases, by other amino acids (Jiang et al., 1997). Agonists had no activity at R128A (TM3) and R310A and S314A (TM7) mutant receptors and a markedly reduced potency at K280A (TM6) and Q307A (TM7) mutant receptors. Previously, positively charged residues of the human P2Y$_2$ receptor (H262, R265, and R292 in TM6 and TM7) were similarly found to be critical for activation (Erb et al., 1995). These results suggest that residues on the exofacial side of TM3 and TM7 are critical determinants of the ATP binding pocket. In contrast, there was no change in the potency or maximal effect of agonists with the S317A mutant receptor, and alanine replacement of F131, H132, Y136, F226, or H277 resulted in mutant receptors that exhibited a seven- to 18-fold reduction in potency compared to that observed with the wild-type receptor. These residues thus appear to subserve a less important modulatory role in ligand binding to the P2Y$_1$ receptor.

Since changes in the potency of 2-MeSADP and HT-AMP paralleled the changes in potency of 2-MeSATP at the various mutant receptors, the β-

and γ-phosphates of the adenine nucleotides appear to be less important than the α-phosphate in ligand/P2Y$_1$ receptor interactions (Jiang et al., 1997). However, T221A and T222A mutant receptors exhibited much larger reductions in triphosphate (89- and 33-fold vs. wild-type receptors, respectively), rather than di- or monophosphate potency. This result may be indicative of a greater role of these TM5 residues in γ-phosphate recognition. Taken together the results suggest that the adenosine and α-phosphate moieties of ATP bind to critical residues in TM3 and TM7 on the exofacial side of the receptor.

Within the extracellular domains

We have investigated the role in receptor activation of all charged amino acids (D, E, K and R) and cysteines in the extracellular loops (EL) 2 and 3 of the human P2Y$_1$ receptor by alanine scanning mutagenesis (Hoffmann et al., 1999, Fig. 1).

Surface detection of most of the mutant receptors by ELISA showed at least 10% expression at the surface of the plasma membrane compared to the wild-type receptor. Control experiments in which COS-7 cells were transfected with lower amounts of P2Y$_1$ wild-type DNA showed full stimulation and no shift for EC$_{50}$ values but surface expression rates dropped to approximately 10%. Therefore, assuming ≥ 10% receptor expression, shifts in the potency of 2-MeSADP and other agonists upon single amino acid replacement directly reflect the structural perturbations of the receptor, rather than insufficient receptor protein reaching the surface. In a few cases (see below), the receptor protein was undetectable at the cell surface. In those cases, the mutation was assumed to interfere with trafficking, and the effect on ligand recognition/activation could not be determined.

Two essential disulfide bridges in the extra-cellular domains have been identified, and several charged residues in the EL 2 (E209) and 3 (R287) have been found to be critical for receptor activation (Hoffmann et al., 1999). The C124A and C202A mutation, located in the upper part of transmembrane helix (TM) 3 and EL 2, prevented PLC stimulation by up to 100 μM 2-MeSADP. These data indicate a disulfide bridge in the P2Y$_1$

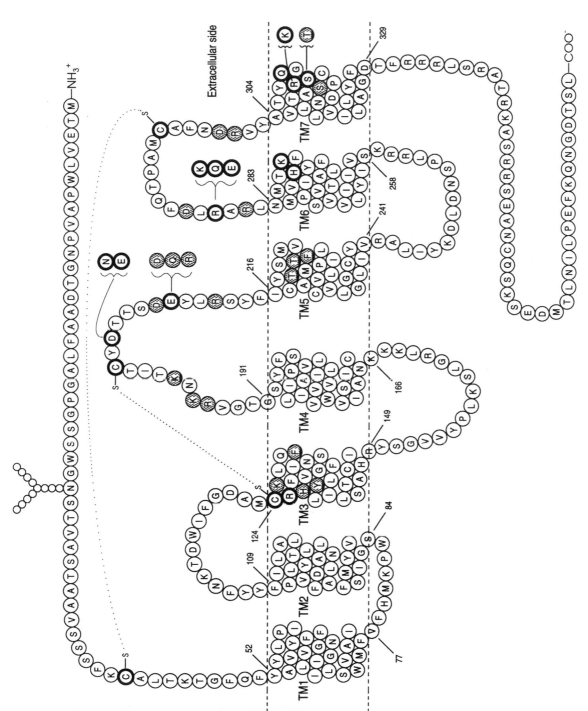

Fig. 1. Human P2Y₁ receptor topology showing transmembrane helical (TM) regions, extracellular loops (EL), and intracellular loops. Single amino acid replacements were made in TM3, 5, 6, and 7, in EL2 and EL3, and in the N-terminal segment. Residues which when mutated to either Ala or the indicated amino acid (brackets), decreased the potency of 2-MeSADP by factors of one-to-five-fold (shaded), five-to-20-fold (shaded and underlined), or >20-fold (heavy circles).

receptor between loop 2 and the upper part of TM 3, as found in many other G protein-coupled receptors. This disulfide bridge seems to be critical for the proper receptor trafficking to the cell surface. Presently it is unknown whether it is also critical for receptor activation. In contrast, the C42A and C296A mutant receptors (located in the N-terminal domain and EL 3) were activated by 2-MeSADP, but the EC_{50} values for stimulation were over 1000-fold greater than for the wild-type receptor. The double mutant receptor C42A/C296A exhibited no additive shift in the dose–response curve for 2-MeSADP. These data indicate that C42 and C296 form another disulfide bridge in the extracellular region which is critical for activation processes.

Upon replacement of charged amino acids in EL 2 and 3, we found only minor deviations from the agonist affinity at wild type receptors, with two remarkable exceptions (Hoffmann et al., 1999). E209 in the EL 2 exhibits a > 1000-fold shift in EC_{50} value when substituted with alanine, while it responds like wild-type if substituted with amino acids capable of H-bonding (D, Q or R). Thus, E209 appears to be required for its ability to form H-bonding alone.

R287 in EL 3 was impaired similarly to E209 when substituted by alanine, i.e. dose–response curves where shifted by > 1000-fold, and the curve shape was identical. Substitution of R287 by lysine, another positively charged residue, only partially restored the potency of 2-MeSADP as a $P2Y_1$ receptor agonist, with an EC_{50} value of 75 nM. The possibility of a required ionic interaction between essential residues R287 and E209, which may be in physical proximity according to preliminary molecular modeling of the loop regions, was considered. This possibility was ruled out since the double mutant E209A/R287A receptor demonstrates a shift that is additive relative to the single mutations. Refined molecular modeling indicates a direct interaction of the negatively charged phosphate chain of the ligand with R287. Moreover, the selective reduction in potency of $3'NH_2$–ATP in activating the E209R mutant receptor is consistent with the hypothesis of direct contact between EL2 and nucleotide ligands. Our findings support ATP binding to at least two distinct domains of the $P2Y_1$

receptor, both outside the TM region ("meta-binding" sites) and within the TM core (Moro et al., 1999).

Novel P2X receptor antagonists

We have developed selective agonists and antagonists for both P2Y and P2X receptors (Table 1). The P2X ligands were developed through an empirical rather than computational approach. Molecular modeling of inotropic P2X receptors is limited by the lack of a template protein, as well as specific knowledge of loop conformation (van Rhee et al., 1997).

Structural modifications of the non-selective P2 antagonist pyridoxal-5'-phosphate-6-azophenyl-2',4'-disulfonate (PPADS, Lambrecht et al., 1992, Fig. 2), have been made through functional group substitution on the sulfophenyl ring and at the phosphate moiety through the inclusion of phosphonates, demonstrating that a phosphate linkage is not required for P2 receptor antagonism (Kim et al., 1998). Phosphonates are not hydrolyzable and thus may be more stable than phosphate analogues in pharmacological studies. Substituted 6-azophenyl and 6-azonaphthyl derivatives were evaluated (Fig. 2). Among the 6-azophenyl derivatives, 5'-methyl, ethyl, propyl, vinyl, and allyl phosphonates were included. The compounds have been tested as antagonists at P2Y receptors in the turkey erythrocyte (Harden et al, 1988) and in the guinea-pig taenia coli and at $P2X_1$ receptors in guinea-pig vas deferens and bladder and at recombinant rat $P2X_2$ receptors expressed in *Xenopus* oocytes. Competitive binding assay at human $P2X_1$ receptors in differentiated HL-60 cell membranes was carried out using [^{35}S]ATP-γ-S. A 2'-chloro analogue of isoPPADS (MRS 2157, Fig. 2), a vinyl phosphonate derivative, MRS 2206, and an azonaphthyl derivative, MRS 2166, were particularly potent in binding at human $P2X_1$ receptors. Potencies of phosphate derivatives at turkey erythrocyte $P2Y_1$ receptors were generally similar to PPADS itself (IC_{50} 18.2 ± 1.7 μM; Fig. 3), except for the *p*-carboxyphenylazo phosphate derivative, MRS 2159, and its *m*-chloro analogue, MRS 2160, which were selective for P2X vs. $P2Y_1$ ($IC_{50} > 100$ μM) receptors. At 30 μM, both MRS 2159 and 2160 significantly antagonized P2X receptor-induced contraction of guinea pig urinary bladder and vas deferens. MRS 2160 was very potent at rat $P2X_2$ receptors expressed in *Xenopus* oocytes with an IC_{50} value of 0.82 ± 0.28 μM, while MRS 2159 was less potent with an IC_{50} value of 11.9 ± 1.4 μM.

Among the phosphonate derivatives (Fig. 2), [4-formyl-3-hydroxy-2-methyl-6-azo-(2′-chloro-5′-sulfonylphenyl)-5-pyridyl]-methylphosphonic acid (MRS 2192) showed high potency at $P2Y_1$ receptors with an IC_{50} of 4.35 ± 0.36 μM (Fig. 3). The corresponding 2′,5′-disulfonylphenyl derivative, MRS 2191, was nearly inactive at turkey erythrocyte $P2Y_1$ receptors. Thus a single ring substitution, sulfo instead of chloro, has a major effect on the selectivity of these methylphosphonates as P2Y receptor antagonists. MRS 2191 was relatively potent at recombinant $P2X_2$ receptors with an IC_{50} value of 1.1 ± 0.2 μM. Also, an ethyl phosphonate derivative, MRS 2142, while inactive at turkey $P2Y_1$ receptors, was particularly potent at recombinant $P2X_2$ receptors (IC_{50} 1.5 ± 0.1 μM).

A pyridoxine cyclic phosphate (cyclic pyridoxine-$\alpha^{4,5}$-monophosphate, MRS 2219; Fig. 2) and its 6-azoaryl derivative (cyclic pyridoxine-$\alpha^{4,5}$-monophosphate-6-azophenyl-2′,5′-disulfonic acid, MRS 2220) selectively potentiate and antagonize, respectively, $P2X_1$ receptors expressed in *Xenopus* oocytes (Jacobson et al., 1998). These derivatives are novel analogues of the P2 receptor antagonists pyridoxal-5′-phosphate and the 6-azophenyl-2′,4′-disulfonate derivative (Lambrecht et al., 1992; PPADS), in which the phosphate group was cyclized by esterification to a CH_2OH group at the 4-position. The cyclic pyridoxine-$\alpha^{4,5}$-monophosphate, MRS 2219, was found to be a selective potentiator of ATP-evoked responses at rat $P2X_1$ receptors with an EC_{50} value of 5.9 ± 1.8 μM, while the corresponding 6-azophenyl-2′,5′-disulfonate derivative, MRS 2220, was a selective antagonist (Fig. 4). The potency of compound MRS 2220 at the recombinant $P2X_1$ receptor (IC_{50} 10.2 ± 2.6 μM) was lower than PPADS (IC_{50} 98.5 ± 5.5 nM) or iso-PPADS (IC_{50} $42.5 \pm$

TABLE 1

Estimates for antagonist potencies (IC_{50} values, nM) in functional assays at P2 receptors.[a]

Compound	$P2X_1$	$P2X_2$	$P2X_3$	$P2X_4$	$P2Y_1$	$P2Y_2$	$P2Y_4$	$P2Y_6$	$P2Y_{AC}$
PPADS	98.5 ± 5.5[e]	1200 ± 200[e]	239.5 ± 38.0,[e] 1700 ± 200 (h)	b	$18,200 \pm 1700$[e]. ~ 4000 (h)	c	c	69%[f] (h)	b
isoPPADS	43 ± 18	398 ± 129	84 ± 4	ND	28.1 ± 8.0	ND	ND	ND	ND
pyridoxal-5-phosphate	~ 3000	$39,500 \pm 19,000$	ND	$219,000 \pm 2400$	$\sim 50,000$ (h)	b	ND	ND	ND
suramin	4900 ± 1000	$10,400 \pm 2000$[e]	~ 3000, $14,900 \pm 1900$ (h)	b	d	$48,000 \pm 17,000$	b	27%[f] (h)	4000 ± 2300
reactive blue 2	ND	360 ± 80	ND	$128,000 \pm 11,800$	d	b	33%[f] (h)	87%[f] (h)	25 ± 7
MRS 2179	1150 ± 200	c	$\sim 10,000$	c	330 ± 59	b	b	b	b
MRS 2220	$10,200 \pm 2600$	b	$58,300 \pm 100$	b	b	b	b	b	b

[a] Effects of antagonists on inward current induced by activation by ATP, at the indicated concentrations, of recombinant rat $P2X_1$ (3 μM), $P2X_2$ (10 μM), $P2X_3$ (1 μM), and $P2X_4$ (30 μM) receptors, expressed in *Xenopus* oocytes, using the twin electrode voltage clamping technique; or at phospholipase C-coupled $P2Y_1$ receptor of turkey erythrocytes, at recombinant human $P2Y_2$ and $P2Y_4$ receptors, and at recombinant rat $P2Y_6$ receptors, unless noted. (h) indicates human clone.
[b] inactive as antagonist at 100 μM.
[c] inactive as antagonist at 10 μM.
[d] right shift at 30 μM and decrease in maximal effect (Boyer et al., 1994).
[e] Similar experiments gave $IC_{50} \sim 1$ μM (Charlton et al., 1996; North and Barnard, 1997).
[f] Percent inhibition at 100 μM (Communi et al., 1996; Robaye et al, 1997)
ND not determined.

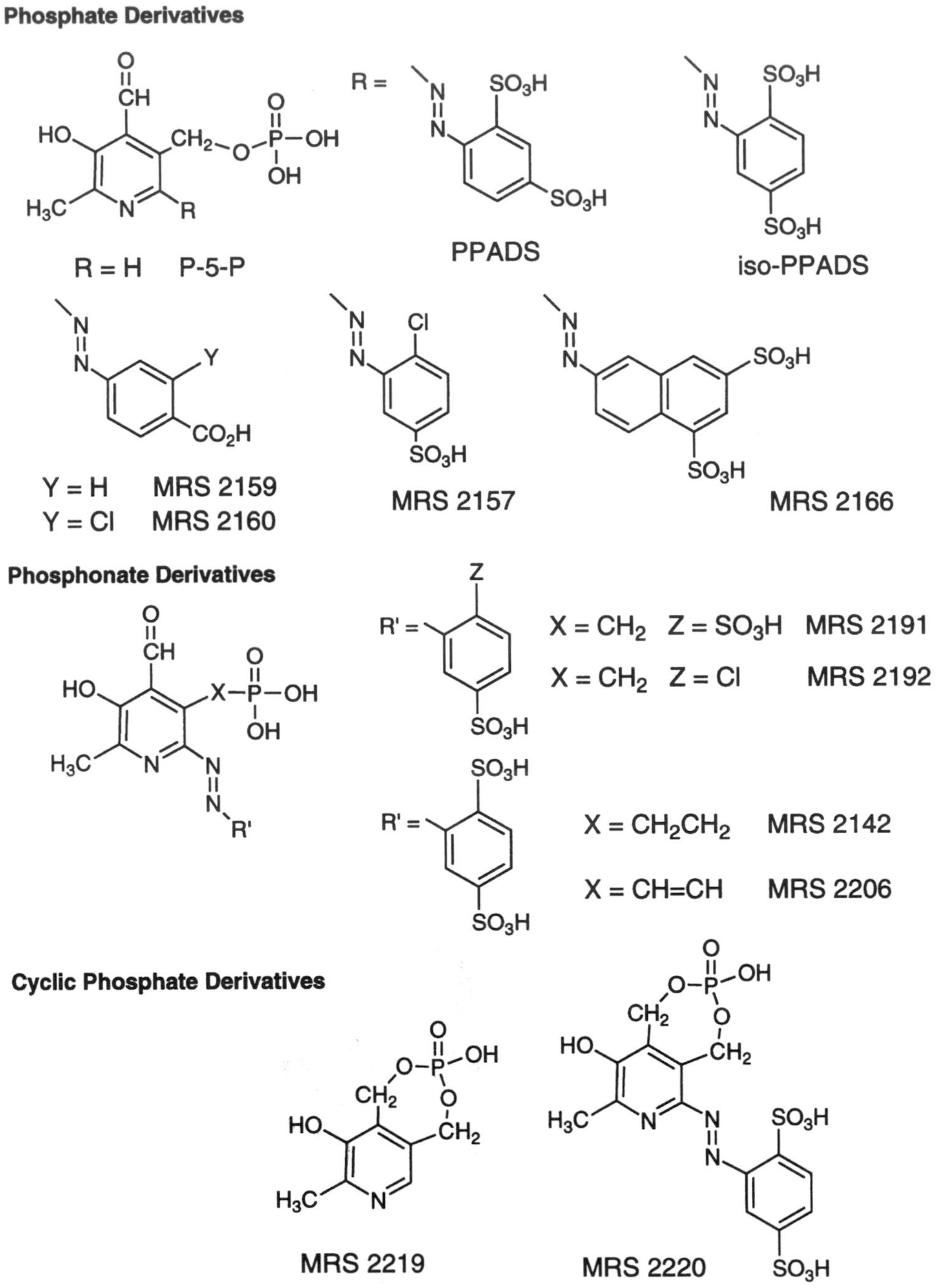

Phosphate Derivatives

R = H   P-5-P

PPADS

iso-PPADS

Y = H   MRS 2159
Y = Cl   MRS 2160

MRS 2157

MRS 2166

Phosphonate Derivatives

X = CH$_2$ Z = SO$_3$H   MRS 2191

X = CH$_2$ Z = Cl   MRS 2192

X = CH$_2$CH$_2$ MRS 2142

X = CH=CH   MRS 2206

Cyclic Phosphate Derivatives

MRS 2219

MRS 2220

Fig. 2. Structures of pyridoxal-5'-phosphate (P-5-P) and other related P2 receptor antagonists, including those recently synthesized in the molecular recognition section (MRS).

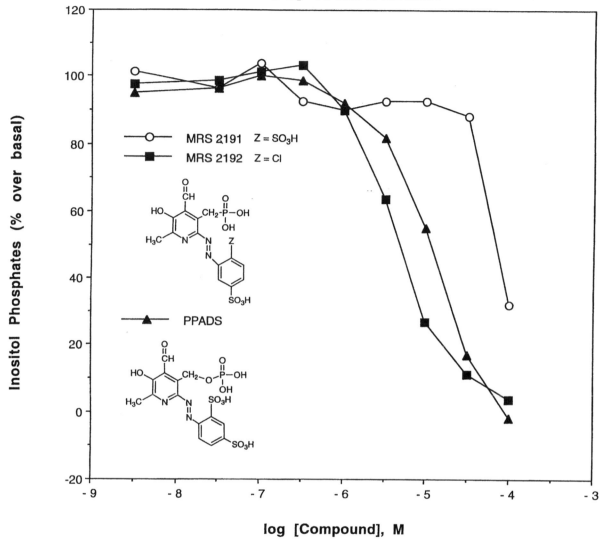

Fig. 3. Stimulation of phospholipase C in turkey erythrocyte membranes (Harden et al., 1988). Membranes were incubated for 30 min at 37°C. Data are presented as percent of maximum accumulation of tritiated inositol phosphates above basal levels in the presence of the antagonist and 10 nM 2-MeSATP, average of two experiments. EC_{50} values were 4.35 μM for MRS 2192 and 18.2 μM for PPADS.

17.5 nM), although unlike PPADS its effect was reversible with washout and surmountable. Compound MRS 2220 also showed weak antagonistic activity at the rat $P2X_3$ receptor (IC_{50} 58.3 ± 0.1 μM), while at recombinant rat $P2X_2$ and $P2X_4$ receptors no enhancing or antagonistic properties were evident. MRS 2219 and MRS 2220

were found to be inactive as either agonists or antagonists at the phospholipase C-coupled $P2Y_1$ receptor of turkey erythrocytes, at recombinant human $P2Y_2$ and $P2Y_4$ receptors, and at recombinant rat P2Y6 receptors. Similarly, neither MRS 2219 nor MRS 2220 at a concentration of 100 μM had measurable affinity at adenosine rat A_1, rat A_{2A},

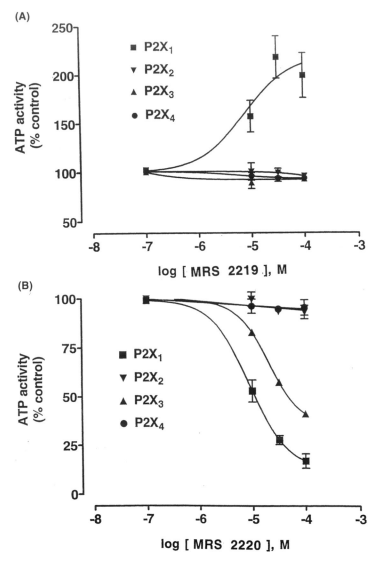

Fig. 4. Effects of MRS 2219 (A) and MRS 2220 (B) on inward current induced by ATP, at the indicated concentrations, of recombinant rat $P2X_1$ (3 μM), $P2X_2$ (10 μM), $P2X_3$ (1 μM), and $P2X_4$ (30 μM) receptors, expressed in *Xenopus* oocytes, using the twin electrode voltage clamping technique. The agonist concentrations correspond approximately to the EC_{70} values. IC_{50} values for MRS 2220 ($n = 4$) at $P2X_1$ and $P2X_3$ receptors were 10.2 ± 2.6 μM and 58.3 ± 0.1 μM, respectively.

or human A3 receptors. The lack of an aldehyde group in these derivatives indicates that Schif's base formation with the $P2X_1$ receptor is not necessarily required for recognition of pyridoxal phosphate derivatives. Thus, MRS 2219 and MRS 2220 are relatively selective pharmacological probes of $P2X_1$ receptors, filling a long-standing need in the P2 receptor field, and may also be important lead compounds for future studies.

Novel P2Y receptor antagonists

Adenosine $3',5'$- and $2',5'$-*bis*phosphates were previously demonstrated to act as competitive antagonists at the $P2Y_1$ receptor (Boyer et al., 1996b). $2'$- and $3'$-Deoxyadenosine *bis*phosphate analogues containing various structural modifications at the 2- and 6-position of the adenine ring, on the ribose moiety, and on the phosphate groups

TABLE 2

Effects of 3′,5′-bisphosphate derivatives on stimulation of PLC in turkey erythrocytes and inhibition of effects of 10 nM 2-MeSATP

Compound	R_1	R_2	R_3	R_4	Agonist Effect (%maximum)	EC_{50} (μM)	Antagonist Effect (%maximum)	IC_{50} (μM)
1	H	NH_2	H	OH	21%	1.28	77%	4.19
2	H	NH_2	H	OCH_3	35%	12.9	65%	12.4
3	H	NH_2	H	H	12%	6.26	87%	5.76
4	Cl	NH_2	H	H	19%	0.651	80%	2.01
5	SCH_3	NH_2	H	H	22%	0.550	77%	2.11
6	H	NH_2	Br	H	0%		100%	36.7
7 (MRS 2179)	H	$NHCH_3$	H	H	0%		99%	0.330
8	H	$NHCH_2CH_3$	H	H	0%		100%	1.08
9	H	$NH(CH_2)_2CH_3$	H	H	0%		<20%	
10	H	$NHCOC_6H_6$	H	H	0%		0%	
11	H	$N(CH_3)_2$	H	H	0%		70%	46.7
12	H	Cl	H	H	<10%		<10%	
13	H	OH	H	H	0%		<20%	
14	H	SCH_3	H	H	0%		78%	29.1

have been synthesized with the goal of developing more potent and selective $P2Y_1$ antagonists (Camaioni et al., 1998). Single-step phosphorylation reactions of adenosine nucleoside precursors were carried out. The activity of each analogue at $P2Y_1$ receptors (Table 2) was determined by measuring its capacity to stimulate phospholipase C in turkey erythrocyte membranes (agonist effect) and to inhibit phospholipase C stimulation elicited by 10 nM 2-MeSATP (antagonist effect). Both 2′- and 3′-deoxy modifications were well tolerated. The carbocyclic analogue of compound 3 proved to be a partial agonist of moderate potency at $P2Y_1$ receptors (Nanadanan et al., 1999). The N^6-methyl modification (2′-deoxy-N^6-methyladenosine-3′,5′-bisphosphate, MRS 2179) both enhanced antagonistic potency of 2′-deoxyadenosine 3′,5′-bisphosphate by 17-fold and eliminated resid-

ual agonist properties observed with the lead compounds (Boyer et al., 1998). In a Schild analysis MRS 2179 was found to be a competitive antagonist with a K_B value 104 nM at $P2Y_1$ receptors in turkey erythrocyte membranes (Fig. 5, Boyer et al., 1998). MRS 2179 did not partially activate either the turkey or human $P2Y_1$ receptor, thus it is a pure antagonist. MRS 2179 was inactive at the P2Y receptor in C6 glioma cells which is coupled to inhibition of adenylate cyclase (J. Boyer et al., unpublished). The N^6-ethyl modification provided intermediate potency as an antagonist, while the N^6-propyl group completely abolished both agonist and antagonist properties (Camaioni et al., 1998). 2-Methylthio and 2-chloro analogues were partial agonists of intermediate potency. A 2′-methoxy group provided intermediate potency as an antagonist while enhancing agonist activity.

An N^1-methyl analogue was a weak antagonist with no agonist activity. An 8-bromo substitution and replacement of the N^6-amino group with methylthio, chloro, or hydroxy groups, greatly reduced the ability to interact with P2Y$_1$ receptors. Benzoylation or dimethylation of the N^6-amino group also abolished the antagonist activity. In summary, our results further define the structure activity of adenosine *bis*phosphates as P2Y$_1$ receptor antagonists and have led to the identification of MRS 2179 as the most potent antagonist reported to date for this receptor. Thus, MRS 2179 was a potent, competitive antagonist, selective for the P2Y$_1$ receptor vs. four other P2Y subtypes. However MRS 2179 has not yet been evaluated at the recently cloned P2Y$_{11}$ receptor (Communi et al., 1997).

Although MRS 2179 is selective for P2Y$_1$ from among five different metabotropic P2 receptors,

caution is advised when using this agent when inotropic P2 receptors are present, since it was found to antagonize one subtype, i.e. the rat P2X$_1$ receptor, expressed in *Xenopus* oocytes (Fig. 6, King et al., unpublished). Ion current induced by 3 μM ATP acting at P2X$_1$ receptors was blocked by MRS 2179 with an IC$_{50}$ value of 1.2 ± 0.2 μM. At the rat P2X$_3$ receptor, MRS 2179 is a much weaker antagonist with an IC$_{50}$ value of ~ 10 μM. The compound at a concentration of 10 μM was inactive at the rat P2X$_2$ receptor, while at the rat P2X$_4$ receptor, a potentiation of ion current by 25% was observed.

Binding model for antagonists at human P2Y$_1$ receptors

An antagonist P2Y$_1$ receptor binding model has been developed (Moro et al., 1998). The structural

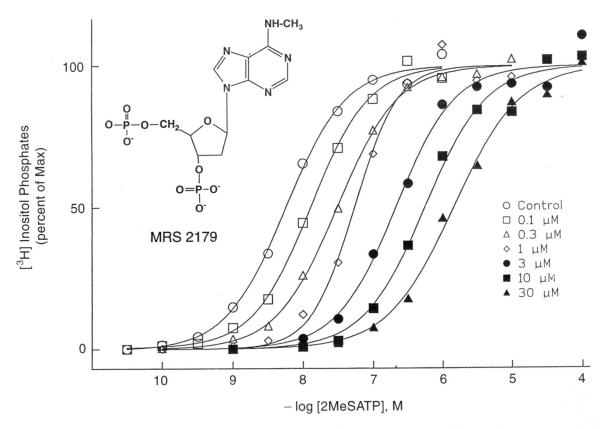

Fig. 5. Competitive inhibition by MRS 2179 of phospholipase C activation (by 10 nM 2-MeSATP) in turkey erythrocyte membranes (Boyer et al., 1998).

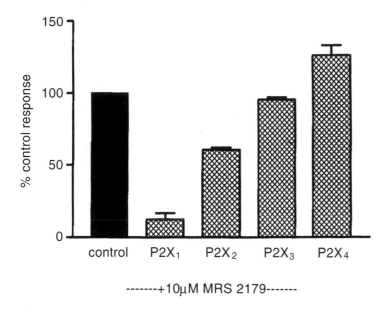

-------+10μM MRS 2179-------

Fig. 6. Effect of 10 μM MRS 2179 on inward current induced by activation by ATP, at the indicated concentrations, of recombinant rat $P2X_1$ (3 μM), $P2X_2$ (10 μM), $P2X_3$ (1 μM), and $P2X_4$ (30 μM) receptors, expressed in *Xenopus* oocytes, using the twin electrode voltage clamping technique ($n=4$). The agonist concentrations correspond approximately to the EC_{70} values. IC_{50} values for MRS 2179 at $P2X_1$ and $P2X_3$ receptors were 1.2 ± 0.2 μM and ~10 μM, respectively. Potentiation of 25% was observed at $P2X_4$ receptors.

similarity between the potent antagonist MRS 2179 and nucleotide agonists, which bind to a single putative binding region within the TM domains, suggests that receptor activation resulting in a specific conformational change may depend on subtle differences between ligands.

The molecular basis for recognition by human $P2Y_1$ receptors of the selective, competitive antagonist MRS 2179 was probed using site-directed mutagenesis and molecular modeling. The potency of this antagonist was measured in mutant receptors in which key residues in TMs 3, 5, 6, and 7 were replaced by Ala or other amino acids. The capacity of MRS 2179 to block stimulation of phospholipase C promoted by 2-MeSADP was lost in $P2Y_1$ receptors having F226A, K280A, or Q307A mutations, indicating that these residues are critical for the binding of the antagonist molecule. Mutation of the residues His132, Thr222, and Tyr136 had an intermediate effect on the capacity of MRS 2179 to block the $P2Y_1$ receptor. These positions therefore appear to have a modulatory role in recognition of this antagonist. F131A, H277A, T221A, R310K, or S317A mutant receptors exhibited an apparent

affinity for MRS 2179 that was similar to that observed with the wild-type receptor. Thus, Phe131, Thr221, His277, and Ser317 are not essential for recognition of the nucleotide antagonist. A computer-generated model of the human $P2Y_1$ receptor was built and analyzed to help interpret these results. The model was derived from using primary sequence comparison, secondary structure predictions, and three-dimensional homology building, using rhodopsin as a template, and was consistent with data obtained from mutagenesis studies. A putative nucleotide binding site was localized, following a *cross docking* procedure to obtain energetically refined 3D structures of the ligand-receptor complexes, and used to predict which residues are likely to be in proximity to agonists and antagonists. According to our computational model TM6 and TM7 are close to the adenine ring, TM3 and TM6 are close to the ribose moiety, and TM3, TM6, and TM7 are near the triphosphate chain.

Initial results indicated that both suramin and PPADS retained antagonist properties at the $P2Y_1$ mutant receptors examined (Jiang et al, 1997).

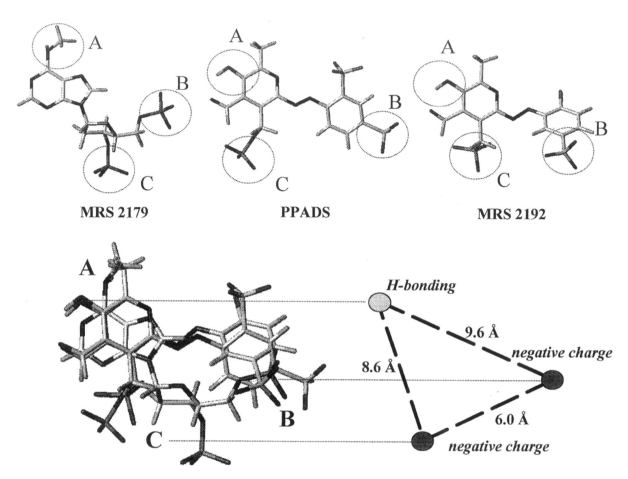

Fig. 7. Lower left: possible superposition of MRS 2179, PPADS, and MRS 2192 using steric and electrostatic alignment methodology. Lower right: scheme of the hypothetical pharmacophore map extrapolated using the superposition of MRS 2179, PPADS, and MRS 2192 structures. At top: Structures of MRS 2179, PPADS, and MRS 2192 showing positions corresponding to the general pharmacophore.

However, recently we have determined with full dose response curves that K280A and Q307A mutations greatly diminish the potency of the pyridoxal phosphate-related antagonists, i.e. 20 μM PPADS is ineffective (Guo et al., 1998). Thus, as for adenosine receptor antagonists, there appears to be a spatial overlap between the binding regions of the receptor for agonist and antagonist ligands, even those of highly divergent structure. Molecular modeling (Moro et al., 1998) using PowerFit (Molecular Simulations, Mahwah NJ) has suggested a possible model of superimposition of two classes of antagonists, nucleotides related to MRS 2179 and non-nucleotides related to pyridoxal phosphate (Fig. 7). In these

energetically minimized conformations, it is possible to demonstrate overlap of negatively charged moieties, two monophosphate groups for MRS 2179 and 5′-phosphate and phenyl-4-sulfonate groups for PPADS. Thus, a preliminary pharmacophore model features a hydrogen bond donor (e.g. the 6-NH_2 of MRS 2179 and 3-OH of PPADS), which is proposed to bind in vicinity of Glu307 of the $P2Y_1$ receptor (Moro et al., 1998), and two anionic groups. The sulfonate group of PPADS (corresponding to 5′-phosphate of MRS 2179) would interact directly with Lys280. By analogy (Moro et al., 1998), this also allows prediction of overlap between PPADS and ATP-related agonists. In conclusion, pyridoxal phosphate antagonists appear to bind in the TM region of the human P2Y$_1$ receptor, since the capacity of PPADS to block stimulation of phospholipase C was lost in receptors having K280A or Q307A mutations (TM6 and 7).

Conclusions

The cloning of at least 13 subtypes of P2 receptors has presented a unique challenge to medicinal chemists: the design of selective agonists and antagonists for this multiplicity of receptors with few existing leads. The human P2Y$_1$ receptor as representative of the P2Y family of metabotropic purine and pyrimidine nucleotide receptors may be modeled based on a rhodopsin template, and the resulting model is highly consistent with pharmacological and mutagenesis results. Charged residues in both the transmembrane and extracellular domains and two disulfide bridges essential for receptor activation have been identified. Selective P2Y$_1$ receptor antagonists such as MRS 2179 are under development. Modeling of P2X receptors has not been achieved, since no template for the extracellular nucleotide binding region exists. Nevertheless, a selective antagonist, MRS 2220, and a potentiator, MRS 2219, of this subtype have been identified. Both are based on pyridoxal-5'-phosphate antagonists (such as PPADS), for which the SAR is being examined at all of the P2 receptor subtypes.

References

Boyer, J.L., Zohn, I.E., Jacobson, K.A. and Harden, T.K. (1994) Differential effects of P$_2$-purinergic receptor antagonists on phospholipase C- and adenylyl cyclase-coupled P$_{2Y}$ purinergic receptors. Br. J. Pharmacol., 113: 614–620.

Boyer, J.L., Siddiqi, S., Fischer, B., Romera-Avila, T., Jacobson, K.A. and Harden, T.K. (1996a) Identification of potent P2Y purinoceptor agonists that are derivatives of adenosine 5'-monophosphate. Br. J. Pharmacol., 118: 1959–1964.

Boyer, J.L., Romero-Avila, T., Schachter, J.B. and Harden, T.K. (1996b) Identification of competitive antagonists of the P2Y$_1$ receptor. Mol. Pharmacol., 50: 1323–1329.

Boyer, J.L., Mohanram, A., Camaioni, E., Jacobson, K.A. and Harden, T.K. (1998) Competitive and selective antagonism of P2Y$_1$ receptors by N^6-methyl 2'-deoxyadenosine 3',5'-bisphosphate. Br. J. Pharmacol., 124: 1–3.

Camaioni, E., Boyer, J.L., Mohanram, A., Harden, T.K. and Jacobson, K.A. (1998) Deoxyadenosine-bisphosphate derivatives as potent antagonists at P2Y$_1$ receptors. J. Med. Chem., 41: 183–190.

Charlton, S.J., Brown, C.A., Weisman, G.A., Turner, J.T., Erb, L. and Boarder, M.R. (1996) PPADS and suramin as antagonists at cloned P2Y- and P2U-purinoceptors. Br. J. Pharmacol., 118: 704–710.

Communi, D., Motte, S., Boeynaems, J.-M. and Pirroton, S. (1996) Pharmacological characterization of the human P2Y$_4$ receptor. Eur. J.. Pharmacol., 317: 383–389.

Communi, D., Govaerts, C., Parmentier, M. and Boeynaems, J.-M. (1997) Cloning of a Human P2 Receptor Coupled to Phospholipase C and Adenylyl Cyclase. J. Biol. Chem., 272: 31969–31973.

Erb, L., Garrad, R., Wang, Y.J., Quinn, T., Turner, J.T. and Weisman, G.A. (1995) Site-directed mutagenesis of P2U purinoceptors – positively charged amino-acids in transmembrane helix-6 and helix-7 affect agonist potency and specificity. J. Biol. Chem., 270: 4185–4188.

Fredholm, B.B., Abbracchio, M.P., Burnstock, G., Daly, J.W., Harden, T.K., Jacobson, K.A., Leff, P. and Williams, M. (1994) Nomenclature and classification of purinoceptors: A report from the IUPHAR subcommittee. Pharmacol. Rev., 46: 143–156.

Guo, D., Moro, S., Von Kügelgen, I., Kim, Y.-C. and Jacobson, K.A. (1999) Recognition of pyridoxal phosphate-related antagonists occurs within transmembrane domains of the human P2Y$_1$ receptor. Faseb J., 12: A465, Abstract 390.9.

Harden, T.K., Hawkins, P.T., Stephens, L., Boyer, J.L. and Downes, P. (1988) Phosphoinositide hydrolysis by guanosine 5'-(γ-thio)triphosphate-activated phospholipase C of turkey erythrocyte membranes. Biochem. J., 252: 583–593.

Hoffmann, C., Moro, S., Nicholas, R.A., Harden, T.K. and Jacobson, K.A. (1999) The role of amino acids of the extracellular loops of the human P2Y$_1$ receptor in surface expression and activation processes. J. Biol. Chem., in press.

Jacobson, K.A., Kim, Y.-C., Wildman, S.S., Mohanram, A., Harden, T.K., Boyer, J.L., King, B.F. and Burnstock, G. (1998) A pyridoxine cyclic-phosphate and its 6-arylazo-derivative selectively potentiate and antagonize activation of P2X$_1$ receptors. J. Med. Chem., 41: 2201–2206.

Jiang, Q., Guo, D., Lee, B. X., van Rhee, A. M., Kim, Y. C., Nicholas, R., Schachter, J., Harden, T. K. and Jacobson, K. A. (1997) Mutational analysis of residues essential for ligand recognition at the human P2Y$_1$ receptor. Mol. Pharmacol., 52: 499–507.

Kim, Y.-C., Camaioni, E., Ziganshin, A.U., Ji, X.-J., King, B.F., Wildman, S.S., Rychkov, A., Yoburn, J., Kim, H., Mohanram, A., Harden, T.K., Boyer, J.L., Burnstock, G. and Jacobson, K.A. (1998) Synthesis and structure activity relationships of pyridoxal-6-azoaryl-5'-phosphate and phosphonate derivatives as P2 receptor antagonists. Drug Devel. Res., 45: 52–66.

Lambrecht, G., Friebe, T., Grimm, U., Windscheif, U., Bungardt, E., Hildebrandt, C., Baumert, H. G., Spatzkumbel,

G. and Mutschler, E. (1992) PPADS, a novel functionally selective antagonist of P2 purinoceptor-mediated responses. *Eur. J. Pharmacol.*, 217: 217–219.

Moro, S., Guo, D., Camaioni, E., Boyer, J.L., Harden, T.K. and Jacobson, K.A. (1998) Human P2Y$_1$ receptor: Molecular modeling and site-directed mutagenesis as tools to identify agonist and antagonist recognition sites. *J. Med. Chem.*, 41: 1456–1466.

Moro, S., Hoffmann, C. and Jacobson, K.A. Role of the extracellular loops of G protein-coupled receptors in ligand recognition: A molecular modeling study of the human P2Y$_1$ receptor. *Biochemistry*, 1999, 38: 3498–3507.

Nandanan, E., Camaioni, E., Jang, S.Y., Kim, Y.-C., Cristalli, G., Herdewijn, P., Secrist, J.A., Tiwari, K.N., Mohanram, A., Harden, T.K., Boyer, J.L. and Jacobson, K.A. (1999) Structure activity relationships of bisphosphate nucleotide derivatives as P2Y$_1$ receptor antagonists and partial agonists. *J. Med. Chem.*, in press.

North, R.A. and Barnard, E.A. (1997) Nucleotide receptors. *Current Opinion in Neurobiol.*, 7: 346–357.

Robaye, B., Boeynaems, J.-M. and Communi, D. (1997) Slow desensitization of the human P2Y$_6$ receptor. *Eur. J.. Pharmacol.*, 329: 231–236.

Schachter, J.B., Li, Q., Boyer, J.L., Nicholas, R.A. and Harden, T.K. (1996) Second messenger cascade specificity and pharmacological selectivity of the human P2Y$_1$-purinoceptor. *Br. J. Pharmacol.*, 118: 167–173.

van Rhee, A.M. and Jacobson, K.A. (1996) Molecular architecture of G protein-coupled receptors. *Drug Devel. Res.*, 37: 1–38.

van Rhee, A.M., Jacobson, K.A., Garrad, R., Weisman, G.A. and Erb, L. (1997) P2 receptor modeling and identification of ligand binding sites. In: J.T. Turner, G. Weisman and J. Fedan, (Eds.), *The P2 Nucleotide Receptors*, in the series "*The Receptors*", Humana Press, Clifton, New Jersey, pp. 135–166.

Release of nucleotides

P. Illes and H. Zimmermann (Eds.)
Progress in Brain Research, Vol 120

CHAPTER 11

Release of ATP and UTP from astrocytoma cells

T. Kendall Harden* and Eduardo R. Lazarowski

Department of Pharmacology, University of North Carolina School of Medicine, Chapel Hill, NC 27599, USA

Introduction

Studies in our laboratory in the 1980s focussed on identification of cell lines that natively express G protein-coupled P2Y receptors. Surprisingly, most cell lines that were examined exhibited second messenger signaling responses to extracellular nucleotides. Indeed, 1321N1 human astrocytoma cells, which we had widely studied as a model for both β-adrenergic (Doss et al., 1981; Waldo et al., 1983) and muscarinic cholinergic receptors (Hughes et al., 1984; Evans et al., 1985; Harden et al., 1985), was one of the few cell lines for which no evidence of P2Y receptor expression could be obtained. Based on the large data base we had accumulated on the G protein-mediated regulation of adenylyl cyclase and phospholipase C in these cells, 1321N1 cells proved ideal for stable expression of cloned P2Y receptors. As such, these cells have become the most widely used cell type for heterologous expression of the P2Y receptors.

Rigorous characterization of the pharmacological selectivity of the five cloned mammalian P2Y receptors (i.e. the $P2Y_1$, $P2Y_2$, $P2Y_4$, $P2Y_6$, and $P2Y_{11}$ receptors) has required that we also address the metabolism and interconversion of nucleotides by the ectoenzymes expressed by 1321N1 cells. For example, we have shown that ectonucleoside diphosphokinase plays an important role in these and other cells in converting nucleotide diphosphates to their corresponding nucleoside

---

*Corresponding author. Tel.: +1 919-966-4816; fax: +919-966-5640; e-mail: tkh@med.unc.edu

triphosphate (Lazarowski et al., 1997b). Our studies of P2Y receptors expressed in 1321N1 cells also have led to the discovery that mild mechanical stimulation of these cells causes non-lytic release of both ATP (Lazarowski et al., 1995) and UTP (Lazarowski et al., 1997a). As such, these cells provide an excellent model for the study of the autocrine/paracrine roles that nucleotides may play. They may also prove invaluable in determining the mechanism whereby release of nucleotides occur from nonexcitable cells. We summarize here some of the results describing nucleotide release from 1321N1 cells.

Results

Our early studies on $P2Y_1$ (Filtz et al., 1994) and $P2Y_2$ (Parr et al., 1994; Lazarowski et al., 1995) receptors expressed in 1321N1 cells suggested the occurrence of release of cellular ATP. That is, stable expression of either of these receptors resulted in basal levels of [^{3}H]inositol phosphates that were much greater than the levels observed with empty vector-transfected 1321N1 cells. Inclusion of the nucleotide hydrolyzing enzyme, apyrase, in the medium of P2Y receptor-expressing cells reduced basal inositol phosphate levels, further suggesting that the endogenous receptor-activating molecules were nucleotides. This possibility was directly examined by several approaches.

Results obtained with P2Y receptor-expressing 1321N1 cells (which were identical to those observed with control cells) are described.

Endogenous ATP pools were labeled by pre-incubation of cells with [³H]adenine. After 3 h, the [³H]adenine-containing medium was replaced with fresh medium and [³H]-labeled species appearing in the extracellular medium were determined by high performance liquid chromatography (HPLC) at various times subsequent to the medium change (Fig. 1). [³H]ATP accumulated rapidly and transiently with maximal levels obtained during the first minute. Longer incubations resulted in a gradual decrease of [³H]ATP and transient formation of [³H]ADP. There was a negligible accumulation of [³H]AMP and little accumulation of [³H]adenosine, but a sustained accumulation of their metabolic products, [³H]inosine, [³H]xanthine, and [³H]hypoxanthine, was observed. Extracellular [³H]ATP was barely detectable after 2 h, and intracellular [³H]ATP levels were essentially unchanged over the 2 h incubation.

The release of ATP from 1321N1 cells was also directly quantified by the luciferin-luciferase assay.

Up to 300 pmol ATP/10^6 cells was released upon replacing the incubation medium with fresh medium (Fig. 2), resulting in a concentration of ATP of approximately 100 nM for confluent cells in a well of a 24-well culture dish (18 mm) containing 0.5 ml of medium. ATP levels subsequently decayed with a $t_{1/2} = 52$ min (Fig. 2). Notably, ATP concentrations reached a steady-state value of 5.6 ± 0.6 nM (50-fold over the detection limit of the luciferase assay) within 3 h. These results suggest that extracellular ATP decayed to a baseline in which constitutive release balanced hydrolysis. Although all of the results described here were from 1321N1 human astrocytoma cells, we have also observed similar mechanically induced ATP release from another neural cell line, C6 rat glioma cells.

Knowledge of ATP release by 1321N1 cells has some practical importance in studies directed at determining the pharmacological selectivities of cloned P2Y receptors. Stable expression of the

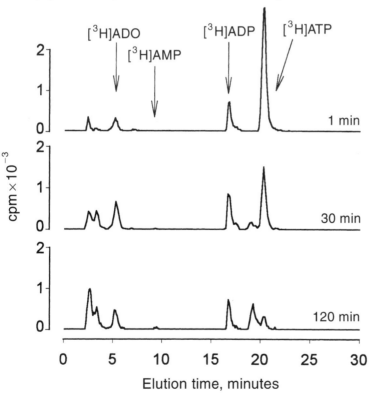

Fig. 1. Release and metabolism of [³H]ATP. Medium samples were collected at the times indicated following a medium change of [³H]adenine-labeled 1321N1 cells . Extracellular [³H]-species were analyzed by HPLC.

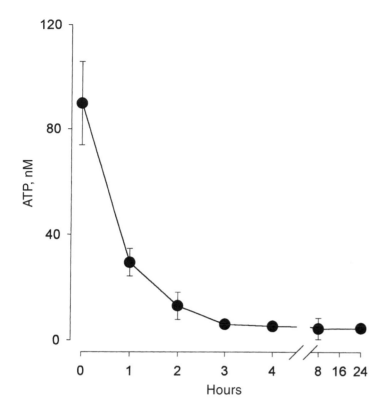

Fig. 2. ATP released from 1321N1 cells by a medium change decayed to steady state. Medium samples were taken at the times indicated after a medium change. ATP concentrations were determined by the luciferine–luciferase reaction.

human $P2Y_2$ receptor in 1321N1 cells resulted in basal levels of [^{3}H]inositol phosphates that were markedly greater than levels obtained with vector-transfected cells (Parr et al., 1994; Lazarowski et al., 1995). Thus, the release of endogenous ATP as illustrated above apparently confounds any quantitative pharmacological analysis of drug specificities at $P2Y_2$ receptors expressed in these cells. To address this problem more directly, experiments were carried out under conditions where extracellular accumulation of ATP was minimized by inclusion of apyrase, a non-specific ATPase, in both the medium used to wash cells and in the assay medium. Addition of apyrase resulted in reduction of basal inositol phosphate levels in $P2Y_2$ receptor-expressing 1321N1 cells to values that were essentially identical to those obtained with vector cells. Only small responses to ATP and UTP were observed in the presence of apyrase. In contrast, a marked inositol phosphate response that was

50–70% higher than accumulation observed with carbachol, which activates an endogenous muscarinic cholinergic receptor, was obtained with the hydrolysis-resistant ATP analogue, ATPγS.

Although incubation with apyrase proved useful for reduction of basal [^{3}H]inositol phosphate levels in $P2Y_2$ receptor-expressing cells, presumably by hydrolyzing any released ATP, the hydrolytic activity of this enzyme introduces another set of concerns for pharmacological studies of $P2Y_2$ receptors. As such, a methodology was pursued that would circumvent the need to use apyrase. The most successful approach was simply to quantitate receptor-promoted inositol phosphate accumulation in cells that received no washes. Thus, after labeling cells for 18 h with [^{3}H]inositol, LiCl was added without a change of medium; after an additional 15 min incubation the cells were challenged with agonists. Under these conditions, extracellular ATP accumulated at concentrations that were far below threshold values for receptor stimulation (Fig. 2), and as with apyrase treatment, low basal levels of inositol phosphates were observed (Lazarowski et al., 1995). More importantly, and unlike the data obtained with apyrase, responses to ATP and UTP were comparable to those obtained with ATPγS. These results illustrate that phosphoinositide hydrolysis is triggered during cell medium changes, apparently by the release of substantial amounts of ATP. Based on these results, inositol phosphate formation is routinely assessed in our laboratory in cells that are not subjected to any washes after the [³H]inositol labeling period.

Our initial evidence for the release of cellular UTP emanated from studies designed to examine the pharmacological action of ATP on the P2Y$_4$ receptor. Time course experiments revealed that 1 μM UTP promoted a rapid and sustained accumulation of inositol phosphates. In marked contrast, ATP produced no effect at 1 μM and the inositol phosphate response observed at 100 μM ATP was preceded by a 5–10 min delay (Lazarowski et al., 1997a).

Surprisingly, accumulation of inositol phosphates occurred after mechanical stimulation of the P2Y$_4$ receptor-expressing 1321N1 cells (P2Y$_4$-1321N1 cells), but not after mechanical stimulation of wild-type cells (Lazarowski et al., 1997a). Although mechanical stimulation of 1321N1 cells results in release of ATP, as stated above ATP added at concentrations (1 μM) in excess of those that occur after mechanical stimulation had no effect on inositol phosphate accumulation in P2Y$_4$-1321N1 cells.

Although no effect was observed in wild type 1321N1 cells, mechanical stimulation by medium displacement, or by applying a flow pulse (70 μl/s for 5 s) also resulted in an immediate Ca^{2+} mobilization in P2Y$_4$-1321N1 cells (Lazarowski et al., 1997a). The Ca^{2+} response to mechanical stimulation but not to carbachol, was completely abolished by apyrase or by the UTP-specific enzyme UDP-glucose pyrophosphorylase in the presence of glucose-1P (Lazarowski et al., 1997a). UTP (1 μM) promoted a rapid Ca^{2+} response in P2Y$_4$-1321N1 cells (Fig. 3A), but no rapid effect was observed with 100 μM ATP, 100 μM CTP, 100 μM UDP, or 100 μM GTP (Lazarowski et al.,

1997a). Taken together these results suggested to us that mechanical stimulation of 1321N1 cells results in release of UTP and that higher levels of accumulation of UTP may occur in the presence of exogenous ATP.

Consistent with the hypothesis that mechanical stimulation of 1321N1 cells results in release of UTP, the kinetics of the Ca^{2+} response following mechanical stimulation of P2Y$_4$ receptor-expressing1321N1 cells were similar to that observed after addition of UTP (Fig. 3A). In contrast, mechanical stimulation of 1321N1 cells expressing the UDP selective P2Y$_6$ receptor resulted in a Ca^{2+} signal that was slower relative to the signal triggered by 1 μM UDP (Fig. 3B). These results suggest that the P2Y$_6$ receptor-mediated Ca^{2+} response to mechanical stimulation was secondary to the release of UTP and its subsequent conversion to UDP.

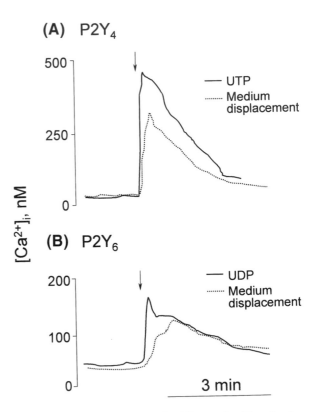

Fig. 3. Autocrine stimulation of uridine nucleotide selective receptors. Changes in intracellular Ca^{2+} were measured with fura-2 loaded 1321N1 cells expressing the P2Y$_4$ (A) or the P2Y$_6$ (B) receptor. The arrows indicate application of stimulus.

Since no sensitive assays were available, we devised methodology to assess directly the release of cellular UTP (Lazarowski et al., 1997a). This assay takes advantage of the selectivity of UDP-glucose pyrophosphorylase for UTP and utilizes [14C]glucose-1P as a cosubstrate. The UTP-dependent conversion of [14C]glucose-1P to [14C]UDPG is illustrated in Fig. 4A and is consistent with previous reports on the selectivity of the enzyme (Steelman and Ebner. 1966). ATP, GTP and CTP were not substrates. Moreover, ATP, GTP, and CTP at concentrations as high as 100 μM had no effect on the quantitation of UTP. Levels as low as 1 nM

UTP can be detected with this assay (Fig. 4A; (Lazarowski et al., 1997a)). UTP could be detected in the medium (0.5 ml in an individual well of a 12-well plate) of resting wild-type 1321N1 cells (Fig. 4B). A 14-fold increase in UTP concentration (70 nM; 35 pmol/10^6 cells) was observed in the extracellular medium 2 min after mechanical stimulation (Fig. 4B). Time course experiments indicated that maximal UTP release occurred within 2 min after mechanical stimulation of the cells and that extracellular UTP decreased gradually thereafter (Fig. 5). A similar pattern was observed for ATP release using a luciferin-luciferase assay, and extracellular ATP levels were typically five times greater than UTP levels. Released ATP and UTP represented approximately 1% of their intracellular pools. No difference in the extent of UTP or ATP release was observed between wild-type cells and P2Y₄-1321N1 cells (Lazarowski et al., 1997a). Addition of 100 μM ATP to the medium of resting wild-type or P2Y₄-1321N1 cells for 10 min resulted in a three-to-five-fold increase in extracellular UTP, indicating that the slow-occurring effects of ATP on Ca²⁺ levels in P2Y₄ receptor-expressing cells follows from conversion of ATP to UTP. We have

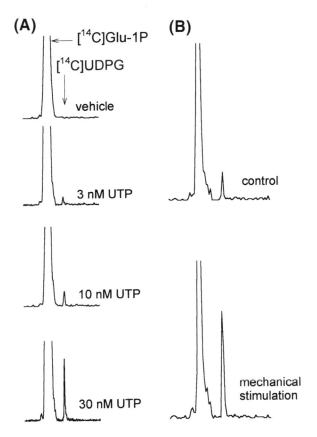

(A) [14C]Glu-1P
[14C]UDPG
vehicle
3 nM UTP
10 nM UTP
30 nM UTP

(B)
control
mechanical stimulation

Fig. 4. Quantification of UTP as the conversion of [14C]glucose-1P to [14C]UDPG. HPLC tracings illustrating (A) UTP-dependent conversion of [14C]glucose-lP to [14C]UDPG, (B) formation of [14C]UDPG by endogenous UTP present in medium samples from control or mechanically stimulated 1321N1 human astrocytoma cells. Incubations were for 1 h at 37°C in the presence of 0.5 U/ml UDPG-pyrophosphorylase, 0.5 U/ml inorganic pyrophosphatase and 1 μM [14C]glucose-lP 0.15 μCi as described (Lazarowski et al., 1997a).

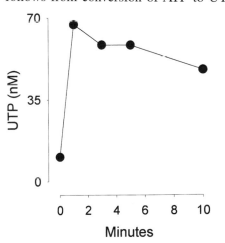

Fig. 5. Time course of UTP release from mechanically stimulated 1321N1 cells. UTP was quantified in the extracellular medium of 1321N1 cells that were stimulated by a medium displacement. Confluent cells on 24-mm plastic dishes were bathed with 0.5 ml serum-free DMEM medium. The medium was stirred with a micro-pipette and samples were collected at the indicated times. UTP was quantified as detailed in the legend of Fig. 4.

now utilized this method to show that UTP can be detected in the medium from a variety of neural and non-neural cells.

Mechanically stimulated release of ATP from 1321N1 cells was not accompanied by detectable release of lactate dehydrogenase (Lazarowski et al., 1995). To further rule out the possibility that cell lysis occurs during mechanical stimulation, P2Y$_4$-1321N1 cells were loaded with calcein, a Mn^{2+}-sensitive fluorescent dye, and changes in fluorescence were examined in the presence of 1 mM extracellular $MnCl_2$ (Lazarowski et al., 1997a). Mechanical stimulation caused no change in the calcein–fluorescence signal. In contrast, the signal obtained from impaling a single cell in a population of 250 cells was approximately 0.4% of the total signal and was approximately four times background. Thus, cell lysis can be observed with a sensitivity of at least 0.2%, and the triggered release of UTP and ATP by mechanical stimulation of 1321N1 cells is not a consequence of cell damage.

Discussion

The original reports by Burnstock and colleagues of release of ATP during nerve transmission (Burnstock, 1972) have been followed by demonstration of non-lytic mechanically promoted ATP release from essentially all tissues. For example, changes in flow rates induce the release of ATP from freshly harvested aortic endothelial cells (Milner et al., 1990), and shear forces trigger ATP release by cultured mouse fibroblasts (Grierson and Meldolesi, 1995). ATP release by mechanical stimulation also has been described for rat basophilic cells (Osipchuk and Cahalan, 1992), hepatocytes (Schlosser et al., 1996), T84 colonic cells and Calu-3 lung adenocarcinoma cells (Grygorczyk and Hanrahan, 1997). We have observed ATP release following mechanical stimulation of NIH 3T3 cells and primary cultures of human nasal epithelial cells (Watt et al., 1998), as well as from 1321N1 human astrocytoma (Lazarowski et al., 1995, 1997a) cells as we describe here.

Unlike the well-documented autocrine/paracrine function of extracellular adenine nucleotides, the potential existence and role of uridine nucleotide release have not been established. A major obstacle in assessing the significance of extracellular UTP has been the lack of a reliable method for accurately detecting the mass of UTP within the low nanomolar range. Although direct evidence for UTP release has not existed, three G protein-linked mammalian P2Y receptors (P2Y$_2$, P2Y$_4$, and P2Y$_6$) have been identified that recognize UTP or UDP as the most potent and/or selective agonist. Thus, whereas the discovery of three different uridine nucleotide-activated receptors has suggested that UTP is an extracellular signaling molecule, our results demonstrating non-lytic release of UTP from 1321N1 human astrocytoma cells in pharmacologically effective concentrations directly support this hypothesis.

UTP potentially can be detected in biological samples by utilizing radioactive precursors to radiolabel the intracellular uridine nucleotide pool (Andersson et al., 1988). The disadvantage of such an approach is that the specific radioactivity of UTP cannot be determined, and therefore, no quantitation of mass can be made. Aspartate is utilized during de novo synthesis of pyrimidines (Andersson et al., 1988). However, incorporation of this non-essential amino acid is difficult to achieve with cultured human cell lines. Alternatively, [^{3}H]uridine potentially can be incorporated into the pyrimidine salvage cycle (Andersson et al., 1988) provided that active RNA synthesis is occurring and an active nucleoside transporter is available in the target cells. Indeed, qualitative evidence for UTP release from intact cells recently was reported by Saiag and coworkers (Saiag et al., 1995) who demonstrated that an increase in the flow rate of perfused [^{3}H]uridine-labeled endothelial cells resulted in the release of a species tentatively identified as [^{3}H]UTP.

The concentration of UTP (up to 70 nM) detected in the extracellular medium of 1321N1 human astrocytoma cells after mechanical stimulation was somewhat lower than expected from receptor response studies. That is, mechanically stimulated P2Y$_4$-1321N1 cells exhibited elevated Ca^{2+} levels comparable to the levels observed with 0.3–1 μM UTP. This difference could not be explained by rapid degradation of UTP since the half-life of [^{3}H]UTP added to 0.5 ml of medium bathing 1321N1 cells (at 0.1 μM) is 10–15 min (Lazarowski et al., 1997b). The simplest explanation is that transient accumulation of released nucleotides at the cell surface can differ significantly from the accumulation measured in the bulk medium. A highly efficient coupling of mechanically released UTP with $P2Y_4$ receptors could also be promoted by undefined cell surface structures, although this has not been proven.

Dubyak and coworkers (Beigi et al., 1998) have recently established methodology that has led to the detection of ATP levels at the surface of human platelets. A chimeric protein that consists of the IgG-binding domain of protein A was fused in-frame with the complete sequence of firefly luciferase. Attachment of the protein to cells was effected by specific interaction with antibodies directed against native surface antigens. Light production by the cell-attached enzyme was linearly related to ATP concentration, and the method was utilized to quantitate a transient cell surface-localized increase in ATP levels during secretion from activated platelets. Application of this methodology to other cell types, including those of neural origin will be important to assess the relationship of cell surface nucleotide levels to those measured in the bulk medium.

Our results suggest that the stimulatory effect of ATP observed in $P2Y_4$-1321N1 cells occurs due to an indirect action of this nucleotide. UTP detected in the medium in the absence of mechanical stimulation likely represents a balance of tonic release and metabolism of nucleotide(s) and provides a potential mechanism for the effects of ATP observed on $P2Y_4$-1321N1 cells. We have previously characterized a non-specific ecto-nucleotidase activity in 1321N1 cells that hydrolyzes both ATP and UTP with similar K_m and V_{max} values (Lazarowski et al., 1997b). We also have identified a very active nucleoside diphosphokinase on the surface of 1321N1 cells that, in the presence of excess ATP, phosphorylates UDP to UTP (Lazarowski et al., 1997b). Thus, addition of ATP to the medium of 1321N1 cells can result in increased accumulation of endogenously released UTP by two means. First, ATP can compete with endogenously released UTP for hydrolysis by ecto-nucleotidases. Second, ATP can serve as a co-substrate with endogenous UDP for nucleoside diphosphokinase-promoted formation of UTP. The increased UTP concentration found in the presence of exogenous ATP supports this hypothesis. A corollary to the latter observation is that UDP, which is not a $P2Y_4$ receptor agonist (Nicholas et al., 1996), can indirectly activate this receptor as a consequence of nucleoside diphosphokinase-promoted formation of UTP from ATP present in the medium (Lazarowski et al., 1997b).

The mechanism whereby UTP and ATP are released from mechanically stimulated cells is unclear. Schwiebert and coworkers (Schwiebert et al., 1995) have proposed that the cystic fibrosis transmembrane conductance regulator (CFTR) acts as an ATP channel. However, CFTR is not expressed in wild-type 1321N1 cells, and heterologous expression of CFTR in 1321N1 cells resulted in no differences in mechanically stimulated ATP release (Watt et al., 1998). Whether other ion channels are involved in the release process is unclear, and involvement of a nucleotide transporter is an obvious possibility. We also need to establish whether release of ATP and UTP is coordinately or independently regulated. The amount of ATP released from 1321N1 cells exceeded that of UTP. Indeed, in preliminary experiments measuring ATP and UTP release from a variety of neural and non-neural cells we have observed that the ratio of ATP release to UTP release is roughly 5:1. This ratio is similar to the ratio of intracellular ATP to UTP that we measured in these cells. These results suggest a mechanical-stimulated release mechanism, e.g. by a transporter, that relatively non-selectively moves intracellular nucleotides from the intracellular to extracellular compartment.

P2Y receptors that are activated by uridine nucleotides, e.g. the $P2Y_2$, $P2Y_4$, and $P2Y_6$ receptors, are present in the brain (Webb et al., 1998). Therefore, the release of neuronal UTP as an extracellular signaling molecule is a strong possibility. Whether mechanical stimulation-promoted release of UTP from human astrocytoma and C6 glioma cells is reflective of a function of glial cells in vivo is not known. However, the description of a uridine nucleotide selective receptor on C6-2B glioma cells (Lazarowski and Harden, 1994), an astrocyte-like cell line, and of uridine nucleotide receptors on primary cultures of astrocytes (Pearce and Langley, 1994; Ho et al., 1995) opens the possibility of a specific target for UTP in glial cells. Astrocytes are known to be active elements in normal brain function and development and respond to various forms of injury with increased cell proliferation and hypertrophy (Abbracchio et al., 1996). Gliotic-like responses can be induced by extracellular nucleotides, which are known to have trophic effects on astrocytes via activation of various P2 receptors (Neary, 1996; Abbracchio et al., 1996). It will be important to determine whether mechanically-stimulated release of UTP and ATP occurs from astrocytes in vivo and whether such a release from 1321N1 human astrocytoma cells is reflective of an autocrine signaling function of these molecules in astrocytes following trauma, stroke, or seizure.

In summary, we have developed an assay for detection of UTP in sub-nanomolar concentrations, and we have shown that mechanically stimulated 1321N1 human astrocytoma and C6 glioma cells release both UTP and ATP into the extracellular medium in the absence of detectable cell lysis. Exogenous ATP, UTP, and UDP are metabolically converted by ecto-nucleotidases and ecto-nucleoside diphosphokinase on the surface of 1321N1 cells, and exogenous ATP, UTP, and UDP activate subsets of P2Y receptors on various types of glial and nerve cells. Thus, autocrine/paracrine activation of P2Y receptors as well as activation of P2Y receptors as a consequence of neuronally released nucleotides likely represent neurobiologically important phenomena.

Acknowledgment

We are indebted to Dr. Lâszló Homolya for assisting during Ca^{2+} measurements.

References

Abbracchio, M.P., Ceruti, S., Bolego, C., Puglisi, L., Burnstock, G. and Cattabeni, F. (1996) Trophic roles of P2 purinoceptors in central nervous system astroglial cells. *Ciba Foundation Symposium*, 198: 142–147.

Andersson, M., Lewan, L. and Stenram, U. (1988) Compartmentation of purine and pyrimidine nucleotides in animal cells. *Int. J. Biochem.*, 20: 1039–1050.

Beigi, R., Kobatake, E., Aizawa, M. and Dubyak, G.R. (1999). Detection of local ATP release from activated platelets using cell surface-attached firefly luciferase. *Am. J. Physiol.*, 276 (Cell Physiol. 45): C267–C275.

Burnstock, G. (1972) Purinergic nerves. *Pharmacol. Rev.*, 24: 509–581.

Doss, R.C., Perkins, J.P. and Harden, T.K. (1981) Recovery of β-adrenergic receptors following long-term exposure of astrocytoma cells to catecholamine: role of protein synthesis. *J. Biol. Chem.*, 256: 12281–12286.

Evans, T., Martin, M.W., Hughes, A.R. and Harden, T.K. (1985) Guanine nucleotide-sensitive, high affinity binding of carbachol to muscarinic cholinergic receptors of 1321N1 astrocytoma cells is insensitive to pertussis toxin. *Mol. Pharmacol.*, 27: 32–37.

Filtz, T.M., Li, Q., Boyer, J.L., Nicholas, R.A. and Harden, T.K. (1994) Expression of a cloned P2Y-purinergic receptor that couples to phospholipase C. *Mol. Pharmacol.*, 46: 8–14.

Grierson, J.P. and Meldolesi, J. (1995) Shear stress-induced $[Ca^{2+}]i$ transients and oscillations in mouse fibroblasts are mediated by endogenously released ATP. *J. Biol. Chem.*, 270: 4451–4456.

Grygorczyk, R. and Hanrahan, J.W. (1997) CFRT-independent ATP release from epithelial cells triggered by mechanical stimuli. *Am. J. Physiol.*, 272: C1058–C1066

Harden, T.K., Petch, L.A., Traynelis, S.F. and Waldo, G.L. (1985) Agonist-induced alteration in the membrane form of muscarinic cholinergic receptors. *J. Biol. Chem.*, 260: 13000–13006.

Ho, C., Hicks, J. and Salter, M.W. (1995) A novel P2-purinoceptor expressed by a subpopulation of astrocytes from dorsal spinal cord of the rat. *Brit. J. Pharmacol.*, 116: 2909–2918.

Hughes, A.R., Martin, M.W. and Harden, T.K. (1984) Pertussis toxin differentiates between two mechanisms of attentuation of cyclic AMP acumulation by muscarinic cholinergic receptors. *Proc. Natl. Acad. Sci. USA*, 81: 5680–5684.

Lazarowski, E.R. and Harden, T.K. (1994) Identification of a uridine nucleotide-selective G-protein-linked receptor that activates phospholipase C. *J. Biol. Chem.*, 269: 11830–11836.

Lazarowski, E.R., Homolya, L., Boucher, R.C. and Harden, T.K. (1997a) Direct demonstration of mechanically induced release of cellular UTP and its implication for uridine nucleotide receptor activation. *J. Biol. Chem.*, 272: 24348–24354.

Lazarowski, E.R., Homolya, L., Boucher, R.C. and Harden, T.K. (1997b) Identification of an ecto-nucleoside diphosphokinase and its contribution to interconversion of P2 receptor agonists. *J. Biol. Chem.*, 272: 20402–20407.

Lazarowski, E.R., Watt, W.C., Stutts, M.J., Boucher, R.C. and Harden, T.K. (1995) Pharmacological selectivity of the cloned human phospholipase C-linked P2U-purinergic receptor: potent activation by diadenosine tetraphosphate. *Br. J. Pharmacol.*, 116: 1619–1627.

Milner, P., Bodin, P., Loesch, A. and Burnstock, G. (1990) Rapid release of endothelin and ATP from isolated aortic endothelial cells exposed to increased flow. *Biochem. Biophys. Res. Commun.*, 170: 649–656.

Neary, J.T. (1996) Trophic actions of extracellular ATP on astrocytes, synergistic interactions with fibroblast growth factors and underlying signal transduction mechanisms. *Ciba Foundation Symposium*, 198: 130–139.

Nicholas, R.A., Watt, W.C., Lazarowski, E.R., Li, Q. and Harden, T.K. (1996) Uridine nucleotide selectivity of three phospholipase C-activating P2 receptors: Identification of a UDP-selective, a UTP-selective, and an ATP- and UTP-specific receptor. *Mol. Pharmacol.*, 50: 224–229.

Osipchuk, Y. and Cahalan, M. (1992) Cell-to-cell spread of calcium signals mediated by ATP receptors in mast cells. *Nature*, 359: 241–244.

Parr, C.E., Sullivan, D.M., Paradiso, A.M., Lazarowski, E.R., Burch, L.H., Olsen, J.C., Erb, L., Weisman, G.A., Boucher, R.C. and Turner, J.T. (1994) Cloning and expression of a human P2U nucleotide receptor, a target for cystic fibrosis pharmacology. *Proc. Natl. Acad. Sci. USA*, 91: 3275–3279.

Pearce, B. and Langley, D. (1994) Purine- and pyrimidine-stimulated phosphoinositide breakdown and intracellular calcium mobilisation in astrocytes. *Brain Res.*, 660: 329–332.

Saiag, B., Bodin, P., Shacoori, V., Catheline, M., Rault, B. and Burnstock, G. (1995) Uptake and flow-induced release of uridine nucleotides from isolated vascular endothelial cells. *Endothelium*, 2: 279–285.

Schlosser, S.F., Burgstahler, A.D. and Nathanson, M.H. (1996) Isolated rat hepatocytes can signal to other hepatocytes and bile duct cells by release of nucleotides. *Proc. Natl. Acad. Sci. USA*, 93: 9948–9953.

Schwiebert, E.M., Egan, M.E., Hwang, T.-H., Fulmer, S.B., Allen, S.S., Cutting, G.R. and Guggino, W.B. (1995) CFTR regulates outwardly rectifying chloride channels through an autocrine mechanism involving ATP. *Cell*, 81: 1063–1073.

Steelman, V.S. and Ebner, K.E. (1966) The enzymes of lactose biosynthesis. I. Purification and properties of UDPG pyrophosphorylase from bovine mammary tissue. *Biochim. Biophys. Acta*, 128: 92–99.

Waldo, G.L., Northup, J.K., Perkins, J.P. and Harden, T.K. (1983) Characterization of an altered membrane form of the β-adrenergic receptor occurring during agonist-induced desensitization. *J. Biol. Chem.*, 258: 13900–13908.

Watt, W.C., Lazarowski, E.R. and Boucher, R.C. (1998) Cystic fibrosis transmembrane regulator-independent release of ATP Its implications for the regulation of $P2Y_2$ receptors in airway epithelia. *J. Biol. Chem.*, 273: 14053–14058.

Webb, T.E., Henderson, D.J., Roberts, J.A. and Barnard, E.A. (1998) Molecular cloning and characterization of the rat $P2Y_4$ receptor. *J. Neurochem.*, 71: 1348–1357.

P. Illes and H. Zimmermann (Eds.)
Progress in Brain Research, Vol 120
© 1999 Elsevier Science BV. All rights reserved

CHAPTER 12

Quantal ATP release from motor nerve endings and its role in neurally mediated depression

E.M. Silinsky*, J.K. Hirsh, T.J. Searl, R.S. Redman and M. Watanabe

Department of Molecular Pharmacology and Biological Chemistry, Northwestern University Medical School, 303 East Chicago Avenue, Chicago, IL 60611, USA

Introduction

Evidence for over two decades has suggested that adenine nucleotides are stored in cholinergic nerve endings (for review see Zimmermann, 1994) and released together with ACh from motor nerve endings when nucleotides are assayed over the time course of many minutes using the firefly luciferase assay (Silinsky and Hubbard, 1973, Silinsky, 1975). In addition, application of either ATP or its hydrolysis products inhibits the release of ACh (Ginsborg and Hirst, 1972; Ribeiro and Walker, 1975; Silinsky, 1980). This inhibitory effect of ATP is dependent upon nucleotide hydrolysis to adenosine (Ribeiro and Sebastiao, 1987; Redman and Silinsky, 1994a) and occurs via an action on specific A_1 adenosine receptors (Silinsky, 1980; Nagano et al., 1992; Redman and Silinsky, 1993).

These observations led to the proposal of the hypothesis represented in cartoon form in Fig. 1. As shown in this figure, in addition to the release of the neurotransmitter ACh and its subsequent activation of nicotinic receptors to generate end-plate potentials (EPPs), ATP is released from nerve terminals as well. This nucleotide is then hydrolyzed by ecto-ATPases and ecto-5′-nucleotidases to form adenosine (Zimmermann, 1996), which acts on a nerve terminal A_1 adenosine receptor to inhibit the

subsequent release of neurotransmitter and to produce neuromuscular depression. Note the decline in ACh release, as represented by a reduction in the size of the EPP with repetitive stimulation. This decline in EPP amplitude is the electrophysiological correlate of neuromuscular depression and is due to events that occur in the nerve ending. Such depression may be severely debilitating in patients with certain neuromuscular diseases such as myasthenia gravis and may also contribute to the neuromuscular fade in patients recovering from paralysis with non-depolarizing neuromuscular blockers (see for example Prior et al., 1997). The numbered sites indicate potential loci for pharmacological intervention and will be discussed below.

Evidence that adenosine derived from endogenous ATP is the mediator of neuromuscular depression in frog and mouse

If the model shown in Fig. 1 is an accurate depiction of the mechanism of neuromuscular depression, then the following three treatments would be predicted to reduce or eliminate neuromuscular depression:

(i) Inhibition of the enzyme ecto 5′-nucleotidase (Fig. 1, site 1) which prevents the creation of endogenous adenosine, should eliminate depression. The inhibitor α,β-methylene ATP, which is such an agent, does indeed prevent

---

*Corresponding author. Tel.: +1 312-503-8287; fax: +312-503-0796; e-mail: e-silinsky@nwu.edu

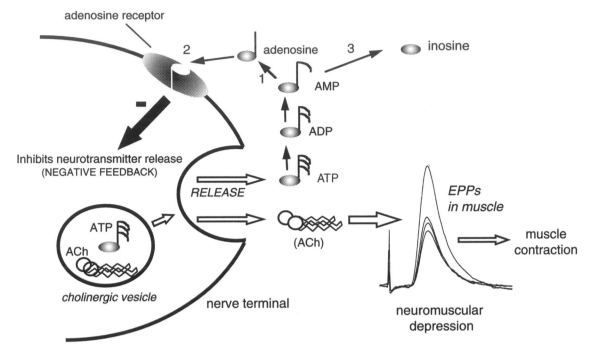

Fig. 1. Model of neuromuscular depression. Nerve stimulation evokes the co-release of both ATP (depicted for convenience as a 32nd note) and ACh from cholinergic vesicles. ACh produces the post-junctional excitation detected electrophysiologically as the EPP. In normal muscle (i.e. muscle that is not treated with tubocurarine or α-bungarotoxin), the EPP is so large that it produces a muscle action potential and a twitch in the muscle fiber. ATP and its hydrolysis products are devoid of significant postjunctional effects on the electophysiological correlates of neuromuscular transmission in the adult. However, after hydrolysis to ADP (16th note), AMP (8th note) and adenosine (quarter note) by ecto-enzymes, adenosine is capable of binding to A1 adenosine receptors in the plasma membrane of the nerve ending and inhibiting the subsequent release of ACh. This depression is manifested as a progressive decrease in the amplitude of the EPP with repetitive stimulation. The sites numbered 1, 2 and 3 are targets for pharmacological intervention and are discussed in the text.

neuromuscular depression in the frog (Fig. 2A).

(ii) The presence of a competitive inhibitor at adenosine A_1 receptors (Fig. 1, site 2) should reversibly eliminate neuromuscular depression. Figure 2B shows that the selective A_1 antagonist 8-cyclopentyltheophylline (CPT) eliminates neuromuscular depression in the frog.

(iii) Addition of adenosine deaminase to the extracellular fluid should reverse neuromuscular depression by rapidly degrading endogenously-released adenosine to the inactive derivative inosine (Fig. 1, site 3). Figure 2B also shows this to be the case.

Finally, the effects of exogenous adenosine should be blocked due to occupation of the A_1 adenosine receptor by endogenously released adenosine during maximal neuromuscular depression. Figure 2C shows this prediction to be borne out by the experimental results. For a complete description of these experiments, the reader is referred to Redman and Silinsky (1994a).

Whilst the data summarized in Fig. 2 were obtained from experiments at frog cutaneous pectoris nerve–muscle junctions, similar results have been observed in the mouse phrenic nerve–hemidiaphragm neuromuscular preparation (Fig. 3). Note that α,β-methylene ATP (Fig 3A) and adenosine deaminase (Fig. 3B) each reverse neuromuscular depression.

As alluded to above, the data shown in Fig. 2 were obtained from the frog cutaneous pectoris nerve–muscle preparation. This is a thin muscle in which conditions may be chosen to minimize the

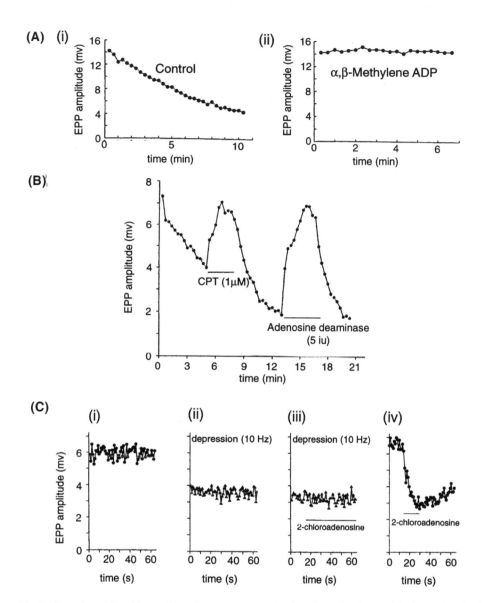

Fig. 2. Experimental evidence from frog neuromuscular junctions for the model shown in Fig 1. In this figure, (A) (i) shows presynaptic neuromuscular depression and (A) (ii) shows its antagonism by α,β-methylene ADP. For both (A) and (B), the motor nerve was stimulated at a frequency of 0.05 Hz. The Ringer solution was modified as described in the legend to Fig. 6 (i.e. modest concentrations of potassium channel blockers, altered concentrations of Ca^{2+} and Mg^{2+}) but similar results were obtained in normal Ringer solution. (B) shows that under similar conditions as (A), neuromuscular depression was eliminated by the addition of 1 μM 8-cyclopentyltheophylline (CPT); depression resumed upon washout of the drug. The application of 5 i.u./ml adenosine deaminase also completely abolished neuromuscular depression. Whilst the data are not shown, perineural Ca^{2+} currents were measured simultaneously with the EPPs in this modified Ringer solution and these did not change during the entire experiment (see below and Redman and Silinsky (1994a) Figs. 8 and 9). (C) Occlusion of the inhibitory effects of exogenous 2-chloroadenosine during neuromuscular depression in normal Ringer solution. The preparation was first stimulated at 1.0 Hz to obtain a control value for the EPP amplitude (i). Continuous stimulation at 10 Hz caused a 40% depression of EPP amplitude (ii) from the control level (i). Application of 25 μM 2-chloroadenosine during this state of depression had no effect (iii). After washout of the drug and subsequent 20 min rest, stimulation at 1.0 Hz produced EPPs of the original amplitude and in this non-depressed state the application of 2-chloroadenosine produced a 45% depression in EPP amplitude (iv). (Reproduced with permission from Redman and Silinsky (1994a)).

148

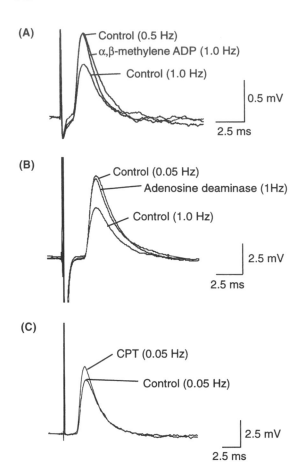

Fig. 3. Endogenous adenosine and neuromuscular depression at mouse motor nerve endings. The mouse phrenic nerve–hemidiaphragm preparation was used in these experiments. (A) Elimination of prejunctional depression by α,β-methylene ADP (50 μM). Depression was produced by increasing the frequency of stimulation from 0.5 Hz to 1 Hz. Responses are the average of 3–6 stimuli. (B) Elimination of neuromuscular depression by adenosine deaminase. (1 i.u./ml) Each response is the average of 16 stimuli. (C) ACh release at low frequencies of nerve stimulation (0.05 Hz) is increased by the adenosine antagonist CPT. Each response is the average of 16 stimuli.

leak of adenosine and isolate the effects of adenosine to that released during neuromuscular depression. While under some circumstances, this is possible in mouse (Fig. 3A,B), often endogenous adenosine is leaking from unknown sources and not related to neuromuscular depression. This is shown in Fig. 3C, in which the adenosine antagonist CPT increases ACh release under basal conditions in which the nerve is stimulated only infrequently. This does not normally occur at neuromuscular

junctions in the frog cutaneous pectoris nerve–muscle preparation, possibly because of the diaphanous nature of this muscle and the ability to apply test solutions by localized fast flow application. However, when endogenous adenosine is elevated artificially by continuous stimulation or by muscle fibre damage due to careless dissection of this frog preparation, then results similar to the mouse experiments shown in Fig. 3C occur in frog as well (see Redman and Silinsky, 1994a). In summary, the evidence presented thus far suggests that ATP is indeed an important mediator of neuromuscular depression in frog and mouse, although it must be mentioned that mouse motor nerve endings appear less sensitive to the effects of exogenous adenosine derivatives (Singh et al., 1990; Nagano et al., 1992; Prior et al., 1997).

Synchronous quantal release of ATP

The earlier results utilizing the firefly luciferin–luciferase assay suggested that ACh is co-released with adenine nucleotides (Silinsky, 1975). However, the time resolution of these experiments were of the order of minutes. If the model of Fig. 1 is true, then ATP must be released concomitantly with ACh by a discrete, temporally isolated nerve impulse and in sufficient concentration to mediate neuromuscular depression. To demonstrate this, one requires an excised patch of membrane that contains both ACh-gated channels (i.e. nicotinic receptors) and channels gated directly by extracellular ATP (i.e. P2X ATP receptors – see North and Barnard, 1997). This patch, when brought to the vicinity of motor nerve endings, should reveal channel openings characteristic of ATP within milliseconds of a nerve impulse. Just such a source of membranes was found by Silinsky and Gerzanich (1993). As part of a project demonstrating that ATP behaves as a fast neurotransmitter substance between mammalian sympathetic neurones cultured from the celiac ganglia of guinea-pig (Silinsky et al. 1992), it was found that the cell bodies of these neurons possess non-selective cation channels gated by extracellular ATP. These channels have a characteristic conductance and opening and closing pattern and can be studied in excised (outside-out) patches (Silinsky and

Gerzanich, 1993). Note the single channel currents induced by extracellular ATP in these excised patches exhibit a profound inward rectification, with a mean conductance at -50 mV of approximately 22 pS (Fig. 4A). Such patches were used in this earlier study to detect ATP release from stimulated neurites in these cultures. These patches also possessed ACh receptors with about twice the conductance and with considerably shorter apparent mean open times than the P2X receptors.

The experimental approach to detect the quantal release of ATP from cholinergic motor nerve endings to frog skeletal muscle is depicted in Fig. 4B. In the experiment shown, the cover slip containing acutely dissociated celiac neurons, possessing both ATP and ACh receptors were placed in a chamber containing a frog cutaneous pectoris nerve–muscle preparation. When excised patches were brought in proximity to the motor nerve ending and the nerve stimulated, responses characteristic of both ATP-gated and ACh-gated ion channels were detected. (see Fig. 4B, in which the predominant openings are caused by ATP). In the remaining experiments (Fig. 5), ACh receptors and their associated ionic channels were blocked by hexamethonium (150 μM), partly to eliminate an intriguing occlusive interaction between these discrete ligand gated channels (Silinsky and Gerzanich, 1993; Searl et al., 1998). Figure 5 shows that these channel openings produced by motor nerve stimulation are completely abolished by the ATP antagonist suramin (Fig. 5A) and occur with a very brief millisecond latency similar to that of the EPP recorded simultaneously (Fig. 5B). The nerve-evoked ATP release, as measured with the patch electrode, was Ca^{2+}-dependent (Fig. 5C), and was mimicked by the local application of 5–25 μM ATP applied to the excised patch (data not shown, see Silinsky and Redman (1996) for more details).

Evoked synaptic events such as those shown in Figs. 4 and 5 that appear with a brief latency after nerve stimulation are generally attributed to the quantal release of the activating substance from vesicular stores. Indeed it was possible in one preliminary experiment to detect quantal fluctuations in ATP release characteristic of such neurotransmitter release from synaptic vesicles. The results of this section thus suggest that ATP is indeed released together with ACh from synaptic vesicles within milliseconds of nerve stimulation and in sufficient concentration to be the mediator of neuromuscular depression.

What nerve terminals components are responsible for the action of exogenous or endogenous adenosine?

Based upon earlier experiments revealing that neurally evoked release mediated by extracellular Sr^{2+}, a partial agonist for evoked ACh release, is inhibited in a non-competitive manner by adenosine receptor agonists in the frog, it was suggested that adenosine inhibited evoked ACh release downstream of alkaline earth cation entry (Silinsky, 1981). Subsequent experiments conducted on the effects of adenosine on ACh release appeared to support this suggestion. First, neither Na^+ nor K^+ currents in the nerve ending are affected during the maximal inhibitory effects of adenosine analogs (Silinsky, 1984; see also Silinsky and Solsona, 1992). Next, when spontaneous ACh release is increased in frequency by methods which bypass calcium channels (e.g. the Ca^{2+} ionophore, ionomycin, or Ca^{2+}-containing liposomes), the inhibitory effect of adenosine on this accelerated ACh release is preserved (Silinsky, 1984; Hunt and Silinsky, 1993). Indeed, in the presence of concentrations of La^{3+} that fully block both Ca^{2+} entry and ACh release associated with Ca^{2+} entry, adenosine derivatives still inhibit spontaneous ACh release. Furthermore, adenosine analogs inhibit neurally evoked asynchronous release mediated by Ba^{2+}, without changing the rate constant of decay of such release (Silinsky, 1984) suggesting that adenosine is not increasing the rate of cation clearance into storage sites.

With regard to changes in cation clearance, the mobile calcium buffer BAPTA and its derivatives have been found to mimic the behaviour of the normal Ca^{2+} clearance mechanism in motor nerve endings in a highly efficient manner, and in some experiments accurately mimicked the inhibitory effect of adenosine on both evoked and spontaneous ACh release (Hunt et al., 1994). Because BAPTA is far more efficacious than any hypothetical Ca^{2+} buffering system linked to adenosine,

(A)

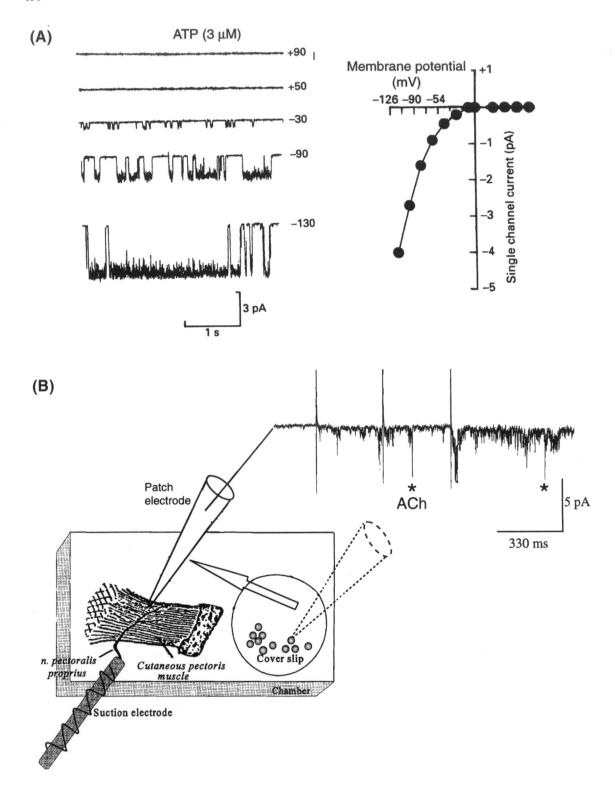

ATP (3 µM)

+90

+50

−30

−90

−130

3 pA

1 s

Membrane potential (mV)

+1

−126 −90 −54

−1

−2

−3

−4

−5

Single channel current (pA)

(B)

*

ACh

*

5 pA

330 ms

Patch electrode

n. pectoralis proprius

Cutaneous pectoris muscle

Cover slip

Chamber

Suction electrode

Fig. 4. Excised (outside-out) patches containing ATP-gated ionic channels (P2X ATP receptors, (A) and their use to detect the quantal release of ATP (B)). (A) Individual traces (left) and current voltage relationships (right) from excised patches (reproduced with permission from Silinsky and Gerzanich, 1993). (B) The method in which an outside-out patch of celiac neurons (situated on a cover slip) was moved to a neuromuscular junction in a cutaneous pectoris muscle subjected to mild protease/collagenase digestion (as described in Grinnell et al. (1989) and Silinsky and Redman (1996)). Preparations were then treated with 100 nM α-bungarotoxin for a time sufficient to block all EPPs in the muscle. Three consecutive stimuli delivered to the motor nerve via a suction electrode evoked currents characteristic of both ATP and ACh as detected with the patch electrode. (For details see Silinsky and Redman (1996) from the example of which Fig. 4B is reproduced.) Asterisks show single channel events produced by ACh.

BAPTA derivatives might be predicted to interfere with the inhibitory actions of adenosine, if adenosine is inhibiting ACh release by increasing cation clearance. However, this is not the case; BAPTA derivatives failed to alter either the potency or the maximum inhibitory effect of adenosine as an inhibitor at frog motor nerve endings (generally a 50% inhibition of both evoked and spontaneous ACh release). Furthermore, adenosine failed to alter facilitation at motor nerve endings; facilitation is another monitor of the extent of calcium clearance (Hunt et al., 1994). Finally, some agents that prevent the uptake and release of Ca^{2+} from storage sites do not impair the ability of adenosine to inhibit ACh release (e.g. TMB-8; see Hunt et al., 1990).

In the 1990s it was still unknown whether *physiologically functional ACh release* reflected as EPPs (Fig. 1) is antagonized by adenosine as a consequence of inhibition of Ca^{2+} entry via nerve terminal Ca^{2+} channels or at a target downstream of calcium entry. Two experimental approaches may be taken in this regard: (i) to make electrophysiological measurements of Ca^{2+} currents simultaneously with ACh release, and (ii) to evoke EPPs in the absence of Ca^{2+} entry through Ca^{2+} channels and determine whether adenosine inhibits ACh release.

With respect to the first experimental approach, it has been found that simultaneous measurements of evoked ACh release (EPPs) may be made together with measurements of the extracellular reflection of nerve terminal Ca^{2+} currents flowing in the perineurium under conditions of modest K^+ channel blockade. These perineural Ca^{2+} currents, while not in the strictest sense membrane ionic currents, are highly correlated with the membrane conductance change that occurs at nerve endings when Ca^{2+} channels are opened (Gunderson et al., 1982, Brigant and Mallart, 1982). These channels

represent the Ca^{2+} conductance pathway that mediates evoked ACh release (see Redman and Silinsky, 1995). The method, which is described in detail in Redman and Silinsky (1995), is depicted in Fig. 6. The Na^+ current, recorded in the regions of the final node of Ranvier produces the depolarisation that opens Ca^{2+} channels and generates a Ca^{2+} current at the nerve ending (upward deflection; see figure legend for caveats in the terminology and additional details). The Ca^{2+} current then initiates ACh release, and the ACh in turn binds to nicotinic receptors to produce the EPPs (for further details, see figure legend).

Figure 6B shows that the selective A_1 receptor agonist N^6-cyclopentyladenosine (CPA) reduces ACh release but has no effect on the perineural Ca^{2+} currents. In addition, neither during depression nor following reversal of depression with the highly selective adenosine receptor antagonist 8-cyclopentyl-1,3, dipropylxanthine (also termed DPCPX or CPX, Fig. 6C) or adenosine deaminase (not shown, see Redman and Silinsky, 1994a) are the perineural Ca^{2+} currents affected by adenosine. These results confirm earlier experiments made in the presence of more complete K^+ channel blockade suggesting that adenosine does not inhibit the Ca^{2+} current responsible for evoked ACh release in the frog (Silinsky and Solsona, 1992). In contrast to these results, other agents such as aminoglycoside antibiotics also reduce ACh release to the same degree as adenosine but do so as a consequence of reducing Ca^{2+} entry. As expected, reduction in EPP amplitudes occurs simultaneously with a reduction in Ca^{2+} currents in the presence of these antibiotics (Redman and Silinsky, 1994b).

The second experimental approach is to exploit the observation that synchronous evoked ACh release reflected as EPPs may be generated by a nerve impulse in the absence of Ca^{2+} entry through Ca^{2+} channels, when Ca^{2+} is delivered to the

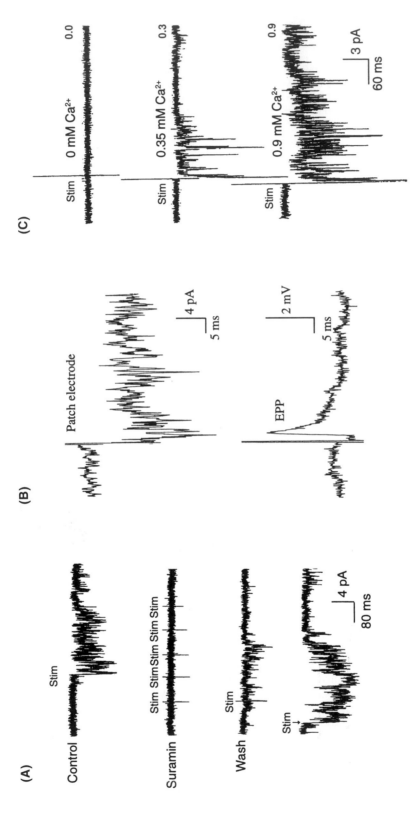

Fig. 5. Properties of ATP release by single motor nerve impulses. In (A)–(C), nicotinic receptors/ion channels were blocked by hexamethonium (200 μM). (A) Sensitivity of evoked ATP release to the ATP antagonist suramin (50 μM). Control shows response to single stimulus (Stim) in the absence of suramin. The remaining traces in (A) show reversible antagonism by suramin (recovery from suramin antagonism was complete after 45 sec (bottom trace)). (B) The ATP response (upper trace-patch electrode) and the EPP (lower trace) share a similar, brief latency in the ms range. Preparation was treated 100 nM α-bungarotoxin for a sufficient time to produce small EPPs. (C) Calcium-dependency of evoked ATP release from motor nerve endings.

secretory apparatus by alternative means (e.g. Ca^{2+}-containing lipid vesicles or liposomes; Silinsky et al., 1995). Note that Ca^{2+} delivered in liposomes supports the generation of EPPs even in the presence of the Ca^{2+} chelator EGTA in the extracellular fluid (Fig. 7A (i) shows the raw data, (ii) shows the averaged data). Similar levels of evoked release were generated by Ca^{2+} liposomes in the presence of the Ca^{2+}-channel blocker Co^{2+} (not shown), in the absence of EGTA (Fig. 7A (iii)) or when 0.25 mM extracellular Ca^{2+} was added to the control, Ca^{2+}-free solution (see Fig. 7A (iv) for a discussion of these results, see Silinsky et al., 1995). Note that adenosine inhibits neurally-evoked ACh release mediated by intracellular Ca^{2+} (delivered via liposomes) in the absence of Ca^{2+} entry (Fig. 7B) to a similar degree to the inhibition of evoked ACh release mediated by Ca^{2+} entry via membrane ionic channels (Silinsky, 1984). Control experiments were also performed in this study using confocal laser scanning microscopy to demonstrate that these liposomes are both interacting with the membrane of the nerve endings and delivering their entrapped contents to the cytoplasmic surface of the nerve ending (see Silinsky et al. (1995) Figs. 5 and 6). In addition, suitable electrophysiological controls were made to exclude the possibility that these liposomes were rupturing locally near the external surface of the nerve ending and delivering Ca^{2+} to the outside surface of the nerve ending (see Silinsky et al. (1995) Fig. 4). The results in the frog thus suggest that evoked ACh release is inhibited by adenosine via an effect on the secretory apparatus rather than by reducing Ca^{2+} entry via Ca^{2+} channels.

Given that a strategic component of the secretory apparatus has been suggested to be the target site for the inhibitory actions of adenosine derivatives (Silinsky, 1981, 1984), it appears of interest to discuss a few possible substrates for this effect. The list of candidates may be narrowed by the observations regarding the effects of the alkaline earth cation Sr^{2+} in the frog; extracellular Sr^{2+} has been shown to be a partial agonist for evoked ACh release (Silinsky, 1981). First, as alluded to above, the relationship between extracellular Sr^{2+} concentration and inhibition by adenosine appears to be non-competitive (Silinsky, 1981), again suggesting

that Ca^{2+} channels are not the target sites for the inhibitory action of adenosine. This is in contrast to the effects of intracellular Sr^{2+}, i.e. Sr^{2+} delivered to the nerve terminal cytoplasm in liposomes, in which the relationship appears to be competitive ($n = 2$; unpublished experiments). Specifically, adenosine produces a greater percent inhibition at lower levels of release (and low intracellular Sr^{2+} concentrations) than at higher levels of release (and higher intracellular Sr^{2+} concentrations). This is characteristic of the mathematics of competitive inhibition and indeed the concentration–response curve for Sr^{2+} and ACh release in the presence of adenosine can be well fit by the mathematical framework for a competitive relationship between adenosine and intracellular Sr^{2+}. This suggests that a presynaptic protein that possesses a binding site for intracellular Ca^{2+} or Sr^{2+}, namely one that has a C2 Ca^{2+} calcium binding domain (see Geppert and Südhof, 1998), could be a viable target at the secretory apparatus for the inhibitory effects of adenosine.

In this regard, based upon studies using phorbol esters and non-selective protein kinase inhibitors, it has been suggested that phospholipase C or protein kinase C (PKC) could be a target for the action of adenosine (Sebastiao and Ribeiro, 1990). Protein kinase C possesses both C1 phorbol ester binding domains and C2 calcium-binding domains. Indeed, we have found that active phorbol esters, which are known to stimulate PKC, produce a parallel increase in neurotransmitter release and a decrease in calcium currents (Redman et al., 1997). These results suggest that the stimulatory effects of phorbol esters are not due to an increase in Ca^{2+} entry via membrane Ca^{2+} channels but rather to an action on a component of the secretory apparatus. In a subsequent study, however, we found to our surprise that a battery of antagonists acting at different sites on PKC failed to inhibit the action of phorbol esters (Searl and Silinsky, 1998). These results thus raise the possibility that a phorbol ester receptor (C1)-containing protein and not the enzyme PKC is the target site for the action of phorbols on ACh release (Searl and Silinsky, 1998). This phorbol receptor, when activated, appears to increase the number of release sites or the efficiency by which Ca^{2+} is able to increase ACh

154

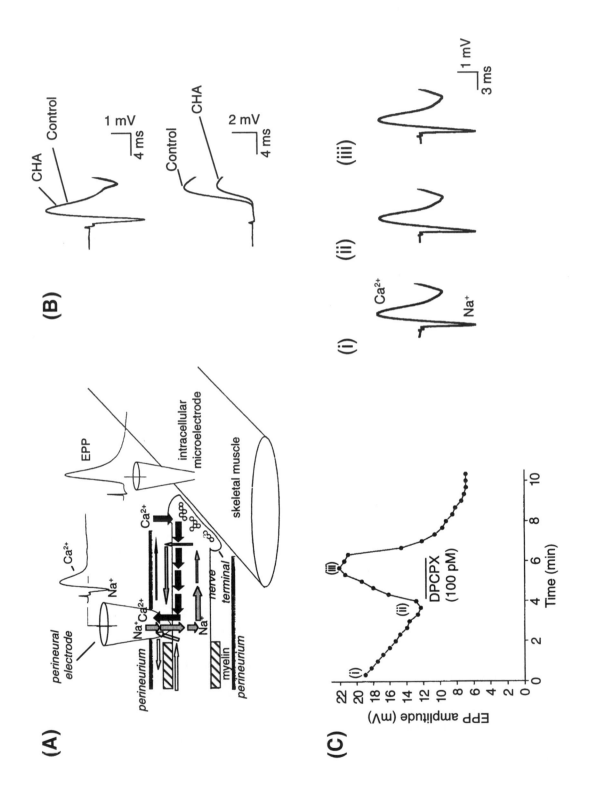

Fig. 6. Simultaneous measurements of EPPs and prejunctional ionic currents during neuromuscular depression and during the action of exogenous or endogenous adenosine. (A) Method used to make simultaneous measurements of perineural Ca^{2+} currents and EPPs from frog motor nerve endings recorded after a blockade of a proportion of the K^+ channels with TEA (250 μM) and DAP (100 μM). The solution also contained normal NaCl, low Ca^{2+} (0.9 mM), and high Mg^{2+} (10 mM) concentrations. The Ca^{2+} channels are concentrated at the nerve ending proper and not upstream in the preterminal unmyelinated axon. The perineural deflections actually represent the voltage change produced across the resistance of the perineural sheath surrounding several small bundles of axons by currents flowing between the nerve terminals and the myelinated axons. These extracellular currents are proportional to the difference in potential between the nerve endings and the nodes of Ranvier. Ion movements initiated by inward Ca^{2+} currents at the nerve terminals (filled arrows) flow across the perineural resistance and are detected as outward currents upstream at the recording microelectrode. The electrophysiological deflections depicted are actual perineural currents and EPPs recorded simultaneously in Ca^{2+} current Ringer. The peak of the upward (largely Ca^{2+}) component of the perineural current when flowing across the extracellular resistance produced a voltage change of 1.8 mV; the EPP was 7.4 mV. Calibration below the perineural current = 2 ms. The downward spike preceding the Ca^{2+} current represents the Na^+ current originating from the junction of the myclinated and non-myelinated axon (for further details and caveats about this method, see text and Redman and Silinsky (1995), from which this figure was reproduced). (B) A highly selective adenosine A1 receptor agonist, N^6-cyclohexyladenosine (CHA, 2 nM), reduces the average EPP without affecting perineural Ca^{2+} currents. Note the 50% inhibition of ACh release is the maximal inhibitory effect for adenosine receptor agonists (see Silinsky, 1984). The effect of adenosine receptor agonists on EPP amplitudes is entirely presynaptic. (C) Simultaneous measurements of ACh release reflected as EPP amplitudes (left) and perineural Ca^{2+} currents (right, (i)–(iii)) recorded during neuromuscular depression and during the actions of DPCPX. The nerve was stimulated at 0.05 Hz for the duration of the experiment. Note the blockade of neuromuscular depression by 100 pM DPCPX without changes in the corresponding Ca^{2+} currents during depression or recovery (i)–(iii)). Only the initial component of the current traces are shown, although the currents from several experiments exhibited repetitive firing (see Redman and Silinsky, 1995). Calibration bar represents a voltage change of 1.0 mV produced by current flow across the perineural resistances.

release (Redman et al., 1997). It is noteworthy that active phorbol esters have been found to decrease the $[Ca^{2+}]$ required for binding to the C2 domain and hence to activate PKC (Newton, 1995). A remarkable parallel to our results has been found in smooth muscle of the canine airway. Specifically, Bremerich et al. (1998) found that phorbol dibutyrate increased the Ca^{2+} sensitivity of permeabilized tracheal smooth muscle by a non-PKC mechanism.

One possible target site for the action of phorbol esters in the nerve terminal is Munc-13, the mammalian homologue of *unc-13*. This protein, which posesses the C1 zinc-finger phorbol ester binding domain, has been found to be localized in the plasma membrane fraction of synaptosomes, is implicated in the Ca^{2+}-dependent docking of synaptic vesicles at neuromuscular junctions (Brose et al., 1995; Betz et al., 1998), and binds to important presynaptic membrane proteins such as syntaxin (Geppert and Sudhof, 1998; see below). In addition to a C1 phorbol receptor, Munc-13 also possesses a C2 Ca^{2+}-binding domain, raising the possibility that a homologue of Munc-13 or a related protein mediates the stimulatory effects of phorbol esters. We thus decided to investigate whether such a phorbol ester receptor could be the target site for

the action of adenosine. Our preliminary results were variable. In some experiments, phorbol esters failed to change the potency or maximal inhibitory effect of adenosine. In others, the effect of adenosine was attenuated during phorbol ester treatment (see e.g. Sebastiao and Ribeiro, 1990). Further experiments on frog and mice are necessary to exclude this phorbol ester receptor as a target for the inhibitory effects of adenosine.

An alternative target for the action of adenosine is the rabphilin/rab3a complex of presynaptic proteins. Specifically, recent results have demonstrated that in cultured mouse hippocampus, synchronous evoked neurotransmitter release is increased in mice deficient in rab3a, a GTP-binding protein that affects Ca^{2+}-dependent fusion of synaptic vesicles with the nerve terminal membrane (Geppert et al., 1997). These authors also demonstrate an enhanced paired-pulse facilitation of transmitter release. As increases in neurotransmitter release are generally associated with reduced facilitation (most likely due to an increased depression at the higher level of transmitter release), and rab3a is also abundant at cholinergic motor nerve endings (Zimmermann et al., 1993), the increased neurotransmitter release and increased facilitation raise the possibility that presynaptic depression is

156

reduced or absent in rab3a knockouts. It would thus be of interest to test the hypothesis that rab3a knockouts demonstrate increased neurotransmitter release and decreased depression due to a reduced sensitivity to adenosine. Such experiments are currently in progress.

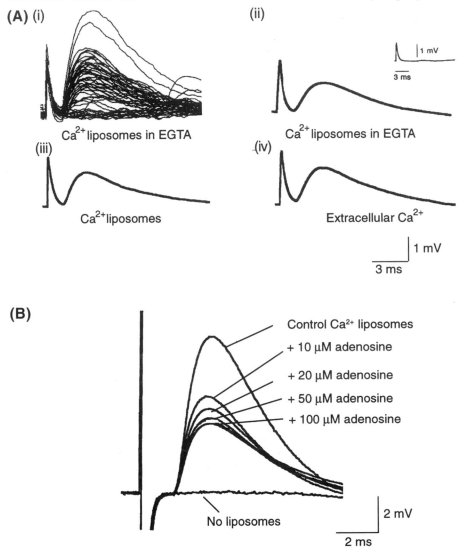

Fig. 7. Evoked ACh release in the absence of Ca^{2+} entry (Ca^{2+} delivered via liposomes) and its antagonism by adenosine. (A) (i) The raw data obtained in the presence of Ca^{2+} liposomes in 1 mM Mg^{2+} solutions containing 2 mM EGTA (MgEGTA solution). (ii) shows the averaged EPP in MgEGTA Ringer from the traces shown in (i) (mean number of ACh quanta released, $m = 1.36$; $n = 45$ stimuli). Absence of ACh release in the absence of Ca^{2+} liposomes is shown as inset (average of 20 stimuli). (A) (iii) The effect of Ca^{2+} liposomes in 1 mM Mg^{2+} solutions without EGTA ($m = 1.34$; $n = 31$ stimuli). (A) (iv) ACh release in the presence of 0.2 mM extracellular Ca^{2+} (1 mM Mg^{2+}) in the absence of Ca^{2+} liposomes ($m = 1.49$; $n = 15$ stimuli). Reproduced with permission from Silinsky et al. (1995). (B) Evoked ACh release supported by Ca^{2+} liposomes is antagonized by adenosine with the same potency and effectiveness that occurs for this nucleoside to inhibit ACh release initiated by Ca^{2+} entry via Ca^{2+} channels. Continuous nerve stimulation (1 Hz) failed to evoke ACh release in the absence of Ca^{2+} liposomes (lowest trace, no liposomes). The effect of Ca^{2+} liposomes (top trace) was similar to that seen when these liposomes were applied in the presence of 2 mM EGTA (not shown). The maximal inhibitory effect of adenosine was about 45% of the control value in this experiment. Each trace is the averaged response to between 227 and 320 stimuli.

It should be mentioned that in some situations, Ca^{2+} currents appear to be depressed by adenosine (for review see Miller, 1998). Given the intimacy between Ca^{2+} channels and the secretory apparatus (see e.g. Südhof, 1995; Goda and Südhof, 1997) this is not surprising. Indeed, it was predicted well before the elegant studies in presynaptic cell biology and biochemistry of the 1990s that Ca^{2+} currents may be affected indirectly by presynaptic modulators as a consequence of an action on a strategic component of the secretory apparatus (see Silinsky (1986), p. 185). It is noteworthy that this prediction has been borne out. For example, it has been found that at cholinergic synapses, cleavage of syntaxin (a presynaptic membrane protein crucial for neurotransmitter release and linked secondarily to Ca^{2+} channels) by botulinum toxin C1, prevented the inhibitory effect of G proteins on Ca^{2+} currents (Stanley and Mirotznik, 1997). More experiments are evidently necessary using simultaneous measurements of Ca^{2+} currents and neurotransmitter release in mice in which specific synaptic proteins have been knocked out to determine the relationship between Ca^{2+} current modulation and specific components of the secretory apparatus.

Conclusions

In this paper, we reviewed the mechanisms of presynaptic neurotransmitter modulation within the framework of a physiologically relevant behaviour, namely the process of neurally mediated prejunctional depression at the skeletal neuromuscular junction. At the frog and mouse neuromuscular junctions, depression appears to be due to the release of endogenous ATP, which after degradation to adenosine, acts back on the nerve ending to inhibit the subsequent release of the neurotransmitter acetylcholine. In support of this, we also presented the evidence that ATP is released synchronously from motor nerve endings within milliseconds of a solitary nerve impulse and in the appropriate concentration range to mediate neuromuscular depression. The data suggest that, in the frog, adenosine mediates the inhibition of *both spontaneous and evoked ACh release* and by an action at the secretory apparatus and hence downstream of calcium entry.

Acknowledgement

This work was supported by a research grant from the NIH (NS12782).

References

Betz, A., Ushery, U., Rickmann, M. Augustin, I., Neher, E., Südhof, T.C., Rettig, J. and Brose, N. (1998) Munc13-1 is a presynaptic phorbol ester receptor that enhances neurotransmitter release. *Neuron*, 21: 123–136.

Bremerich, D.H., Kai, T., Warner, D.O. and Jones, K.A. (1998) Effect of phorbol esters on Ca^{2+} sensitivity and myosin light-chain phosphorylation in airway smooth muscle. *Am. J. Physiol.*, 274: C1253–C1260.

Brigant, J.L. and Mallart, A. (1982) Presynaptic currents in mouse motor endings. *J. Physiol.*, 333: 619–636.

Brose, N., Hoffmann, K., Hata, Y. and Südhof, T.C. (1995) Mammalian homologues of *Caenorhabditis elegans unc-13* gene define a novel family of C2 domain proteins. *J. Biol. Chem.*, 270: 25273–25280.

Geppert, M., Goda, Y., Stevens, C.F. and Südhof, T.C. (1997). The small GTP-binding protein rab3a regulates a late step in synaptic vesicle fusion. *Nature*, 387: 810–814.

Geppert, M. and Südhof, T.C. (1998) Rab3 and Synaptotagmin: The yin and yang of synaptic membrane fusion. *Ann Rev. Neurosci.*, 21: 75–95.

Ginsborg, B.L. and Hirst, G.D.S. (1972). The effect of adenosine on the release of the transmitter from the phrenic nerve of the rat. *J. Physiol.*, 224: 629–645.

Goda, Y. and Südhof, T.C. (1997). Calcium regulation of neurotransmitter release: reliably unreliable? *Curr. Opin. Cell Biol.*, 9: 513–515.

Grinnell, A.C., Gundersen, C.B., Meriney, S.D. and Young, S.H. (1989) Direct measurement of ACh release from exposed frog nerve terminals: Constraints on interpretation of non-quantal release. *J. Physiol.*, 419: 225–251.

Gunderson, C.B., Katz, B. and Miledi, R. (1982) The antagonism between botulinum toxin and calcium in motor nerve terminals. *Proc. R. Soc. (Lond)*, 216: 369–376.

Hunt, J.M., Silinsky, E.M., Hirsh, J.K., Ahn, D. and Solsona, C. (1990). The effects of an antagonist of intracellular calcium translocation (TMB-8) on transmitter release and the action of adenosine at motor nerve endings. *Eur. J. Pharmacol.*, 178: 259–266.

Hunt, J.M., Redman, R.S. and Silinsky, E.M. (1994) Reduction by intracellular calcium chelation of acetylcholine secretion without occluding the effects of adenosine at frog motor nerve endings. *Br. J. Pharmacol.*, 111: 753–758.

Hunt, J.M. and Silinsky, E.M. (1993) Ionomycin-induced acetylcholine release and its inhibition by adenosine at frog motor nerve endings. *Br. J. Pharmacol.*, 110: 828–832.

Miller, R.J. (1998) Presynaptic receptors. *Ann. Rev. Pharmacol. Toxicol.*, 38: 201–227.

Nagano, O., Foldes, F.F., Natasuka, H., Reich, D., Ohta, Y., Sperlagh, B. and Vizi, E.S. (1992) Presynaptic A_1-purinoceptor-mediated inhibitory effects of adenosine and its

158

stable analogues on the mouse hemidiaphragm preparation. *Naunyn-Schmiedeberg's Archiv. Pharmacol.*, 346: 197–202.

Newton, A.C. (1995) Protein kinase C: Structure, function and regulation. *J. Biol Chem.*, 270: 28495–28498.

North, R.A. and Barnard, E. (1997) Nucleotide receptors. *Curr. Opin. Neurobiol.*, 7: 346–357.

Prior, C., Breadon, E.L. and Lindsay, K.E. (1997) Modulation by presynaptic adenosine A_1 receptors of nicotinic receptor antagonist-induced neuromuscular block in the mouse. *Eur. J. Pharmacol.*, 327: 103–108.

Ribeiro, J.A. and Sebastiao, A.M. (1987) On the role, inactivation and origin of endogenous adenosine at the frog neuromuscular junction. *J. Physiol.*, 384: 571–585.

Ribeiro, J.W. and Walker, J. (1975) The effects of adenosine triphosphate and adenosine diphosphate on transmission at the rat and frog neuromuscular junction. *Br. J. Pharmacol.*, 54: 213–218.

Redman, R.S. and Silinsky, E.M. (1993) A selective adenosine antagonist (8-cyclopentyl-1,3,-dipropylxanthine) eliminates both neuromuscular depression and the action of exogenous adenosine by an effect on A_1 receptors. *Mol. Pharmacol.*, 45: 835–840.

Redman, R.S. and Silinsky, E.M. (1994a) ATP released together with acetylcholine as the mediator of neuromuscular depression at frog motor nerve endings. *J. Physiol.*, 477.1: 117–127.

Redman, R.S. and Silinsky, E.M. (1994b) Aminoglycoside antibiotics decrease calcium currents in frog motor nerve endings. *Br. J. Pharmacol.*, 113: 375–378.

Redman, R.S. and Silinsky, E.M. (1995) On the simultaneous electrophysiological measurements of neurotransmitter release and perineural calcium currents from frog motor nerve endings. *J. Neurosci. Methods*, 57: 151–159.

Redman, R.S., Searl, T.J., Hirsh, J.K. and Silinsky, E.M. (1997) Opposing effects of phobol esters on transmitter release and calcium currents at frog motor nerve endings. *J. Physiol.*, 501.1: 41–48.

Searl, T.J. and Silinsky, E.M. (1998) Increases in acetylcholine release produced by phorbol esters are not mediated by protein kinase C at motor nerve endings. *J. Pharmacol. Exper, Ther.*, 285: 247–251.

Searl, T.J., Redman, R.S. and Silinsky, E.M. (1998) Mutual occlusion of ATP-gated channels and nicotinic receptors on sympathetic neurons of the guinea-pig. *J. Physiol.*, 510.3: 783–791.

Sebastiao, A.M. and Ribeiro, J.A. (1990) Interactions between adenosine and phorbol esters or lithium at the frog neuromuscular junction. *Br. J. Pharmacol.*, 100: 55–62.

Silinsky, E.M. (1975) On the association between transmitter secretion and the release of adenine nucleotides from mammalian motor nerve terminals. *J. Physiol.*, 247: 145–162.

Silinsky, E.M. (1980) Evidence for specific adenosine receptors at cholinergic nerve endings. *Br. J. Pharmacol.*, 71: 191–194.

Silinsky, E.M. (1981) On the calcium receptor that mediates depolarization-secretion coupling at cholinergic motor nerve endings. *Br. J. Pharmacol.*, 73: 413–429.

Silinsky, E.M. (1984) On the mechanism by which adenosine receptor activation inhibits the release of acetylcholine from motor nerve endings. *J. Physiol.*, 346: 243–256.

Silinsky, E.M. (1986) Inhibition of transmitter release by adenosine: Are calcium currents depressed or are the intracellular effects of calcium impaired? *Trends in Pharmacol. Sci.*, 7: 180–185.

Silinsky, E.M., Gerzanich, V. and Vanner, S.M. (1992) ATP mediates excitatory synaptic transmission in mammalian neurones. *Br. J. Pharmacol.*, 106: 762–763.

Silinsky, E.M. and Gerzanich, V. (1993) On the excitatory effects of ATP and its role as a neurotransmitter in coeliac neurons of the guinea-pig. *J. Physiol.*, 464: 197–212.

Silinsky, E.M. and Hubbard, J.I. (1973) Release of ATP from rat motor nerve terminals. *Nature*, 243: 404–405.

Silinsky, E.M. and Redman, R.S. (1996) Synchronous release of ATP and neurotransmitter within milliseconds of a motor nerve impulse in the frog. *J. Physiol.*, 492.3: 815–822.

Silinsky, E.M. and Solsona, C.S. (1992) Calcium currents at motor nerve endings: Absence of effects of adenosine receptor agonists in the frog. *J. Physiol.*, 457: 315–328.

Silinsky, E.M., Watanabe, M., Redman, R.S., Qiu, R., Hirsh, J.K., Hunt, J.M., Solsona, C.S., Alford, S. and Macdonald, R.C. (1995) Neurotransmitter release evoked by nerve impulses without calcium entry through calcium channels in frog motor nerve endings. *J. Physiol.*, 482.3: 511–520.

Singh, Y.N., Dryden, W.F. and Chen, H. (1990) The inhibitory effects of some adenosine analogues on transmitter release at the mammalian neuromuscular junction. *Can. J. Physiol. Pharmacol.*, 64: 1446–1450.

Stanley, E.F. and Mirotznik, R.R. (1997) Cleavage of syntaxin prevents G-protein regulation of presynaptic calcium channels. *Nature*, 385: 340–343.

Südhof, T.C. (1995) The synaptic vesicle cycle: A cascade of protein–protein interactions. *Nature*, 375: 645–653.

Zimmermann, H. (1994) Signalling via ATP in the nervous system. *Trends Neurosci.*, 17: 420–426.

Zimmermann, H., Volknandt, W., Wittich, B. and Hausinger, A. (1993) Synaptic vesicle life cycle and turnover. *J. Physiol. (Paris)*, 87: 159–170.

Zimmermann H. (1996) Biochemistry, localization and functional roles of ecto-nucleotidases in the nervous system. *Prog. Neurobiol.*, 49: 589–618.

P. Illes and H. Zimmermann (Eds.)
Progress in Brain Research, Vol 120
© 1999 Elsevier Science BV. All rights reserved

CHAPTER 13

Receptor- and carrier-mediated release of ATP of postsynaptic origin: Cascade transmission

E. Sylvester Vizi* and Beáta Sperlágh

Department of Pharmacology, Institute of Experimental Medicine, Hungarian Academy of Sciences, POB 67, H-1450 Budapest, Hungary

Introduction

The concept of cotransmitters has been discussed by Burnstock (1990). Two transmitters to fulfill the expectation to be cotransmitters should be released from the same terminals by means of exocytosis and should synergistically activate the receptors on the target cell. ATP has been described as a typical example of a cotransmitter (Burnstock, 1990; Von Kügelgen and Starke, 1991b). Electrophysiological evidence was obtained that ATP activates an inward current on many neuronal and smooth muscle preparations via P2X receptors expressed in the brain (cf. Illes et al., 1996) including the rat medial habenula nucleus (Edwards et al., 1992) the dorsal horn of the rat spinal cord (Bardoni et al., 1997), the locus coeruleus (Nieber et al., 1997) in sympathetic ganglia (Silinsky, et al., 1992; Evans, et al., 1992; Cloues, et al., 1993; Nakazawa and Inoue, 1993; Nakazawa, 1994) as well as in enteric neurons (Galligan and Bertrand, 1994; Zhou and Galligan, 1998). In addition, it has been shown that ATP is released in response to neuronal activity from vas deferens (Kirkpatrick and Burnstock, 1987; Kasakov et al., 1988; Vizi and Burnstock, 1988; Von Kügelgen and Starke, 1991a; Sperlágh and Vizi, 1992; Vizi, et al., 1992; Todorov et al., 1996) from medial habenula (Sperlágh et al., 1995;

Sperlágh et al., 1998a), hypothalamus (Sperlágh et al., 1998b), cervical ganglia (Vizi et al., 1997), and hippocampus (Cunha et al., 1996). Depolarization-induced purely neuronal release of ATP has been demonstrated from cultured neuronal cells, such as sympathetic neurons (von Kügelgen et al., 1994) or from synaptosomes (Potter and White, 1980; Richardson and Brown, 1987). However, due to the action of rapid and highly effective degradation of extracellular ATP by Ca^{2+}/Mg^{2+} dependent ectoATPase (Dunwiddie et al., 1997; Zimmermann et al., 1998) the real biophase concentration of endogenous ATP in the vicinity of the postsynaptic receptors is difficult to estimate.

Nevertheless, neurochemical evidence is available that ATP is co-stored with transmitters (cf. Zimmermann, 1994; cf. Sperlágh and Vizi, 1996) and released together with noradrenaline (Kirkpatrick and Burnstock, 1987; Kasakov et al., 1988; Vizi and Burnstock, 1988; Von Kügelgen and Starke, 1991a; Vizi et al., 1992; Sperlágh et al., 1998b) and acetylcholine (Vizi et al., 1997), and it was, therefore, strongly suggested that it is a cotransmitter (Burnstock, 1990; Von Kügelgen and Starke, 1991b).

Receptor and carrier-mediated release of ATP

It has generally been thought that ATP and noradrenaline (NA) are coreleased from the same

*Corresponding author. Tel.: +36-1-210-0819; fax: +36-1-210-0813; e-mail: Esvizi@koki.hu

vesicles. However, the results of recent studies (e.g. Sperlágh and Vizi, 1992; Westfall et al., 1996; cf. Von Kügelgen, 1996) on the presynaptic modulation of chemical transmission suggest that ATP may not be released from the same terminals as transmitters. It was even suggested that it is released from the postsynaptic site in response to the effect of endogenous ligands on their receptors (Vizi et al., 1992).

Since ATP is present not only in the synaptic vesicles but also in the in the cytoplasm of every living cells, its extracellular presence does not necessarily suggest a neuronal origin, the activated target cells, as well as glial cells are also potential sources of ATP. Therefore in previous studies we (Vizi and Burnstock, 1988; Vizi, et al., 1992) and others (Westfall et al., 1987; Katsuragi et al., 1990; Katsuragi et al., 1991; Von Kügelgen and Starke, 1991a; Tokunaga et al., 1993; Kirkpatrick and Burnstock, 1994; Von Kügelgen and Starke, 1994) have found that endogenous ATP is released to the extracellular space in response to the action of various agonists on their specific receptors.

Noradrenergic transmission and ATP

Thus, in the guinea-pig vas deferens preparation, where the co-transmitter function of ATP with noradrenaline is supported by electrophysiological and pharmacological evidence (Meldrum and Burnstock, 1983; Sneddon and Burnstock, 1984), and ATP elicits the twitch component of the biphasic response of the smooth muscle in response to field stimulation, both α_1-adrenoceptor and P2X-purinoceptor agonists were able to release ATP in a concentration-dependent manner (Fig. 1A,C). Furthermore, the action of noradrenaline to release ATP persisted after sympathetic denervation of the vasa deferentia by 6-hydroxidopamine pretreatment, indicating that the source of α_1-adrenoceptor-mediated release of ATP is the postsynaptic target cell (Fig. 1B). Electrical stimulation-induced release of ATP was reduced to approximately 50% by the α_1-adrenoceptor antagonist, prazosin, showing that the origin of ATP in this tissue is heterogeneous, partly neuronal (presynaptic), and partly non-neuronal, (postsynaptic).

Role of ATP released from postsynaptic site: cascade transmission

On the basis of these results, a new type of transmission, the cascade transmission has been proposed (Vizi et al., 1992), where a primary neurotransmitter (e.g. noradrenaline) releases a secondary transmitter (ATP) in a receptor-mediated manner from postsynaptic sites. ATP released in this way may amplify or modulate the secretory response and participate in chemical transmission in the following ways:

(a) it may contribute to fast synaptic transmission;
(b) it may modulate the action of primary transmitters on postsynaptic receptors;
(c) by itself, or after degradation to adenosine it may be involved in the prejunctional modulation of neurotransmitter release via A_1 or P_2 purinoceptors.

The functional significance of cascade transmission could be particularly important in those systems, where the primary transmitter diffuses to a long distance from its release sites to reach postsynaptic receptors, i.e. when non-synaptic interaction occurs (Vizi and Lábos, 1991).

The contribution of α_1-adrenoceptor-mediated release of ATP to extracellular ATP accumulation has also been shown in other preparations, innervated by the sympathetic nervous system, such as rat tail artery (Westfall et al., 1987; Shinozuka et al.,1991), rabbit aorta (Sedaa et al.,1989) and human saphenous vein (Rump and Von Kügelgen, 1994). In addition to smooth muscle, the postsynaptic target cells in blood vessels could also be the endothelial cells, which have been shown to be able to release ATP in response to mechanical stimuli via a glibenclamide-sensitive mechanism (Hassessian et al., 1993) or by agonists (Yang et al., 1994; Shinozuka et al., 1994). The relative contribution of α_1-adrenoceptor-mediated ATP release to the overall ATP outflow varies between tissues, and may reach 90–97% of the total stimulation-induced overflow (Sedaa et al., 1989; Von Kügelgen and Starke, 1994). Interestingly, α_1-adrenoceptor activation and subsequent adenosine accumulation has been found to be responsible for the beneficial effect of ischemic preconditioning on infarct size in *in vivo* canine and rat hearts (Kitakaze et al., 1994; Sato et al., 1997).

Cholinergic transmission and ATP

The receptor-operated ATP release has been shown to occur not only in noradrenergic tissues, but also in tissues innervated by cholinergic nerves. Thus, in the longitudinal muscle strip of guinea-pig ileum, ATP is co-released with acetylcholine from the

varicosities of the myenteric plexus (White, 1978), and acts as a prejunctional modulator of the neurally released acetylcholine (Vizi and Knoll, 1976; Sperlágh and Vizi, 1991). In the same preparation, carbamylcholine, a muscarinic-type acetylcholine receptor agonist has been shown to release ATP, an effect reversed by atropine, and the M_3 selective muscarinic receptor antagonist 4-DAMP (Table 1). Furthermore, atropine and 4-DAMP considerably reduced nerve stimulation-induced ATP overflow, indicating that ATP released

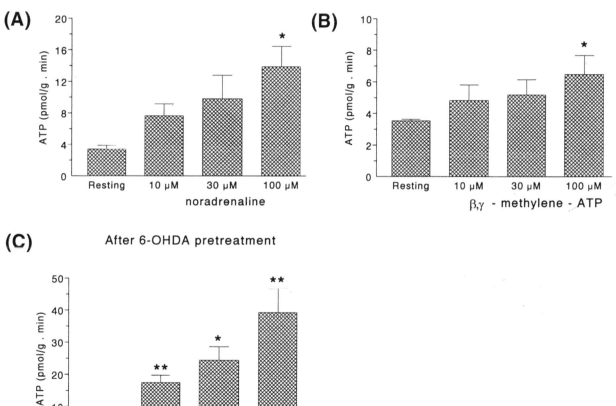

Fig. 1. Receptor-operated ATP release from isolated guinea-pig vas deferens preparation. Vasa deferentia were superfused with Krebs' solution and exposed to bath application of increasing concentrations of the α_1-adrenoceptor agonist noradrenaline (A) or the P2X-receptor agonist β,γ-methyleneATP (C). ATP release by noradrenaline was also studied in vasa deferentia of rats pretreated with 6-hydroxidopamine (100 mg/kg + 250 mg/kg i.p.) (B). This pretreatment reduced almost completely the endogenous noradrenaline content in the preparations (from 15.32 ± 1.89 to 1.32 ± 0.28 μg/g ($P < 0.01$, $n = 4$). The release of ATP is measured by luciferin–luciferase assay and is expressed in nmol/g.min. The amount of ATP released in three consecutive samples (9 min) was assayed in the absence or presence of noradrenaline or β,γ-methyleneATP. Results show mean ± SEM of 4–6 identical experiments. Asterisks indicate significant differences from resting release (*$P < 0.05$, **$P < 0.01$).

TABLE 1

Release of endogenous ATP from isolated longitudinal strips of guinea-pig ileum by electrical stimulation and by carbamylcholine (100 μM)

	ATP release (pmol/g)	
	control	+ Atropine (1 μM)
1. Electrical stimulation	750.3 ± 187.9	22.79 ± 16.64*
2. Carbamylcholine	10126 ± 4336	540.5 ± 186.7**

Preparations were placed to thermoregulated organ baths and perfused continuously with Krebs' solution with the rate of 1 ml/min. After 60 min preperfusion period electrical field stimulation (2 Hz, 1 ms, 240 pulses) were applied or carbamylcholine (100 μM) was perfused. Atropine (1 μM) was administered 15 min before electrical stimulation or carbamylcholine application. ATP release was assayed by the luciferin–luciferase assay. The release of ATP evoked by electrical stimulation or carbamylcholine was calculated by the area-under-the-curve method and expressed in pmol/g. Statistical comparison was carried out by Student t-test and asterisks indicate significant differences between atropine treated and control preparations ($n = 7$, *$P < 0.001$, **$P < 0.05$). Atropine alone did not change the outflow of ATP.

by neural activity also derived partly from post-synaptic sources, in response to the action of endogenous acetylcholine on M_3-receptors (Nita-hara et al.,1995). Our results are in agreement with those of Katsuragi et al. (1992, 1993) who also found that muscarinic agonists are able to release ATP from innervated smooth muscle preparations.

In addition to muscarinic receptor activation, other stimuli were also effective to release ATP in the myenteric plexus–ileal smooth muscle preparation. Thus, when different concentrations of ADP were added to the tissues, and the amount of nucleotides in the incubation media were analyzed by the HPLC–UV technique, the amount of ADP declined after its addition, and simultaneously appeared in the medium in consequence to the activity of ectonucleotidases (Figs. 2 and 3). However, in addition to AMP, ATP also appeared in the extracellular fluid after addition of ADP, and its amount was dependent on the initial concentration of ADP, indicating that ADP, acting on P2X purinoceptors, elicited the release of ATP (Fig. 3). To rule out the possibility that ADP rephosphory-lated to ATP by ectokinases, the sum of nucleotides ([ATP + ADP + AMP]) was also calculated, which verified a net release of nucleotides (most likely ATP) in response to the addition of ADP.

Similar results have been obtained in another preparation, containing exclusively neuro-neuronal synapses, i.e. in the rat sympathetic ganglion. In addition to acetylcholine (ACh) a classical fast transmitter, it has been suggested that ATP is also involved in neurotransmission in sympathetic ganglia (Evans et al., 1992): it accelerates the decay of the current produced by ACh (Nakazawa and Inoue, 1993). In contrast, Inokuchi and McLachlan (1995) failed to observe such an effect. Nevertheless, Nakazawa (1994) found that there is an interaction between the ATP- and ACh-evoked conductances in rat sympathetic neurons; that is, ATP activates a subpopulation of the nicotinic receptor channels. Acutely isolated superior cervical ganglia of the rat release the multiple amount of ATP to low frequency electrical stimulation, in comparison to other preparations, which might be explained by tissue structure, i.e. the ganglion is a pure neuronal tissue, where nerve terminals are packaged tightly to each other (Vizi et al., 1997). Again, carba-mylcholine, the muscarinic receptor agonist was found to release ATP and this release was mediated by the activation of nicotinic acetylcholine receptors, because hexamethonium, a nicotinic receptor antagonist inhibited it (Vizi et al., 1997). Since carbamylcholine-evoked release of ATP did not change after 10 days of surgical denervation of the ganglia, the source of this type of release is identified at the postsynaptic target neuron (Table 2).

Inhibition of carrier function by cooling prevents receptor-mediated release of ATP

A further intriguing question is that if ATP is released in a receptor-operated manner, how does it enter to the extracellular space. Being a highly charged molecule, ATP cannot pass the cell membrane by simple diffusion. Since agonist-evoked ATP release in general proved to be independent of extracellular Ca^{2+} (Vizi et al., 1992), apart from cytolytic release, carrier-mediated release seemed to be the most likely mechanism responsible for receptor-operated ATP release. Autocrine signaling by transmembrane release of ATP through membrane transporters has already been implicated in non-neuronal cells (Wang et al., 1996; Mitchell et al., 1998) but not in neuronal interactions. Another indication for the carrier mediated release of ATP came from studies, where carrier-mediated membrane transport systems for nucleotides have been reported in crayfish motor nerves and brain synaptosomes (Luqmani 1981; Chaudry et al., 1985; Sun and Lee, 1985). Although the temperature-dependent function of membrane carriers has been

known for a long time, it has only recently recognized that unlike exocytotic transmitter release, neurotransmitter uptake systems are inhibited by low temperature, and 12°C is the cut-off point where vesicular and carrier-mediated release could be separated (Table 3) (Vizi, 1998). By the use of the FRIGOMIX cooling system (Braun Biotech International, Germany), the perfusion fluid could be quickly and gradually cooled and the

(A)   60 nmol ADP

(C)   1500 nmol ADP

(B)   300 nmol ADP

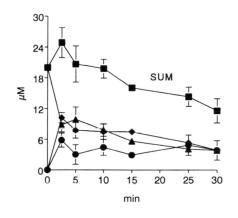
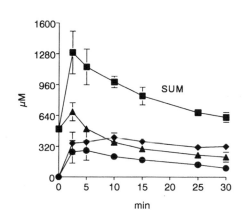
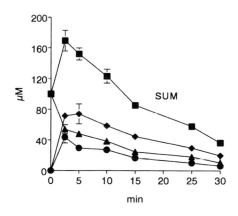

Fig. 2. ADP-evoked release of nucleotides in the longitudinal muscle strip of guinea-pig ileum. (A) 60 nmol (B) 300 nmol and (C) 1500 nmol ADP was added to the preparations, incubated in 3 ml Krebs solution at 37°C. Subsequently, 70 µl aliquots were taken from the tubes and analyzed for ATP (diamonds) ADP (triangles) and AMP (circles) by HPLC–UV. Squares show the sum of nucleotides (SUM = [ATP + ADP + AMP]) in the aliquots after the addition of ADP. Results are expressed in µM and plotted as a function of time ($n = 3$). Note that the y axis are differently scaled, i.e. the action of ADP to release nucleotides was concentration-dependent.

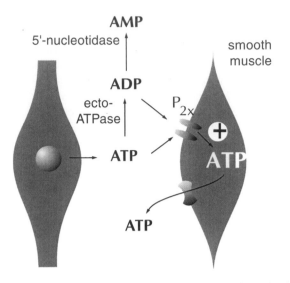

Fig. 3. Schematic representation of the mechanism of ADP evoked-nucleotide release in the guinea-pig ileum longitudinal smooth muscle preparation. ATP released from the varicosities of myenteric plexus sequentially dephosphorylated by the ectoATPase and 5′-nucleotidase enzyme. ADP, acting on P2X-purinoceptors, present on smooth muscle, gives rise to further release of ATP.

temperature of the preparations could be precisely controlled. In the guinea-pig ileum–myenteric plexus preparation we found that ATP release by carbamylcholine was strictly temperature dependent, indicating the involvement of a yet unidentified membrane carrier in this effect (Fig. 4). Therefore, we propose that acetylcholine, co-released with ATP from the varicosities of myenteric plexus upon neuronal activity, releases ATP from the smooth

muscle via a carrier-mediated mechanism acting on M_3-subtype muscarinic receptors. The action of acetylcholine on M_3 receptors is prevented by atropine, a muscarine receptor antagonist, whereas the carrier whereby ATP is released upon M_3 receptor activation is blocked by cooling the temperature to 12°C or less (Fig. 5). Furthermore, ATP release evoked by other agonists, acting on receptors located on smooth muscle, such as β, γ-methyleneATP a P2X receptor-agonist, and histamine also exhibited temperature-sensitivity in this preparation.

As evidence for a similar mechanism, in the rat superior cervical ganglion, where the field stimulation induced-release of ATP exhibited frequency dependence, lowering the bath temperature to 7°C partially inhibited the evoked-release of ATP, when higher frequencies were applied (Table 4). In contrast, acetylcholine release remained unaffected by the low temperature at any frequency showing that acetylcholine is released entirely in an exocytotic fashion (Vizi et al., 1997). Since the amount of acetylcholine released by stimulation with a constant shocks number at different frequencies remained unchanged, it is reasonable to assume that the biophase concentration of acetylcholine in the vicinity of the postsynaptic terminals was higher, when the same number of shocks stimulated the nerve terminal within a shorter period of time, i.e. at high frequency stimulation. This higher biophase concentration of acetylcholine in turn could result in a higher postsynaptic release of ATP (Table 4).

TABLE 2

Release of ATP by carbamylcholine from isolated superior cervical ganglion

	ATP release (pmol/g.5 min)	
	Control	Decentralized
Resting	187.5 ± 18.3	50 ± 14.4
Carbamylcholine 1 μM	$462.5 \pm 29.5*$	$162.5 \pm 24*$
Carbamylcholine 1 μM + Hexamethonium 100 μM	190.1 ± 12.5	60.4 ± 10.5

In 10 rats the preganglionic nerve of the SCG was cut in one side (decentralized) and the ganglion prepared from the other side was used as control. The ganglia were taken 10 days after cutting of preganglionic nerve. Significance, unpaired Student t-test. $*P < 0.01$ compared with the resting release of ATP in the absence of drugs and assayed from the same ganglion. The content of radioactivity of control and decentralized ganglia after the isolated ganglia had been loaded with 2 μCi/ml [^{3}H]choline was also determined: control = $1,587,098 \pm 43708$ Bq/g (n = 8); decentralized = $148,071 \pm 11,477$ Bq/g ($n = 8$). The difference is significant, $P < 0.01$. Carbamylcholine was added to the perfusion fluid at the concentration of 1 μM ($n = 5$), hexamethonium at concentration of 100 μM 10 min prior to exposure of ganglion to carbamylcholine ($n = 5$).

TABLE 3

Differential temperature sensitivity of vesicular and carrier mediated transmitter release

	Temperature dependence	Cold (12°C)
Vesicular release	No	Increased
Carrier-mediated release	Yes	Inhibited

Conclusions

One of the main features of the nervous system is its plasticity, achieved mainly by pre- and post-synaptic modulation of chemical transmission. It is becoming increasingly clear that ATP subserves important extracellular functions in addition to its well-established role in cell metabolism.

The finding that low temperature inhibits ATP release evoked by carbamylcholine suggests that the receptor-operated release of ATP is a carrier-mediated process. Nevertheless, it is still unclear whether a cellular compartmentalization available for this kind of postsynaptic release of ATP exists. A possible candidate for the postsynaptically-releasable ATP stores could be the so called *caveolae*, which are multiple invaginations on the membrane of the smooth muscle, present in various tissues, including the smooth muscle of vas deferens and guinea-pig ileum (Anderson, 1993). Interestingly, ectoATPase and 5′-nucleotidase precipitates are heavily localized on the surface of

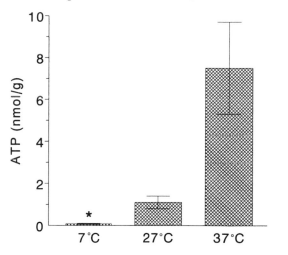

Fig. 4. Temperature-dependence of the release of ATP evoked by carbamylcholine (100 μM) in the longitudinal muscle strip of guinea-pig ileum. Preparations were placed to thermoregulated organ baths and perfused continuously with Krebs' solution with the rate of 1 ml/min. After 60 min preperfusion period carbamylcholine (100 μM) was perfused at different temperatures. Quick cooling of the preparations were obtained by a controlled thermoelectric device (FRIGOMIX) according to method of Vizi (1997). The release of ATP evoked by carbamylcholine was calculated by the area-under-the-curve method and is expressed in nmol/g. Results show mean ± SEM of four-to-six identical experiments. Asterisks indicate significant differences between 37°C and lower temperature ($*P < 0.05$, $**P < 0.01$).

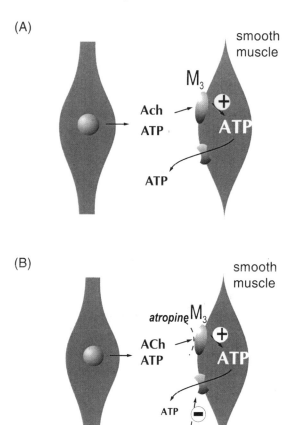

Fig. 5. Schematic representation of the mechanism of ATP release in the guinea-pig ileum longitudinal smooth muscle preparation. (A) ATP is co-released with acetylcholine from the varicosities of myenteric plexus, upon neuronal activity. Acetylcholine, acting on M_3 type muscarinic receptors, in turn releases ATP from the smooth muscle via a carrier-mediated mechanism. (B) The action of acetylcholine on M_3 receptors is prevented by atropine, a muscarine receptor antagonist, whereas the carrier whereby ATP is released upon M_3 receptor activation is blocked by cooling the temperature to 12°C or less.

caveolae observed by enzymecytochemistry (Fig. 6). In neurons, vesicular exocytosis is also a potential mechanism for receptor-operated post-synaptic release of ATP, since Zaidi and Matthews (1997) observed in rat superior cervical ganglion that exocytotic release of transmitters occurs from neuronal cell bodies and from dendrites upon stimulation of ganglia by potassium depolarization or cholinergic agonists. A further direction for future investigation would be the molecular identification of ATP transporters in different neuronal tissues, where the functional significance of purinergic neurotransmission has already been established.

In summary, it has been found that ATP can be released not only from the presynaptic nerve

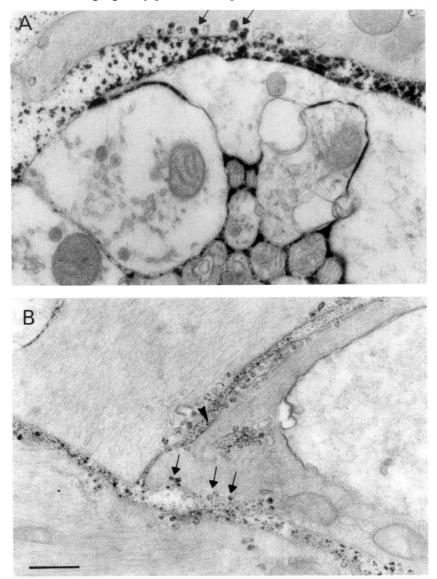

Fig. 6. EctoATPase enzymecytochemistry in ileal smooth muscle (A) and 5'-nucleotidase enzyme hystochemistry in vas deferens (B) of the guinea pig (from A. Kittel and E. Bácsy with permission). (A) The inside of caveolae (arrow) is strongly reactive. Heavy reaction deposit is also seen on or between the membranes of nerve terminals and sometimes in the intercellular matrix. (B) 5'-nucleotidase activity in caveolae (arrows) and on the cell membrane (arrowhead) of vas deferens smooth muscle cells.

TABLE 4

Effect of low temperature on release of ATP and [³H]acetylcholine at rest and in response to stimulation in rat superior cervical ganglion ($n = 8$)

	Release at 37°C		Release at Release at 7°C	
	[³H]ACh(%)	ATP(nmol/g)	[³H]ACh(%)	ATP(nmol/g)
Resting in 5 min	1.83 ± 0.14	0.125 ± 0.020	1.89 ± 0.21	0.025 ± 0.004*
Stimulation evoked (360 shocks)				
2 Hz	9.5 ± 1.1	0.73 ± 0.08	9.87 ± 1.63	0.72 ± 0.13
6 Hz	10.66 ± 1.83	7.3 ± 0.31	10.32 + 0.75	3.4 ± 0.49*
10 Hz	11.82 ± 2.14	30.1 ± 2.77	11.64 ± 0.64	6.45 ± 0.8*

Note that while the release of acetylcholine was not changed by lowering the temperature, that of ATP was significantly reduced both at rest and in response to stimulation. *Significant ($p > 0.01$) difference from values measured at 37°C.

terminals but also from the postsynaptic target cells and this is a receptor- and carrier-mediated process. Therefore, we suggest that ATP of postsynaptic origin, released from the postsynaptic cell (e.g. from the smooth muscle, sympathetic neuron) in response to the effect of endogenous ligands (e.g. noradrenaline, ACh), is involved in signal transmission. Thus, there is evidence for a cascade transmission, where a primary transmitter (e.g. noradrenaline, ACh) releases a secondary transmitter (ATP) in a receptor- and transporter-mediated manner. On the basis of these results, we propose that ATP acts locally and non-synaptically to amplify or modulate the secretory response elicited by a more traditional signal.

Acknowledgments

This work was supported by the grants of Hungarian Research Foundation (OTKA T025614) and Hungarian Medical Research Council (ETT). The authors are grateful to Dr. Ágnes Killel and Dr. Ernö Bácsy for providing electromicrographs and to Ms. Zsuzsanna Körössy, Ms. Éva Szénássy and Ms. Katalin Lengyel for expert technical assistance.

References

Anderson, R.G.W. (1993) Caveloae: Where incoming and outgoing messengers meet. *Proc. Natl. Acad. Sci. USA*, 90: 10909–10913.

Bardoni, R., Goldstein, P.A., Lee, C.J., Gu, G.J. and MacDermott, A.B. (1997) ATP P2X receptors mediate fast synaptic transmission in the dorsal horn of the rat spinal cord. *J. Neurosci.*, 1797: 5297–5304.

Burnstock, G. (1990) Noradrenaline and ATP as cotransmitters in sympathetic nerves. *Neurochem. Int.*, 17: 357–368.

Chaudry, I.H., Clemens, M.G. and Baue, A.E. (1985) Uptake of ATP by tissues. In: T.W. Stone (Ed.), *Purines: Pharmacology and Physiological Roles*. Macmillan, London, pp. 115–124.

Cloues, R., Jones, S. and Brown, D. (1993) Zn^{2+} potentiates ATP-activated currents in rat sympathetic neurones. *Eur. J. Physiol.*, 424: 152–158.

Cunha, R.A., Vizi, E.S., Ribeiro, J.A. and Sebastiao, A.M. (1996) Preferential release of ATP and its extracellular catabolism as a source of adenosine upon high- but not low-frequency stimulation of rat hippocampal slices. *J. Neurochem.*, 67: 2180–2187.

Dunwiddie, T.W., Diao, L.H. and Proctor, W.R. (1997) Adenosine nucleotides undergo rapid quantitative conversion to adenosine in the extracellular space in rat hippocampus. *J. Neurosci.*, 17: 7673–7682.

Edwards, F.A., Gibb, A.J. and Colquhoun, D. (1992) ATP receptor-mediated synaptic currents in the central nervous system. *Nature*, 359: 144–147.

Evans, R.J., Derkach, V. and Suprenant, A. (1992) ATP mediates fast synaptic transmission in mammalian neurons. *Nature*, 357: 503–505.

Galligan, J.J. and Bertrand, P.P. (1994) ATP mediates fast synaptic potentials in enteric neurons. *J. Neurosci.*, 14: 7563–7571.

Hassessian, H., Bodin, P. and Burnstock, G. (1993) Blockade by glibenclamide of the flow-evoked endothelial release of ATP that contributes to vasodilatation in the pulmonary vascular bed of the rat. *Br. J. Pharmacol.*, 109: 466–472.

Illes, P., Nieber, K. and Nörenberg, W. (1996) Electrophysiological effects of ATP on brain neurones. *J. Auton. Pharm.*, 16: 407–411.

Inokuchi, H. and McLachlan, E.M. (1995) Lack of evidence for P2X-purinoceptor involvement in fast synaptic responses in

168

intact sympathetic ganglia isolated from guinea-pigs. *Neurosci.*, 69: 651–659.

Kasakov, L., Ellis, J., Kirkpatrick, K., Milner, P. and Burnstock, G. (1988) Direct evidence for concomitant release of noradrenaline, adenosine 5'-triphosphate and neuropeptide Y from sympathetic nerve supplying the guinea-pig vas deferens. *J. Auton. Nerv. Syst.*, 22: 75–82.

Katsuragi, T., Tokunaga, T., Usune, S. and Furukawa, T. (1990) A possible coupling of postjunctional ATP release and transmitter's receptor stimulation in smooth muscles. *Life Sci.*, 46: 1301–1307.

Katsuragi, T., Tokunaga, T., Ogawa, S., Soejima, O., Sato, C. and Furukawa, T. (1991) Existence of ATP-evoked ATP release system in smooth muscle. *J. Pharmacol. Exp. Ther.*, 259: 513–518.

Katsuragi, T., Soejima, O., Tokunaga, T. and Furukawa, T. (1992) Evidence for postjunctional release of ATP evoked by stimulation of muscarinic receptors in ileal longitudinal muscles of guinea-pig. *J. Pharmacol. Exp. Ther.*, 260: 1309–1313.

Katsuragi, T., Shirakabe, K., Soejima, O., Tokunaga, T., Matsuo, K., Sato, C. and Furukawa, T. (1993) Possible transsynaptic cholinergic neuromodulation by ATP released from ileal longitudinal muscles of guinea-pigs. *Life Sci.*, 53: 911–918.

Kirkpatrick, K. and Burnstock, G. (1987) Sympathetic nerve-mediated release of ATP from the guinea-pig vas deferens is unaffected by reserpine. *Eur. J. Pharmacol.*, 138: 207–214.

Kirkpatrick, K.A. and Burnstock, G. (1994) Release of endogenous ATP from the vasa deferentia of the rat and guinea-pig by the indirect sympathomimemtic amine tyramine. *J. Auton. Pharmacol.*, 14: 325–335.

Kitakaze, M., Hori, M., Morioka, T., Minamino, T., Takashima, S., Okazaki, Y., Node, K., Komamura, K., Iwakura K., Itoh, T., Inoue, M. and Kamada, T. (1994) Alpha$_1$-adrenoceptor activation mediates the infarct size-limiting effect of ischemic preconditioning through augmentation of 5'-nucleotidase activity. *J. Clin. Invest.*, 93: 2197–2205.

Luqmani, Y.A. (1981) Nucleotide uptake by isolated cholinergic synaptic vesicles: evidence for a carrier of adenosine 5'-triphosphate. *Neurosci.*, 6: 1011–1021.

Meldrum, L.A. and Burnstock, G. (1983) Evidence that ATP acts as a co-transmitter with noradrenaline in sympathetic nerves supplying the guinea-pig vas deferens. *Eur. J. Pharmacol.*, 92: 161–163.

Mitchell, C.H., Carre, D.A., McGlinn, A.M., Stone, R.A. and Civan, M.M. (1998) A release mechanism for stored ATP in ocular ciliary epithelial cells. *Proc. Natl. Acad. Sci. USA*, 95: 7174–7178.

Nakazawa, K. (1994) ATP-activated current and its interaction with acetylcholine-activated current in rat sympathetic neurones. *J. Neurosci.*, 14: 740–750.

Nakazawa, K. and Inoue, K. (1993) ATP and acetylcholine activated channels co-existing in cell-free membrane patches from rat sympathetic neuron. *Neurosci. Lett.*, 163: 97–100.

Nieber, K., Poelchen, P. and Illes, P. (1997) Role of ATP in fast excitatory synaptic potentials in locus coeruleus neurones of the rat. *Br. J. Pharmacol.*, 122: 423–430.

Nitahara, K., Kittel, A., Liang, S.D. and Vizi, E.S. (1995) A$_1$-receptor mediated effect of adenosine on the release of acetylcholine from the myenteric plexus: role and localization of ectoATPase and 5'-nucleotidase. *Neurosci.*, 67: 159–168.

Potter, P. and White, T.D. (1980) Release of adenosine 5'-triphosphate from synaptosomes from different regions of rat brain. *Neurosci.*, 5: 1351–1356.

Richardson, P.J. and Brown, S.J. (1987) ATP release from affinity-purified rat cholinergic nerve terminals. *J. Neurochem.*, 48: 622–630.

Rump, L.C. and von Kügelgen, I. (1994) A study of ATP as a sympathetic cotransmitter in human saphenous vein. *Br. J. Pharmacol.*, 111: 65–72.

Sato, T., Obata, T., Yamanaka, Y. and Arita, M. (1997) Stimulation of alpha$_1$-adrenoceptors and protein kinase C mediated activation of ecto-5'-nucleotidase in rat hearts in vivo. *J. Physiol. (Lond.)*, 503: 119–127.

Sedaa, K.O., Bjur, R.A., Shinozuka, K. and Westfall, D.P. (1989) Nerve and drug-induced release of adenine nucleosides and nucleotides from rabbit aorta. *J. Pharmacol. Exp. Ther.*, 252: 1060–1067.

Shinozuka, K., Sedaa, K.O., Bjur, R.A. and Westfall, D.P. (1991) Participation by purines in the modulation of norepinephrine release by methoxamine. *Eur. J. Pharmacol.*, 192: 431–434.

Shinozuka, K., Hashimoto, M., Masumura, S., Bjur, R.A., Westfall, D.P. and Hattori K. (1994) In vitro studies of release of adenine nucleotides and adenosine from rat vascular endothelium in response to α$_1$-adrenoceptor stimulation. *Br. J. Pharmacol.*, 113: 1203–1208.

Silinsky, E.M., Gerzanich, V. and Vanner, S.M. (1992) ATP mediates excitatory synaptic transmission in mammalian neurones. *Br. J. Pharmacol.*, 106: 762–763.

Sneddon, P. and Burnstock, G. (1984) Inhibition of excitatory junction potentials in guinea-pig vas deferens by alpha-beta-methyleneATP: further evidence for ATP and noradrenaline as cotransmitters. *Eur. J. Pharmacol.*, 100: 85–90.

Sperlágh, B. and Vizi, E.S. (1991) Effect of presynaptic P$_2$ receptor stimulation on transmitter release. *J. Neurochem.*, 56: 1466–1470.

Sperlágh, B. and Vizi, E.S. (1992) Is the neuronal ATP release from guinea-pig vas deferens subject to α$_2$-adrenoceptor-mediated modulation? *Neurosci.*, 51: 203–209.

Sperlágh, B. and Vizi, E.S. (1996) Neuronal synthesis, storage and release of ATP. *Semin. in Neurosci.*, 8: 175–186.

Sperlágh, B., Kittel, A., Lajtha, A. and Vizi, E.S. (1995) ATP acts as fast neurotransmitter in rat habenula: Neurochemical and enzymecytochemical evidence. *Neurosci.*, 66: 915–920.

Sperlágh, B., Maglóczky, Zs., Vizi, E.S. and Freund, T. (1998a) The triangular septal nucleus as the major source of ATP release in the rat habenula: A combined neurochemical and morphological study. *Neurosci.*, 86: 1195–1207.

Sperlágh, B., Sershen, H., Lajtha, A. and Vizi, E.S. (1998b) Co-release of endogenous ATP and [³H]noradrenaline from rat hypothalamic slices: origin and modulation by α_2-adrenoceptors. *Neurosci.*, 82: 511–520.

Sun, A.Y. and Lee, D.Z. (1985) Synaptosomal ADP uptake. *J. Neurochem.*, 44: S90.

Todorov, L., Mihaylova-Todorova, S., Craviso, G.L., Bjur, R.A. and Westfall, T.D. (1996) Evidence for the differential release of the cotransmitters ATP and noradrenaline from sympathetic nerves of the guinea-pig vas deferens. *J. Physiol. (Lond.)*, 496: 731–748.

Tokunaga, T., Katsuragi, T., Sato, C. and Furukawa, T. (1993) ATP release evoked by isoprenaline from adrenergic nerves of guinea-pig atrium. *Neurosci. Lett.*, 186: 95–98.

Vizi, E.S. (1998) Different temperature dependence of carrier-mediated (cytoplasmic) and stimulus-evoked (exocytotic) release of transmitter: A simple method to separate the two types of release. *Neurochem. Int.*, 33: 359–366.

Vizi, E.S. and Burnstock, G. (1988) Origin of ATP release in the rat vas deferens: Concomitant measurement of ³H-noradrenaline and ¹⁴C-ATP. *Eur. J. Pharmacol.*, 158: 69–77.

Vizi, E.S. and Knoll, J. (1976) The inhibitory effect of adenosine and related nucleotides on the release of acetylcholine. *Neurosci.*, 1: 391–398.

Vizi, E.S. and Lábos, E. (1991) Non-synaptic interactions at presynaptic level. *Progr. in Neurobiol.*, 37: 145–163.

Vizi, E.S., Sperlágh, B. and Baranyi, M. (1992) Evidence that ATP, released from the postsynaptic site by noradrenaline, is involved in mechanical responses of guinea-pig vas deferens: cascade transmission. *Neurosci.*, 50: 455–465.

Vizi, E.S., Liang, S.D., Sperlágh, B., Kittel, A. and Jurányi, Z. (1997) Studies on the release and extracellular metabolism of endogenous ATP in rat superior cervical ganglion: support for neurotransmitter role of ATP. *Neurosci.*, 79: 893–903.

Von Kügelgen, I., Allgaier, C., Schobert, A. and Starke, K. (1994) Co-release of noradrenaline and ATP from cultured sympathetic neurons. *Neurosci.*, 61: 199–202.

Von Kügelgen, I. and Starke, K. (1991a) Release of noradrenaline and ATP by electrical stimulation and nicotine in guinea-pig vas deferens. *Naunyn-Schmiedeberg's Arch. Pharmacol.*, 344: 419–429.

Von Kügelgen, I. and Starke, K. (1991b) Noradrenaline-ATP co-transmission in the sympathetic nervous system. *Trends Pharmacol. Sci.*, 12: 319–324.

Von Kügelgen, I. and Starke, K. (1994) Corelease of noradrenaline and ATP by brief pulse trains in guinea-pig vas deferens. *Naunyn-Schmiedeberg's Arch. Pharmacol.*, 350: 123–129.

Von Kügelgen, I. (1996) Modulation of neural ATP release through presynaptic receptors. *Semin. in Neurosci.*, 8: 247–257.

Wang, Y., Roman, R., Lidofsky, S.D. and Fitz, J.G. (1996) Autocrine signaling through ATP release represents a novel mechanism for cell volume regulation. *Proc. Natl. Acad. Sci. USA*, 93: 12020–12025.

Westfall, D.P., Sedaa, K. and Bjur, R.A. (1987) Release of endogenous ATP from rat caudal artery. *Blood Vessels*, 4: 125–127.

Westfall, D.P Todorov,L.D., Mihaylova-Todorova, S.T. and Bjur, R.A. (1996) Differences between the regulation of noradrenaline and ATP release. *J. Auton. Pharmacol.*, 16: 393–395.

White, T.D. (1978) Release of ATP from a synaptosomal preparation by elevated extracellular K and by veratridine. *J. Neurochem.*, 40: 1069–1075.

Yang, S.Y., Cheek, D.J., Westfall, D.P. and Buxton, I.L.O. (1994) Purinergic axis in cardiac blood vessels: Agonist-mediated release of ATP from cardiac endothelial cells. *Circ. Res.*, 74: 401–407.

Zaidi, Z.F. and Matthews, M.R. (1997) Exocytotic release from neuronal cell bodies, dendrites and nerve terminals in sympathetic ganglia of the rat, and its differential regulation. *Neurosci.*, 80: 861–891.

Zhou, X. and Galligan, C. (1998) P2X-purinoceptors in cultured myenteric neurons of guinea-pig small intestine. *J. Physiol. (Lond.)*, 496: 719–729.

Zimmermann, H. (1994) Signalling via ATP in the nervous system. *Trends in Neurosci.*, 17: 420–426.

Zimmermann, H., Braun N., Kegel, B. and Heine P. (1998) New insights into molecular structure and function of ecto-nucleotidases in the nervous system. *Neurochem. Int.* 32: 421–425.

Presynaptic modulation by nucleotides of neurotransmitter release

P. Illes and H. Zimmermann (Eds.)
Progress in Brain Research, Vol 120

P2-receptors controlling neurotransmitter release from postganglionic sympathetic neurones

Ivar von Kügelgen[1,*], Wolfgang Nörenberg[1], Helga Koch[1], Angelika Meyer[1], Peter Illes[2] and Klaus Starke[1]

[1]*Department of Pharmacology, University of Freiburg, Hermann-Herder-Strasse 5, D-79104 Freiburg, Germany*
[2]*Department of Pharmacology, University of Leipzig, Härtelstrasse 16–18, D-04107 Leipzig, Germany*

Introduction

Extracellular nucleotides control the function of many cells including neurones by an action on membrane-bound P2-receptors (for reviews see Burnstock, 1990; Illes and Nörenberg, 1993; Zimmermann, 1994; North and Barnard, 1997; Silinsky et al., 1997). This article discusses the properties and the functional roles of P2-receptors present on postganglionic sympathetic neurones. A physiological role is well established for release-inhibiting P2-receptors located at the axon terminals of postganglionic sympathetic neurones. These presynaptic P2-receptors are activated by cotransmitter ATP and operate as a kind of autoreceptor mediating a feedback inhibition of ongoing, action-potential evoked sympathetic transmitter release (Fuder and Muth, 1993; Von Kügelgen et al., 1993). In addition to release-modulating P2-receptors, postganglionic sympathetic neurones possess P2-receptors, activation of which causes the generation of action potentials and subsequently transmitter release. These excitatory P2-receptors have been proposed to be involved in fast synaptic neuro-neural transmission in sympathetic ganglia (Evans et al., 1992; Silinsky et al., 1992).

*Corresponding author. Tel.: +49 228 735445; fax: +49 228 735404; e-mail: kugelgen@uni-bonn.de

Modulation of action-potential evoked transmitter release

Presynaptic release-inhibiting P2-receptors

In many tissues adenine nucleotides such as ATP inhibit the action-potential evoked release of noradrenaline. Figure 1a shows an example with the nucleotide analogue ADPβS. ADPβS given at increasing concentrations inhibited the electrically evoked release of [³H]-noradrenaline in segments of the rat pancreas in a concentration-dependent manner (continuous line in Fig. 1a). A part of this inhibitory effect was mediated by the well-known adenosine A_1-receptors (Koch et al., 1998) either directly or after breakdown of the nucleotide to AMP or adenosine (see Ross et al., 1998). A second component of inhibition, however, was mediated by separate P2-receptors. The occurrence of two separate release-inhibiting purinoceptors (A_1 and P2) has been demonstrated in several tissues by the selective interaction of adenine nucleotides with P2-receptor antagonists (for a discussion of the concept of P3-purinoceptors see Silinsky et al., 1997). In the rat pancreas and the rat heart atrium, for example, the P2-receptor antagonist reactive blue 2 and its isomer cibacron blue 3GA selectively diminished the release-inhibiting action of ATP, ATPγS and ADPβS, but not that of the pure adenosine A_1-receptor agonist N^6-cyclopentyladenosine (Von Kügelgen et al., 1995a; Koch et al.,

174

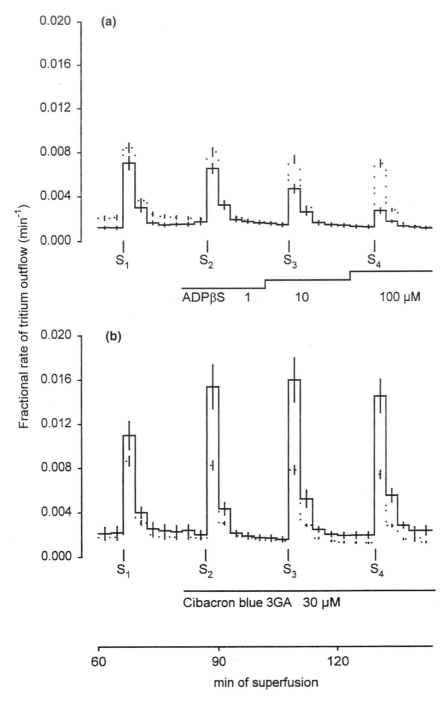

Fig. 1. Tritium outflow from segments of the rat pancreas preincubated with [³H]-noradrenaline: effects of electric stimulation, ADPβS (a) and cibacron blue 3GA (b). After preincubation with [³H]-noradrenaline the tissue segments (see Koch et al., 1998) were superfused with [³H]-noradrenaline-free buffer containing desipramine (1 μM) and yohimbine (1 μM). At $t = 66$, 87, 108 and 129 min of superfusion they were stimulated by 60 pulses at 1 Hz (S1 to S4). ADPβS or cibacron blue 3GA was added as indicated (continuous lines in (a) and (b). Control experiments in the absence of ADPβS and cibacron blue 3GA are presented in broken lines. The figure shows efflux-versus-time curves. Tritium outflow is expressed as fractional rate (min^{-1}). Means ± SEM of five to 10 experiments.

1998). Similarly, the P2-receptor antagonist suramin preferentially attenuated the release-inhibiting effects of adenine nucleotides but not of adenosine or its analogues in the mouse vas deferens (Von Kügelgen et al., 1989, 1994a; Kurz et al., 1993). The operation of inhibitory P2-receptors at postganglionic sympathetic axons seems to be widespread; they have so far been demonstrated in several species and in tissues innervated by sympathetic neurones located in paravertebral, pelvic as well as prevertebral ganglia (upper part of Table 1).

The involvement of extracellular mediators in the effects of the P2-receptor agonists seems to be unlikely since their release-inhibiting actions persisted almost unchanged in the presence of blockade of several other receptor systems (including α_1- and α_2-adrenoceptors and muscarinic receptors) as well as in the presence of blockade of prostaglandin synthesis (e.g., Von Kügelgen et al., 1995a). Therefore, the findings suggest that the release-inhibiting P2-receptors are located at the sympathetic axon terminals, i.e. that they are presynaptic receptors. (With the exception of the preparation of chick sympathetic neurones all preparations summarised in the upper part of Table 1 contain only sympathetic axons but no sympathetic nerve cell bodies.)

In most tissues, ATP and ATPγS were the most potent agonists in inhibiting transmitter release. In addition, ADPβS has recently been recognised to preferentially activate the presynaptic P2-receptor when used at low concentrations (Koch et al., 1998). Reactive blue 2 and its isomer cibacron blue 3GA acted as antagonists at the presynaptic P2-receptor. This pharmacological profile differs from the properties of all cloned P2-receptors (see North and Barnard, 1997). Nevertheless, the presynaptic P2-receptor is likely to belong to the group of G-protein-coupled P2Y-receptors, as indicated by the attenuation of P2-receptor-mediated effects on noradrenaline release after pretreatment of mouse vas deferens and heart atrium segments with pertussis toxin or N-ethylmaleimide (Von Kügelgen et al., 1993, 1995a).

In synapses with ATP as a cotransmitter the presynaptic P2-receptors can act as autoreceptors, i.e. receptors activated by (co)transmitter ATP itself. Blockade of the presynaptic P2-receptors by P2-receptor antagonists then causes an enhancement in transmitter release. Figure 1b shows as an example the marked increase in the electrically evoked release of [³H]-noradrenaline in segments of the rat pancreas after blockade of presynaptic P2-receptors by cibacron blue 3GA (30 μM; continuous line in Fig. 1b). The increase was observed after incubation with cibacron blue 3GA for only 6 min and remained approximately unchanged after 27 and 48 min of incubation (Fig. 1b). Under identical conditions, the P2-receptor antagonists

TABLE 1

Sympathetically innervated tissues (or preparations of cultured sympathetic neurones) in which P2-receptors modulate transmitter release

Release-inhibiting P2-receptors	
Mouse heart atrium	Von Kügelgen et al., 1995b
Mouse vas deferens	Von Kügelgen et al., 1989, 1993, 1994a; Kurz et al., 1993
Rat iris	Fuder and Muth, 1993
Rat heart atrium	Von Kügelgen et al., 1995a
Rat tail artery	Gonçalves and Queiroz, 1996
Rat kidney	Bohmann et al., 1997
Rat pancreas	Koch et al., 1998
Rat vas deferens	Kurz et al., 1993; Von Kügelgen et al., 1994a
Rabbit vas deferens	Grimm et al., 1994
Guinea-pig saphenous artery	Fujioka and Cheung, 1987
Chick sympathetic neurones	Allgaier et al., 1994, 1995a
Release-enhancing P2-receptors	
Rabbit ear artery	Miyahara and Suzuki, 1987 (but see Ishii et al., 1995)
Guinea-pig ileum	Sperlagh and Vizi, 1993
Chick sympathetic neurones	Allgaier et al., 1994, 1995a

reactive blue 3GA (3 to 30 μM) and reactive red 2 (3 and 10 μM) also markedly increased the release of [³H]-noradrenaline in the rat pancreas (up to 114% of control; Koch et al., 1998). In contrast, blockade of adenosine A₁-receptors caused a much smaller increase, even after a prolonged time of incubation (Koch et al., 1998). This confirms the view that the increase induced by the P2-receptor antagonists was due to a blockade of presynaptic P2-receptors (and not due to an interaction with adenosine receptors or the metabolism of adenine nucleotides). Similarly as illustrated in Fig. 1b, P2-receptor antagonists increased the action-potential evoked transmitter release in all preparations summarised in the upper part of Table 1 (in cultured chick sympathetic neurones, however, the increase in [³H]-noradrenaline release was very small; Allgaier et al., 1995a). Hence, release-inhibiting P2-receptors at postganglionic sympathetic axons generally seem to be activated by endogenous nucleotides, most likely by ATP.

By analysing the effects of P2-receptor antagonists on sympathetic transmitter release in more detail, it was shown that adenine nucleotides released by nerve activity activate the presynaptic P2-receptors. In the rat heart atrium, rat kidney, rat pancreas and the mouse vas deferens, P2-receptor antagonists increased the release of noradrenaline evoked by long pulse trains (train duration several seconds to 1 min) very markedly, but the release evoked by short pulse trains (train duration less than 100 ms) not at all or only slightly (Von Kügelgen et al., 1993, 1995a; Bohmann et al., 1997; Koch et al., 1998). These operational conditions indicate that the release of adenine nucleotides by ongoing nerve activity is a prerequisite for the function of the negative feedback mediated by the presynaptic P2-receptors (similarly known for other autoreceptor mechanisms; see Starke et al., 1989). Since the increase in transmitter release due to blockade of presynaptic P2-receptors largely persisted in preparations with blocked smooth muscle contraction (to prevent non-neuronal ATP release; see Von Kügelgen et al., 1993) cotransmitter ATP itself is likely to activate the presynaptic P2-receptors in agreement with an autoreceptor role of these P2-receptors.

The operation of presynaptic P2-receptors is not restricted to the postganglionic sympathetic nervous system. The noradrenergic and serotonergic axons in the rat occipito-parietal cortex and the noradrenergic axons in the rat hippocampus also possess inhibitory P2-receptors most likely of the group of P2Y-receptors (Von Kügelgen et al., 1994b, 1997a; Koch et al., 1997). However, in contrast to the peripheral nervous system the presynaptic P2-receptors in the rat occipito-parietal cortex and hippocampus do not seem to be activated by endogenous ligands. The role of these presynaptic P2-receptors remains to be studied.

Release-enhancing P2-receptors

In some tissues or preparations (summarised in the lower part of Table 1), adenine nucleotides also caused a facilitation of stimulation-evoked transmitter release. α,β-Methylene-ATP, for example, increased the electrically evoked release of [³H]-noradrenaline (and [³H]-acetylcholine) in the guinea-pig ileum (Sperlagh and Vizi, 1991). The involved presynaptic receptor has been described as a P2X-receptor (Sperlagh and Vizi, 1991). It is not yet known whether these release-enhancing (modulatory) P2-receptors are identical with or differ from the excitatory P2-receptors, activation of which directly induces transmitter release (see below).

Generation of action-potentials and induction of transmitter release

Extracellular adenine and uracil nucleotides have been shown to excite postganglionic sympathetic neurones of the rat (Cloues et al., 1993; Connolly et al., 1993; Boehm, 1994; Boehm et al., 1995; Khakh et al., 1995; Von Kügelgen et al., 1997b), the guinea pig (Evans et al., 1992; Silinsky and Gerzanich, 1993; Reekie and Burnstock, 1994), the chicken (Allgaier et al., 1995b) and the bullfrog (Siggins et al., 1977; Akasu et al., 1983). In many of these neurones, the nucleotides induced not only membrane depolarisation but also the generation of action potentials (see references above and Nörenberg et al., 1999). The generation of action potentials will subsequently cause transmitter release. In agreement with the electrophysiological observations, a nucleotide-induced release of noradrenaline has been demonstrated in rat and chick sympathetic neurones (Boehm 1994; Allgaier et al., 1995b; Boehm et al., 1995; Von Kügelgen et al., 1997b). It is now evident that at least two different types of P2-receptors are involved in these excitatory effects. One receptor is sensitive to adenine nucleotides whereas the other receptor is predominantly activated by uracil nucleotides (Connolly et al., 1993; Boehm et al., 1995; Von Kügelgen et al., 1997b). Figure 2 illustrates the operation of two separate excitatory P2-receptors in cultured sympathetic neurones derived from thoracolumbal ganglia of new-born rats. Electrical stimulation as well as ATP and UDP induced a transient release of [³H]-noradrenaline (broken lines in Fig. 2; see Von Kügelgen et al., 1997b). The P2-receptor antagonist suramin diminished only the response to ATP, but not to UDP (continuous lines in Fig. 2).

Excitatory P2X-receptors

The receptors mediating the fast excitatory actions of adenine nucleotides in postganglionic sympathetic neurones have been identified as P2X-receptors (Evans et al., 1992; Silinsky and Gerzanich, 1993; Khakh et al., 1995; Von Kügelgen et al., 1997b; Nörenberg et al., 1999). These receptors are ligand-gated ion channels with a relatively high permeability for sodium and calcium ions (Bean et al., 1990; Fieber and Adams, 1991; Surprenant et al., 1995; Evans et al., 1996; Rogers et al., 1997). In sympathetic neurones derived from rat thoracolumbal ganglia, activation of P2X-receptors has been shown to cause the influx of cations followed by cell membrane depolarisation, the generation of action potentials, calcium influx through voltage-sensitive calcium-channels and exocytotic transmitter release (Von Kügelgen et al., 1997b, 1998). In agreement with

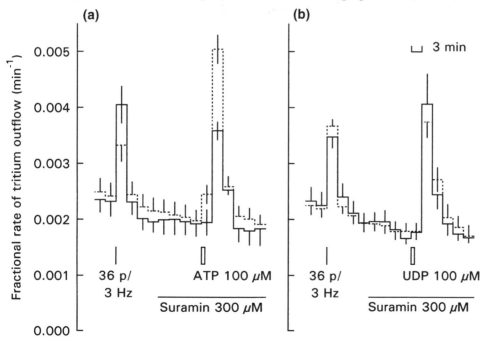

Fig. 2. Tritium outflow from dissociation cultures of rat sympathetic neurones preincubated with [³H]-noradrenaline: effects of electric stimulation, ATP (a), UDP (b), and suramin. Thoracolumbal sympathetic ganglia from newborn rats were dissociated and the cells were cultured for 6 to 7 days on a coverslip (see Von Kügelgen et al., 1997). After preincubation with [³H]-noradrenaline the cultures were superfused with [³H]-noradrenaline-free buffer containing desipramine (1 μM). At $t = 66$ min of superfusion they were stimulated electrically by 36 pulses (p) at 3 Hz. Then they were stimulated by addition of ATP (100 μM) or UDP (100 μM) at $t = 90$ min for 1 min. Suramin was added to the superfusion buffer 12 min before nucleotide addition (continuous lines in (a) and (b)). Control experiments in the absence of suramin are presented in broken lines. The figure shows efflux-versus-time curves. Tritium outflow is expressed as fractional rate (min⁻¹). Means ± SEM of five to 12 experiments.

this chain of events leading to transmitter release blockade of sodium channels by tetrodotoxin (0.5 µM) and blockade of N-type calcium channels by ω-conotoxin (0.1 µM) markedly reduced the ATP-induced release of [^{3}H]-noradrenaline from rat thoracolumbal ganglion neurones. Similarly, the ATP-induced increase in intraaxonal calcium concentration (measured by fura-2 microfluorimetry over neuronal processes) was significantly reduced by both tetrodotoxin and ω-conotoxin. Omission of calcium totally abolished both responses to ATP (Von Kügelgen et al., 1997b, 1998). The portion of ATP-induced increase in intraaxonal calcium concentration and ATP-induced noradrenaline release remaining in the presence of tetrodotoxin and ω-conotoxin (about 40 to 50%) is probably due to calcium influx through the P2X-receptor itself.

Findings of several electrophysiological studies (e.g. ATP-induced single channel currents in excised outside-out patches; Silinsky and Gerzanich, 1993) indicate that excitatory P2X-receptors occur at the cell bodies of postganglionic sympathetic neurones. The occurrence of fast ATP-gated ion channels at the somata or dendrites of these neurones is one basis for the assumption that ATP might act as a neurotransmitter in fast synaptic transmission in sympathetic ganglia (Evans et al., 1992; Silinsky et al., 1992). A second piece of evidence is the observation that ATP is in fact released in sympathetic ganglia due to stimulation of preganglionic nerve fibres (Vizi et al., 1997). However, it remains to be demonstrated that ATP plays this role in ganglionic transmission in vivo (see Inokuchi and McLachlan, 1995).

The failure of adenine nucleotides to induce noradrenaline release (at least to any major extent) in the isolated, sympathetically innervated tissues summarised in the upper part of Table 1 suggests that the sympathetic axons in these tissues possess no (or only a small number of) P2X-receptors. Hence, the occurrence of P2X-receptors may be restricted to the cell bodies or dendrites of the sympathetic neurones innervating these tissues. Similarly, excitatory P2X-receptors operate at the noradrenergic nerve cell bodies in the rat locus coeruleus (Tschöpl et al., 1992; Shen and North, 1993) but not at the noradrenergic axons in the rat occipito-parietal cortex or hippocampus (Von

Kügelgen et al., 1994b; Koch et al., 1997). As described above, the axons in these tissues possess release-inhibiting P2-receptors. This differential distribution of excitatory P2X-receptors at nerve cells bodies and inhibitory P2-receptors at axon terminals may be widespread in the nervous system.

However, under some conditions postganglionic sympathetic axons may also possess excitatory P2X-receptors. Excitatory P2X-receptors might be involved in the release-enhancing effects of adenine nucleotides observed in the rabbit ear artery and the guinea-pig ileum (see above). Moreover, in cultures of rat thoracolumbal sympathetic ganglion cells P2X-receptors occur not only at somata but also at axons. The experiments shown in Fig. 3 demonstrate a localisation at sympathetic axons. ATP induced the release of [^{3}H]-noradrenaline not only in dissociation cultures of rat thoracolumbal ganglion cells (Fig. 2a) and in cultures of whole sympathetic ganglia with outgrown processes (Fig. 3c) but also in preparations containing almost exclusively outgrown processes (neurites) but no nerve cell bodies (Fig. 3a; prior to the superfusion experiment the ganglia with the nerve cell bodies but not the processes were removed from the coverslip used for culturing and superfusion; see Przywara et al., 1993). Moreover, the resistance of a part of the ATP-induced [^{3}H]-noradrenaline release and a part of the ATP-induced increase in intraaxonal calcium level after blockade of the propagation of action potentials by tetrodotoxin strongly confirms the view that under our experimental conditions P2X-receptors are present at the axons of the cultured sympathetic neurones. The isolated tissues in which the lack of a release-inducing effect of adenine nucleotides indicated the absence of P2X-receptors at the sympathetic axons (see above) had been obtained from adult animals. In contrast, the thoracolumbal ganglia used for culturing were derived from new-born rats. Hence, the difference in the occurrence of P2X-receptors at the axons may indicate a developmental change in the expression of neuronal P2-receptors.

Excitatory P2-receptors for uracil nucleotides

In addition to P2X-receptors, a separate P2-receptor mediating the excitatory actions of uracil

179

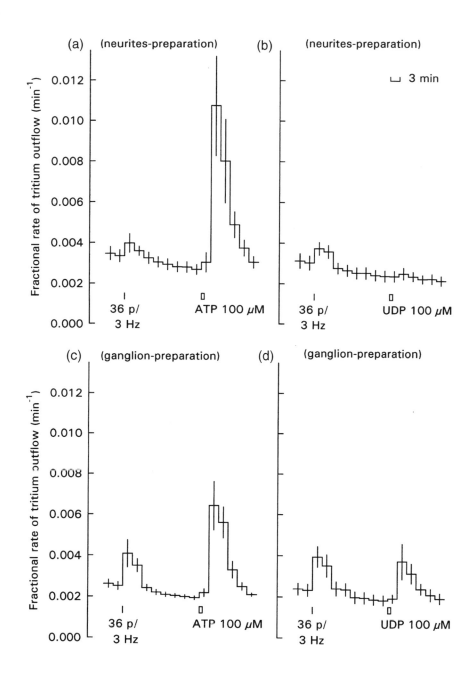

Fig. 3. Tritium outflow from axon preparations or whole ganglion cultures of rat sympathetic neurones preincubated with [³H]-noradrenaline: effects of electric stimulation, ATP (a) and (c) and UDP (b) and (d). Whole thoracolumbal sympathetic ganglia from newborn rats (not dissociated) were cultured for 6 to 7 days on a coverslip (for culture conditions see Von Kügelgen et al., 1997b). During these 6 to 7 days of culturing, a network of neuronal processes grew over the coverslip. In some preparations, the ganglia with the nerve cell bodies were mechanically removed immediately before a subsequent superfusion experiment (similarly as described by Przywara et al., 1993). After preincubation with [³H]-noradrenaline the preparations were then superfused with [³H]-noradrenaline-free buffer containing desipramine (1 μM). At t = 66 min of superfusion they were stimulated electrically by 36 pulses (p) at 3 Hz. Then they were stimulated by addition of ATP 100 μM or UDP 100 μM at t = 90 min for 1 min. The figure shows efflux-versus-time curves. Tritium outflow is expressed as fractional rate (min⁻¹). Means ± SEM of five to 9 experiments.

nucleotides has recently been identified on rat postganglionic sympathetic neurones. As shown in Fig. 2b activation of this receptor also induces the release of [^{3}H]-noradrenaline. In rat sympathetic neurones P2X-receptor blockade by suramin, PPADS and pyridoxal-phosphate attenuated the release response to ATP but not that to UDP and UTP demonstrating the involvement of a separate receptor (Fig. 2b and Boehm et al., 1995; Von Kügelgen et al., 1997b). These findings are in agreement with previous observations that uracil nucleotides caused a depolarisation of rat superior cervical ganglia by an action on a receptor distinct from a P2X-receptor (Connolly et al., 1993).

The uracil nucleotide-sensitive receptor is not a ligand-gated ion channel but most likely a member of the group of G-protein-coupled P2Y-receptors (Von Kügelgen et al., 1997b; Nörenberg et al., 1999). Several uracil nucleotide-sensitive P2Y-receptors have recently been cloned (see Nicholas et al., 1996; Communi and Boeynaems, 1997). However, due to the lack of selective antagonists a more detailed characterisation of the subtype of the uracil nucleotide-sensitive P2-receptor found in sympathetic neurones is not yet possible. Uracil nucleotides also induce the generation of action potentials in sympathetic neurones (see Nörenberg et al., 1999). In accord with this mechanism of action, tetrodotoxin totally abolished the uracil nucleotide-induced release of [^{3}H]-noradrenaline in cultures of rat sympathetic neurones (Boehm et al., 1995; Von Kügelgen et al., 1998). The chain of events leading to the generation of action potentials after activation of the uracil nucleotide-sensitive receptors remains doubtful. An inhibition of the M-current has been proposed to be involved in the excitatory action of uracil nucleotides in rat superior cervical ganglion neurones (Boehm and Bofill-Cardona 1998). However, as discussed by Nörenberg et al. (1999) several contradictory findings argue against the involvement of this mechanism in rat thoracolumbal ganglion neurones.

The uracil nucleotide-sensitive receptors might exclusively be located at the nerve cell bodies of the sympathetic neurones. UDP induced a release of [^{3}H]-noradrenaline in dissociation cultures (Fig. 2b) and cultures of whole ganglia (Fig. 3d) but not in preparations containing only neuronal processes (neurites; Fig. 3b). In agreement with a somato-dentritic localisation of the uracil nucleotide-sensitive receptors is the total blockade of the release response to uracil nucleotides by tetrodotoxin (see above). Whether the uracil nucleotide-sensitive P2-receptors on postganglionic sympathetic neurones play a physiological or pathophysiological role is not known. Uracil nucleotides are stored in several cell types and cellular elements including platelets (e.g. Goetz et al., 1971) and have recently been demonstrated to be released from astrocytoma cells (Lazarowski et al., 1997).

Conclusions

Postganglionic sympathetic neurones possess excitatory as well as inhibitory P2-receptors differing in function, signal transduction mechanisms and anatomical location. A physiological role is well established for inhibitory P2-receptors located at sympathetic axon terminals. These presynaptic P2-receptors are activated by cotransmitter ATP and inhibit further sympathetic transmitter release. Their occurrence seems to be widespread; they have been found in tissues innervated by neurones located in paravertebral, pelvic as well as prevertebral ganglia. The receptors are likely to belong to the group of G-protein-coupled P2Y-receptors. Excitatory P2-receptors located at the nerve cell bodies or dendrites of postganglionic sympathetic neurones have been proposed to be involved in fast synaptic neuro-neural transmission in sympathetic ganglia. Sympathetic neurones cultured from rat paravertebral ganglia possess two different subtypes of excitatory P2-receptors. Activation of both receptors induced the generation of action potentials and transmitter release. One receptor is an adenine nucleotide sensitive P2X-receptor. Activation of this receptor induced cation influx followed by membrane depolarisation, calcium entry through voltage-sensitive calcium-channels and exocytosis. The other receptor is activated by uracil nucleotides and belongs most likely to the group of G-protein-coupled P2Y-receptors. The chain of events leading to transmitter release upon activation of the uracil nucleotide-sensitive P2-receptor remains unclear.

Acknowledgements

The work was supported by grants of the Deutsche Forschungsgemeinschaft (SFB 505; Il 20/7-1; Ku 1173/1-1) and the European Commission (BMH4 CT96-0676).

References

Akasu, T., Hirai, K. and Koketsu, K. (1983) Modulatory actions of ATP on membrane potentials of bullfrog sympathetic ganglion cells. *Brain Res.*, 258: 313–317.

Allgaier, C., Pullmann, F., Schobert, A., Von Kügelgen, I. and Hertting, G. (1994) P_2 purinoceptors modulating noradrenaline release from sympathetic neurons in culture. *Eur. J. Pharmacol.*, 252: R7–R8.

Allgaier, C., Wellmann, H., Schobert, A. and Von Kügelgen, I. (1995a) Cultured chick sympathetic neurons: modulation of electrically evoked noradrenaline release by P2-purinoceptors. *Naunyn-Schmiedebergs Arch. Pharmacol.*, 352: 17–24.

Allgaier, C., Wellmann, H., Schobert, A., Kurz, G. and Von Kügelgen, I. (1995b) Cultured chick sympathetic neurons: ATP-induced noradrenaline release and its blockade by nicotinic receptor antagonists. *Naunyn-Schmiedebergs Arch. Pharmacol.*, 352: 25–30.

Bean, B.P., Williams, C.A. and Ceelen, P.W. (1990) ATP-activated channels in rat and bullfrog sensory neurons: current-voltage relation and single-channel behavior. *J. Neurosci.*, 10: 11–19.

Boehm, S. (1994) Noradrenaline release from rat sympathetic neurons evoked by P2-purinoceptor activation. *Naunyn-Schmiedeberg's Arch. Pharmacol.*, 350: 454–458.

Boehm, S. and Bofill-Cardona, E. (1998) Uridine nucleotide receptors modulate M-type K^+ channels of rat sympathetic neurons. *Naunyn-Schmiedebergs Arch. Pharmacol.*, 357: R33.

Boehm, S., Huck, S. and Illes, P. (1995) UTP- and ATP-triggered transmitter release from rat sympathetic neurones via separate receptors. *Br. J. Pharmacol.*, 116: 2341–2343.

Bohmann, C., Von Kügelgen, I. and Rump, L.C. (1997) P2-receptor modulation of noradrenergic neurotransmission in rat kidney. *Br. J. Pharmacol.*, 121: 1255–1262.

Burnstock, G. (1990) Co-transmission. *Arch. Int.Pharmacodyn. Ther.*, 304: 7–33.

Cloues, R., Jones, S. and Brown, D.A. (1993) Zn2+ potentiates ATP-activated currents in rat sympathetic neurons. *Pflügers Arch.*, 424: 152–158.

Communi, D. and Boeynaems, J.M. (1997) Receptors responsive to extracellular pyrimidine nucleotides. *Trends Pharmacol. Sci.*, 18: 83–86.

Connolly, G.P., Harrison, P.J. and Stone, T.W. (1993) Action of purine and pyrimidine nucleotides on the rat superior cervical ganglion. *Br. J. Pharmacol.*, 110: 1297–1304.

Evans, R.J., Derkach, V. and Surprenant, A. (1992) ATP mediates fast synaptic transmission in mammalian neurons. *Nature*, 357: 503–505.

Evans, R.J., Lewis, C., Virginio, C., Lundström, K., Buell, G., Surprenant, A. and North, R.A. (1996) Ionic permeability of, and divalent cation effects on, two ATP gated cation channels (P2X receptors) expressed in mammalian cells. *J. Physiol.*, 497: 413–422.

Fieber, L.A. and Adams, D.J. (1991) Adenosine triphosphate-evoked currents in cultured neurones dissociated from rat parasympathetic cardiac ganglia. *J. Physiol.*, 434: 239–256.

Fuder, H. and Muth, U. (1993) ATP and endogenous agonists inhibit evoked [^{3}H]-noradrenaline release in rat iris via A_1- and P_{2Y}-like purinoceptors. *Naunyn-Schmiedeberg's Arch. Pharmacol.*, 348: 352–357.

Fujioka, M. and Cheung, D.W. (1987) Autoregulation of neuromuscular transmission in the guinea-pig saphenous artery. *Eur. J. Pharmacol.*, 139: 147–153.

Goetz, U., Da Prada, M. and Pletscher, A. (1977) Adenine-, guanine- and uridine-5′-phosphonucleotides in blood platelets and storage organelles of various species. *J. Pharmacol. Exp. Ther.*, 178: 210–215.

Gonçalves, J. and Queiroz, M.G. (1996) Purinoceptor modulation of noradrenaline release in rat tail artery: tonic modulation mediated by inhibitory P2Y- and facilitatory A2A-purinoceptors. *Br. J. Pharmacol.*, 117: 156–160.

Grimm, U., Fuder, H., Moser, U., Bäumert, H.G., Mutschler, E. and Lambrecht, G. (1994) Characterization of the prejunctional muscarinic receptors mediating inhibition of evoked release of endogenous noradrenaline in rabbit isolated vas deferens. *Naunyn-Schmiedeberg's Arch. Pharmacol.*, 349: 1–10.

Illes, P. and Nörenberg, W. (1993) Neuronal ATP receptor and their mechanism of action. *Trends Pharmacol. Sci.*, 14: 50–54.

Inokuchi, H. and McLachlan, E.M. (1995) Lack of evidence for P2X-purinoceptor involvement in fast synaptic responses in intact sympathetic ganglia isolated from guinea-pigs. *Neuroscience*, 69: 651–659.

Ishii, R., Shinozuka, K., Kunitomo, M., Hashimoto, T. and Takeuchi, K. (1995) Characterization of the facilitatory prejunctional purinoceptor on adrenergic nerves of the rabbit ear artery. *J. Pharmacol. Exp. Ther.*, 273: 1390–1395.

Khakh, B.S., Humphrey, P.P.A. and Surprenant, A. (1995) Electrophysiological properties of P2X-purinoceptors in rat superior cervical, nodose and guinea-pig coeliac neurones. *J. Physiol.*, 484: 385–395.

Koch, K., Von Kügelgen I. and Starke, K. (1997) P2-Receptor-mediated inhibition of noradrenaline release in the rat hippocampus. *Naunyn-Schmiedebergs Arch. Pharmacol.*, 355: 707–715.

Koch, K., Von Kügelgen I. and Starke, K. (1998) P2-Receptor-mediated inhibition of noradrenaline release in the rat pancreas. *Naunyn-Schmiedebergs Arch. Pharmacol.*, 357: 431–440.

Kurz, K., Von Kügelgen, I. and Starke, K. (1993) Prejunctional modulation of noradrenaline release in mouse and rat vas deferens: contribution of P_1- and P_2-purinoceptors. *Br. J. Pharmacol.*, 110: 1465–1472.

Lazarowski, E.R., Homolya, L., Boucher, R.C. and Harden, T.K. (1997) Direct demonstration of mechanically induced release of cellular UTP and its implication for uridine nucleotide receptor activation. *J. Biol. Chem.*, 272: 24348–24354.

Miyahara, H. and Suzuki, H. (1987) Pre- and post-junctional effects of adenosine triphosphate on noradrenergic transmission in the rabbit ear artery. *J. Physiol.*, 389: 423–440.

Nicholas, R.A., Watt, W.C., Lazarowski, E.R., Li, Q. and Harden, K. (1996) Uridine nucleotide selectivity of three phospholipase C-activating P2 receptors: identification of a UDP-selective, a UTP-selective, and an ATP- and UTP-specific receptor. *Mol. Pharmacol.*, 50: 224–229.

Nörenberg, W., von Kügelgen, I., Meyer, A. and Illes, P. (1999) Electrophysiological analysis of P2-receptor mechanisms in rat sympathetic neurones. *Prog. Brain Res.*, 120: 203–222.

North, R.A. and Barnard, E.A. (1997) Nucleotide receptors. *Curr. Opin. Neurobiol.*, 7: 346–357.

Przywara, D.A., Bhave, S.V., Chowdhury, P.S., Wakade, T.D. and Wakade, A.R. (1993) Sites of transmitter release and relation to intracellular Ca2 + in cultured sympathetic neurons. *Neuroscience*, 52: 973–986.

Reekie, F.M. and Burnstock, G. (1994) Some effects of purines on neurones of guinea-pig superior cervical ganglia. *Gen. Pharmacol.*, 25: 143–148.

Rogers, M., Colquhoun, L.M., Patrick, J.W. and Dani, J.A. (1997) Calcium flux through predominantly independent purinergic ATP and nicotinic acetylcholine receptors. *J. Neurophysiol.*, 77: 1407–1417.

Ross, F.M., Brodie, M.J. and Stone, T.W. (1998) Adenosine monophosphate as a mediator of ATP effects at P1 purinoceptors. *Br. J. Pharmacol.*, 124: 818–824.

Siggins, G.R., Gruol, D.L., Padjen, A.L. and Forman, D.S. (1977) Purine and pyrimidine mononucleotides depolarise neurones of explanted amphibian sympathetic ganglia. *Nature*, 270: 263–265.

Silinsky, E.M. and Gerzanich, V. (1993) On the excitatory effects of ATP and its role as a neurotransmitter in coeliac neurons of the guinea-pig. *J. Physiol.*, 464: 197–212.

Silinsky, E.M., Gerzanich, V. and Vanner, S.M. (1992) ATP mediates excitatory synaptic transmission in mammalian neurones. *Br. J. Pharmacol.*, 106: 762–763.

Silinsky, E.M., Von Kügelgen, I., Smith, A. and Westfall, D.P. (1997) Functions of extracellular nucleotides in peripheral and central neuronal tissues. In: J.T. Turner, G.A. Weisman, and J.S. Fedan (Eds.), *The P2 nucleotide receptors*. Humana, Totowa, pp. 259–290.

Shen, K.Z. and North, R.A. (1993) Excitation of rat locus coeruleus neurons by adenosine 5′-triphosphate: ionic mechanism and receptor characterization. *J. Neurosci.*, 13: 894–899.

Sperlagh, B. and Vizi, E.S. (1991) Effect of presynaptic P_2 receptor stimulation on transmitter release. *J. Neurochem.*, 56: 1466–1470.

Starke, K., Göthert, M. and Kilbinger, H. (1989) Modulation of neurotransmitter release by presynaptic autoreceptors. *Physiol. Rev.*, 69: 864–989.

Surprenant, A., Buell, G. and North, R.A. (1995) P2X receptors bring new structure to ligand-gated ion channels. *Trends Neurosci.*, 18: 224–229.

Tschöpl, M., Harms, L., Nörenberg, W. and Illes, P. (1992) Excitatory effects of adenosine 5′-triphosphate on rat locus coeruleus neurones. *Eur. J. Pharmacol.*, 17: 71–77.

Vizi, E.S., Liang, S.D., Sperlagh, B., Kittel, A. and Juranyi, Z. (1997) Studies on the release and extracellular metabolism of endogenous ATP in rat superior cervical ganglion: support for neurotransmitter role of ATP. *Neuroscience*, 79: 893–903.

Von Kügelgen, I., Schöffel, E. and Starke, K. (1989) Inhibition by nucleotides acting at presynaptic P2-receptors of sympathetic neuro-effector transmission in the mouse isolated vas deferens. *Naunyn-Schmiedeberg's Arch. Pharmacol.*, 340: 522–532.

Von Kügelgen, I., Kurz, K. and Starke, K. (1993) Axon terminal P_2-purinoceptors in feedback control of sympathetic transmitter release. *Neuroscience*, 56: 263–267.

Von Kügelgen, I., Kurz, K. and Starke, K. (1994a) P2-purinoceptor-mediated autoinhibition of sympathetic transmitter release in mouse and rat vas deferens. *Naunyn-Schmiedeberg's Arch. Pharmacol.*, 349: 125–132.

Von Kügelgen, I., Späth, L. and Starke, K. (1994b) Evidence for P2-purinoceptor-mediated inhibition of noradrenaline release in rat brain cortex. *Br. J. Pharmacol.*, 113: 815–822.

Von Kügelgen, I., Stoffel, D. and Starke, K. (1995a) P2-purinoceptor-mediated inhibition of noradrenaline release in rat atria. *Br. J. Pharmacol.*, 115: 247–254.

Von Kügelgen, I., Stoffel, D. and Starke, K. (1995b) Presynaptic P2-purinoceptors at sympathetic axons in mouse atria. *Naunyn-Schmiedeberg's Arch. Pharmacol.*, 351: R138.

Von Kügelgen, I., Koch, H. and Starke, K. (1997a) P2-receptor-mediated inhibition of serotonin release in the rat brain cortex. *Neuropharmacology*, 36: 1221–1227.

Von Kügelgen, I., Nörenberg, W., Illes, P., Schobert, A. and Starke, K. (1997b) Differences in the mode of stimulation of cultured rat sympathetic neurons between ATP and UDP. *Neuroscience*, 78: 935–941.

Von Kügelgen, I., Nörenberg, W., Meyer, A., Illes, P. and Starke, K. (1998) Mechanisms of nucleotide-induced noradrenaline release from rat sympathetic neurons. *Drug Devel. Res.*, 43: 61.

Zimmermann, H. (1994) Signalling via ATP in the nervous system. *Trends Neurosci.*, 17: 420–426.

P. Illes and H. Zimmermann (Eds.)
Progress in Brain Research, Vol 120
© 1999 Elsevier Science BV. All rights reserved

CHAPTER 15

Adenine nucleotides as inhibitors of synaptic transmission: Role of localised ectonucleotidases

Ana M. Sebastião[1,*], Rodrigo A. Cunha[1,2], J. Francisco Cascalheira[3] and J. Alexandre Ribeiro[1]

[1]*Laboratory of Neurosciences, Faculty of Medicine, University of Lisbon, 1600 Lisbon, Portugal*
[2]*Department of Chemistry and Biochemistry, Faculty of Sciences, University of Lisbon, 1649-028 Lisbon, Portugal*
[3]*Laboratory of Biochemistry, Department of Chemistry, University of Beira Interior, 6200 Covilhã, Portugal*

Introduction

To be or not to be ATP or adenosine the active substance to modulate neuronal activity has been a matter of debate for several years, and in some cases is still an open question partly addressed in this paper. Since Burnstock's proposal of purinergic nerves (Burnstock, 1972), evidence is growing to show that this nucleotide can behave as a fast neurotransmitter at the peripheral (see e.g. Burnstock and Kennedy, 1985) as well as at the central (Edwards et al., 1992; Silinsky and Gerzanich, 1993; Nieber et al., 1997) nervous systems. ATP can also behave as a modulator of synaptic transmission, either through activation of metabotropic P_{2y} receptors or through activation of ionotropic P_{2x} receptors (see e.g. Inoue et al., 1996). On the other hand, the release of ATP into the extracellular space is followed by a cascade of enzymatic events leading to a fast production of adenosine (see Zimmermann, 1996). The enzymes involved in this process, the ecto-nucleotidases, may have a dual function: (1) terminating the action of ATP as a neurotransmitter/neuromodulator and/ or (2) forming another neuromodulator, adenosine. Indeed, adenosine, by activating A_1 (see e.g.

Ribeiro, 1995), A_2 (see Sebastião and Ribeiro, 1996) and A_3 (see Jacobson, 1998) receptors is an efficient modulator of neuronal functioning.

To explore the role of ATP and adenosine as neuromodulators, it is critical to understand whether any action of ATP is mediated by the intact nucleotide or mediated by adenosine through activation of metabotropic P_1 purinoceptors (A_1, A_{2A}, A_{2B} and A_3 subtypes). Since no selective tools are so far available to block ecto-ATPase activities or P_2 receptor activation, attempts to deal with this ambiguity have found two main difficulties: the first is related to the transient nature of the ATP responses, due either to P_2 receptor desensitisation or to fast hydrolysis of the nucleotide to adenosine, which may have hidden nucleotide effects, and the second is the possibility that the hydrolysis of ATP to adenosine with subsequent activation of adenosine receptors and adenosine re-uptake occur in restricted areas of the synapse, leading to virtually undetectable ATP metabolites. In this paper, evidence for localised hydrolysis of adenine nucleotides at the synaptic level will be reviewed.

Neuromuscular junction

The first evidence that adenosine inhibits neurotransmitter release was obtained by Ginsborg and Hirst (1971) at the neuromuscular junction.

*Corresponding author. Tel.: 351-1-7936787; fax: 351-1-7936787; e-mail: anaseb@neurociencias.pt

Because ATP is released from motor nerve terminals (Silinsky and Hubbard, 1973; Cunha and Sebastião, 1993; Silinsky and Redman, 1996) and inhibits the release of acetylcholine at the neuromuscular junction (Ribeiro and Walker, 1973), ATP instead of its metabolite, adenosine, would apparently be the most interesting substance to investigate. Since no post-junctional effects of ATP were detected (Ribeiro and Walker, 1973), it was proposed that ATP and/or their metabolites are negative feedback neuromodulators of the release of neurotransmitters, and that ATP could be responsible for the 'Wedenski inhibition' occurring at the neuromuscular junction in consequence of high frequency nerve stimulation (Ribeiro, 1979). The development of tools that selectively affect adenosine responses or extracellular adenosine accumulation allowed a better discrimination between nucleotide or nucleoside effects. By using these tools subsequent studies concluded that the inhibitory effect of adenine nucleotides on neuromuscular transmission is likely mediated by adenosine (Ribeiro and Sebastião, 1987; Redman and Silinsky, 1994; but see Giniatullin and Sokolova, 1998). We now know that ATP and adenosine itself can be released from stimulated motor nerve endings as well as from contracting muscle fibres and that the hydrolysis of released adenine nucleotides contributes to about a half to the total amount of extracellular adenosine (Cunha and Sebastião, 1993). Accordingly, removal of this extracellular adenosine (i.e. with adenosine deaminase) facilitates neuromuscular transmission, this increase being about twice that caused by preventing the formation of adenosine from released adenine nucleotides (Ribeiro and Sebastião, 1987).

The enzymatic machinery responsible for the formation of adenosine from adenine nucleotides at an innervated skeletal muscle has been fully characterised (Cunha and Sebastião, 1991; Cascalheira and Sebastião, 1992), and includes an ecto-ATPase, converting ATP to ADP, an ecto-adenylate kinase, converting ADP to AMP and ATP, an ecto-AMP deaminase, converting AMP to IMP, and an ecto-5'-nucleotidase, responsible for the formation of adenosine from AMP as well as inosine from IMP. In addition, an ecto-nucleoside diphosphate kinase activity was detected at the innervated sartorius muscle (Cascalheira and Sebastião, 1992). The ecto-adenylate kinase activity is inhibited by diadenosine pentaphosphate (AP$_5$A), the ecto-AMP deaminase activity is inhibited by coformycin and the ecto-5'-nucleotidase activity is inhibited by α,β-methylene-ADP (AOPCP) either when using AMP or IMP as substrate (Cunha and Sebastião, 1991; Cascalheira and Sebastião, 1992). Extracellular adenosine is inactivated by a dipyridamole-sensitive (Ribeiro and Sebastião, 1987) and nitrobenzylthioinosine-sensitive (Sebastião and Ribeiro, 1988) uptake system. At the rat (Sebastião and Ribeiro, 1988), but not at the frog (Cunha and Sebastião, 1991) neuromuscular junction, an ecto-adenosine deaminase activity also seems to be present.

Taking advantage of the existence of ATP analogues that are poor substrates for the ecto-nucleotidases, we investigated their action on neuromuscular transmission. It was observed that, in contrast with ATP or adenosine, the ATP analogue substituted in the α,β position of the phosphate chain, α,β-methylene ATP, is virtually devoid of effect on neuromuscular transmission (Ribeiro and Sebastião, 1987). However, α,β-methylene ATP is dephosphorylated to AOPCP at the innervated muscle (Cascalheira and Sebastião, 1992), making it difficult to conclude whether the absence of effect of α,β-methylene ATP is due to its inability to form adenosine or to its conversion to an ecto-5'-nucleotidase inhibitor, which by itself has a small excitatory action on neuromuscular transmission (Ribeiro and Sebastião, 1987). As assessed by HPLC analysis of bath samples, β,γ-methylene ATP is a very stable ATP analogue at the innervated sartorius muscle of the frog. Indeed, when β,γ-methylene ATP (10 μM) was added to this preparation, its concentration remained stable for 2 hours and no metabolites could be detected in the bath (Cascalheira and Sebastião, 1992). Surprisingly, β,γ-methylene ATP inhibited the amplitude of evoked endplate potentials without affecting the resting membrane potential, indicating that it inhibits neuromuscular transmission (Ribeiro and Sebastião, 1987). The effects of this ATP analogue and of ATP were compared in the same endplate and it was observed that 10 μM of β,γ-methylene ATP caused an inhibition similar to that caused by 2.5 μM ATP. Also, the time course of the effects of equipotent concentrations of ATP and of β,γ-methylene ATP do not differ appreciably (Ribeiro and Sebastião, 1987).

The inability to detect extracellular catabolism of β,γ-methylene ATP taken together with the observation that this ATP analogue inhibits neuromuscular transmission, could suggest the presence of inhibitory P_2 receptors. If so, the action of β,γ-methylene ATP should not be affected by inhibition of ecto-5'-nucleotidase. To investigate this, the action of the ATP analogue in the absence and in the presence of AOPCP was compared. Surprisingly, this ecto-5'-nucleotidase inhibitor could prevent the inhibitory effect of β,γ-methylene ATP on neuromuscular transmission, as well as preventing the inhibitory effect of ATP (Ribeiro and Sebastião, 1987). As expected, no modification of the effect of adenosine by AOPCP was observed. Though AOPCP has proved to be a very useful tool to inhibit ecto-5'-nucleotidase, and thus to prevent adenosine formation from adenine nucleotides (Cunha and Sebastião, 1991), one has to have in mind that: (1) AOPCP can be phosphorylated to α,β-methylene ATP when in the presence of ATP (Cascalheira and Sebastião, 1992) and high amounts of ATP may be released into the endplate upon stimulation (Silinsky, 1975; Cunha and Sebastião, 1993); and (2) AOPCP either directly or due to its conversion to α,β-methylene ATP may prevent the activation of P_2 receptors (Burnstock and Kennedy, 1985; Khakh et al., 1995).

Another possibility to test potential adenosine-mediated effects caused by adenine nucleotides is to use drugs that interfere with adenosine action, and to see whether the adenine nucleotide responses are equally affected. One of the drugs is dipyridamole, an adenosine uptake blocker, which induces a clear potentiation of the inhibitory effect of adenosine on evoked endplate potentials (Ribeiro and Sebastião, 1987) as well as reduces the rate of disappearance of adenosine from the bath when this nucleoside is added to innervated muscles (Cunha and Sebastião, 1991). Dipyridamole potentiated by a similar degree the inhibitory actions of ATP, adenosine and β,γ-methylene ATP on neuromuscular transmission (Fig. 1a). Also, when using an A_1 adenosine

receptor antagonist, 1,3-dipropyl-8-cyclopentyl-xanthine (DPCPX), it was observed that this xanthine antagonises the inhibitory effects of submaximal concentrations of β,γ-methylene ATP on neuromuscular transmission, as it does in relation to adenosine (Cascalheira and Sebastião, 1992) or to selective adenosine receptor agonists (Sebastião and Ribeiro, 1989).

Two of the observations described above, i.e., the apparent absence of metabolism of β,γ-methylene ATP and its inhibitory action on neuromuscular transmission, would suggest the presence of inhibitory P_2 receptors at the frog sartorius neuromuscular junction. However, three other observations (potentiation by dipyridamole, antagonism by DPCPX and prevention by AOPCP of the effect of β,γ-methylene ATP) render this conclusion unbearable. A possibility to explain all the data is that at the synaptic level there is an enzyme able to catabolise β,γ-methylene ATP, but the restricted localisation of this enzyme does not allow detection of catabolism in the whole preparation. Thus, local catabolism of the nucleotide may produce sufficient amounts of adenosine in the vicinity of the adenosine receptor, adenosine being quickly taken up, and neither the produced adenosine levels are high enough to be detected in the bath nor the amounts of the nucleotide decrease sufficiently to detect a change in the bath.

The conclusion that inhibitory effects of adenine nucleotides are mediated by hydrolysis to adenosine and subsequent activation of A_1 adenosine receptors does not necessarily preclude the existence and activation of P_2 receptors at the neuromuscular junction. It is known that ATP enhances the spontaneous release of acetylcholine from developing neuromuscular synapses, an action probably mediated by P_2 receptor activation (Fu and Huang, 1994). Also, at the skate electric organ (a model of the neuromuscular junction), P_2 receptor activation triggers the release of calcium from intracellular stores (Green et al., 1997). More difficult to reconcile with previously published evidence is the recent report that ATP inhibits transmitter release at the frog neuromuscular junction by activating its own presynaptic receptor, which appears different from that operated by adenosine (Giniatullin and Sokolova, 1998). The

186

(a) innervated sartorius (frog)

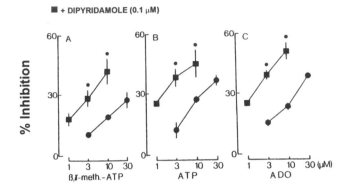

(b) hippocampus (rat)

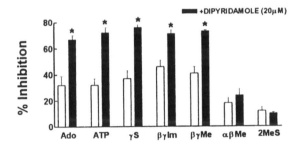

Fig. 1. Potentiation by the adenosine uptake blocker, dipyridamole, of the inhibitory action of γ-phosphate substituted ATP analogues on neurotransmission. In (a) the ordinates represent percentage inhibition of nerve-evoked twitch responses caused by β,γ-methylene ATP (β,γ-meth-ATP, (A)), by ATP (B) or by adenosine (ADO, (C)), in the absence (●) or in the presence (■) of dipyridamole (0.1 μM) at the innervated sartorius muscle of the frog. Note that the dipyridamole-induced shift to the left of the concentration response curve of the nucleotides and of the nucleoside is similar. In (b) the ordinates represent the percentage inhibition of the slope of field excitatory post-synaptic potentials recorded from the CA1 area of rat hippocampal slices upon stimulation of the Schaffer collaterals, in the absence (open bars) and in the presence (filled bars) of dipyridamole (20 μM). Ado: adenosine; γS: ATPγS; βγIm: βγ-imido-ATP; βγMe: βγ-methylene-ATP; αβMe: αβ-methylene-ATP; 2MeS: 2-methylthio-ATP, with all these purines tested at 10 μM. Note that, due to the lower sensitivity of the adenosine transporters to dipyridamole in the rat than in the frog, higher concentrations of the inhibitor were used in the rat preparation, but as in (a) dipyridamole potentiated by a similar degree the inhibitory effects of adenosine, of ATP and of the γ-substituted ATP analogues. αβ-methylene-ATP and 2-methylthio-ATP caused a nearly negligible inhibition of synaptic transmission in the hippocampus, which was not affected by dipyridamole. Adapted from Cascalheira and Sebastião (1992) and from Cunha et al. (1998).

conclusion was based upon the observation that the effect of ATP is prevented by α,β-methylene-ATP (used as a desensitising agent on P₂ receptors) and by suramin (used as a P₂ antagonist), is not antagonised by theophylline (an adenosine receptor antagonist) and not prevented by concavalin A (used as a 5′-nucleotidase inhibitor). Whether the ability of α,β-methylene ATP to prevent the ATP effect (Giniatullin and Sokolova, 1998) was due to

its conversion to the 5′-nucleotidase inhibitor, AOPCP (Fig. 2a; see also Cascalheira and Sebastião, 1992) and whether the ability of suramin to prevent the effect of ATP (Giniatullin and Sokolova, 1998) was due to its ability to prevent ATP hydrolysis (Ziganshin et al., 1995; Martí et al., 1996), remains to be determined. It would also be important to test whether concavalin A does inhibit ecto-5′-nucleotidase activity at the neuromuscular

(a) innervated sartorius (frog)

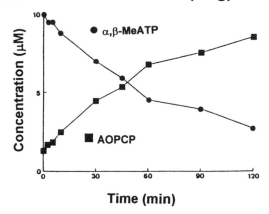

(b) hippocampus (rat)

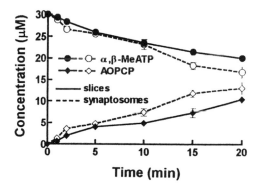

Fig. 2. Progress curves of αβ-methylene-ATP catabolism at innervated sartorius muscle of the frog (a) and hippocampal slices or hippocampal synaptossomes of the rat (b). The nucleotide was incubated with the preparations at zero time; samples were collected from the bath at the times indicated in the abscissa and analysed by HPLC. Note that in all cases αβ-methylene-ATP (αβ-MeATP) was dephosphorylated to αβ-methylene-ADP (AOPCP), a well known inhibitor of ecto-5'-nucleotidase. Adapted from Cascalheira and Sebastião (1992) and from Cunha et al. (1998).

junction, as has been observed with AOPCP (Cunha and Sebastião, 1991), to try to solve why the effect of ATP is prevented by AOPCP (Ribeiro and Sebastião, 1987) but not by concavalin A (Giniatullin and Sokolova, 1998). It is also intriguing that the inhibitory effect of ATP was not antagonised by theophylline (Giniatullin and Sokolova, 1998), in contrast with what has been previously observed by others using a more selective adenosine A_1 receptor

antagonist, DPCPX (Cascalheira and Sebastião, 1992; Redman and Silinsky, 1994) also at the frog neuromuscular junction. Another discrepancy between the results obtained by Giniatullin and Sokolova (1998) and those previously published (Ribeiro and Walker, 1975) is the additivity of the effects of high concentrations of ATP and adenosine observed by Giniatullin and Sokolova (1998) at the frog neuromuscular junction and the non-additivity of both actions observed by Ribeiro and Walker (1975) at the rat neuromuscular junction. All these discrepancies, leading to opposite conclusions separated in time by two decades, besides being stimulating at least for the scientists, if not for science, illustrate quite well the difficulties found whenever trying to elucidate whether an effect of ATP also shared by adenosine is due to an effect mediated by ATP itself or if it depends upon its hydrolysis to adenosine. We do believe that only the conjunction of functional, pharmacological and enzymatic assays will succeed in the elucidation of this old but now renewed question.

Besides inhibitory actions, mediated by A_1 receptors, adenosine also has excitatory actions on motor nerve endings, facilitating acetylcholine release by activation of A_{2A} receptors (Correia-de-Sá et al., 1991). High frequency motor nerve stimulation facilitates activation of excitatory A_{2A} receptors by endogenous adenosine (Correia-de-Sá et al., 1996), whereas at low frequencies of stimulation the tonic inhibition by endogenous adenosine predominates (Sebastião and Ribeiro, 1985; Correia-de-Sá et al., 1996). It is interesting that adenosine formed from released adenine nucleotides preferentially activates excitatory A_{2A} receptors of motor nerve endings (Cunha et al., 1996a), which could be anticipated from the intrinsic properties of the ecto-nucleotidase pathway (James and Richardson, 1993). Whether the tonic preponderance of A_{2A} receptors under high frequency stimulation (Correia-de-Sá et al., 1996) is related to a preferential release of ATP at high frequency stimulation (Cunha and Sebastião, 1993), remains to be elucidated. It is of interest that at the hippocampus, which shares many similarities with the neuromuscular junction with respect to adenosine neuromodulation (Ribeiro and Sebastião, 1991; Sebastião et al., 1995), high-frequency stimulation favours ATP release (Wieraszko et al., 1989; Cunha et al., 1996b).

Hippocampus

Due to its relative simplicity, the neuromuscular junction has several advantages to investigate synaptic phenomena, namely (1) transmission is mediated by only one neurotransmitter type, acetylcholine; and (2) each synaptic potential reflects the activity of only one nerve terminal. However, the presence of muscle fibres, which are rich in ectonucleotidase activities and have a much larger surface area than the nerve terminals, makes it difficult to extrapolate some of the data obtained at the neuromuscular junction to central nervous system synapses. In spite of possessing a greater synaptic complexity than the neuromuscular junction, the hippocampal slices, with well defined synaptic circuits, have been proved to be a useful and probably the best in vitro model to study how synaptic transmission is affected by neuromodulators in the central nervous system.

ATP is released upon stimulation of hippocampal pathways in a frequency dependent manner (Wieraszko et al., 1989; Cunha et al., 1996b). Release of ATP, as well as of adenosine, occurs from neuronal and non-neuronal structures, including glial cells (for a review see Cunha, 1997). Immunocytochemical studies revealed that ecto-5'-nucleotidase is present in astrocytes throughout the hippocampus, whereas the neuronal distribution of the enzyme appears predominant in the mossy fibre synapses of the CA3 area of the hippocampus (Zimmermann et al., 1993). However, using histochemical and enzymatic approaches it was observed that ecto-enzymatic activities able to metabolise extracellular AMP to adenosine are present in nerve terminals throughout the hippocampus (Bernstein et al., 1978; Cunha et al., 1992). Ecto-5'-nucleotidase activity is particularly associated with cholinergic nerve terminals of the hippocampus (Cunha et al., 1992), where ATP inhibits acetylcholine release by a process dependent upon its hydrolysis to adenosine (Cunha et al., 1994b). It is interesting that cholinergic nerve terminals of the cerebral cortex do not possess ecto-5'-nucleotidase activity (Cunha et al., 1992), and ATP by itself decreases acetylcho-

line release (Cunha et al., 1994b). This reinforces the idea of a key role for ecto-5'-nucleotidase in structures where the main purinergic modulator is adenosine.

ATP exerts neuromodulatory actions in the hippocampus (for reviews see Illes and Nörenberg, 1993; Wieraszko, 1996; Inoue et al., 1996), and some of these actions are due to activation of P_2 receptors, which are expressed in the hippocampus (Kidd et al., 1995; Collo et al., 1996). Adenosine, by activating inhibitory A_1 (Sebastião et al., 1990), excitatory A_{2A} (Sebastião and Ribeiro, 1992) or A_3 receptors (Fleming and Mogul, 1997) affects synaptic transmission and/or neuronal excitability in the hippocampus. In addition, cross-talk between adenosine receptors subtypes, including A_{2A}/A_1 (Cunha et al., 1994a) and A_3/A_1 (Dunwiddie et al., 1997b) occurs in the hippocampus. All this diversity of purinergic functions and receptors in the hippocampus implies a delicate balance between ATP release and extracellular adenosine formation. In fact, several lines of evidence suggest a well controlled gradient of purines in the synaptic cleft (see Cunha, 1997).

By using ATP analogues derived from the inactive enantiomer of adenosine, L-adenosine, Stone and Cusack (1989) concluded that the inhibitory effect of ATP on synaptic transmission in the CA1 area of hippocampus depends upon its hydrolysis to adenosine. Recently, it was proposed that the inhibitory action of ATP on epileptiform activity in the hippocampus is dependent upon its hydrolysis to AMP, which could act as a P_1 receptor ligand (Ross et al., 1998). Unfortunately, in this study the protocols that successfully prevented the inhibitory effect of AMP (exogenous addition of AMP deaminase or exogenous addition of ecto-5'-nucleotidase plus adenosine deaminase) do not allow a clear distinction between the consequences of accelerating AMP metabolism and prevention of adenosine formation. At least the post-synaptic A_1 adenosine receptors in the hippocampus are apparently insensitive to AMP (Dunwiddie et al., 1997a).

ATP analogues with substitutions in the γ-phosphate inhibit synaptic transmission in the CA1 area of the hippocampus, but surprisingly, these ATP analogues are apparently resistant to ecto-nucleotidase activity in the hippocampal slices of the rat (Cunha et al., 1998). In an attempt to reconcile these observations, that could suggest that the effect of ATP on synaptic transmission is mediated by P_2 receptors, with previous indications that the effect of the nucleotide depends upon previous hydrolysis to adenosine (Stone and Cusack, 1989), functional assays with drugs that affect adenosine production or adenosine removal were performed. It was observed that at the hippocampus (Cunha et al., 1998), as it occurs at the neuromuscular junction (Cascalheira and Sebastião, 1992), the inhibitory effects of the usually considered 'stable ATP analogues' are potentiated by drugs that prevent adenosine removal, such as dipyridamole (Fig. 1b) and are attenuated by inhibition of ecto-5′-nucleotidase with AOPCP. Furthermore, their effects are inhibited by extracellular adenosine removal with adenosine deaminase, are antagonised by the A_1 receptor antagonist, DPCPX, and are not antagonised by the P_2-receptor antagonists, suramin, PPADS or reactive-blue (Cunha et al., 1998). One should then conclude that the inhibitory action of ATP and of γ-phosphate substituted ATP analogues on synaptic transmission in the hippocampus is mediated by activation of inhibitory A_1 adenosine receptors, after extracellular catabolism of these nucleotides to adenosine. The discrepancy between the stability of the γ-substituted ATP analogues at the hippocampal slices and their action on synaptic transmission is once more suggestive of localised catabolism of adenine nucleotides. Although low amounts of adenosine are probably formed, due to the low rate of catabolism, substrate channelling might allow a high local concentration of adenosine to be reached in the surroundings of the A_1 receptor. Interestingly, catabolism of γ-substituted ATP analogues could be observed in hippocampal synaptosomes (Cunha et al., 1998), though at a much lower rate than that of ATP. The virtual absence of catabolism of the ATP analogues in the slices and its catabolism in the synaptosomes is another strong indication that these nucleotides can be catabolised at restrict locations. Fast and localised catabolism of ATP also occurs at the post-synaptic level, i.e. in the dendrites of CA1 pyramids, where ATP can be converted within

miliseconds into adenosine to activate A_1 receptors (Dunwiddie et al., 1997a).

The extracellular catabolism of the γ-substituted ATP analogues in the synaptosomes is different from that of ATP, because the only phosphorylated metabolite detected in the bath is AMP, with ADP being undetectable (Cunha et al., 1998). Since p-nitrophenylphosphate did not modify the catabolism of the ATP analogues, the involvement of nonspecific phosphatases can be precluded. A possible candidate to hydrolyse the γ-substituted ATP analogues at the nerve endings is an ecto-ATP pyrophosphatase, but the lack of selective inhibitors of this enzyme makes it difficult to directly test this hypothesis.

It is of interest that a long (30 min) pre-incubation with α,β-methylene ATP reduces the inhibitory effect of ATP in the hippocampus, as well as inhibits extracellular AMP catabolism (Cunha et al., 1998). Since α,β-methylene ATP is catabolised to AOPCP both in hippocampal slices and synaptosomes (Fig. 2b, see also Cunha et al., 1998), the ability of this nucleotide to attenuate the ATP action might be attributed to inhibition of ecto-5′-nucleotidase, rather than to prevention of P_2 receptor functioning.

Excitatory synaptic transmission in the hippocampus is mediated by glutamate, and as mentioned above, the inhibitory action of ATP on hippocampal synaptic transmission is mediated by hydrolysis to adenosine (Cunha et al., 1998). This excitatory circuit is highly regulated by a network of inhibitory interneurones, which use GABA as the neurotransmitter, but neither ATP nor γ-substituted ATP analogues affect GABA release in the hippocampus (Cunha and Ribeiro, 1998). Noradrenaline release in the rat hippocampus is under inhibitory control of P_2-receptors (Koch et al., 1997). This contrasts with what occurs in the rabbit cerebral cortex where the adenine nucleotide-dependent inhibition of noradrenaline release might be mediated by direct activation of A_1 adenosine receptors (Von Kügelgen et al., 1992). Whether these differences are related to species differences awaits further investigation. In fact, it was recently observed that in the rat cerebral cortex presynaptic P_{2y} receptors mediate inhibition of serotonin release (Von Kügelgen et al., 1997). It is

190

interesting that in the rat cerebral cortex, but not in the rat hippocampus, ATP itself can inhibit acetylcholine release (Cunha et al., 1994b). Taken together, these studies suggest that adenine nucleotides are more important modulators of neurotransmitter release in cerebral cortex than in the hippocampus.

Concluding remarks

By using a cholinergic model from the peripheral nervous system, the neuromuscular junction, as well as a more complex preparation from the central nervous system, the hippocampus, where excitatory transmission is mainly glutamatergic, we obtained evidence for localised catabolism of adenine nucleotides followed by substrate channelling to A_1 receptors. This localised catabolism may mask some adenosine-mediated ATP effects. Thus, to establish an effect of ATP as such it seems essential to show that substances that prevent adenosine formation, or affect adenosine action do not modify the nucleotide effect (see also Harden et al., 1997). These substances include ecto-5'-nucleotidase inhibitors (e.g. AOPCP), adenosine uptake blockers (e.g. dipyridamole), selective adenosine receptor antagonists (e.g. DPCPX, for A_1 receptors) and adenosine deaminase itself to remove extracellular adenosine. When using adenosine deaminase one has to be aware that the rate of adenosine formation from AMP, through endogenous ecto-5'-nucleotidase, may be faster than the rate of adenosine deamination, through exogenously applied adenosine deaminase. If this occurs, an adenosine-dependent nucleotide action will probably be incompletely prevented by adenosine deaminase. Ideally, several tools designed to change each of the steps should be used to minimise the side effects. We consider that α,β-methylene ATP, due to its ability to form AOPCP (Fig. 2) and to inhibit nucleotide catabolism, is of limited usefulness when used as a tool to prevent P_2 receptor activation. Also, the P_2 antagonist, suramin, has to be used with caution, due to its ability to inhibit ecto-nucleotidases. Our proposal is that only the conjunction of functional, pharmacological and enzymatic assays and the convergence of the data will lead to a successful answer to the

quite old but still alive dichotomy of ATP/P_2 vs. adenosine/P_1 inhibitory neuromodulation.

Acknowledgements

This work has been supported by grants from the European Union (BIOMED II-BMH4-CT98-0676), and Fundação para a Ciência e Tecnologia (FCT).

References

Bernstein, H.-G., Weiß, J. and Luppa, H. (1978) Cytochemical investigations of the localization of 5'-nucleotidase in the rat hippocampus with special reference to synaptic regions. *Histochemistry*, 55: 261–267.

Burnstock, G. (1972) Purinergic nerves. *Pharmacol. Rev.*, 24: 509–581.

Burnstock, G. and Kennedy, C. (1985) Is there a basis for distinguishing two types of P_2-purinoceptors? *Gen. Pharmacol.*, 16: 433–440.

Cascalheira, J.F. and Sebastião, A.M. (1992) Adenine nucleotide analogues, including γ-phosphate-substituted analogues, are metabolised extracellularly in innervated frog sartorius muscle. *Eur. J. Pharmacol.*, 222: 49–59.

Collo, G., North, R.A., Kawashima, E., Merlo-Pich, E., Neidhart, S., Surprenant, A. and Buell, G. (1996) Cloning of $P2x_5$ and $P2x_6$ receptors and the distribution and properties of an extended family of ATP-gated ion channels. *J. Neurosci.*, 16: 2495–2507.

Correia-de-Sá, P., Sebastião, A.M. and Ribeiro, J.A. (1991) Inhibitory and excitatory effects of adenosine receptor agonists on evoked transmitter release from phrenic nerve endings of the rat. *Br. J. Pharmacol.*, 103: 1614–1620.

Correia-de-Sá, P., Timóteo, M.A. and Sebastião, J.A. (1996) Presynaptic A_1 inhibitory/A_{2A} facilitatory adenosine receptor activation balance depends on motor nerve stimulation paradigm at the rat hemidiaphragm. *J. Neurophysiol.*, 76: 3910–3919.

Cunha, R.A. (1997) Release of ATP and adenosine and formation of extracellular adenosine in the hippocampus. In: Y. Okada (Ed.), *The Role of Adenosine in the Nervous System*. Elsevier, Amsterdam, pp. 135–142.

Cunha, R.A. and Ribeiro, J.A. (1998) In search of a neuromodulatory role for ATP in the hippocampus. *J. Neurochem.*, 71, (suppl.): S33.

Cunha, R.A. and Sebastião, A.M. (1991) Extracellular metabolism of adenine nucleotides and adenosine in the innervated skeletal muscle of the frog. *Eur. J. Pharmacol.*, 197: 83–92.

Cunha, R.A. and Sebastião, A.M. (1993) Adenosine and adenine nucleotides are independently released from both the nerve terminals and the muscle fibres upon electrical stimulation of the innervated skeletal muscle of the frog. *Pflügers Arch.*, 424: 503–510.

Cunha, R.A., Sebastião, A.M. and Ribeiro, J.A. (1992) Ecto-5'-nucleotidase is associated with cholinergic nerve terminals in the hippocampus but not in the cerebral cortex of the rat. *J. Neurochem.*, 59: 657–666.

Cunha, R.A., Johansson, B., van der Ploeg, I., Sebastião, A.M., Ribeiro, J.A. and Fredholm, B.B. (1994a) Evidence for functionally important adenosine A_{2a} receptors in the rat hippocampus. *Brain Res.*, 649: 208–216.

Cunha, R.A., Sebastião, A.M. and Ribeiro, J.A. (1994b) Purinergic modulation of the evoked release of [^{3}H]acetylcholine from the hippocampus and cerebral cortex of the rat: role of the ectonucleotidases. *Eur. J. Neurosci.*, 6: 33–42.

Cunha, R.A., Correia-de-Sá, P., Sebastião, A.M. and Ribeiro, J.A. (1996a) Preferential activation of excitatory adenosine receptors at rat hippocampal and neuromuscular synapses by adenosine formed from released adenine nucleotides. *Br. J. Pharmacol.*, 119: 253–260.

Cunha, R.A., Vizi, E.S., Ribeiro, J.A. and Sebastião, A.M. (1996b) Preferential release of ATP and its extracellular catabolism as a source of adenosine upon high- but not low-frequency stimulation of rat hippocampal slices. *J. Neurochem.*, 67: 2180–2187.

Cunha, R.A., Sebastião, A.M. and Ribeiro, J.A. (1998) Inhibition by ATP of hippocampal synaptic transmission requires localized extracellular catabolism by ecto-nucleotidases into adenosine and channeling to adenosine A_1 receptors. *J. Neurosci.*, 18: 1987–1995.

Dunwiddie, T.V., Diao, L. and Proctor, W.R. (1997a) Adenine nucleotides undergo rapid, quantitative conversion to adenosine in the extracellular space in rat hippocampus. *J. Neurosci.*, 17: 7673–7682.

Dunwiddie, T.V., Diao, L., Kim, H.O., Jiang, J. and Jacobson, K.A. (1997b) Activation of hippocampal adenosine A_3 receptors produces desensitization of A_1 receptor-mediated responses in rat hippocampus. *J. Neurosci.*, 17: 607–614.

Edwards, F.A., Gibb, A.J. and Colquhoun, D. (1992) ATP receptor-mediated synaptic currents in the central nervous system. *Nature*, 359: 144–147.

Fleming, K.M. and Mogul, D.J. (1997) Adenosine A_3 receptors potentiate hippocampal calcium current by a PKA-dependent/PKC independent pathway. *Neuropharmacology*, 36: 353–362.

Fu, W.M. and Huang, F.L. (1994) Potentiation by endogenously released ATP of spontaneous transmitter secretion at developing neuromuscular synapses in Xenopus cell cultures. *Br. J. Pharmacol.*, 111: 880–886.

Giniatullin, R.A. and Sokolova, E.M. (1998) ATP and adenosine inhibit transmitter release at the frog neuromuscular junction through distinct presynaptic receptors. *Br. J. Pharmacol.*, 124: 839–844.

Ginsborg, B.L. and Hirst, G.D.S. (1971) Cyclic AMP, transmitter release and the effect of adenosine at the neuromuscular junction. *Nature New Biol.*, 232: 63–64.

Green, A.C., Dowdall, M.J. and Richardson, C.M. (1997) ATP acting on P_{2y} receptors triggers calcium mobilization in Schwann cell at the neuroelectrocyte junction in the skate. *Neuroscience*, 80: 635–651.

Harden, T.K., Lazarowski, E.R. and Boucher, R.C. (1997) Release, metabolism and interconversion of adenine and uridine nucleotides: implications for G-protein coupled P2 receptor agonist selectivity. *Trends Pharmacol. Sci.*, 18: 43–46.

Inoue, K., Koizumi, S. and Ueno, S. (1996) Implications of ATP receptors in brain functions. *Prog. Neurobiol.*, 50: 483–492.

Illes, P. and Nörenberg, W. (1993) Neuronal ATP receptors and their mechanism of action. *Trends Pharmacol. Sci.*, 14: 50–54.

Jacobson, K.A. (1998) Adenosine A_3 receptors: novel ligands and paradoxical effects. *Trends Pharmacol. Sci.*, 19: 184–191.

James, S. and Richardson, P.J. (1993) Production of adenosine from extracellular ATP at the striatal cholinergic synapse. *J. Neurochem.*, 60: 219–227.

Khakh, B.S., Humphrey, P.P.A. and Suprenant, A. (1995). Electrophysiological properties of P_{2x}-purinoceptors in rat superior cervical, nodose and guinea-pig coeliac neurones. *J. Physiol. Lond.*, 484: 385–395.

Kidd, E.J., Grahames, C.B., Simon, J., Michel, A.D., Barnard, E.A. and Humphrey, P.P. (1995) Localization of P_{2x} purinoceptor transcripts in the rat nervous system. *Mol. Pharmacol.*, 48: 569–573.

Koch, H., Kügelgen, I., and Starke, K. (1997) P_1-receptor-mediated inhibition of noradrenaline release in the rat hippocampus. *Naunyn-Schmiedeberg's Arch. Pharmacol.*, 355: 707–715.

Martí, M.E., Cantí, C., Gómez de Aranda, I., Miralles, F., Solsona, C. (1996) Action of suramin upon ecto-apyrase activity and synaptic depression of Torpedo electric organ. *Br. J. Pharmacol.*, 118: 1232–1236.

Nieber, K., Polchen, W and Illes, P. (1997) Role of ATP in fast excitatory synaptic potentials in locus coeruleus of the rat. *Br. J. Pharmacol.*, 122: 423–430.

Redman, R.S. and Silinsky, E.M. (1994) ATP released together with acetylcholine as the mediator of neuromuscular depression at frog motor nerve endings. *J. Physiol. Lond.*, 477: 117–127.

Ribeiro, J.A. (1979) Purinergic modulation of transmitter release. *J. Theor. Biol.*, 80, 259–270.

Ribeiro, J.A. (1995) Purinergic inhibition of neurotransmitter release in the central nervous system. *Pharmacol. Toxicol.*, 77: 299–305.

Ribeiro, J.A. and Sebastião, A.M. (1987) On the role, inactivation and origin of endogenous adenosine at the frog neuromuscular junction. *J. Physiol. Lond.*, 384: 571–585.

Ribeiro, J.A. and Sebastião, A.M. (1991) Purinergic modulation of neurotransmitter release in the peripheral and central nervous systems. In: J. Feigenbaum and M. Hanani (Eds.), *Presynaptic Regulation of Neurotransmitter Release, A Handbook.* Freund, London, pp. 451–495.

Ribeiro, J.A. and Walker, J. (1973) Action of adenosine triphosphate on end-plate potentials recorded from muscle fibres of the rat diaphragm and frog sartorius. *Br. J. Pharmacol.*, 49: 724–725.

Ribeiro, J.A. and Walker, J. (1975) The effects of adenosine triphosphate and adenosine diphosphate on transmission at the rat and frog neuromuscular junctions. *Br. J. Pharmacol.*, 54: 213–218.

Ross, F., Brodie, M.J. and Stone, T.W. (1998). Adenosine monophosphate as a mediator of ATP effects at P1 purinoceptors. *Br. J. Pharmacol.*, 124: 818–824.

Sebastião, A.M. and Ribeiro, J.A. (1985) Enhancement of transmission at the frog neuromuscular junction by adenosine deaminase: Evidence for an inhibitory role of endogenous adenosine on neuromuscular transmission. *Neurosci. Letts.*, 62: 267–270.

Sebastião, A.M. and Ribeiro, J.A. (1988) On the adenosine receptor and adenosine inactivation at the rat diaphragm neuromuscular junction. *Br. J. Pharmacol.*, 94: 109–120.

Sebastião, A.M. and Ribeiro, J.A. (1989) 1,3,8- and 1,3,7-substituted xanthines: Relative potency as adenosine receptor antagonists at the frog neuromuscular junction. *Br. J. Pharmacol.*, 96: 211–219.

Sebastião, A.M. and Ribeiro, J.A. (1992) Evidence for the presence of excitatory A_2 adenosine receptors in the rat hippocampus. *Neurosci. Letts.*, 138: 41–44.

Sebastião, A.M. and Ribeiro, J.A. (1996) Adenosine A_2 receptor-mediated excitatory actions in the nervous system. *Prog. Neurobiol.*, 48: 167–189.

Sebastião, A.M., Stone, T.W. and Ribeiro, J.A. (1990) On the inhibitory adenosine receptor at the neuromuscular junction and hippocampus of the rat: Antagonism by 1,3,8-substituted xanthines. *Br. J. Pharmacol.*, 101: 453–459.

Sebastião, A.M., Cunha, R.A., Correia-de-Sá, P., de Mendonça, A. and Ribeiro, J.A. (1995) Role of A_{2a} receptors in the hippocampus and motor nerve endings. In: L. Belardinelli and A. Pelleg (Eds.), *Adenosine and Adenine Nucleotides: From Molecular Biology to Integrative Physiology.* Kluwer Academic Publishers, Boston, pp. 251–261.

Silinsky, E.M. (1975) On the association between transmitter secretion and the release of adenine nucleotides from mammalian motor nerve terminals. *J. Physiol. Lond.*, 247: 145–162.

Silinsky, E.M. and Gerzanich, V. (1993) On the excitatory effects of ATP and its role as a neurotransmitter in coeliac neurones of the guinea-pig. *J. Physiol. Lond.*, 464: 197–212.

Silinsky, E.M. and Hubbard, J.I. (1973) Release of ATP from rat motor nerve terminals. *Nature*, 243: 404-405.

Silinsky, E.M. and Redman, R.S. (1996) Synchronous release of ATP and neurotransmitter within milliseconds of a motor nerve impulse in the frog. *J. Physiol. Lond.*, 492: 815-822.

Stone, T.W. and Cusack, N.J. (1989) Absence of P2-purinoceptors in hippocampal pathways. *Br. J. Pharmacol.*, 97: 631-635.

Von Kügelgen, I., Spath, L. and Starke, K. (1992) Stable adenine nucleotides inhibit [^{3}H]noradrenaline release in rabbit brain cortex slices by direct action at presynaptic adenosine A_1-receptors. *Naunyn-Schmiedeberg's Arch. Pharmacol.*, 346: 187-196.

Von Kügelgen, I., Koch, H. and Starke, K. (1997) P_2-receptor-mediated inhibition of serotonin release in the rat brain cortex. *Neuropharmacology*, 36: 1221-1227.

Wieraszko, A. (1996) Extracellular ATP as a neurotransmitter: Its role in synaptic plasticity in the hippocampus. *Acta Neurobiol. Exp.*, 56: 637-648.

Wieraszko, A., Goldsmith, G. and Seyfried, T.N. (1989) Stimulation-dependent release of adenosine triphosphate from hippocampal slices. *Brain Res.*, 485: 244-250.

Ziganshin, A., Ziganshina, L.E., Bodin, P., Bailey, D. and Burnstock, G. (1995) Effects of P_2-purinoceptor antagonists on ecto-nucleotidase activity of guinea-pig vas deferens cultured smooth muscle cells. *Biochem. Mol. Biol. Int.*, 36: 863-869.

Zimmermann, H. (1996) Biochemistry, localization and functional roles of ecto-nucleotidases in the nervous system. *Prog. Neurobiol.*, 49: 589-618.

Zimmermann, H., Vogel, M. and Laube, U. (1993) Hippocampal localization of 5'-nucleotidase as revealed by immunocytochemistry. *Neuroscience*, 55: 105-112.

P. Illes and H. Zimmermann (Eds.)
Progress in Brain Research, Vol 120
© 1999 Elsevier Science BV. All rights reserved

CHAPTER 16

The functions of ATP receptors in the synaptic transmission in the hippocampus

Kazuhide Inoue*, Schuichi Koizumi, Shinya Ueno, Aya Kita and Makoto Tsuda

Division of Pharmacology, National Institute of Health Sciences, 1-18-1 Kamiyoga, Setagaya, Tokyo 158, Japan

Introduction

ATP has joined the growing list of compounds shown to function as neurotransmitters in various tissues including smooth muscle (Burnstock and Kennedy, 1985) peripheral neurons (Bleehen and Keele, 1977; Bean and Friel, 1990) and the central nervous system (CNS) (Inoue et al., 1992; Edwards et al., 1992; Harms et al., 1992; Shen and North, 1993). Extracellular ATP produces responses through two subclasses of ATP receptors, $P2X_n$ and $P2Y_n$ (Abbracchio and Burnstock, 1994). Several laboratories have identified ATP receptors by cDNA cloning (Webb et al., 1993; Lustig et al., 1993; Brake et al., 1994; Valera et al., 1994; Bo et al., 1995; Chen et al., 1995; Lewis et al., 1995; Buell et al., 1996; Surprenant et al., 1996). The $P2X_n$ subclass has so far been divided into seven subtypes and been shown to be composed of ligand-gated ion channels. The $P2Y_n$ subclass has been reported to couple to PLC_β or adenylyl cyclase via GTP-binding proteins. The investigation of the molecular biology of purinoceptors has grown rapidly in recent years resulting in significant amounts of new information (Burnstock and King, 1996; North and Barnard, 1997). Since there are no selective agonists and antagonists, it is very difficult to examine the functions of these receptors in the CNS. This review aims to discuss the implications of ATP receptors in the function of the hippocampus in spite of this paucity of data.

ATP is released from the hippocampus (Wieraszko et al., 1989) and several mRNAs for certain types of ATP receptors are present in this area (Collo et al., 1996). Although ATP induces fast synaptic currents in cultured hippocampal neurons (Inoue et al., 1992) and long-lasting enhancement of the population spikes (Wieraszko et al., 1989; Nishimura et al., 1990 ; Fujii et al., 1995) and ATP produces an increase in intracellular Ca^{2+} ($[Ca^{2+}]i$) in hippocampal cells (Inoue et al., 1995), it has been shown by an electrophysiological method, Ca^{2+} imaging (Koizumi and Inoue, 1997) and direct detection of glutamate that ATP inhibits the glutamate release from cultured rat hippocampus. In the meanwhile ATP stimulates the release of GABA, an inhibitory neurotransmitter which inhibits the excitable function of glutamate. Besides these actions of ATP on the regulation of neurotransmitters, ATP also activates microglia cells to release plasminogen which has been reported to promote the development of mesencephalic dopaminergic neurons and to enhance neurite outgrowth from explants of neocortical tissue (Nakajima et al., 1992a, 1992b). These data suggest that ATP may protect the brain function from excess stimulation.

The effect of ATP in hippocampal neurons; the study using Ca^{2+} imaging

The hippocampus is well known to be involved in memory and learning and to be particularly sensi-

*Corresponding author. Tel. and fax: +81-3-3700-9698; e-mail: inoue@nihs.go.jp

tive to ischemic insults. Glutamate, the major excitatory neurotransmitter in the hippocampus has been extensively studied in relation to neuronal cell death and long term potentiation (LTP) which may underlie the process of memory and learning (Bliss and Lynch, 1988; Malenka et al., 1989). Although the activation of glutamate receptors is an event considered to be necessary for the induction of LTP in the hippocampus, many other factors, such as arachidonic acid (Bliss et al., 1991), nitric oxide (NO) and carbon monoxide (Zhuo et al., 1993) as well as ATP are thought to be involved in synaptic plasticity.

ATP is released from brain synaptosomal preparations by stimulation with KCl (White, 1978) and from Schaffer collateral–comissural afferents of hippocampal slices by electrical stimulation (Wieraszko and Seyfried, 1989). ATP induces fast synaptic currents in cultured neurons from the hippocampus (Inoue et al., 1992) as well as in slices from the medial habenula (Edwards et al., 1992, 1997). In addition, exogenously applied ATP is able to induce long-lasting enhancement of the population spikes recorded from mouse (Wieraszko and Seyfried, 1989) and guinea pig (Nishimura et al., 1990; Fujii et al., 1995) hippocampal slices, and can potentiate LTP in the hippocampus (Wieraszko and Ehrlich, 1994). These data suggested the excitatory action of ATP in the hippocampus. Indeed, ATP can produce glutamate release and an increase in $[Ca^{2+}]i$ in some neurons (less than 20% of cells tested) (Inoue et al., 1992, 1995).

We then examined the effect of ATP on the release of glutamate from more than one million cells of primary cultured rat hippocampus using a biochemical analytical method. The results however, were opposite to our expectation, i.e. ATP did not stimulate glutamate release but inhibited the spontaneous and high K^+-evoked release of glutamate from these cells (Fig. 1A). Long term depression (LTD), which is an important cellular mechanism for memory and learning in the hippocampus (Artola and Singer, 1993; Bear and Malenka, 1994) is thought to be due to a long term decrease in glutamate release from presynaptic terminals in the hippocampus (Bolshakov and Siegelbaum, 1994). Therefore, it is very important to determine whether ATP does inhibit or stimulate the release of glutamate.

Kudo's group found that Ca^{2+} oscillations in the cultured hippocampus can be used as an indication of glutamate release (Ogura et al., 1987). Therefore we can examine the effect of ATP on the glutamate release in individual neurons by the measurement of the effect on Ca^{2+} oscillations. Figure 1(B) shows typical Ca^{2+} oscillations in 7-day-cultured hippocampal neurons (Koizumi and Inoue, 1997). The Ca^{2+} oscillations were observed in cells cultured for 5–14 days. Traces (a)–(d) demonstrate synchronous Ca^{2+} oscillations whereas trace (e) shows non-synchronous oscillations, suggesting that cells (a)–(d) are coupled via synaptic connections. Removal of extracellular Ca^{2+} or the application of tetrodotoxin (TTX, 3 μM) abolished the Ca^{2+}-oscillations (Fig. 1(C)(a)). In addition, D-2-amino-phosphonovalerate (APV, 100 μM), an N-methyl-D-aspartate (NMDA) glutamate receptor antagonist, and 6-cyano-7-nitro quinoxalline-2,3-dione (CNQX, 30 μM), a non-NMDA receptor antagonist, also inhibited the Ca^{2+} oscillations (Fig. 1(C)(b)). When the effects of ATP on the Ca^{2+} oscillations were evaluated, at least two types of responses were observed. In some cells (about 20% of cells tested) ATP induced a transient rise in $[Ca^{2+}]i$ followed by an inhibition of the Ca^{2+} oscillations (Fig. 1(D)(a) and (b)). In other cells, only the inhibition of the Ca^{2+} oscillations was observed (Fig. 1(D)(c)). Thus, ATP-induced inhibition of the Ca^{2+} oscillation was observed in all neurons tested. ATP is metabolized to adenosine which is well known to inhibit several responses in the central nervous system (Phillis et al., 1975; Okada and Kuroda, 1980). Indeed, we demonstrate here that adenosine (10 μM) inhibited the Ca^{2+} oscillations in our cultured hippocampal preparation as results previously reported (Kudo et al., 1991). Inhibition of the Ca^{2+} oscillations by adenosine was completely blocked by 100 μM aminophylline, a non-selective adenosine receptor antagonist, and by 100 μM 8-CPT, an adenosine A_1 receptor antagonist, but not by CP66713, an antagonist of adenosine A_{2A} receptors (data not shown). The inhibition was mimicked by CHA (10 μM), an adenosine A_1 receptor agonist, but not by CGS-22494 (10 μM), an adenosine A_{2A} receptor agonist. These results support the theory that the

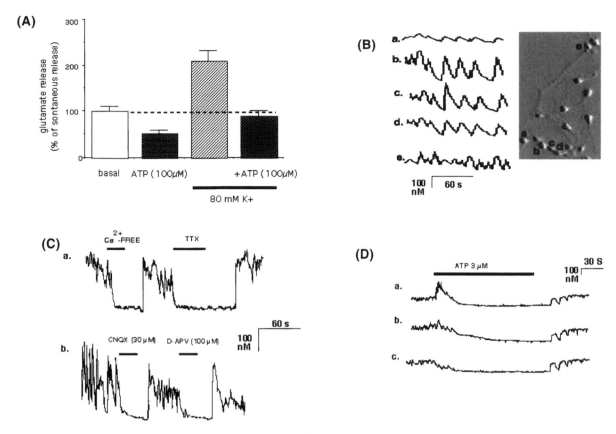

Fig 1 Effects of ATP on the glutamate release and on the Ca^{2+} oscillations in the cultured rat hippocampal neurons. (A) Effect of ATP (100 μM) on glutamate release. Glutamate was measured by HPLC-ECD using L-glutamate oxidase column. (B) Typical periodical Ca^{2+}-oscillation in the neurons. (a)–(e) correspond to a–e shown in phase-contrast image of the hippocampal neurons (200). Horizontal and vertical bars show 60 s and 100 nM, respectively. (C) (a) The effects of removal of extracellular Ca^{2+} or application of TTX on the Ca^{2+} oscillations. (b) The effects of CNQX or D-APV on the Ca^{2+} oscillations. CNQX (30 μM) or D-APV (100 μM) were added to the cells for 1 min (horizontal bars) and each application was separated by 1 min. (D) Typical traces of the effect of 3 μM ATP on Ca^{2+} oscillations. ATP was applied to the cells for 2 min (horizontal bar). a. ATP produced a transient rise in [Ca^{2+}]i followed by a suppression of the oscillations. (c) ATP inhibited the Ca^{2+} oscillations without the transient [Ca^{2+}]i rise. Horizontal and vertical scales show 30 s and 100 nM, respectively. Four such separate experiments were performed ($n = 128$). ((B)–(D) were reproduced with the permission from Koizumi and Inoue, 1997)

adenosine-induced inhibition of the Ca^{2+} oscillations is mediated by A$_1$ adenosine receptors. Although 100 μM aminophylline completely blocked the inhibitory action of adenosine, it did not affect ATP-dependent inhibition (Fig. 2(a)). Similarly, 8-CPT, at 100 μM, did not abolish the ATP-induced inhibition although it did show partial attenuation of the ATP-dependent inhibitory action (data not shown). Figure 3 shows the effects of various ATP analogues on the Ca^{2+} oscillations. ATP was inhibitory over a concentration range of 0.01 to 10 μM. In some preparations, ATPγS and

α,β-MeATP, which are non-hydrolyzable ATP analogues, also mimicked the inhibitory action. Moreover, UTP, which is not an adenosine derivative, mimicked the inhibition. The potency rank order of inhibition is: 2MeSATP > ATP > ATPγS > UTP > α,β-MeATP. Since the Ca^{2+} responses to glutamate were almost the same with or without ATP treatments, ATP (10 μM) did not inhibit the sensitivity of postsynaptic glutamate receptors. This means that the site of action of ATP was thought not to be in postsynapse. We conclude that ATP-induced inhibition was mediated via ATP

196

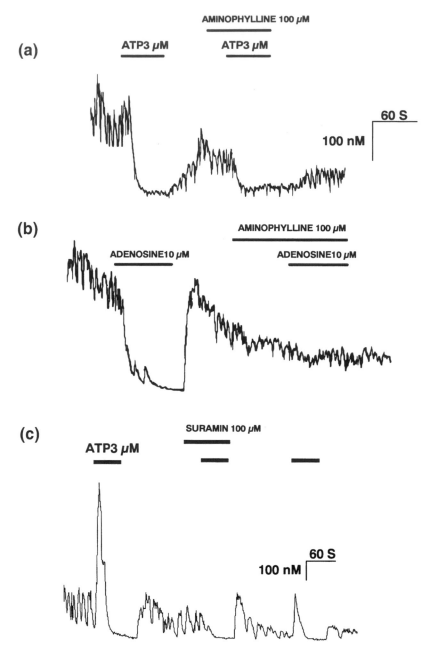

Fig. 2. The effects of aminophylline (100 μM) or suramin (100 μM) on the inhibition by ATP (3 μM) or adenosine (10 μM) of the Ca²⁺ oscillations. Aminophylline was applied to the cells 60 s before and during an application of drugs. (a) The effects of aminophylline on the inhibition by ATP (100 μM) of the Ca²⁺ oscillations. (b) The effects of aminophylline on the inhibition by adenosine (10 μM) of the Ca²⁺ oscillations. (c) The effects of suramin on the stimulation and inhibition by ATP (3 μM) of the Ca²⁺ oscillations. (Reproduced with permission from Koizumi and Inoue, 1997)

197

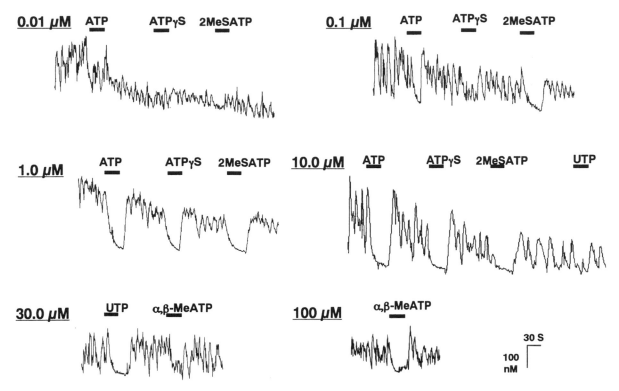

Fig. 3. The effects of various analogues of ATP on the Ca²⁺ oscillations in the hippocampal neurons. Traces show the representative results. Each compound was applied to the cells for 30 s (horizontal bars) at a concentration shown in the left side of each trace. Horizontal and vertical scales show 30 s and 100 nM, respectively. At least three such separate experiments were performed ($n = 96 - 128$). (Reproduced with permission from Koizumi and Inoue, 1997)

receptors since (1) ATP exhibited the inhibitory action even in the presence of P1-purinoceptor antagonists, and (2) non-hydrolyzable ATP analogues or (3) UTP inhibited the oscillation. We thus demonstrated that ATP has an inhibitory role via presynaptic ATP receptors in the suppression of the release of glutamate in hippocampal neurons.

Another inhibitory neurotransmitter, γ-aminobutyric acid (GABA), may affect the Ca²⁺ oscillations. Indeed, Ca²⁺ oscillations were inhibited by GABA, and were amplified by bicuculline (an antagonist to GABA_A receptors) in some neurons (data not shown). This indicates the continuous inhibition by GABA of the glutamatergic system in the hippocampus. The inhibition was dependent on the number of days the cells were in culture as the amplification by bicuculline was greater in neurons cultured for more than 10 days than those which were cultured for less than seven days. The data is in agreement with the physio-

logical data mentioned below and in previous reports, i.e. the inhibition by GABA of Ca²⁺ oscillations in various brain regions including the hippocampus was observed only in mature neurons (Obrietan and Van den Pol, 1995), and a GABA-mediated inhibition was found to be developmentally dependent in the hippocampus (Wagner and Alger, 1995). The inhibition by ATP of glutamate release is not through GABA because of the data obtained in the physiological examination. The detail of this are explained below.

We have previously reported that ATP produces a rise in [Ca²⁺]i in cultured rat hippocampal neurons (Inoue et al., 1995) where we observed Ca²⁺ responses to ATP in the presence of tetrodotoxin (TTX, 3 μM), hexamethonium (C6, 100 μM), APV (100 μM), CNQX (30 μM), bicuculline (10 μM), and cadmium (Cd²⁺, 300 μM), indicating that ATP directly stimulated excitatory P₂-purinoceptors in the postsynaptic neurons to promote the increase in

[Ca^{2+}]i. The finding that ATP produced a transient rise in [Ca^{2+}]i in some cells (Fig. 1(C)(a)), and that this increase was inhibited by suramin (Fig. 2(c)), suggests that the response is mediated by excitatory ATP receptors in the postsynaptic neurons. In addition to the excitatory ATP receptors, inhibitory ATP receptors also seem to be present in the hippocampal neurons. In other words, ATP receptors can play reciprocal roles, i.e. inhibitory and stimulatory in the functions of the hippocampus. Such reciprocal roles have been reported in other neurotransmitter systems in the CNS. Exogenously applied ATP inhibits the release of norepinephrine from rat brain cortex (Von Kügelgen et al., 1994) and rat hippocampus (Koch et al., 1997) whereas ATP causes depolarization (Harms et al., 1992) and inward currents (Shen and North, 1993) in the noradrenergic neurons of the nucleus locus coeruleus. Also, ATP inhibits dopamine release from the neostriatum (Von Kügelgen et al., 1997) whereas ATP increases the release of dopamine from the striatum (Zhang et al., 1995; Kittner et al., 1997). Therefore, the effect of ATP appears to be dependent on the site of the action and subtype of ATP receptors stimulated. One possible mechanism of this inhibition is the inhibition of presynaptic Ca^{2+} channels which are responsible for neurotransmitter release. ATP has been reported to inhibit L-type voltage-gated Ca^{2+} channels (VGCCs) in PC12 cells (Nakazawa and Inoue, 1992) and N- and P/Q-type VGCCs in adrenal chromaffin cells (Currie and Fox, 1996). The release of glutamate in the hippocampal neurons is regulated by both presynaptic N- and Q-type VGCCs (Scholz and Miller, 1995). These findings raise the possibility that ATP may exhibit its inhibitory action by suppressing presynaptic VGCCs thereby reducing the release of glutamate, which may in turn result in inhibition of the Ca^{2+} oscillations. One possible mechanism of the facilitation of glutamate release is the increase of Ca^{2+} influx through presynaptic P2X receptor/channels which can evoke neurotransmitter release as previously described in PC12 cells (Inoue et al., 1989; Nakazawa and Inoue, 1992). More recently, it has been reported as supporting evidence that ATP co-released with glutamate from central terminals of dorsal root ganglia (DRG) neurons activates ionotropic P2X receptors in the terminal, resulting in the increase of glutamate release (Gu and MacDermott, 1997).

The effect of ATP in hippocampal neurons; the electrophysiological study

The hippocampal pyramidal neurons are innervated by glutamatergic neurons and GABAergic interneurons. In the cultured rat hippocampal neurons, we can detect spontaneous synaptic currents which are blocked by TTX (3 μM) or extracellular Ca^{2+} free condition. These currents are composed of currents through glutamate receptor/channels (I_{glu}) and those through GABA$_A$ receptor/channels (I_{GABA}). Therefore, we need to separate these currents clearly for the examination of the effect of ATP on the release of glutamate or GABA in electrophysiological method. According to the Nernst equation, ionic equilibrium potentials vary linearly with the absolute temperature and logarithmically with the ionic concentration ratio. Since I_{GABA} is equal to Cl$^-$ current and I_{glu} is non-selective cation currents, we can adjust the equilibrium potential of I_{GABA} to around $-30\,mV$ and that of I_{glu} to around 0 mV by setting the concentration of Cl$^-$ or cations of extracellular buffer and internal pipette solution (Fig. 4, upper right). Figure 4(a)–(c) shows the currents detected in the hippocampal neurons cultured for seven days. At $-20\,mV$ of holding potential, both outward and inward currents appeared as shown in Fig. 4(b). The outward currents, which is theoretically I_{Cl^-} (I_{GABA}), were blocked by bicuculline (10 μM). The inward currents, which is theoretically I_{glu}, were blocked by CNQX (30 μM). At 0 mV of holding potential, only outward currents appeared and these currents were blocked by bicuculline (10 μM) as shown in Fig. 4(a). At $-30\,mV$ of holding potential in the Mg^{2+} free extracellular buffer, only inward currents appeared and these currents were blocked by APV (100 μM) and CNQX (30 μM) as shown in Fig. 4(c). Amplitude and frequency of both currents increased with the age of the cultures. I_{glu} was detected in all neurons cultured more than 10 days but not less than seven days in a physiological buffer (containing Mg^{2+}). Figure 5 shows the effect of ATP on the I_{glu} in 11 days cultured hippocampal

I_{GABA} and $I_{glutamate}$ in hippocampal neurons

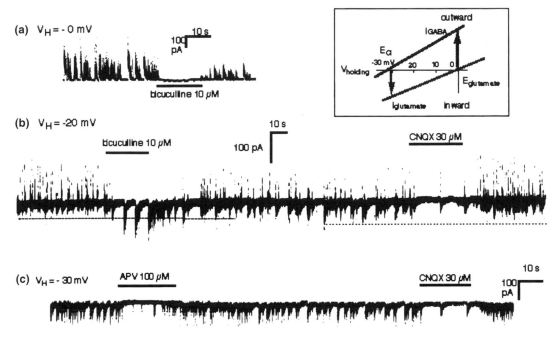

Fig. 4. Spontaneous synaptic currents through glutamate- and GABA-receptors in cultured rat hippocampal neurons. We adjusted the equilibrium potential of currents via GABAA receptor/channels (IGABA, equal to Cl^- current) to around -30 mV and that of currents via ionotropic glutamate receptors (I_{glu}) to around 0 mV by setting the concentration of Cl^- or cations of extracellular buffer and internal pipette solution according to the Nernst equation (upper right). (a)–(c) Currents detected in the hippocampal neurons cultured for seven days. (a) At 0 mV of holding potential, only outward currents appeared and these currents were blocked by bicuculline (10 μM). (b) At -20 mV of holding potential, both outward and inward currents appeared. The outward currents, which is theoretically I_{Cl^-} (I_{GABA}), were blocked by bicuculline (10 μM). The inward currents, which is theoretically I_{glu}, were blocked by CNQX (30 μM). (c) At -30 mV of holding potential in the Mg^{2+} free extracellular buffer, only inward currents appeared and these currents were blocked by APV (100 μM) and CNQX (30 μM).

neurons in the buffer. CNQX (30 μM) inhibited the current (Fig. 5(b)) but APV (100 μM) did not (Fig. 5(a)). ATP (30 μM) strongly inhibited the current (Fig. 5(c)). The inhibitory action of ATP was not blocked by PPADS (Fig. 5(d)), a blocker of several subtypes of ATP receptors. Drug application was done by Y-tube method (Ueno et al., 1997) which is able to change the solution containing the drug in 50 ms. Thus ATP is able to stimulate the neurons directly without delay before the chance of enzymatic degradation. Since ATP (30 μM) did not affect the glutamate-evoked current in the postsynaptic neurons (data not shown), the inhibition of I_{glu} by ATP is thought to be via presynaptic PPADS-insensitive ATP receptor(s) by the decrease of the glutamate release.

The release of glutamate may be under the continuous inhibitory regulation by GABA through $GABA_A$ receptors because bicuculline (10 μM) strongly enhanced I_{glu} for long period (Fig. 5(e)). A similar effect can be observed in Fig. 4(b) and in the Ca^{2+}-imaging study mentioned above. The enhanced I_{glu} was completely blocked by CNQX (10 μM) (data not shown). The idea that the inhibitory action of ATP to glutamate release may be due to the enhancement of the GABA release is not likely because ATP (30 μM) was able to inhibit the enhanced I_{glu} under the application of bicuculline (10 μM) shown in Fig. 5(f). Recently, it has been reported that GABAergic interneurons are the major postsynaptic targets of mossy fibres in the rat hippocampus. Therefore, CNQX can block the

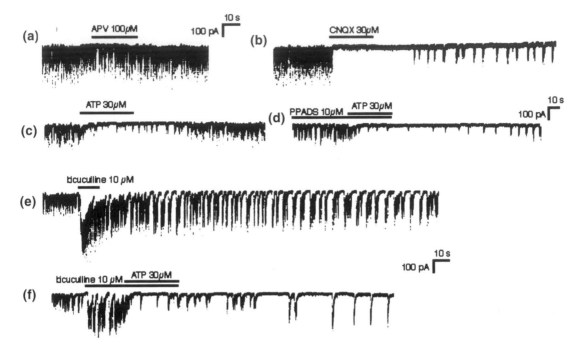

Fig. 5. The effect of ATP on the I_{glu} in 11 days cultured hippocampal neurons. Holding potential was around -30 mV. (a) and (b) Effect of APV (100 μM) or CNQX (30 μM). (c) Effect of ATP (30 μM) on the current. (d) The inhibitory action of ATP was not blocked by PPADS (10 μM), a blocker of several subtypes of ATP receptors. Drug application was done by Y-tube method which is able to change the solution containing the drug during 50 ms. Thus ATP is able to stimulate the neurons without delay or enzymatic degradation. (e) Effect of bicuculline (10 μM) on the currents. (f) The inhibitory effect of ATP (30 μM) on the enhanced I_{glu} under the application of bicuculline (10 μM).

input by glutamate to interneurons resulting in the inhibition of GABA release. Indeed, CNQX (30 μM) strongly inhibited the frequency of I_{GABA} as shown in Fig. 4(b), right-hand-side.

ATP stimulates plasminogen release from microglia

Recently, it has been suggested that microglia, a type of glial cell, is activated in response to pathological changes in CNS and can play a variety of important roles of CNS (Nakajima and Kohsaka, 1993; Verkhratsky and Kettenmann, 1996; Kreutzberg, 1996). Activated microglia are mainly scavenger cells but also perform various other functions in tissue repair and neural regeneration. These cells are able to release several cytotoxic substances in vitro, such as NO, as well as arachidonic-acid derivatives, excitatory amino acids, quinolinic acid and cytokines (Colton and Gilbert, 1987; Banati et al., 1993). Free oxygen

radicals released by microglia have a neurotoxic effect in co-cultures of neurones and microglia (Thery et al., 1993). Activated microglia may also play a protective role. They produce the urokinase-type plasminogen activator and plasminogen, which promotes the development of mesencephalic dopaminergic neurons and enhances neurite outgrowth from explants of neocortical tissue (Nakajima et al., 1992a, 1992b). The activation and proliferation of microglia occurring after axotomy of the facial nerve are accompanied by an increase in urokinase activity in the facial nucleus (Nakajima et al., 1996). These properties of microglia can be modulated by cytokines and neurotransmitters acting through receptors for CNS signalling molecules including ATP.

It has been reported that ATP is released from hippocampal slices by electrical stimulation of Schaffer collateral–comissural afferents (Wieraszko et al., 1989), and that ATP mediates synaptic transmission in cultured rat hippocampal neurons

(Inoue et al., 1992, 1995). It has also been reported that exogeneously applied plasminogen enhances synaptic transmission via NMDA-glutamate receptors in cultured rat hippocampal neurons (Inoue et al., 1994). It is well known that microglia express receptors for ATP, G-protein coupled-type ATP receptors such as P2Y and ionotropic ATP receptors such as P2Z (recently named P2X$_7$) (Neary et al., 1996; Surprenant et al., 1996). ATP induces the release of interleukin(IL)-1β from mouse brain microglia through P2Z (Ferrari et al., 1996), produces currents in rat brain (Walz et al., 1993) or forebrain microglia (Nörenberg et al., 1994; Langosch et al., 1994) and stimulates an increase in intracellular Ca^{2+} in rat brain (Walz et al., 1993). These data suggest that ATP mediates signals from neurons to microglia in hippocampus, resulting in the stimulation of the release of plasminogen. Therefore we examined the effect of ATP on the release of plasminogen from microglia with the aim of being able to speculate on the physiological significance of this release (Inoue et al., 1998). Figure 6(A) shows the effect of ATP (10–100 μM) and glutamate (100 μM) on the release of plasmi-

nogen from microglia using an immunoblotting method with DAB staining. Fig. 6(B) shows quantitative data from the immunoblotting experiments using a high sensitivity ECL method. These data demonstrate that ATP stimulated the release of plasminogen in a concentration-dependent manner (the estimated EC50 is approximately 15 μM), whereas glutamate (100 μM) did not induce the release of plasminogen at 10 min after the stimulation. It is reported that NO is released from various cells following the cascade of Ca^{2+} influx, activation of Ca^{2+}/calmoduline-dependent protein kinase, and activation of NO synthase (Palmer et al., 1987). It is also known that ATP stimulates NO release from endothelial cells, presumably through an increase in [Ca^{2+}]i (Olsson and Pearson, 1990). However, we did not detect the presence of NO after addition of ATP (10–100 μM) while the addition of lipopolysaccharide (LPS, 0.5 μg/ml), as a positive control, did produce NO from the microglia (Inoue et al., 1998).

We examined the intracellular Ca^{2+} dependency of the plasminogen release using BAPTA-AM, which is converted in the cytosol to BAPTA, an

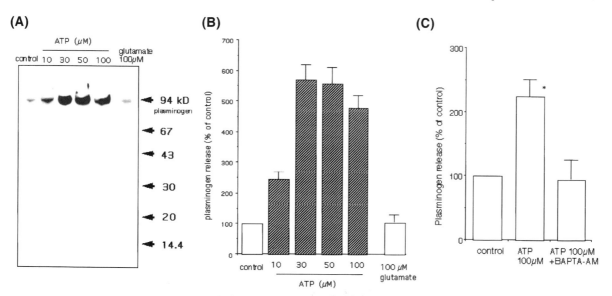

Fig. 6. The effect of ATP (10–100 μM) and glutamate (100 μM) on the release of plasminogen from microglia 10 min after stimulation. (A) Immunoblotting with 0.02% 3,3'-diaminobenzidine staining. (B) Quantitative data by immunoblotting with enhanced chemiluminescent staining. Data are mean ± SEM of three-to-five blots measured. (C) The effect of 1 h pretreatment of cells with BAPTA-AM (200 μM) on the plasminogen release evoked by ATP(100 μM). Data are mean ± SEM of three-to-five blots measured. An asterisk indicates significant difference from the BAPTA–AM-treated group (*P < 0.05). (Reproduced with permission from Inoue et al., 1998)

intracellular Ca^{2+} chelator (Negulescu et al., 1989) since neurotransmitter release from synaptic vesicles is triggered by a transient increase in the internal Ca^{2+} concentration ($[Ca^{2+}]i$) through voltage-gated Ca^{2+} channels after priming (Sudhof, 1995). Figure 6(C) shows the effect of a 1 h pretreatment of cells with BAPTA-AM (200 μM) on the plasminogen release induced by ATP stimulation. BAPTA-AM treatment inhibited the plasminogen release completely. The data suggests that ATP-evoked plasminogen release is dependent on the increase in $[Ca^{2+}]i$.

The next question we asked was how ATP increased the $[Ca^{2+}]i$. It has been reported that microglia express the P2Y receptor (a G-protein coupled-type P_2 receptor) as well as the P_{2Z} ($P2X_7$) receptor (a Ca^{2+}-permeable ligand-gated channel) (Nörenberg et al., 1994; Langosch et al., 1994; Neary et al., 1996; Surprenant et al., 1996). Application of ATP (100 μM) induced an increase in $[Ca^{2+}]i$ (41 out of 41 cells), whereas glutamate (100 μM) did not produce any increase in $[Ca^{2+}]i$ (70 out of 70 cells). This is consistent with the report that microglia express ATP receptors but not glutamate receptors (Verkhratsky and Kettenmann, 1996). The $[Ca^{2+}]i$ increased quickly during a 30 s application of ATP and then decreased to control levels by 2 min after the ATP application. The increase in $[Ca^{2+}]i$ by ATP was concentration-dependent from 1 to 100 μM (Fig. 7(A)) with the estimated EC50 of 12 μM, and was dependent on extracellular Ca^{2+} (12 out of 12 cells). These data indicate that the ionotropic ATP receptor, presumably $P2X_7$, is responsible for the increase in $[Ca^{2+}]i$ in microglia. Then, we examined the effect of BzATP, a selective agonist of $P2X_7$ (Surprenant et al., 1996), and the effect of oxidized ATP, a selective antagonist of $P2X_7$ (Surprenant et al., 1996), on the ATP-induced increase of $[Ca^{2+}]i$. BzATP produced a long-lasting increase in $[Ca^{2+}]i$ even at 1 μM, a concentration at which ATP did not evoke the increase (Fig. 7(B)). Moreover, 1h pretreatment with oxidized ATP (100 μM) completely blocked the increase by ATP (10 μM) as shown in Fig. 7(C). These data, taken together with the fact that the concentration-response curve for the increase in $[Ca^{2+}]i$ by ATP was shifted to the left in the Mg^{2+}-free condition (Fig. 7(A)), indicate

that the ATP-induced $[Ca^{2+}]i$ increase is due to the activation of $P2X_7$ receptors.

It has been reported that activated microglia can display both neuroprotective and neurotoxic functions in the CNS. Plasminogen is thought to be a key molecule for the neurotrophic effect because it promotes the development of mesencephalic dopaminergic neurons and enhances neurite outgrowth from explants of neocortical tissue (Nakajima et al., 1992a, 1992b). NO is thought to be neurotoxic via a mechanism involving free oxygen radicals and it has been demonstrated that NO released from microglia has a neurotoxic effect in co-cultures of neurones and microglia (Thery et al., 1993). Here, we present evidence that ATP produces an increase in $[Ca^{2+}]i$ which stimulates plasminogen release from cultured microglia through $P2X_7$. On the other hand, ATP does not stimulate NO secretion from the same preparations. It is possible that ATP released from nerve endings transmits information to microglia and stimulates them to release plasminogen in the hippocampus. As mentioned above, the released plasminogen may modulate the function of the neurons. Thus, ATP from purinergic neurons can transmit information to glutamate-related neurons via microglia. In other words, a triangular connection of neuron-microglia-neuron may be completed by ATP and plasminogen in the hippocampus. ATP also exists in the cytosol at a concentration of several mM and is able to leak out of cells damaged by ischemia or trauma. It is reported that microglia respond to ischemia in the CA1 area of hippocampus (Yamashita et al., 1994). ATP leaking from damaged cells as well as ATP from nerve endings may be able to activate hippocampal microglia.

Conclusion

The possible function of ATP receptors in the synaptic transmission in the hippocampus was reviewed. This review presents information that ATP inhibits presynaptically the release of glutamate, an excitatory neurotransmitter from the hippocampus. Meanwhile, it has been reported that ATP stimulates the release of glycine, an inhibitory neurotransmitter from interneurons of dorsal horn (see Ree et al., personal communication, 1998). In

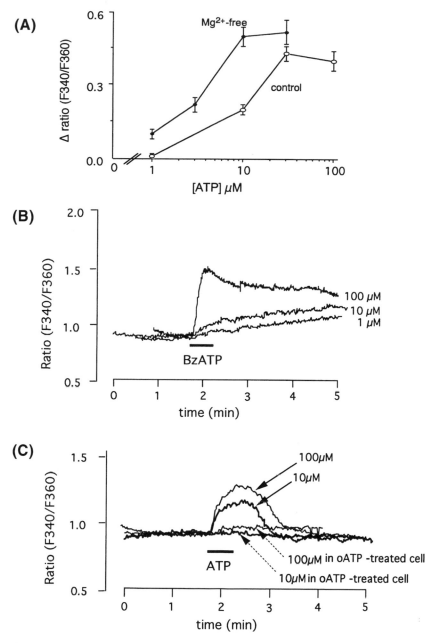

Fig. 7. The effect of ATP (10–100 μM) on the intracellular Ca^{2+} in microglia. (A) The concentration-dependency of the increase in $[Ca^{2+}]i$ by ATP from 1 to 100 μM. Data are mean ± SEM of the responses in 7–10 cells. (B) and (C) Effects of BzATP and oxidizedATP (1 h pretreatment) in microglia. Data represent the mean of the responses in 14 cells. (Reproduced with permission from Inoue et al., 1998)

hippocampus, ATP may stimulate the release of GABA from some interneurons (unpublished data). The inhibition of glutamate release and the stimulation of GABA and glycine release have the same effect as a negative feedback loop, i.e. turning-down the activity of neurons stimulated by glutamate. ATP also activates microglia to release plasminogen, a type of neurotrophic factor, which promotes the development of mesencephalic dopaminergic neurons and enhances neurite outgrowth from explants of neocortical tissue. ATP may thus stimulate the reconstruction of synaptic networks through activating microglia. Moreover, there are many reports on the trophic action of ATP itself in CNS (Neary et al., 1996). Thus, ATP may have a role in the protection of the function of neurons in hippocampus from over-stimulation by glutamate.

Acknowledgements

We would like to express our gratitude to Drs. K. Nakajima, K.Nakazawa, T. Morimoto, P. Illes and S. Kohsaka for collaborations.

References

Abbracchio, M.P. and Burnstock, G. (1994) Purinoceptors: Are there families of P2X and P2Y purinoceptors? *Pharmacol. Ther.*, 64: 445–475.

Artola, A. and Singer, W. (1993) Long-term depression of excitatory synaptic transmission and its relationship to long-term potentiation. *Trends Neurosci.*, 16: 480–487.

Banati, R.B., Gehrmann, J., Schubert, P. and Kreutzberg, G.W. (1993) Cytotoxicity of microglia. *Glia*, 7: 111–118.

Bean, B.P. and Friel, D.D. (1990) ATP-activated channels in excitable cells. In: T. Narahashi (Ed.), *Ion channels,* Vol. 2. New York: Plenum Publishing Corp. pp. 169–203.

Bear, M.F. and Malenka, R.C. (1994) Synaptic plasticity: LTP and LTD. *Curr. Opin. Neurobiol.*, 4: 389–399.

Bleehen, T. and Keele, C.A. (1977) Observations on the algogenic actions of adenosine compounds on the human blister base preparation. *Pain*, 3: 367–377.

Bliss, T.V.P., Clements, M.P., Errington, M.L., Lynch, M.A. and Williams, J.H. (1991) Presynaptic changes associated with long-term potentiation in the dentate gyrus. In: M. Baudry, and J.L. Davis (Eds.), *Long Term Potentiation.* MIT Press, Cambridge, Massachusetts, pp. 3–18,.

Bliss, T.V.P. and Lynch, M.A. (1988) Long term potentiation of synaptic transmission in the hippocampus: properties and mechanisms. In: S.A. Deadwyler and P. Landfield (Eds.). In: *Long Term Potentiation from Biophysics to Behavior.* New York: Alan R. Liss Inc., pp. 3–72.

Bo, X., Zhang, Y., Nassar, M., Burnstock, G. and Schöepfer, R. (1995) A P2X purinoceptor cDNA conferring a novel pharmacological profile. *FEBS Lett.*, 375: 129–133.

Bolshakov, V.Y. and Siegelbaum, S.A. (1994) Postsynaptic induction and presynaptic expression of hippocampal long-term depression. *Science*, 264: 1148–1152.

Brake, A.J., Wagenbach, M.J. and Julius, D. (1994) New stractural motif for ligand-gated ion channels defined by an ionotropic ATP receptor. *Nature*, 371: 519–522.

Buell, G.L., Lewis, C., Collo, G., North, R.A. and Surprenant, A. (1996) An antagonist-insensitive P2X receptor expressed in epithelia and brain. *EMBO J.*, 15: 55–62.

Burnstock, G. and Kennedy, C. (1985) A dual function for adenosine 5′-triphosphate in the regulation of vascular tone; excitatory cotransmitter with noradrenaline from perivascular nerves and locally released inhibitory intravascular agent. *Cir. Res.*, 58: 319–391.

Burnstock, G. and King, B.F. (1996) Numbering of cloned P2 purinoceptors. *Drug Dev. Res.*, 38: 67–71.

Chen, C., Akopian, A.N., Sivilotti, L., Colquhoun, D., Burnstock, G. and Wood, J.N. (1995) A P2X receptor expressed by a subset of sensory neurons. *Nature*, 377: 428–430.

Collo, G., North, R.A., Kawashima, E., Merlo-Pich, E., Neidhart, S., Surprenant, A. and Buell, G. (1996) Cloning of $P2X_5$ and $P2X_6$ receptors and the distribution and properties of an extended family of ATP-gated ion channels. *J. Neurosci.*, 16: 2495–2507.

Colton, C.A. and Gilbert, D.L. (1987) Production of superoxide anions by a CNS macrophage, the microglia. *FEBS Lett.*, 223: 284–288.

Currie, K.P.M. and Fox, A.P. (1996) ATP serves as a negative feedback inhibitor of voltage-gated Ca^{2+} channel currents in cultured bovine adrenal chromaffin cells. *Neuron*, 16: 1027–1036.

Edwards, F.A., Gibb, A.J. and Colquhoun, D. (1992) ATP receptor-mediated synaptic current in the central nervous system. *Nature*, 359: 145–147.

Edwards, F.A., Robertson, S.J. and Gibb, A.J. (1997) Properties of ATP receptor-mediated synaptic transmission in the rat medial habenula. *Neuropharmacol.*, 36: 1253–1268.

Ferrari, D., Villalba, M., Chiozzi, P., Falzoni, S., Ricciardi-Castagnoli, P. and Di Virgilio, F. (1996) Mouse microglial cells express a plasma membrane pore gated by extracellular ATP. *J. Immunol.*, 156: 1531–1539.

Fujii, S., Kato, H., Furuse, H., Ito, K., Osada, H., Hamaguchi, T. and Kuroda, Y. (1995) The mechanism of ATP-induced long-term potentiation involves extracellular phosphorylation of membrane proteins in guinea-pig hippocampal CA1 neurons. *Neurosci. Lett.*, 187: 130–132.

Gu, J.G. and MacDermott, A.B. (1997) Activation of ATP P2X receptors elicits glutamate release from sensory neuron synapses. *Nature*, 389: 749–753.

Harms, L., Finta, E.P., Tschöpl, M. and Illes, P. (1992) Depolarization of rat locus coeruleus neurons by adenosine 5′-triphosphate. *Neuroscience*, 48: 941–952.

Inoue, K., Koizumi, S., Nakajima, K., Hamanoue, M. and Kohsaka, S. (1994) Modulatory effect of plasminogen on NMDA-induced increase in intracellular free calcium concentration in rat cultured hippocampal neurons. *Neurosci. Lett.*, 179: 87–90.

Inoue, K., Koizumi, S. and Nakazawa, K. (1995) Glutamate-evoked release of adenosine 5′-triphosphate causing an increase in intracellular calcium in hippocampal neurons. *NeuroReport*, 6: 437–440.

Inoue, K., Nakajima, K., Morimoto, T., Kikuchi, Y., Koizumi, S., Illes, P. and Kohsaka, S. (1998) ATP stimulates Ca^{2+}-dependent plasminogen release from cultured microglia. *Br. J. Pharmacol*, 123: 1304–1310.

Inoue, K., Nakazawa, K., Fujimori, K. and Takanaka, A. (1989) Extracellular adenosine 5′-trisphosphate-evoked norepinephrine release coupled with receptor-operated Ca permeable channels in PC12 cells. *Neurosci. Lett.*, 106: 294–299.

Inoue, K., Nakazawa, K., Fujimori, K., Watano, W. and Takanaka, A. (1992) Extracellular adenosine 5′-triphosphate-evoked glutamate release in cultured rat hippocampal neurons. *Neurosci. Lett.*, 134: 215–218.

Kittner, H., Krügel, U., Graf, D. and Illes, P. (1997) Influence of 2-methylthio ATP (2-MeSATP) on the mesolimbic dopaminergic system of the rat: Microdialysis, electroencephalography and behavior. *Naunyn-Schmiedeberg's Arch. Pharmacol.*, 355: R36.

Koch, H., Von Kügelgen, I. and Starke, K. (1997) P2-receptor-mediated inhibition of noradrenaline release in the rat hippocampus. *Naunyn-Schmiedeberg's Arch. Pharmacol.*, 355: 707–715.

Koizumi, S. and Inoue, K. (1997) Inhibition by ATP of calcium oscillations in cultured rat hippocampal neurons. *Br. J. Pharmacol.*, 122: 51–58.

Kreutzberg, G.W. (1996) Microglia: A sensor for pathological events in the CNS. *Trends. Neurosci.*, 19: 312–318.

Kudo, Y., Akita, K., Nakazawa, M., Kuroda, Y. and Ogura, A. (1991) The presynaptic effects of adenosine and adenine nucleotides monitored by the modulation of periodic changes in intracellular calcium concentration in cultured hippocampal neurons. In: S. Imai and M. Nakazawa (Eds.), *Role of Adenosine and Adenine Nucleotides in the Biological System.* Elsevier Science Publishers BV, Amsterdam, pp. 235–244.

Langosch, J.M., Gebicke-Haerter, J.P., Nörenberg, W. and Illes, P. (1994) Characterization and transduction mechanisms of purinoceptors in activated rat microglia. *Br. J. Pharmacol.*, 113: 29–34.

Lewis, C., Neidhart, S., Holy, C., North, R.A., Buell, G. and Surprenant, A. (1995) Coexpression of P2X(2) and P2X(3) receptor subunits can account for ATP-gated currents in sensory neurons. *Nature*, 377: 432–435.

Lustig, K.D., Shiau, A.K., Brake, A.J. and Julius, D. (1993) Expression cloning of an ATP receptor from mouse neuroblastoma cells. *Proc. Natl. Acad. Sci. USA*, 90: 5113–5117.

Malenka, R., Kauer, J.A., Perkel, D. and Nicoll, A. (1989) The impact of postsynaptic calcium on synaptic transmission- its role in long-term potentiation. *Trends Neurosci.*, 12: 444–450.

Nakajima, K. and Kohsaka, S. (1993) Functional roles of microglia in the brain. *Neurosci. Res.*, 17: 187–203.

Nakajima, K., Reddington, M., Kohsaka, S. and Kreutzberg, G.W. (1996) Induction of urokinase-type plasminogen activator in rat facial nucleus by axotomy of the facial nerve. *J. Neurochem.*, 66: 2500–2505.

Nakajima, K., Tsuzaki, N., Nagata, K., Takemoto, N. and Kohsaka, S. (1992a) Production and secretion of plasminogen in cultured rat brain microglia. *FEBS Lett.*, 308: 179–182.

Nakajima, K., Tsuzaki, N., Shimojo, M., Hamanoue, M. and Kohsaka, S. (1992b) Microglia isolated from rat brain secrete a urokinase-type plasminogen activator. *Brain Res.*, 577: 285–292.

Nakazawa, K. and Inoue, K. (1992) Roles of Ca^{2+} influx through ATP-activated channels in catecholamine release from pheochromocytoma PC12 cells. *J. Neurophysiol.*, 68: 2026–2032.

Neary, J.T., Rathbone, M.P., Cattabeni, F., Abbracchio, M.P. and Burnstock, G. (1996) Trophic actions of extracellular nucleotides and nucleosides on glial and neuronal cells. *Trends. Neurosci.*, 19: 13–18.

Negulescu, P.A., Reenstra, W.W. and Machen, T.E. (1989) Intracellular Ca requirements for stimulus-secretion coupling in parietal cell. *Am. J. Physiol.*, 256: C241–C251.

Nishimura, S., Mohri, M., Okada, Y. and Mori, M. (1990) Excitatory and inhibitory effects of adenosine on the neurotransmission in the hippocampal slices of guinea pig. *Brain Res.*, 525: 165–169.

Nörenberg, W., Langosch, J.M., Gebicke-Haerter, J.P. and Illes, P. (1994) Characterization and possible function of adenosine 5′-triphosphate receptors in activated rat microglia. *Br. J. Pharmacol.*, 111: 942–950.

North, R.A. and Barnard, E.A. (1997) Nucleotide receptors. *Curr. Opin. Neurobiol.*, 7: 346–357.

Obrietan, K. and Van den Pol, A. (1995) GABA neurotransmission in the hypothalamus: developmental reversal from Ca^{2+} elevating to depressing. *J. Neurosci.*, 15: 5065–5077.

Ogura, A., Iijima, T., Amano, T. and Kudo, Y. (1987) Optical monitoring of excitatory synaptic activity between cultured hippocampal neurons by a multi-site Ca^{2+} fluorometry. *Neurosci. Lett.*, 78: 69–74.

Okada, Y. and Kuroda, Y. (1980) Inhibitory action of adenosine and adenosine analogs on neurotransmission in the olfactory cortex slice of guinea pig-structure–activity relationships. *Eur. J. Pharmacol.*, 61: 137–146.

Olsson, R.A. and Pearson, J.D. (1990) Cardiovascular purinoceptors. *Physiol. Rev.*, 70: 761–845.

Palmer, R.M., Ferrige, G.A. and Moncada, S. (1987) Nitric oxide release accounts for the biological activity of endothelium-derived relaxing factor. *Nature*, 327: 524–526.

Phillis, J.W., Kostopoulos, G.K. and Limacher, J.J. (1975) A potent depressant action of adenosine derivatives on cerebral cortical neurons. *Eur. J. Pharmacol.*, 30: 125–129.

Scholz, K.P. and Miller, J.R.. (1995) Developmental changes in presynaptic calcium channels coupled to glutamate release in cultured rat hippocampal neurons. *J. Neurosci.*, 15: 4612–4617.

Shen. K.-Z. and North, R.A. (1993) Excitation of rat locus coerleus neurons by adenosine 5′-triphosphate: Ionic mechanism and receptor characterization. *J. Neurosci.*, 13: 894–899.

Sudhof, C.T. (1995) The synaptic vesicle cycle: a cascade of protein–protein interactions. *Nature*, 375: 645–653.

Surprenant, A., Rassendren, F., Kawashima, E., North, R.A. and Buell, G. (1996) The cytolytic P_{2Z} receptor for extracellular ATP identified as a P2X receptor ($P2X_7$). *Science*, 272: 735–738.

206

Thery, C., Chamak, B. and Mallat, M. (1993) Neurotoxicity of brain macrophages. *Clin. Neuropathol.*, 12: 288–290.

Ueno, S., Bracamontes, J., Zorumski, C., Weiss, D.S. and Steinbach, J.H. (1997) Bicuculline and gabazine are allosteric inhibitors of channel opening of theGABA$_A$ receptor. *J Neurosci.*, 17: 625–634.

Valera, S., Hussy, N., Evans, R.J., Adami, N., North, R.A., Surprenant, A. and Buell, G. (1994) A new class of ligand-gated ion channel defined by P2X receptor for extracellular ATP. *Nature*, 371: 516–519.

Verkhratsky, A. and Kettenmann, H. (1996) Calcium signalling in glial cells. *Trends. Neurosci.*, 19: 346–352.

Von Kügelgen, I., Koch, H. and Starke, K. (1997) P2-receptor-mediated inhibition of serotonin release in the rat brain cortex. *Neuropharmacology*, 36: 1221–1227.

Von Kügelgen, I., Späth, L. and Starke, K. (1994) Evidence for P2-purinoceptor-mediated inhibition of noradrenaline release in rat brain cortex. *Br. J. Pharmacol.*, 113: 815–822.

Wagner, J.J. and Alger, B.E. (1995) GABAergic and developmental influences on homosynaptic LTD and depotentiation in rat hippocampus. *J. Neurosci.*, 15: 1577–1586.

Walz, W., Ilschner, S., Ohlemeyer, C., Banati, R. and Kettenmann, H. (1993) Extracellular ATP activates a cation conductance and a K+ conductance in cultured microglial cells from mouse brain. *Neurosci.*, 13: 4403–4411.

Webb, T.E., Simon, J., Krishek, B.J., Bateson, A.N., Smart, T.G., King, B.F. and Burnstock, G. (1993) Cloning and functional expression of a brain G-protein-coupled ATP receptor. *FEBS Lett.*, 324: 219–225.

White, T.D. (1978) Release of ATP from a synaptosomal preparation by elevated extracellular K+ and by veratridine. *J. Neurochem.*, 30: 329–336.

Wieraszko, A. and Ehrlich, Y.H. (1994) On the role of extracellular ATP in the induction of long-term potentiation in the hippocampus. *J. Neurochem.*, 63. 1731–1738.

Wieraszko, A., Goldsmith, G. and Seyfried, T.N. (1989) Stimulation-dependent release of adenosine triphosphate from hippocampal slices. *Brain Res.*, 485: 244–250.

Wieraszko, A. and Seyfried, T.N. (1989) ATP-induced synaptic potentiation in hippocampal slices. *Brain Res.*, 491: 356–359.

Xiao, M.Y., Karpefors, M., Gustafsson, B. and Wigstrom, H. (1995) On the linkage between AMPA and NMDA receptor-mediated EPSPs in homosynaptic long-term depression in the hippocampal CA1 region of young rats. *J. Neurosci.*, 15: 4496–4506.

Yamashita, K., Niwa, M., Kataoka, Y., Shigematsu, K., Himeno, A., Tsutsumi, K., Nakano-Nakashima, M., Sakurai-Yamashita, Y., Shibata, S. and Taniyama, K. (1994) Microglia with an endothelin ETB receptor aggregate in rat hippocampus CA1 subfields following transient forebrain ischemia. *J. Neurochem.*, 63: 1042-1051.

Zhang, Y.X., Yamashita, H., Ohshita, T., Sawamoto, K. and Nakamura, S. (1995) ATP increases extracellular dopamine level through stimulation of P2Y purinoceptors in the rat striatum. *Brain Res.*, 691: 205–212.

Zhuo, M., Small, S.A., Kandel, E.R. and Hawkins, R.D. (1993) Nitric oxide and carbon monoxide produce activity-dependent long-term synaptic enhancement in hippocampus. *Science*, 260: 1946–1950.

SECTION VII

Physiology of nucleotide function

P. Illes and H. Zimmermann (Eds.)
Progress in Brain Research, Vol 120
© 1999 Elsevier Science BV. All rights reserved

CHAPTER 17

Electrophysiological analysis of P2-receptor mechanisms in rat sympathetic neurones

Wolfgang Nörenberg[1,*], Ivar von Kügelgen[1], Angelika Meyer[1] and Peter Illes[2]

[1]*Department of Pharmacology, University of Freiburg, Hermann-Herder-Strasse 5, D-79104 Freiburg, Germany*
[2]*Department of Pharmacology, University of Leipzig, Härtelstrasse 16-18, D-04107 Leipzig, Germany*

Introduction

Adenosine-5'-triphosphate (ATP) is an excitatory transmitter/cotransmitter in the central nervous system, as well as periphery (Burnstock, 1990; Von Kügelgen and Starke, 1991a; Edwards et al., 1992; Evans et al., 1992; Silinsky and Gerzanich, 1993; Zimmermann, 1994; Illes et al., 1996; Nieber et al., 1997). In the sympathetic tree of the autonomic nervous system, ATP is stored together with noradrenaline in synaptic vesicles (Zimmermann, 1994) and released upon nerve stimulation (Lew and White, 1987; Von Kügelgen and Starke, 1991b).

Sympathetic neurones are not only a source of ATP, but also respond to the nucleotide. ATP and some of its structural analogues induce depolarisation, e.g. in guinea-pig coeliac ganglion cells and rat superior cervical ganglion (SCG) cells, respectively (Silinsky and Gerzanich, 1993; Connolly et al., 1993). One type of receptor mediating these excitatory actions in mammalian sympathetic neurones belong to the P2X-receptor subclass of nucleotide receptors (Evans et al., 1992; Silinsky and Gerzanich, 1993; Kakh et al., 1995). P2X-receptors are ligand-gated channels which when activated allow cationic inward currents leading to

*Corresponding author. Tel.: +49-761-203-5317; fax: +49-761-203-5318; e-mail: noerenbe@ruf.uni-freiburg.de

membrane depolarisation and eventually to action potential firing. To date, seven subtypes (P2X$_1$ to P2X$_7$) of P2X-receptors have been cloned and functionally expressed (North and Barnard, 1997). It is noticeable that two of this subtypes were extracted originally from sympathetic ganglia cDNA libraries (P2X$_5$ and P2X$_6$ from rat coeliac cells and SCG cells, respectively; Collo et al., 1996) and one from nerve growth factor differentiated PC12 cells (P2X$_2$; Brake et al., 1994) which closely resemble postganglionic sympathetic neurones.

Besides P2X-receptors, sympathetic neurones may also posses excitatory metabotropic nucleotide receptors (P2Y-receptors). An indication for this is provided by findings in sympathetic neurones from SCG, as well as thoracolumbal ganglia of the rat. In these preparations, the pyrimidine nucleotides uridine-5'-triphosphate (UTP) and uridine-5'-diphosphate (UDP) induced the release of previously taken-up [³H]-noradrenaline but did not elicit P2X-receptor mediated inward currents (Boehm et al., 1995; Von Kügelgen et al., 1997). It is noticeable that several of the hitherto cloned and functionally expressed P2Y-receptor subtypes are either equally sensitive to ATP and UTP (P2Y$_2$) or predominantly activated by pyrimidine nucleotides (P2Y$_3$, P2Y$_4$ and P2Y$_6$; for a review see North and Barnard, 1997).

In the following sections, we will present the outcome of some patch clamp investigations aimed to better understand excitatory mechanisms of extracellular nucleotides in thoracolumbal sympathetic neurones of the rat. The scope was twofold: firstly to pharmacologically characterise P2X-receptors present in these cells (Von Kügelgen et al., 1997) and, secondly, to search for electrophysiological correlates of the actions of pyrimidine nucleotides. A possible answer to the latter question may come from a finding in amphibia: UTP depolarised lumbar sympathetic neurones of bullfrogs (Siggins et al., 1977) by inhibition of a potassium current also attenuated by muscarinic agonists, the M-current ($I_{K(M)}$; Adams et al., 1982). However, at present it is not clear whether excitatory effects of pyrimidine nucleotides in mammalian sympathetic neurones are also mediated by the M-current mechanism. This was proposed recently for cultured SCG cells (Boehm and Bofill-Cardona, 1998). However, conflicting results exist. In the same preparation, UTP inhibited M-currents only in cells where $P2Y_2$-receptors were heterologously expressed by cRNA injection, but was completely ineffective in non-injected cells (Filippov et al., 1998).

Membrane potential responses of cultured rat thoracolumbal sympathetic neurones to extracellular nucleotides

In the first step, we tried to characterise electrophysiologically the cell types found in our preparation. To this end, cultures of sympathetic neurones were prepared from new-born rats as described previously and investigated after 5–7 days in culture, i.e. at a time when UDP/UTP-induced [^{3}H]-noradrenaline release could be routinely measured (Von Kügelgen et al., 1997). The cultures contained at least three morphologically different cell types (Fig. 1A): (1) round or oval cells, bright in phase contrast microscopy, with two and more dendrite-like processes, having a greater diameter of 20–40 μm, assumed to be sympathetic neurones, (2) oval cells with a dark soma, having a greater diameter around 30 μm and bearing two long processes which may represent the neuroglial satellite cells, and (3) large polygonal cell bodies devoid of processes, having a greater diameter of 30–50 μm, the fibroblasts. The effects of current injection, ATP (100 μM), UDP (100 μM) and the classic ganglionic stimulant nicotine (100 μM) on the membrane potential were determined in the current clamp mode of the whole-cell patch clamp technique. The phase contrast–bright cells had a membrane potential of -56.7 ± 1.1 mV ($n = 7$). Injection of positive current elicited depolarising electrotonic potentials and, at sufficient strength (25–80 pA) action potentials, thereby confirming that these cells were neurones (Fig. 1B). ATP and nicotine evoked depolarisations, again accompanied by action potential firing. UDP lacked any major effect (Fig. 1B). UDP was also ineffective when applied for 60 s instead of the usual 3–5 s (not shown). The presumed satellite cells (Fig. 1C; $n = 7$) and the fibroblasts (Fig 1D; $n = 5$) had membrane potentials of -42.7 ± 2.7 mV and -66.6 ± 1.3 mV, respectively. In these cells injection of depolarising current of up to 300 pA failed to elicit action potentials and neither UDP nor nicotine caused any significant change. ATP had a biphasic effect in most satellite cells, a depolarisation (seven cells) followed by a hyperpolarisation (five out of seven cells). Fibroblasts were only depolarised by ATP.

It should be noted that the electrophysiological observations confirm that the phase contrast–bright cells are neurones. It is noticeable that the same cells also showed tyrosinehydroxylase-like immunoreactivity (Von Kügelgen et al., 1997). Hence, the phase contrast–bright cells are sympathetic neurones. UDP had no effect in whole-cell recordings. This lack of effect could have been due to the washout of intracellular constituents during whole cell dialysis since the UDP concentration used (100 μM) had been shown previously to have profound stimulatory effects on the release of [^{3}H]-noradrenaline in the same cultures (Von Kügelgen et al., 1997; see also Von Kügelgen et al., 1999). In other words, this observation can be taken as an indication for UDP (and most probably also for UTP) acting via metabotropic P2Y-receptors. Vice versa, ATP was effective, i.e. most likely activates ionotropic P2X-receptors. The effects of ATP on satellite cells and fibroblasts were not investigated further but indicate the presence of P2-receptors also on these cell types. Accordingly, an ATP-induced rise in intracellular calcium was reported

(A)

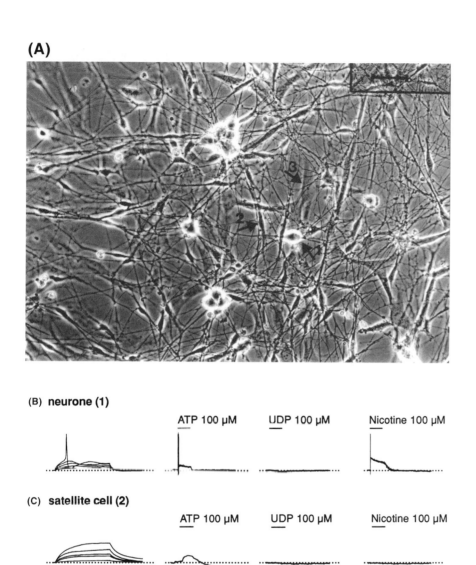

(B) neurone (1)

ATP 100 µM　　UDP 100 µM　　Nicotine 100 µM

(C) satellite cell (2)

ATP 100 µM　　UDP 100 µM　　Nicotine 100 µM

⌐ 20 mV
50 ms

⌐ 20 mV
5 s

(D) fibroblast (3)

ATP 100 µM　　UDP 100 µM　　Nicotine 100 µM

Fig. 1. Effects of current injection, ATP, UDP and nicotine in neurones (B), satellite cells (C) and fibroblasts (D).(A) Microphotograph taken from a culture prepared from thoracolumbal sympathetic ganglia of the rat (bar in the upper right corner = 50 µM). Arrows point to the three major cell types encountered in this preparation: phase contrast–bright neuronal cells (1), darker satellite cells (2) and polygonal fibroblasts (3). Note, that neurones could be easily identified solely by visual inspection. The membrane potential of cells was recorded under current-clamp in the whole-cell configuration of the patch-clamp technique at room temperature. Quasi-physiological bath- (Na^+-rich) and pipette (K^+-rich) solutions were used. Application of drugs was accomplished by a solenoid-valve driven pressurised fast flow superfusion system allowing complete solution exchange in the vicinity of the cell under investigation in approximately 150 ms.

to occur in fibroblasts (Gonzalez et al., 1989; Fine et al., 1989), as well as in Schwann cells (Lyons et al., 1994; Berti-Mattera et al., 1996; Mayer et al., 1997) which belong as the ganglionic satellite cells to peripheral neuroglia.

Membrane current responses of cultured rat thoracolumbal sympathetic neurones to extracellular nucleotides

P2X-receptors can be found in a variety of smooth muscle, neuronal and glial cells, as well as in immunocytes (for reviews see Bean, 1992; Illes and Nörenberg, 1993; Zimmermann, 1994; Surprenant et al, 1995; Illes et al., 1996). In the sympathetic nervous system, the properties of these receptors are well described in rat SCG and guinea-pig coeliac ganglion neurones (Evans et al., 1992; Silinsky and Gerzanich, 1993; Cloues et al., 1993; Nakazawa, 1994; Kakh et al., 1995).

All of the hitherto known P2X-receptors, native-receptors as well as the cloned and expressed subtypes, can be classified into four distinct functional groups according to (1) their kinetic behaviour, (2) their differential sensitivity to the structural analogue α,β-methyleneadenosine-5′-triphosphate (α,β-MeATP), (3) their differential sensitivity against P2-receptor antagonists like pyridoxalphosphate-6-azophenyl-2′-4′-disulphonic acid (PPADS) and suramin, and (4) their ability to form cytolytique pores upon prolonged contact with ATP (Collo et al., 1996; Evans and Surprenant, 1996; Surprenant, 1996; Surprenant et al., 1996). However, until now it is not possible to apply this classification to P2X-receptors in thoracolumbal sympathetic neurones of the rat because the properties of these receptors have been only incompletely characterised. The following section summarises our electrophysiological studies undertaken in order to provide a basis for such a comparison.

At a holding potential of -60 mV, ATP (100 μM) induced inward currents (Fig. 2A). The activating current could be fitted to a single exponential function (inset Fig. 2A). Time constants of activation (τ_{onset}) were in the range of 22 to 37 ms ($n=5$). This fast time course is typical for effects mediated by the rapid opening of ligand gated channels and is very similar to values found in rat SCG neurones or in PC12 cells (Surprenant, 1996; Evans and Surprenant, 1996). From inspection of the current trace in Fig. 2A, it is obvious that the receptor showed only minor desensitisation during the period (1 s) of ATP application. In five cells the current decayed during that time to $73\pm9\%$ of the peak amplitude. This lack of profound desensitisation again closely resembles P2X-receptors in SCG and PC12 cells (Surprenant, 1996; Evans and Surprenant, 1996). As aforementioned, the observation that ATP effects could be elicited under whole-cell patch clamp conditions argues per se in favour of an effect mediated by ionotropic receptors. This was proven by the experiments illustrated in Fig. 2B. No differences were found between current amplitudes evoked in cells under standard conditions or in cells micro-dialysed with the enzymatically stable guanosine-5′-triphosphate analogue Guanosine-5′-O-(2-thiodiphosphate) (GDPβS; 300 μM; $n=5$ each). GDPβS is known to prevent G protein-mediated effects of agonists by precluding the binding of GTP (Dunlap et al., 1987). Hence, this observation unequivocally demonstrates that excitatory effects of ATP on thoracolumbal sympathetic neurones of the rat are mediated by P2X-receptors.

In order to elucidate the sensitivity of P2X-receptors in rat thoracolumbal sympathetic neurones to α,β-MeATP the actions of ATP (100 μM) were compared with those of its structural analogue (100 μM; Fig. 3A). This concentration of ATP elicited 90% of the maximal obtainable effect (EC_{90}) in experiments where complete concentration-response curves (0.3-1000 μM ATP) were obtained. The half-maximal effective concentration (EC_{50}) was 14 μM and curves levelled off at 300 μM of the nucleotide ($n=6$). Given at 100 μM to the same set of cells, ATP-evoked current amplitudes amounted to 122 ± 29 pA, those evoked by α,β-MeATP were only 16 ± 8 pA, i.e. 13% of the ATP effect ($n=7$). Since we did not determine complete concentration-response curves of α,β-MeATP in the present experiments, we do not know whether α,β-MeATP behaves as a full agonist at P2X-receptors of rat thoracolumbal sympathetic neurones; in SCG neurones a partial agonistic behaviour of this compound had been described (Kakh et al., 1995). However, it is clear that the affinity for the P2X-receptor on thoracolumbal neurones is low and one can expect that the EC_{50} value would be in the high μmolar range. Such low sensitivity to α,β-MeATP is also typical for SCG neurones as well as PC12 cells (Surprenant, 1996; Evans and Surprenant,

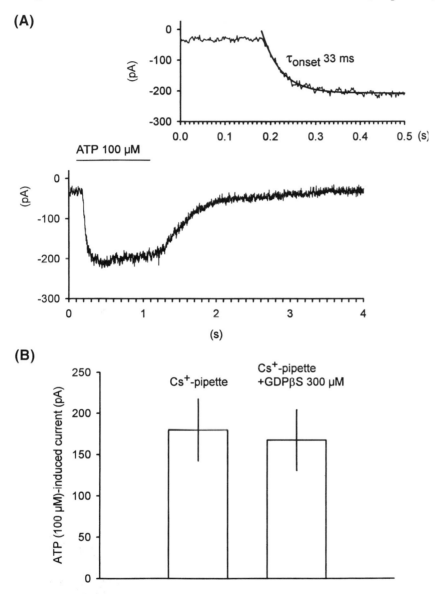

Fig. 2. (A) Sustained inward current (downward deflection) elicited in a rat thoracolumbal sympathetic neuron by ATP (100 μM) pressure applied for 1 s. The inset shows the same recording with an expanded time scale. Current activation could be fitted by a single exponential function revealing an activation time constant of 33 ms in this case. Membrane currents were recorded under voltage-clamp at a holding potential of −60 mV. For isolation of P2X-receptor mediated currents, a Cs^+-rich pipette solution was used. The bath contained tetrodotoxin (0.5 μM). (B) ATP-induced currents are not mediated via G protein coupled receptors. Cells from the same batch were alternatively voltage-clamped either with standard pipette solution or a pipette solution also containing GDPβS (300 μM; n = 5 each). Currents were elicited 20 min after gaining whole-cell access in order to allow for an equilibrium between cell interior and pipette solution.

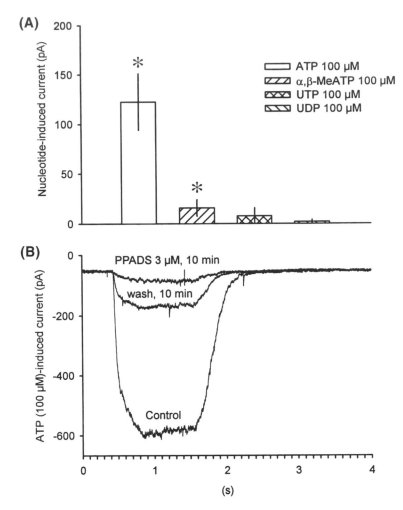

Fig. 3. (A) Comparison between the effects of ATP, α,β-MeATP, UTP and UDP at P2X-receptors of rat thoracolumbal sympathetic neurones. Nucleotides were pressure applied at the concentration of 100 μM for 1 s and the peak amplitude of inward currents at a holding potential of −60 mV was measured. *$P < 0.001$–0.005, significant differences from zero (Student's t-test); $n = 7$ each. (B) Interaction between ATP (100 μM) and the P2-receptor antagonist PPADS (3 μM). The graph shows three superimposed current amplitudes elicited when ATP (for 1 s) was given alone (control) after 10 min of contact and in the continued presence of PPADS, and after an additional 10 min of antagonist washout.

1996). In addition to ATP and α,β-MeATP the pyrimidne nucleotides UTP and UDP (each at 100 μM) were also tested and found to be ineffective ($7.7 ± 7.9$ and $2.0 ± 2.1$ pA, respectively; $n = 7$; Fig. 3A). This is in striking contrast to release experiments, where the same concentration of UTP and UDP elicited near maximal transmitter release (Von Kügelgen et al., 1997).

Finally, we investigated possible interactions between ATP and the P2-receptor antagonists PPADS (Windscheif et al., 1994) and suramin

(Dunn and Blakeley, 1988). Both antagonists, when given at the concentration of 3 μM, greatly diminished ATP-induced effects as illustrated for PPADS in Fig. 3B. Expressed as percentage inhibition, PPADS and suramin depressed current amplitudes evoked by ATP (100 μM) by $84 ± 7$ and $49 ± 3\%$, respectively ($n = 6$ each). From this, in all likelihood, IC_{50} values in the low μmolar range are to be expected. Hence, P2X-receptors in rat thoracolumbal sympathetic neurones show a similar sensitivity to both antagonists as P2X-receptors in

TABLE 1

Comparison between cloned and functionally expressed P2X-receptor subtypes and P2X-receptors in rat thoracolumbal sympathetic neurones

	$P2X_1$	$P2X_2$	$P2X_3$	$P2X_4$	$P2X_5$	$P2X_6$	$P2X_7$	$P2X_{(?)}$
Localisation								
	Smooth muscle, heart, sensory ganglia	Autonomic ganglia, brain, sensory ganglia	Sensory ganglia	Brain, pancreas, glands, other	Autonomic ganglia, heart	Autonomic ganglia, brain	Immunocytes	Thoracolumbal sympathetic ganglia (rat)
Agonist EC_{50} (μM)								
ATP	~1	~10	~1	~10	~10	~10	~300	14
α,βMeATP	~1	>300	~1	>300	~300	~300		high μmolar range
Antagonist IC_{50} (μM)								
Suramin	~1	~1	~1	100–500	>1	100	300	~3
PPADS	~1	~1	~1	30–200	>1	>100	~50	<3
Desensitisation								
	0%	78%	9%	64%	87%	60%	slow	73%

Data for the cloned receptors were taken from Collo et al. (1996) and North and Barnard (1997). Data for rat thoracolumbal sympathetic neurones are from the present study. Sympathetic neurones were voltage-clamped at a holding potential of -60 mV in the whole-cell configuration of the patch clamp technique. ATP concentration-response relations were determined by pressure application of increasing nucleotide concentrations (0.3–1000 μM; 1 s every 90 s). The half-maximal active ATP concentration (EC_{50}) was estimated from a logistic curve fitted to the weighted mean peak current amplitudes. α,β-MeATP (100 μM) was pressure applied for 1 s. After 90 s of washout, ATP at 100 μM (~EC_{90}) was also applied to the same set off cells. α,β-MeATP-evoked current amplitudes amounted to 13% of ATP-evoked effects. Hence, EC_{50} values in the high μmolar range are to be expected. In order to investigate interactions between ATP and the P2-receptor antagonists suramin and PPADS, ATP (100 μM) was pressure applied every 5 min for 1s seven times. In the absence of antagonists, ATP-induced currents were stable. In interaction experiments, antagonists were added for 15 min immediately after the first application of ATP. After that time, suramin and PPADS (3 μM each) inhibited ATP-evoked currents by 49% and 84%, respectively. Hence, half-maximal inhibitory concentrations (IC_{50}) in the low μmolar range are to be expected. Desensitisation was measured from currents induced by 1–2 s application of ATP (30–100 μM) and is expressed as the current remaining at the end of application.

SCG neurones and PC12 cells (Surprenant, 1996). The inhibitory action of both antagonists showed different degrees of recovery upon washout. Ten minutes after antagonist removal the remaining inhibition was in experiments with PPADS $75 \pm 2\%$ (Fig. 3B) and $10 \pm 3\%$ with suramin. Similar differences in the time course of action between PPADS and suramin have also been observed in SCG neurones (Kakh et al., 1995) as well as in cloned and heterologously expressed P2X-receptors from PC12 cells (Evans et al., 1995).

Taken together, the present results confirm the existence of P2X-receptors in thoracolumbal sympathetic neurones of the rat. Moreover, judged from their kinetic and pharmacological properties, these receptors seem to be closely related if not identical to native P2X-receptors found in other autonomic ganglia particularly in the superior cervical gan-

glion of the rat. In addition, a comparison of our receptors (right side of Table 1) with the hitherto known molecular subtypes (Table 1) is now possible. As already mentioned, P2X-receptors can be classified into four functional groups. $P2X_1$ and $P2X_3$ show a high sensitivity to α,β-MeATP as well as rapid desensitisation and belong to the first group. $P2X_2$ and $P2X_5$ receptors, the second group, are rather insensitive to α,β-MeATP, highly sensitive to PPADS and suramin and show only little desensitisation. $P2X_4$ and $P2X_6$ belong to the third group by virtue of their low sensitivities to the antagonists (Collo et al., 1996; North and Barnard, 1997). The $P2X_7$ receptor, found in immunocytes, is the only known member of the fourth group and unique in several respects; it exhibits only poor sensitivity to ATP as compared with $2',2'-O-$ (4-benzoyl-benzoyl)-ATP, the responses to ATP are potentiated by a reduction of extracellular divalent cation concentrations, and upon prolonged contact with agonists the receptor forms large diameter cytolytique pores (Surprenant et al., 1996). Inspection of Table 1 indicates that the P2X-receptor found in thoracolumbal sympathetic neurones of the rat has properties closely related to group two, i.e. is functionally a P2X$_2$/P2X$_5$-like receptor.

Effects of muscarinic agonists and extracellular nucleotides on M-currents in rat thoracolumbal sympathetic neurones

In the last series of experiments, we tried to solve the riddle of the mechanisms behind pyrimidine nucleotide-induced excitation. The secretagogue action of UTP and UDP in combination with their lack of effect at P2X-receptors (Boehm et al., 1995; Von Kügelgen et al., 1997) strongly suggests that sympathetic neurones constitutively express a pyrimidine-sensitive group of P2-receptors distinct from ionotropic P2X-receptors. It is anticipated that these receptors are P2Y-receptors. Indications for this may be derived from the observations that all cloned UTP/UDP-sensitive receptors are metabotropic P2-receptors (North and Barnard, 1997) and, moreover, that UTP/UDP were ineffective under conditions of whole-cell dialysis, i.e. when a significant washout of intracellular constituents is likely to occur (own findings). A possible target for the signalling of these presumed P2Y-receptors comes from a survey of the literature. UTP inhibited a time- and voltage-dependent potassium current, the M-current in bullfrog sympathetic neurones (Adams et al, 1982; Lopez and Adams, 1989) as well as NG108-15 neuroblastoma x glioma cells (Filippov et al., 1994). $I_{K(M)}$ is activated by depolarisation and, due to its lack of inactivation, clamps the neuronal membrane potential back to negative levels thereby impeding neuronal excitability (Marrion, 1997). Pharmacological interventions which inhibit this current, therefore, may lead to membrane depolarisation, calcium influx and eventually transmitter release. It should be noted, that $I_{K(M)}$ is also found in mammalian sympathetic neurones of rats (Brown et al., 1982; Freschi, 1983).

Two approaches were used to investigate possible pyrimidine nucleotide actions on $I_{K(M)}$ in rat

thoracolumbal sympathetic neurones. An indirect one used the cell response to current injection as a parameter of neuronal excitability. In addition, M-current deactivation relaxations were recorded directly. In both cases, effects of UTP and UDP were compared to those of oxotremorine-methiodide (oxo-M), a muscarinic agonist and well-known inhibitor of M-channels (Marrion et al., 1989). In all these experiments we used the amphotericin B perforated patch technique (Rae et al., 1991). Amphotericin forms pores which allow the flux of small ions but exclude molecules of the size of glucose or larger (Cass et al., 1970; Holz and Finkelstein, 1970); intracellular second messenger systems remain relatively undisturbed under these conditions. It is noticeable that M-currents show a considerable rundown under whole-cell conditions (Brown et al., 1989) but remain stable in perforated patch recordings (Caulfield et al., 1994).

The results are illustrated in Fig. 4. Action potentials were evoked by injection of positive current (Fig. 4(A)(a), (B)(a)). Current strength was adjusted to a level, which evoked just a single spike. Both UTP (100 μM) and oxo-M (10 μM) induced repetitive firing. However, the mechanisms behind this increased excitability were different for the two compounds as indicated by the following experiments. In order to investigate possible drug effects on M-currents an experimental protocol described previously for SCG neurones (Marrion et al., 1989) was employed. Cells were voltage-clamped at the relatively depolarised potential of −30 mV to preactivate the M-current, and the current was then intermittently deactivated by hyperpolarising voltage steps to −60 mV. During the hyperpolarising steps the current records showed slow inward relaxations which reflect deactivating tail currents for $I_{K(M)}$ (Fig. 4(A)(b), (B)(b)). Effects of UTP, UDP and oxo-M on $I_{K(M)}$ were then assessed by comparing the amplitudes of these deactivation tails elicited either in the absence or in the presence of the respective compounds (Marrion et al., 1989). Neither UTP (100 μM), nor UDP (100 μM) had any effect on $I_{K(M)}$ (Fig. 4(A)(b), (C)). Expressed as percentage inhibition, UTP inhibited the M-current by $3.8 \pm 5.1\%$, and UDP inhibited by $-1.7 \pm 3.2\%$ ($p > 0.05$; $n = 5$

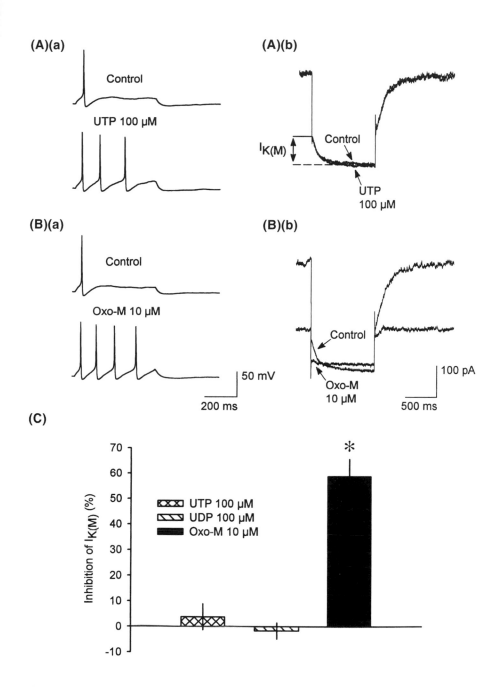

Fig. 4. Effects of UTP (100μM), UDP (100 μM) and oxo-M (10 μM) on the action potential frequency ((A)(a), (B)(a)) and M-current deactivation tails ((A)(b), (B)(b)) in rat thoracolumbal sympathetic neurones. (A)(a), (B)(a), depolarising current pulses, 500 ms in duration, reaching just threshold for a single action potential were applied under current-clamp two times: in the absence of UTP or oxo-M (control) and after 3 min of contact with the respective drug. (A)(b), (B)(b), hyperpolarising voltage pulses from a holding potential of −30 mV to a command potential of −60 mV, 1 s in duration, were applied under voltage-clamp two times: in the absence of UTP or oxo-M (control) and after 3 min of contact with the respective drug. (C) gives the statistical evaluation of experiments similar to those shown in (A)(b) and (B)(b); *$P < 0.001$, significant difference from zero (Student's t-test); $n = 5$ each. Amphotericin B perforated patch method and quasi-physiological bath-(Na^+-rich) and pipette (K^+-rich) solutions were used throughout.

each). The situation was different with oxo-M (10 µM). The muscarinic agonist decreased $I_{K(M)}$ deactivation tails by $58.8 \pm 6.9\%$ (Fig. 4(B)(b), (C); $p < 0.001$; $n = 5$), i.e. to a level very similar to that found in rat SCG neurones (Marrion et al., 1989).

Taken together, rat thoracolumbal sympathetic neurones possess excitatory muscarinic receptors. The mechanism behind these receptors is inhibition of the M-current. UTP (and UDP, not shown) also increased neuronal excitability. This observation together with the known secretagogue action of pyrimidine nucleotides in these cells (Von Kügelgen et al., 1997) strongly suggests the additional presence of UTP/UDP sensitive P2-receptors, most likely of the P2Y-class. The transduction mechanism of these receptors in rat thoracolumbal sympathetic neurones, however, does not include M-channels.

Conclusions

The aim of the present study was to analyse the mechanisms by which ATP, UTP and UDP release noradrenaline from rat postganglionic sympathetic neurones. In visually identified sympathetic neurones, ATP elicited membrane depolarisations as well as inward currents. Intracellular dialysis with GDPβS did not prevent the action of ATP. The effects of ATP were inhibited by the P2-receptor antagonists PPADS and suramin. UTP and UDP did not evoke inward currents. The muscarine receptor agonist oxo-M, as well as UTP increased neuronal excitability. However, only oxo-M but neither UTP nor UDP depressed the M-current.

These results (summarised in Fig. 5) support the conclusion that two distinct types of excitatory P2-receptors coexist in rat thoracolumbal sympathetic neurones. The first type is a P2X-receptor, a ligand-gated cation channel, with properties similar to native P2X-receptors in rat SCG neurones and PC12 cells as well as to the cloned $P2X_2/P2X_5$ subtypes. Activation of this receptor directly induces membrane depolarisation without the need for an intracellular signal transduction machinery. The second type is, in all likelihood, a metabotropic P2Y-receptor sensitive to UTP and UDP. Activation of this P2Y-receptor increases neuronal excitability as does the activation of muscarine receptors also

Rat thoracolumbal sympathetic neuron

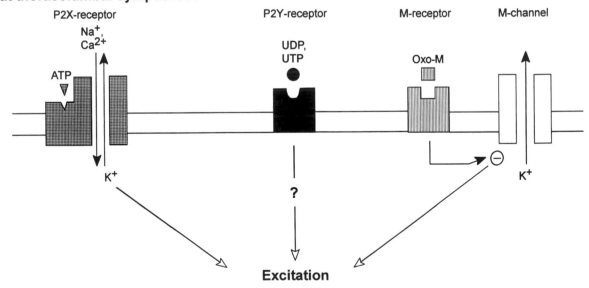

Fig. 5. Transduction mechanisms of coexisting excitatory P2X- and P2Y-receptors as well as muscarine (M) receptors in rat thoracolumbal sympathetic neurones. Activation by ATP of P2X-receptors allows cation flux eventually leading to membrane depolarisation. Activation by oxo-M of muscarinic receptors is excitatory via a mechanism including the closure of M-type potassium channels. The mode of action of UTP and UDP at excitatory P2Y-receptors waits for elucidation. It does not involve, however, the M-channel mechanism.

present in thoracolumbal sympathetic neurones. Muscarine receptors exert their excitatory effects by closure of M-channels. The mechanism behind the presumed P2Y-receptors is still unclear. From our results, however, it can be excluded that M-channels are involved in the transduction pathway of these receptors in rat thoracolumbal sympathetic neurones.

Several open questions remain. UTP either inhibited M-currents in native rat sympathetic neurones (Boehm and Bofill-Cardona, 1998) or had no effect (Filippov et al., 1998; present study). At present, the reason for these conflicting observations is not known. However, it is possible that the neurones were investigated at different stages of cellular differentiation. If so, this obvious discrepancy may reflect developmental changes either in P2-receptor expression or transduction mechanisms. Such shifts in P2X- as well as P2Y-receptor function during ontogenesis have already been described, for example in smooth muscle and neuronal preparations (Brownhill et al., 1997; Wirkner et al., 1998). A second point of uncertainty concerns the physiological role of P2-receptors in sympathetic neurones. Biochemical studies indicate that ATP is co-released with acetylcholine from preganglionic sympathetic nerves (Vizi et al., 1997). Electrophysiological evidence for a role of ATP in sympathetic ganglionic neurotransmission, however, is still lacking or contradictory (Inokuchi and McLachlan, 1995). Continued efforts using combined electrophysiological, biochemical and molecular biological approaches will be necessary to clarify these open questions.

Acknowledgements

We are grateful to Dr. K. Starke for many helpful discussions. The study was supported by the Deutsche Forschungsgemeinschaft (SFB 505; Il 20/7-1).

References

Adams, P.R., Brown, D.A. and Constanti, A. (1982) Pharmacological inhibition of the M-current. *J. Physiol.*, 332: 223–262.

Bean, B.P. (1992) Pharmacology and electrophysiology of ATP-activated ion channels. *Trends Pharmacol. Sci.*, 13: 87–90.

Berti-Mattera, D.N., Wilkins, P.L., Madhun, Z. and Suchovsky, D. (1996) P_2-purinergic receptors regulate phospholipase C and adenylate cyclase activities in immortalized Schwann cells. *Biochem. J.*, 314: 555–561.

Boehm, S., Huck, S. and Illes, P. (1995) UTP- and ATP-triggered transmitter release from rat sympathetic neurones via separate receptors. *Br. J. Pharmacol.*, 116: 2241–2243.

Boehm, S. and Bofill-Cardona, E. (1998) Uridine nucleotide receptors modulate M-type K^+ channels of rat sympathetic neurons. *Naunyn-Schmiedeberg's Arch. Pharmacol.*, 357: R33.

Brake, A.J., Wagenbach, M.J. and Julius, D. (1994) New structural motif for ligand-gated ion channels defined by an ionotropic ATP receptor. *Nature*, 371: 519–523.

Brown, D.A., Adams, P.R. and Constanti, A. (1982) Voltage-sensitive K-currents in sympathetic neurons and their modulation by neurotransmitters. *J. Autonomic Nervous System*, 6: 23–35.

Brown, D.A., Marrion, N.V. and Smart, T.G. (1989) On the transduction mechanism for muscarine-induced inhibition of M-currents in cultured rat sympathetic neurones. *J. Physiol.*, 413: 469–488.

Brownhill, V.R., Hourani, S.M. and Kitchen, I. (1997) Ontogeny of P2-purinoceptors in the longitudinal muscle and muscularis mucosae of the isolated rat duodenum. *Br. J. Pharmacol.*, 122: 225–232.

Burnstock, G. (1990) Co-transmission. *Arch. Int. Pharmacodyn.*, 304: 7–33.

Cass, A., Finkelstein, A. and Krespi, V. (1970) The ion permeability induced in lipid membranes by the polyene antibiotics nystatin and amphotericin B. *J. Gen. Physiol.*, 56: 100–124.

Caulfield, M.P., Jones, S., Vallis, Y., Buckley, N.J., Kim, G.-D., Milligan, G. and Brown, D.A. (1994) Muscarinic M-current inhibition via $G_{\alpha q/11}$ and α-adrenoceptor inhibition via $G_{\alpha 0}$ in rat sympathetic neurones. *J. Physiol.*, 477: 415–422.

Cloues, R., Jones, S. and Brown, D. (1993) Zn^{2+} potentiates ATP-activated currents in rat sympathetic neurones. *Pflüger's Arch.*, 424: 152–158.

Collo, G., North, R.A., Kawashima, E., Merlo-Pich, E., Neidhart, S., Surprenant, A. and Buell, G. (1996) Cloning of $P2X_5$ and $P2X_6$ receptors and the distribution and properties of an extended family of ATP-gated ion channels. *J. Neurosci.*, 16: 2495–2507.

Connolly, G.P., Harrison, P.J. and Stone, T.W. (1993) Action of purine and pyrimidine nucleotides on the rat superior cervical ganglion. *Br. J. Pharmacol.*, 110: 1297–1304.

Dunlap, K., Holz, G.G. and Rane, S.G. (1987) G proteins as regulators of ion channel function. *Trends Neurosci.*, 10: 241–244.

Dunn, P.M. and Blakeley, A.G.H. (1988) Suramin: A reversible P_2-purinoceptor antagonist in the mouse vas deferens. *Br. J. Pharmacol.*, 93: 243–245.

Edwards, F.A., Gibb, A.J. and Colquhoun, D. (1992) ATP-receptor-mediated synaptic currents in the central nervous system. *Nature*, 359: 144–147.

Evans, R.J., Derkach, V. and Surprenant, A. (1992) ATP mediates fast synaptic transmission in rat sympathetic neurons. *Nature*, 357: 503–505.

Evans, R.J., Lewis, C., Buell, G., Valera, S., North, R.A. and Surprenant, A. (1995) Pharmacological characterization of heterologously expressed ATP-gated cation channels (P_{2X} purinoceptors). *Mol. Pharmacol.*, 48: 178–183.

Evans, R.J. and Surprenant, A. (1996) P2X receptors in autonomic and sensory neurons. *Seminars in The Neurosciences*, 8: 217–223.

Filippov, A.K., Selyanko, A.A., Robbins, J. and Brown, D.A. (1994) Activation of nucleotide receptors inhibits M-type K current $[I_{K(M)}]$ in neuroblastoma x glioma hybrid cells. *Pflügers Arch.*, 429: 223–230.

Filippov, A.K., Webb, T.E., Barnard, E.A. and Brown, D.A. (1998) $P2Y_2$ nucleotide receptors expressed heterologously in sympathetic neurons inhibit both N-type Ca^{2+} and M-type K^+ currents. *J. Neuroscience*, 18: 5170–5179.

Fine, J., Cole, P. and Davidson, J.S. (1989) Extracellular nucleotides stimulate receptor-mediated calcium mobilization and inositol phosphate production in human fibroblasts. *Biochem. J.*, 263: 371–376.

Freschi, J.E. (1983) Membrane currents of rat sympathetic neurons under voltage clamp. *J. Neurophysiol.*, 50: 1460–1478.

Gonzalez, F.A., Bonapace, E., Belzer, I., Friedberg, I. and Heppel, L.A. (1989) Two distinct receptors for ATP can be distinguished in Swiss 3T6 mouse fibroblasts by their desensitization. *Biochem. Biophys. Res. Com.*, 164: 706–713.

Holz, R. and Finkelstein, A. (1970) The water and non-electrolyte permeability induced in thin lipid membranes by the polyene antibiotics nystatin and amphotericin B. *J. Gen. Physiol.*, 56: 125–145.

Illes, P. and Nörenberg, W. (1993) Neuronal ATP receptors and their mechanism of action. *Trends Pharmacol. Sci.*, 14: 50–54.

Illes, P., Nieber, K., Fröhlich, R. and Nörenberg, W. (1996) P_2-purinoceptors and pyrimidinoceptors of catecholamine-producing cells and immunocytes. In: G. Burnstock (Ed.), *P_2 purinoceptors: Localization, Function and Transduction Mechanisms*, John Wiley, Chichester, pp. 110–129.

Inokuchi, H. and McLachlan, E.M. (1995) Lack of evidence for P2X-purinoceptor involvement in fast synaptic response in intact ganglia isolated from guinea-pigs. *Neuroscience*, 69: 651–659.

Khakh, B.S., Humphrey, P.P.A. and Surprenant, A. (1995) Electrophysiological properties of P_{2X}-purinoceptors in rat superior cervical, nodose, and guinea-pig coeliac neurones. *J. Physiol.*, 484: 385–395.

Lew, M.J. and White, T.D. (1987) Release of endogenous ATP during sympathetic nerve stimulation. *Br. J. Pharmacol.*, 92: 349–355.

Lopez, H.S. and Adams, P.R. (1989) A G protein mediates the inhibition of the voltage-dependent potassium M current by muscarine, LHRH, substance P and UTP in bullfrog sympathetic neurons. *Eur. J. Neurosci.*, 1: 529–542.

Lyons, S.A., Morell, P. and McCarthy, K.D. (1994) Schwann cells exhibit P_{2Y} purinergic receptors that regulate intracellular calcium and are up-regulated by cyclic AMP analogues. *J. Neurochem.*, 63: 552–560.

Marrion, N.V. (1997) Control of M-current. *Annu. Rev. Physiol.*, 59: 483–504.

Marrion, N.V., Smart, T.G., Marsh, S.J. and Brown, D.A. (1989) Muscarinic suppression of the M-current in the rat sympathetic ganglion is mediated by receptors of the M_1-subtype. *Br. J. Pharmacol.*, 98: 557–573.

Mayer, C., Wächter, J., Kamleiter, M. and Grafe, P. (1997) Intracellular calcium transients mediated by P2 receptors in the paranodal Schwann cell region of myelinated rat spinal root axons. *Neurosci. Lett.*, 224: 49–52.

Nakazawa, K. (1994) ATP-activated current and its interaction with acetylcholine-activated current in rat sympathetic neurons. *J. Neurosci.*, 14: 740–750.

Nieber, K., Poelchen, W. and Illes, P. (1997) Role of ATP in fast excitatory potentials in locus coeruleus neurones of the rat. *Br. J. Pharmacol.*, 122: 423–430.

North, R.A. and Barnard, E.A. (1997) Nucleotide receptors. *Current Opinion in Neurobiology*, 7: 346–357.

Rae, J., Cooper, K., Gates, P. and Watsky, M. (1991) Low access resistance perforated patch recordings using amphotericin B. *J. Neurosci. Meth.*, 37: 15–26.

Siggins, G.R., Gruol, D., Padjen, A. and Formand, D. (1977) Purine and pyrimidine mononucleotides depolarize neurones of explanted amphibian sympathetic ganglia. *Nature*, 270: 263–265.

Silinsky, E.M. and Gerzanich, V. (1993) On the excitatory effects of ATP and its role as a neurotransmitter in coeliac neurons of the guinea-pig. *J. Physiol.*, 464: 197–212.

Surprenant, A. (1996) Functional properties of native and cloned P2X receptors. In: G. Burnstock, (Ed.), *P_2 Purinoceptors: Localization, Function and Transduction Mechanisms*. John Wiley, Chichester, pp. 208–222.

Surprenant, A., Buell, G. and North, R.A. (1995) P2X receptors bring new structure to ligand-gated ion channels. *Trends Neurosci.*, 18: 224–229.

Surprenant, A., Rassendren, F., Kawashima, E., North, R.A. and Buell, G. (1996) The cytolytic P_{2Z} receptor for extracellular ATP identified as a P_{2X} receptor (P2X7). *Science*, 272: 735–737.

Vizi, E.S., Liang, S.-D., Sperlágh, B., Kittel, Á. and Jurányi Z. (1997) Studies on the release and extracellular metabolism of endogenous ATP in rat superior cervical ganglion: support for neurotransmitter role of ATP. *Neuroscience*, 79: 893–902.

Von Kügelgen, I., Nörenberg, W., Illes, P., Schobert, A. and Starke, K. (1997) Differences in the mode of stimulation of cultured rat sympathetic neurons between ATP and UDP. *Neuroscience*, 78: 935–941.

Von Kügelgen, I. and Starke, K. (1991a) Noradrenaline-ATP cotransmission in the sympathetic nervous system. *Trends Pharmacol. Sci.*, 12: 319–324.

Von Kügelgen, I. and Starke, K., (1991b) Release of noradrenaline and ATP by electrical stimulation and nicotine in guinea-pig vas deferens. *Naunyn-Schmiedebergs Arch. Pharmacol.*, 344: 419–429.

Windscheif, U., Ralevic, V., Bäumert, H.G., Mutschler, E., Lambrecht G. and Burnstock, G. (1994) Vasoconstrictor and vasodilator responses to various agonists in the rat perfused mesenteric arterial bed: Selective inhibition by PPADS of contractions mediated via P_{2X}-purinoceptors. *Br. J. Pharmacol.*, 113: 1015–1021.

Wirkner, K., Franke, H., Inoue, K. and Illes, P. (1998) Differential age-dependent expression of α_2-adrenoceptor and P2-purinoceptor function in rat locus coeruleus neurones. *Naunyn-Schmiedebergs Arch. Pharmacol.*, 357: 186–189.

Zimmermann, H. (1994) Signalling via ATP in the nervous system. *Trends Pharmacol. Sci.*, 17: 420–426.

P. Illes and H. Zimmermann (Eds.)
Progress in Brain Research, Vol 120
© 1999 Elsevier Science BV. All rights reserved

P2 receptor-mediated activation of noradrenergic and dopaminergic neurons in the rat brain

Holger Kittner[1], Ute Krügel[1], Wolfgang Poelchen[1], Dirk Sieler[1], Robert Reinhardt[1], Ivar von Kügelgen[2] and Peter Illes[1,*]

[1]*Department of Pharmacology and Toxicology, University of Leipzig, Härtelstrasse 16-18, D-04107 Leipzig, Germany*
[2]*Department of Pharmacology and Toxicology, University of Freiburg, Hermann-Herder-Strasse 5, D-79104 Freiburg, Germany*

Introduction

Adenosine 5′-triphosphate (ATP) has been suggested to be a transmitter in purinergic neurons of the peripheral nervous system as well as a co-transmitter with noradrenaline in the axon terminals of postganglionic sympathetic neurons (Burnstock, 1986; von Kügelgen and Starke, 1991; Zimmermann, 1994). The cell bodies of these neurons are endowed with ATP-sensitive receptors of the P2X-type (Cloues et al., 1993; Nakazawa, 1994; Khakh et al., 1995). While P2X receptors are ligand-activated cationic channels and mediate fast excitatory neurotransmission, P2Y receptors are coupled to G proteins and mediate slow synaptic events or release Ca^{2+} from intracellular stores (Abbracchio and Burnstock, 1994; Fredholm et al., 1994). Guinea-pig coeliac neurons form a fibre network in cell culture systems (Evans et al., 1992). Focal electrical stimulation of these fibres evokes fast excitatory synaptic currents due to the activation of P2X receptors. Hence, ATP is a neuro-neuronal transmitter of coeliac neurons although whether ATP is co-released with noradrenaline has not been investigated.

*Corresponding author. Tel.: +49-341-9724600; fax: +49-341-9724609; e-mail: ILLP@server3.medizin.uni-leipzig.de

Evidence that ATP stimulates central noradrenergic neurons

In the central nervous system, a major group of noradrenergic neurons is concentrated in a pontine nucleus, the locus coeruleus (LC). LC cells project into various areas of the brain, for example the hippocampus, neocortex, hypothalamus, cerebellum and spinal cord (Loughlin and Fallon, 1985). The LC is involved in a number of cognitive and emotional processes including attention and anxiety (Foote et al., 1983) and participates in the persistent adaptation of brain function during drug addiction (Rasmussen et al., 1990).

As measured by extracellular micro-electrodes in a midpontine brain slice preparation, rat LC neurones fire at a constant rate (Tschöpl et al., 1992). ATP caused no consistent effect when given alone, but increased the firing rate when given in the presence of the adenosine A_1 receptor antagonist DPCPX, indicating a balance between a direct excitatory P2 effect and an indirect (mediated by the degradation product, adenosine) inhibitory A_1 effect. Structural analogues of ATP, such as α,β-methylene ATP (α,β-meATP) and 2-methylthio ATP (2-MeSATP) which are rather stable against enzymatic degradation or do not yield the inhibitory metabolite adenosine, unequivocally facilitated the firing of LC neurons (Tschöpl et al., 1992; Fröhlich et al., 1996). The P2 receptor antagonists suramin, reactive blue 2 (RB2) and pyridoxalphosphate-6-azophenyl-2,4-disulphonic acid (PPADS) all depressed the effects of α,β-meATP and 2-MeSATP (Tschöpl et al., 1992; Fröhlich et al., 1996).

Intracellular recordings in LC neurons suggested that α,β-meATP inhibits a resting K^+ conductance (probably via G protein activation) and, at the same time, opens cationic channels (Harms et al., 1992). Direct measurements of ion currents confirmed these findings (Shen and North, 1993) and led to the suggestion that LC cells are endowed both with P2X and P2Y receptors (Illes et al., 1996). It has been shown that functional P2X receptors are not present after birth and become expressed at some later age, reaching maturity only in rats older than 18 days (Wirkner et al., 1998).

Evidence that ATP stimulates central dopaminergic neurons

Midbrain dopaminergic neurons are localized in two major nuclei, the substantia nigra pars compacta and the neighbouring ventral tegmental area (VTA); the main projection areas of these neurons are the striatum and the nucleus accumbens (NAc), respectively (Ungerstedt, 1971). VTA cells comprise the so-called mesolimbic-mesocortical projection system which plays important roles in the regulation of goal directed behaviour and adaptive functions. It has been hypothesized that this system is involved in gating the translation of motivationally-relevant environmental stimuli into adaptive behavioural response (Mogenson, 1987; Le Moal and Simon, 1991). Thereby, dopaminergic transmission is proposed to be the primary determinant in regulating the initiation of a behavioural response to environmental stimuli (Mogenson et al., 1993). In addition, the mesolimbic-mesocortical pathway has important implications in psychiatric disorders such as schizophrenia and attention deficits as well as in mediating the rewarding and reinforcing effects of addictive drugs (Ashby, 1996).

ATP has been shown by microdialysis experiments to increase the basal release of dopamine in the rat striatum (Zhang et al. 1995; 1996). ATP appears to act by stimulating a P2Y receptor, since

the ribosylation of G proteins by intracerebroventricularly applied pertussis toxin decreased the effect of ATP (Zhang et al., 1995). In PC12 cells, various D2 dopamine receptor antagonistic antipsychotic drugs interfered with the ATP-induced increase in intracellular Ca^{2+} (Koizumi et al., 1995) and the subsequent secretion of dopamine (Courtney et al., 1991). These results suggest that P2X receptor-mediated responses may be altered by dopamine D2 receptors (Inoue et al., 1996).

Results and discussion

Noradrenergic mechanisms in the locus coeruleus

Electrophysiological studies

Focal electrical stimulation evokes in the rat LC biphasic synaptic potentials consisting of early depolarizing (p.s.p.) and late hyperpolarizing (i.p.s.p.) components (Fig. 1A). It has been found that the p.s.p. is due to the release of glutamate from afferent fibres predominantly onto non-N-methyl-D-aspartate (non-NMDA) receptors and of γ-aminobutyric acid (GABA) onto GABA$_A$ receptors (Cherubini et al., 1988; Williams et al., 1991). Therefore, kynurenic acid and picrotoxin were used to exclude those fractions of the p.s.p. which are mediated by excitatory amino acid receptors and GABA$_A$ receptors, respectively (Fig. 1; Nieber et al., 1997). In some experiments, CNQX and AP5 were applied instead of kynurenic acid to block non-NMDA and NMDA receptors of LC neurons, respectively. A further application of the P2 receptor antagonist suramin (30–100 μM) caused a concentration-dependent and reversible inhibition, suggesting that the release of ATP is responsible for the residual p.s.p. (Fig. 1). Another, more selective antagonist of P2 receptors, PPADS acted like suramin. Since both suramin and PPADS depressed the p.s.p. in a range of concentrations similar to that which inhibited the effect of pressure-applied α,β-meATP (Nieber et al., 1997), the involvement of P2 receptors appeared to be unequivocally confirmed.

An adenosine deaminase-containing pathway has been identified, projecting from the posterior hypothalamus to the mesencephalic trigeminal nucleus (MNV) in the immediate neighbourhood of

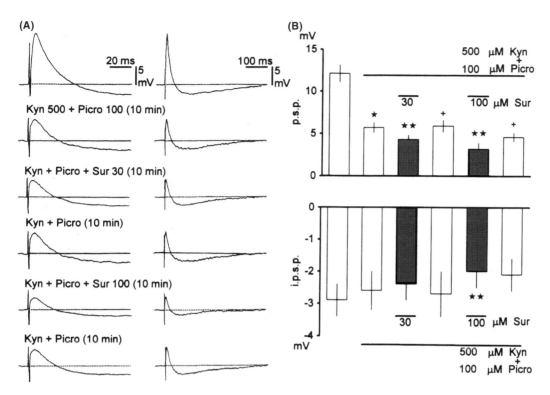

Fig. 1. Effect of kynurenic acid, picrotoxin and suramin on p.s.p./i.p.s.p. sequences evoked by focal electrical stimulation in rat LC neurons. Representative tracings recorded with two different time scales (A). The dotted line indicates the membrane potential. All concentrations are expressed in μM. The superfusion times of antagonists are in parentheses. Means ± S.E.M. of seven experiments similar to those shown in (A) (B). Effects on p.s.p. amplitudes ((B), upper panel). Effects on i.p.s.p. amplitudes ((B), lower panel). Antagonists were present in the superfusion medium over the periods marked by the horizontal bars. *$P < 0.05$; significant difference from synaptic potentials measured before the application of kynurenic acid. **$P < 0.05$; significant difference from synaptic potentials measured before the application of suramin. +$P < 0.05$; significant difference from synaptic potentials measured before the washout of suramin.

the LC (Nagy et al., 1986). It is conceivable that this pathway releases ATP as a neurotransmitter onto its target cells. While MNV neurons responded to ATP with an inward current (Khakh et al., 1997), they were insensitive to the degradation product adenosine (Regenold et al., 1988). LC neurons responded to ATP and adenosine with inward (Harms et al., 1992) and outward currents (Regenold and Illes, 1990), respectively. Hence, ATP may be involved in neurotransmission from a hitherto unidentified afferent purinergic pathway to the LC.

Alternatively, ATP and noradrenaline may be co-released from dendrites or recurrent axon collaterals onto LC neurons producing excitation (p.s.p.) and inhibition (i.p.s.p.), respectively. In fact, idazoxan (1 μM) an α$_2$ adrenoceptor antago-

nist inhibited the i.p.s.p., without altering the p.s.p. (Fig. 2A). In additional experiments, kynurenic acid (500 μM) and picrotoxin (100 μM) were used to isolate the ATPergic fraction of the p.s.p. pharmacologically (Fig. 2B). Under these conditions, the selective noradrenergic neurotoxin 6-hydroxydopamine (500 μM) and the noradrenergic neuron blocker guanethidine (10 μM) depressed the i.p.s.p. (Fig. 2C), confirming that this synaptic potential is due to the release of noradrenaline either from recurrent axon collaterals or dendrites of the LC neurons themselves or from afferent fibres originating in the nucleus para-gigantocellularis (Egan et al., 1983; Williams et al., 1991). However, the p.s.p. was also inhibited by these compounds, indicating that the destruction of noradrenergic nerve terminals by 6-hydroxydopa-

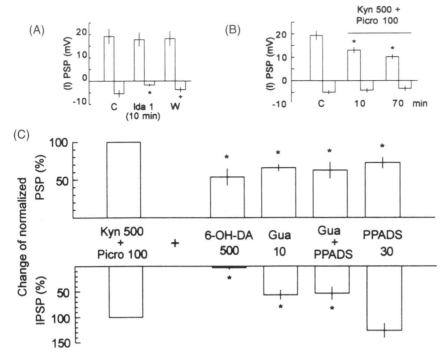

Fig. 2. Effect of 6-hydroxydopamine, guanethidine and PPADS on the pharmacologically isolated ATPergic fraction of the p.s.p. Selective inhibition of the i.p.s.p. by idazoxan (A). Synaptic potentials were recorded before (C) and 10 min after superfusion with idazoxan (Ida; 1 µM), as well as after a subsequent washout (W). Selective inhibition of the p.s.p. by kynurenic acid and picrotoxin (B). Synaptic potentials were recorded before (C), as well as 10 and 70 min after superfusion with kynurenic acid (Kyn; 500 µM) and picrotoxin (Picro; 100 µM). Percentage change of normalized p.s.p. and i.p.s.p. amplitudes by 6-hydroxydopamine (6-OH-DA), guanethidine (Gua), PPADS, and PPADS plus guanethidine with respect to the time-matched controls (C), evaluated after 60 min of superfusion (C). Kynurenic acid (500 µM) and picrotoxin (100 µM) were present in the superfusate for 10 min alone and then for another 60 min together with the drugs under investigation. Means ± SEM of four (A), five (B) and five–seven (C) experiments. *P < 0.05; significant difference from synaptic potentials measured before the application of idazoxan (A), kynurenic acid and picrotoxin (B), or from 100% (C).

mine or the selective blockade of action potential propagation to the axon terminals by guanethidine may equally inhibit the release of noradrenaline and ATP. Further support for this view was supplied by the finding that the application of PPADS (30 µM) to the guanethidine (10 µM)-containing medium failed to cause any further inhibition (Fig. 2C). Therefore, it is most likely that guanethidine and PPADS inhibit the same ATPergic fraction of the p.s.p. Since the ratio of ATP to noradrenaline varies with different stimulation conditions (von Kügelgen and Starke, 1991), a consequence of co-transmission in the LC may be a finely tuned regulation of the neuronal firing rate. This situation is different from that observed in postganglionic sympathetic nerves, where both ATP and noradrenaline are excitatory transmitters (Stjärne,

1989; von Kügelgen and Starke, 1991). Such additive effects of co-transmitters are the usual pattern, although subtractive effects also occur (Kupfermann, 1991).

Dopaminergic mechanisms in the ventral tegmental area

Electrophysiological studies

Intracellular recordings were made in a midbrain slice preparation of the rat brain containing the VTA (Poelchen et al., 1998). Dopaminergic principal cells were identified by their electrophysiological properties and a hyperpolarizing response to dopamine (Fig. 3A, upper panel). Superfusion with dopamine (100 µM) additionally

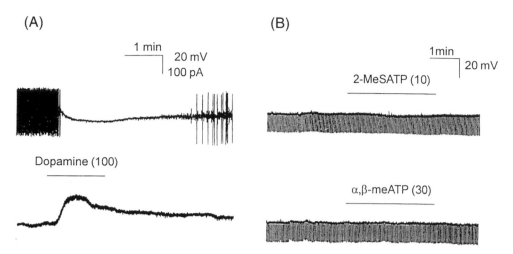

Fig. 3. Effects of dopamine, 2-MeSATP and α,β-meATP on the membrane potential and input resistance of rat VTA principal neurons. Dopamine caused hyperpolarization in the current-clamp mode ((A), upper panel) and outward current at a holding potential of −60 mV in the single electrode voltage-clamp mode ((A), lower panel). Neither 2-MeSATP nor α,β-meATP altered the membrane potential or input resistance (B). Constant current was passed through the electrode to hyperpolarize the cell by about 20 mV. Downward deflections represent electronic potentials caused by hyperpolarizing pulses of constant amplitude. The superfusion times of all drugs are indicated by horizontal lines. The concentrations of dopamine, 2-MeSATP and α,β-meATP appear in brackets and are expressed in μM. Each agonist was tested on a separate cell.

caused a reduction of the spontaneous firing rate and a decrease of the apparent input resistance. At the same time dopamine (100 μM) caused outward current at a holding potential of −60 mV (Fig. 3A, lower panel). In contrast, 2-MeSATP (10 μM) as well as α,β-meATP (30 μM) had no effect when added to the superfusion medium (Fig. 3B). Therefore, it was concluded that dopaminergic neurons of the VTA do not possess somatic P2 receptors.

Microdialysis experiments

In order to investigate the influence of ATP on the somatodendritic release of dopamine from meso-limbic neurons, a microdialysis probe was positioned into the rat VTA, and 2-MeSATP (100 μM; about 13% recovery in the extramembrane space) was applied via the probe (Krügel et al., 1998). Thereby, the level of extracellular dopamine was increased up to 550%, followed by a transient decrease (Fig. 4A). The 2-MeSATP-induced dopamine release was inhibited by the selective P2 receptor antagonist PPADS (30 μM). Moreover, PPADS alone depressed the extracellular level of dopamine to about 55% of its pre-drug value (Fig. 4A).These findings suggest that stim-

ulatory P2 receptors are targets of endogenously released ATP imposing a tonic stimulation on dopamine release. However, it is unclear whether P2X or P2Y receptors are involved, since 2-MeSATP activates both receptor types (Kennedy and Leff, 1995) and high concentrations of PPADS also fail to prefer P2X over P2Y receptors (Humphrey et al., 1995).

The physiological relevance of somatodendritic dopamine release is still poorly understood. Dopamine released by dendrites interacts both with the D_2 autoreceptors localized on the somata and dendrites of dopaminergic cells and with D_1 receptors localized on neighbouring non-dopaminergic cells. It has been assumed that dopamine in the VTA plays a modulatory function mediating an adjustment of dopamine release under various conditions (Kiyatkin, 1995).

Behavioural studies

Unilateral microinjection of 2-MeSATP (5 pmol) into the rat VTA caused enhanced locomotion in the open field situation accompanied by an increased ambulation in the inner areas during the whole observation period. Figure 5A demonstrates a typical pattern of locomotion. There was a clear

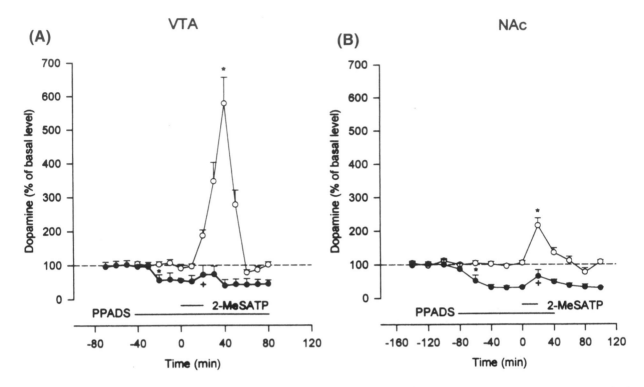

Fig. 4. Effects of 2-MeSATP and PPADS, or their combination on the concentration of extracellular dopamine in the ventral tegmental area (VTA) and nucleus accumbens (NAc). Collection of superfusate and application of drugs was by microdialysis; the probes were positioned into the VTA (A) and NAc (B). 2-MeSATP (100 μM) increased the release of dopamine in normal aCSF (○) to a larger extent than in aCSF containing from the third collection period onwards PPADS (30 μM; ●). PPADS (30 μM) alone depressed the basal release of dopamine. Means ± SEM of five experiments. *$P < 0.05$; significant difference from the dopamine level before superfusion with 2-MeSATP. +$P < 0.05$; significant difference from the dopamine level measured without superfusion with PPADS.

dose-dependency of this effect; while 0.5 pmol 2-MeSATP was inactive, 2-MeSATP both at 5 and 50 pmol caused similar increases in locomotion. The behavioural response to the application of 2-MeSATP into the VTA may be the result of an enhanced activity of mesolimbic dopaminergic neurons with widespread projection targets. Beside the projection to the NAc there are also projections to the caudate-putamen, the olfactory tubercle, the amygdala, the lateral septum and various cortical structures such as the prefrontal or cingulate cortex (Paxinos, 1995). This contrasts with more circumscribed effects after injection of 2-MeSATP into the NAc and may explain differences between the respective behavioural responses initiated (see later).

As a whole, both microdialysis and behavioural studies suggest a P2 receptor-mediated somatodendritic release of dopamine in the VTA. On the other hand, electrophysiological recordings failed to demonstrate a change in membrane potential or input resistance in response to 2-MeSATP. These divergent results may be explained by the fact that P2 receptors are possibly localized at the dendrites, while similar receptors are missing at the cell bodies. Hence, electrophysiological changes leading to dopamine release occur in the dendritic region but are not conducted to more remote somatic sites.

Dopaminergic mechanisms in the nucleus accumbens

Microdialysis experiments

When the microdialysis probe was positioned into the NAc, application of 2-MeSATP (0.1, 1, 10 mM) via the probe enhanced the concentration of

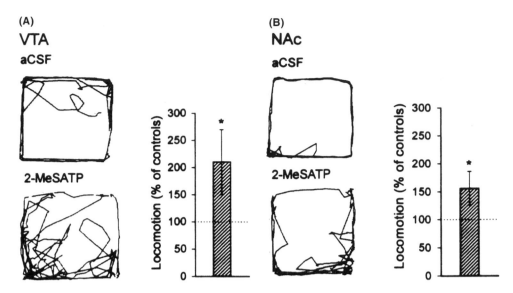

Fig. 5. Examples for typical patterns of movement evoked by the injection of 2-MeSATP (5 pmol) into the VTA (A) or NAc (B) compared with the patterns of movement evoked by the injection of aCSF into the respective brain areas. Measurements were made for 10 min after the injection of aCSF (upper panels) or 2-MeSATP (lower panels). Hatched bars represent the mean locomotor activity calculated as a percentage of the controls indicated by the dotted line ($n = 6$). *$P < 0.05$; significant difference from the effect of aCSF.

dopamine in the extracellular space in a dose-dependent manner (Kittner et al., 1997). Figure 4B shows the elevation of the intra-accumbal dopamine concentration by 2-MeSATP (1 mM) and a subsequent, transiently decreased dopamine level. Pre-treatment with PPADS (30 μM) abolished the 2-MeSATP-induced increase of dopamine release. When PPADS was applied alone, the basal concentration of dopamine declined to about 35% of the control level indicating that endogenous ATP is present in the extracellular space exerting a stimulatory effect on dopamine release via P2 receptors. It is suggested that ATP originating, for example from the axon terminals of VTA neurons, storing dopamine together with this co-transmitter, modulates the basal tonic release of dopamine in the NAc.

The physiological P2 receptor-mediated response to endogenously released ATP is terminated by a fast successive degradation by ectonucleotidases to adenosine (Humphrey et al., 1995; Zimmermann, 1996). Adenosine is known to depress the release of dopamine in the striatum via A_1 receptor activation (Zetterstrom and Fillenz,

1990) and to facilitate it via A_{2A} receptor activation (Okada et al., 1997; Fuxe et al., 1998). Therefore, a functional balance between ATP and its metabolite adenosine may determine the intra-accumbal concentration of dopamine. The mechanisms to maintain or restore the balance between these transmitters/modulators remain to be investigated.

Behavioural studies

Injection of 2-MeSATP (0.5, 5, 50 pmol) into the NAc enhanced locomotion in the open field situation dose-dependently in comparison with the aCSF-treated controls (Fig. 5B). Although 2-MeSATP was without effect at 0.5 pmol, it increased locomotion at 5 pmol maximally, without a further increase of this response at a still higher concentration of 50 pmol. The maximal effect (about 150% of the controls) was observed between 5 to 10 min after the injection of 5 pmol 2-MeSATP. Taken together, the open field behaviour caused by 2-MeSATP is characterized by a prolongation of the motor response to a new environmental stimulus. A similar effect of ATP was also demonstrated in frog embryos (Dale and Gilday, 1996). The pattern of movement in the open field showed an increase of locomotion in the peripheral areas of the open field without an enhancement of the central ambulation (Fig. 5B).

It is noteworthy that no circling behaviour and stereotypies were observed after unilateral injection of 2-MeSATP into the NAc. The open field behaviour after 2-MeSATP application showed close similarities with the motor activation produced by amphetamine which is associated with increased dopaminergic transmission and attenuated by mesoaccumbens dopamine depletion (Clarke et al., 1988; Di Chiara and Imperato, 1988; Kuszenski and Segal, 1989). There is considerable evidence indicating that the alteration of locomotor activity appears to be mediated by mesolimbic structures, whereas stereotypic behaviour is more closely related to the nigrostriatal system in rats (Costall and Naylor, 1975; Kelley et al., 1975). Therefore, the failure of stereotypies suggests that effects are restricted to the mesolimbic system.

Furthermore, after a short resting period (at least 10 min after injection of 2-MeSATP) the behaviour of the animals changed to a qualitatively new pattern of increased locomotion in the central open field areas indicating anxiolytic effects (data not shown). The pattern of movement at this period showed marked similarities to the open field behaviour after the injection of 2-MeSATP into the VTA (Fig. 5A). Considering the enzymatic inactivation of 2-MeSATP, these behavioural changes seem to be a rebound phenomenon mediated by feedback mechanisms in a system with extraordinary complexity rather than a direct 2-MeSATP-mediated effect. It has been proposed that for the locomotor response not only specific nuclei such as the NAc, the VTA or the ventral pallidum are important, but that feedback connections between these nuclei are especially critical in initiating pharmacologically-induced locomotion (Hooks and Kalivas, 1995).

As shown in Fig. 6A, the enhanced locomotor activity 5 to 10 min after the application of 2-MeSATP (5 pmol) was inhibited by pre-treatment with PPADS (5 pmol). It is interesting to note that while at this time period the P2 receptor agonist was stimulatory, and its antagonist had no effect when given alone, their combination resulted in less locomotion than that caused by the injection of aCSF itself. The increased central ambulation in the subsequent observation period was also inhibited by PPADS (data not shown).

Pre-treatment with the D_1 receptor antagonist SCH 23390 (50 pmol) did not alter the 2-MeSATP (5 pmol)-induced locomotion, while the D_2 receptor antagonist sulpiride (50 pmol) decreased it. Only a combination of both compounds which was inactive when given alone, abolished the effect of 2-MeSATP; under these conditions locomotor activity was even decreased below the level caused by the injection of aCSF (Fig. 6B). The results suggest that the behavioural changes produced by 2-MeSATP after injection into the NAc are mainly mediated by dopamine release and subsequent activation of postsynaptic D_1 and D_2 receptors. However, the contribution of further neurotransmitters such as glutamate could not be excluded.

Electroencephalographic investigations

The EEG signals from an electrode implanted into the NAc were telemetrically received, transformed into real time by means of fast Fourier analysis and displayed as continuous spectra of power density. The obtained spectra were divided into six frequency bands (Hz): 0.6–4.0 (δ-band), 4.0–8.0 (v-band), 8.0–9.5 (α_1-band), 9.5–13 (α_2-band), 13–19 (β_1-band) and 19–30 (β_2-band).

Figure 7A shows a typical example for the changes of qualitative EEG pattern after the injection of 2-MeSATP (5 pmol) into the NAc. An apparent decrease in magnitude and increase in frequency of the EEG signal occurred, both typical for EEG desynchronization. The quantitative EEG (EEG power spectrum) was characterized by a selective elevation of the alpha-1 band power (Fig. 7B). Expressed as a percentage of the drug-free baseline EEG, the alpha-1 band power was increased to 150% of its pre-drug level. Similar changes were reported in rats after the systemic application of the dopaminergic agonists apomorphine and quinpirole in relatively large doses (0.5 mg/kg, s.c.; Kropf and Kuschinsky, 1991; Kropf et al. 1992) or after the s.c. application of high d-amphetamine (4 mg/kg, i.p.) or cocaine (30 mg/kg)

(A)

5 - 10 min

(B)

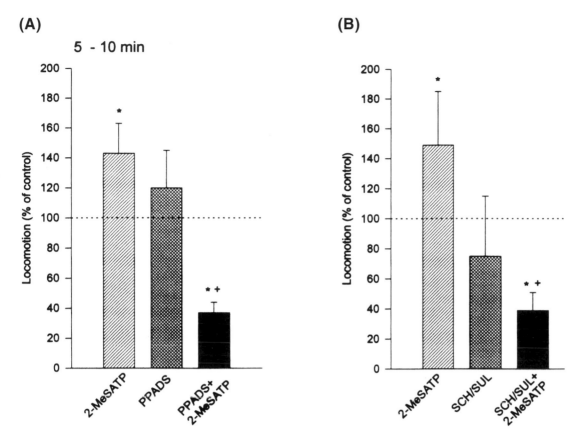

Fig. 6. Effect of intra-accumbal injection of 2-MeSATP on the locomotor activity of rats. Interaction of 2-MeSATP with the P2 receptor antagonist PPADS and the DA$_1$ and DA$_2$ receptor antagonists SCH 23390 and sulpiride. Measurements were made 5 to 10 min after 2-MeSATP application into the NAc. 2-MeSATP, PPADS (5 pmol each) or the combination of both was microinfused in order to investigate the involvement of P2 receptors (A). An equivalent quantity of the solvent (aCSF) was also applied. SCH 23390 and sulpiride (50 pmol each) in combination, or in separate experiments together with 2-MeSATP (5 pmol) were microinfused in order to investigate the involvement of D$_1$ and D$_2$ receptors (B). The locomotor activity of aCSF was taken as 100%. Means ±SEM of 6 animals. *$P < 0.05$; significant difference from the aCSF-treated control group; $^+P < 0.05$; significant difference from the 2-MeSATP group.

doses which increase the synaptic concentration of dopamine (Ferger et al., 1994).

Microdialysis studies with simultaneously recorded EEG activity showed that there is a close relationship between the increase in the alpha-1 power and the enhanced release of dopamine 10–15 min after 2-MeSATP application (Fig. 7B, upper inset). Ten minutes later, a decrease of the extracellular dopamine concentration developed (Fig. 7B, lower inset), accompanied by a marked elevation of the delta power (Fig. 7B). Separate experiments, where the EEG activity and open field behaviour were simultaneously measured, demonstrated a parallel increase in the alpha-1 power and

locomotion (Kittner et al., 1997). Consistent with these data are observations of in vivo microdialysis and voltammetry studies revealing that exposure to environmental stimuli results in an elevated extracellular dopamine concentration in the NAc (Mitchell and Gratton, 1992; Feenstra et al., 1995). There are also interesting correlations to human EEG which report the association of an increased alpha activity with enhanced intentional behaviour and hedonistic emotions (Machleidt et al., 1994).

Additional experiments of our group indicated that the elevation of the alpha-1 power is not uniformly accompanied by an increased accumbal dopamine level (not shown). Compounds such as

232

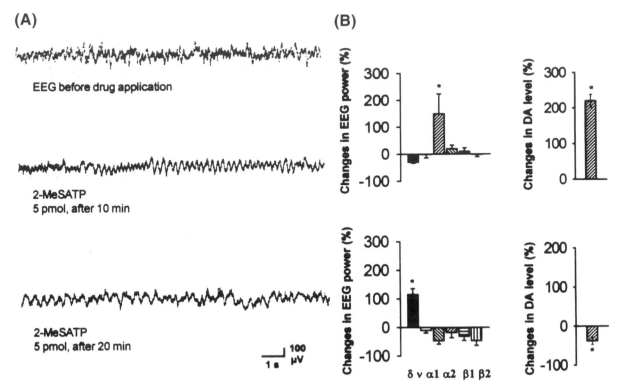

Fig. 7. Effect of intra-accumbens injection of 2-MeSATP on the EEG power spectrum of rats. Typical EEG recordings (A) and the corresponding power spectra ((B), left panels) are shown. The hatched bars represent percentage changes of the basal concentration of dopamine at the respective time periods ((B), right panels). Collection was performed by microdialysis; 2-MeSATP (100µM) was perfused via the microdialysis probe. Means ± SEM of three to four animals. *$P < 0.05$; significant difference from the respective pre-drug frequency band ((B), left panels) or dopamine concentration ((B), right panels).

the D_2 dopamine receptor agonist quinpirole, the P2 receptor antagonist PPADS or the ionotropic glutamate receptor antagonists CPP and CNQX (all applied via the microdialysis probe into the NAc) also evoked a concurrent increase in alpha-1 power and locomotion. The extracellular dopamine level was slightly increased after perfusion with CPP and CNQX, two antagonists selective for NMDA and non-NMDA receptors, respectively, but decreased after perfusion with quinpirole or PPADS. These results suggest that the elevation of the alpha-1 power is not directly associated with the extracellular dopamine level, but probably inversely related to the activity of GABAergic projection neurons which are most likely the locomotor component of limbically initiated behaviour (Kiyatkin, 1995). The inhibition of these neurons either by dopamine receptor agonists or glutamate

receptor antagonists can mediate rewarding functions (Carlsson and Carlsson, 1990; Carlezon and Wise, 1996).

Conclusions

Our investigations as a whole demonstrate the presence of release stimulatory P2 receptors in the central noradrenergic and dopaminergic systems. Hence, exogenously applied or endogenously released ATP may excite noradrenergic neurons of the LC and dopaminergic neurons of the VTA. ATP and its structural analogue 2-MeSATP depolarized LC neurons in a slice preparation. In addition, there is strong evidence for the contribution of ATP to excitatory synaptic potentials recorded from the neurons themselves; this ATP appears to be co-released with noradrenaline from recurrent axon collaterals or dendrites. A combined methodological approach was used to prove the presence of P2 receptors in the VTA and its main projection target in the NAc. Although electrophysiological recordings from VTA neurons in a slice preparation failed to show a depolarizing response to 2-MeSATP, the application of 2-MeSATP into the VTA via a microdialysis probe was a powerful stimulus of dopamine release in vivo. The concomitant behavioural response consisted of increased locomotion and ambulation in the inner areas of an open field system. Microdialysis of the NAc with 2-MeSATP demonstrated a similar although less pronounced release of dopamine. The coincident behavioural (increased locomotion) and EEG activities (selective elevation of the alpha-1 band power) support the involvement of dopaminergic mechanisms. All effects of 2-MeSATP were blocked by the selective P2 receptor antagonist PPADS. In addition, PPADS when given alone decreased the basal release of dopamine both from the VTA and NAc, indicating the existence of a tonic excitatory control by endogenous ATP of dopamine secretion. The divergent in vitro and in vivo results suggest that P2 receptors are probably situated at the dendrites and axon terminals of VTA neurons but not at their cell somata. In concusion, there is direct evidence for the co-release of ATP with noradrenaline and indirect evidence for a similar co-release of ATP with dopamine. ATPergic mechanisms may have important functions related to central catecholaminergic systems; implications for psychiatric disorders or drug dependence are likely to occur.

Acknowledgements

This work was supported by the Bundesministerium für Bildung, Forschung und Technologie, "Biologische und psychosoziale Faktoren von Drogenmissbrauch und Drogenabhängigkeit" (01EB9425) and "Interdisziplinäres Zentrum für Klinische Forschung an der Medizinischen Fakultät Leipzig" (01KS9504, C4).

References

Abbracchio, M.P. and Burnstock, G. (1994) Purinoceptors: Are there families of P2X and P2Y purinoceptors. *Pharmacol. Ther.*, 64: 445–475.

Ashby, C.R. (1996) *The Modulation of Dopaminergic Neurotransmission by Other Neurotransmitters*. CRC Press, Boca Raton, p. 223.

Burnstock, G. (1986) The changing face of autonomic neurotransmission. *Acta Physiol. Scand.*, 126: 67–91.

Carlezon, W.A. and Wise, R.A. (1996) Rewarding actions of phencyclidine and related drugs in nucleus accumbens shell and frontal cortex. *J. Neurosci.*, 16: 3112–3122.

Carlsson, M. and Carlsson, A. (1990) Interactions between glutamatergic and monoaminergic systems within the basal ganglia: Implications for schizophrenia and Parkinson's disease. *Trends Neurosci.*, 13: 272–276.

Cherubini, E., North, R.A. and Williams, J.T. (1988) Synaptic potentials in rat locus coeruleus neurones. *J. Physiol.*, 496: 431–442.

Clarke, P.B.S., Jakubovic, A. and Fibiger, H.C. (1988) Anatomical analysis of the involvement of mesolimbocortical dopamine in the locomotor stimulant actions of d-amphetamine and apomorphine. *Psychopharmacology* 96: 511–520.

Cloues, R., Jones, S. and Brown, D.A. (1993) Zn^{2+} potentiates ATP-activated currents in rat sympathetic neurons. *Pflügers Arch.*, 424: 152–158.

Costall, B. and Naylor, R.J. (1975) The behavioral effects of dopamine applied intracerebrally to the areas of the mesolimbic system. *Eur. J. Pharmacol.*, 32: 87–92.

Courtney, N.D., Howlett, A.C. and Westfall, T.C. (1991) Dopaminergic regulation of dopamine release from PC12 cells via a pertusis toxin-sensitive G protein. *Neurosci Lett.*, 122: 261–264.

Dale, N. and Gilday, D. (1996) Regulation of rythmic movements by purinergic neurotransmitters in frog embryos. *Nature*, 383: 259–263.

Di Chiara, G. and Imperato, A. (1988) Drugs abused by humans preferentially increase synaptic dopamine concentrations in the mesolimbic system of freely moving rats. *Proc. Natl. Acad. Sci. U.S.A.*, 85: 5274–5278.

Egan, T.M., Henderson, G., North, R.A. and Williams, J.T. (1983) Noradrenaline-mediated synaptic inhibition in rat locus coeruleus neurones. *J. Physiol.*, 345: 477–488.

Evans, R.J., Derkach, V. and Surprenant, A. (1992) ATP mediates fast synaptic transmission in mammalian neurons. *Nature*, 357: 503–505.

Feenstra, M.G., Botterblom, M.H. and van Uum, J.F. (1995) Novelty-induced increase in dopamine release in the prefrontal cortex *in vitro*, inhibition by diazepam. *Neurosci. Lett.*, 189: 81–94.

Ferger, B., Kropf, W. and Kuschinsky, K. (1994) Studies on electroencephalogram (EEG) in rats suggest that moderate doses of cocaine or d-amphetamine activate D_1 rather than D_2 receptors. *Psychopharmacology*, 114: 297–308.

Foote, S.L., Bloom, F.E. and Aston-Jones, G. (1983) Nucleus locus coeruleus: New evidence of anatomical and physiological specificity. *Physiol. Rev.*, 63: 844–915.

Fredholm, B.B., Abbracchio, M.P., Burnstock, G., Daly, J.W., Harden, T.K., Jacobson, K.A., Leff, P. and Williams, M. (1994) Nomenclature and classification of purinoceptors. *Pharmacol. Rev.*, 46: 143–156.

234

Fröhlich, R., Boehm, S. and Illes, P. (1996) Pharmacological characterization of P_2 purinoceptor types in rat locus coeruleus neurons. *Eur. J. Pharmacol.*, 315: 255–261.

Fuxe, K., Ferre, S., Zoli, M. and Agnati, L.F. (1998) Integrated events in central dopamine transmission as analyzed at multiple levels. Evidence for intramembrane adenosine A_{2A}/dopamine D_2 and adenosine A_1/dopamine D_1 receptor interactions in the basal ganglia. *Brain Res. Rev.*, 26: 258–273.

Harms, L., Finta, E.P., Tschöpl, M. and Illes, P. (1992) Depolarization of rat locus coeruleus neurons by adenosine 5′-triphosphate. *Neuroscience*, 48: 941–952.

Hooks, M.S. and Kalivas, P.W. (1995) The role of mesoaccumbens-pallidal circuitry in novelty-induced behavioral activation. *Neuroscience*, 64: 587–597.

Humphrey, P.P.A., Buell, G., Kennedy, I., Khakh, B.S., Michel, A.D., Suprenant, A. and Trezise, D.J. (1995) New insights on P_{2X} purinoceptors. *Naunyn-Schmiedeberg's Arch. Pharmacol.*, 352: 585–596.

Illes, P., Nieber, K., Fröhlich, R. and Nörenberg, W. (1996) P2 purinoceptors and pyrimidinoceptors of catecholamine-producing cells and immunocytes. In: D.J. Chadwick and J.A. Goode, (Eds.), *P2 Purinoceptors: Localization, Function and Transduction Mechanisms*. John Wiley, Chichester, pp. 110–129.

Inoue, K., Koizumi, S. and Ueno, S. (1996) Implications of ATP receptors in brain functions. *Prog. Neurobiol.*, 50: 483–492.

Kelley, P.H., Seviour, P.W. and Iversen, S.D. (1975) Amphetamine and apomorphine responses in the rat following 6-OHDA lesions of the nucleus accumbens septi and corpus striatum. *Brain Res.*, 94, 507–526.

Kennedy, C. and Leff, P. (1995) How should P2X purinoceptors be classified pharmacologically? *Trends Pharmacol. Sci.*, 16: 168–174.

Khakh, B.S., Humphrey, P.P.A. and Henderson, G. (1997) ATP-gated cation channels (P2X purinoceptors) in neurons of the trigeminal mesencephalic nucleus (MNV). *J. Physiol.*, 498: 709–715.

Khakh, B.S., Humphrey, P.P.A. and Surprenant, A. (1995) Electrophysiological properties of P_{2X} purinoceptors in rat superior cervical, nodose and guinea-pig coeliac neurones. *J. Physiol.*, 484: 385–395.

Kittner, H., Krügel, U. and Illes, P. (1997) 2-Methylthio ATP enhances dopaminergic mechanisms in the mesolimbic mesocortical system of the rat: an *in vivo* study. *Naunyn Schmiedeberg's Arch. Pharmacol.*, 356: R52.

Kiyatkin, E.A. (1995) Functional significance of mesolimbic dopamine. *Neurosci. Biobehav. Rev.*, 19: 573–598.

Koizumi, S., Ikeda, M., Nakazawa, K., Inoue, K., Ito, K. and Inoue, K. (1995) Inhibition by haloperidol of adenosine 5′ triphosphate-evoked responses in rat pheochromocytoma cells. *Biochem. Biophys. Res. Commun.*, 210: 624–630.

Kropf, W., Kriegelstein, J. and Kuschinsky, K. (1992) Effects of stimulation of putative dopamine autoreceptors on electroencephalographic power spectrum in comparison with effects produced by blockade of postsynaptic dopamine receptors in rats. *Eur. Neuropsychopharmacol.*, 2: 467–474.

Kropf, W. and Kuschinsky, K. (1991) Electroencephalographic correlates of the sedative effects of dopamine agonists presumably acting on autoreceptors. *Neuropharmacology*, 30: 953–960.

Krügel, U., Franke, H., Kittner, H. and Illes, P. (1998) In vivo effects of 2-methylthio ATP after application into the ventral tegmental area: a microdialysis, behavioural and immunohistochemical study. *Eur. J. Neurosci.*, 10: 286.

Kupfermann, I. (1991) Functional studies of cotransmission. *Physiol. Rev.*, 71: 683–732.

Kuszenski, R. and Segal, D. (1989) Concomitant characterization of behavioral and striatal neurotransmitter response to amphetamine using *in vivo* microdialysis. *J. Neurosci.*, 9: 2051–2065.

Le Moal, M. and Simon, H. (1991) Mesocorticolimbic dopaminergic network: Functional and regulatory roles. *Physiol. Rev.*, 71: 155–234.

Loughlin, S.E. and Fallon, J.H. (1985) Locus coeruleus. In: G. Paxinos (Ed.), *The Rat Nervous System, Vol. 2, Hindbrain and Spinal Cord*. Academic Press, Sidney, pp. 79–93.

Machleidt, W., Gutjahr, L. and Hinrichs, H. (1994) Die EEG Spektralmuster der Grundgefühle: Hunger, Angst, Aggression, Trauer und Freude. *EEG-EMG*, 25: 81–97.

Mitchell, J.B. and Gratton, A. (1992) Partial dopamine depletion of the prefrontal cortex leads to enhanced mesolimbic dopamine release elicited by repeated exposure to natural reinforcing stimuli. *J. Neurosci.*, 12: 3609–3618.

Mogenson, G.J. (1987) Limbic-motor integration. In: A.N. Epstein and A.R. Morrison (Eds.), *Progress in Psychobiology and Physiology*. Academic Press, New York, pp. 117–170.

Mogenson, G.J., Brudzynski, S.M., Wu, M., Yang, C.R. and Yim, C.C.Y. (1993) From motivation to action: A review of dopaminergic regulation of limbic-nucleus accumbens-ventral pallidum-pedunculopontine nucleus circuitries involved in limbic-motor integration. In: P.W. Kalivas and C.D. Barnes (Eds.), *Limbic Motor Circuits and Neuropsychiatry*. CRC Press, Boca Raton, pp. 193–236.

Nagy, J.I., Buss, M. and Daddona, P.E. (1986) On the innervation of trigeminal mesencephalic primary afferent neurons by adenosine deaminase-containing projections from the hypothalamus in the rat. *Neuroscience*, 17: 141–156.

Nakazawa, K. (1994) ATP-activated current and its interaction with acetylcholine-activated current in rat sympathetic neurons. *J. Neurosci.*, 14: 740–750.

Nieber, K., Poelchen, W. and Illes, P. (1997) Role of ATP in fast excitatory synaptic potentials in locus coeruleus neurones of the rat. *Br. J. Pharmacol.*, 122: 423–430.

Okada, M., Kiryu, K., Kawata, Y., Mizuno, K., Wada, K., Tasaki, H. and Kaneko, S. (1997) Determination of the effects of caffeine and carbamazepine on striatal dopamine release by *in vivo* microdialysis. *Eur. J. Pharmacol.*, 321: 181–188.

Paxinos, G. (1995) *The Rat Nervous System*. 2nd edn., Academic Press, San Diego.

Poelchen, W., Sieler, D., Inoue, K. and Illes, P. (1998) Effect of extracellular adenosine 5′-triphosphate on principal neurons of the rat ventral tegmental area. *Brain Res.*, 800: 170–173.

Rasmussen, K., Beitner-Johnson, D.B., Krystal, J.H., Agha-janian, G.K. and Nestler, E.J. (1990) Opiate withdrawal and the rat locus coeruleus: Behavioral, electrophysiological, and biochemical correlates. *J. Neurosci.*, 10: 2308–2317.

Regenold, J.T. and Illes, P. (1990) Inhibitory adenosine A_1-receptors on rat locus coeruleus neurones. An intracellular electrophysiological study. *Naunyn-Schmiedeberg's Arch. Pharmacol.*, 341: 225–231.

Regenold, J.T., Haas, H.L. and Illes, P. (1988) Effects of purinoceptor agonists on electrophysiological properties of rat trigeminal mesencephalic neurones *in vitro*. *Neurosci. Lett.*, 92: 347–350.

Shen, K.-Z. and North, R.A. (1993) Excitation of rat locus coeruleus neurons by adenosine 5'-triphosphate: ionic mechanisms and receptor characterization. *J. Neurosci.*, 13: 894–899.

Stjärne, L. (1989) Basic mechanisms and local modulation of nerve impulse-induced secretion of neurotransmitters from individual sympathetic nerve varicosities. *Rev. Physiol. Biochem. Pharmacol.*, 112: 1–37.

Tschöpl, M., Harms, L., Nörenberg, W. and Illes, P. (1992) Excitatory effects of adenosine 5'-triphosphate on rat locus coeruleus neurones. *Eur. J. Pharmacol.*, 213: 71–77.

Ungerstedt, U. (1971) Stereotaxic mapping of the monoamine pathways in the rat brain. *Acta Physiol. Scand.*, 367: 1–48.

von Kügelgen, I. and Starke, K. (1991) Noradrenaline-ATP co-transmission in the sympathetic nervous system. *Trends Pharmacol. Sci.*, 12: 319–324.

Williams, J.T., Bobker, D.H. and Harris, G.C. (1991) Synaptic potentials in locus coeruleus neurons in brain slices. *Prog. Brain Res.*, 88: 167–172.

Wirkner, K., Franke, H., Inoue, K. and Illes, P. (1998) Differential age-dependent expression of α_2 adrenoceptor- and P2 purinoceptor-functions in rat locus coeruleus neurons. *Naunyn-Schmiedeberg's Arch. Pharmacol.*, 357: 186–189.

Zetterstrom, T. and Fillenz, M. (1990) Adenosine agonists can both inhibit and enhance *in vivo* striatal dopamine release. *Eur J. Pharmacol.*, 3: 137–143.

Zhang, Y.-X., Yamashita, H., Ohshita, T., Sawamoto, N. and Nakamura, S. (1996) ATP induces release of newly synthesized dopamine in the rat striatum. *Neurochem. Int.*, 28: 395–400.

Zhang, Y.-X., Yamashita, H., Ohshita, T., Sawamoto, K. and Nakamura, S. (1995) ATP increases extracellular dopamine level through stimulation of P_{2y} purinoceptors in the rat striatum. *Brain Res.*, 691: 205–212.

Zimmermann, H. (1994) Signalling via ATP in the nervous system. *Trends Neurosci.*, 17: 420–426.

Zimmermann, H. (1996) Extracellular purine metabolism. *Drug Develop. Res.*, 39: 168–174.

P. Illes and H. Zimmermann (Eds.)
Progress in Brain Research, Vol 120
© 1999 Elsevier Science BV. All rights reserved

ATP receptor-mediated component of the excitatory synaptic transmission in the hippocampus

Yuri Pankratov[1], Ulyana Lalo[2], Enrique Castro[3], Maria Teresa Miras-Portugal[3] and Oleg Krishtal[1,*]

[1]*Department of Cellular Membranology, Bogomoletz Institute of Physiology, Bogomoletz St. 4, 252024 Kiev, Ukraine*
[2]*Department of General Physiology of Nervous System, Bogomoletz Institute of Physiology, Bogomoletz St. 4, 252024 Kiev, Ukraine*
[3]*Departamento de Bioquimica, Facultad de Veterinaria, Universidad Complutense de Madrid, 28040 Madrid, Spain.*

Introduction

ATP is a neurotransmitter in the peripheral nervous system. Excitatory post-synaptic currents (EPSCs) mediated by ATP have been recorded in ganglionic neurons in culture (Evans et al., 1992) and in myenteric neurons (Zhou and Galligan, 1996). Recent data suggest that ATP may be a fast neurotransmitter also in the central nervous system (Edwards and Gibb, 1993; Zimmermann, 1994). ATP is released exocytotically from isolated nerve ending preparations (Potter and White, 1980) and activates depolarizing inward currents in neurons of spinal ganglia (Krishtal et al., 1983), in neurones of spinal cord (Jahr and Jessel, 1983), cerebral cortex (Sun et al., 1992), locus coeruleus (Harms et al., 1992; Shen and North, 1993) and hippocampus (Inoue et al., 1992; Balachandran and Bennett, 1996). The action of ATP as a fast neurotransmitter is mediated by ionotropic P2X-type ATP receptors (Surprenant et al., 1995), of which seven subtypes have been cloned so far. All of them are expressed in the central nervous system, notably $P2X_4$ and $P2X_6$ subunits (Buell et al., 1996). Localization of ATP receptors by ligand-binding techniques (Bo and Burnstock, 1994; Balcar et al., 1995) and in situ hybridization (Kidd et al., 1995; Collo et al.,

1996) show a broad distribution of P2X receptors in the brain, particularly in the cerebellum and hippocampus. The $P2X_4$ and $P2X_6$ ATP-receptors are abundant in CA1–CA3 hippocampal pyramidal cells from adult rats (Collo et al., 1996; Le et al., 1998), although wide distribution of mRNA for $P2X_3$ subunits has also been reported (Seguela et al. 1996). In young (P5) animals $P2X_1$ and several isoforms of $P2X_7$ are also present in the hippocampus (Kidd et al., 1995; Simon et al., 1997). The presence of heteropolymers of $P2X_3$ and $P2X_2$ or $P2X_4$ (Lewis et al.,1995) cannot be excluded as well. Since different subtypes of P2X receptors have different affinities to agonists and antagonists and exhibit different kinetics and permeability (Buell et al, 1996), their co-expression or hetero-polymerization could provide diverse properties of ATP-mediated neurotransmission.

Despite the widespread distribution of P2X receptors across the brain, EPSCs mediated by ATP have been demonstrated so far only at a particular localization, namely in the medial habenula (Edwards et al., 1992). Here we report the pharmacological identification of a purinergic component in the excitatory synaptic input to CA1 neurons of the rat hippocampus.

Methods

Experiments were carried out on transverse 200 to 400 μm thick hippocampal slices of Wistar rat

*Corresponding author. Tel.: + 38-044-293-21-42; fax: + 38-044-293-21-42; e-mail: krishtal@biph.serv.kiev.ua

(17–19 day-old animals) brains, removed rapidly after decapitation on a guillotine and placed into ice-cold artificial cerebrospinal fluid (ACSF-I), containing (mM): NaCl 130, KCl 3, $MgCl_2$ 2.0, NaH_2PO_4 1, $NaHCO_3$ 26, glucose 15, gassed with 95% O_2–5% CO_2 to obtain a final pH of 7.4. The CA1 region was functionally isolated from CA2 and CA3 areas. "Minislices" were prepared by making a cut orthogonal to the *stratum pyramidale* up to distal edge of mossy fiber layer. After dissection, slices were placed in a holding chamber and incubated at 32°C for 40 min, then they were maintained at a room temperature (22–24°C) for 3–6 h until they were placed in the recording chamber, also at room temperature. During incubation and recording slices were kept in ASCF-II of the following composition (in mM): NaCl 135, KCl 2.7, $CaCl_2$ 2.5, $MgCl_2$ 0.75, NaH_2PO_4 1, $NaHCO_3$ 26, glucose 15, picrotoxin 0.1, gassed with 95% O_2/5% CO_2 to obtain a final pH of 7.4.

Whole-cell voltage clamp recordings were obtained from CA1 neurons using the conventional patch clamp technique. The patch pipette (1–2.5 MΩ) was filled with the following intracellular solution: CsF 120 mM, Tris-Cl 20 mM, pH 7.3. The series and the input resistances were 6–8 MΩ and 60–300 MΩ, respectively and varied by less than 20% in the cells accepted for analysis. To record EPSCs the Schaffer collateral/commisural pathway was stimulated at 0.05–0.08 or 0.2 Hz with a 50 μm thick Ni/Cr bipolar electrode positioned on the slice surface in the stratum radiatum. Currents were monitored using a RK400 (BioLogic, France) amplifier, filtered at 1.0 kHz with 5-pole Bessel filter, digitized at 2.5 kHz and stored on the computer disk.

To examine responses to ATP in pyramidal cells, ATP (250 μM) was applied by pressure ejection ($P = 5–10$ kPa) from the pipette with 200–300 μm internal diameter. The pressure was switched by a solenoid valve controlled by a computer. The pipette was positioned perpendicularly to the flow of the bath perfusing solution near the edge of the stratum radiatum. The agonist was applied in the presence of 10 μM CNQX and 50 μM D-APV at 2–5 min intervals using 3 s pulses for registration of ATP-evoked currents and 10 s pulses for registration of intracellular Ca^{2+}-transients.

For the measurements of intracellular Ca^{2+} concentration, hippocampal neurons were loaded with the calcium indicator fura-2 by 25 min incubation of the hippocampal slices with a Tyrode solution supplemented with fura-2 acetoxymethylester (fura-2/AM) (10 μM, diluted in DMSO) and pluronic F-127 detergent (0.02%) at 35°C. For the period of dye-loading, slices were kept under a controlled air environment (5% CO_2 + 95% O_2). Subsequently the slices were incubated in ASCF-II for additional 40 min to ensure Fura-2/AM deetherification. For Fura-2 excitation, the cells were alternately illuminated at wavelengths 360 ± 5 nm and 390 ± 5 nm. Excitation filters were mounted in a filter wheel set at five revolutions per second. The emitted light was collected at 530 ± 10 nm by a photomultiplier. The filter wheel and photomultiplier outputs were controlled by the Fura-2 Data Acquisition System (Luigs and Neumann, Rattingen, Germany). Signals corresponding to both excitation wavelengths were fed to a computer via a TIDA interface (Batelle, Germany). $[Ca^{2+}]_i$ values were calculated off-line. To reduce the background fluorescence and select the region of interest, the UV illumination was attenuated by an adjustable diaphragm installed in the light path. Dye-loaded neurons were positioned in such a way that the fluorescent signal was collected from their somata. The system was calibrated in vitro, the actual $[Ca^{2+}]_i$ was calculated from the ratio of fluorescence recorded at 360 and 390 excitation wavelengths as described elsewhere (Grynkiewicz et al., 1985).

The following compounds were purchased from RBI: 6-cyano-7-nitroquinoxaline-2,3-dione (CNQX), 2-amino-5-phosphovaleric acid (D-APV), pyridoxal phosphate-6-azophenyl-2'-4'-disulfonic acid (PPADS), α,β-methylene ATP (α,β-meATP), ATP_{gS}, mecamylamine and tropisetron. Fura-2/AM was delivered by Molecular Probes (Eugene, OR, USA). The other chemicals were from Sigma.

Results

Our investigation was initiated by the observation that the specific P2-purinoceptor antagonist PPADS had a small, but consistent and reversible, inhibitory effect on the EPSC recorded in CA1 pyramidal neurons in response to stimulation of Schaffer collateral at especially low stimulation frequencies (0.1–0.05 Hz). This effect was not significant when evaluated for the total EPSC, but it became well manifested when the larger, AMPA component of the EPSC was blocked by CNQX (Fig. 1A). Under these conditions the EPSC was reduced by an average of $24 \pm 12\%$ (at -50 mV holding potential) in 11 out of 14 cells tested, although the cell to cell variability was quite high (ranging from 12% to 43% of inhibition). The onset and recovery from inhibition was quite slow, taking about 10 and 20 min respectively.

The inhibitory action of PPADS on EPSCs in hippocampus has already been reported (Motin and Bennet, 1995) and attributed to the inhibition of glutamate release from presynaptic terminals. However, the data obtained in the present work argue against this conclusion. As shown in Fig. 1B, PPADS also altered the voltage dependence of the EPSC. It was more effective at negative potentials

and thus the PPADS-sensitive component of the EPSC increases monotonically with hyperpolarization unlike the bell-shaped I/V curve characteristic NMDA-activated currents. The change in the shape of the current–voltage relationship cannot be attributed to the changes in the voltage clamping conditions since PPADS does not appreciably modify the input resistance of the cells or the series resistance error. In addition, the reduction of the EPSC elicited by PPADS was not proportional to the control current indicating that the effect is not due to inhibition of excitatory transmitter release and represents a component which cannot be attributed to the action of either AMPA or NMDA receptors.

In fact, in 51 out of 65 cells tested the EPSC was not completely blocked after bath application of CNQX (10 μM) and D-APV (50 μM) and a small residual EPSC remained (rEPSC, see Fig. 2). This rEPSC averaged up to $25 \pm 11\%$ of the total EPSC (at -50 mV). The rEPSC could not be further

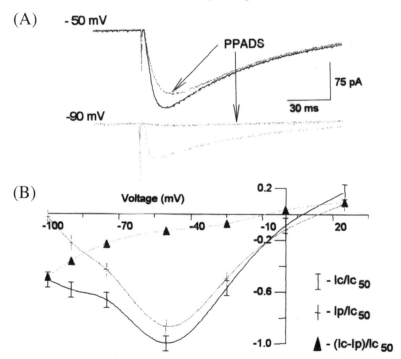

Fig. 1. Effect of bath application of PPADS on EPSC in CA1 hippocampal neuron in the presence of 10 μM CNQX. Schaffer collaterals were stimulated at 0.05 Hz. (A) EPSCs before and after application of 10 μM PPADS at two holding potentials (all current traces are the mean of four sequential records). (B) Current–voltage relationship for the amplitude of EPSC in control (I_c), and under 10 μM PPADS (I_p). Currents were normalized to the value in control at -50 mV (I_{c50}) and all measurements were made in the presence of 10 μM CNQX. Each point is mean ± SEM for seven cells.

240

reduced by a two-fold increase in glutamatergic antagonists, indicating that rEPSC was not due to incomplete inhibition of glutamate receptors (Fig. 2A). These concentrations of CNQX and D-APV exceed the reported IC50 for corresponding receptors by a factor larger than 60 (Honoré et al., 1988; Olverman et al., 1988). In contrast, the rEPSC was strongly suppressed by PPADS in 13 out 17 cells tested. PPADS (10 μM) reduced the rEPSC by 60 ± 15% (at −75 mV). At this concentration

PPADS is expected to be selective for ATP receptors and does not inhibit glutamatergic, serotoninergic or cholinergic receptors (Lambrecht et al., 1992; Motin and Bennett, 1995; Buell et al., 1996). Conversely, the rEPSC was not altered by the 5-HT3 antagonist tropisetron (10 μM) or the nicotinic antagonist mecamylamine (10 μM) (data not shown).

It should be mentioned that stimulation of Schaffer collaterals at frequencies of 0.2 Hz or

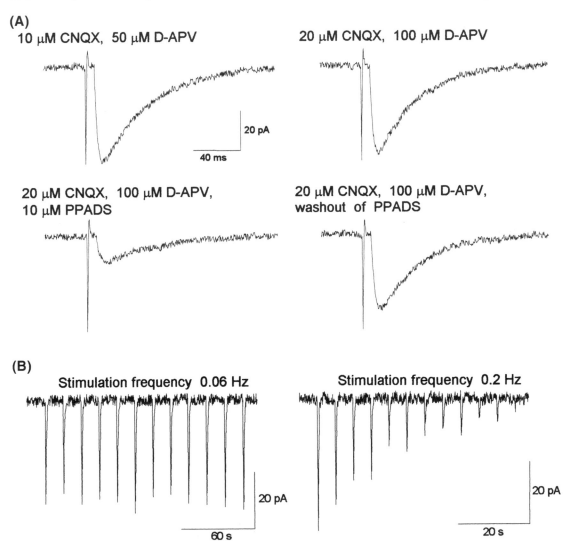

Fig. 2. EPSC recorded in the presence of CNQX and D-APV. (A) Changes in EPSC amplitude following bath application of 10 and 20 μM CNQX, 50 and 100 μM D-APV and 10 mM PPADS, holding potential −75 mV, stimulation frequency 0.05 Hz, each point is mean ± s.e.m for four sequential trials. (B) The time course of rEPSC amlitude at different stimulation frequencies. Holding potential was −75 mV and the measurements were performed in the presence of 10 μM CNQX and 50 μM D-APV.

higher led to quick and irreversible inhibition of the rEPSC (Fig. 2B). This phenomenon was observed in 11 cells. In view of this circumstance, the stimulation at 0.05 Hz was used in all experiments reported. The effect of modulators of P2X receptors was studied only in the cells with a rEPSC remaining stable for more than 15 min.

In the majority of cells the rEPSCs could be well approximated with exponential rise and decay times of 3.1 ± 1.2 ms and 32 ± 13 ms, respectively ($n = 25$), whereas the analogous parameters for pharmacologically isolated AMPA and NMDA components of EPSCs were 1.5 ± 0.3 ms, 14 ± 3 ms

and 7.9 ± 4.8 ms, 63 ± 7.4 ms, respectively ($n = 20$). Hence, residual EPSCs are slower than the AMPA component but faster than the NMDA component.

The voltage dependence of the rEPSC is demonstrated in Fig. 3A and B. The residual current exhibits inward rectification and has a reversal potential of -7 ± 5 mV ($n = 7$). These properties are similar to those of ATP-elicited currents demonstrated in other types of cells (Krishtal et al., 1983; Edwards et al., 1992, Seguela et al., 1996). All the described observations allowed to hypothesize that rEPSC is mediated by PPADS-sensitive P2X receptors.

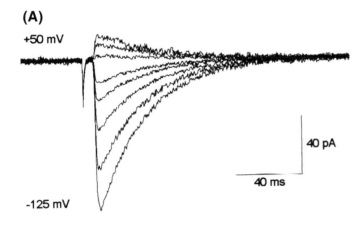

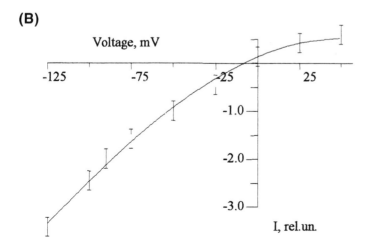

Fig. 3. Voltage dependance of rEPSC. (A) rEPSCs (mean of four records) measured in the presence of 10 μM CNQX and 50 μM D-APV at 25 mV interval between holding potentials -125 and $+40$ mV; (B) mean current-voltage relationship for rEPSC. Currents were normalized to the value at -50 mV. Each point is mean $\pm$ SEM of 11 cells.

To test this hypothesis we investigated the effect of some known modulators of P2X receptors. It has been shown that Zn^{2+} potentiates P2X-mediated responses by increasing the receptor affinity to ATP (Li et al., 1993; Brake et al., 1994; Soto et al., 1996). We have found that rEPSC was markedly potentiated (by $250 \pm 170\%$) by 10 μM of extracellular Zn^{2+} in 13 out 18 cells (Fig. 4A). In the rest of the cells Zn^{2+} was causing insignificant inhibition of rEPSC (see below). The potentiated current

was also sensitive to blockade by PPADS. In the other five cells (28% of total number of 18 cells) Zn^{2+} inhibited the rEPSC. This effect was not so large as potentiation and amounted to $25 \pm 11\%$. As already mentioned, the PPADS effect was highly variable from cell to cell. Figure 4B and C depicts a cell showing the rEPSC only slightly inhibited by PPADS. Independent of the sensitivity to PPADS, the rEPSC could be almost totally blocked by adding to the superfusing solution the non-hydro-

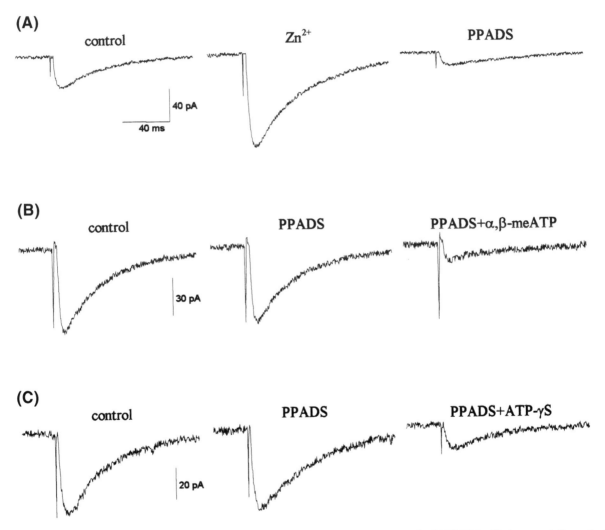

Fig. 4. The pharmacological properties of rEPSCs. (A) Effects of 10 μM $ZnCl_2$ and 10 μM PPADS added sequentially to the bath solution. (B) Effect of bath application of 20 μM α,β-meATP on rEPSC in the cell which exhibits weak sensitivity to PPADS. (C) Effect of bath application of 50 μM ATP-γ-S on rEPSC in the cell which exhibits weak sensitivity to PPADS. The cells in (A), (B) and (C) are different. All measurements were in the presence of 10 μM CNQX and 50 μM D-APV, at holding potential of −75 mV. Each trace is the mean of four consecutive responses.

lizable ATP analogs, α,β-meATP (20 μM, $n = 3$) or ATPγS (50 μM, $n = 7$). This blocking effect developed 5–10 min after application and was irreversible for the rest of the experiment up to 2–3 h.

Although the expression of P2X subunits in the hippocampus has been described by now, the presence of ATP-elicited inward currents has been observed only in cultured hippocampal neurons (Inoue et al., 1992; Balachandran and Bennett, 1996). Thus, we tested the sensitivity of pyramidal cells in situ to exogenous ATP. Local application of a solution with high concentrations of ATP was used for this purpose. Since ATP may elicit glutamate release from hippocampal neurons (Inoue et al., 1992) or presynaptic terminals (Motin and Bennett, 1995), ATP was applied in the presence of picrotoxin, CNQX and D-APV. In this way we avoided glutamatergic and GABA-ergic signals that could result from possible ATP-evoked neurotransmitter release. Under these conditions clear inward currents in response to ATP were observed in 13 out of 15 cells (Fig. 5). In the same way as the rEPSC, the ATP-evoked currents were markedly and reversibly inhibited by PPADS and potentiated by Zn^{2+} in four out of five cells tested.

The application of ATP in a concentration of 250 μM proved to produce transient increases in $[Ca^{2+}]_i$ in the majority of hippocampal neurones (in 19 out of 25 cells tested). The amplitude of ATP-induced $[Ca^{2+}]_i$ transients reached 72 ± 28 nM ($n = 19$; see Fig. 6). Inhibition of synaptic transmission in the slices by 5 min-incubation with 1 μM tetrodotoxin did not alter the agonist-induced $[Ca^{2+}]_i$ transients suggesting that they are mediated by direct activation of receptors expressed in the plasma membrane of pyramidal neurons (data not shown). In order to determine the sources of ATP-induced $[Ca^{2+}]_i$ increase we applied ATP in a Ca^{2+}-free bath solution. The response was strongly, but incompletely depressed (see Fig. 6A) indicating that the observed $[Ca^{2+}]_i$ increase has arisen from the activation of both ionotropic and metabotropic purinoreceptors. The ATP-induced $[Ca^{2+}]_i$ transients were inhibited by PPADS (Fig. 6B) and facilitated by Zn^{2+} (Fig. 6C). The effect of Zn^{2+} on $[Ca^{2+}]_i$ transients was not as large as for the rEPSC and amounted to $70 \pm 30\%$. This discrepancy can be

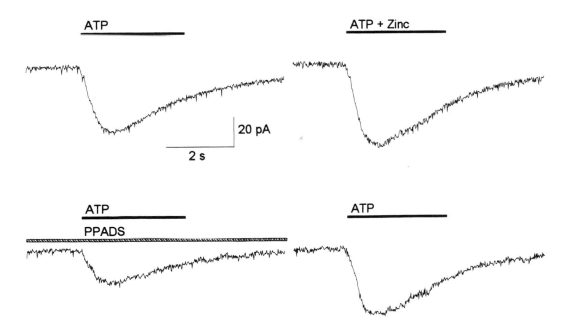

Fig. 5. Effect of 10 μM of PPADS and 10 μM of $ZnCl_2$ on the inward currents evoked by the application of 250 μM ATP in CA1 region of hippocampal slice. PPADS was added 5 min before ATP application. All measurements were in the presence of 10 μM CNQX and 50 μM D-APV, at holding potential of −75 mV. Each trace is the mean of four consecutive responses.

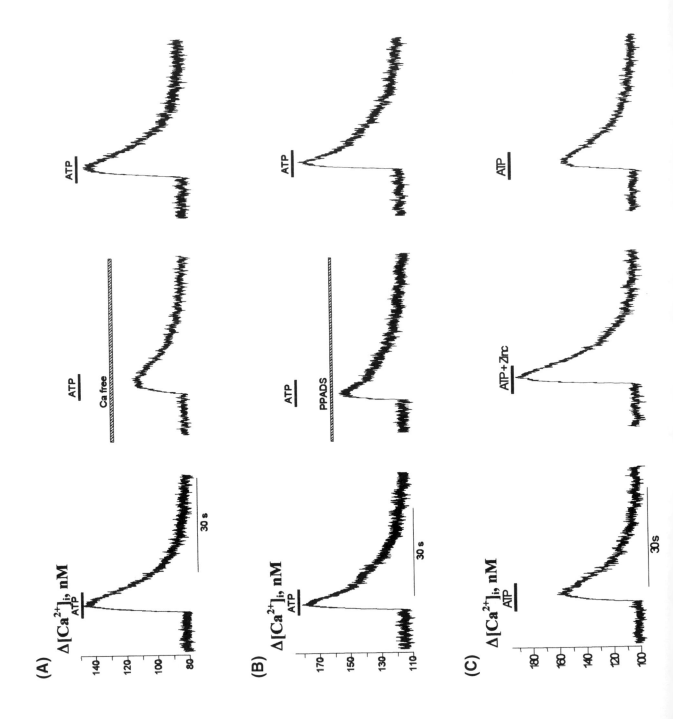

attributed to the partial contribution of Ca^{2+} influx through ionotropic receptors to the general increase in the intracellular Ca^{2+}.

Modulatory effects of Zn^{2+} have been reported not only for P2X, but also for GABA and NMDA receptors (Mayer and Vyklicky 1989; Martina et al., 1996; Paolette et al., 1997). However, the potentiation of receptor activity by zinc is a distinctive property of ATP-receptors. By now this potentiation has been demonstrated only for recombinant $P2X_2$ and $P2X_3$ receptor subtypes. Correspondingly, it is still unknown which effect Zn^{2+} has on other subtypes. In the dorsal root ganglion neurons of bullfrog, it was shown (Li et al., 1997) that zinc can inhibit ATP-activated currents and this inhibition can be reversed by sulfhydryl group reducing agent dithiothreitol (DTT). On the other hand, the sulfhydryl reducing reagents, such as DTT or glutathione, reverse the action of Zn^{2+} on the NMDA (Tang and Aizenman, 1993; Paoletti et al., 1997) and GABA receptors (Pan, Z.-H. et al., 1995; Martina et al., 1996) receptors. On the basis of these data, we tried to elucidate whether the reducing reagent DTT can reverse the action of Zn^{2+} on the ATP-mediated EPSCs.

As mentioned above, Zn^{2+} failed to potentiate the rEPSC in a fraction of neurons tested, but caused its inhibition. We have found that such rEPSCs were increased by $51 \pm 22\%$ ($n = 5$) by the application of 2 mM DTT (Fig. 7A). Alternatively, in the cells, where rEPSC was potentiated by Zn^{2+}, DTT decreased its amplitude by $47 \pm 21\%$ (in eight cells out of eight tested). Furthermore, in three cells we observed a simultaneous inhibition of the purinergic component and potentiation of the NMDA component of synaptic transmission, as shown in Fig. 7B. For this purpose, D-APV was applied in a concentration lower than usual to obtain incomplete block of NMDA receptors. At a holding potential of -50 mV, the EPSC represented the sum of a more rapid purinergic component and slower sustained NMDA component while at a holding potential of -90 mV, the EPSC consisted only of the purinergic one. As the result of DTT application the purinergic component became smaller, whereas the NMDA component became larger and slower, in agreement with the observation of Tang and Aizenman (1993). The effects of Zn^{2+} and DTT were opposite in all the cells tested, but it is worth noting that DTT can increase, while Zn^{2+} can decrease neurotransmitter release, presumably via presynaptic NMDA autoreceptors (Woodward and Blair, 1991; Breukel et al., 1998) and this effect can interfere with the action of DTT and Zn^{2+} on postsynaptic ATP-receptors.

Discussion

The current paradigm establishes that excitatory synaptic transmission in hippocampal CA1/CA3 synapses is mediated solely by excitatory amino acid receptors. Here we demonstrate the presence of a clear non-glutamatergic component in the EPSC measured in CA1 pyramidal neurons. Based on the selectivity of PPADS and the sensitivity to Zn^{2+} and ATP analogs we conclude that the transmitter underlying this component is most likely ATP acting on ionotropic P2X receptors. Our data suggest that about one fifth of the excitatory input to CA1 neurons is purinergic. The rapid inhibition of the purinergic component of the postsynaptic current by stimulation at the normally used frequencies may explain why this component remained undetected so far. ATP released from damaged cells or its build-up in the synaptic clefts at higher stimulation frequencies may ultimately suppress P2X activity. In addition, one cannot exclude the inhibitory action of ATP itself on the synaptic transmission via presynaptic purinoreceptors (Koizumi et al., 1997; Koch et al.,1997).

Fig. 6. ATP-induced Ca^{2+}-transients in CA1 pyramidal neurons. (A) $[Ca^{2+}]_i$ elevation evoked by application of 250 μM of ATP measured in control conditions and 5 min after slice incubation in Ca-free solution. (B) Inhibition of the ATP-induced Ca^{2+}-transients in hippocampal pyramidal neurons by PPADS. Traces represent the Ca^{2+}-transients induced by 250 μM ATP in control and in the presence of 10 μM PPADS added 5 min before ATP application. (C) Potentiation of the ATP-induced Ca^{2+}-transients by Zn^2. Traces represent the Ca^{2+}-transients induced by 250 μM ATP in control and in the presence of 10 μM $ZnCl_2$ added 5 min before ATP application.

In fact, the increased stimulation frequency in medial habenula slices increases the number of failures in ATP-mediated EPSCs (Edwards et al., 1997). The pronounced cell-to-cell variability in the sensitivity of the rEPSC to PPADS as well as the diversity of the modulatory effect of Zn^{2+}, can be accounted for by the multiplicity of P2X receptor subunits within the hippocampus. Non-

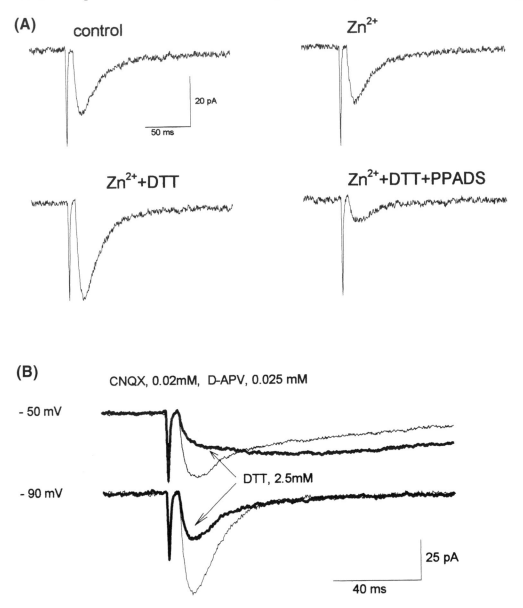

Fig. 7. Zn^2+ and the reducing agent DTT may have reciprocally different effects on rEPSCs in different cells. (A) Example of the neuron in which Zn^2 had a weak inhibitory effect on the rEPSC. rEPSC (mean of five records) was measured in the presence of 20 μM CNQX and 100 μM D-APV, before (control) and after application of 10 μM $ZnCl_2$; subsequently 2 mM DTT and 10 μM PPADS were added. (B) Example of the neuron in which Zn^2 had a facilitatory effect on rEPSC, a low concentration of D-APV allows to record the rEPSC and a sustained NMDA component simultaneously. EPSC (mean of five records) was measured in the presence of 20 μM CNQX and 25 μM D-APV before (thin line) and after (thick line) application of 2mM DTT at indicated holding potentials. See text for explanations.

inactivating (P2X$_2$(a)), moderately- (P2X$_2$(b)) and rapidly-inactivating (P2X$_1$ and P2X$_3$) subtypes of P2X receptors are known to be present in the hippocampal formation of rats (Kidd et al., 1995; Collo et al., 1996; Seguela et al., 1996; Simon et al., 1997). Among these, P2X$_2$(a) and P2X$_1$ are blocked by PPADS, while P2X$_2$(b) are less sensitive to these antagonists (Simon et al., 1997). The P2X$_2$ and P2X$_3$ subtypes can be potentiated by Zn^{2+}, while for other subtypes the effect of zinc remains unclear. On the other hand, PPADS-insensitive P2X$_4$ subunits are localized in a subpopulation of CA1 pyramidal neurons in adult rats (Le et al., 1998). The cellular distribution of other P2X subunits in the hippocampus is not known. Thus, co-expression or heteropolymarization of several subtypes of P2X receptors could provide diverse properties of the action of ATP as a fast neurotransmitter.

What is the source of synaptically active ATP? Some studies have claimed that ATP may be co-released with glutamate by presynaptic terminals of Schaffer collateral (Wieraszko et al., 1989; Cunha et al., 1996), however this point has been discussed (Hamann and Attwell, 1996). Recently it has been shown that noradrenergic collaterals release ATP at the excitatory synaptic input to locus coeruleus neurons (Nieber et al., 1997). The noradrenergic system (and other aminergic and cholinergic fibers that co-store ATP) project to the hippocampal formation and thus may provide a supply of partly purinergic fibers with ATP mediating synaptic responses (von Kügelgen et al., 1994).

Conclusions

The excitatory synaptic transmission in the CA1 area of hippocampal slices of 17–19 day-old rats has been investigated in situ by using patch clamp and intracellular calcium concentration ($[Ca^{2+}]_i$) measurements. Excitatory post-synaptic currents (EPSCs) were elicited by stimulating Schaffer collaterals at frequencies below 0.2 Hz. A small component of the EPSC remains uninhibited after full inhibition of glutamatergic transmission by 6-cyano-7-nitroquinoxaline-2,3-dione (CNQX, 20 μM) and 2-amino-5-phosphovaleric acid (D-APV, 100 μM). The amplitude of this residual EPSC (rEPSC) comprised $25 \pm 11\%$ of the total EPSC when measured at a holding potential of -50 mV. The rEPSC was blocked by bath application of the selective P2 purinoreceptor blocker pyridoxal phosphate-6-azophenyl-2′-4′-disulfonic acid (PPADS, 10 μM) or the non-hydrolizable ATP analogs, ATP-γ-S and α,β-methylene-ATP at 50 and 20 micromolar concentrations respectively. The rEPSC was considerably potentiated (to $250 \pm 170\%$ of control) by external Zn^{2+}(10 μM) in 70% of the cells tested and was inhibited for $25 \pm 11\%$ in the rest of the cells. Dithiothreitol (DTT) reversed both effects of Zn^{2+} in all cells investigated. In another series of experiments, exogenous ATP was applied to the CA1 neurons in situ and elicited the inward current as well as the changes in $[Ca^{2+}]_i$. Inward currents as well as $[Ca^{2+}]_i$-transients were inhibited by PPADS to the same extent as the rEPSCs. It is concluded that about one fifth of the postsynaptic current generated at the excitatory input to CA1 neurons of rats is due to purinergic transmission which uses various subtypes of P2X receptors.

Acknowledgement

The present research was supported in part by an International Research Scholar's award from the Howard Hughes Medical Institute (HHMI 75195-5480001).

References

Balachandran, C. and Bennett, M.R. (1996) ATP-activated cationic and anionic conductances in cultured rat hippocampal neurons. *Neurosci. Lett.*, 204: 73–76.

Balcar, V.J., Li, Y., Killinger, S. and Bennett, M.R. (1995) Autoradiography of P2x ATP receptors in the rat brain. *Br. J. Pharmacol.*, 115: 302–306.

Bo, X. and Burnstock, G. (1994) Distribution of [^{3}H]alpha,beta-methylene ATP binding sites in rat brain and spinal cord. *Neuroreport*, 5: 1601–1604.

Brake, A.J., Wagenbach, M.J. and Julius, D. (1994) New structural motif for ligand-gated ion channels defined by an ionotropic ATP receptor. *Nature*, 371: 519–523.

Breukel, A.I.M., Besselsen, E., Lopes, F.H., da Silva, W.E. and Ghijsen J.M. (1998) A presynaptic N-methyl-D-aspartate autoreceptor in rat hippocampus modulating amino acid release from a cytoplasmic pool. *Eur. J. Neurosci.*, 10(1): 106–115.

Buell, G., Collo, G. and Rassendren, F. (1996) P2X receptors: An emerging channel family. *Eur. J. Neurosci.*, 8:

248

2221–2228.

Collo, G., North, R.A., Kawashima, E., Merlo-Pich, E., Neidhart, S., Surprenant, A. and Buell, G. (1996) Cloning of P2X5 and P2X6 receptors and the distributions and properties of an extended family of ATP-gated ion channels. *J. Neurosci.*, 16: 2495–2507.

Cunha, R.A., Vizi, E.S., Ribeiro, J.A. and Sebastiao, A.M. (1996) Preferential release of ATP and its extracellular catabolism as a source of adenosine upon high- but not low-frequency stimulation of rat hippocampal slices. *J. Neurochem.*, 67: 2180–2187.

Edwards, F.A., Gibb, A.J. and Colquhoun, D. (1992) ATP receptor-mediated currents in the central nervous system. *Nature*, 359: 144–147.

Edwards, F.A. and Gibb, A.J. (1993) ATP-a fast neurotransmitter. *FEBS. Lett.*, 325: 86–89.

Edwards, F.A., Robertson, S.J. and Gibb, A.J. (1997) Properties of ATP receptor-mediated synaptic transmission in the rat medial habenula. *Neuropharmacology*, 36: 1253–1268.

Evans, R.J., Derkach, V. and Surprenant, A. (1992) ATP mediates fast synaptic transmission in mammalian neurons. *Nature*, 357: 503–505.

Grynkiewicz, G., Poenie, M. and Tsien, R.Y. (1985) A new generation of Ca^{2+} indicators with greatly improved fluorescent properties. *J. Biol Chem.*, 260: 3440–3450.

Hamann, M. and Attwell, D. (1996) Non-synaptic release of ATP by electrical stimulation in slices of rat hippocampus, cerebellum and habenula. *Eur. J. Neurosci.*, 8: 1510–1515.

Harms, L., Finta E.P., Tscöpl M. and Illes, P. (1992) Depolarization of rat locus coeruleus by adenosine 5′-triphoshate. *Neuroscience*, 48: 941–952.

Honoré, T., Davies, S.N., Drejer, J., Fletcher, E.J., Jacobsen, P., Lodge, D. and Nielsen, F. (1988) Quinoxalinediones: Potent competitive non-NMDA glutamate receptor antagonists. *Science*, 241: 701–703.

Inoue, K., Nakazawa, K., Fujimori, K., Watano, T. and Takanaka, A. (1992) Extracellular adenosine 5′-triphosphate-evoked glutamate release in cultured hippocampal neurons. *Neurosci. Lett.*, 134: 215–218.

Jahr, C.E. and Jessel, T.M. (1983) ATP excites a subpopulation of rat dorsal horn neurones. *Nature*, 304: 730–733.

Kidd, E.J., Grahames, C.B., Simon, J., Michel, A.D., Barnard, E.A. and Humphrey, P.P. (1995) Localization of P2X purinoceptor transcripts in the rat nervous system. *Mol. Pharmacol.*, 48: 569–573.

Koizumi, S., Inoue, K. and Koch, H. (1997) Inhibition by ATP of calcium oscillations in rat cultured hippocampal neurones. *Br. J. Pharmacol.*, 122: 51–58.

Koch, H., Kugelgen, I. And Starke, K. (1997) P2-receptor-mediated inhibition of noradrenaline release in the rat hippocampus. *Naunyn. Schmiedebergs. Arch. Pharmacol.*, 355: 707–715.

Krishtal, O.A., Marchenko, S.M. and Pidoplichko, V.I. (1983) Receptor for ATP in the membrane of mammalian sensory neurones. *Neurosci. Lett.*, 35: 41–45.

Lambrecht, G., Friebe, T., Grimm, U., Windscheif, U., Bungardt, E., Hildebrandt, C., Baumert, H.G., Spatz-Kum-bel, G. and Mutschler, E. (1992) PPADS, a novel functionally selective antagonist of P2 purinoceptor-mediated responses. *Eur. J. Pharmacol.*, 217: 217–219.

Le, K.T., Villeneuve, P., Ramjaun, A.R., McPherson, P.S., Beaudet, A. and Seguela, P. (1998) Sensory presynaptic and widespread somatodendritic immunolocalization of central ionotropic P2X ATP receptors. *Neuroscience*, 83: 177–190.

Lewis, C., Neidhart, S., Holy, C., North, R.A., Buell, G and Surprenant, A. (1995) Coexpression of P2X2 and P2X3 receptor subunits can account for ATP-gated currents in sensory neurons. *Nature*, 377: 432–435.

Li, C., Peoples, R.W., Li, Z. and Weight, F.F. (1993) Zn^{2+} potentiates excitatory action of ATP on mammalian neurons. *Proc. Natl. Acad. Sci. USA*, 90: 8264–8267.

Li, C., Peoples, R.W. and Weight, F.F. (1997) Inhibition of ATP-activated current by zinc in dorsal root ganglion neurones of bullfrog. *J. Physiol. (Lond.)*, 505 (Pt.3): 641–653.

Martina, M., Mozrzymas, J.W., Strata, F. and Cherubini, E. (1996) Zinc modulation of bicuculline-sensitive and -insensitive GABA receptors in the developing rat hippocampus. *Eur. J. Neurosci.*, 8(10): 2168–2176.

Mayer, M.L. and Vyklicky, L. Jr. (1989) The action of zinc on synaptic transmission and neuronal excitability in cultures of mouse hippocampus. *J. Physiol. (Lond.)*, 415: 351–365.

Motin, L. and Bennett, M.R. (1995) Effect of P2-purinoceptor antagonists on glutamatergic transmission in the rat hippocampus. *Br. J. Pharmacol.*, 115: 1276–1280.

Nieber, K., Poelchen, W. and Illes, P. (1997) Role of ATP in fast excitatory synaptic potentials in locus coeruleus neurones of the rat. *Br. J. Pharmacol.*, 122: 423–431.

Olverman, H.J., Jones, A.W. and Watkins, J.C. (1988) [^{3}H]D-2-Amino-5-phosphonopentanoate as a ligand for N-methyl-D-aspartate receptors in the mammalian central nervous system. *Neuroscience*, 26: 1–15.

Pan Z.-H., Bahring, R., Grantyn, R. and Lipton, S. (1995) Differential modulation by sulfhydryl redox agents and glutathione of GABA- and glycine-evoked currents in rat retinal ganglion cells. *J. Neurosci.*, 15(2): 1384–1391.

Paoletti, P., Ascher, P. and Neyton, J. (1997) High-affinity zinc inhibition of NMDA NR1–NR2 receptors. *J. Neurosci.*, 17(15): 5711–5725.

Potter, P. and White, T.D. (1980) Release of adenosine 5′-triposphate from synaptosomes from different regions of rat brain. *Neuroscience*, 5: 1351–1356.

Seguela, P., Haghighi, A., Soghomonian, J. and Cooper, E. (1996) A novel neuronal P2X ATP receptor ion channel with wide distribution in the brain. *J. Neurosci.*, 16: 448–455.

Shen, K.Z. and North, R.A. (1993) Excitation of rat locus coeruleus neurons by adenosine 5′-triphosphate: ionic mechanism and receptor characterization. *J. Neurosci.*, 13: 894–899.

Simon, J., Kidd, E.J., Smith, F.M., Chessel, I.P., Murrel-Lagnado, R., Humphrey, P.P.A. and Barnard, E.A. (1997) Localization and functional expression of splice variants of the P2X2 receptor. *Mol. Pharmacol.*, 52: 237–248.

Soto, F., Garcia-Guzman, M., Gomez-Hernandez, J.M., Hollmann, M., Karschin, C. and Stuehmer, W. (1996) P2X4: An

ATP-activated ionotropic receptor cloned from rat brain. *Proc Natl Acad Sci USA*, 93: 3684–3688.

Sun, M.K., Wahlstedt, C. and Reis, D.J. (1992) Actions of externally applied ATP on rat reticulospinal vasomotor neurons. *Eur. J. Pharmacol.*, 224: 93–96.

Surprenant, A., Buell, G. and North, R.A. (1995) P2X receptors bring new structure to ligand-gated ion channels. *Trends. Neurosci.*, 18: 224–229.

Tang, L.H and Aizenman, E. (1993) The modulation of *N*-methyl-D-aspartate receptors by redox and alkylating reagents in rat cortical neurones in vitro. *J. Physiol. (Lond.)*, 465: 303–325.

von Kügelgen, I., Kurz, K., Bultmann, R., Driessen, B. and Starke, K. (1994) Presynaptic modulation of the release of the co-transmitters noradrenaline and ATP. *Fundam. Clin. Pharmacol.*, 8: 207–213.

Wieraszko, A., Goldsmith, G. and Seyfried, T.N. (1989) Stimulation-dependent release of adenosine triphosphate from hippocampal slices. *Brain Res.*, 485: 244–250.

Woodward, J.J. and Blair, R. (1991) Redox modulation of *N*-Methyl-D-Aspartate-stimulated neurotransmitter release from rat brain slices. *Journ. Neurochem.*, 57: 2059–2064.

Zhou, X. and Galligan, J.J., (1996) P2X purinoceptors in cultured myenteric neurons of guinea-pig small intestine. *J. Physiol. (Lond.)*, 496: 719–729.

Zimmermann, H. (1994) Signalling via ATP in the nervous system. *Trends. Neurosci.*, 17: 420–426.

P. Illes and H. Zimmermann (Eds.)
Progress in Brain Research, Vol 120

Nucleotide and dinucleotide effects on rates of paroxysmal depolarising bursts in rat hippocampus

Fiona M. Ross[1], Martin J. Brodie[2] and Trevor W. Stone[1],*

[1]*Institute of Biomedical and Life Sciences,
Division of Neuroscience and Biomedical Systems, West Medical Building,
University of Glasgow, Glasgow, G12 8QQ, UK*
[2]*Department of Medicine, University of Glasgow, Western Infirmary, Glasgow G12 8QQ, UK*

Introduction

The proposals that ATP could be a fast excitatory neurotransmitter in the medial habenula and locus coeruleus (Edwards et al., 1992; Nieber et al., 1997) has rekindled interest in the possible location, mechanism and physiological function of purine nucleotide receptors in the CNS. ATP can induce inward current and depolarisation in the cerebellum (Ikeuchi and Nishizaki, 1996), nucleus tractus solitarius (Ueno et al., 1992) hypothalamus (Chen et al., 1994), locus coeruleus (Shen and North, 1993; Frohlich et al., 1996) and hippocampus (Inoue et al., 1992, 1995; Balachandran and Bennett, 1996; Dave and Mogul, 1996).

However, ATP is susceptible to metabolism by a variety of enzymes, which can produce adenosine 5'-monophosphate (AMP) as well as adenosine (Stone and Simmonds, 1991; Zimmerman, 1996). It was, therefore, of interest to further investigate the modulation of epileptiform activity by ATP in the CA3 region of the hippocampus taking into account the possible actions of these metabolites.

In addition, several reports exist describing the actions of dinucleotides in different regions of the central nervous system. AP$_4$A and AP$_5$A raised the level of excitation in nodose ganglion neurones (Marchenko et al., 1988), and within the hippocampus AP$_4$A and AP$_5$A depressed postsynaptic field potentials and intracellular postsynaptic currents (Klishin et al., 1994). AP$_5$A produced a reversible increase in the current through calcium channels (Panchenko et al., 1996) in hippocampal neurones. The present chapter, therefore, summarises work designed to determine the effects of adenine nucleotides and dinucleotides on epileptiform bursts of activity.

Effects of adenine nucleotides

Of a series of nucleotide analogues, ATP, ADP and AMP were able to depress the frequency of epileptiform bursts. The stable compound α,β-methylene ATP produced an increase of burst frequency, which was prevented by suramin or pyridoxalphosphate-6-azophenyl-2',4'-disulphonic acid (PPADS). 2-MethylthioATP was inactive up to 50 μM. The A1 receptor blocker 1,3-dipropyl-8-cyclopentylxanthine (CPX), however, blocked the inhibitory activity of both adenosine and ATP.

Adenosine deaminase

A concentration of adenosine deaminase of 0.2 U ml^{-1} was able to prevent responses to adenosine at 50 μM, but had little effect on the

*Corresponding author. Tel.: +44 41-330-4481; fax:
+44 41-330-4100; e-mail: T.W.Stone@bio.gla.ac.uk

inhibitory effect of ATP (50 µM) (Fig. 1). Adenosine deaminase only partially reduced the effect of AMP at the concentration which eliminated responses to adenosine. This was more apparent towards the end of the 10 min perfusion period, when the depression of control burst rate was reduced from approximately 40% to less than 20%, than when the maximum inhibition of rate was reduced from around 65% to near 40%. The depressant effect of AMP remained significant at both times. The concentration of adenosine deaminase used in this experiment was likely to have been near maximal; the same profile of change was observed, with a roughly 50% reduction of the maximal AMP response when AMP deaminase was employed at a 10-fold higher dose.

AMP deaminase (E.C. 3.5.4.6)

AMP deaminase (0.2 U ml^{-1}) tended to decrease the rate of epileptiform activity to a small extent, but over the 10 min perfusion period this did not reach significance. AMP at 50 µM depressed the rate of epileptiform activity by approximately 60%. This effect was rapid in onset and had a slight tendency to drift back towards control during the perfusion time. AMP deaminase metabolises AMP to inosine 5'-monophosphate (IMP) which was inactive. The depression of activity caused by AMP (50 µM) was totally inhibited when AMP deaminase (0.2 U ml^{-1}) was co-perfused with AMP (Fig. 2).

ATP is metabolised through a number of stages by ecto-ATPases, one of the resulting metabolites being AMP. In order to neutralise any AMP produced during the perfusion of ATP, AMP deaminase was used at the concentration which inhibited the effect of AMP at 50 µM. The responses to ATP at both 50 and 200 µM were reduced by the enzyme (Fig. 3). There was a tendency for the rate of bursting to decline progressively throughout the application of the enzyme but, by the end of the ATP application, AMP deaminase had still reduced the effect sufficiently that burst frequency was not significantly below control levels. Indeed, the small degree of inhibition produced by ATP plus AMP deaminase after 10 min of perfusion was similar to that produced by the enzyme alone and may therefore be attributed to an action of the enzyme itself. The maximum amount of inhibition produced by ATP was reduced by AMP deaminase from 96% ± 3.45 to 20% ± 11.98.

→

Fig. 1. (A) Graph of the frequency of generation of spontaneous burst discharges with time. Perfusion of the slices with ATP (50 µM) during the period indicated by the bar above the graph depressed discharge rate. Superfusion with a combination of ATP and adenosine deaminase (0.2 U ml^{-1}) did not prevent the depression by ATP. (B) Histogram indicating the comparison between the effect of ATP alone and with adenosine deaminase. Each point represents the mean ± SEM for $n = 6$ slices (ATP) or $n = 4$ slices (ATP + deaminase). ## $P < 0.01$ relative to the control rate in the absence of any added agent. All the work summarised in this and subsequent figures was performed on transverse hippocampal slices, 450–500 µm thick, prepared from male Wistar rats (180–250 g). The slices were kept within an interface chamber containing artificial cerebrospinal fluid (aCSF) gassed with 95%O$_2$–5%CO$_2$ for at least 1 h prior to use. The composition of the aCSF was as follows (mM): NaCl 115; NaHCO$_3$ 25; KCl 2; KH2PO$_4$ 2.2; CaCl$_2$ 2.5; MgSO$_4$ 1.2; glucose 10; saturated with 95%O$_2$–5%CO$_2$. After incubation individual slices were transferred to a 1 ml submersion chamber which was continually perfused with aCSF or modified aCSF at a rate of 3.5–4 ml min^{-1}. The temperature of the chamber was maintained at approximately 34°C. A bipolar stimulation electrode was placed in the hippocampal CA3 region to allow orthodromic stimulation of the mossy fibres. The response was recorded via a glass capillary electrode in the pyramidal cell layer of the CA3 region. Stimulation (0.2 Hz.) was applied briefly to check the viability of the slice and the correct positioning of the recording electrode, after which stimulation was halted and the perfusing medium changed from normal aCSF to magnesium-free aCSF containing 4-aminopyridine (4AP) at 50 µM. After approximately 5–20 min spontaneous bursts of population spikes occurred which were continuously recorded on a Gould storage oscilloscope and a Grass pen recorder and subsequently plotted as frequency against time. Drugs were perfused for a minimum of 10 min. The control frequency (bursts per minute) was calculated as the mean of the three observations immediately preceding start of drug perfusion. The effect of added agents was measured both as the mean of the final three observations made during the 10 min period of perfusion and as the maximum amount of inhibition produced during the period of application. Results are expressed as a percentage of the control rate ± standard error of the mean (SEM) for n slices. Statistical analysis of control against test rate was carried out using a paired Student's t-test. Multiple comparisons were made using a Student's t-test (paired or unpaired) or analysis of variance (ANOVA) followed by a Student-Newman-Keuls pos$t$-test. $P < 0.05$ was taken to indicate significance. Reproduced with permission from Ross et al. (1998a).

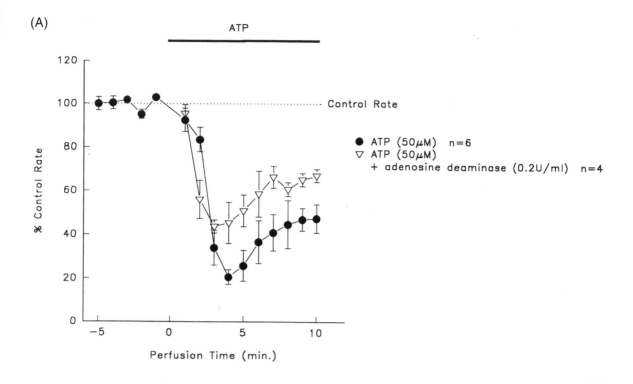

(A)

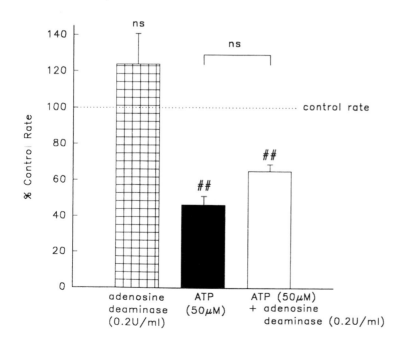

(B)

(A)

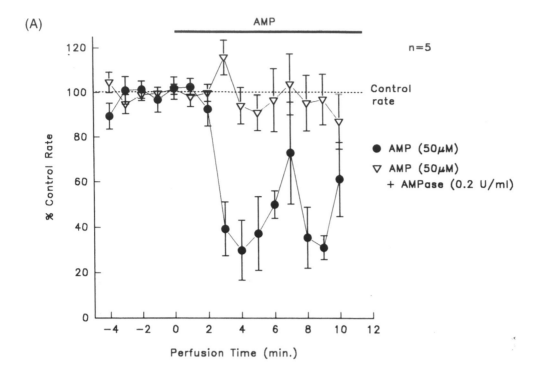

(B)

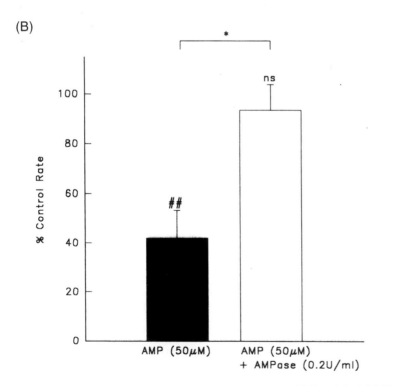

Fig. 2. The time course of the decrease in discharge rate by AMP and the inhibition by AMP deaminase is shown in (A). The mean effect at the end of a 10 min perfusion is summarised in (B). AMP significantly alters the rate from control (## $P < 0.01$). AMP deaminase significantly reduces the effect of AMP (* $P < 0.05$). (Reproduced with permission from Ross et al. (1998b)).

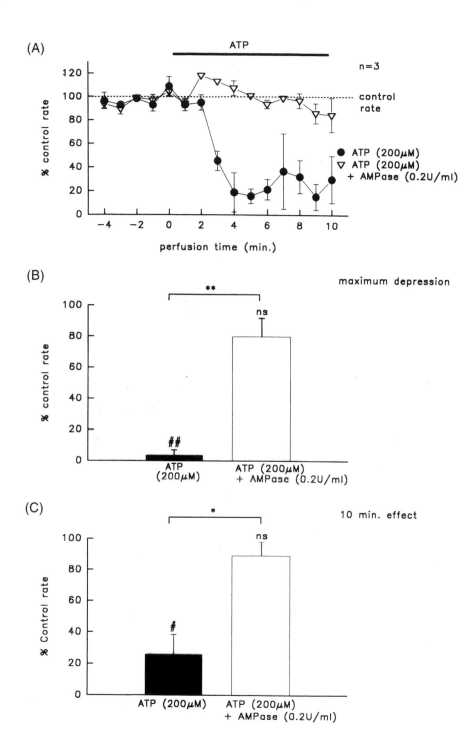

Fig. 3. (A) represents the time course of the effect of ATP (200 μM) and ATP + AMP deaminase on the rate of activity. The maximum inhibition produced by ATP is summarised in (B) and the net effect at the end of 10 min superfusion is shown in (C). Only ATP alone significantly changes the rate from control. AMP deaminase significantly inhibits the effect of ATP. The maximum effect results are calculated as the mean ± SEM of the maximum effect in individual slices, not necessarily at the same time point. Hence the maximum and terminal values differ slightly between the time course graph and the histograms. # $P < 0.05$, ## $P < 0.01$ relative to control levels, * $P < 0.05$, ** $P < 0.01$ for difference between the columns. (Reproduced with permission from Ross et al. (1998b)).

5'-nucleotidase (E.C. 3.1.3.5)

The effects of AMP and ATP were also examined during superfusion of the brain slices with 5'-nucleotidase (E.C.3.1.3.5). Adenosine deaminase was included to remove adenosine formed from the nucleotides. The inhibitory effect of AMP was completely prevented (Fig. 4). ATP (50 μM) showed an initial depression, although the inhibitory effect of ATP was reduced to a level which was not significantly different from control values. By the end of the 10 min superfusion period the discharge rate had returned to control rates.

Results

Effects of adenine dinucleotides

At a concentration of 10 μM, adenosine, diadenosine tetraphosphate (AP$_4$A) and diadenosine pentaphosphate (AP$_5$A) significantly decreased the activity rate. Adenosine deaminase did not change the maximum degree of depression produced by AP$_4$A or AP$_5$A at 10 μM. However, when the mean effect at the end of a 10 min perfusion was analysed in the presence of adenosine deaminase, AP$_4$A and AP$_5$A (10 μM) no longer significantly altered the rate from control (Fig. 5).

In the presence of cyclopentyltheophylline, AP$_4$A (10 μM) no longer depressed the rate of epileptiform activity (Fig. 6). AMPase (0.2 U ml^{-1}), did not significantly change the discharge rate from control during the perfusion of agonists. At this concentration AMPase significantly inhibited the depression in discharge rate produced by AMP.

Discussion

Nucleotides

Clearly ATP can depress epileptiform activity in vitro. The excitation by α,β-methyleneATP and blockade by suramin and pyridoxal-phosphate-6-azopheny-2'-4'-disulphonic acid (PPADS) suggests the involvement of P2X receptors. However, the depressant effect of ATP was prevented by a P1 (adenosine) receptor antagonist, 8-cyclo-pentyl-1,3-dimethylxanthine, but not by adenosine deaminase. This combination of results strongly implies that ATP was acting on a P1 receptor or that the P2 inhibitory receptors could be blocked by xanthines. Evidence has been obtained by others for nucleotides producing response mediated by P1 receptors (von Kugelgen et al., 1992; Cunha et al, 1994; Barajas-Lopez et al., 1995; King et al., 1996). However, the metabolism of ATP by ectoenzymes has been reported in numerous brain regions including vestibular neurones (Cummins and Hyden, 1962), hippocampal neurones (Cunha et al., 1992, 1994) and synaptosomes from a number of brain regions including the cortex (Lin and Way, 1982) and the hippocampus (Nagy et al., 1986). These enzymes are able to terminate the action of ATP released endogenously (Richardson and Brown, 1987; Terrian et al., 1989; Kennedy et al., 1996). The formation of AMP from ADP or ATP results from the action of either ectoATPase, ectoADPase or adenylate kinase which is also thought to be an ecto-enzyme (Nagy et al., 1989). 5'-nucleotidase is also located on the external face of most cell membranes (Lee et al., 1986) and dephosphorylates AMP to form adenosine.

The role of AMP has not been considered by us or several previous groups (von Kügelgen et al., 1992; Kurz et al., 1993; Cunha et al., 1994; Barajas-Lopez et al., 1995; King et al., 1996). In the present study, AMP exerted a depressant effect in our model of epileptiform activity, and ATP depressed spontaneous activity in a similar manner. However, AMP deaminase inhibited the action of ATP (50 or 200 μM). Some groups failed to block ATP responses with AMP deaminase, thus substantiating their suggestion that ATP was able to activate P1 receptors in some tissues (Griese et al., 1991; Bo et al., 1993; Côte et al., 1993).

The conclusion from the work described here, however, is clearly that the depressant activity of ATP is mediated largely via metabolism to AMP, rather than adenosine.

Dinucleotides

The depressant nature of the dinucleotides in this study is in agreement with previous work in which AP$_4$A and AP$_5$A at 5 μM inhibited extracellular

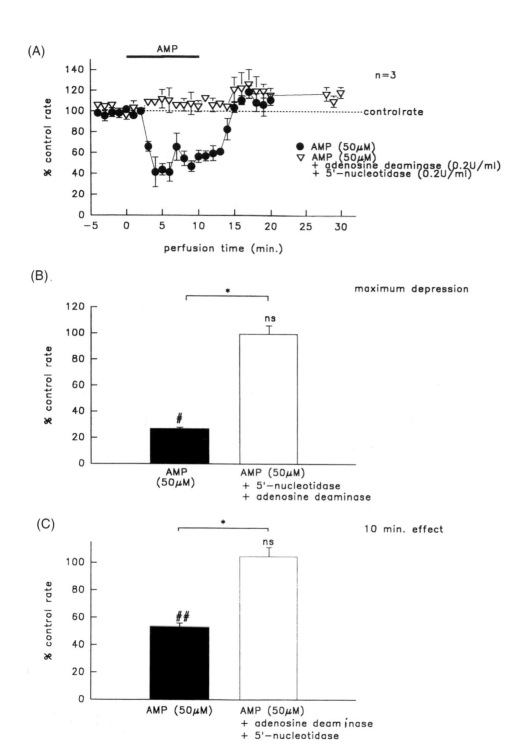

Fig. 4. The time course of the effect of AMP alone and with 5′-nucleotidase and adenosine deaminase is shown in (A). The maximum extent of inhibition and the 10 min. effect are analysed in (B) and (C) respectively. The maximum effect results are calculated as the mean ± SEM of the maximum effect in individual slices, not necessarily at the same time point. Hence the maximum and terminal values differ slightly between the time course graph and the histograms. # $P < 0.05$, ## $P < 0.01$ relative to control levels, * $P < 0.05$, for difference between the columns. (Reproduced with permission from Ross et al. (1998b)).

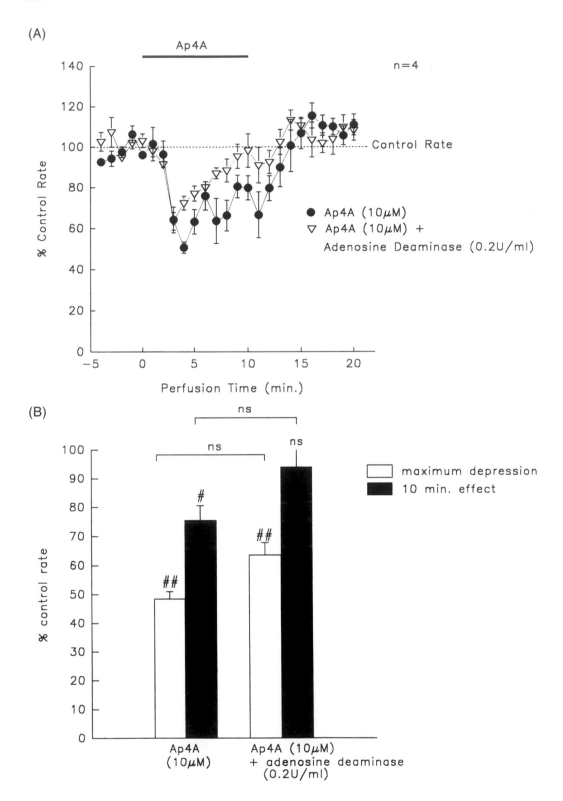

postsynaptic field potentials and intracellular post-synaptic currents. In isolated hippocampal CA3 neurones the application of AP_5A at 5 μM enhanced the amount of inward current through calcium channels in a rapid and reversible manner (Panchenko et al., 1996). The use of the antagonist ω-conotoxin GVIA suggested the involvement of N-type calcium channels. In rat midbrain synaptosomes AP_4A and AP_5A induced calcium entry in a voltage-independent process, and AP_5A potentiated calcium movement through N-type channels (Panchenko et al., 1996), although in the former case known calcium antagonists were unable to block the effect.

The nucleotides formed following the action of hydrolases on AP_nA can be metabolised by ecto-nucleotidases and 5′-nucleotidase producing adenosine. In agreement with earlier reports, adenosine reduced the frequency of discharge rate in this model of epileptiform activity. Adenosine deaminase (0.2 U ml^{-1}) totally prevented the effect of adenosine as well as AP_4A and AP_5A. At low concentrations AP_nA may increase the cellular secretion or release of adenosine which could then depress epileptiform activity. Adenosine deaminase did not block the peak responses to higher concentration of 10 μM AP_4A and AP_5A. However, if the extent of the inhibition was analysed as the mean change in rate at the end of a 10 min perfusion then adenosine deaminase did reduce the response to AP_4A and AP_5A. The results imply that AP_4A and AP_5A have direct inhibitory effects, but that these may decline, perhaps due to desensitisation. This early direct action of the dinucleotides is then continued by the formation or release of adenosine.

5′-adenylic acid deaminase (AMPase) removes an amino group from AMP to form IMP which is inactive. AMPase did not change the activity of AP_4A, suggesting that AMP was not responsible for this effect. A similar conclusion was reached by Gu and Geiger (1994) in rat cerebellum.

Diadenosine polyphosphates are stored in brain synaptosomes (Pintor et al., 1992) and this location, their release and their potent actions on peripheral as well as central excitable tissue are consistent with a possible neurotransmitter function for adenine dinucleotides.

Much debate has existed surrounding the receptors activated by adenine dinucleotides since they were first shown to reproduce the contractile effects of adenine nucleotides on a variety of smooth muscle preparations (Stone, 1981). Inhibition by the P2X receptor antagonist pyridoxal-6-azophe-nyl-2,4-disulphonic acid (PPADS), and desensitisation by α,β-methyleneATP suggest the involvement of receptors of the P2X subclass (Hoyle et al., 1989; Tepel et al;., 1995). Displacement studies and competition studies using the displacement of [^{35}S]ADP-β-S by diadenosine polyphosphates in rat brain synaptic terminals found a binding profile which suggested a novel receptor designated as P2D. Functional studies in rat brain synaptosomes have also shown that ATP and AP_4A display different response profiles with regard to antagonism and cross desensitisation, thereby supporting the existence of a P2D receptor (Pintor and Miras-Portugal, 1995; Pivorun and Nordone 1996) now reclassified as a P2Y receptor (Fredholm et al., 1997).

The depression of hippocampal field potentials and excitatory postsynaptic currents by AP_5A was inhibited by the A1 receptor antagonist 8-cyclo-pentyl-1,3-dimethylxanthine (CPT) (Klishin et al., 1994) suggesting that adenine dinucleotides can directly activate P1 receptors. Similarly theophylline, a P1 receptor antagonist, reduced the inhibition of cortical neurones by dinucleotides

Fig. 5. (a) Graph of the frequency of generation of spontaneous burst discharges with time. Perfusion of the slices with AP_4A (10 μM) during the period indicated by the bar above the graph depressed discharge rate. Superfusion with a combination of AP_4A and adenosine deaminase (0.2 U ml^{-1}) did not change the maximum depression by the dinucleotide. (b) Histogram indicating the comparison between the effect of AP_4A alone and with adenosine deaminase measured either as the maximum depression of discharge rate or as the depression seen after 10 min of perfusion. In neither cases was there any significant difference between the effect of AP_4A alone and with adenosine deaminase, although AP_4A no longer produced a depression of discharge rate after 10 min. Each point represents the mean ± s.e.mean for $n = 4$ slices. # $P < 0.05$, ## $P < 0.01$ relative to the control rate in the absence of any added agent. (Reproduced with permission from Ross et al. (1998c)).

(Stone and Perkins, 1981) and CPT prevented the depression of epileptiform activity by AP₄A. Overall, therefore, the data indicate that adenine dinucleotides may be able to activate P1 receptors directly.

Summary and conclusions

Slices of rat hippocampus can be induced to generate spontaneous interictal-like bursts of action potentials when perfused with a medium containing

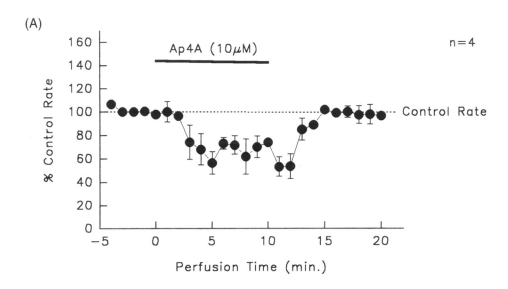

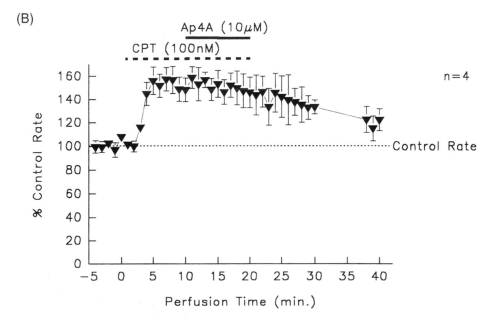

Fig. 6. The effect of CPT on the depression of activity produced by AP₄A. The control response to AP₄A is seen in (a), but this was prevented by superfusion with CPT. CPT was perfused for 10 min. prior to the addition of AP₄A (b). Each point represents the mean ± SEM for $n = 4$ slices. (Reproduced with permission from Ross et al. (1998c)).

no added magnesium and 4-aminopyridine (4AP). The frequency of these bursts is depressed by adenosine 5'-triphosphate (ATP) and this effect can be prevented by cyclopentyltheophylline but not by adenosine deaminase. AMP (50 μM) had a similar action to reduce discharge rate. At 10 μM, adenosine, diadenosine tetraphosphate and diadenosine pentaphosphate all decreased the burst frequency. Adenosine deaminase (0.2 U ml^{-1}) totally annulled the inhibition of epileptiform activity produced by 10 μM adenosine but reduced only the later components of the inhibition by 10 μM diadenosine tetraphosphate and diadenosine pentaphosphate. Cyclopentyltheophylline prevented the depression of burst discharges by diadenosine tetraphosphate. 5'-adenylic acid deaminase (AMP deaminase, AMPase) did not significantly alter the discharge rate over the 10 min superfusion period used for drug application but did prevent the depressant effect of AMP and ATP. AMP deaminase did not prevent the inhibitory effects of diadenosine tetraphosphate. The results suggests that in the CA3 region of the hippocampus, diadenosine tetraphosphate and diadenosine pentaphosphate act partly by stimulating xanthine sensitive receptors directly and partly via metabolism to adenosine, and that AMP may be responsible for the inhibitory effects of ATP on epileptiform activity.

Acknowledgement

This work was supported by the Epilepsy Research Association of Scotland.

References

Balachandran, C. and Bennett, M.R. (1996) ATP-activated cationic and anionic conductances in cultured rat hippocampal neurons. *Neurosci. Lett.*, 204: 73–76.

Barajas-Lopez, C., Muller, M.J., Prietogomez, B. and Espinosaluna, R. (1995) ATP inhibits the synaptic release of acetylcholine in submucosal neurons. *J. Pharmacol. Exp. Ther.*, 3: 1238–1245.

Bo, H., Altschuld, R.A. and Hohl, C.M. (1993) Adenosine stimulation of AMP deaminase activity in adult-rat cardiac myocytes. *Am. J. Physiol.*, 264: C48–C35.

Chen, Z.P., Levy, A. and Lightman, S.L. (1994) Activation of specific ATP receptors induces a rapid increase in intracellular calcium ions in rat hypothalamic neurons. *Brain Res.*, 641: 249–256.

Côte, S., van Sande, J. and Boeynaems, J.M. (1993) Enhancement of endothelial camp accumulation by adenine nucleotides: Role of methylxanthine-sensitive sites. *Am. J. Physiol.*, 264: H1498–H1503.

Cummins, J. and Hyden, H. (1962) Adenosine triphosphate levels and adenosine triphosphatases in neurons, glia and neuronal membranes of the vestibular nucleus. *Biochim. Biophy. Acta.*, 60: 271–283.

Cunha, R.A., Sebastião, A.M. and Ribeiro, J.A. (1992) Ecto-5'-nucleotidase is associated with cholinergic nerve terminals in the hippocampus but not in the cerebral cortex. *J. Neurochem.*, 59: 657–666.

Cunha, R.A., Ribeiro, J.A. and Sebastiao, A.M. (1994) Purinergic modulation of the evoked release of (3H)acetylcholine from the hippocampus and cerebral cortex of the rat: Role of the ectonucleotidases. *Eur.J. Pharmacol.*, 6: 33–42.

Dave, S. and Mogul, D.J. (1996) ATP receptor activation potentiates a voltage-dependent Ca channel in hippocampal neurons. *Brain Res.*, 715: 208–216.

Edwards, F.A., Gibb, A.J. and Colquhoun, D. (1992) ATP receptor mediated synaptic currents in the central nervous system. *Nature*, 359: 144–147.

Fredholm, B.B., Abbracchio, M.P., Burnstock, G., Dubyak, G.R., Harden, T.K., Jacobson, K.A., Schwabe, U. and Williams, M. (1997) Towards a new nomenclature for P1 and P2 receptors. *Trends in Pharmacol. Sci.*, 18: 79–82.

Frohlich, R., Boehm, S. and Illes, P. (1996) Pharmacological characterisation of P2 purinoceptor types in rat locus coeruleus neurons. *Eur. J. Pharmacol.*, 315: 255–261.

Griese, M., Gobran, L.I. and Rooney, S.A. (1991) A2 and P2 purine receptor interactions and surfactant secretion in primary cultures of type II cells. *Am. J. Physiol.*, 261: L140–L147.

Gu, J.G. and Geiger, J.D. (1994) Effects of diadenosine polyphosphates on sodium nitroprusside-induced soluble guanylate cyclase activity in rat cerebellum. *Neurosci. Lett.*, 169: 185–187.

Hoyle, C., Hoyle, C.H.V., Chapple, C. and Burnstock, G. (1989) Isolated human bladder: evidence for an adenine dinucleotide acting on P_{2x} purinoceptors and for purinergic transmission. *Eur. J. Pharmacol.*, 174: 115–118.

Ikeuchi, Y. and Nishizaki, T. (1996) P2 purinoceptor-operated potassium channel in rat cerebellar neurons. *Biochem. Biophys. Res. Comm.*, 218: 67–71.

Inoue, K., Nakazawa, K., Fujimori, K., Watano, T. and Takanaka, A. (1992) Extracellular adenosine 5'-triphosphate-evoked glutamate release in cultured hippocampal neurons. *Neurosci. Lett.*, 134: 215–218.

Inoue, K., Koizumi, S. and Nakazawa, K. (1995) Glutamate-evoked release of adenosine 5'-triphosphate causing an increase in intracellular calcium in hippocampal neurons. *Neuroreport*, 6: 437–440.

Kennedy, C., Westfall, T.D. and Sneddon, P. (1996) Modulation of purinergic neurotransmission by ecto-ATPase. *Seminars in the Neurosciences*, 8: 195–199.

King, B.F., Pintor, J., Wang, S., Ziganshin, A.U., Ziganshina, L.E. and Burnstock, G. (1996) A novel P1 purinoceptor activates an outward K+ current in follicular oocytes of Xenopus laevis. *J. Pharmacol. Exp. Ther.*, 276: 93–100.

Klishin, A., Lozovaya, N., Pintor, J., Miras-Portugal, M.T. and Krishtal, O. (1994) Possible functional role of diadenosine polyphosphates: Negative feedback for excitation in hippocampus. *Neuroscience*, 58: 235–236.

Kurz, K., Von Kugelgen, I and Starke, K. (1993) Prejunctional modulation of noradrenaline release in mouse and rat vas deferens: Contribution of P1 and P2-purinoceptors. *Br. J. Pharmacol.*, 110: 1465–1472.

Lee, K.S., Schubert, P., Reddington, M. and Kreutzberg, G.W. (1986) The distribution of adenosine A1 receptors and 5'-nucleotidase in the hippocampal formation of several mammalian species. *J. Comparative Neurology*, 246: 427–434.

Lin, S.C. and Way, L. (1982) A high affinity Ca^{2+}-ATPase in enriched nerve-ending plasma membranes. *Brain Res.*, 235: 387–392.

Marchenko, S.M., Obukhov, A.G., Volkova, T.M. and Tarussova, N.B. (1988) Bis(adenosyl-5')tetraphosphate as a partial agonist of ATP receptors in rat sensory neurons. *Neirofiziologiya*, 20: 427–431.

Nagy, A.K., Shuster, T.A. and Delgado-Escueta, A.V. (1986) Ecto-ATPase of mammalian synaptosomes: identification and enzymatic characterisation. *J. Neurochem.*, 47: 976–986.

Nagy, A.K., Shuster, T.A. and Delgado-Escueta, A.V. (1989) Rat brain synaptosomal ATP : AMP-phosphotransferase activity. *J. Neurochem.*, 53: 1166–1172.

Nieber, K., Poelchen, W. and Illes, P. (1997) Role of ATP in fast excitatory synaptic potentials in locus coeruleus of the rat. *Br. J. Pharmacol.*, 122: 423–430.

Panchenko V.A., Pintor J., Tsyndrenko, A.YA., Miras-Portugal, M.T. and Krishtal, O.A. (1996) Diadenosine polyphosphates selectively potentiate N-type Ca^{2+} channels in rat central neurons. *Neurosci.*, 70: 353–360.

Pintor, J., Díaz-Ray, A., Torres, M. and Miras-Portugal, M.T. (1992) Presence of diadenosine polyphosphates-Ap_4A and Ap_5A-in rat brain synaptic terminals. Ca^{2+} dependent release evoked by 4-aminopyridine and veratridine. *Neurosci. Lett.*, 136: 141–144.

Pintor, J. and Miras-Portugal, M.T. (1995) A novel receptor for diadenosine polyphosphates coupled to calcium increase in rat midbrain synaptosomes. *Br. J. Pharmacol.*, 115: 895–902.

Pivorun, E.B. and Nordone, A. (1996) Brain synaptosomes display a diadenosine tetraphosphate (Ap_4A)-mediated Ca^{2+} influx distinct from ATP-mediated influx. *J. Neurosci. Res.*, 44: 478–489.

Richardson, P.J. and Brown, S.J. (1987) ATP release from affinity-purified rat cholinergic nerve terminals. *J. Neurochem.*, 48: 622–630.

Ross, F.M., Brodie, M.J. and Stone, T.W. (1998a) Modulation by adenine nucleotides of epileptiform activity in the CA3 region of rat hippocampal slices. *Br. J. Pharmacol.*, 123: 71–80.

Ross, F.M., Brodie, M.J. and Stone, T.W. (1998b) AMP as a mediator of ATP effects at P1 purinoceptors. *Br. J. Pharmacol.*, 124: 818–824.

Ross, F.M., Brodie, M.J. and Stone, T.W. (1998c) The effects of adenine dinucleotides on epileptiform activity in the CA3 region of rat hippocampal slices. *Neuroscience*, 85: 217–228.

Shen, K.Z. and North, R. A. (1993) Excitation of rat locus coeruleus neurons by ATP-ionic mechanisms and receptor characterisation. *J. Neurosci.*, 13: 894–899.

Stone, T.W. (1981) Actions of adenine dinucleotides on the vas deferens, guinea-pig taenia caeci and bladder. *Eur. J. Pharmacol.*, 75: 93–102.

Stone, T.W. and Perkins, M.N. (1981) Adenine dinucleotide effects on rat cortical neurones. *Brain Res.*, 229: 241–245.

Stone, T. W. and Simmonds, H.A. (1991) *Purines: Basic and Clinical Aspects*. Kluwer Academic Press, Dordrecht.

Tepel, M., Bachmann, J., Schlüter, H. and Zidek, W. (1995) Diadenosine polyphosphate-induced increase in cytosolic free calcium in vascular smooth muscle cells. *J. Hypertension*, 13: 1686–1688.

Terrian, D.M., Herandez, P.G., Rea, M.A. and Peters, R.I. (1989) ATP release, adenosine formation, and modulation of dynorphin and glutamic acid release by adenosine analogues in rat hippocampal mossy fiber synaptosomes. *J. Neurochem.*, 53: 1390–1399.

Ueno, S., Harata, N., Inoue, K. and Akaike, N. (1992) ATP-gated current in dissociated rat nucleus solitarii neurons. *J. Neurophysiol.*, 68: 778–785.

von Kügelgen, I., Späth, L. and Starke, K. (1992) Stable adenine nucleotides inhibit [^{3}H]-noradrenaline release in rabbit brain cortex slices by direct action at presynaptic adenosine A1-receptors. *Naunyn-Schmied. Arch. Pharmacol.*, 346: 187–196.

Zimmermann, H. (1996) Biochemistry, localisation and functional roles of ecto-nucleotidases in the nervous system. *Progr. Neurobiol.*, 49: 589–618.

Physiology of
nucleoside function

P. Illes and H. Zimmermann (Eds.)
Progress in Brain Research, Vol 120
© 1999 Elsevier Science BV. All rights reserved

CHAPTER 21

The function of A_2 adenosine receptors in the mammalian brain: Evidence for inhibition vs. enhancement of voltage gated calcium channels and neurotransmitter release

Frances A. Edwards* and Susan J. Robertson

Department of Physiology, University College London, Gower St., London WC1E 6BT, UK

Introduction

Adenosine is best known as an inhibitory neurotransmitter in the central nervous system (CNS). The specific adenosine receptors (P_1 receptors) fall into various classes (A_1, $A_{2(A \text{ and } B)}$ and A_3) with different functions. The inhibitory actions are most commonly mediated by the widespread A_1 receptor which, amongst other mechanisms, act via G_i to decrease cyclic AMP levels and to decrease activation of Ca^{2+} channels (e.g. Dolphin et al., 1986; Fredholm and Dunwiddie, 1988; for reviews Dunwiddie and Fredholm, 1989; Stiles, 1992). In addition to the A_1 receptor, A_2 receptors are widespread in the brain. Of these the A_{2B} receptor is more widely distributed, being evenly spread throughout almost all brain tissues (for review see Feoktistov and Biaggioni, 1997). In contrast, the A_{2A} receptor was thought to have a very narrow distribution in the brain, mostly in the basal ganglia, but more recent studies, using high resolution techniques, have revealed a wider distribution in hippocampus, cortex, nucleus tractus solitarius and elsewhere (Cunha et al., 1994a; Sebastiao and Ribeiro, 1996; Ongini et al., 1997). These receptors have generally been considered to mediate excitatory effects in the brain partly

because they have the opposite effect to A_1 receptors in that they act via G_s to increase cyclic AMP levels. Another class of adenosine receptor, A_3, is much more sparsely distributed in the central nervous system. This receptor is methylxanthine insensitive and hence is clearly distinguishable from the other P_1 receptors. It will not be considered here. (For review of adenosine receptor nomenclature see Fredholm et al., 1994.)

Finally a receptor known as the P_3 receptor has been described (Shinozuka et al., 1988) which, unlike the other adenosine receptors, has been reported to be activated by ATP as well as adenosine. Unfortunately the effects of the various pharmacological agents which are active at the various P_1 receptors have largely not been tested against the reported P_3 receptors. Hence, unless activation by ATP has clearly been tested (which is usually not the case), the possibility of a contribution from the P_3 receptor cannot be completely ruled out (including Robertson and Edwards, 1998). However as this, so far uncloned receptor, has not yet been reported to occur in the brain, we will not consider it further here and will assume that identification of P_1 receptors stands.

Recently we reported that neurotransmission mediated by ATP via P2X receptors in synapses in the medial habenula nucleus of 3-week-old rats was inhibited by specific activation of adenosine A_2 receptors (Robertson and Edwards, 1998). This

---

*Corresponding author. Tel.: +44-171-4193286; fax: +44-171-8132807; e-mail: f.a.edwards@ucl.ac.uk

direct demonstration of inhibition of transmitter release via A_2 receptors has caused some surprise in purinergic circles as the perception of the previous literature is that activation of A_2 receptors enhances, rather than inhibits, transmitter release.

In this paper we look at the results we have reported, in the context of previous reports on the role of A_2 receptors in synaptic transmission in the mammalian brain. We come to the conclusion that there are very few cases in which A_2 receptors have been shown to be excitatory in the sense of directly increasing the release of a specific CNS neurotransmitter. Although in several previous studies the overall effect of activation of A_2 receptors has been to increase network excitability, most of these results would be compatible with an indirect effect via inhibition of γ-aminobutyric acid (GABA) release.

The two systems which have been directly studied most clearly in this context have been voltage gated Ca^{2+} channels and GABAergic transmission. Activation of A_2 receptors has been shown to result in inhibition of both calcium channels and GABAergic transmission. Presumably localisation of A_2 receptors on GABAergic terminals results in inhibition of calcium channels and hence inhibition of GABA release. Note that depending on the circuitry, this could result overall in increased excitability in an intact network.

A_2 receptor cascade: stimulates an increase in cAMP

It seems clear from biochemical studies that stimulation of A_2 adenosine receptors is coupled to G-proteins and can result in stimulation of G_s and hence activation of adenylate cyclase resulting in an increase in cyclic AMP (cAMP) formation (for review see Sebastiao and Ribeiro, 1996; Fredholm et al., 1994; Ongini and Fredholm, 1996). More recently it has been suggested that in addition to this pathway, stimulation of A_{2B} receptors in HEK 293 cells can lead to stimulation of G_q and via this pathway activation of phospholipase C and hence intracellular calcium mobilisation. Note that there has so far been no evidence that A_{2A} receptors can activate this pathway. In addition A_{2B} receptors

have been suggested to couple directly to calcium channels via G_s (for review see Feoktistov and Biaggioni, 1997).

A_{2A} receptor activation results in inhibition of calcium channels

Several recent reports have linked A_2 receptor activation to inhibition of voltage gated calcium channels. For example, Chen and Van den Pol (1997) reported that both specific agonists of A_1 and specific agonists of A_2 receptors resulted in a decrease in calcium channel current amplitude without a change in voltage dependence (charge carried by barium) in cultured GABAergic neurones from the suprachiasmic and arcuate nucleus. They also confirmed that these effects were specifically inhibited by the appropriate antagonists.

Recently in another interesting paper, Park et al. (1998) showed that A_2 receptor activation results in inhibition of both L- and N-type calcium channels in PC12 cells via increase in cAMP and activation of protein kinase A. Also of interest is an earlier report from the same group which showed an inhibition of calcium influx through P2X receptors (Park et al., 1997) which we discuss further below in relation to our data on inhibition of ATP release. Other recent reports also suggested inhibition of calcium channel activity. For example, in a subpopulation of striatal medium spiny neurones, Svenningsson et al. (1998) report phosphorylation of DARPP-32, a protein which when phosphorylated is a potent and selective inhibitor of protein phosphatase-1 and part of a signalling cascade which would be expected to result in inhibition of voltage dependent calcium channels (Surmeier et al., 1995).

Interestingly, in the hippocampus, several studies have also reported activation of P-type calcium channels (but not N- or L-type) by A_2 receptor activation. (Note that though they may have some influence, P-type channels are not the primary channels on which evoked neurotransmitter release is usually dependent.) In a study using freshly dissociated hippocampal neurones from the CA3 region of the guinea pig, Mogul et al. (1993) showed enhancement of ω-agatoxin sensitive calcium channels (but not ω-conotoxin sensitive channels) by receptors A_{2B} receptor activation.

Two other studies were performed on hippocampal synaptosomes. A similar effect was reported by Gonçalves et al. (1997) who demonstrated increases in $[^{45}Ca^{2+}]$ uptake due to activation of A_2 receptor which were inhibited by ω-agatoxin in the CA3 but not the CA1 region of the hippocampus. This may indicate that the effect of A_2 receptor activation is different on different types of voltage gated channels which could allow differential effects depending on the channel most important in neurotransmitter release in a particular terminal. Note, however, that in such synaptosomal studies of mixed synaptic terminals, various interactions may occur which lead to indirect effects and so the interpretation must be treated with caution (see below).

It would not be surprising, in the light of: (1) the relatively recent finding of many new neuronal cells expressing A_{2A} receptors; (2) the improving pharmacology in the field; and (3) the lack of reports on biochemical pathways involved in A_2 receptor activation in the CNS, if other G-protein coupled pathways were also in the future discovered linked to this receptor.

A_2 effects measured on single identified synapses

Inhibition of GABA release

Two groups have studied the effects of specific A_2 adenosine receptor agonists on GABAergic inhibitory postsynaptic currents. Both observed similar results.

Mori et al. (1996) (Fig. 1), recorded evoked and spontaneous miniature ipscs in striatal medium spiny neurones. They clearly showed that activation of A_{2A} receptors by bath application of the specific agonist CGS 21680 caused an inhibition of GABA release resulting in a 30% decrease in evoked transmission. They confirmed that the effect was presynaptic by demonstrating a decrease in the frequency but not the amplitude of miniature currents as well as no change in the whole cell

response to applied GABA. It is of particular interest to note that in this study, it was demonstrated that the inhibition of transmitter release was mimicked by application of the membrane permeable cAMP analogues 8-Bromo-cAMP and dibutyryl cAMP and hence was consistent with A_{2A} receptors acting via its established pathway of increasing cAMP via activation of adenylate cyclase .

GABA release was also clearly shown to be inhibited by both A_1 and A_2 receptors in neurones cultured from the suprachiasmic and arcuate nuclei. This study used patch clamp of neurones which were in some cases synaptically connected in networks and in others were single GABAergic grown on microislands which formed autapses onto themselves (Chen and Van den Pol, 1997). Like Mori et al. (1996), they demonstrated a decrease in evoked GABAergic transmission as well as a decrease in miniature frequency without a change in miniature distribution. As outlined above, in the same cells they also demonstrated a decrease in calcium currents in response to both A_1 and A_2 receptor stimulation. Thus they concluded that activation of both A_1 and A_2 receptors decreased release of GABA by decreasing the influx of calcium presynaptically.

Inhibition of ATP release

Recently, using monosynaptic stimulation of purinergic fibres in the medial habenula nucleus and recording responses in individual cells using patch clamp techniques, we found that synaptic release of ATP could be modulated via a variety of receptors (Robertson and Edwards, 1998) (Fig. 2). In particular, we showed that bath application of specific agonists for either A_1 or A_2 receptors leads to an inhibition of ATP release. Moreover these effects were respectively inhibited by lower or higher concentrations of 8-cyclopentyltheophylline (1 vs. 10 μM, 8CPT), as would be predicted from the K_is for this antagonist for A_1 vs. A_2 receptors. It was of interest to note that endogenous adenosine, (effective at low stimulation frequencies (0.5–1 Hz) but increasingly effective at higher stimulation frequencies (2–10 Hz)), only activated A_2 receptors at

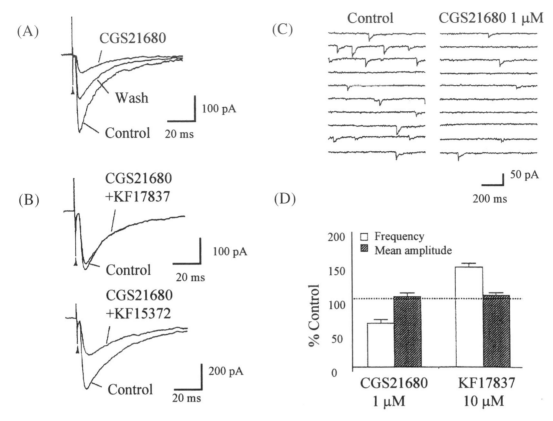

Fig. 1. Activation of A_{2A} receptors inhibits GABAergic synaptic transmission using patch clamp recording from striatal medium spiny neurons in striatal slices (Mori et al. 1996). (A) Average of 15 consecutive evoked inhibitory postsynaptic currents (ipscs) show that ipscs are reversibly inhibited by the specific A_{2A} receptor agonist CGS 21680 (1 μM). (B) Average of 12 consecutive ipscs (in each panel) show that the A_2 adenosine receptor antagonist KF17837 (0.1 μM) completely inhibits the effect of A_2 receptor activation (upper panel) while the A_1 receptor antagonist KF15372 has no effect (lower panel). (C) Raw data traces showing miniature ipscs recorded in tetrodotoxin (1 μM) with (right panel) and without (left panel) CGS 21680. (D) Pooled data from six cells showing that the frequency of ipscs decreased as a result of A_2 receptor activation without a change in amplitude. Note that alone the A_2 receptor antagonist KF17837 caused an increase in frequency, again without a change in amplitude, suggesting background activation of the A_2 receptor by endogenous adenosine. (All recordings were made in the presence of antagonists of AMPA, NMDA and muscarinic acetylcholine receptors.) (Adapted from Mori et al. (1996)).

these synapses and despite their higher affinity for adenosine did not reach the A_1 receptors. These studies were carried out in the presence of bicuculline to block $GABA_A$ receptors as well as CNQX and 7-chlorokynurenate to block AMPA and NMDA glutamate receptors respectively.

In contrast to the effects on ATP release, glutamate release (onto the same cell-type, in the presence of bicuculline), was inhibited by background or exogenously applied adenosine, acting on A_1 receptors. Glutamatergic synapses were not measurably affected by activation of A_2 receptors. Hence in the medial habenula nucleus both A_1 and

A_2 receptors mediate inhibition of transmitter release but A_2 receptors are only effective at purinergic and not glutamatergic synapses.

Note that we have not yet determined the subtype of A_2 receptor involved in inhibition of ATP-mediated transmission. However, it is interesting to note that there are a variety of mechanisms amongst the studies outlined above which could mediate this effect. In particular, release would be inhibited by inhibition of voltage gated calcium channel. Moreover the effects reported on inter-action of A_2 receptors with the ATP-gated cation channels themselves (Chen and Van den Pol, 1997)

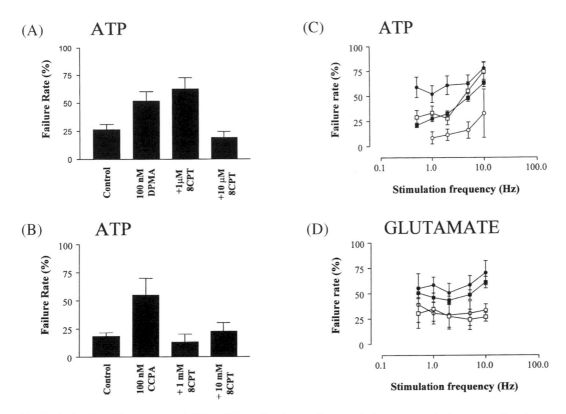

Fig. 2. Activation of A_{2A} receptors inhibits ATP-mediated synaptic transmission using patch clamp recording from neurones in slices of medial habenula nucleus (Robertson and Edwards, 1998). All data refer to the percentage failure to release of evoked ATP currents, recorded in the presence of antagonists of $GABA_A$ receptors, AMPA receptors and NMDA receptors. (A) Inclusion of DPMA in the recording solution increases the percentage failure rate. This effect is inhibited by 8CPT but only at high enough concentrations to inhibit A_2 receptors. 1 μM 8CPT (which would be expected to be specific to A_1 receptors) had no effect. (B) The A_1 agonist CCPA also increases failure rate of ATP release and the effect is completely inhibited by 8CPT (1 μM), indicating that the drug effectively penetrates the slice. (C) Effects of background adenosine on ATP release is mediated by A_2 receptors. Under control conditions, as the stimulation frequency is increased from 0.5–10 Hz, the failure rate increases (solid squares). The failure rate is increased by the inclusion of DPMA in the solution (closed circles) and this effect occludes with the effect of increasing stimulation frequency. Inclusion of 8CPT (1 μM) alone has no effect at any stimulation frequency (open squares), while increasing 8CPT to 10 μM (open circles) where it would be expected to block A_2 receptors results in a decrease in failure rate throughout the frequency range and statistically no significant effect of frequency remains. Note that even at low frequencies, the high concentration of 8CPT has a substantial effect on decreasing failure rate, indicating considerable activation of A_2 receptors by background adenosine. The lack of effect of the low concentration indicates that background adenosine does not seem to reach A_1 receptors, despite their higher affinity for adenosine. (D) Effects of background adenosine on glutamate release is mediated by A_1 receptors. The effects of stimulation frequency and adenosine agonists and antagonists on glutamatergic synaptic currents in the same cell-type. Control currents are not significantly affected by changing stimulation frequency (solid squares). Inclusion of DPMA (closed circles) has no significant effect on failure rate of glutamatergic currents. 8CPT alone (open symbols; squares 1 μM; circles 10 μM) decreases failure rate throughout the frequency range at both concentrations. This indicates background activation by endogenous adenosine acting via an A_1 receptor while A_2 receptors appear not to be present on these terminals. All differences noted were tested by one or two factor Analysis of Variance (as appropriate) and were at least significant at $P < 0.05$. (Adapted from Robertson and Edwards (1998) (see for details of statistics)).

is also of interest. An inhibition of ATP-gated channels mediated by A_2 receptors could contribute to pre- or postsynaptic changes in transmission at this synapse.

Finally we can also not completely rule out an interaction between ATP release and inhibition of GABA release as we have not yet investigated the effect of blocking $GABA_B$ receptors. However, this would be expected to give the opposite effect to that seen here and hence is unlikely to be the cause of A_2 receptor mediated inhibition of ATP-mediated transmission.

Excitation of glycine release

One study has investigated the effects of A_2 receptor activation on glycine release in the rat brainstem (Umemiya and Berger, 1994). In contrast to ATP and GABA, this study showed an increase of evoked transmission and a decrease in failure rate in response to application of the A_2 specific agonist DPMA or by applying adenosine in the presence of 8CPT (1 μM) (at which concentration it would be expected to be specific in antagonising A_1 adenosine receptors). The study also investigates the effects of A_1 vs. A_2 receptor activation on a range of different calcium channels and, as outlined above, reported that activation of A_2 receptors specifically results in an increase in P-type calcium channels while A_1 receptors were particularly effective in inhibiting N-type channels.

Thus the increase in glycinergic transmission seen in this study may indicate that the glycine synapses studied are particularly sensitive to calcium influx through P-type calcium channels, whereas the GABAergic synapses described above may be more strongly dependent on N-type channels.

Extracellullar recording and release studies

No conclusions about release of specific transmitters can be made without isolating a single type of synapse from its network

The general perception of A_2-receptors as causing excitation and increased transmitter release comes from the observation reported by various groups that activation of A_2 receptors increases excitation as measured by the size of the population spike, slope of the field epsp or changes in the release of a variety of neurotransmitters in the presence of A_2 agonists (e.g. Cunha et al., 1994b; Barraco et al., 1996; Cunha et al., 1996; Phillis and O'Leary, 1997; for review see Sebastiao and Ribeiro, 1996).

These studies have usually been performed in vitro, in slices or other preparations in which networks of neurones remain connected. On the other hand, some in vivo studies have reported that activation of A_2 receptors results in inhibition of firing (e.g. Lin and Phillis, 1991; Thomas and Spyer, 1999). Such network studies give considerable information on the overall picture of whether activation of a particular receptor type will be excitatory or inhibitory on the final outflow of the network tested. However, unless antagonists are included to simplify the possible interactions occurring, no conclusions can be reached on the specific question addressed here of whether A_2 receptors directly act on a particular type of terminal to cause increase or decrease of release of the transmitter(s) from that type of terminal.

For example, as soon as a network includes any feedback or feedforward GABAergic inhibition, where the GABA release is inhibited by activation of A_2 receptors, as observed in the studies outlined above, the expected result would be disinhibition and thus an increase in transmitter release from various other types of synapse (Fig. 3a). On the other hand, if there are multiple types of GABAergic neurones included in the network as is seen in the striatum (see Ferré et al., 1997, for review), depending on whether any or all of the subtypes feature A_2 receptors, the result could be an excitation or inhibition of release of any transmitter (Fig. 3b), including GABA itself. Such complexities are illustrated in the various studies of the interaction of A_2 receptor activation and D_2 dopamine receptor activation. For example, Mayfield et al. (1996) illustrate that a selective A_2 agonist can increase or decrease GABA release depending on the concentration at which it is applied and that at low concentrations the activation of A_2 receptors can prevent the inhibition of GABA released caused by activation of D_2 receptors. All these effects only occur if release is stimulated electrically (i.e. if the network is activated), while spontaneous release is not affected. This may be partly a problem of detection but it could equally be a question of dependence on interactions within the network which may not be seen in the case of spontaneous release. Similar complex interaction between these systems were recently illustrated and

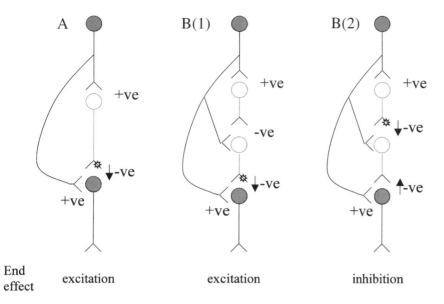

End effect   excitation   excitation   inhibition

Fig. 3. Schematic diagram representing the different end effects that A_2 receptor-mediated inhibition of GABA release could have under fairly simple network conditions. Line drawings of neurones represent glutamatergic excitatory neurones (unbroken lines) and GABAergic inhibitory neurones (broken lines). The star symbol on a synapse represents existence of an A_2 receptor. The label '+ve' represents excitation (i.e. glutamate as transmitter) and '−ve' represents inhibition. Arrows represent changes which will occur as a direct or indirect result of A_2 receptor activation causing inhibition of GABA release from the terminal on which it is situated. The end effect refers to excitation or inhibition of release of transmitter from the last neurone in the network shown. (A) If only one GABAergic neurone is connected in feed forward inhibition then A_2 receptor activation will result in disinhibition and hence an excitatory end effect. (B) If two inhibitory neurones are connected in series within a similar network then dependent on which has A_2 receptors the end effect will vary. Note these are extremely simple networks compared to those in existence in most networks studied.

reviewed by Fuxe et al. (1998). In their studies they used in vivo microdialysis and measured increases in GABA release in response to A_2 receptor activation, which acted synergistically with D_2 receptor inhibition. Again this is an end effect of a complex system and cannot be assumed to be a direct effect on GABAergic terminals.

While this problem causes the most profound ambiguity in slice or in vivo preparations, similar problems also arise in studies using synaptosomes. For example, if a preparation contains a variety of terminals, stimulated release will expose all terminals to all released transmitters. Hence if some terminals feature A_2 receptors, any terminal which has receptors for the transmitter released by the A_2 sensitive terminal, could in turn be affected by A_2 receptor agonists. Thus we will only include here studies where at least the known effects of GABAergic inhibition were excluded by inclusion of antagonists of $GABA_A$ and $GABA_B$ receptors. This still does not exclude other interactions

however, such as those known to occur with dopamine receptors in the striatum (Lepiku et al., 1997; Mayfield et al., 1996).

Despite recording from single cells, slice studies of synaptic potentials can also suffer from similar problems of interpretation. In a recent study Li and Henry (1998) used intracellular recording of epsps and reported that activation of A_2 receptors resulted in an increase in the amplitude of the synaptic potentials associated with an increase in input resistance of the membrane and a depolarisation. Note that because they were recording synaptic potentials (rather than currents under voltage clamp), the increase in input resistance could itself have caused an increase in the potential amplitude. The depolarisation, on the other hand, by decreasing the driving force, would have decreased the amplitude. Hence though it is clear that the end effect was to increase excitability, the authors point out that the effect was probably postsynaptic and in fact it would depend on the balance between change in input resistance and membrane potential. Though it was not the given interpretation, these results would again be completely compatible with a decreased level of background GABA in the slice, resulting in a decreased tonic activation of GABA$_A$ receptors.

Thus the results found in studies involving networks are often found to be contradictory, as clearly outlined in an interesting review by Latini et al. (1996). The disagreement between different studies presumably occurs because of slight difference in preparations, which allow somewhat different parts of the network or different subsets of the total neuronal population to remain intact under different conditions.

One release study in which the interpretation was clarified by use of a variety of antagonists, reported the effects of very low concentrations of the A$_{2A}$ agonist CGS 21680. They found that CGS 21680 inhibited K$^+$-evoked release of GABA and enhanced K$^+$-evoked release of acetylcholine from synaptic terminals isolated from synaptosomes (Kurokawa et al., 1994). They demonstrated that these effects were inhibited by a specific A$_2$ receptor antagonist but not affected by bicuculline, saclofen, naloxone, mecamylamine, atropine or DCPCX. Thus the effects were not indirect due to the changes in acetylcholine release affecting GABA release or vice versa, nor were they mediated by A$_1$-adenosine receptors nor naloxone sensitive opioid receptors. It is interesting to note that in this study where the system is relatively simplified, having no connections intact and many receptors being discounted, that again GABA release was shown to be inhibited.

Conclusion

Thus the literature to date seems to indicate that activation of A$_2$ receptors, usually acting through G$_s$ and accumulation of cAMP, results in down-regulation of N- and L-type calcium channels and up-regulation of P-type calcium channels. GABA release tends to be inhibited by activation of A$_2$ receptors which would be consistent with the inhibition of N- and L-type calcium channels and also with the frequent observation of increased excitability of neuronal networks. The observation

that glycinergic transmission was increased may reflect an unusual dominance of P-type calcium channels in this system, but further investigation is needed in this area. In terms of our own results showing inhibition of ATP-mediated synaptic currents, this seems to be completely compatible with the literature and could reflect inhibition of N-type calcium channels and/or interaction of A$_2$ receptors with P2X receptors.

References

Barraco, R.A., Clough-Helfman, C. and Anderson, G.F. (1996) Augmented release of serotonin adenosine A$_2$a receptor activation and desensitization by CGS 21680 in the rat nucleus tractus solitarius. *Brain Res.*, 733: 155–161.

Chen, G. and Van den Pol, A.N. (1997) Adenosine modulation of calcium currents and presynaptic inhibition of GABA release in suprachiasmatic and arcuate nucleus neurons. *J. Neurophysiol.*, 77: 3035–3047.

Cunha, R.A., Correia-De-Sá, P., Sebastiao, A.M. and Ribeiro, J.A. (1996) Preferential activation of excitatory adenosine receptors at rat hippocampal and neuromuscular synapses by adenosine formed from released adenine nucleotides. *Br. J. Pharmacol.*, 119: 253–260.

Cunha, R.A., Johansson, B., Ploeg, I., Sebastiao, A.M., Ribeiro, J.A. and Fredholm, B.B. (1994a) Evidence for functionally important adenosine A$_2$a receptors in the rat hippocampus. *Brain Res.*, 649: 208–216.

Cunha, R.A., Milusheva, E., Vizi, E.S., Ribeiro, J.A. and Sebastiao, A.M. (1994b) Excitatory and inhibitory effects of A$_1$ and A$_{2A}$ adenosine receptor activation on the electrically evoked [^{3}H]acetylcholine release from different areas of the rat hippocampus. *J. Neurochem.*, 63: 207–214.

Dolphin, A.C., Forda, S.R. and Scott, R.H. (1986) Calcium-dependent currents in cultured rat dorsal root ganglion neurones are inhibited by an adenosine analogue. *J. Physiol. (Lond.)*, 373: 47–61.

Dunwiddie, T.V. and Fredholm, B.B. (1989) Adenosine A 1 receptors inhibit adenylate cyclase activity and neurotransmitter release and hyperpolarize pyramidal neurons in rat hippocampus. *J. Pharmacol. Exp. Ther.*, 249: 31–37.

Feoktistov, I. and Biaggioni, I. (1997) Adenosine A$_{2B}$ receptors. *Pharmacol. Rev.*, 49: 381–402.

Ferré, S., Fredholm, B.B., Morelli, M., Popoli, P. and Fuxe, K. (1997) Adenosine-dopamine receptor–receptor interactions as an integrative mechanism in the basal ganglia. *Trends Neurosci.*, 20: 482–487.

Fredholm, B.B., Abbracchio, M.P., Burnstock, G., Daly, J.W., Harden, T.K., Jacobson, K.A., Leff, P. and Williams, M. (1994) Nomenclature and classification of purinoceptors. *Pharmacol. Rev.*, 46: 143–156.

Fredholm, B.B. and Dunwiddie, T.V. (1988) How does adenosine inhibit transmitter release? *Trends Pharmacol. Sci.*, 9: 130–134.

Fuxe, K., Ferré, S., Zoli, M. and Agnati, L.F. (1998) Integrated events in central dopamine transmission as analyzed at multiple levels. Evidence for intramembrane adenosine A_{2A} dopamine D_2 and adenosine A_1 dopamine D_1 receptor interactions in the basal ganglia. *Brain Res. Rev.*, 26: 258–273.

Gonçalves, M.L., Cunha, R.A. and Ribeiro, J.A. (1997) Adenosine A_{2A} receptors facilitate $^{45}Ca^{2+}$ uptake through class A calcium channels in rat hippocampal CA3 but not CA1 synaptosomes. *Neurosci. Lett.*, 238: 73–77.

Kurokawa, M., Kirk, I.P., Kirkpatrick, K.A., Kase, H. and Richardson, P.J. (1994) Inhibition by KF17837 of adenosine A_{2A} receptor-mediated modulation of striatal GABA and ACh release. *Br. J. Pharmacol.*, 113: 43–48.

Latini, S., Pazzaglia, M., Pepeu, G. and Pedata, F. (1996) A_2 Adenosine receptors: Their presence and neuromodulatatory role in the central nervous system. *Gen. Pharmacol.*, 27: 925–933.

Lepiku, M., Rinken, A., Järv, J. and Fuxe, K. (1997) Modulation of [^{3}H]quinpirole binding to dopaminergic receptors by adenosine A_{2A} receptors. *Neurosci. Lett.*, 239: 61–64.

Li, H. and Henry, J.L. (1998) Adenosine A_2 receptor mediation of pre- and postsynaptic excitatory effects of adenosine in rat hippocampus in vitro. *Eur. J. Pharmacol.*, 347: 173–182.

Lin, Y. and Phillis, J.W. (1991) Characterization of the depression of rat cerebral cortical neurons by selective adenosine agonists. *Brain Res.*, 540: 307–310.

Mayfield, R.D., Larson, G., Orona, R.A. and Zahniser, N.R. (1996) Opposing actions of adenosine A_{2A} and dopamine D_2 receptor activation on GABA release in the basal ganglia: Evidence for an A_{2A}/D_2 receptor interaction in globus pallidus. *Synapse*, 22: 132–138.

Mogul, D.J., Adams, M.E. and Fox, A.P. (1993) Differential activation of adenosine receptors decreases N-type but potentiates P-type Ca^{2+} current in hippocampal CA3 neurons. *Neuron*, 10: 327–334.

Mori, A., Shindou, T., Ichimura, M., Nonaka, H. and Kase, H. (1996) The role of adenosine A_{2a} receptors in regulating GABAergic synaptic transmission in striatal medium spiny neurons. *J. Neurosci.*, 16: 605–611.

Ongini, E., Adami, M., Ferri, C. and Bertorelli, R. (1997) Adenosine A_{2A} receptors and neuroprotection. *Ann. NY Acad. Sci.*, 825: 30–48.

Ongini, E. and Fredholm, B.B. (1996) Pharmacology of adenosine A_{2A} receptors. *Trends Pharmacol. Sci.*, 17: 364–372.

Park, T.J., Chung, S.K., Han, M.K., Kim, U.H. and Kim, K.T. (1998) Inhibition of voltage-sensitive calcium channels by the A_{2A} adenosine receptor in PC12 cells. *J. Neurochem.*, 71: 1251–1260.

Park, T.J., Song, S.K. and Kim, K.T. (1997) A_{2A} adenosine receptors inhibit ATP-Induced Ca^{2+} influx in PC12 cells by involving protein kinase A. *J. Neurochem.*, 68: 2177–2185.

Phillis, J.W. and O'Leary, D.S. (1997) Purines and the nucleus tractus solitarius: Effects on cardiovascular and respiratory function. *Clin. Exp. Pharmacol. Physiol.*, 24: 738–742.

Robertson, S.J. and Edwards, F.A. (1998) ATP and glutamate are released from separate neurones in the rat medial habenula nucleus: Frequency dependence and adenosine-mediated inhibition of release. *J. Physiol. (Lond.)*, 508: 691–701.

Sebastiao, A.M. and Ribeiro, J.A. (1996) Adenosine A_2 receptor-mediated excitatory actions on the nervous system. *Prog. Neurobiol.*, 48: 167–189.

Shinozuka, K., Bjur, R.A. and Westfall, D.P. (1988) Characterization of prejunctional purinoceptors on adrenergic nerves of the rat caudal artery. *Naunyn Schmiedebergs Arch. of Pharmacol.*, 338: 221–227.

Stiles, G.L. (1992) Adenosine receptors. *J. Biol. Chem.*, 267: 6451–6454.

Surmeier, D.J., Bargas, J., Hemmings, H.C. Jr., Nairn, A.C. and Greengard, P. (1995) Modulation of calcium currents by a D_1 dopaminergic protein kinase/phosphatase cascade in rat neostriatal neurons. *Neuron*, 14: 385–397.

Svenningsson, P., Lindskog, M., Rognoni, F., Fredholm, B.B., Greengard, P. and Fisone, G. (1998) Activation of adenosine A_{2A} and dopamine D1 receptors stimulates cyclic AMP dependent phosphorylation of DARPP-32 in distinct populations of striatal projection neurons. *Neurosci.*, 84: 223–228.

Thomas, T. and Spyer, K.M. (1999) A novel influence of adenosine on ongoing activity in rat rostral ventrolateral medulla. *Neurosci.*, 88(4): 1213–1223.

Umemiya, M. and Berger, A.J. (1994) Activation of adenosine A_1 and A_2 receptors differentially modulates calcium channels and glycinergic synaptic transmission in rat brainstem. *Neuron*, 13: 1439–1446.

P. Illes and H. Zimmermann (Eds.)
Progress in Brain Research, Vol 120

CHAPTER 22

An adenosine A$_3$ receptor-selective agonist does not modulate calcium-activated potassium currents in hippocampal CA1 pyramidal neurons

Thomas V. Dunwiddie[1,2,3,*], Kenneth A. Jacobson[4] and Lihong Diao[2]

[1]*Neuroscience Program, University of Colorado Health Sciences Center, Denver, CO, USA*
[2]*Department of Pharmacology, University of Colorado Health Sciences Center, Denver, CO, USA*
[3]*Veterans Administration Medical Research Service, Denver, CO, USA*
[4]*Molecular Recognition section, Laboratory of Bioorganic Chemistry, National Institute of Diabetes, Digestive and Kidney Diseases,
National Institutes of Health, Bethesda, MD 20892, USA*

Introduction

The most recently discovered adenosine receptor, the A$_3$ receptor, belongs to the general class of G-protein coupled receptors, and has been linked to several putative effector mechanisms. One distinguishing characteristic of this receptor is that many adenosine agonists have affinities for the A$_3$ receptor that are much lower than their corresponding affinities at the adenosine A$_1$ receptor (Zhou et al., 1992), and this is true as well for the endogenous ligand, adenosine. It has been reported that A$_3$ receptors are coupled in an inhibitory fashion to adenylyl cyclase (Zhou et al., 1992; Zhao et al., 1997), although in other systems, A$_3$ receptor activation activates phospholipase C and leads to an elevation in inositol phosphate levels (Ali et al., 1990; Ramkumar et al., 1994; Auchampach et al., 1997), and this occurs in brain as well (Abbracchio et al., 1995). If this is the case, then increases in intracellular Ca^{2+} and activation of protein kinase C should occur as a consequence of agonist occupation of the A$_3$ receptor. Recent studies in brain have suggested that A$_3$ receptors can inhibit the function of presynaptic modulatory receptors (Dunwiddie et al., 1997; Macek et al., 1998), and that this effect is mediated via PKC (Macek et al., 1998; Diao and Dunwiddie, unpublished observations). However, other reports have proposed that postsynaptic A$_3$ receptors in hippocampus may elicit responses that are linked to activation of PKA but not PKC (Fleming and Mogul, 1996), suggesting that a further consideration of effector pathways for this receptor is warranted. One response in hippocampal CA1 pyramidal neurons that is affected by increases in intracellular Ca^{2+}, as well as activation of either PKC or PKA, is the K$^+$-mediated afterhyperpolarization that follows activation of voltage dependent Ca^{2+} channels by a depolarizing stimulus. Previous studies have demonstrated that activation of β-adrenergic receptors will nearly abolish the AHP (Madison and Nicoll, 1982; Haas and Konnerth, 1983), an effect that can be mimicked directly by activation of PKA (Abdul-Ghani et al., 1996). Activation of PKC with phorbol esters will also inhibit this response (Malenka et al., 1986), and activation of either muscarinic receptors (Cole and Nicoll, 1984) or metabotropic glutamate receptors (Abdul-Ghani et al., 1996) reduce the AHP via increases in intracellular Ca^{2+} and subsequent activation of CaM-K. Furthermore,

---

*Corresponding author. Tel.: + 1 303-315-4222; fax: + 1 303-315-4814; e-mail: Tom.Dunwiddie@UCHSC.edu

adenosine itself has been previously reported to either increase (Haas and Greene, 1984) or decrease (Dunwiddie, 1985) this current. Thus, this response appeared to be a likely candidate that would be sensitive to activation of multiple second messenger systems, but an effect of A_3 receptor activation on this conductance has not been described.

The recent development of CI-IB-MECA, an agonist that is approximately 2500-fold selective for the A_3 vs. the A_1 receptor, and 1400-fold selective for the A_3 vs. the A_{2a} receptor (Jacobson et al., 1995), has facilitated the study of the physiological consequences of selective activation of this receptor. Behavioral studies have demonstrated an A_3 receptor-mediated depression of locomotor activity (Jacobson et al., 1993), but the effects of A_3 receptor activation on neuronal activity at the cellular level are only poorly understood. A recent report has suggested that activation of A_3 receptors can increase Ca^{2+} currents in hippocampal CA3 neurons (Fleming and Mogul, 1996), but apart from this, there have been few reports of specific actions of A_3 receptor activation on neuronal activity.

Because the hippocampus and cerebellum show the highest levels of A_3 mRNA in brain (De et al., 1993), and because the responses to A_1 and A_2 receptor activation have been well-characterized in this brain region (Dunwiddie, 1985; Greene and Haas, 1991; Cunha et al., 1994, 1995), we have investigated the electrophysiological actions of Cl-IB-MECA in this brain region.

Materials and methods

Slice preparation

Hippocampal slices were obtained from 6- to 8-week-old, male Sprague-Dawley rats (Sasco Animal Laboratories, Omaha, NE) using standard techniques (Dunwiddie and Lynch, 1978; Dunwiddie and Hoffer, 1980). Animals were decapitated, the hippocampus was dissected free of surrounding tissue, and 400 μm slices were cut from the middle portion of the hippocampus with a TC-2 tissue chopper (Sorvall, Norwalk, CT). Slices were immediately placed in ice-cold artificial cerebral spinal fluid (aCSF) consisting of (in mM): NaCI, 124; KCl, 3.3; KH_2PO_4, 1.2; $MgSO_4$, 2.4; $CaCl_2$, 2.5; D-glucose, 10; and $NaHCO_3$ 25.7, pH 7.4, pregassed with 95% O_2–5% CO_2, and were then transferred to an incubation chamber maintained at 33°C to equilibrate for at least 1 h before electrophysiological recording. When recording, the slices were transferred to a submersion recording chamber (1 ml volume) where they were placed on a nylon net and superfused with medium at a rate of 2 ml/min. The superfusion medium was gassed with humidified 95% O_2–5% CO_2 and maintained at a temperature of 33°C to 34°C.

Electrophysiological recording

Whole-cell patch recording experiments were performed using patch pipettes pulled from borosilicate glass (O.D. 1.5 mm, I.D. 0.86 mm, with filament; Sutter Instrument Co., Novato, CA). Electrodes had tip resistance of 5–8 MΩ when filled with a solution containing (in mM): K-Gluconate, 125; KCl, 15; HEPES, 10; $CaCl_2$, 0.1; KBAPTA, 1; K-ATP, 2; Tris-GTP, 0.3; pH adjusted to 7.3 with KOH, osmolarity adjusted to 280–290 mOsm. Whole-cell electrophysiological recordings were made using the "blind" patch recording technique (Blanton et al., 1989). The access resistance was continually monitored by observation of the cell membrane capacitative transients in response to a brief voltage step and was below 25 MΩ in all experiments. Pyramidal neurons were differentiated from glial cells by their ability to fire action potentials, and from interneurons by their high membrane capacitance and accommodation of cell firing in response to depolarizing current injection. Cells were voltage-clamped using an Axoclamp-2A amplifier (Axon Instruments, Burlingame, CA) in the single-electrode voltage-clamp mode. The membrane potentials were corrected by −11 mV for the liquid junction potential between the electrode filling solution and the bath (Neher, 1992).

I_{AHP} was evoked every 40 s by briefly (150 ms) stepping to 0 mV from holding potentials of −55 mV. Membrane resistance was determined by measuring the current change in response to a −4 mV command step every 40 s, and holding current was measured every 20 s. Responses were recorded using an R.C. Electronics ISC-16 A/D board and software developed in our laboratory.

At least 10 to 15 min of stable baseline responses were obtained in each experiment before drug applications began. Following acquisition of the baseline data, the effects of drugs were then tested by adding them directly to the superfusion medium with a calibrated syringe pump (Razel Scientific Instruments, Inc., Stamford, CT). The net change in flow rate using this approach was never more than 1%. The peak amplitude of the AHP current (I_{AHP}) was determined for each response and then averaged during the pre-drug control, drug, and the post-drug washout periods. At least 10 responses were included in each average. The data were analyzed as mean percent change in the response amplitude when compared to responses obtained during the control period.

Statistical analysis

All data were analyzed between groups, using the unpaired Student's t-test and nonparametric test (Mann-Whitney test) with a $P < 0.05$ criterion for statistical significance.

Chemicals

Adenosine, K-ATP, and Tris-ATP were obtained from Sigma (St. Louis, MO); baclofen, carbachol, 8-cyclopentyl-1,3-dimethylxanthine (CPT), and 3,7-dimethyl-1-propargylxanthine (DMPX) were purchased from Research Biochemicals (Natick, MA); 2-chloro-N^6-(3-iodobenzyl)-adenosine-5-N-methyluronamide (Cl-IB-MECA) was provided by Research Biochemical International (Natick, MA) as part of the Chemical Synthesis Program of the National Institute of Mental Health, contract N01MH30003. The K-gluconate, KCl, $CaCl_2$ and HEPES used in the patch electrode recording solution were from Fluka Chemika-Biochemika (Ronkonkoma, New York), and K-BAPTA was from Molecular Probes (Eugene, Oregon). All drugs were dissolved in distilled water except CPT and Cl-IB-MECA, which were initially dissolved in 100% dimethylsulfoxide (DMSO) at greater than 1000 times final concentration and then diluted with aCSF to the desired concentration. The final

concentration of DMSO in the superfusion medium never exceeded 0.05%.

Results

When hippocampal brain slices were superfused with adenosine (50 µM), a rapid outward current developed that showed little desensitization during a 10 min superfusion period, and reversed readily upon return to control buffer (Fig. 1A). This response was completely blocked by the selective adenosine A_1 receptor antagonist N^6-cyclopentyl-theophylline (CPT; 1 µM), and by CPT + DMPX (a moderately selective A_2 antagonist; 50 µM). In either case, there was no significant effect of adenosine superfusion on the holding current (Fig. 1A). These latter conditions are nearly identical to those under which both adenosine and N^6-2-(4-aminophenyl)ethyl-adenosine (APNEA) have been reported to increase Ca^{2+} currents in isolated hippocampal CA3 pyramidal neurons, an effect attributed to activation of A_3 receptors (Fleming and Mogul, 1997). These results suggested that under resting conditions, activation of A_3 receptors by adenosine did not have any significant effect upon the holding current in CA1 pyramidal neurons voltage clamped at –55 mV. This observation was confirmed by superfusion of slices with either 100 nM or 1 µM Cl-IB-MECA, also in the presence of CPT + DMPX to block any possibility of activation of A_1 or A_2 receptors; in this situation, there was also no significant change in the holding current required to clamp neurons to –55 mV (Fig. 1B).

It has been suggested that activation of protein kinase C (PKC) in hippocampal pyramidal neurons is linked to a reduction in the magnitude of afterhyperpolarizations (AHPs) induced by depolarizing current injection (Malenka et al., 1986; Engisch et al., 1996). Because previous studies have shown that A_3 receptors can also be linked to activation of phospholipase C (PLC), and hence activation of PKC, we characterized the effects of adenosine and the selective A_3 agonist Cl-IB-MECA on I_{AHP}. A 150 ms command step to 0 mV typically elicited a slow, outward current that lasted 10 s or more, and which was usually between 50–500 pA in amplitude (Fig. $2A_2$, $2B_2$). The time course of this response was consistent with the time

(A)

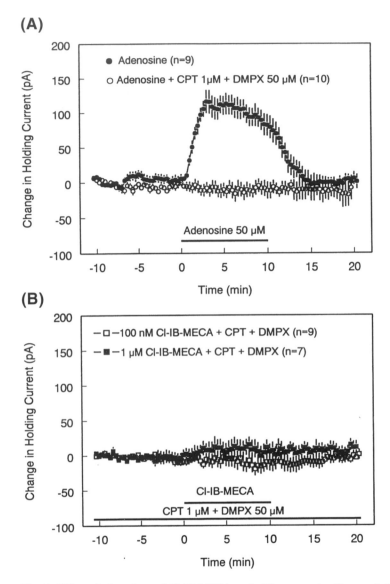

(B)

Fig. 1. Effect of adenosine and Cl-IB-MECA on holding currents. Changes induced in current required to clamp CA1 hippocampal pyramidal neurons to -55 mV are indicated for groups of cells treated with identical drug superfusion protocols. Although not indicated, CPT + DMPX superfusion was maintained for at least 20 min prior to superfusion with adenosine or Cl-IB-MECA. CPT is an A_1 selective antagonist with approximately 130X selectivity for A_1 vs. A_{2A} receptors and a K_I for A_1 receptors of 11 nM (Bruns et al., 1986); DMPX has a K_I of 11 μM at A_{2A} receptors and is 4–11 $\times$ selective for A_{2A} vs. A_1 (Ukena et al., 1986; Seale et al., 1988). Adenosine induced a highly significant outward change in the holding current (A) and this effect was blocked by CPT + DMPX. Neither concentration of Cl-IB-MECA produced any change in the holding current in the presence of the A_1/A_2 receptor antagonists.

course of the slow AHP activated by depolarizing current injection in current clamped neurons (Madison and Nicoll, 1982; Haas and Greene, 1984). To confirm the identity of this current, and to verify its previously reported sensitivity to other pharmacological agents, we examined the effects of carbachol (10 μM; Fig. 2A), and isoproterenol (500 nM; Fig. 2B), both of which have been reported to inhibit this current. Bath superfusion with either of these agents virtually abolished the I_{AHP}, suggesting that under voltage clamp conditions, this current retained the pharmacological

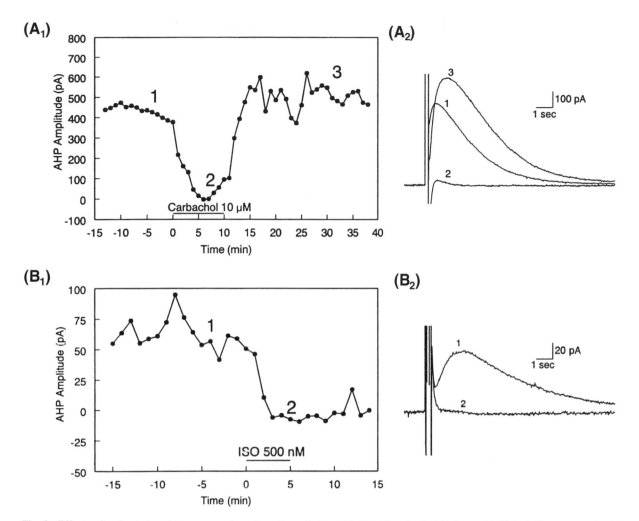

Fig. 2. Effects of carbachol and isoproterenol on I_{AHP}. Superfusion with 10 μM carbachol (A$_1$) or with 500 nM isoproterenol (ISO; B$_1$) rapidly blocked the currents evoked by a 150 ms depolarizing step to 0 mV. The amplitudes shown at the left correspond to the peak amplitude of the I_{AHP}, or in the case of the inhibited responses, to the holding current at the time corresponding to the peak amplitude of I_{AHP} prior to drug superfusion. The actual currents elicited by the depolarizing step are indicated at right (A$_2$ and A$_1$); the numbers in the figures at the left indicate the times at which the actual waveforms at the right were acquired.

sensitivity that has been previously described for the slow AHP.

In subsequent experiments, slices were superfused with either 100 nM (Fig. 3A) or 1 μM (Fig. 3B) Cl-IB-MECA, and the amplitude of I_{AHP} was monitored. Neither concentration of this A$_3$ selective agonist had any significant effect on AHP amplitude. On the other hand, as we have previously noted (Dunwiddie, 1985), adenosine itself produced a rapid and highly significant reduction in the amplitude of I_{AHP}, and this effect reversed readily upon washing (Fig. 3C). The effect of

adenosine, which was highly significant by itself ($t = 4.92$, $P < 0.01$), did not appear to be mediated via A$_3$ receptors, because this action was successfully antagonized by the combination of CPT + DMPX ($t = 1.66$, n.s.; Fig. 4A). On the other hand, neither concentration of Cl-IB-MECA had a significant effect when tested on seven to 10 cells under the same conditions (Fig. 4B).

Thus, in summary, adenosine at a concentration of 50 μM induced a significant outward change in the holding current, and significantly attenuated I_{AHP}, and both of these effects were completely

280

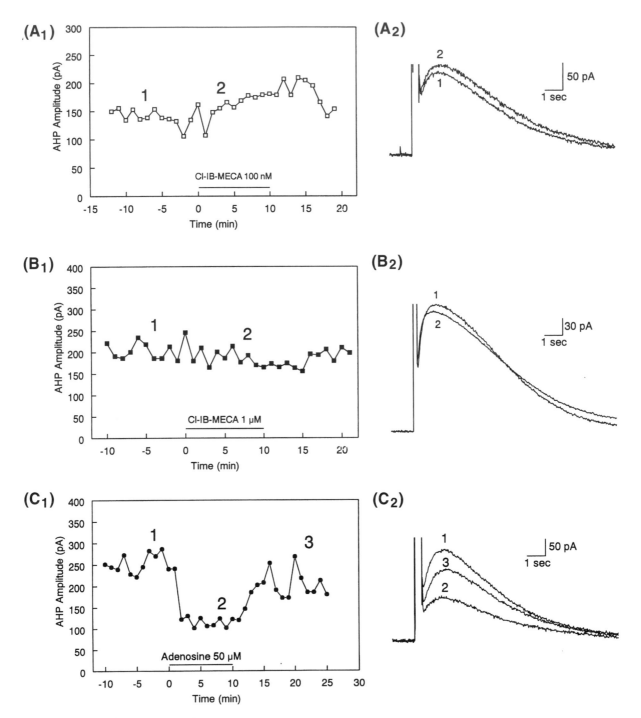

Fig. 3. Effects of Cl-IB-MECA and adenosine on I_{AHP} amplitude. The time courses of changes in the amplitude of I_{AHP} are shown at the left and averaged responses obtained at the indicated times at the right, as in Fig. 2. Neither concentration of Cl-IB-MECA had any significant effect on the response, but 50 μM adenosine depressed I_{AHP} amplitude in a readily reversible fashion.

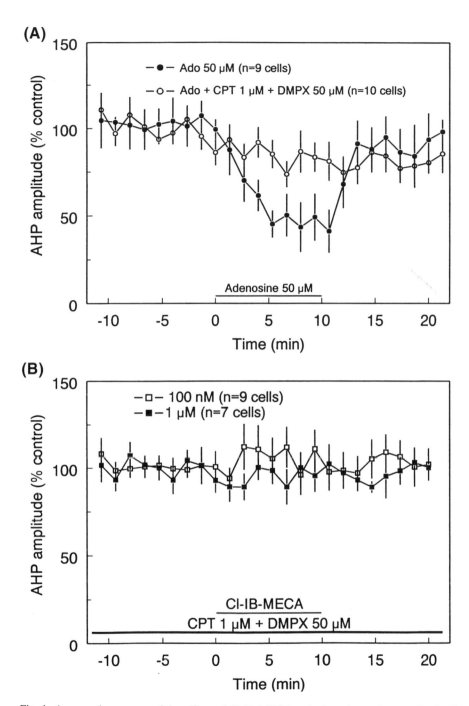

Fig. 4. Average time course of the effects of Cl-IB-MECA and adenosine on I_{AHP} amplitude. Groups of slices were superfused with drugs using identical treatment protocols, and the mean ± SEM response as a percentage of pre-drug response amplitude is illustrated. Adenosine induced a reversible depression of this response that was blocked by CPT + DMPX (A), whereas Cl-IB-MECA had no effect, either in the absence of inhibitors (Fig. 3A and B), or in slices pretreated with the A_1/A_2 inhibitors (B).

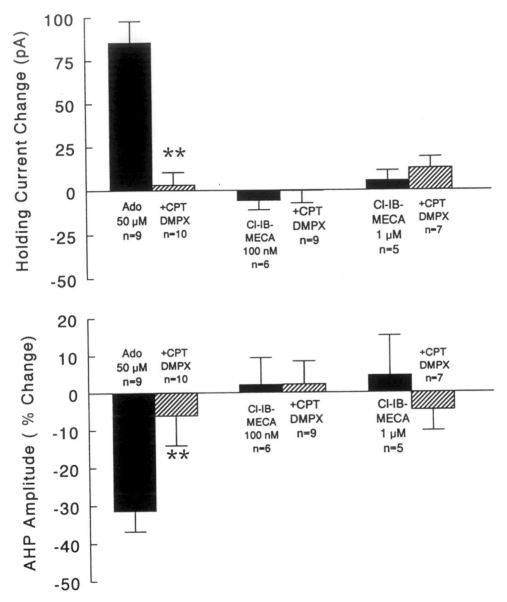

Fig. 5. Average effects of drug superfusion on holding current and I_{AHP}. The mean ± SEM response for changes in the holding current (upper) and I_{AHP} (lower) are shown. All of the effects observed with adenosine were completely antagonized by CPT + DMPX (**$P < 0.01$; hatched bars), whereas Cl-IB-MECA had no significant effect upon either measure, either alone or in combination with the combined antagonists.

blocked by the combination of the A₁ selective antagonist CPT and the somewhat A₂ selective antagonist DMPX (Fig. 5); these observations are consistent with previous work suggesting that A₁ receptor agonists can elicit both of these effects (Dunwiddie, 1985; Gerber et al., 1989). On the other hand, Cl-IB-MECA had no significant effect

on either one of these parameters, and this was true both when it was tested alone, and in the presence of CPT + DMPX (Fig. 5).

Discussion

The slow AHP that can be elicited in hippocampal pyramidal neurons by brief depolarization is a response that has been shown to be quite sensitive to activation of multiple second messenger systems, and appears to be inhibited by activation of PKA, PKC, and CaM-K. For these reasons, it appeared to be an appropriate candidate to determine whether activation of A_3 receptors was linked to changes in the activity of any of these kinases. However, the results of these studies demonstrated that neither the selective A_3 receptor agonist, Cl-IB-MECA, nor the non-selective ligand adenosine (in the presence of A_1- and A_2-selective blockers), had any significant effect on the amplitude of I_{AHP}. Furthermore, these results also suggest that A_3 receptor activation does not inhibit the voltage gated Ca^{2+} channels that mediate the Ca^{2+} influx that activates this conductance. In the absence of antagonists, adenosine elicited a reduction in the amplitude of I_{AHP}, but these effects were blocked by the combination of the A_1 (CPT) and A_2 (DMPX) selective antagonists. This is most probably the result of a direct reduction in Ca^{2+} channel activity mediated via A_1 receptors, or the activation of inwardly rectifying K^+ channels via A_1 receptors, which could indirectly reduce the Ca^{2+} influx. In an intact system such as the brain slice, it is not possible to effectively clamp CA1 pyramidal neurons during a depolarizing step to 0 mV; the effectiveness of the clamp, and hence the net Ca^{2+} influx, can be affected by the activation of other conductances, such as the inwardly-rectifying K^+ channels that are linked to activation of A_1 receptors.

At least three kinds of responses might have been expected based upon the known actions of A_3 receptors. First, the enhancement of Ca^{2+} currents that has been associated with A_3 receptor activation in CA3 neurons (Fleming and Mogul, 1997) might have been expected to lead to an enhancement of I_{AHP}. Second, activation of PLC would have been expected to lead to a reduction in I_{AHP}, either via activation of PKC, or via the IP3-dependent release of intracellular Ca^{2+}. Finally, because it has been proposed that A_3 receptors may activate PKA in hippocampal neurons (Fleming and Mogul, 1997), this might also have been expected to lead to a decrease in I_{AHP}.

The fact that none of these effects were observed suggests several possibilities. It is possible that

there were offsetting effects, such that an enhancement of Ca^{2+} influx combined with second messenger mediated down-regulation of the I_{AHP} led to no significant change in the current. Although possible, it seems unlikely that such precisely offsetting responses could occur in every slice; the lack of change even in the variability of the response (Fig. 4B) is not consistent with this conclusion. A somewhat more likely alternative perhaps is that the effects on Ca^{2+} currents are confined solely to CA3 neurons; little work has been done on the localization of A_3 receptors, and it is certainly possible that the CA1 and CA3 neurons differ in terms of receptor expression and/or coupling. Alternatively, A_3 receptors may be present on CA1 pyramidal neurons, but localized in such a way that they do not directly affect I_{AHP}. Much recent work on adenylyl cyclase has emphasized the localization or anchoring of this enzyme near appropriate substrates, and it is possible that while β-receptors activate adenylyl cyclase in a subcellular compartment with ready access to the channels that mediate I_{AHP}, A_3 receptors activate a cyclase that lacks such access. Future experiments will be required to determine which of these alternatives is the case.

Conclusions

The adenosine A_3 receptor is found in brain, and has been linked to a number of physiological responses, but the cellular mechanisms that underlie these effects are still unclear. In the present experiments, the effects of a selective A_3 agonist, 2 - chloro - N^6 - (3 - iodobenzyl) - adenosine - 5′ - N - methyluronamide (Cl-IB-MECA), were characterized on hippocampal CA1 pyramidal neurons. More specifically, we examined the possibility that activation of A_3 receptors could lead to activation of protein kinase A (PKA), protein kinase C (PKC), or calcium/calmodulin dependent protein kinase II (CaM-K), which might then induce significant changes in the activity of hippocampal pyramidal neurons. CA1 pyramidal neurons were recorded in the whole cell mode using patch electrodes, and the effects of adenosine agonists were determined on the holding current, and on the amplitude of the slow outward calcium-activated K^+ currents activated following a depolarizing step in the command voltage (I_{AHP}). As has been previously reported, a variety of agents reduced the I_{AHP}, including the β-adrenoceptor agonist isoproterenol, the muscarinic agonist carbachol, and adenosine itself; however, Cl-IB-MECA had no significant effects on this current. Cl-IB-MECA by itself also had no effect on the holding current, and had a weak but not statistically significant antagonistic effect on the outward currents induced by activation of A_1 receptors on pyramidal neurons. These results suggest that either A_3 receptors are not present on the cell bodies of hippocampal CA1 pyramidal neurons, that they are present but not linked to PKA, PKC, or CaM-K activation, or that they only activate these kinases in cell compartments that are not accessible to the ion channels that underlie the AHP response.

Acknowledgements

This work was supported by grant R01 NS 29173 from the National Institute of Neurological Disorders and Stroke, and by the Veterans Administration Medical Research Service.

References

Abbracchio, M.P., Brambilla, R., Ceruti, S., Kim, H.O., von Lubitz, D.K.J.E., Jacobson, K.A. and Cattabeni, F. (1995) G protein-dependent activation of phospholipase C by adenosine A_3 receptors in rat brain. *Mol. Pharmacol.*, 48: 1038–1045.

Abdul-Ghani, M.A., Valiante, T.A., Carlen, P.L. and Pennefather, P.S. (1996) Metabotropic glutamate receptors coupled to IP3 production mediate inhibition of IAHP in rat dentate granule neurons. *J. Neurophys.*, 76: 2691–2700.

Ali, H., Cunha-Melo, J.R., Saul, W.F. and Beaven, M.A. (1990) Activation of phospholipase C via adenosine receptors provides synergistic signals for secretion in antigen-stimulated RBL-2H3 cells. Evidence for a novel adenosine receptor. *J. Biol. Chem.*, 265: 745–753.

Auchampach, J.A., Rizvi, A., Qiu, Y., Tang, X.L., Maldonado, C., Teschner, S. and Bolli, R. (1997) Selective activation of A3 adenosine receptors with N^6-(3- iodobenzyl)adenosine-5′-N-methyluronamide protects against myocardial stunning and infarction without hemodynamic changes in conscious rabbits. *Circ. Res.*, 80: 800–809.

Blanton, M.G., Lo Turco, J.J. and Kriegstein, A.R. (1989) Whole cell recording from neurons in slices of reptilian and mammalian cerebral cortex. *J. Neurosci. Meth.*, 30: 203–210.

Bruns, R.F., Daly, J.W. and Pugsley, T.A. (1986) Characterization of A2 receptor labeled by [^{3}H]-NECA in rat striatal membranes. *Mol. Pharmacol.*, 29: 331–346.

Cole, A.E. and Nicoll, R.A. (1984) Characterization of a slow cholinergic post-synaptic potential recorded in vitro from rat hippocampal pyramidal cells. *J. Physiol. (Lond.)*, 352: 173–188.

Cunha, R.A., Johansson, B., Fredholm, B.B., Ribeiro, J.A. and Sebastiáo, A.M. (1995) Adenosine A2A receptors stimulate acetylcholine release from nerve terminals of the rat hippocampus. *Neurosci. Lett.*, 196: 41–44.

Cunha, R.A., Johansson, B., Van der Ploeg, I., Sebastiao, A.M., Ribeiro, J.A. and Fredholm, B.B. (1994) Evidence for functionally important adenosine A2a receptors in the rat hippocampus. *Brain Res.*, 649: 208–216.

De, M., Austin, K.F. and Dudley, M.W. (1993): Differential distribution of A_3 receptor in rat brain. *Soc. Neurosci. Abstr.*, 19: 42.11 (Abstract).

Dunwiddie, T.V. (1985) The physiological role of adenosine in the central nervous system. *Int. Rev. Neurobiol.*, 27: 63–139.

Dunwiddie, T.V., Diao, L.H., Kim, H.O., Jiang, J.L. and Jacobson, K.A. (1997) Activation of hippocampal adenosine A_3 receptors produces a desensitization of A1 receptor-mediated responses in rat hippocampus. *J. Neurosci.*, 17: 607–614.

Dunwiddie, T.V. and Hoffer, B.J. (1980) Adenine nucleotides and synaptic transmission in the in vitro rat hippocampus. *Br. J. Pharmacol.*, 69: 59–68.

Dunwiddie, T.V. and Lynch, G.S. (1978) Long-term potentiation and depression of synaptic responses in the rat hippocampus: Localization and frequency dependency. *J. Physiol. (Lond.)*, 276: 353–367.

Engisch, K.L., Wagner, J.J. and Alger, B.E. (1996) Whole-cell voltage-clamp investigation of the role of PKC in muscarinic inhibition of IAHP in rat CA1 hippocampal neurons. *Hippocampus*, 6: 183–191.

Fleming, K.M. and Mogul, D.J. (1996): Adenosine A_3 receptors potentiate hippocampal calcium current by a PKA-dependent/PKC-independent pathway. *Drug Dev. Res.*, 37: 121 (Abstract).

Fleming, K.M. and Mogul, D.J. (1997) Adenosine A_3 receptors potentiate hippocampal calcium current by a PKA-dependent/PKC-independent pathway. *Neuropharmacol.*, 36: 353–362.

Gerber, U., Greene, R.W., Haas, H.L. and Stevens, D.R. (1989) Characterization of inhibition mediated by adenosine in the hippocampus of the rat in vitro. *J. Physiol. (Lond.)*, 417: 567–578.

Greene, R.W. and Haas, H.L. (1991) The electrophysiology of adenosine in the mammalian central nervous system. *Prog. Neurobiol.*, 36: 329–341.

Haas, H.L. and Greene, R.W. (1984) Adenosine enhances afterhyperpolarization and accommodation in hippocampal pyramidal cells. *Pflugers Arch.*, 402: 244–247.

Haas, H.L. and Konnerth, A. (1983) Histamine and noradrenaline decrease calcium-activated potassium conductance in hippocampal pyramidal cells. *Nature*, 302: 432–434.

Jacobson, K.A., Kim, H.O., Siddiqi, S.M., Olah, M.E., Stiles, G.L. and von Lubitz, D.K.J.E. (1995) A_3-adenosine receptors: design of selective ligands and therapeutic prospects. *Drugs of the Future*, 20: 689–699.

Jacobson, K.A., Nikodijevic, O., Shi, D., Gallo-Rodriguez, C., Olah, M.E., Stiles, G.L. and Daly, J.W. (1993) A role for central A3 adenosine receptors: Mediation of behavioral depressant effects. *FEBS Lett.*, 336: 57–60.

Macek, T.A., Schaffhauser, H. and Conn, P.J. (1998) Protein kinase C and A3 adenosine receptor activation inhibit presynaptic metabotropic glutamate receptor (mGluR) function and uncouple mGluRs from GTP-binding proteins. *J. Neurosci.*, 18: 6138–6146.

Madison, D.V. and Nicoll, R.A. (1982) Noradrenaline blocks accommodation of pyramidal cell discharge in the hippocampus. *Nature*, 299: 636–638.

Malenka, R.C., Madison, D.V., Andrade, R. and Nicoll, R.A. (1986) Phorbol esters mimic some cholinergic actions in hippocampal pyramidal neurons. *J. Neurosci.*, 6: 475–480.

Neher, E. (1992) Ion channels for communication between and within cells. *Science*, 256: 498–502.

Ramkumar, V., Stiles, G.L., Beaven, M.A. and Ali, H. (1994) The A3 adenosine receptor is the unique adenosine receptor which facilitates release of allergic mediators in mast cells. *J. Biol. Chem.*, 268: 16887–16890.

Seale, T.W., Abla, K.A., Shamim, M.T., Carney, J.M. and Daly, J.W. (1988) 3,7-Dimethyl-1-propargylxanthine: A potent and selective in vivo antagonist of adenosine analogs. *Life Sci.*, 43: 1671–1684.

Ukena, D., Shamim, M.T., Padgett, W. and Daly, J.W. (1986) Analogs of caffeine: Antagonists with selectivity for A2 adenosine receptors. *Life Sci.*, 39: 743–750.

Zhao, Z.H., Francis, C.E. and Ravid, K. (1997) An A_3-subtype adenosine receptor is highly expressed in rat vascular smooth muscle cells: Its role in attenuating adenosine-induced increase in cAMP. *Microvasc. Res.*, 54: 243–252.

Zhou, Q.-Y., Li, C., Olah, M.E., Johnson, R.A., Stiles, G.L. and Civelli, O. (1992) Molecular cloning and characterization of an adenosine receptor: The A3 adenosine receptor. *Proc. Natl. Acad. Sci. (USA)*, 89: 7432–7436.

P. Illes and H. Zimmermann (Eds.)
Progress in Brain Research, Vol 120

CHAPTER 23

Brain hypoxia: Effects of ATP and adenosine

K. Nieber*, D. Eschke and A. Brand

Institut für Pharmazie der Universität Leipzig, Lehrstuhl Pharmakologie für Naturwissenschaftler, Brüderstr. 34, D-04103 Leipzig, Germany

Introduction

Neurons in the mammalian central nervous system are highly sensitive to the availability of oxygen. Hypoxia and ischemia of sufficient duration or severity can alter neuronal function and cell morphology and lead to cell injury or death. The vulnerability to cerebral ischemia is strikingly dependent upon the age of an individual. Young rats have a better chance of survival than adult rats (Duffy et al., 1975; Nabetani et al., 1995). This increased survivability is associated with a reduced rate of energy utilization during ischemia in very young animals possibly due to less-developed functional activity (Kass and Lipton, 1986). Many studies have been performed in order to understand the mechanisms underlying hypoxia-induced synaptic and extrasynaptic disturbances in the brain in vivo and in vitro. During the last decade several new aspects became the focus of interest. Gene expression, gene therapy, apoptosis, growth factors and inflammatory mechanisms during and after ischemia were topics which were intensively discussed (Krieglstein, 1996). Despite the enormous progress during the past years our knowlegde about processes which initiate the pathophysiological cascade of cell death after ischemia or prevent its further development, is limited and many open questions exist. To study hypoxia-induced changes in the membrane properties of single neurons

humerous electrophysiological investigations have been carried out in vitro in slice preparations obtained from various brain areas. Neocortical (Luhmann and Heinemann, 1992; Luhmann, 1996) or hippocampal pyramidal cells in the CA1 region (Hansen et al., 1982; Fujiwara et al., 1987; Grigg and Anderson, 1989; Leblond and Krnjevic, 1989; Zhang and Krnjevic, 1993) and CA3 region (Ben-Ari and Lazdunski, 1989, Ben-Ari et al., 1990) as well as neurons of the locus coeruleus (Nieber et al., 1995; Yang et al., 1997), brain stem (Morawietz et al., 1995; Ballanyi et al., 1996) and substantia nigra (Watts et al., 1995) were frequently used for such investigations.

Short-term hypoxia induced typical changes to membrane potential consisting of hypoxic depolarization followed by hypoxic hyperpolarization, and eventually by a second hyperpolarization during reoxygenation (Fujiwara et al., 1987; Leblond and Krnjevic, 1989; Nieber et al., 1995). It is evident that a complex interaction of several factors is involved in the early and in the late phase of the neuronal dysfunction. Adenosine 5′-triphosphate (ATP) has been shown to store and supply energy in neurons, but it also acts as an extracellular signal molecule under physiological and pathophysiological conditions. ATP is able to control the excitability from the inner and outer side of the neuronal membrane. ATP released from metabolically damaged cells may activate purinoceptors of the P2-type. Furthermore, ATP dependent potassium (K_{ATP}) channels are highly interesting, because these channels are regulated by the intra-

---
*Corresponding author. Tel.: +49-341-9736812; fax: +49-314-9736709; e-mail: nieber@rz.uni-leipzig.de

cellular ATP concentration. During hypoxia the intracellular concentration of ATP decreases. Subsequently, the K_{ATP} channels open (Ashcroft and Ashcroft, 1990). The resulting outflow of potassium hyperpolarizes the membrane and lowers the neuronal excitability. Extracellular ATP can be rapidly degraded by ecto-5′-nucleotidases to adenosine. Adenosine is an inhibitory signal molecule acting on P1-purinoceptors. The extracellular level of adenosine increases dramatically following hypoxia.

The purine nucleoside adenosine influences numerous physiological processes, including effects on neuronal communication within the central nervous system. These observations suggest that adenosine may play a role as neurotransmitter or neuromodulator. Adenosine and its analogues inhibit the release of other neurotransmitters and depress neuronal depolarization. The involvement of adenosine in the cascade of hypoxic changes is complex. There is evidence for a transient neuroprotective potential of adenosine both at presynaptic and postsynaptic terminals (Rudolphi et al., 1992). This overview describes the involvement of ATP on neuronal hypoxic changes in membrane potential and discusses the contribution of adenosine as a protective agent in cerebral hypoxia.

Effects of hypoxia on membrane potenial

Brief periods of hypoxia elicit a characteristic sequence of changes in membrane potential that causes a reversible loss of neuronal function. Hippocampal pyramidal cells of the CA1 region show typical triphasic changes of the membrane potential. An early hypoxic depolarization and an increase in excitability is followed by a hypoxic hyperpolarization, associated with a decrease in excitability. During reoxygenation an even more pronounced posthypoxic hyperpolarization develops (Fujiwara et al., 1987; Leblond and Krnjevic, 1989; Hsu and Huang, 1997).

Locus coeruleus neurons respond to brief hypoxic stimuli with an initial transient increase in the spontaneous firing rate that is followed by depression of the action potentials. When spontaneous firing is blocked by hyperpolarizing current

injection, hypoxia induces effects comparable to those described for hippocampal pyramidal cells: an early hypoxic depolarization, followed by a hyperpolarization and a posthypoxic hyperpolarization (Nieber et al., 1995).

Similarly, neurons of the substantia nigra (Watts et al., 1995) and the dorsal vagal nucleus (Trapp and Ballanyi, 1995; Ballanyi et al., 1996) respond to hypoxia with a depolarization and hyperpolarization. In neocortical pyramidal neurons hypoxia exhibits a hypoxic hyperpolarization (Rosen and Morris, 1991; O'Reilly et al., 1995) or a more complex response including hypoxic and posthypoxic hyperpolarization is observed (Luhmann and Heinemann, 1992). In another group of neocortical neurons the only response to hypoxia is a depolarizing shift in membrane potential (Fig. 1A). The divergent results may be explained by the variable resting membrane potential values of the individual pyramidal cells and by differences between experimental procedures (Aitken et al., 1995). A monophasic reversible depolarization during hypoxia is also observed in striatal neurons (Calabresi et al., 1995; Pisani et al. 1997).

The hypoxic hyperpolarization is due to the activation of a potassium current and its reversal potential is near -90 mV (Luhmann and Heinemann, 1992; Nieber et al., 1995). The hypoxic depolarization is supposed to be due to the blockade of the K^+,Na^+-ATPase and to the activation of excitatory amino acid receptors by endogenous glutamate. The reversal potential of this response was found to be near -40 mV (Rosen and Morris, 1991). Furthermore, the appearance of the hypoxic depolarization depends on glucose concentration in the superfusion medium (Fowler, 1989). A removal of glucose from the medium results in a stronger hypoxic depolarization (Fowler, 1992).

Hypoxia-induced decrease of intracellular ATP-concentration and inactivation of Na⁺,K⁺-ATPase

The energetic state of hippocampal, cortical and striatal tissues has been investigated in vivo (Hisanaga et al., 1986; Onodera et al., 1986) and in vitro (Yoneda and Okada, 1989; Paschen and Djuricic,

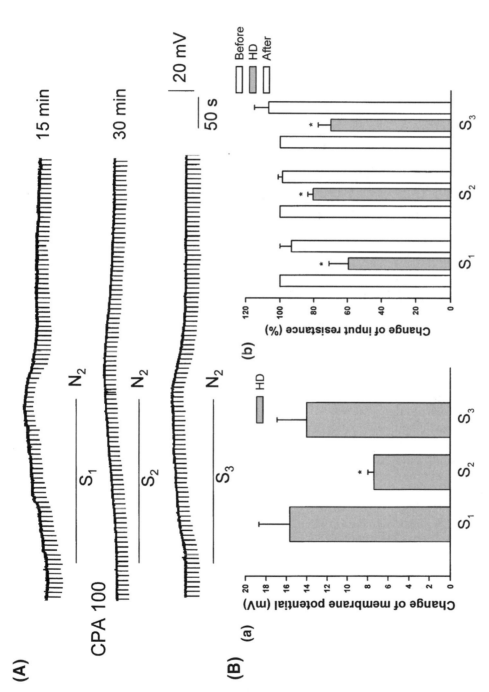

Fig. 1. Effects of cyclopentyl adenosine (CPA) on hypoxic changes of the membrane potential and input resistance on neocortical pyramidal cells. (A) Superfusion with a hypoxic medium (95%N_2–5%CO_2) led to depolarization (HD) of the neurons. Constant current was passed through the recording electrode to keep the cell at −90 mV. Hypoxia was induced three times for five minutes (S_1–S_3; marked by horizontal bars). CPA (100 µM) was applied for five minutes before, during and fifteen minutes after S_2. The intervals between the traces are shown. Hypoxic changes were measured in six experiments similar to that indicated in (A). (B) Means ± SEM are presented. The magnitude of the HD during the consecutive hypoxic stimuli is indicated by filled columns (a). The control HD was depressed in the presence of CPA by 47.1 ± 6.9%. The depression was reversed by washout. The percentage change of input resistance before (first set of open columns in each group, designated as 100%), during (filled columns in each group) and after the HD (second set of open columns in each group) is indicated (b). Hypoxia depressed the input resistance in the mean of six experiments but CPA (100 µM) did not influence this alteration. In addition, CPA (100 µM) failed to influence both the membrane potential and input resistance of the neurons. *$P < 0.05$ significant difference from the control.

1995; Milusheva et al. 1996; Pissarek et al., 1998). Hypoxia or deprivation of glucose results in a marked decline of ATP levels in neocortical, hippocampal or striatal tissue. Detailed studies on neocortical brain slices indicate that if pyramidal neurons respond to short-term hypoxia (5 min) with a marked depolarization, nucleoside triphosphate levels of pyramidal neurons decrease (GTP, ATP, UTP, CTP), while the respective diphosphate levels are transiently increased (GDP, ADP) or remain constant (UDP) (Pissarek et al., 1998). In slices subjected to 30 min of hypoxia the triphosphate levels continued to decrease, while the levels of GDP and ADP returned to control values. However, tri- to diphosphate ratios progessively declined for ATP/ADP and GTP/GDP, but not for UTP/UDP when the duration of hypoxia was increased from five to 30 min.

In the brain, about 40% of the energy production is required for Na^+,K^+-ATPase activity under resting conditions (Skou, 1992). Na^+,K^+-ATPase is a surface membrane protein responsible for translocating sodium and potassium ions across the cell membrane, utilizing ATP as the driving force. The transport produces both a chemical and an electrical gradient across the cell membrane. The electrical gradient is essential for maintaining the resting membrane potential. Hypoxia-induced fall in the intracellular ATP-concentration blocks the Na^+,K^+-ATPase. Such a blockade causes a rapid increase in extracellular K^+-concentrations ($[K^+]_o$). The increase in $[K^+]_o$ starts within the first minutes after the onset of hypoxia. The magnitude depends on the age of the animal, the CNS region, tissue temperature, presence or absence of glucose and severity of O_2 deprivation (Haddad and Donnelly 1990; Jiang and Haddad, 1991; Donnelly et al., 1992; Perez-Pinzon et al., 1995). A rapid reduction of ATP-driven Na^+,K^+-transport and a corresponding progressive increase in $[K^+]_o$ have been made responsible for the early phase of hypoxia-induced changes of membrane parameters (Fujiwara et al., 1987; Nieber et al., 1995). This assumption is supported by intracellular recordings from hippocampal CA1 neurons indicating that prolonged application of ouabain causes an initial small hyperpolarization, a subsequent slow depolarization, and a rapid depolarization (Tanaka et al.,

1997). Furthermore, it is shown that the rapid depolarization is voltage-independent and is probably due to a non-selective increase in permeability to different ions und the underlying conductance change is primarily the result of an inhibition of Na^+,K^+-ATPase activity.

Involvement of K_{ATP}-channels

There is a general agreement that the hypoxic hyperpolarization seen in different brain regions is caused by the depletion of ATP following by an increase in K^+ conductance (Grigg and Anderson, 1989; Leblond and Krnjevic, 1989; Zhang and Krnjevic, 1993). There is disagreement whether ATP-sensitive K^+ (K_{ATP}) channels or Ca^{2+}-dependent K^+ (K_{Ca}) channels are involved. A group of authors argue that hypoxia slows down processes that maintain a very low cytosolic Ca^{2+} concentration (Leblond and Krnjevic, 1989). This leads to activation of K_{Ca} conductance. However, there is much data indicating the involvement of K_{ATP} channels. K_{ATP} channels are an important class of ionic channels linking the bioenergetic metabolism to membrane excitability (Lazdunski, 1994). They were first described in cardiac myocytes (Ashcroft, 1988), and later found in pancreatic β-cells, skeletal and smooth muscle cells, but also in various neurons in the brain. These channels are blocked by intracellular ATP, and this inhibitory effect does not require ATP hydrolysis (Ashcroft and Ashcroft, 1990). A certain class of drugs, the potassium channel openers (cromakalim, lemakalim, RO 31-6930, diazoxide), activates a K_{ATP} current while sulfonylurea antidiabetics (glibenclamide, gliquidone, tolbutamide) inhibit it, thereby antagonizing the effect of the channel openers. Autoradiographic techniques and binding studies with ^{3}H-glibenclamide, a blocker of K_{ATP} channels, indicate that in newborn and adult rats specific glibenclamide binding is saturable (Haddad and Jiang, 1994). These binding sites are heterogenously distributed throughout the brain. The highest distribution has been observed in the cortex and the substantia nigra. Moderate densities were found in the basal ganglia and the medullary hypoglossal nucleus. The lowest density is found in the hypothalamus and the brain stem (Haddad and

Jiang, 1994). The presence of functional K_{ATP} channels was proven by a variety of methods including the measurement of K^+ efflux (Lazdunski, 1994), the determination of transmitter release (Amoroso et al., 1990) and electrophysiological recordings (Illes et al., 1994).

The activation of K_{ATP} channels may become relevant during hypoxia, glucose deprivation or the combination of both when cellular ATP-levels decline (Tromba et al., 1992; Paschen and Djuricic, 1995; Espanol et al., 1996; Milusheva et al., 1996). It seems reasonable to assume that purine and pyrimidine nucleotides might allosterically regulate K_{ATP} channels (Lazdunski, 1994). It is interesting that preincubation of the brain slices with creatine, which markedly increases tissue phosphocreatine levels and prevents depletion of intracellular ATP, diminishes the failure of synaptic transmission in hippocampal (Carter et al., 1995) and neocortical neurons (Luhmann and Heinemann, 1992) or the inhibition of firing in locus coeruleus neurons (Illes et al., 1994). Additionally, when ATP is injected into hippocampal CA1 neurons, hypoxia does not result in hyperpolarization (Fujimura et al., 1997).

There is further pharmacological evidence supporting the hypothesis that K_{ATP} channels are opened during hypoxia. K_{ATP} channel agonists mimic the hypoxic hyperpolarization or an outward current and the respective antagonists reverse these effects in CA1 hippocampal pyramidal neurons (Erdemli and Krnjevic, 1996; Fujimura, 1997), neocortical pyramidal cells (Ashcroft, 1988), neurons of the locus coeruleus (Finta et al., 1993), the substantia nigra (Watts et al., 1995) or the dorsal vagal nucleus (Trapp and Ballanyi, 1995). Glibenclamide and tolbutamide prevent half of the K^+ loss usually observed during hypoxia in a dose dependent way (Haddad and Jiang, 1994). Moreover, the release of glutamate in hippocampal slices, evoked by hypoxia combined with glucose deprivation is reduced in the presence of K_{ATP} channel agonists, while it was enhanced by K_{ATP} channel antagonists (Zini et al., 1993).

Another interesting series of observations focuses on the effect of K_{ATP} channel antagonists on hypoxia-induced hyperpolarization. Tolbutamide practically eliminates the hypoxic hyperpolarization in various areas of the brain, e.g. hippocampus (Grigg and Anderson, 1989; Godfraind and Krnjevic, 1993), neocortex (Luhmann and Heinemann, 1992), substantia nigra (Murphy and Greenfield, 1992), locus coeruleus (Nieber et al., 1995) and dorsal vagal neurons (Trapp and Ballanyi, 1995). Additionally, the hypoxic depolarization in hippocampal CA3 neurons can be reduced by application of the K_{ATP} channel agonist diazoxid (Krnjevic and Ben-Ari, 1989). These findings support the hypothesis that K_{ATP} channels contribute to the hypoxic hyperpolarization in various brain areas. However, there are also data confirming the suggestion that these channels might not have a protective function during hypoxia. In neocortical neurons which respond only with depolarization to hypoxia, K_{ATP} channels are not opened by diazoxide or hypoxia alone; apparently only the combination of these two manipulations become effective (Pissarek et al., 1998). In the presence of glucose, blockade of K_{ATP} channels by sulfonylurea antidiabetics does not lead to a progessive hypoxic depolarization in a group of dorsal vagal neurons in the brainstem (Ballanyi et al., 1996). All these results point out that the importance of the K_{ATP} channels in particular for providing an adaptive central nervous mechanism varies between the brain regions and depends on the experimental conditions.

Role of adenosine

Adenosine is an endogenous nucleoside which is released under pathophysiological conditions including hypoxia (Rudolphi et al., 1992; Wallman-Johansson and Fredholm, 1994; Latini et al., 1995). It has been shown that adenosine plays an important role in the regulation of neuronal functions. The main source of adenosine is the dephosphorylation of 5'-AMP (Meghji, 1993). Adenosine, released by energy depletion, is formed predominantly intracellularly (Lloyd et al., 1993). Another way to gain adenosine is the hydrolysis of S-adenosylhomocysteine. However, investigations on hippocampal neurons exclude a significant contribution of the transmethylation pathway to adenosine accumulation during hypoxia (Latini et al., 1995).

Adenosine exerts its action via membrane bound receptors. To date, four G-protein coupled adenosine receptors have been described: A_1-, A_{2A}-, A_{2B}- and A_3-receptors (Collis and Hourani, 1993; Fredholm et al., 1994; Palmer and Stiles, 1995). The A_1- and A_2-receptors, which couple negatively (A_1) or positively (A_2) to adenylate cyclase (Fredholm et al., 1994), were characterized pharmacologically (van Calker et al., 1979). The adenosine A_3-receptor was defined and characterized after its cloning (Meyerhof et al., 1991; Zhou et al., 1992). The stimulation of phospholipase C activity represents a principal transduction mechanism for this receptor in the brain (Abbracchio et al., 1995). Adenosine receptor subtypes have been shown to exist in different areas of the central nervous system (Fastbom et al., 1987; Sebastiao and Ribeiro, 1996). Adenosine is involved in several steps along the pathophysiological cascade resulting from hypoxia. Various studies support the idea that adenosine is a neuroprotective agent in cerebral hypoxia or ischemia (Rudolphi et al., 1992; Schubert and Kreutzberg, 1993). In vivo, the administration of adenosine analogues (Daval et al., 1989), an adenosine deaminase antagonist (Phillis and O'Regan, 1989), or an adenosine uptake blocker (Andine et al., 1990) offers neuroprotection in the hippocampal CA1 region after transient forebrain ischemia. The upregulation of adenosine A_1-receptors by chronic administration of caffein also protects CA1 region from ischemic damage (Rudolphi et al., 1989). Thus, it appears that adenosine increases the threshold for postischemic brain damage. In vitro, adenosine contributes to the hypoxic hyperpolarization and immediately depresses synaptic transmission in the central nervous system in different brain regions (Schubert and Mitzdorf, 1979; Fowler, 1989; Gribkoff and Bauman, 1992; Katchman and Hershkowitz, 1993; Phillis et al., 1997). Adenosine prevents the hypoxia-induced membrane depolarization and this leads to a reduction of Ca^{2+} influx. However, adenosine is not only responsible for the initial phase of synaptic suppression but also for the sustained suppression of synaptic transmission during prolonged moderate hypoxia (Arlinghaus and Lee, 1996). Several mechanisms were found by which adenosine acts neuroprotectively. The inhibitory effect of adenosine is mainly due to the inhibition of neurotransmitter release, especially glutamate from presynaptic nerve terminals (Correia-de-Sa et al., 1991) and hyperpolarization by opening K^+ channels on the postsynaptic side in hippocampal (Croning et al., 1995) and locus coeruleus (Nieber et al., 1995) neurons. Additionally, adenosine reduces the Ca^{2+} influx and facilitates the resynthesis of ATP during recovery (Mori and Mishino, 1995). By inhibiting neuronal Ca^{2+} influx, adenosine counteracts the presynaptic release of the excitatory neurotransmitter glutamate, which enhances the intracellular Ca^{2+} concentration by activation of metabotropic glutamate receptors and induces membrane depolarization by activation of ion channel-linked glutamate receptors.

The mechanisms underlying the effects of adenosine seem to involve adenosine A_1- and A_2-receptors, but to date the A_1-receptors may be of special importance. Activation of A_1-receptors by selective receptor agonists prevents or blocks the electrophysiolgical correlates of hypoxia in hippocampal cells (Domenici et al., 1996) or has a protective function in hypoxia-induced decrease in glucose metabolism (Tominaga et al., 1992). In pyramidal cells of the somatosensory cortex the stable adenosine A_1-receptor analogue cyclopentyl-adenosine (CPA) decreases the hypoxic depolarization (Fig. 1). The effect of CPA is antagonized by the adenosine A_1-receptor antagonist DPCPX (Fig. 2). Furthermore, adenosine A_1-receptor antagonists can attenuate hypoxia-induced increase in $[K^+]_o$ and prevent transmission failure but do not attenuate hypoxia-induced hyperpolarization (Croning et al., 1995).

Very little is known about the role of adenosine A_2-receptors during hypoxia. An excitatory effect of a selective A_2-receptor agonist (CGS 21680) has been reported for rat hippocampus slices (Sebastiao and Ribeiro, 1992). Another group did not find any effect on synaptic transmission by this agonist (Longo et al., 1995). Together with in vivo studies (von Lubitz et al., 1995) it seems likely that adenosine A_2-receptors play a limited role in hypoxic changes. The A_2-receptor mediated adenosine action might be of interest in other neurodegenerative processes such as schizophrenia or Parkinsoń's disease (Sebastiao and Ribeiro, 1996).

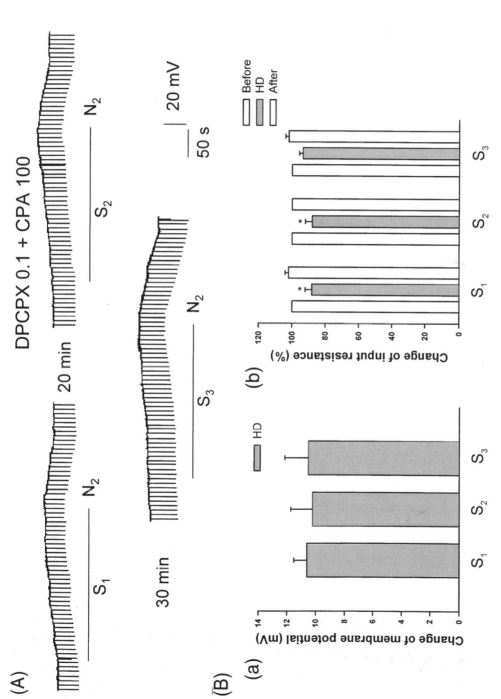

Fig. 2. Effects of cyclopentyl adenosine (CPA) in the presence of DPCPX on hypoxic changes of the membrane potential and input resistance of neocortical pyramidal cells. (A) The experimental procedure was similar to that shown in Fig. 1. Additionally, DPCPX (0.1 μM) was applied ten minutes before S₂ and was present during and fifteen minutes after S₂. Hypoxic changes were measured in five experiments similar to that indicated in (A). (B) Means SEM are presented. The magnitude of the HD during the consecutive hypoxic stimuli is indicated by filled columns (a). The control HD was not changed when CPA was applied in presence of DPCPX. The percentage change of input resistance before (first set of filled columns in each group, designated as 100%), during (filled columns in each group) and after the HD (second set of open columns in each group) is indicated (b). Hypoxia depressed the input resistance in the mean of five experiments. CPA (100 μM) applied in presence of DPCPX (0.1 μM) did not influence this alteration. In addition, DPCPX (0.1 μM) failed to influence both the membrane potential and input resistance of the neurons. *P < 0.05 significant difference from the control.

A perspective tool for further investigations on the role of adenosine might be the adenosine A_3-receptor. Adenosine A_3-receptor agonists have been shown to stimulate the phospholipase C hydrolysis and IP_3 generation in rat brain slices. An increase of inositol triphosphate triggers the Ca^{2+}-release from an inositole triphosphate sensitive internal Ca^{2+}-store (Berridge and Irvine, 1989) and contributes to the cell injury by increasing $[Ca^{2+}]_i$ (Abbracchio et al., 1995). This hypothesis is supported by in vivo experiments in which acute stimulation of the A_3-receptor by selective agonists causes an impaired postischemic cerebral blood flow, enhanced mortality and extensive neuronal changes in the hippocampus (von Lubitz et al., 1994). In vitro experiments indicate that high concentrations of the adenosine A_3-receptor agonist 2-chloro-IB-MECA induce neurotoxicity in neurons and subcytotoxic concentrations augmented glutamate neurotoxicity in rat cerebellar granule neurons (Sei et al., 1997). Further studies with selective adenosine A_3-receptor antagonists are necessary to investigate the role of this receptor subtype during and after hypoxia.

Conclusions

It is well known that the presence of oxygen and glucose are indispensible to the maintenance of neural function in the mammalian brain. In these animals, hypoxia causes irreversible brain damage and neuronal activity cannot recover even after relatively short periods of hypoxia. Experiments using brain slices enable us to evaluate the relationship between membrane parameters and energy metabolism during hypoxia. In addition to the lack of high energy phosphates, during hypoxia the release of excitatory amino acid transmitters is enhanced and depolarization of neuronal membranes due to accumulation of K^+ in extracellular space as well as that of intracellular Ca^{2+} can lead to the damage of neurons. However, in some neurons the early depolarization is followed by a hyperpolarization. This phenomenon has been considered to be due to the opening of K^+ channels. Pharmacological experiments on different brain regions indicate the involvement of K_{ATP} channels. The K_{ATP} channels might be functionally relevant for providing an adaptive mechanism, which serves to regulate membrane excitability during hypoxia.

Recent experimental data indicate a probable role of adenosine as an endogenous neuroprotective substance in brain hypoxia. This nucleoside is rapidly formed during hypoxia as a result of breakdown of ATP. It acts on adenosine receptors that are present in most neuronal cells and that produce cellular effects that tend to antagonize some processes which contribute to the cell death.

Acknowledgement

The work in the laboratory of Dr. K. Nieber was supported by the Bundesministerium für Bildung, Forschung und Technologie (BMB + F), Interdisziplinäres Zentrum für Klinische Forschung (IZKF) at the University of Leipzig (01KS9504, C4/C12).

References

Abbracchio, M.P., Brambilla, R., Ceruti, S., Kim, H.O., von Lubitz, D.K.J.E., Jacobson, K.A. and Cattabeni, F. (1995) G protein-dependent activation of phospholipase C by adenosine A_3 receptors in rat brain. *Mol. Pharmacol.*, 48: 1038–1045.

Aitken, P.G., Breese, F.R., Dudek, F.F., Edwards, F., Espanol, M.T., Larkman, P.M., Lipton, P., Newman, G.C., Nowak Jr., T. S., Panizzon, K.L., Raley-Susman, K.M., Reid, K.H., Rice, M.E., Sarvey, J.M., Schoepp, D.D., Segal, M., Taylor, C.P., Teyler, T.J. and Voulalsa, P.J. (1995) Preparative methods of brain slices: A discussion. *J. Neurosci. Meth.*, 59: 139–149.

Amoroso, S., Schmid-Antomarchi, H., Fosset, M. and Lazdunski, M. (1990) Glucose, sulfonylureas, and neurotransmitter release: Role of ATP-sensitive K^+ channels. *Science*, 247: 852–854.

Andine, P., Rudolphi, K.A., Fredholm, B.B. and Hagberg, H. (1990) Effect of propentofylline (HWA 285) on extracellular purines and excitatory amino acids in CA1 of rat hippocampus during transient ischaemia. *Br. J. Pharmacol.*, 100: 814–818.

Arlinghaus, L. and Lee, K.S. (1996) Endogenous adenosine mediates the sustained inhibition of excitatory synaptic transmission during moderate hypoxia. *Brain Res.*, 724: 265–268.

Ashcroft, F.M. (1988) Adenosine 5′-triphosphate-sensitive potassium channels. *Annu. Rev. Neurosci.* 11: 97–118.

Ashcroft, S.J.H. and Ashcroft, F.M. (1990) Properties and functions of ATP-sensitive K-channels. *Cell. Signalling*, 2: 197–214.

Ballanyi, K., Doutheil, J. and Brockhaus, J. (1996) Membrane potentials and microenvironment of rat dorsal vagal cells in vitro during energy depletion. *J. Physiol. (Lond.)*, 495: 769–784.

Ben-Ari, Y. and Lazdunski, M. (1989) Galanin protects hippocampal neurons from the functional effect of anoxia. *Eur. J. Pharmacol.*, 165: 331–332.

Ben-Ari, Y., Krnjevic, K. and Crepel, V. (1990) Activators of ATP-sensitive K$^+$ channels reduce anoxic depolarization in CA3 hippocampal neurons. *Neuroscience*, 37: 55–60.

Berridge, M.J. and Irvine, R.F. (1989) Inositol phosphates and cell signalling. *Nature*, 341: 197–205.

Calabresi, P., Pisani, A., Mercuri, N.B. and Bernardi, G. (1995) On the mechanisms underlying hypoxia-induced membrane depolarization in striatal neurons. *Brain*, 118: 1027–1038.

Carter, A.J., Müller, E., Pschorn, U. and Stransky, W. (1995) Preincubation with creatine enhances levels of creatine phosphate and prevents anoxic damage in rat hippocampal slices. *J. Neurochem.*, 64: 2691–2699.

Collis, M.G. and Hourani, S.M.O. (1993) Adenosine receptor subtypes. *Trends Neurosci.*, 14: 360–366.

Correia-de-Sa, P., Sebastiao, A.M. and Ribeiro, J.A. (1991) Inhibitory and excitatory effects of adenosine receptor agonists on evoked transmitter release from phrenic nerve endings of the rat. *Br. J. Pharmacol.*, 103: 1614–1620.

Croning, M.D., Zetterstrom, T.S., Grahame-Smith, D.G. and Newberry, N.R. (1995) Action of adenosine receptor antagonists on hypoxia-induced effects in the rat hippocampus in vitro. *Br. J. Pharmacol.*, 116: 2113–2119.

Daval, J.L., von Lubitz, D.K., Deckert, J., Redmons, D.J. and Marangos, P.J. (1989) Protective effect of cyclohexyladenosine on adenosine A1-receptors, guanine nucleotide and forskolin binding sites following transient brain ischemia: A quantitative autoradiografic study. *Brain Res.*, 491: 212–226

Domenici, M.R., de Carolis, A.S. and Sagratella, S. (1996) Block by N6-L-phenylisopropyl-adenosine of the electrophysiological and morphological correlates of hippocampal ischaemic injury in the gerbil. *Br. J. Pharmacol.*, 118: 1551–1557.

Donelly D.F., Jiang, C. and Haddad, G.G. (1992) Comparative responses of brainstem and hippocampal neurons to O$_2$ deprivation: In vitro intracellular studies. *Am. J. Physiol.*, 261: L549–L554.

Duffy, T.E., Kohle, S.J. and Vannucci, R.C. (1975) Carbohydrate and energy metabolism in perinatal rat brain: Relation to survival in anoxia. *J. Neurochem.* 24: 271–276.

Erdemli, G. and Krnjevic, K. (1996) Is activation of metabotropic glutamate receptors responsible for acute hypoxic changes in hippocampal neurons? *Brain Res.*, 723: 1–7.

Espanol, M.T., Litt, L., Chang, L.H., Weinstein, P.R. and Chan, P.H. (1996) Adult rat brain slice preparation for nuclear magnetic resonance spectroscopy studies of hypoxia. *Anesthesiology*, 84: 201–210.

Fastbom, J., Pazos, A. and Palacios, J.M. (1987) The distribution of adenosine A$_1$ receptors and 5′-nucleotidase in the brain of some commonly used experimental animals. *Neuroscience*, 22: 813–826.

Finta, E.P., Harms, L., Sevcik, J., Fischer, H.D. and Illes, P. (1993) Effect of potassium channel openers and their antagonists on rat locus coeruleus neurones. *Br. J. Pharmacol.*, 109: 308–315.

Fowler, J.C. (1989) Adenosine antagonists delay hypoxia-induced depression of neuronal activity in hippocampal brain slice. *Brain Res.*, 490: 378–384.

Fowler, J.C. (1992) Escape from inhibition of synaptic transmission during in vitro hypoxia and hypoglycemia in the hippocampus. *Brain Res.*, 573: 169–173.

Fredholm, B.B., Abbracchio, M.P., Burnstock, G. Daly, J.W., Harden, T.K., Jacobson, K.A., Leff, P. and Williams, M. (1994) Nomenclature and classification of purinoceptors. *Pharmacol. Rev.*, 46: 143–156.

Fujimura, N., Tanaka, E., Yamamoto, S., Shigemori, M. and Higashi, H. (1997) Contribution of ATP-sensitive potassium channels to hypoxic hyperpolarization in rat hippocampal CA1 neurons in vitro. *J. Neurophysiol.*, 77: 378–385.

Fujiwara, N., Higashi, H., Shimoji, K. and Yoshimura, M. (1987) Effects of hypoxia on rat hippocampal neurones in vitro. *J. Physiol. (Lond.)*, 384: 131–151.

Godfraind, J.M. and Krnjevic, K. (1993) Tolbutamide suppresses anoxic outward current of hippocampal neurons. *Neurosci. Lett.*, 162: 101–104.

Gribkoff, V.K. and Bauman, L.A. (1992) Endogenous adenosine contributes to hypoxic synaptic depression in hippocampus from young and aged rats. *J. Neurophysiol.* 68: 620–628.

Grigg, J.J. and Anderson, E.G. (1989) Glucose and sulfonylureas modify different phases of the membrane potential change during hypoxia in rat hippocampal slices. *Brain Res.*, 489: 302–310.

Haddad, G.G. and Donnelly, D.F. (1990) O$_2$ deprivation induces a major depolarization in brain stem neurons in the adult but not in the neonatal rat. *J. Physiol. (Lond.)*, 429: 4–11–428.

Haddad, G.G. and Jiang, C. (1994) Mechanisms of neuronal survival during hypoxia: ATP-sensitive K$^+$ channels. *Biol. Neonate*, 65: 160–165.

Hansen, A.J., Hounsgaard, J. and Johnsen, H. (1982) Anoxia increases potassium conductance in hippocampal nerve cells. *Acta Physiol. Scand.*, 115: 301–310.

Hisanaga, K., Onodera, H. and Kogure, K. (1986) Changes in levels of purine and pyrimidine nucleotides during acute hypoxia and recovery in neonatal rat brain. *J. Neurochem.*, 47: 1344–1350.

Hsu, K.S. and Huang, C.C. (1997) Characterization of the anoxia-induced long-term synaptic potentiation in area CA1 of the rat hippocampus. *Br. J. Pharmacol.*, 122: 671–681.

Illes, P., Sevcik, J., Finta, E.P., Fröhlich, R., Nieber, K. and Nörenberg, W. (1994) Modulation of locus coeruleus neurons by extra- and intracellular adenosine 5′-triphosphate. *Brain Res. Bull.*, 35: 513–519.

Jiang, C. and Haddad, G.G. (1991) Effect of anoxia on intracellular and extracellular potassium activity in hypoglossal neurons in vitro. *J. Neurophysiol.*, 66: 103–111.

Kass, I.S. and Lipton, P. (1986) Protection of hippocampal slices from young rats against anoxic transmission damage is due to better maintenance of ATP. *J. Physiol. (Lond.)*, 413: 1–11.

Katchman, A.N. and Hershkowitz, N. (1993) Adenosine antagonists prevent hypoxia-induced depression of excitatory but not inhibitory synaptic currents. *Neurosci. Lett.*, 159: 123–126.

Krieglstein, J. (1996) *Pharamacology of cerebral ischemia 1996*. Medipharm Scientific Publischers, Stuttgart.

Krnjevic, K. and Ben-Ari, Y. (1989) Anoxic changes in dentate granule cells. *Neurosci. Lett.*, 107: 89–93.

Latini, S., Corsi, C., Pedata, F. and Pepeu, G. (1995) The source of brain adenosine outflow during ischemia and electrical stimulation. *Neurochem. Int.*, 27: 239–244.

Lazdunski, M. (1994) ATP-sensitive potassium channels: An overview. *J. Cardiovasc. Pharmacol.*, 24 (Suppl 4): S1–S5.

Leblond, J. and Krnjevic, K. (1989) Hypoxic changes in hippocampal neurons. *J. Neurphysiol.*, 62: 1–14.

Lloyd, H.G.E., Lindström, K. and Fredholm, B.B. (1993) Intracellular formation and release of adenosine from rat hippocampal slices evoked by electrical stimulation or energy depletion. *Neurochem. Int.*, 23: 173–185.

Longo, R., Zeng, Y.C. and Sagratell, S. (1995) Opposite modulation of 4-aminopyridine and hypoxic hyperexcitability by A_1 and A_2 adenosine receptor ligands in rat hippocampal slices. *Neurosci. Lett.*, 200: 21–24.

Luhmann, H.J. (1996) Ischemia and lesion induced imbalances in cortical function. *Prog. Neurobiol.*, 48: 131–166.

Luhmann, H.J. and Heinemann, U. (1992) Hypoxia-induced functional alterations in adult rat neocortex. *J. Neurophysiol.*, 67: 798–811.

Meghji, P. (1993) Storage, release, uptake, and inactivation of purines. *Drug Dev. Res.*, 28: 214–219.

Meyerhof, V., Müller-Brechlin, R. and Richter, D. (1991) Molecular cloning of a novel putative G-protein coupled receptor expressed during rat spermiogenesis. *FEBS Lett.*, 2884: 155–160.

Milusheva, E.A., Doda, M. Baranyi, A. and Vizi, E.S. (1996) Effect of hypoxia and glucose deprivation on ATP level, adenylate energy charge and $[Ca^{2+}]_0$-dependent and independent release of [^{3}H]-dopamine in rat striatal slices. *Neurochem. Int.*, 28: 501–507.

Morawietz, G., Ballanyi, K., Kuwana, S. and Richter, D.W. (1995) Oxygen supply and ion homeostasis of the respiratory network in the in vitro perfused brainstem of adult rats. *Exp. Brain Res.*, 106: 265–274.

Mori, H. and Mishina, M. (1995) Neurotransmitter receptors VIII: Structure and function of the NMDA receptor channel. *Neuropharmacology*, 34: 1219–1237.

Murphy, K.P. and Greenfield, S.A. (1992) Neuronal selectivity of ATP-sensitive potassium channels in guinea-pig substantia nigra revealed by responses to anoxia. *J. Physiol. (Lond.)*, 453: 167–183.

Nabetani, M., Okada, Y., Kawai, S. and Nakamura, H. (1995) Neural activity and the levels of high energy phosphates during deprivation of oxygen and/or glucose in hippocampal slices of immature and adult rats. *Int. J. Devi. Neuroscience*, 13: 3–13.

Nieber, K., Sevcik, J. and Illes, P. (1995) Hypoxic changes in rat locus coeruleus neurons in vitro. *J. Physiol. (Lond.)*, 486: 33–46.

O'Reilly, J.P., Jiang, C. and Haddad, G.G. (1995) Major differences in response to graded hypoxia between hypoglossal and neocortical neurons. *Brain Res.*, 683: 179–186.

Onodera, H., Iijima, K. and Kogure, K. (1986) Mononucleotide metabolism in the rat brain after transient ischemia. *J. Neurochem.*, 46: 1704–1710.

Palmer, T.M. and Stiles, G.L. (1995) Adenosine receptors. *Neuropharmacology*, 34: 683–694.

Paschen W. and Djuricic, B. (1995) Comparison of in vitro ischemia-induced disturbances in energy metabolism and protein synthesis in the hippocampus of rats and gerbils. *J. Neurochem.*, 65: 1692–1697

Perez-Pinzon, M.A., Tao, L. and Nicholson, C. (1995) Extracellular potassium, volume fraction, and tortuosity in rat hippocampal CA1, CA3, and cortical slices during ischemia. *J. Neurophysiol.*, 74: 565–573.

Phillis, J.W., Edstrom, J.P. Kostopoulos, G.K. and Kirkpatrick, J.R. (1997) Effects of adenosine and adenine nucleotides on synaptic transmission in the cerebral cortex. *Can. J. Physiol. Pharmacol.*, 57: 1289–1312.

Phillis, J.W. and O'Regan, M.H. (1989) Deoxycoformycin antagonizes ischemia-induced neuronal degeneration. *Brain Res. Bull.*, 22: 537–540.

Pisani, A., Calabresi, P. and Bernardi, G. (1997) Hypoxia in striatal and cortical neurones: membrane potential and Ca^{2+} measurements. *Neuroreport*, 8: 1143–1147.

Pissarek, M., Garcia de Arriba, S., Schäfer, M., Sieler, D., Nieber, K. and Illes, P. (1998) Changes by short-term hypoxia in the membrane properties of pyramidal cells and the levels of purine and pyrimidine nucleotides in slices of rat neocortex; effects of agonists and antagonists of ATP-dependent potassium channels. *Naunyn-Schmiedeberg's Arch. Pharmacol.*, 358: 430–439.

Rosen, A.S. and Morris, M.E. (1991) Depolarizing effects of anoxia on pyramidal cells of the rat neocortex. *Neurosci. Lett.*, 124: 169–173.

Rudolphi, K.A., Schubert, P., Parkinson, F.E. and Fredholm, B.B. (1992) Neuroprotective role of adenosine in cerebral ischaemia. *Trends Pharmacol. Sci.*, 13: 439–445.

Rudolphi, K.A., Keil, M., Fastbom, J., Fredholm, B.B. (1989) Ischemic damage in gerbil hippocampus is reduced following upregulation of adenosine (A1) receptors by caffeine treatment. *Neurosci. Lett.*, 103: 275–288.

Schubert, P. and Kreutzberg, G.W. (1993) Cerebral protection by adenosine. *Acta Neurochir. Suppl. (Wien)*, 57: 80–88.

Schubert, P. and Mitzdorf, U. (1979) Analysis and quantitative evaluation of the depressive effect of adenosine on evoked potentials in hippocampal slices. *Brain Res.*, 172: 186–190.

Sebastiao, A.M. and Ribeiro, J.A. (1992) Evidence for the presence of excitatory A_2 adenosine receptors in the rat hippocampus. *Neuropharmacology*, 26: 1181–1985.

Sebastiao, A.M. and Ribeiro, J.M. (1996) Adenosine A_2 receptor-mediated excitatory actions on the nervous system. *Prog. Neurobiol.*, 48: 167–189.

Sei, Y., von Lubitz, D.K.J.E., Abbracchio, M.P., Ji, X.D. and Jacobson, K.A. (1997) Adenosine A_3 receptor agonist-induced neurotoxicity in rat cerebellar granule neurons. *Drug Dev. Res.*, 40: 267–273.

Skou, J.C. (1992) The Na,K-ATPase. *J. Bioenerg. Biomembr.*, 24: 249–261.

Tanaka, E., Yamamoto, S. Kudo, Y., Mihara, S. and Higashi, H. (1997) Mechanisms underlying the rapid depolarization produced by deprivation of oxygen and glucose in rat hippocampal CA1 neurons in vitro. *J. Neurophysiol.*, 78: 891–902

Tominaga, K., Shibata, S. and Watanabe, S. (1992) A neuroprotective effect of adenosine A_1-receptor agonist on ischemia-induced decrease in 2-deoxyglucose uptake in rat hippocampal slices. *Neurosci. Lett.*, 145: 67–70.

Trapp S. and Ballanyi K. (1995) K_{ATP} channel mediaton of anoxia-induced outward current in rat dorsal vagal neurons in vitro. *J. Physiol. (Lond.)*, 487: 37–50.

Tromba, C., Salvaggio, A., Racagni, G. and Volterra, A. (1992) Hypoglycemia-activated K^+ channels in hippocampal neurons. *Neurosci. Lett.*, 143: 185–189.

van Calker, D., Müller, M. and Hamprecht, B. (1979) Adenosine regulates via two different types of receptors, the accumulation of cyclic AMP in cultured brain cells. *J. Neurochem.*, 33: 999–1005.

von Lubitz, D.K., Lin, R.C. and Jacobson, K.A. (1995) Cerebral ischemia in gerbils: effects of acute and chronic treatment with adenosine A_{2A} receptor agonist and antagonist. *Eur. J. Pharmacol.*, 287: 295–302.

von Lubitz, D.K., Lin, R.C., Popik, P., Carter, M.F. and Jacobson, K.A. (1994) Adenosine A_3 receptor stimulation and cerebral ischemia. *Eur. J. Pharmacol.*, 263: 59–67.

Wallman-Johansson, A. and Fredholm, B.B. (1994) Release of adenosine and other purines from hippocampal slices stimulated electrically or by hypoxia/hypoglycemia. Effect of chlormethiazole. *Life Sci.*, 55: 721–728.

Watts, A.E., Hicks, G.A. and Henderson, G. (1995) Putative pre- and postsynaptic ATP-sensitive potassium channels in the rat substantia nigra in vitro. *J. Neurosci.*, 15: 3065–3074.

Yang, J.J., Chou, Y.C., Lin, M.T. and Chiu, T.H. (1997) Hypoxia-induced differential electrophysiological changes in rat locus coeruleus neurons. *Life Sci.*, 61: 1763–1773.

Yoneda, K. and Okada, Y. (1989) Effects of anoxia and recovery on the neurotransmission and level of high-energy phosphates in thin hippocampal slices from the guinea-pig. *Neuroscience*, 28: 401–407.

Zhang, L. and Krnjevic, K. (1993) Whole-cell recording of anoxic effects on hippocampal neurons in slices. *J. Neurophysiol.*, 69: 118–127.

Zhou, Q.Y., Ci, C., Olah, M.E., Johnson, R.A., Stiles, G.L. and Civelli, O. (1992) Molecular cloning and characterization of an adenosine receptor: The A_3 adenosine receptor. *Proc. Natl. Acad. Sci. USA*, 89: 7432–7436.

Zini, S., Roisin, M.P., Armengaud, C. and Ben-Ari, Y. (1993) Effect of potassium channel modulators on the release of glutamate induced by ischaemic-like conditions in rat hippocampal slices. *Neurosci. Lett.*, 153: 202–205.

Nucleotide effects on neuronal differentiation and glial proliferation

P. Illes and H. Zimmermann (Eds.)
Progress in Brain Research, Vol 120
© 1999 Elsevier Science BV. All rights reserved

CHAPTER 24

Adenosine and P2 receptors in PC12 cells. Genotypic, phenotypic and individual differences

Giulia Arslan and Bertil B. Fredholm*

Department of Physiology and Pharmacology, Karolinska Institutet, 171 77 Stockholm, Sweden

Introduction

PC12 cells, derived from a spontaneous rat pheochromocytoma (Greene and Tischler, 1976), are widely used to study neuronal differentiation, effects of growth factors and the regulation of neurotransmitter release. Wild-type PC12 cells differentiate to a more neuronal phenotype upon treatment with, for example, nerve growth factor (NGF) and basic fibroblast growth factor (bFGF) (Greene et al., 1991). But other types of signals may also be of critical importance in the outgrowth of nerve fibres and in their finding of their correct targets. ATP and its breakdown product adenosine could provide such a signal.

Adenosine, acting on so-called A_{2A} adenosine receptors, increases cyclic AMP formation in PC12 cells (Hide et al., 1992; van der Ploeg et al., 1996). Adenosine analogues and forskolin can induce neurite outgrowth in PC12 cells (Roth et al., 1991). It has been shown that adenosine can act synergistically with NGF to induce neurite outgrowth in PC12 cells (Rathbone et al., 1992), whereas it antagonises NGF effects on MAP-kinase activity (Arslan et al., 1997).

The effects of ATP and related nucleotides on PC12 cells have been extensively investigated. In PC12 cells ATP is a powerful stimulator of catecholamine release (Inoue et al., 1989) probably via an increase in the intracellular levels of Ca^{2+} (Fasolato et al., 1990). The ATP stimulated increase in Ca^{2+} appears to be due to several actions— possibly mediated via different receptors. There are, however, marked quantitative as well as qualitative differences between the results of different studies. The culture conditions of PC12 cells as reported differ between laboratories and PC12 cells have been shown to be able to change their phenotype, not only when cultured with different differentiating factors such as neurotrophins, but also spontaneously (Greene et al., 1991; Clementi et al., 1992b).

We decided therefore to investigate the possible sources of variability, which can be schematically subdivided into three levels: first, genetic differences within the same cell clone due to spontaneous mutations of some cells leading to the coexistence of different cell types; second, phenotypic differences due to differentiation induced by specific substances such as NGF; third, differences that can be revealed even in the same clone in the same phenotypic phase. We used four PC12 clones, isolated based on neomycin resistance, which show a relatively stable phenotype and have been employed for studies of calcium movements (Fasolato et al., 1991; Zacchetti et al., 1991; Clementi et al., 1992c).

*Corresponding author. Tel.: +46-8-728-79-40; fax: +46-8-34-12-80; e-mail: Bertil.Fredholm@fyfa.ki.se

Materials and methods

Materials

Cell culture media, foetal calf and horse serum, and cell culture flasks were from NordCell (Bromma, Sweden). Mouse 2.5S nerve growth factor (NGF) was from Promega (Madison, WI, USA). D-myo-2-[^{3}H] Inositol 1,4,5-trisphosphate and [^{3}H]-cAMP were from Amersham and 2,8-[^{3}H] adenosine 3′,5′-cyclic monophosphate (44.5 Ci/mM) was from New England Nuclear. Fura-2 free acid, fura 2-AM, ionomycin, bradykinin, ADP, UTP, α,β-methylene ATP, nifedipine, inositol 1,4,5-trisphosphate, EGTA, and forskolin were all from Sigma (St. Louis, MO, USA). Polynucleotide-kinase and γ-^{32}P-ATP were from Amersham. ATP was from Fluka, and suramin was from Calbiochem. 5′-N-ethylcarboxamidoadenosine (NECA) and 2-[p-(2-carbonylethyl) phenylethylamino-5′-N-ethylcarboxamidoadenosine (CGS 21680) and PPADS were from Research Biochemical IS International (Wayland, MA, USA). [^{3}H]-SCH 58 261 ([^{3}H]-5 - amino - 7 - (2 - phenylethyl) - 2 - (2-furyl)-pyrazolo-[4,3-e]-1,2,4 triazolo[1,5-c]pyrimidine) was given to us by Ennio Ongini, Schering Plough (Milan, Italy). The anti-A$_{2A}$ antibody was a kind gift from Dr. Joel Linden (Charlottesville, VA, USA). Antibodies against G proteins, vinculin and phospho-tyrosine were from Santa Cruz Biotechnology Inc. (Santa Cruz, CA, USA). Cy3-labelled donkey anti-rabbit, LRSC-labelled goat anti-rabbit and FITC donkey anti-mouse antibodies were purchased from Jackson Immunochemicals (West Grove, PA, USA).

Cell culture

The PC12 cells were subcultured two times weekly at a density of 400,000/ml. The cells were induced to differentiate by treatment with nerve growth factor (50 ng/ml) for three to seven days or by addition of 10 μM forskolin together with 30 μM rolipram for three days.

Biochemical methods

A$_{2A}$ receptor binding

Binding of [^{3}H]-SCH 58261 to A$_{2A}$ receptors was studied as previously described (Zocchi et al., 1996) using adenosine deaminase treated membranes of PC12 cells. Assays were carried out in GF/B filter Millipore plates in triplicates in a final volume of 0.3 ml. Non-specific binding was determined in the presence of 50 μM NECA.

Protein analysis by Western blotting

Proteins were run on a 12% polyacrylamide, denaturing gel, blotted onto a membrane (Immobilon, Millipore, Bedford, MA, USA) and non-specific binding blocked by an overnight incubation in Tris-buffered saline (TBS; Tris-HCl, pH 8, 150 mM NaCl) containing Tween-20 (0.5%) and 5% dry milk. Membranes were incubated with primary antibodies for 1 hour, rinsed several times in Tris-buffered saline, incubated with a secondary antibody (horseradish peroxidase conjugated anti-rabbit) and developed by an enhanced chemiluminescence system (Amersham, GB).

cAMP measurements

The method described earlier (Nordstedt and Fredholm, 1990), using a cell harvester (Skatron AS) to separate free and bound [^{3}H]-cAMP over glass fibre filters was used.

Inositol 1,4,5-trisphosphate measurements

InsP$_3$ was measured in deproteinised extracts of PC12 cells using a modified competitive binding assay as described previously (Gerwins, 1993).

RT-PCR

Total RNA was isolated using the Ultraspec™ RNA Isolation system (BIOTECH BULLETIN NO:27, 1992) and resuspended in diethyl pyrocarbonate treated water. cDNA was synthesised by reverse transcription of total RNA. Oligonucleotide primers for the rat adenosine receptors were: [5′ - CGAATTCAACCTGCAGAACGTCACC - 3′, sense] and [5′-TCGAATTCGCGGTC(G/A)-ATGGCGAT(A/G)-3′, antisense] for the A_{2A} receptors, [5′-CAGAC(G/C)CCCACCAACTACTT-3′, sense] and [5′-GCCACCA(G/T)GAAGAT(C/T)-TT(A/G)ATG-3′, antisense] for the A_{2B} receptor, [5′-TCGAATTCCAGGCTGCCTACATTGGCAT-3′, sense] and [5′-TCGAATTCCAGAAAGGT-GACCCGGAACT-3′, antisense] and for the A_1 receptor, [5′-AGGTTTGAGTCCAAAGAAT-CCG-3′, sense] and [5′-AATACCACGAC-GAGTGCCTTGT-3′, antisense] for the A_3 receptor. They are expected to code for products of 301 (A_1), 216 (A_{2A}), 513 (A_{2B}) and 922 (A_3) bp.

Primers for the rat P2 receptors were: [5′-GGTGGGTGTTTGTCTATGAA-3′], [3′-AGT-GTTCTAAGTTGTGGGGG-5′] for the $P2X_1$ receptor, [5′-TACCCCTTCCCAGACCTT-3′], [3′-CTTGACCGTGTGTTCCCG-5′] for the $P2X_2$ receptor, [5′-GGGGTACCGGTTGCGAACTG-GAAAGATGTGGTT-3′] [3′-TATCCGCGGGG-TTGATGTTGGGGAGGATGTT-5′] for the $P2X_4$ receptor and [5′-ACGCCTCCACCACCTACA-3′] [3′-CAACACACCCACGACCAC-5′] for the $P2Y_2$ receptor. These primers are expected to code for products of 540 ($P2X_1$), 430 ($P2X_2$), 201 ($P2X_4$) and 287 ($P2Y_2$) bp. Primers were chosen using Oligo primer Analysis Software (4.0) and were synthesised by Scandinavian Gene Synthesis, Köping, Sweden. The resulting single stranded cDNA was submitted to 35 cycles of PCR under standard conditions.

Calcium measurements using microfluorimetry or imaging

For calcium measurements cells were plated at a density of $1.5-2 \times 10^5$ in six well plates for 48–72 h. On the day of experiment the culture medium was replaced with a solution of the following composition: 120 mM NaCl, 5 mM KCl, 1.6 mM Mg_2SO_4, 1.2 mM KH_2PO_4, 1.2 mM $CaCl_2$, 20 mM HEPES, 10 mM D-glucose and 0.1% bovine serum albumin (pH adjusted to 7.4 with NaOH). Fura 2 was loaded in cytosol by incubation with 2 μM Fura 2-AM, for 40 min, in the dark, at room temperature. Under these conditions the compartmentalisation of the dye is minimal, as judged by remaining fluorescence ($4.8 \pm 1.1\%$, $n = 4$) after selective permeabilisation of plasmalemma with 10-15 μg/ml digitonin. At the end of incubation period, cells were washed several times with Fura free solution and allowed at least 15 min for complete de-esterification of the dye.

The coverslip was then mounted in a special designated chamber (capacity 0.150 ml) in the stage of an inverted fluorescence microscope (Zeiss, Axiovert 35) equipped for fura 2 microfluorimetry. The temperature of the chamber was maintained at 35°C and the cells were perfused at a rate of 0.9 ml/min. The entire bath solution was changed in less than six seconds.

Excitation at the two wavelengths 340 nm and 380 nm was acquired at an emission of 510 nm every 2.5 s. The ratio 340 to 380 was converted to calcium values using the classical equation described by Grynkiewicz et al. (1985), after calculation of R_{min} value as minimal fluorescence reached after perfusion with Krebs solution without calcium and with 20 mM EGTA, 5 mM Tris and 10 μM ionomycin (pH = 8.2, no bovine serum albumin added) and R_{max} as maximal fluorescence reached in presence of 10 mM calcium and 10 μM ionomycin. The autofluorescence of the cell, defined as the remaining fluorescence in the presence of 5 mM $MnCl_2$ and 10 μM ionomycin was subtracted.

In each experiment, fluorescence measurements were limited to one cell, by means of an adjustable external diaphragm. However, the minimal opening of diaphragm available (1 mm) was larger than dimensions of the non-differentiated PC12 cells. Thus part of the signal was obtained from outside the cell and hence our estimates of the resting calcium value in these cells is probably low. In the case of differentiated cells we selected each time one cell with at least one neurite longer than the maximal diameter of the cell. The fluorescence signal collected from one cell was stable for more than 20 min, and the bleaching was minimal.

Immunohistochemistry

PC12 cells were plated on poly-L-lysine coated glasses in a 24-well plate (20,000 per well), synchronised for 48 h with quiescence medium and treated with NGF. After stimulation, the cells were

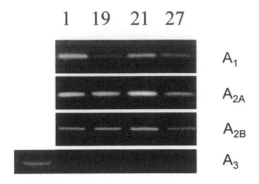

Testis

Fig. 1. Expression of mRNA adenosine receptor subtypes in four different subclones of PC12 cells (1, 19, 21 and 27) using RT-PCR. In the case of the A_3 receptor a positive control (rat testis) was included.

fixed for 10 min by 4% paraformaldehyde in phosphate buffer (125 mM, pH 7.4). Incubation with the primary antibody was performed at 4°C overnight in a wash buffer (phosphate buffer, 20 mM, pH 7.4, 500 mM NaCl, 0.3% Triton X-100) supplemented with 15% goat serum. The cells were then rinsed in wash buffer six times and incubated with the secondary antibody for 1 h at room temperature.

Results and discussion

Genotypic variations in adenosine receptors

We first examined signal transduction via A_{2A} adenosine receptors in four (denoted: clone 1, 19,

21 and 27) of the several clones isolated by Meldolesi and coworkers (Zacchetti et al., 1991; Clementi et al., 1992a). As seen in Fig. 1, mRNAs for the adenosine A_1, A_{2A}, and A_{2B} receptors were present in all four clones. Primers for the adenosine A_3 receptor were also tested, but no transcripts were detected (panel 4, Fig. 1). No attempt was made to quantify the reactions.

Since the cAMP increase elicited by adenosine analogues in PC12 cells seems to be mostly mediated by activation of A_{2A} receptors (van der Ploeg et al., 1996), A_{2A} receptor expression was examined first. A_{2A} receptors were quantitated using the antagonist radioligand [^{3}H]-SCH 58261. The estimated B_{max} value was highest in clone 1 followed by clones 21, 19 and 27 (Fig. 2). This pattern was mimicked by the expression of receptor protein as quantitated using western blot (not shown). We next examined some of the components of the downstream pathway leading to cAMP production, namely G_s and adenylyl cyclase. The amount of G_s protein was estimated by western blotting and appeared to be higher in clones 21 and 27 than in the other two clones (not shown). Responses to forskolin and to CGS 21680, a selective adenosine A_{2A} receptor agonist, are shown in Fig. 2. Forskolin was employed in order to test adenylyl cyclase activity and revealed that, in accordance with the binding data, responses were higher in clones 1 and 21 than in clones 19 and 27. CGS 21680 further confirmed also that direct

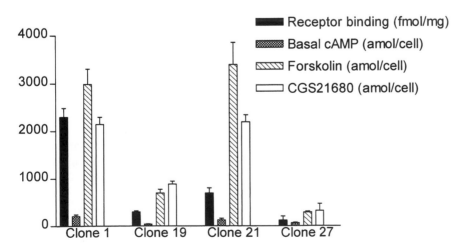

Fig. 2. Differences in receptor number (B_{max} of [^{3}H]-SCH 58261 binding) and in cAMP responses in four different PC12 cell clones.

activation of the adenosine A_{2A} receptor led to higher concentration of cAMP in clone 1 and 21 than in clones 19 and 27. The non-selective adenosine analogue NECA was next employed to find out whether a component of the cAMP response could be due to activation of A_{2B} receptors as well (Fig. 2). The responses induced by NECA were similar to those observed with CGS 21680, being approximately equal in clones 1 and 21 but lower in clones 19 and very low in clone 27.

These results show several things: (1) there are major differences in receptor and G protein expression between PC12 cell clones; (2) the functional response depends only partly on the number of receptors, but the abundance of other components of the signalling cascade plays a major role, and (3) the presence of mRNA for a receptor does not mean that there is a sufficient amount of functional receptor expressed to allow a biological response.

Phenotypic variations in adenosine receptors and variations between individual cells

Once it was established that clone 1 was an appropriate and reliable tool for studies on adenosine receptors in PC12 cells, differentiation in this clone was induced by one week of NGF treatment in order to examine whether this process could affect adenosine A_{2A} receptor expression and activity. Undifferentiated PC12 cells expressed very high levels of adenosine A_{2A} receptors and exhibited strong cAMP responses when stimulated with adenosine agonists (see above). NGF induced differentiation was accompanied by a down-regulation of adenosine A_{2A} receptor: receptor binding decreased to 500 fmol/mg, immunoreactive A_{2A} receptor protein was decreased to about half and cAMP production was reduced by 60% (Arslan et al., 1997). The drastic reduction of the cAMP response in differentiated cells was not due to a decrease in G_s proteins or adenylyl cyclase. Instead the fact that the decreases in receptor number (75%) and receptor response (up to 60%) were very similar suggests that the decrease in the number of receptors is at least a major component in the decreased functional response. There were also major differences in the amount of receptor and receptor mRNA between individual PC12 cells of

the same clone and after the same treatment with NGF (not shown).

P2 receptors in PC12 cells

PC12 cells from clone 1 expressed $P2X_1$, $P2X_2$, $P2X_4$ and $P2Y_2$ receptors as judged by PCR. It has previously been shown that they do not express $P2X_3$ receptors. By functional and pharmacological criteria these cells do not express $P2X_5$, $P2X_6$ or $P2X_7$ receptors. Neither do they appear to express $P2Y_1$, $P2Y_6$ or $P2Y_{11}$ receptors (Arslan et al., 1996). Judging by RT-PCR there were differences in the expression of P2 receptors between different clones. However, the absence of good radioligands and antibodies made it impossible for us to explore such differences further.

Ca^{2+} responses in PC12 cells and phenotypic variations in Ca^{2+} responses

In single PC12 cells ATP induced a rapid, transient increase in $[Ca^{2+}]_i$. The magnitude of the response decreased upon repeated stimulation, indicating some desensitisation. In undifferentiated cells the increase in $[Ca^{2+}]_i$ was completely dependent upon the presence of extracellular Ca^{2+} (Fig. 3A), indicating that in these cells ATP could not mediate any release of Ca^{2+} from intracellular stores. By contrast, in cells treated for one week with NGF ATP did induce a Ca^{2+} transient even in the absence of Ca^{2+} ions in the medium (Fig. 3B).

Upon differentiation with NGF the peak increase in Ca^{2+} increased from 351 ± 26 nM to 580 ± 5 nM. The increase in Ca^{2+}-free medium was 0 ± 6 nM in control cells and 82 ± 23 nM in NGF-differentiated cells. We also found that a small component of the Ca^{2+} transient in differentiated cells was due to entry of Ca^{2+} via so-called store-operated Ca^{2+}-channels. Thus, the increase in the magnitude of the Ca^{2+} response in differentiated PC12 cells seemed to be due to an increase in the contribution by P2Y receptors. Indeed, ATP- and UTP-induced formation of inositol trisphosphate was increased in NGF-differentiated cells (Fig. 4). This provides additional evidence that in differentiated, but not in undifferentiated PC12 cells there are functional P2Y receptors. Furthermore, the fact that the

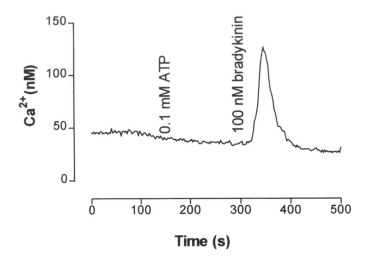

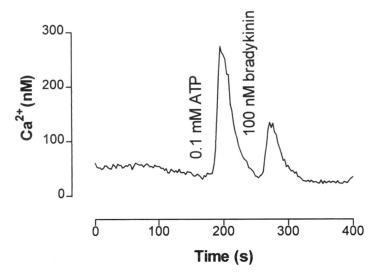

Fig. 3. Comparison of the effect of ATP (100 μM) and bradykinin (100 nM) in a Ca^{2+}-free medium on the levels of intracellular Ca^{2+} in individual control (upper panel) or NGF-treated (lower panel) PC12 cells as determined using the Fura-2 technique in a microfluorimetric setup. Note that ATP was completely ineffective in control cells, but at least as effective as bradykinin in NGF-treated cells.

response was equal for UTP and ATP (not shown) suggests that the response is most probably attributable to a $P2Y_2$ or a $P2Y_4$ receptor.

It has been reported that ATP can raise cAMP in PC12 cells (Yakushi et al., 1996). This could indicate that these cells express a novel form of the P2Y receptor, called the $P2Y_{11}$ receptor (Communi et al., 1997). In apparent agreement with this we found that ATP could indeed increase cAMP formation, but only in NGF differentiated cells

(Fig. 5). We also showed that this effect was not due to hydrolysis of ATP to adenosine that would in turn act on adenosine A_{2A} receptors to raise cAMP levels (Arslan et al., 1997). Adenosine causes an almost 10-fold increase in cAMP accumulation in these cells, and this is blocked by the selective antagonist SCH 58261. The failure to see any effect of the latter compound therefore suggests that under our experimental conditions hydrolysis of ATP to active adenosine is a minor problem. By

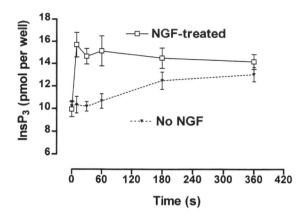

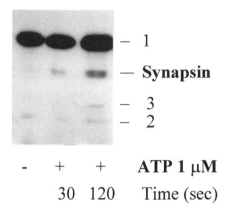

ATP-induced synapsin phosphorylation

Fig. 4. Stimulation of endogenous PLC-β as determined by the generation of inositol 1,4,5-trisphosphate (InsP₃) by UTP (100 μM) in NGF-treated (unfilled symbols—unbroken line) and control (filled symbols—broken line) PC12 cells. Mean-±SEM of two experiments in triplicate.

contrast, the response to ATP was completely abolished by indomethacin, a known inhibitor of cyclooxygenases. Thus, our tentative conclusion is that the effect of ATP is an indirect one mediated via the release of a prostaglandin.

The P2 receptor antagonist suramin could not be used since it also blocked responses to bradykinin. This was not seen with another antagonist, PPADS (North and Barnard, 1997). This compound was completely ineffective in control PC12 cells, indicating that the P2 receptor responsible for the Ca-transients was the P2X₄ receptor. In NGF-

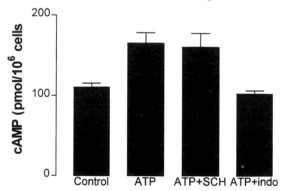

Fig. 5. Effect of ATP (100 μM) on cAMP formation in NGF-differentiated PC12 cells. Note that the increase was completely abolished by the addition of 3 μM indomethacin (ATP + INDO) to block cyclooxygenase. By contrast, the response was not blocked by an effective concentration (300 nM) of a selective inhibitor of adenosine A₂ₐ receptors, SCH 58261 (ATP + SCH).

Fig. 6. Time dependent phosphorylation of synapsin 1 induced by ATP in NGF-differentiated PC12 cells. The figure shows a Western-blot using an antibody (RU-19) directed against phosphorylated synapsin. An ATP (1 μM) and time dependent phosphorylation of synapsin is seen. In addition a major band that is not affected (1) and two minor bands (2, 3) that are partly affected by ATP can be seen.

treated cells, however, a slight response was observed. This could be due to an effect on either P2X₂ or P2Y₂ receptors.

Phenotypic variation in phosphoproteins including synapsin phosphorylation

Synapsins were first discovered as a major neuronal substrate for cyclic AMP-dependent phosphorylation, but its role is now considered to be much wider, ranging from regulation of synaptic vesicle disposition in nerve terminals to neuronal differentiation and synaptogenesis (Greengard et al., 1993). It is known that synapsin I can be rapidly phosphorylated upon neuronal calcium entry by calcium/calmodulin-dependent protein kinase II. A peptide antibody, RU 19, directed to the consensus phosphorylation site in the synapsin I molecule was recently generated (Matovcik et al., 1994). The PC12 cell clone 1 cells are known to possess synapsin I (Clementi et al., 1992a). Using a Western blot it could be demonstrated that the synapsin band was phosphorylated, as detected by RU 19, in an ATP stimulated manner (Fig. 6). The ATP-induced synapsin phosphorylation was abol-

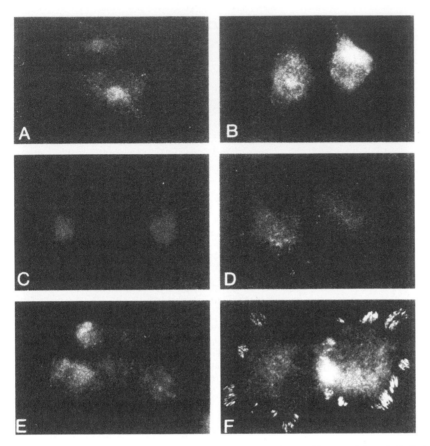

Fig. 7. Immunolabelling of control (panels A, C, E) or 3 days NGF-treated (panels B, D, E) PC12 cells. The upper two panels (A,B) show synapsin. The lower panels (C–F) show P-synapsin antibody: The two middle panels without ATP, the two lower panels with 1 μM ATP. Note that in the NGF-treated cells ATP induced an increase in immunofluorescence with the P-synapsin antibody that was evident both in the locations (cytosolic c.f. panels A, B) where synapsin is found, and at a very peripheral location.

ished by omitting Ca^{2+} from the medium, indicating that the phosphorylation is indeed Ca-dependent (not shown). However, there were also other major bands, not altered by ATP stimulation of the PC12 cells, as well as bands, unrelated to synapsin, that were altered by ATP.

Using immunofluorescence an increase in immunoreactivity to the phosphoprotein-specific antibody was found after ATP-treatment, but only in cells that had been allowed to differentiate into a more neuronal phenotype by treatment with NGF for three days (Fig. 7). The precise location of the immunofluorescence differed with the time of treatment with NGF (not shown). The labelled sites were in growth zones, and seemed to accompany the tips of the extending neurites: the further the neurites extended the more removed these sites

were from the cell body. It could also be seen that not only ATP but also carbachol and potassium-depolarisation induced the specific labelling (Fig. 8). The effect of ATP was dependent on calcium (Fig. 8E) in the medium. This was true also for the other agents (not shown), indicating that the relevant factor for inducing phosphoprotein immunofluorescence is an increase in intracellular calcium, secondary to calcium influx.

Judging from studies on co-localisation of synapsin and RU-19 immunofluorescence much of the latter was not identical with synapsin. This conclusion was further enforced by the finding that the more immunopurified the antibody the less consistent labelling of the structures in growth zones (not shown). Since the labelled phosphoprotein appeared to locate to sites at the very end of actin

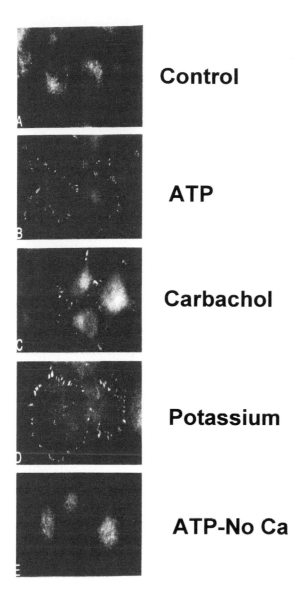

Control

ATP

Carbachol

Potassium

ATP-No Ca

Fig. 8. In PC12 cells treated for 3 days with NGF an increase in immunofluorescence with the P-synapsin antibody can be induced by ATP, carbachol and potassium depolarisation and it is calcium-dependent. The five panels show immunofluorescence using the phospho-synapsin antibody as the primary antibody. Under control conditions (panel A) and after 2 min treatment with 1 μM ATP in the absence of Ca^{2+} in the medium (panel E; ATP-No Ca) background staining over nuclei and part of cytosol is found. Treatment with ATP for 2 min in the presence of Ca^{2+} in the medium (panel B) induces specific labelling of peripheral structures. A similar effect is seen after stimulation with the cholinergic drug carbachol for the same amount of time (panel C) or after potassium (45 mM) depolarisation (Panel D).

filaments, as judged by double staining with RU-19 antibody and phalloidin, the protein may be related to adhesion structures related to growth zones. However, there was not a complete overlap with vinculin, in that RU-19 appeared to label structures that were more peripheral. There was, however, a gross correspondence between the structures labelled with RU-19 and those that stained with a phosphotyrosine antibody. We therefore tentatively conclude that ATP, via a calcium-dependent mechanism, causes phosphorylation of synapsin I and in addition of a protein localised to growth zones in differentiating PC12 cells. The identity of this protein remains uncertain.

Conclusions

We have further demonstrated that there are large variations between individual PC12 cells in their ability to respond to purines. Part of this variability can be explained by a clonal heterogeneity. We also show that differentiation of PC12 into a more neuron-like phenotype with NGF alters the expression of the receptors for purines. Adenosine A_{2A} receptors decrease whereas the expression of P2Y receptors increases. Thus, after differentiation a substantial part of the Ca^{2+} response of PC12 cells to ATP can be attributed to stimulation of a P2Y receptor, probably a $P2Y_2$ receptor. We also show that after treatment with NGF, ATP induces calcium-dependent phosphorylation of synapsin and an as yet unidentified protein located close to focal adhesions in areas of neurite extension. The results are compatible with the view that ATP and adenosine can modify the effect of such factors that regulate neuronal phenotype.

Acknowledgements

We want to thank Drs. Emilio Clementi and Jacopo Meldolesi for providing the PC12 cell clones and Dr. Andrew Cernik for providing us with the RU19 antibody for phosphosynapsin. Dr. Catalin Filipeanu is gratefully acknowledged for his help with some of the early experiments regarding calcium responses of PC12 cells and Drs. Clemens Allgaier and Björn Kull for their help with developing the PCR probes and some of the initial PCR analyses.

Mrs. Eva Irenius gave valuable technical assistance in several parts of these studies. They were supported by grants from the Strategic Fund for Scientific Research and Wallenbergstiftelsen.

References

Arslan, G., Irenius, E., Kull, B., Filipeanu, C., Czernik, A., Clementi, E. and Fredholm, B.B. (1996) NGF differentiation induces changes in ATP evoked calcium responses and protein phosphorylation in PC12 cells. *Purines '96, Milano, Italy, July 6–9.*

Arslan, G., Kontny, E. and Fredholm, B.B. (1997) Downregulation of adenosine A_{2A} receptors upon NGF-induced differentiation of PC12 cells. *Neuropharmacology*, 36: 1319–1326.

Clementi, E., Racchetti, G., Zacchetti, D., Panzeri, M.C. and Meldolesi, J. (1992a) Differential expression of markers and activities in a group of PC12 nerve cell clones. *Eur. J. Neurosci.*, 4: 944–953.

Clementi, E., Scheer, H., Raichman, M. and Meldolesi, J. (1992b) ATP-induced Ca^{2+} influx is regulated via a pertussis toxin-sensitive mechanism in a PC12 cell clone. *Biochem. Biophys. Res. Commun.*, 188: 1184–1190.

Clementi, E., Scheer, H., Zacchetti, D., Fasolato, C., Pozzan, T. and Meldolesi, J. (1992c) Receptor-activated Ca^{2+} influx. Two independently regulated mechanisms of influx stimulation coexist in neurosecretory PC12 cells. *J. Biol. Chem.*, 267: 2164–2172.

Communi, D., Govaerts, C., Parmentier, M. and Boeynaems, J.M. (1997) Cloning of a human purinergic P2Y receptor coupled to phospholipase C and adenylyl cyclase. *J. Biol. Chem.*, 272: 31969–31973.

Fasolato, C., Pizzo, P. and Pozzan, T. (1990) Receptor-mediated calcium influx in PC12 cells. ATP and bradykinin activate two independent pathways. *J. Biol. Chem.*, 265: 20351–20355.

Fasolato, C., Zottini, M., Clementi, E., Zacchetti, D., Meldolesi, J. and Pozzan, T. (1991) Intracellular Ca^{2+} pools in PC12 cells. Three intracellular pools are distinguished by their turnover and mechanisms of Ca^{2+} accumulation, storage, and release. *J. Biol. Chem.*, 266: 20159–21067.

Gerwins, P. (1993) Modification of a competitive protein binding assay for determination of inositol 1,4,5-trisphosphate. *Anal. Biochem.*, 210: 45–49.

Greene, L.A., Sobeih, M.M. and Teng, K.K. (1991). Methodologies for the culture and experimental use of the PC12 rat pheochromocytoma cell line. In: G. Banker and K. Goslin (Eds.), *Culturing Nerve Cells*. MIT Press, Cambridge, pp. 178–212.

Greene, L.A. and Tischler, A.S. (1976) Establishment of a noradrenergic clonal line of rat adrenal pheochromocytoma cells which respond to nerve growth factor. *Proc. Natl. Acad. Sci. USA*, 73: 2424–2428.

Greengard, P., Valtorta, F., Czernik, A.J. and Benfenati, F. (1993) Synaptic vesicle phosphoproteins and regulation of synaptic function. *Science*, 259: 780–785.

Grynkiewicz, G., Poenie, M. and Tsien, R.Y. (1985) A new generation of Ca^{2+} indicators with greatly improved fluorescence properties. *J. Biol. Chem.*, 260: 3440–3450.

Hide, I., Padgett, W.L., Jacobson, K.A. and Daly, J.W. (1992) A_{2A} adenosine receptors from rat striatum and rat pheochromocytoma PC12 cells: characterization with radioligand binding and by activation of adenylate cyclase. *Mol. Pharmacol.*, 41: 352–359.

Inoue, K., Nakazawa, K., Fujimori, K. and Takanaka, A. (1989) Extracellular adenosine 5′-triphosphate-evoked norepinephrine secretion not relating to voltage-gated Ca channels in pheochromocytoma PC12 cells. *Neurosci. Lett.*, 106: 294.

Matovcik, L.M., Karapetian, O., Czernik, A.J., Marino, C.R., Kinder, B.K. and Gorelick, F.S. (1994) Antibodies to an epitope on synapsin I detect a protein associated with the endocytic compartment in non-neuronal cells. *Eur. J. Cell Biol.*, 65: 327–340.

Nordstedt, C. and Fredholm, B.B. (1990) A modification of a protein-binding method for rapid quantification of cAMP in cell-culture supernatants and body fluid. *Analyt. Biochem.*, 189: 231–234.

North, R.A. and Barnard, E.A. (1997) Nucleotide receptors. *Curr. Opin. Neurobiol.*, 7: 346–357.

Rathbone, M.P., Deforge, S., Deluca, B., Gabel, B., Laurenssen, C., Middlemiss, P. and Parkinson, S. (1992) Purinergic stimulation of cell division and differentiation: mechanisms and pharmacological implications. *Med. Hypotheses*, 37: 213–219.

Roth, J.A., Marcucci, K., Lin, W.H., Napoli, J.L., Wagner, J.A. and Rabin, R. (1991) Increase in beta-1,4-galactosyltransferase activity during PC12 cell differentiation induced by forskolin and 2-chloroadenosine. *J. Neurochem.*, 57: 708–713.

van der Ploeg, I., Ahlberg, S., Parkinson, F.E., Olsson, R.A. and Fredholm, B.B. (1996) Functional characterization of adenosine A_2 receptors in Jurkat cells and PC12 cells using adenosine receptor agonists. *Naunyn Schmiedebergs Arch. Pharmacol.*, 353: 250–260.

Yakushi, Y., Watanabe, A., Murayama, T. and Nomura, Y. (1996) P2 purinoceptor-mediated stimulation of adenylyl cyclase in PC12 cells. *Eur. J. Pharmacol.*, 314: 243–248.

Zacchetti, D., Clementi, E., Fasolato, C., Lorenzon, P., Zottini, M., Grohovaz, F., Fumagalli, G., Pozzan, T. and Meldolesi, J. (1991) Intracellular Ca^{2+} pools in PC12 cells. A unique, rapidly exchanging pool is sensitive to both inositol 1,4,5-trisphosphate and caffeine-ryanodine. *J. Biol. Chem.*, 266: 20152–20158.

Zocchi, C., Ongini, E., Conti, A., Monopoli, A., Negretti, A., Baraldi, P.G. and Dionisotti, S. (1996) The non-xanthine heterocyclic compound, SCH 58261, is a new potent and selective A_{2A} adenosine receptor antagonist. *J. Pharmacol. Exp. Ther.*, 276: 398–404.

P. Illes and H. Zimmermann (Eds.)
Progress in Brain Research, Vol 120

Nucleotide receptor signalling in spinal cord astrocytes: Findings and functional implications

Conor J. Gallagher[1,2] and Michael W. Salter[1,2,3,*]

[1]*Programme in Brain and Behaviour, Hospital for Sick Children, University of Toronto, Toronto, Ont., Canada*
[2]*Institute of Medical Science, University of Toronto, Toronto, Ont., Canada*
[3]*Department of Physiology, University of Toronto, Toronto, Ont., Canada*

Introduction

The central nervous system is made up of both neuronal and non-neuronal cell types. Astrocytes are non-neuronal cells and comprise the largest group of CNS cells, outnumbering neurons by approximately 10 to 1 (Kuffler, 1984). Emerging evidence suggests that astrocytes may play a crucial role in information processing in the CNS. Astrocytes are non-excitable cells and the primary mode of signalling in astrocytes is via fluctuations in intracellular Ca^{2+} concentration ($[Ca^{2+}]_i$) (Verkhratsky and Kettenmann, 1996). Ca^{2+} signals regulate many intracellular processes in these cells, including growth, differentiation and the secretion of neuroactive substances. Astrocytes express a variety of cell surface receptors, including ligand-gated ion channels and G-protein coupled receptors, which render them capable of responding to extracellular stimuli. Thus, extracellular ligands that induce changes in $[Ca^{2+}]_i$ may play important roles in the functions of astrocytes.

Synaptic junctions in the CNS are ensheathed by astrocytes, which may serve to regulate K^+ homeostasis and to maintain the isolation of individual synapses through the uptake of neurotransmitter diffusing out of the synaptic cleft (Barres, 1991). Astrocyte processes are within the diffusion range of synaptically released transmitters. If astrocytes express the appropriate receptors, they may be stimulated by released neurotransmitters. The expression of neurotransmitter receptors on astrocytes raises the possibility that astrocytes might play a much more dynamic and important role than simply being the housekeepers of the synapse. A corollary to the intimate localization of astrocytes with synapses is that substances released from synaptically located astrocytes may reach and act on neurons with the potential to modify synaptic transmission.

There is abundant evidence for the presence of extracellular ATP in the CNS (Burnstock, 1997). For example, ATP and other nucleotides may be released from neurons and other cells under both normal and pathological conditions (Born and Kratzer, 1984; Edwards et al., 1992). It is known that astrocytes respond to extracellular ATP with a rise in $[Ca^{2+}]_i$ (Neary et al., 1988, Bruner and Murphy, 1993; Salter and Hicks, 1994). The present paper focuses on the characterization of Ca^{2+} signals evoked by extracellular purine and pyrimidine nucleotides in astrocytes from the spinal cord.

Sources of extracellular nucleotides in the nervous system

Release of nucleotides from neurons was first described by Holton and Holton (1954) who

*Corresponding author. Tel.: + 1 416-813-6272; fax: + 1 416-813-7921; e-mail: mike.salter@utoronto.ca

showed that ATP was released from the peripheral endings of primary sensory neurons. From the results of their experiments it was inferred, by Dale's principle, that release of ATP may occur at the central terminals of primary afferent fibres in the CNS. Since then, ATP has been shown to be present in synaptic vesicles of the autonomic and sensory systems where it co-localizes with noradrenaline, acetylcholine, and various other neuroactive compounds (for review see Burnstock, 1997).

The first evidence that ATP released from nerve terminals in the CNS may participate in synaptic transmission was obtained through the measurement of miniature and evoked post-synaptic currents using the whole-cell patch clamp technique in brain slices taken from the medial habenula (Edwards et al., 1992). In this preparation fast synaptic currents that were resistant to blockers of glutamate, GABA and nicotinic receptors were found to be inhibited by P2 receptor blockers.

Release from purinergic terminals is not, however, the only source of extracellular nucleotides. Cytosolic ATP concentration has been reported to be greater than 5 mM (Gordon, 1986) and under pathological conditions, where cell lysis occurs, extrusion of ATP can result in high concentrations of nucleotides in the extracellular milieu. Born and Kratzer (1984) found that the extracellular concentration of ATP in the vicinity of damaged tissue can reach 20 μM, which is sufficient to stimulate all known metabotropic purine receptors.

Although neurons are the primary source of exocytotically released nucleotides, endothelial cells are also capable of releasing nucleotides (El Moatassim et al., 1992). Endothelial cells might serve as a source of extracellular nucleotides under both normal and pathological conditions. The blood-brain barrier is formed by both astrocytic foot processes and endothelial cells. ATP released from vascular endothelial cells or autonomic nerve terminals and acting on adjacent astrocytes or endothelial cells may alter the permeability of the blood-brain barrier. Furthermore, astrocytes are themselves capable of releasing ATP in a Ca^{2+}-dependent manner, as has been shown in response to stimulation with glutamate (Queiroz et al., 1997).

Pharmacological characteristics of P2Y receptors on spinal cord astrocytes

ATP has been found to be released from synaptosomes prepared from the dorsal horn of the spinal cord (White et al., 1985) and it has been proposed that ATP is released at synapses in the spinal cord dorsal horn (for review see Salter et al., 1993). As synapses are intimately surrounded by astrocytes we questioned what effects extracellular ATP would exert on astrocytes in this region.

In order to investigate the effects of nucleotides on spinal cord astrocytes we have made use of primary cultures prepared from the dorsal horn of the spinal cord. In these cultures astrocytes are readily distinguishable morphologically from other cells in the culture (Salter and Hicks, 1994). Individual astrocytes can, therefore, be studied using optical and electrophysiological methods. We found that astrocytes cultured from the spinal cord dorsal horn respond to extracellular ATP by elevating $[Ca^{2+}]_i$ (Salter and Hicks, 1994). The elevation in $[Ca^{2+}]_i$ was due to release of Ca^{2+} from intracellular stores. Adenosine and AMP produced no change in $[Ca^{2+}]_i$ indicating that the responses are not mediated by P1 receptors. Moreover, because ATP did not cause an influx of extracellular Ca^{2+}, and application of the P2X selective agonist α,β-methyleneATP had no effect on $[Ca^{2+}]_i$ we suggested that purine-induced elevations of $[Ca^{2+}]_i$ in astrocytes are mediated by metabotropic nucleotide (P2Y) receptors (Salter and Hicks, 1994).

As ATP may activate several types of metabotropic receptor (O'Connor et al., 1991) it was possible that more than one receptor mediated the responses of spinal cord astrocytes to ATP. In order to investigate which subtypes of nucleotide receptors are on spinal cord astrocytes we used the agonists 2-methylthioATP (2MeSATP) and UTP which had been shown to differentially activate distinct subtypes of P2Y receptors (O'Connor et al., 1991).

Application of 2MeSATP was found to produce an elevation in $[Ca^{2+}]_i$ in 99% of astrocytes in the spinal cord dorsal horn. On the other hand, UTP induced Ca^{2+} responses in only 72% of spinal cord astrocytes (Ho et al., 1995). That there is a population of 2MeSATP-sensitive astrocytes unresponsive to UTP suggested that it is unlikely that these compounds are activating a single type of receptor. No receptor-specific antagonists were available at the time of these experiments, therefore to investigate whether a single receptor mediated both responses we employed a cross-desensitization strategy. We found that desensitization of the Ca^{2+} response evoked by 2MeSATP did not prevent astrocytes from responding to UTP and vice versa. Therefore, it was proposed that at least two subtypes of nucleotide receptor are expressed by spinal cord astrocytes (Ho et al., 1995); one receptor is selectively activated by 2MeSATP and the other is selectively activated by UTP.

In characterizing the responses of astrocytes it was found that Ca^{2+} responses to 2MeSATP are blocked by the competitive P2 antagonists suramin and PPADS (Salter and Hicks, 1994; Ho et al., 1995). Recently an antagonist, A3P5PS, which is reportedly selective for the $P2Y_1$ receptor has been described (Boyer et al., 1996). We have found that this compound blocks responses of spinal cord astrocytes to 2MeSATP but does not affect responses to UTP (Fam and Salter, 1998). It has been reported that ADP and 2MeSADP are ligands for the cloned $P2Y_1$ receptor expressed in a heterologous system (Leon et al., 1997). We have found that spinal cord astrocytes also respond to 2MeSADP and ADP and that these responses can be blocked by A3P5PS (Fam and Salter, 1998). These observations suggest that the receptor activated by 2MeSATP on spinal cord astrocytes may be identical to the cloned $P2Y_1$ receptor.

More than one subtype of P2Y receptor may be activated by UTP. We employed pharmacological methods to determine which of these receptors was mediating the response to UTP in spinal cord astrocytes. While 72% of spinal cord astrocytes respond to UTP less than 10% respond to UDP (Shing and Salter, unpublished observations). Moreover, in the presence of the $P2Y_1$ antagonist A3P5PS these cells are insensitive to ADP but maintain their responsiveness to ATP. ATP and UTP are approximately equipotent at activating spinal cord astrocytes (Ho et al., 1995). Responses to UTP are blocked by suramin (Ho et al., 1995) but are found to be unaffected by A3P5PS. Ca^{2+} responses of spinal cord astrocyte to UTP are blocked by

PPADS and are not attenuated by inhibition of $G_{i/o}$ G-proteins with pertussis toxin (Ho et al., 1995). Of the cloned UTP-sensitive P2Y receptors only the $P2Y_2$ subtype is activated equally well by UTP and ATP. However, the Ca^{2+} response generated by stimulation of $P2Y_2$ is attenuated by pertussis toxin and is insensitive to the P2Y antagonist PPADS (Erb et al., 1993). The $P2Y_4$ receptor is primarily a UTP receptor, however, it is insensitive to suramin (Boarder and Hourani, 1998) while the $P2Y_6$ receptor is specifically a UDP receptor (Communi et al., 1996). The pharmacological characteristics of the spinal cord astrocyte UTP response do not correspond to those of any of the cloned pyrimidinoceptors. Taken together our pharmacological data suggest that spinal cord astrocytes express a novel subtype of P2Y receptor.

The evidence described above, and that from other labs, indicates that astrocytes in culture respond to exogenous ATP via P2Y receptors (Neary et al., 1988, Bruner and Murphy, 1993; Salter and Hicks, 1994; Centemeri et al., 1997). However, whether P2Y receptors are expressed on astrocytes in situ was, for a time, controversial (Kimelberg et al., 1997). It was reported that, in hippocampal slices, the effects of extracellular ATP on astrocytes were due to hydrolysis of the trinucleotide by ecto- or endo-nucleotidases and the subsequent activation of P1, rather than P2, receptors (Porter and McCarthy, 1995). However, the presence of P2Y receptors was subsequently identified on astrocytes acutely isolated from both rat cortex and hippocampus (Cai and Kimelberg, 1997) and also from astrocytes in an acute preparation of the rat retina (Newman and Zahs, 1997). Therefore, it is now clear that metabotropic P2 receptors are expressed by astrocytes in a physiologically relevant context.

Signal transduction pathways activated by astrocyte P2Y receptor stimulation

Elucidating the pathways through which metabotropic P2 receptors signal is an important step towards understanding the function of these receptors in astrocytes. For this purpose we used the selective agonists 2MeSATP and UTP, characterized above, to investigate the signal transduction pathways activated by P2Y receptor stimulation in spinal cord astrocytes. In order to manipulate the signal transduction pathway directly, we used the whole-cell patch clamp method to perfuse reagents directly into astrocytes (Salter and Hicks, 1995).

As a first step, we found that Ca^{2+} responses to 2MeSATP or UTP were blocked by intracellular administration of a G-protein function blocker, GDPβS (Idestrup and Salter, 1998) indicating that each receptor subtype is coupled to release of intracellular Ca^{2+} via a G-protein. To elucidate the G-protein mediating the Ca^{2+} response we used an inhibitor of the $G_{i/o}$ G-proteins, pertussis toxin (Yamane and Fung, 1993). It was found that neither the response to 2MeSATP nor that to UTP was affected by pertussis toxin indicating that the G-proteins mediating the responses are pertussis toxin insensitive G-proteins, likely G_q (Idestrup and Salter, 1998).

As mentioned above, responses to brief stimulation of spinal cord astrocytes with 2MeSATP or UTP are unaffected by the absence of extracellular Ca^{2+} suggesting that activation of these receptors induces the release of Ca^{2+} from intracellular stores. There are two principal types of releasable intracellular Ca^{2+} store in cells: ryanodine/caffeine-sensitive and IP_3-sensitive. All cells possess one or both of these types of Ca^{2+} store (for review see Berridge, 1993). We found that extracellular application of caffeine to spinal cord astrocytes, which is known to cause release of Ca^{2+} from ryanodine/caffeine-sensitive stores, produces no increase in $[Ca^{2+}]_i$. Thus, it was concluded that spinal cord astrocytes do not possess such stores (Salter and Hicks, 1994). This observation implicated IP_3-sensitive stores as the source of the nucleotide-induced $[Ca^{2+}]_i$ elevation. To determine whether these cells contain IP_3-sensitive stores, IP_3 was perfused intracellularly in the absence of extracellular Ca^{2+}. IP_3 perfusion caused an increase in $[Ca^{2+}]_i$ confirming that spinal cord astrocytes possess an IP_3 releasable pool of Ca^{2+} (Salter and Hicks, 1995). This Ca^{2+} pool was depleted by the endoplasmic reticulum Ca^{2+}/ATPase blocker thapsigargin, which was found to eliminate the response to intracellularly applied IP_3 (Salter and Hicks, 1995) and to extracellularly applied 2MeSATP and

UTP (Idestrup and Salter, 1998). Thus, we concluded that Ca^{2+} release in response to these nucleotides occurs from IP_3-sensitive stores.

IP_3 is a product of the hydrolysis of phosphatidylinositol bisphosphate (PIP_2) by the enzyme phospholipase C (PLC), and most cells express several isoforms of this enzyme (Rhee and Bae, 1997). In order to examine which isoform of this enzyme is involved in the signal transduction pathway for spinal cord P2Y receptors we applied an inhibitor of PLCβ, U-73122. This compound was found to block responses to 2MeSATP and to UTP. In contrast, the inactive isomer, U-73343, had no effect on 2MeSATP or UTP responses indicating that each receptor signals via the activation of PLCβ (Salter and Hicks, 1995).

Taking the evidence together it was concluded that spinal cord astrocytes express two distinct subtypes of nucleotide receptors. Each of these receptors signals via phospholipase Cβ with the subsequent production of IP_3 and release of Ca^{2+} from intracellular stores (Fig. 1).

Calcium response patterns evoked by extracellular nucleotides

As described above, elevation of extracellular nucleotide concentration in the CNS can occur as a result of synaptic release or through diffusion of nucleotides out of damaged cells. Under physiological conditions synaptic release of ATP may produce intermittent, brief elevations of extracellular ATP. On the other hand, under pathological conditions, lysis of cells may result in the diffusion of nucleotides into the extracellular space, resulting in persistent elevation of the ATP level in the extracellular milieu. We considered it possible that the signalling pathways engaged by the two receptors expressed on spinal cord astrocytes produce differing responses to these distinct patterns of ATP release. For this reason we have compared the Ca^{2+} responses evoked by prolonged stimulation of spinal cord astrocytes with 2MeSATP and UTP.

We have found that when spinal cord astrocytes are stimulated with brief (1–10 s) applications of 2MeSATP or UTP the peak and time course of the Ca^{2+} responses are not distinguishably different from each other: within 5 s of the start of the application $[Ca^{2+}]_i$ reaches a rapid peak and then decays back to baseline, typically by 2 min. However, when spinal cord astrocytes are subjected to prolonged stimulation with agonist a profound difference in the temporal pattern of signalling was uncovered. Prolonged stimulation of spinal cord astrocytes with 2MeSATP results in a rapid spike of $[Ca^{2+}]_i$ and in the continued presence of agonist $[Ca^{2+}]_i$ decays to the baseline level over a period of 4 min. This response was unaffected by removal of extracellular Ca^{2+} which suggests that this response is generated wholly by release of Ca^{2+} from intracellular stores.

In contrast, while extracellular UTP produces a similar initial Ca^{2+} response to that evoked by 2MeSATP, during prolonged application of UTP the $[Ca^{2+}]_i$ level does not fully decline to the baseline value. Rather $[Ca^{2+}]_i$ is maintained at a plateau level at approximately 25% of the peak of the response (Fig. 2). In the absence of extracellular Ca^{2+} the peak of the UTP-evoked Ca^{2+} response is

unaffected but the plateau phase is eliminated. This suggests that the plateau of elevated $[Ca^{2+}]_i$ requires extracellular Ca^{2+} and might be due to an influx of extracellular Ca^{2+}. If agonist application is stopped during the plateau phase $[Ca^{2+}]_i$ returns to baseline indicating that maintenance of this influx requires continued activation of the receptor.

In most cell types influx of Ca^{2+} stimulated by release of stored intracellular Ca^{2+} has been reported to occur by means of a Ca^{2+}-release activated Ca^{2+} current, I_{CRAC}, which effects refilling of the depleted stores (for review see Parekh and Penner, 1997). I_{CRAC} may be stimulated by thapsigargin (Lytton et al., 1991), and the persistent stimulation of G_q coupled receptors, such as the carbachol-sensitive muscarinic acetylcholine receptor, may also stimulate I_{CRAC} (Boulay at al., 1997). The molecular identity of the Ca^{2+} channel which carries I_{CRAC} has remained elusive, however, the resulting influx has several characteristics: it is permeable to the divalent cations Mn^{2+} and Ba^{2+}

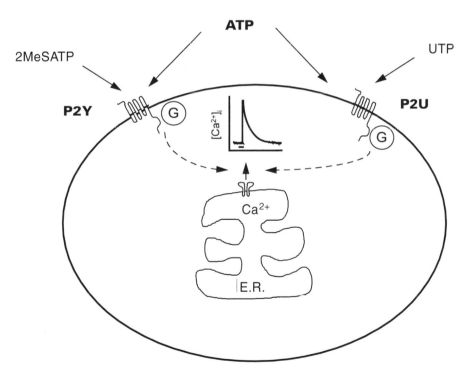

Fig. 1. Spinal cord astrocytes express two pharmacologically distinct subtypes of G-protein coupled receptors that are activated by ATP. Each receptor subtype can be stimulated independently using selective ligands: one subtype is selectively activated by 2meSATP but not by UTP, while the other is activated by UTP and not by 2meSATP. Brief stimulation of either receptor subtype triggers the release of Ca^{2+} from IP$_3$-sensitive intracellular stores in the endoplasmic reticulum (E.R.) via a PLCβ/IP$_3$ pathway.

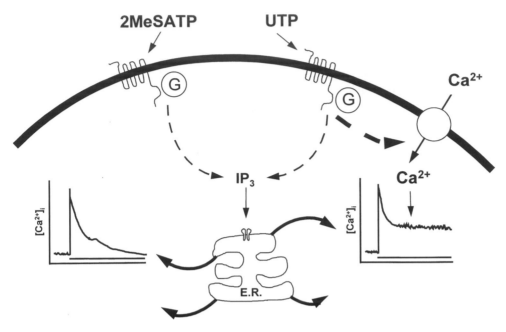

Fig. 2. Prolonged stimulation of spinal cord astrocytes exposes a difference in the temporal patterns of Ca^{2+} signalling evoked by receptor-selective ligands. Prolonged exposure of spinal cord astrocytes to 2MeSATP produces a Ca^{2+} response which decays back to baseline while prolonged stimulation of the UTP-sensitive receptor induces an agonist-dependent plateau of elevated $[Ca^{2+}]_i$.

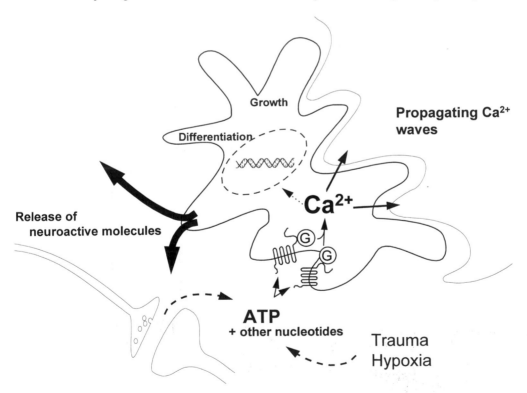

Fig. 3. Possible physiological and pathophysiological roles for extracellular nucleotides on spinal cord astrocytes.

and is blocked by submicromolar concentrations of the trivalent cation La^{3+} (Aussel et al., 1996; Parekh and Penner, 1997). We have found that Ca^{2+} stores depletion with thapsigargin in spinal cord astrocytes stimulates a Ca^{2+} influx pathway that is blocked by nM concentrations of La^{3+} and by 10 μM Gd^{3+} and is permeable to Mn^{2+}. Therefore, it appears that I_{CRAC} in spinal cord astrocytes may be generated by the application of thapsigargin.

To investigate whether the Ca^{2+} influx stimulated by UTP is I_{CRAC} we applied known inhibitors of I_{CRAC} to astrocytes during UTP stimulation. In contrast to the astrocyte I_{CRAC}, we have found that the UTP-induced $[Ca^{2+}]_i$ plateau was not inhibited by low concentrations of La^{3+} but was blocked by 10 μM Gd^{3+}. Furthermore, the UTP-induced Ca^{2+} influx pathway was not permeable to Mn^{2+}. Thus, the Ca^{2+} influx triggered by UTP application cannot be accounted for by I_{CRAC}. We found that the plateau phase was blocked by the receptor-operated Ca^{2+} channel blocker, SKF-96365, indicating that UTP stimulation does indeed open a membrane channel permeable to Ca^{2+}. We thus hypothesize that the UTP-triggered Ca^{2+} influx is through a G-protein coupled receptor activated channel distinct from I_{CRAC}.

Interestingly, our results indicate that the influx channel activated by UTP is not activated when astrocytes are stimulated with 2MeSATP. These findings, therefore, demonstrate that there is divergence in the signalling pathways activated by the two receptors and that the UTP-activated receptor engages an additional pathway. Thus, the receptor activated by 2MeSATP, putatively the $P2Y_1$ receptor, signals the onset of astrocyte stimulation by extracellular nucleotides and then desensitizes. In that the $P2Y_1$ receptor detects the onset of new events, $P2Y_1$ receptor signalling functions as a signal differentiator. In contrast, the spinal cord astrocyte UTP receptor detects the onset and, through the plateau of elevated $[Ca^{2+}]_i$, the continued presence of extracellular nucleotides. In this way the pyrimidinoceptor functions like a signal integrator. Thus, two spinal cord astrocyte nucleotide receptors encode differing temporal Ca^{2+} signals in response to persistent stimulation.

It is possible, therefore, that through Ca^{2+} signals, astrocytes may be capable of responding to and encoding patterns of synaptic input or of persistent nucleotide receptor stimulation in the CNS. Such signals may effect actions within the same cell or may be integrated and transmitted via regenerating Ca^{2+} waves throughout an astrocyte syncytium.

Decoding Ca^{2+} signals in astrocytes

It has been known for many years that, in many types of cell, increasing $[Ca^{2+}]_i$ can modify the expression of various genes (Hardingham et al., 1997). Recently, it has been reported that it is not simply raised $[Ca^{2+}]_i$, but the pattern, persistence and frequency of $[Ca^{2+}]_i$ signals that are the critical determinants of which genes are expressed (Dolmetsch et al., 1997, 1998). Dolmetsch et al. have studied the activation of transcription factors, including NF-AT and NF-κB, the expression of which was known to be stimulated by increasing $[Ca^{2+}]_i$. It was found that transient and prolonged elevations of $[Ca^{2+}]_i$ differentially activate distinct transcription factors in B-lymphocytes which, like astrocytes, are non-excitable cells (Dolmetsch et al., 1997). In these experiments intracellular Ca^{2+} signals were induced by varying the concentration of extracellular Ca^{2+} in the presence of the Ca^{2+} ionophore, ionomycin. A transient Ca^{2+} signal was found to induce the transcription factors NF-κB and JNK but not NF-AT, whereas prolonged elevation in $[Ca^{2+}]_i$ induced NF-AT. In subsequent experiments it was shown that the frequency of $[Ca^{2+}]_i$ oscillations confers specificity on the genes induced by the Ca^{2+} signals (Dolmetsch et al., 1998). It was found that Ca^{2+}-activated nuclear transcription factors are tuned to particular frequencies of Ca^{2+} signals. Thus, the temporal pattern of elevations in $[Ca^{2+}]_i$ induced by specific metabotropic receptor activation can confer specific downstream effects of this ubiquitous intracellular messenger. In astrocytes, the two nucleotide receptors we have studied share common elements in their signalling pathways but different patterns of $[Ca^{2+}]_i$ signals are produced depending on the subtype of receptor activated and the duration of stimulation. It is therefore possible that the different patterns of Ca^{2+} signals might be differentially decoded by means of activating distinct sets of transcription factors and expression of differing genes.

Possible roles for extracellular nucleotides in spinal cord astrocytes

Ca²⁺ waves in astrocytes

Within the CNS, each astrocyte is apposed to many other astrocytes which are coupled together via gap junctions (Giaume and McCarthy, 1996). In this way astrocytes form a functional syncytium which permits communication between cells in the syncytium. Propagation of Ca^{2+} waves has been demonstrated through syncytia of astrocytes in hippocampal slices (Dani et al., 1992), in confluent astrocyte cultures (Finkbeiner, 1992) and in acute preparations of the mammalian retina (Newman and Zahs, 1997). The mechanism of coupling between astrocytes may be via gap junctional communication whereby Ca^{2+}, or some other diffusable intracellular messenger, may pass though the junction and trigger release of Ca^{2+} from intracellular stores in the coupled cells (Venance et al., 1997). However, it is known that gap junctions are disabled by elevated $[Ca^{2+}]_i$ and it has been found that Ca^{2+} waves can be propagated, in vitro, across regions devoid of cells (Hassinger et al., 1996). Thus, there is evidence for an alternative mechanism of propagating Ca^{2+} waves, possibly by a diffusable extracellular messenger. Glutamate has been proposed as this extracellular messenger(Parpura et al., 1994), however it is unlikely to be the signal as glutamate receptor antagonists have no effect on the propagation of Ca^{2+} waves (Hassinger et al., 1996, Newman and Zahs, 1998).

Release of ATP from cultured astrocytes has been demonstrated in response to stimulation of cells with glutamate (Queiroz et al., 1997) and stimulation of retinal glia with ATP induces Ca^{2+} waves. This demonstrates that astrocytes are capable of both releasing ATP and of being stimulated by ATP. It is conceivable that astrocytic Ca^{2+} waves can be triggered by a number of stimuli but are propagated through the release of ATP and activation of P2Y receptors on nearby astrocytes. In mast cells ATP has been shown to be the diffusable extracellular signal in cell to cell transmission of Ca^{2+} signals and is capable of inducing Ca^{2+}-dependent vesicular secretion in these cells (Osipchuk and Cahalan, 1992).

Calcium waves in astrocytes may function as a signal amplifier; Ca^{2+} signals originating at one point in the astrocyte network may spread throughout the syncitium inducing effects in a large number of cells. It is possible that these waves are propagated by ATP acting as a diffusable extracellular messenger. (See note added in proof).

Astrocyte-neuron interactions

The modulation of neuronal function by astrocytes is of potential physiological importance. Astrocytes may modulate synaptic transmission on two levels. The first is through moment-to-moment regulation of synaptic efficacy and the second may be through the provision of longer term trophic support required for the maintenance of existing, or the genesis of new, synaptic connections.

It has been shown that when neurons are cocultured with glial cells the amplitude and frequency of miniature EPSCs is greatly enhanced relative to a glia-free neuronal culture, without altering the number of synaptic contacts. The presence of astrocytes in the culture exerts a trophic effect on the efficiency of synaptic transmission (Pfrieger and Barres, 1997).

Calcium waves in astrocytes have been found to modulate synaptic transmission in a retinal preparation (Newman and Zahs, 1998). The mechanism by which this modulation occurs is currently unclear but it is possible that signalling from astrocytes to neurons may occur either via direct communication or by a diffusable messenger. Direct communication from astrocytes to neurons was proposed by Nedergaard (1994). She demonstrated the transmission of electrically-induced Ca^{2+} signals from astrocytes to overlying neurons which were blocked by inhibitors of gap junctional communication, halothane and octanol, implying a direct astrocyte–neuron communication pathway via gap junctions. There is also evidence for diffusable signalling between astrocytes and neurons. It has been demonstrated that stimulation of a G-protein coupled receptor on cortical astrocytes induced the Ca^{2+}-dependent release of glutamate (Bezzi et al.,

1998; Parpura et al., 1994) and, as discussed above, ATP can be released from astrocytes following stimulation with glutamate (Queiroz et al., 1997).

In addition to releasing classical neurotransmitters, astrocytes are known to synthesize and secrete various neurotrophic factors (Rudge, 1993). The neurotrophins BDNF, NT-3 and NT-4/5 were identified in astrocytes by Friedman et al. (1998). It was concluded by Rubio (1997) that astrocytes play an important role in regulating the extracellular availability of neurotrophins in the CNS and that secretion of neurotrophins by astrocytes is enhanced during CNS inflammation. The secretion of such factors may be important in the maintenance or plasticity of synaptic function and heightened release of such factors, such as may occur during CNS inflammation, may induce phenotypic or morphological changes in adjacent neurons.

As discussed in the previous section, neuroactive substances released from astrocytes may act on astrocytes themselves but these substances may also stimulate adjacent neurons, both pre-synaptically and post-synaptically to modify synaptic transmission. Thus astrocytes may provide ongoing trophic support for neurons while at the same time they are also capable of the modulation of neuronal function through the consequences of Ca^{2+} wave propagation.

Potential role for astrocyte nucleotide receptors in CNS injury

Mechanical or ischaemic trauma to the CNS may result in cell death by necrosis and/or apoptosis. Cell death by necrosis involves the lysis of the cell and the diffusion of intracellular contents into the extracellular space. Since the concentration of intracellular nucleotides is in the millimolar range, levels of ATP and its degradation products may reach high concentrations in the extracellular space after cell necrosis. Following trauma, astrocytes undergo injury-induced proliferation and differentiation, termed "astrogliosis", which results in what is known as a "glial scar" (Norenberg, 1994). The glial scar is thought to be a limiting factor in CNS regeneration following trauma (Bovolenta et al., 1992). Several authors have shown that extrac-

ellular nucleotides may induce astrogliosis (Abbracchio et al., 1994; Neary et al., 1994). A link between P2Y receptor activation and the induction of astrogliosis is supported by evidence that stimulation of $P2Y_1$ and $P2Y_2$ receptors on cultured astrocytes can stimulate the proliferative MAP kinase pathway, which is known to couple to cell differentiation (Neary and Zhu, 1994; Soltoff et al., 1998). Furthermore, stimulation of astrocyte $P2Y_1$ receptors with ADPβS induces the expression of the immediate early genes Fos and Jun (Priller et al., 1998), transcription factors that induce gene expression and which have been found to be upregulated in the CNS following injury (Morgan, 1987). Thus, it appears that P2Y receptor stimulation may activate intracellular pathways that induce cell differentiation and proliferation.

At the microscopic level it has been shown that the exogenous application of ATP to astrocytes can induce morphological and constitutive changes that mimic the changes that occur in gliotic astrocytes. Such changes include a transformation of cells to a stellate morphology and increases in the GFAP content of these cells (Abbracchio et al., 1994; Neary et al., 1994). It is thought that these effects are mediated through P2Y receptors (Bolego et al., 1997) although this is still controversial, as there is evidence to suggest that these effects are due to the hydrolysis of tri-and di-nucleotides to AMP and adenosine which then activate astrocytic P1 receptors (Porter and McCarthy, 1995). However, inasmuch as astrogliosis depends on nucleotide-evoked elevation of $[Ca^{2+}]_i$, our failure to detect P1 receptors coupling to increased $[Ca^{2+}]_i$ on astrocytes from the spinal cord (Salter and Hicks, 1994), suggests that P2Y receptors may mediate the induction of astrogliosis in this region.

Although most studies have focused on purine nucleotides, release of UTP was recently demonstrated from 1321N1 astrocytoma cells following mechanical stimulation (Lazarowski et al., 1997). This raises the possibility that endogenous UTP may be released from astrocytes in response to mechanical trauma in vivo.

The morphological changes induced by ATP stimulation of astrocytes require prolonged periods of stimulation. Thus, in terms of possible involvement of the two metabotropic P2 receptors we have been studying, involvement of the P2Y$_1$ receptor seems unlikely as the Ca^{2+} signal evoked by stimulating this receptor subtype fades rapidly with sustained stimulation. On the other hand, the pyrimidinoceptor-triggered Ca^{2+} signal persists during prolonged elevations of extracellular nucleotide. MacVicar (1987) showed that morphological changes in astrocytes are prevented in the presence of the Ca^{2+} channel blockers Co^{2+} and Cd^{2+}, and therefore concluded that Ca^{2+} entry was required for astrocyte differentiation. The work of Dolmetsch et al. (1997), discussed above, leads us to consider that each receptor likely activates a distinct pattern of transcription factors and induces the expression of differing sets of genes. Therefore, it is conceivable that it is the prolonged stimulation of the spinal cord astrocyte pyrimidinoceptor that is the trigger for astrogliosis in the spinal cord.

Conclusions

We have determined that spinal cord astrocytes express two functionally distinct subtypes of nucleotide receptor. One receptor appears to be the cloned P2Y$_1$ receptor while the other represents a novel P2Y receptor subtype activated by UTP. Each receptor couples to the release of Ca^{2+} from IP$_3$-sensitive intracellular stores but when subjected to sustained stimulation the two subtypes of receptor induce distinctive Ca^{2+} response patterns. We suggest that the differing Ca^{2+} signals evoked by the different receptors may have distinct functional consequences in terms of signal discrimination, gene expression, growth and proliferation.

Note added in proof: Since the submission of this chapter two groups (Cotrina et al. (1998); Guthrie et al. (1999)) have demonstrated that ATP mediates the propagation of Ca^{2+} waves between astrocytes.

Acknowledgements

This work was supported by grants from the Medical Research Council of Canada and the Nicole Fealdman Memorial Fund. C.J.G. is supported by the Rick Hansen Man in Motion Foundation and the Physiotherapy Foundation of Canada. M.W.S. is an MRC Scientist.

References

Abbracchio, M.P., Saffrey, M.J., Höpker, V. and Burnstock, G. (1994) Modulation of astroglial cell proliferation by analogues of adenosine and ATP in primary cultures of rat striatum. *Neuroscience*, 59: 67–76.

Aussel, C., Marhaba, R., Pelassy, C. and Breittmayer, J.P. (1996) Submicromolar La^{3+} concentrations block the calcium release-activated channel, and impair CD69 and CD25 expression in CD3- or thapsigargin-activated Jurkat cells. *Biochem. J.*, 313: 909–913.

Barres, B.A. (1991) New roles for glia. *J. Neurosci.*, 11: 3685–3694.

Berridge, M.J. (1993) Inositol trisphosphate and calcium signalling. *Nature*, 361: 315–325.

Bezzi, P., Carmignoto, G., Pasti, L., Vesce, S., Rossi, D., Rizzini, B.L., Pozzan, T. and Volterra, A. (1998) Prostaglandins stimulate calcium-dependent glutamate release in astrocytes. *Nature*, 391: 281–285.

Boarder, M.R. and Hourani, S.M. (1998) The regulation of vascular function by P2 receptors: Multiple sites and multiple receptors. *Trends. Pharmacol. Sci.*, 19: 99–107.

Bolego, C., Ceruti, S., Brambilla, R., Puglisi, L., Cattabeni, F., Burnstock, G. and Abbracchio, M.P.(1997) Characterization of the signalling pathways involved in ATP and basic fibroblast growth factor-induced astrogliosis. *Br. J. Pharmacol.*, 121: 1692–1699.

Born, G.V. and Kratzer, M.A. (1984) Source and concentration of extracellular adenosine triphosphate during haemostasis in rats, rabbits and man. *J. Physiol. (Lond.)*, 354: 419–429.

Boulay, G., Zhu, X., Peyton, M., Jiang, M., Hurst, R., Stefani, E. and Birnbaumer, L. (1997) Cloning and expression of a novel mammalian homolog of Drosophila transient receptor potential (Trp) involved in calcium entry secondary to activation of receptors coupled by the G$_q$ class of G protein. *J. Biol. Chem.*, 272: 29672–29680.

Bovolenta, P., Wandosell, F. and Nieto-Sampedro, M. (1992) CNS glial scar tissue: a source of molecules which inhibit central neurite outgrowth. *Prog. Brain Res.*, 94: 367–379.

Boyer, J.L., Romero-Avila, T., Schachter, J.B. and Harden, T.K. (1996) Identification of competitive antagonists of the P2Y$_1$ receptor. *Mol. Pharmacol.*, 50: 1323–1329.

Bruner, G. and Murphy, S. (1993) Purinergic P2Y receptors on astrocytes are directly coupled to phospholipase A2. *Glia*, 7: 219–224.

Burnstock, G. (1997) The past, present and future of purine nucleotides as signalling molecules. *Neuropharmacology*, 36: 1127–1139

Cai, Z. and Kimelberg, H.K. (1997) Glutamate receptor-mediated calcium responses in acutely isolated hippocampal astrocytes. *Glia*, 21: 380–389.

Centemeri, C., Bolego, C., Abbracchio, M.P., Cattabeni, F., Puglisi, L., Burnstock, G. and Nicosia, S. (1997) Characterization of the Ca^{2+} responses evoked by ATP and other nucleotides in mammalian brain astrocytes. *Br. J. Pharmacol.*, 121: 1700–1706.

Communi, D., Parmentier, M. and Boeynaems, J.M. (1996) Cloning, functional expression and tissue distribution of the human P2Y$_6$ receptor. *Biochem. Biophys. Res. Commun.*, 222: 303–308.

Cotrina, M.L., Lin, J.H. and Nedergaard, M. (1998) Cytoskeletal assembly and ATP release regulate astrocytic calcium signaling. *J. Neurosci.*, 18: 8794–8804.

Dani, J.W., Chernjavsky, A. and Smith, S.J. (1992) Neuronal activity triggers calcium waves in hippocampal astrocyte networks. *Neuron*, 8: 429–440.

Dolmetsch, R.E., Lewis, R.S., Goodnow, C.C. and Healy, J.I. (1997) Differential activation of transcription factors induced by Ca^{2+} response amplitude and duration. *Nature*, 386: 855–858.

Dolmetsch, R.E., Xu, K. and Lewis, R.S. (1998) Calcium oscillations increase the efficiency and specificity of gene expression. *Nature*, 392: 933–936.

Edwards, F.A., Gibb, A.J. and Colquhoun, D. (1992) ATP receptor-mediated synaptic currents in the central nervous system. *Nature*, 359: 144–147.

El-Moatassim, C., Dornand, J. and Mani, J.-C. (1992) Extracellular ATP and cell signalling. *Biochim. Biophys. Acta Mol. Cell Res.*, 1134: 31–45.

Erb, L., Lustig, K.D., Sullivan, D.M., Turner, J.T. and Weisman, G.A. (1993) Functional expression and photoaffinity labeling of a cloned P2U purinergic receptor. *Proc. Natl. Acad. Sci. USA*, 90: 10449–10453.

Fam, S.R. and Salter, M.W. (1998) *Abstr. Soc. Neurosci.* 24: 1601.

Finkbeiner, S. (1992) Calcium waves in astrocytes—Filling in the gaps. *Neuron*, 8: 1101–1108.

Friedman, W.J., Black, I.B. and Kaplan, D.R. (1998) Distribution of the neurotrophins brain-derived neurotrophic factor, neurotrophin-3, and neurotrophin-4/5 in the postnatal rat brain: an immunocytochemical study. *Neuroscience*, 84: 101–114.

Giaume, C. and McCarthy, K.D. (1996) Control of gap-junctional communication in astrocytic networks. *Trends. Neurosci.*, 19: 319–325.

Gordon, J.L. (1986) Extracellular ATP: Effects, sources and fate. *Biochem. J.*, 233: 309–319.

Guthrie, P.B., Knappenberger, J., Segal, M., Bennett, M.V.L., Charles, A.C. and Kater, S.B. (1999) ATP released from astrocytes mediates glial calcium waves. *J. Neurosci.*, 19: 520–528.

Hardingham, G.E., Chawla, S., Johnson, C.M. and Bading, H. (1997) Distinct functions of nuclear and cytoplasmic calcium in the control of gene expression. *Nature*, 385: 260–265.

Hassinger, T.D., Guthrie, P.B., Atkinson, P.B., Bennett, M.V. and Kater, S.B. (1996) An extracellular signaling component in propagation of astrocytic calcium waves. *Proc. Natl. Acad. Sci. USA*, 93: 13268–13273.

Ho, C., Hicks, J. and Salter, M.W. (1995) A novel P2-purinoceptor expressed by a subpopulation of astrocytes from the dorsal spinal cord of the rat. *Br. J. Pharmacol.*, 116: 2909–2918.

Holton, F.A. and Holton, P. (1954) The capillary dilator substances in dry powders of spinal roots; a possible role of adenosine triphosphate in chemical transmission from nerve endings. *J. Physiol. (Lond.)*, 126: 124–140.

Idestrup, C.P. and Salter, M.W. (1998) P2Y and P2U receptors differentially release intracellular Ca^{2+} via the phospholipase C/inositol 1,4,5-triphosphate pathway in astrocytes from the dorsal spinal cord. *Neuroscience*, 86: 913–923.

Kimelberg, H.K., Cai, Z., Rastogi, P., Charniga, C.J., Goderie, S., Dave, V. and Jalonen, T.O. (1997) Transmitter-induced calcium responses differ in astrocytes acutely isolated from rat brain and in culture. *J. Neurochem.*, 68: 1088–1098.

Kuffler, S.W. (1984) In: S.W. Kuffler, J.G. Nicholls and A.R. Martin (Eds.), *From Neuron to Brain*, seconnd ed, Sinauer, Sunderland, Massachusetts. p. 324.

Lazarowski, E.R., Homolya, L., Boucher, R.C. and Harden, T.K. (1997) Direct demonstration of mechanically induced release of cellular UTP and its implication for uridine nucleotide receptor activation. *J. Biol. Chem.*, 272: 24348–24354.

Leon, C., Hechler, B., Vial, C., Leray, C., Cazenave, J.P. and Gachet, C. (1997) The P2Y$_1$ receptor is an ADP receptor antagonized by ATP and expressed in platelets and mega-karyoblastic cells. *FEBS Lett.*, 403: 26–30.

Lytton, J., Westlin, M. and Hanley, M.R. (1991) Thapsigargin inhibits the sarcoplasmic or endoplasmic reticulum Ca-ATPase family of calcium pumps. *J. Biol. Chem.*, 266: 17067–17071.

MacVicar, B.A. (1987) Morphological differentiation of cultured astrocytes is blocked by cadmium or cobalt. *Brain Res.*, 420: 175–177.

Morgan, J.I., Cohen, D.R., Hempstead, J.L. and Curran, T. (1987) Mapping patterns of c-fos expression in the central nervous system after seizure. *Science*, 237: 192–197.

Neary, J.T., Van Breemen, C., Forster, E., Norenberg, L.O.B. and Norenberg, M.D. (1988) ATP stimulates calcium influx in primary astrocyte cultures. *Biochem. Biophys. Res. Commun.*, 157: 1410–1416.

Neary, J.T., Baker, L., Jorgensen, S.L. and Norenberg, M.D. (1994) Extracellular ATP induces stellation and increases glial fibrillary acidic protein content and DNA synthesis in primary astrocyte cultures. *Acta Neuropathol.(Berl)*, 87: 8–13.

Neary, J.T. and Zhu, Q. (1994) Signaling by ATP receptors in astrocytes. *NeuroReport*, 5: 1617–1620.

Nedergaard, M. (1994) Direct signaling from astrocytes to neurons in cultures of mammalian brain cells. *Science*, 263: 1768–1771.

Newman, E.A. and Zahs, K.R. (1997) Calcium waves in retinal glial cells. *Science*, 275: 844–847.

Newman, E.A. and Zahs, K.R. (1998) Modulation of neuronal activity by glial cells in the retina. *J. Neurosci.*, 18: 4022–4028.

Norenberg, M.D. (1994) Astrocyte responses to CNS injury. *J. Neuropathol. Exp. Neurol.*, 53: 213–220.

322

O'Connor, S.E., Dainty, I.A. and Leff, P. (1991) Further subclassification of ATP receptors based on agonist studies. *Trends Pharmacol. Sci.* 12: 137–141

Osipchuk, Y. and Cahalan, M. (1992) Cell-to-cell spread of calcium signals mediated by ATPreceptors in mast cells. *Nature*, 359: 241–244.

Parekh, A.B. and Penner, R. (1997) Store depletion and calcium influx. *Physiol. Rev.*, 77: 901–930.

Parpura, V., Basarsky, T.A., Liu, F., Jeftinija, K., Jeftinija, S. and Haydon, P.G. (1994) Glutamate-mediated astrocyte-neuron signalling. *Nature*, 369: 744–747.

Pfrieger, F.W. and Barres, B.A. (1997) Synaptic efficacy enhanced by glial cells in vitro. *Science*, 277: 1684–1687.

Porter, J.T. and McCarthy, K.D. (1995) Adenosine receptors modulate $[Ca^{2+}]_i$ in hippocampal astrocytes in situ. *J. Neurochem.*, 65: 1515–1523.

Priller, J., Reddington, M., Haas, C.A. and Kreutzberg, G.W. (1998) Stimulation of P2Y-purinoceptors on astrocytes results in immediate early gene expression and potentiation of neuropeptide action. *Neuroscience*, 85: 521–525.

Queiroz, G., Gebicke-Haerter, P.J., Schobert, A., Starke, K. and von Kuegelgen, I. (1997) Release of ATP from cultured rat astrocytes elicited by glutamate receptor activation. *Neuroscience*, 78: 1203–1208.

Rhee, S.G. and Bae, Y.S. (1997) Regulation of phosphoinositide-specific phospholipase C isozymes. *J. Biol. Chem.*, 272: 15045–15048.

Rubio, N. (1997) Mouse astrocytes store and deliver brain-derived neurotrophic factor using the non-catalytic gp95trkB receptor. *Eur. J. Neurosci.*, 9: 1847–1853.

Rudge, J.S. (1993) Astrocyte-Derived Neurotrophic Factors. In: S. Murphy (Ed.) *Astrocytes: Pharmacology and Function.* Academic Press, San Diego. pp. 267–305.

Salter, M.W., De Koninck, Y. and Henry, J.L. (1993) Physiological roles for adenosine and ATP in synaptic transmission in the spinal dorsal horn. *Prog. Neurobiol.*, 41: 125–156.

Salter, M.W. and Hicks, J.L. (1994) ATP-evoked increases intracellular calcium in cultured neurones and glia from the dorsal spinal cord. *J. Neurosci.*, 14: 1563–1575.

Salter, M.W. and Hicks, J.L. (1995) ATP causes release of intracellular Ca^{2+} via the phospholipase $C\beta/IP_3$ pathway in astrocytes from the dorsal spinal cord. *J. Neurosci.*, 15: 2961–2971.

Soltoff, S.P., Avraham, H., Avraham, S. and Cantley, L.C. (1998) Activation of $P2Y_2$ receptors by UTP and ATP stimulates mitogen-activated kinase activity through a pathway that involves related adhesion focal tyrosine kinase and protein kinase C. *J. Biol. Chem.*, 273: 2653–2660.

Venance, L., Stella, N., Glowinski, J. and Giaume, C. (1997) Mechanism involved in initiation andpropagation of receptor-induced intercellular calcium signaling in cultured rat astrocytes. *J. Neurosci.*, 17: 1981–1992.

Verkhratsky, A. and Kettenmann, H. (1996) Calcium signalling in glial cells. *Trends. Neurosci.*, 19: 346–352.

White, T.D., Downie, J.W. and Leslie, R.A. (1985) Characteristics of K^+- and veratridine-induced release of ATP from synaptosomes prepared from dorsal and ventral spinal cord. *Brain Res.*, 334: 372–374.

Yamane, H.K. and Fung, B.K.K. (1993) Covalent modifications of G-proteins. *Annu. Rev. Pharmacol. Toxicol.*, 33: 201–241.

P. Illes and H. Zimmermann (Eds.)
Progress in Brain Research, Vol 120

Trophic signaling pathways activated by purinergic receptors in rat and human astroglia

Joseph T. Neary[1,2,*], Micheline McCarthy[1,3], Ann Cornell-Bell[4] and Yuan Kang[1]

[1]*Research Service, VA Medical Center, Miami, FL 33125, USA*
[2]*Departments of Pathology, Biochemistry and Molecular Biology, University of Miami School of Medicine, Miami, FL 33125, USA*
[3]*Department of Neurology, University of Miami School of Medicine, Miami, FL 33125, USA*
[4]*Cognetix/Viatech Imaging Laboratories, Ivoryton, CT, 06442, USA*

Introduction

Extracellular nucleotides and nucleosides exert mitogenic and morphogenic effects on neural cells (Neary et al., 1996). These trophic actions may be important in development and synaptogenesis as well as in brain injury and repair. In particular, extracellular nucleotides and nucleosides may play a role in gliosis, the hypertrophic and hyperplastic response of astrocytes to brain injury (Eng et al., 1987; Norenberg, 1994). Such cells, which are often referred to as reactive astrocytes, are characterized by the generation of cellular processes (stellation) and by an increase in expression of GFAP; cellular proliferation is also frequently observed. Recent studies from several laboratories have demonstrated that these characteristics can be mimicked by the application of ATP or ATP analogs to astrocytes in culture (Rathbone et al., 1992; Neary et al., 1994; Abbracchio et al., 1994; Neary et al., 1994a; Bolego et al., 1997). In vivo, GFAP staining was increased following infusion of an adenosine analog into rat brain (Hindley et al., 1994). Thus, the release of purines, including ATP, from damaged or dying cells following injury

(Gordon, 1986) such as hypoxia (Bergfeld and Forrester, 1992; Fredholm, 1997) may contribute to the gliosis associated with many common neurological conditions such as trauma, stroke, seizure, and degenerative and demyelinative disorders.

After CNS injury, there are increases in extracellular levels of a number of growth factors, cytokines, and other mitogens and morphogens which have been implicated in gliosis (Eddleston and Mucke, 1993; Ridet et al., 1997). Polypeptide growth factors such as fibroblast growth factors (FGFs) are of particular interest because of their potent effects on differentiation and proliferation. Interestingly, nucleotides and nucleosides can act synergistically with polypeptide growth factors to enhance trophic effects (Huang et al., 1989; Schafer et al., 1995; Gysbers and Rathbone, 1996). In astrocytes, ATP synergistically enhances FGF-induced mitogenesis (Neary et al., 1994b). These findings suggest a role for extracellular nucleotides and nucleosides, acting alone or in concert with polypeptide growth factors, in the formation of reactive astrocytes.

While much is known about factors that can initiate gliotic-like responses, very little is known about the molecular mechanisms that mediate the formation of reactive astrocytes. To understand these mechanisms, we believe it will be important to delineate the signal transduction pathways that

---
*Corresponding author. Tel.: +1 305-324-4455; fax: +1 305-324-3126; e-mail: jneary@mednet.med.miami.edu

mediate cell growth. Mitogen-activated protein kinase (MAPK) cascades are intriguing candidates because emerging evidence now points to a crucial role for these signaling pathways in cellular proliferation and differentiation (reviewed in Seger and Krebs, 1995; Neary, 1997) as well as in tissue injury and recovery (e.g., Hu and Wieloch, 1994). These cascades consist of at least three cytoplasmic protein kinases which are activated sequentially: MAPK kinase kinase → MAPK kinase → MAPK (Fig. 1). At least three parallel MAPK pathways have been identified; these are frequently referred to as ERK (extracellular signal regulated protein kinase), SAPK (stress-activated protein kinase, also known as JNK for c-Jun N-terminal kinase) and p38/MAPK. The latter two cascades have been implicated in growth arrest and apoptosis (Kyriakis and Avruch, 1996). MAPK cascades are stimulated by growth or stress signals, and the activated MAPKs can target proteins in the cytoplasm, membrane, or cytoskeleton, or they can translocate to the nucleus where they activate or induce transcription factors, thereby leading to the expression of genes important in cell growth or death. Some stimuli are highly selective for one pathway, and this has led to the concept that the ERK pathway mediates proliferation and differentiation

whereas the SAPK and p38 pathways are involved in growth arrest and apoptosis. However, it is now clear that some stimuli activate two or more pathways, thereby suggesting that highly-coordinated signaling between pathways is needed to mediate a specific cellular response.

Signaling from P2Y receptors to the ERK cascade in rat astrocytes

Because the ERK cascade plays a crucial role in regulating cellular proliferation and differentiation, we postulated that the ATP-induced morphogenic and mitogenic responses in astrocytes are mediated by the ERK cascade. To test this hypothesis, we first demonstrated that extracellular ATP rapidly stimulates ERK in primary cultures of rat cortical astrocytes (Neary and Zhu, 1994c). This stimulation could have been due to breakdown of ATP to adenosine with subsequent activation of P1 receptors, but experiments with P1 and P2 agonists and antagonists revealed that P2 receptors rather than P1 receptors stimulate ERK in rat astrocytes. In collaboration with Prof. G. Burnstock, B. King, and colleagues, we found that rat cortical astrocytes express both P2y- and P2u-like receptors and that these receptors are coupled to calcium and ERK

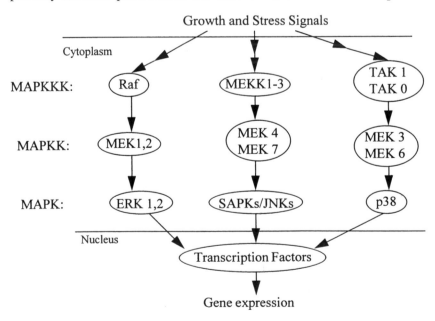

Fig. 1. Diagram of activation of MAPK cascades.

signaling pathways (King et al., 1996). Regarding ERK activation, we found that UTP and 2-MeSATP were approximately as effective as ATP, ADP was less effective than ATP, whereas α,β-meATP was ineffective. We also noted that the astroglia P2Y receptors were readily desensitized upon subsequent applications of ATP. Pertussis toxin treatment reduced ATP- and UTP-evoked ERK activation by approximately 50% (Neary et al., 1997), thereby suggesting that signaling from P2Y receptors to ERK is mediated at least partly by Gα subunits such as Gi or Go. RT-PCR analysis has revealed that P2Y$_1$ (Webb et al., 1996), P2Y$_2$ and P2Y$_4$ (J.T. Neary and W.-J. Nie, unpublished observations) receptor subtypes are expressed in primary cultures of rat cortical astrocytes. To test

for the expression of purinergic receptors in rat brain, we have conducted experiments with unrolled hippocampal preparations. Preliminary results indicate that functional P2u-like receptors are expressed in astrocytes in vivo (Fig. 2); studies with other P2Y agonists are currently under way.

To investigate the signaling pathway from P2Y receptors to ERK, we started by examining protein kinase C (PKC) because PKC links some types of G protein-coupled receptors to ERK, although the mechanism of coupling is not well understood. We found that signaling from P2Y receptors to ERK involves PKC because down-regulation of PKC blocked activation of ERK by ATP (Neary, 1996). Because P2Y receptors are also coupled to phosphatidylinositol-specific phospholipase C (PI-PLC)

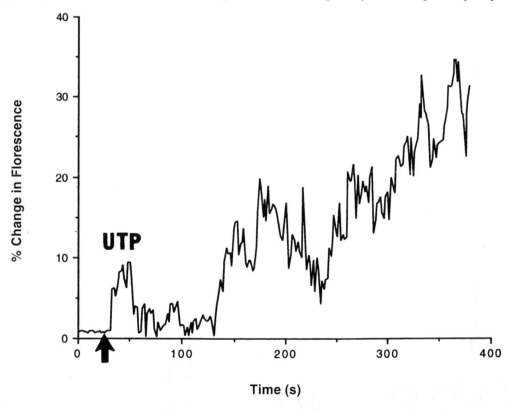

Fig. 2. Functional astrocytic P2Y receptors in neonatal hippocampus. After dissection, the hippocampus was unrolled to expose the dentate gyrus, CA3–CA1 fields, and margin of cortex as a flat sheet which was incubated with the Ca^{2+} indicator Fluo3-AM (10 μM) for 45 min. Tissue was placed in a flow-through chamber and perfused with physiological salt solution that was constantly bubbled with 95% air/5% CO$_2$. Images were collected on Nikon PCM200 confocal scanning laser microscope with the fluorescein excitation filter. After collecting baseline images, UTP (100 uM) was added. Data is presented as fluorescence intensity vs. time of a cell with characteristic astrocyte morphology (larger cell body, absence of axons and dendrites). UTP brought about a rapid increase in intracellular Ca^{2+} followed by oscillations of Ca^{2+}, indicative of P2Y receptors sensitive to UTP.

TABLE 1

Signaling from P2Y receptors to ERK is not dependent on calcium

Treatment	ERK Activty (fold stimulation; mean ± s.e.m.)
ATP	2.57 ± 0.25
EGTA + ATP	2.47 ± 0.20 (n.s.)
BAPTA + ATP	2.68 ± 0.12 (n.s.)

Prior to addition of ATP (100 μM, 5 min), astrocytes were treated with EGTA (3 mM, 5 min) or BAPTA-AM (30 μM, 30 min) to chelate extracellular or intracellular calcium, respectively. Note that ATP-evoked activation of ERK was not appreciably reduced by either EGTA or BAPTA, thereby indicating that signaling from P2Y purinergic receptors to ERK proceeds independently of calcium. (n.s.; not significant)

leading to diacylglycerol (DG) and inositol phosphate formation with subsequent calcium mobilization and activation of PKC, we reasoned that PI-PLC, calcium, PKC and ERK are part of a common pathway. But surprisingly, inhibition of PI-PLC did not block signaling from P2Y receptors to ERK, although it did block inositol phosphate formation (Kang et al., 1996). Moreover, chelation of intra- or extracellular calcium did not block signaling from P2Y receptors to ERK (Table 1). Thus, our results indicate that P2Y receptors are coupled independently to the ERK cascade and the PI-PLC/calcium pathway.

If signaling from P2Y receptors to ERK is independent of PI-PLC and calcium, but dependent on PKC, we reasoned that a calcium-independent isoform of PKC may be involved in the P2Y/ERK pathway. In support of this hypothesis, we found that an inhibitor of calcium-dependent PKCs, Gö 6976, was significantly less effective in blocking ATP-evoked ERK activation than Ro 31-8220, an inhibitor of both calcium-dependent and calcium-independent PKC isoforms (Fig. 3); similar results have been obtained with another inhibitor of calcium-dependent and -independent PKCs, GF102903X (J.T. Neary and Y. Kang, unpublished observations). Moreover, ATP stimulated a rapid translocation of a calcium-independent PKC isoform, PKCδ, but not a calcium-dependent isoform, PKCγ (Kang et al., 1997). Time course studies rcvealed maximum PKCδ translocation at 15–30 s. This rapid translocation of PKCδ occurs prior to

ATP stimulation of ERK and thus is consistent with PKCδ serving as an upstream activator of the ERK cascade.

We have also addressed the question of how P2Y receptors activate the PKC linked to ERK. Our experiments suggest the diacylglycerol needed to activate PKC is generated from phosphatidylcholine (PC) rather than PI because an inhibitor of PC hydrolysis, D609, blocked signaling from P2Y receptors to ERK (Neary et al., 1997). Diacylglycerol can be formed from PC by either PC-PLC or phospholipase D (PLD). Our experiments indicate a role for PLD because ATP rapidly stimulated an increase in choline, rather than phosphocholine (J.T. Neary and Y. Kang, unpublished observations). The rapid time course (15–30 s peak response) is consistent with the rapid translocation of PKCδ and subsequent activation of the ERK cascade. Thus, our studies to date suggest that PLD and PKCδ are key components of the signaling pathway from P2Y receptors to ERK.

The ERK cascade mediates mitogenic signaling by ATP/P2Y receptors

The evidence that P2Y receptors are linked to the ERK cascade is not sufficient to determine whether the ERK cascade mediates mitogenic signaling by P2Y receptors because, in addition to activating and inducing transcription factors involved in gene expression needed for proliferation and differentiation, ERK has other targets including proteins in the cytoplasm, membrane, and cytoskeleton. Thus, ERK may serve other cellular functions besides regulating cell growth. To investigate the role of the ERK cascade in P2Y receptor-induced mitogenesis, we utilized the selective MEK inhibitor, PD 098059 (Alessi et al., 1995). If the ATP-evoked activation of ERK and the ATP-induced mitogenic response are causally related, PD 098059 should block the ability of ATP to stimulate DNA synthesis. Indeed, this was observed (Neary et al., 1997), thereby indicating that the ERK cascade mediates mitogenic signaling by ATP/P2 receptors in astrocytes. Prior to these experiments, we conducted dose-response and time course studies with PD 098059 to determine the optimum conditions for inhibition of serum-stimu-

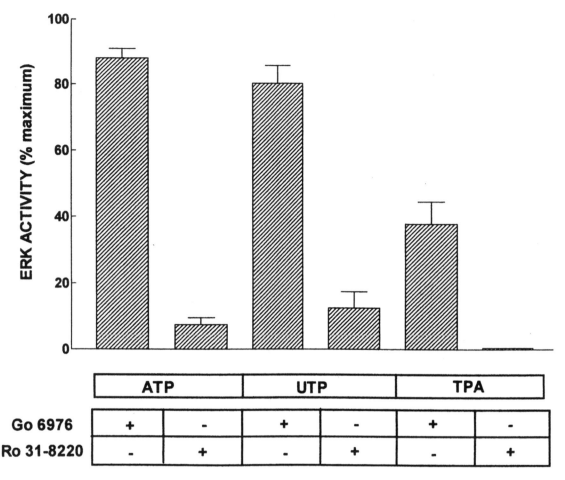

Fig. 3. Signaling from P2Y receptors to ERK involves calcium-independent PKC. Primary rat astrocyte cultures were treated with Gö 6976 (5 uM, 20 min), an inhibitor of calcium-dependent PKC isoforms, or with Ro 31-8220 (5 μM, 20 min), an inhibitor of both calcium-dependent and -independent PKC isoforms, prior to addition of ATP (100 μM, 5 min), UTP (100 μM, 5 min), or TPA (100 nM, 5 min). ERK activity data were obtained from a minimun of three experiments, each conducted with duplicate culture plates. Inhibition of ATP, UTP, and TPA stimulated ERK activity was expressed as percent of maximum activation; fold-stimulation (mean ± SEM) of ERK activity by ATP, UTP, and TPA was 3.70 ± 0.15, 4.06 ± 0.28, and 2.95 ± 0.13, respectively.

lated mitogenesis (50 uM, 30 min). Then, by means of ERK activity assays as well as immunoblots with an antibody that recognizes phosphorylated ERK1 and ERK2, we demonstrated that under these conditions PD 098059 inhibits ATP-evoked ERK activation.

Differences between signaling from ATP/P2Y and FGF-2 receptors to the ERK cascade

FGF-2 also signals to the ERK cascade, and because ATP and FGF-2 act synergistically to

stimulate astroglial mitogenesis, we compared the ATP and FGF-2 signaling pathways. As a first approach, we took advantage of the ability of protein kinase A (PKA) pathways to modulate the ERK cascade by interacting with Raf-1, the first kinase in the cascade. We found that activation of PKA blocked signaling from FGF-2 to ERK but not from ATP to ERK (Neary and Zhu, 1994c). This suggested that FGF-2 receptors recruit Raf-1, but ATP/P2Y receptors do not. As another approach, in collaboration with Dr. J. Avruch, Massachusetts General Hospital, we directly measured Raf-1 activation. In confirmation of our PKA experiments, we found that while FGF-2 activates Raf-1, extracellular ATP does not (Lenz et al., 1998). Thus, while ATP activates ERK, it does so by a pathway distinct from that of FGF-2. However, UTP did stimulate Raf-1, thus indicating that different P2Y receptors are coupled to ERK by distinct pathways (Lenz et al., 1998). Further studies are needed to determine the nature of the MEK activator recruited by ATP/P2Y receptors.

Stimulation of ATP and FGF receptors leads to activation of AP-1 transcriptional complexes

The ERK cascade can relay trophic signals to the nucleus by activating or inducing transcription factors or complexes, thereby leading to changes in gene expression. Activator protein (AP)-1 is a functional transcriptional complex that has been implicated in the expression of a wide variety of genes, including GFAP (Masood et al., 1993), and activation of MAPK cascades leads to increased AP-1 activity (Karin, 1995). In addition, AP-1 complexes are increased following brain injury (Dash et al., 1995). In view of the trophic actions of extracellular ATP and its ability to activate ERK, we hypothesized that stimulation of P2Y receptors would lead to AP-1 complex formation. By means of electrophoretic mobility shift assays, we found that both ATP and UTP induced AP-1 binding activity in rat astrocytes (Neary et al., 1996b). The AP-1 complex is a heterodimer consisting of members of Fos and Jun families of transcription factors (Angel and Karin, 1991), and we found that c-Fos was present in the ATP-induced AP-1 complexes. This finding is in very good agreement with the work of Abbracchio and colleagues who showed by immunostaining that c-Fos and c-Jun are increased following activation of P2 receptors in striatal primary cultures (Bolego et al., 1997). FGF-2 also stimulated AP-1 formation, although again not as quickly as ATP (Neary et al., 1996a), a finding consistent with the time course of FGF-induced ERK activation. As with ERK, the ATP-evoked AP-1 activity was due to P2Y rather than P1 receptors and thus was not a result of the breakdown of ATP to adenosine. Signaling from P2Y receptors to AP-1 was also dependent on PKC

(Neary et al., 1996b), and inhibition of the ERK cascade by the selective MEK inhibitor PD 098059 (Alessi et al., 1995) reduced the ability of ATP and UTP to stimulate AP-1 binding activity by approximately 50% (data not shown). This indicates that activation of this transcriptional complex is dependent in part on the ERK cascade and suggests that additional cascades may be involved in AP-1 formation. These findings demonstrate that P2Y receptors can signal to the nucleus; such signaling may contribute to the changes in gene expression which underlie the trophic actions of extracellular ATP.

ATP and FGF-2 activate p38/MAPK in astrocytes

Because inhibiting the ERK cascade did not completely block AP-1 formation, we investigated the possibility that ATP and FGF-2 activate other MAPK pathways. We found that ATP and FGF-2 stimulated phosphorylation of p38/MAPK (Fig. 4). To determine whether p38 is involved in mitogenesis, we utilized a selective inhibitor of p38/MAPK, SB 203580 (Cuenda et al., 1995), and found that this inhibitor completely blocked ATP-mediated mitogenic signaling and reduced FGF-mediated mitogenic signaling by at least 50% (Neary et al., 1998). These preliminary results indicate that parallel MAPK pathways are involved in purinergic and FGF signaling and suggest that coordinated signaling among the ERK and p38 cascades may underlie gliotic responses.

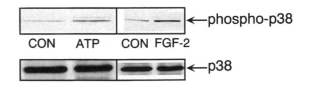

Fig. 4. ATP and FGF-2 activate p38/MAPK. Rat primary astrocyte cultures were treated with FGF-2 (25 ng/ml) or ATP (100 μM) for 15 min, and lysates containing equivalent amounts of protein were subjected to SDS-electrophoresis. Immunoblots were probed with antibodies which recognizephosphorylated p38 or total p38/MAPK (New England BioLabs) (CON, control). Similar results were obtained in two additional experiments.

Mitogenic signaling from purinergic and FGF-2 receptors to MAPK cascades in human astrocytes

To extend our results to a system more relevant to human neurological disorders, we have begun to investigate mitogenic signaling pathways activated by purines and FGF-2 in human astrocytes. These cells were obtained from first trimester rostral CNS tissue as previously described (McCarthy et al., 1995). Procedures for procurement and use of human fetal CNS tissue were approved and monitored by the University of Miami School of Medicine's Subcommittee for the Protection of Human Subjects. Interestingly, we found differences in purine-evoked mitogenic signaling between rat and human astrocytes (Neary et al., 1998a). Mitogenesis in the human astrocytes was stimulated not only by ATP, but also by adenosine analogs. P1 and P2 agonists also stimulated ERK; this is in contrast to rat astrocytes in which P2, but not P1, receptors are linked to ERK (Neary and Zhu, 1994c). Further studies are needed to identify the subtypes of P1 and P2 receptors coupled to ERK in human astrocytes, but in preliminary studies we have found that a selective A3-receptor agonist, IB-MECA (kindly provided by Dr. K. Jacobson, National Institutes of Health, Bethesda, MD, USA) is partially effective in stimulating ERK activity.

Studies with the MEK inhibitor PD 098059 indicated that mitogenic signaling by P1 and P2 receptors was mediated by ERK (Neary et al., 1998a). Signaling from P1 and P2 receptors to ERK is also dependent on PKC in human astrocytes. These findings suggest a role for both P1 and P2 receptors in the proliferation of human astrocytes. As in rat astrocytes, ATP enhances FGF-induced mitogenesis in human astrocytes (Fig. 5). This effect is reduced by SB 203580 (Fig. 5), thereby indicating a role for p38/MAPK in ATP- and FGF-mediated mitogenic signaling in human astrocytes.

Conclusions

Our findings indicate that MAPK cascades mediate mitogenic signaling by purinergic receptors in both rat and human astrocytes. In rat astrocytes, P2 receptors are linked to the ERK cascade and mitogenesis, whereas in human astrocytes, both P1

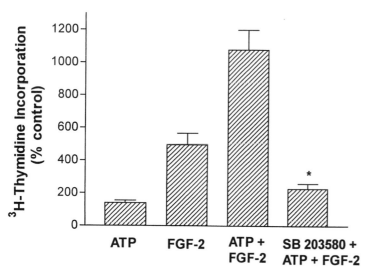

Fig. 5. Inhibition of p38/MAPK reduces ATP-enhanced, FGF-2-induced mitogenesis in human fetal astrocytes. Human fetal astrocytes were treated with or without SB 203580 (10 μM) for 60 min prior to addition of ATP (100 μM) and/or FGF-2 (25 ng/ml); DNA synthesis was measured 22 hr later. Data (mean ± SEM) were obtained from three independent experiments, each conducted in quadruplicate. ATP significantly ($p < 0.01$) enhanced FGF-2-stimulated mitogenesis; this effect was inhibited by SB 203580 (*, $p < 0.001$). SB 203580 also reduced mitogenic signaling by either ATP or FGF-2 alone (data not shown). These findings suggest a role for p38/MAPK in ATP- and FGF-2-induced mitogenic signaling.

and P2 receptors are coupled to ERK. The signaling pathway from rat and human purinergic receptors to the ERK cascade involves PKC. Studies with rat astrocytes reveal that P2Y receptors are coupled to ERK by a pathway which is independent of the PI-PLC/inositol phosphate/calcium pathway. In these cortical astrocytes, signaling from P2Y receptors to ERK involves PLD and a calcium-independent

PKC, PKCδ (Fig. 6). Following activation of ERK by P2Y receptor occupancy, functional AP-1 transcriptional complexes are induced which can, in turn, lead to changes in gene expression underlying the trophic effects of extracellular nucleotides. ATP, in addition to its trophic actions, can also enhance mitogenesis induced by FGF-2, a key growth factor implicated in neuronal survival. Our findings

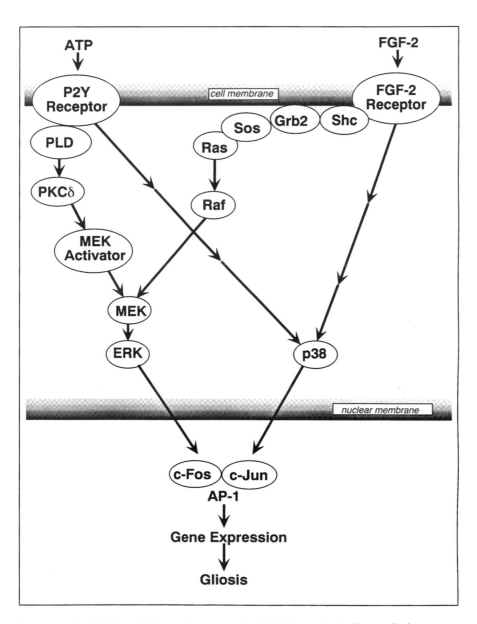

Fig. 6. Model of ATP and FGF signaling coupled to MAPK cascades leading to gliosis.

suggest that ATP, acting alone or in concert with FGF-2, plays an important role in development and synaptogenesis as well as in brain injury and repair. Astroglial purinergic receptors and components of their signaling pathways may serve as targets for the development of novel therapeutic strategies for the management of brain injury and other neurological conditions.

Acknowledgments

We are grateful to Dr. J. Avruch, Massachusetts General Hospital, Boston, MA, USA, for providing c-Raf antibodies and recombinant, inactive MEK and ERK used in the c-Raf coupled assay, Dr. K. Jacobson, National Institutes of Health, Bethesda, MD, USA, for providing IB-MECA, Dr. G. Lawton, Roche Research Centre, Hertfordshire, UK, for providing Ro 31-8220, and Dr. J.C. Lee, Smith-Kline Beecham Pharmaceuticals, King of Prussia, PA, USA, for providing SB 203580. We also gratefully acknowledge the assistance of Dr. Scott Whittemore, University of Miami School of Medicine, Miami, FL, USA, in obtaining and culturing human fetal astrocytes. This work was supported by the Department of Veterans Affairs.

References

Abbracchio, M.P., Saffrey, M.J., Hopker, V. and Burnstock, G. (1994) Modulation of astroglial cell proliferation by analogues of adenosine and ATP in primary cultures of rat striatum. *Neurosci.*, 59: 67–76.

Alessi, D.R., Cuenda, A., Cohen, P., Dudley, D.T. and Saltiel, A.R. (1995) PD 098059 is a specific inhibitor of the activation of mitogen-activated protein kinase kinase in vitro and in vivo. *J. Biol. Chem.*, 270: 27489–27494.

Angel, P. and Karin, M. (1991) The role of Jun, Fos and the AP-1 complex in cell-proliferation and transformation. *Biochim. Biophys. Acta*, 1072: 129–157.

Bergfeld, G.R. and Forrester, T. (1992) Release of ATP from human erythrocytes in response to a brief period of hypoxia and hypercapnia. *Cardio. Res.*, 26: 40–47.

Bolego, C., Ceruti, S., Brambilla, R., Puglisi, L., Cattabeni, F., Burnstock, G. and Abbracchio, M.P. (1997) Characterization of the signalling pathways involved in ATP and basic fibroblast growth factor-induced astrogliosis. *Br. J. Pharmacol.*, 121: 1692–1699.

Cuenda, A., Rouse, J., Doza, Y.N., Meier, R., Cohen, P., Gallagher, T.F., Young, P.R. and Lee, J.C. (1995) SB 203580 is a specific inhibitor of a MAP kinase homologue which is stimulated by cellular stresses and interleukin-1. *FEBS Lett.*, 364: 229–233.

Dash, P.K., Moore, A.N. and Dixon, C.E. (1995) Spatial memory deficits, increased phosphorylation of the transcription factor CREB, and induction of the AP-1 complex following experimental brain injury. *J. Neurosci.*, 15: 2030–2039.

Eddleston, M. and Mucke, L. (1993) Molecular profile of reactive astrocytes: implications for their role in neurologic disease. *Neurosci.*, 54: 15–36.

Eng, L.R., Reier, P.J. and Houle, J.D. (1987) Astrocyte activation and fibrous gliosis: Glial fibrillary acidic protein immunostaining of astrocytes following intraspinal cord grafting of fetal CNS tissue. *Prog. Brain Res.*, 71: 439–455.

Fredholm, B.B. (1997) Adenosine and neuroprotection. *Int. Rev. Neurobiol.*, 40: 259–280.

Gordon, J.L. (1986) Extracellular ATP: Effects, sources and fate. *Biochem. J.*, 233: 309–319.

Gysbers, J.W. and Rathbone, M.P. (1996) GTP and guanosine synergistically enhance NGF-induced neurite outgrowth from PC12 cells. *Int. J. Devl. Neurosci.*, 14: 19–34.

Hindley, S., Herman, M.A.R. and Rathbone, M.P. (1994) Stimulation of reactive gliosis in vivo by extracellular adenosine diphosphate or an adenosine A2 receptor agonist. *J. Neurosci. Res.*, 38: 399–406.

Hu, B.-R. and Wieloch, T. (1994) Tyrosine phosphorylation and activation of mitogen-activated protein kinase in the rat brain following transient cerebral ischemia. *J. Neurochem.*, 62: 1357–1367.

Huang, N., Wang, D. and Heppel, L.A. (1989) Extracellular ATP is a mitogen for 3T3, 3T6, and A431 cells and acts synergistically with other growth factors. *Proc. Natl. Acad. Sci. USA*, 86: 7904–7908.

Kang, Y., Zhu, Q., Yu, E. and Neary, J.T. (1996) Signal transduction pathways coupled to P2Y purinoceptors in astrocytes. *Soc. Neurosci. Abstr.*, 22: 1734.

Kang, Y., Zuniga, S. and Neary, J.T. (1997) Signalling from ATP/P2Y receptors to MAP kinase in astrocytes involves a calcium-independent PKC isoform. *J. Neurochem.*, 69 (Suppl.): S207.

Karin, M. (1995) The regulation of AP-1 activity by mitogen-activated protein kinases. *J. Biol. Chem.*, 270: 16483–16486.

King, B.F., Neary, J.T., Zhu, Q., Wang, S., Norenberg, M.D. and Burnstock, G. (1996) P2 purinoceptors in rat cortical astrocytes: Expression, calcium-imaging and signalling studies. *Neurosci.*, 74: 1187–1196.

Kyriakis, J.M. and Avruch, J. (1996) Sounding the Alarm: Protein kinase cascades activated by stress and inflamation. *J. Biol. Chem.*, 271: 24313–24316.

Lenz, G., Luo, Z., Avruch, J., Rodnight, R. and Neary, J.T. (1998) P2Y purinergic receptor subtypes recruit different MEK activators in astrocytes. *FASEB J.*, 12: A1469.

Masood, K., Bresnard, F., Su, Y. and Brenner, M. (1993) Analysis of a segment of the human glial fibrillary acidic protein gene that directs astrocyte-specific transcription. *J. Neurochem.*, 61: 160–166.

McCarthy, M., Wood, C., Fedoseyeva, L. and Whittemore, S.R. (1995) Media components influence viral gene expression

332

assays in human fetal astrocyte cultures. *J. NeuroVirol.*, 1: 275–285.

Neary, J.T. (1996) Trophic actions of extracellular ATP on astrocytes, synergistic interactions with fibroblast growth factors, and underlying signal transduction mechanisms. *Ciba Foundation Symposium*, 198: 130–141.

Neary, J.T. (1997) MAPK cascades in cell growth and death. *News in Physiol. Sci.*, 12: 286–293.

Neary, J.T., Baker, L., Jorgensen, S.L. and Norenberg, M.D. (1994) Extracellular ATP induces stellation and increases GFAP content and DNA synthesis in primary astrocyte cultures. *Acta Neuropathol.*, 87: 8–13.

Neary, J.T., Dash, P.K. and Zhu, Q. (1996a) Signaling by extracellular ATP and bFGF in astrocytes. *Adv. Gene Technol.*, 7: 13.

Neary, J.T., Kang, Y. and Mehta, H. (1998) Mitogenic signaling and MAPK cascades activated by fibroblast growth factors and extracellular ATP in astrocytes. *Miami Nature Biotechnology Short Reports*, 9: 119–120.

Neary, J.T., Kang, Y. and Zuniga, S. (1997) The MAPK/ERK cascade mediates mitogenic signaling by ATP/P2 receptors in astrocytes. *Soc. Neurosci. Abstr.*, 23: 70.

Neary, J.T., McCarthy, M., Kang, Y. and Zuniga, S. (1998a) Mitogenic signaling from P1 and P2 purinergic receptors to mitogen-activated protein kinase in human fetal astrocytes. *Neurosci. Lett.*, 242: 159–162.

Neary, J.T., Rathbone, M.P., Cattabeni, F., Abbracchio, M.P. and Burnstock, G. (1996) Trophic actions of extracellular nucleotides and nucleosides on glial and neuronal cells. *Trends Neurosci.*, 19: 13–18.

Neary, J.T., Whittemore, S.R., Zhu, Q. and Norenberg, M.D. (1994a) Destabilization of glial fibrillary acidic protein mRNA by ammonia and protection by extracellular ATP. *J. Neurochem.*, 63: 2021–2027.

Neary, J.T., Whittemore, S.R., Zhu, Q. and Norenberg, M.D. (1994b) Synergistic activation of DNA synthesis in astrocytes by fibroblast growth factor and extracellular ATP. *J.Neurochem.*, 63: 490–494.

Neary, J.T. and Zhu, Q. (1994c) Signaling by ATP receptors in astrocytes. *NeuroReport*, 5: 1617–1620.

Neary, J.T., Zhu, Q., Kang, Y. and Dash, P.K. (1996b) Extracellular ATP induces formation of AP-1 complexes in astrocytes via P2 purinoceptors. *NeuroReport*, 7: 2893–2896.

Norenberg, M.D. (1994) Astrocyte responses to CNS injury. *J. Neuropath. Exp. Neurol.*, 53: 213–220.

Rathbone, M.P., Middlemiss, P.J., Kim, J.-L., Gysbers, J.W., DeForge, S.P., Smith, R.W. and Hughes, D.W. (1992) Adenosine and its nucleotides stimulate proliferation of chick astrocytes and human astrocytoma cells. *Neurosci. Res.*, 13: 1–17.

Ridet, J.L., Malhotra, S.K., Privat, A. and Gage, F.H. (1997) Reactive astrocytes: cellular and molecular cues to biological function. *Trends Neurosci.*, 20: 570–577.

Schafer, K.-H., Saffrey, M.J. and Burnstock, G. (1995) Trophic actions of 2-chloroadenosine and bFGF on cultured myenteric neurones. *NeuroReport*, 6: 937–941.

Seger, R. and Krebs, E.G. (1995) The MAPK signaling cascade. *FASEB J.*, 9: 726–735.

Webb, T.E., Feolde, E., Vigne, P., Neary, J.T., Runberg, A., Frelin, C. and Barnard, E.A. (1996) The P2Y purinoceptor in rat brain microvascular endothelial cells couples to inhibition of adenylate cyclase. *Br. J. Pharmacol.*, 119: 1385–1392.

P. Illes and H. Zimmermann (Eds.)
Progress in Brain Research, Vol 120

CHAPTER 27

Signalling mechanisms involved in P2Y receptor-mediated reactive astrogliosis

Maria P. Abbracchio*, Roberta Brambilla, Stefania Ceruti and Flaminio Cattabeni

Institute of Pharmacological Sciences, University of Milan, Via Balzaretti 9, 20133 Milan, Italy

Reactive astrogliosis: good or bad for CNS repair?

The central nervous system (CNS) is made up of several cell populations, mainly neurons, oligodendrocytes, astrocytes and microglial cells.

Many functions have been attributed to astrocytes, including support during CNS development, maintenance of ion homeostasis, uptake of chemical mediators and modulation of neurotransmission, as well as contribution to CNS immune response (for review, see: Eddleston and Mucke, 1993; Ridet et al., 1997). Another role of astrocytes is to respond to various type of noxius stimuli (e.g., hypoxia/ischemia and trauma) with vigorous and rapid astrogliosis, a reaction characterized by both increased astrocytic proliferation and cellular hypertrophy, as shown by increased cellular size and presence of longer and thicker astrocytic processes which intensively stain for the astrocytic-specific marker glial fibrillary acidic protein (GFAP). Astrocytic hypertrophy accompanied by increased GFAP expression seems to be a common characteristic in all types of astrogliosis, even in the absence of glial cell proliferation as is the case for neonatal anoxia (Hatten et al., 1991). The origin of reactive astrocytes is still an open question. There seems to be a remarkable similarity between reactive astrocytes and embryonic radial

*Corresponding author. Tel.: +39-2-20488316; fax.: +39-2-29404961; e-mail: ABBRACCH@imiucca.csi. unimi.it

glia, the type of glia supporting guidance to migrating neuroblasts during neurogenesis (Cameron and Rakic, 1991; see also, Abbracchio et al., 1995a) and believed to differentiate into mature astrocytes later during brain development, possibly as a consequence of completion of neuronal migration (ibidem). Reactive astrocytes can indeed re-express embryonic markers, such as vimentin and PSA-NCAM, the polysialyated form of neural cell adhesion molecule, whose expression has been related to the ability of astrocytes to support neuritic extension (Ridet et al., 1997). This has suggested that reactive astrocytes may arise from de-differentiation of already existing astrocytic cells. However, the hypothesis that a CNS lesion may lead to a recapitulation of ontogenic events is quite a difficult one to assess experimentally. Multipotential stem cells are probably widely present in the adult CNS, and it can also be hypothesized that, after a lesion, quiescent precursor cells could give rise to reactive astrocytes under the stimulation of local trophic factors (for review, see Ridet et al., 1997).

There is still debate as to the exact significance of reactive astrogliosis. Although this phenomenon has long been considered as the major impediment to axonal regrowth after an injury, the formation of a glial barrier around the lesion is also an advantage, because it isolates the still intact CNS tissue from secondary lesions. In addition, some data strongly suggest that in certain conditions reactive astrocytes could provide a permissive substrate for neuritic extension and hence for regeneration of damaged axons (permissive versus inhibitory reactive gliosis). A beneficial functional role for activated astrocytes is suggested by the demonstration that these are needed for axonal regrowth and guidance in vitro (Hatten et al., 1991) and by the fact that suppression of GFAP expression by antisense mRNA also obliterates the formation of stable astrocytic processes in response to neuronal signals (Weinstein et al., 1991). These findings are corroborated by the demonstration that during the post-ischemic and post-traumatic period, astroglial cells produce various neurotrophins and pleiotrophins as well as several adhesion molecules which may favour axonal regeneration (for review, see Neary et al., 1996a; Ridet et al., 1997).

It is, however, clear that in several cases astrogliosis is detrimental to neuronal recovery, likely due to an abnormal evolution towards a non-permissive gliotic scar. Interestingly, ultrastructural analysis has revealed that the non-permissive scar contains strikingly more gap junctions compared to the permissive gliotic scar (Alonso and Privat, 1993). Moreover, it is becoming increasingly clear that activated astrocytes may hyperproduce inflammatory mediators, including various cytokines and products of arachidonic acid metabolites (Pearce et al., 1989; O'Banion et al., 1996; Blom et al., 1997). Excessive and/or chronic astrocytic activation may hence lead to increased inflammatory events, as seems to be the case for various neurological disorders such as Alzheimer's disease (Rogers et al., 1993; Breitner et al., 1994; Anderson et al., 1995), prion-related disorders (Brown et al., 1996) and acute ischemia or trauma (Planas et al., 1995; Ohtsuki et al., 1996). To confirm a role for inflammation in these neurological disorders, epidemiological studies have shown that steroidal as well as non-steroidal anti-inflammatory drugs lower the risk of developing Alzheimer's disease (Rogers et al., 1993; Breitner et al., 1994; Anderson et al., 1995). Hence, an emerging requirement in the field of brain repair is the development of strategies to block post-traumatic gliosis. X-irradiation, local injection of antibodies against reactive astrocytic markers or of chemicals interfering with the physiology of activated astrocytes are among the different strategies that have been proposed for the experimental manipulation of the glial scar (Ridet et al., 1997). Of course, a key issue in this field concerns the identification of the endogenous molecules involved in the formation and maintenance of reactive astrocytes. We need to characterize the crucial triggers of reactive astrogliosis, to define their role in the different phases of the astrogliotic reaction, and, more importantly, to establish their role in permissive versus inhibitory reactive gliosis. Only by exploiting this information we will be able to develop a correct strategy to the in vivo manipulation of the astrocytic reaction.

Role of nucleotides and nucleosides in modulation of astrocytic function

Several factors have been implicated in the initiation of astrogliosis, including growth factors (epidermal growth factor; fibroblast growth factor, FGF; platelet-derived growth factor), cytokines and myelin basic proteins (Norton et al., 1992). In addition to these factors, we and others have been exploring the hypothesis that nucleotides and nucleosides may as well play a role in astrocytic activation following trauma and ischemia. This hypothesis is based on the fact that large amounts of nucleotides and nucleosides are released following brain trauma or ischemia. Under such pathological conditions, injured or hypoxic cells release their contents: these include not only the soluble nucleotide and nucleoside pools, but also the products of RNA and DNA degradation. Thus, at the site of injury, cells are exposed to significantly elevated concentrations of ATP, GTP and adenosine for prolonged periods of time (for review, see Neary et al., 1996a). Therefore, it is likely that, under such pathological conditions, nucleotides and nucleosides may serve as signals that act in concert with "classic" polypeptidic growth factors to initiate brain repair mechanisms, including astrogliosis, activation of microglial cells and regeneration of damaged neuronal axons. A possible role in the initiation of astrogliosis is also confirmed by the presence on astroglial cells (Abbracchio et al., 1995a) of specific receptors for both adenosine and ATP/UTP (the P1 and P2 receptors, respectively, Fredholm et al., 1994, 1997). In particular, the presence of G-protein-coupled P2Y receptors has been widely demonstrated on astrocytes obtained from either mammalian brain or spinal cord (Bruner and Murphy, 1990; Abbracchio et al., 1995b; Salter and Hicks, 1995; Neary, 1996; Bolego et al., 1997; see also below). To characterize the functional effects evoked by these receptors, we have investigated the effects induced by the relatively hydrolysis-resistant ATP analogue α,β-methyleneATP (α,β-meATP) on rat striatal astrocytes. A 2 h challenge of these cells with this purine analogue resulted, three days later, in marked elongation of GFAP-positive astrocytic processes, an event that reproduces in vitro the astrocytic hypertrophy known to occur during in vivo reactive astrogliosis. Similar effects on astrocytic elongation were also observed with ATP and other P2 receptor agonists, such as β,γ-meATP, ADPβS, 2methylthioATP and, to a lesser extent, UTP (Bolego et al., 1997). Interestingly, the maximal effect evoked by α,β-meATP was qualitatively and quantitatively similar to that induced by a classic trigger of reactive astrogliosis, such as basic FGF (bFGF, Abbracchio et al., 1995b). However, specific independent receptors seemed to be involved in purine- and bFGF-induced reactive astrogliosis. In fact, the P2 receptor antagonist suramin concentration-dependently reversed α,β-meATP- but not bFGF-induced reactive astrogliosis. Vice versa, the specific tyrosine kinase inhibitor genistein prevented bFGF- but not α,β-meATP-induced astrocytic changes, confirming that separate receptors and transductional mechanisms mediate growth factor- and purine-induced astrogliosis. However, we also noticed that the addition of both agents to cells did not result in an additive effect on the elongation of astrocytic processes (Abbracchio et al., 1995b), suggesting that, despite the involvement of separate receptors, the two transductional cascades activated by the growth factor and the purine analogue may merge at some common intracellular target crucially involved in mediating the long-term functional changes typical of reactive astrogliosis (see also below).

Transduction mechanisms responsible for P2Y receptor-mediated reactive astrogliosis

In a subsequent study, in an attempt to characterize the short-term events responsible for the long-term

actions of ATP analogues on astroglial cells, we have characterized in detail the transduction pathways involved in ATP-induced astrogliosis.

It is known that extracellular ATP exerts its effects through either ligand-gated ion channels (P2X receptors) or G-protein-coupled receptors (P2Y receptors) (Abbracchio and Burnstock, 1994). At least seven distinct P2X and 11 P2Y receptors have been cloned so far (for review, see Turner et al. (Eds.), 1997). In brain, ATP has been implicated both in fast neuro-neuronal communication via P2X receptors, and in regulation of slower metabotropic responses via P2Y receptors. Hence our first step was to verify whether the gliotic P2 receptor activated by α,β-meATP belonged to the ligand-gated or G-protein-coupled family. Previous evidence suggested that α,β-meATP may act as a potent agonist at some of the ligand-gated P2X ($P2X_1$ and $P2X_3$) receptors, which are channels for Na^+/K^+ and Ca^{2+} (Evans et al., 1997). However, at concentrations which effectively induced reactive astrogliosis, neither α,β-meATP nor β,γ-meATP elicited any significant increase of the intracellular Ca^{2+} concentration ($[Ca^{2+}]_i$) (Centemeri et al., 1997), ruling out a role for ligand-gated P2X receptors. On the contrary, pertussis toxin (PTX), which inactivates Gi/Go proteins, completely abolished α,β-meATP but not bFGF-induced elongation of astrocytic processes (Bolego et al., 1997), suggesting that the gliotic P2 receptor does indeed belong to the P2Y receptor family. Hence, we believe that the gliotic response to ATP analogues in our experimental system is mediated by an "atypical" P2Y receptor sensitive to α,β-meATP (see also below).

Cloning of the various members of the P2Y receptor family has enabled the characterization of their transduction mechanisms in transfected systems. Basically, these receptors have been linked to either stimulation of phospholipase C (PLC) and generation of inositol-tris-phosphates (IP_3) or to inihibition of adenylyl cyclase activity (Harden et al., 1997). Despite the fact that the prototypical response to P2Y receptor activation in many native systems is stimulation of PLC (for review, see Abbracchio, 1997), previously identified astroglial P2Y receptors coupled to this transduction system (Salter and Hicks, 1995) did not seem to play a role in our experimental model. Such a conclusion is based on the above reported lack of effect of α,β-meATP on $[Ca^{2+}]_i$, and also on the fact that neomycin, an inhibitor of PLC, had no effect on α,β-meATP-induced elongation of astrocytic processes (Bolego et al., 1997). P2Y receptors that inhibit adenylyl cyclase (Boyer et al., 1993), were also ruled out based on lack of effect of α,β-meATP on forskolin-stimulated adenylyl cyclase activity (Bolego et al., 1997). Several lines of evidence instead suggested that the gliotic P2Y receptor is linked to activation of phospholipase A_2 via a PTX-sensitive G-protein. This hypothesis was corroborated by the following data: (1) challenge of cells with α,β-meATP resulted in a rapid (within 15 min) and concentration-dependent release of [^{3}H] arachidonic acid (AA) from [^{3}H]-AA pre-labelled cells; (2) PLA$_2$ inhibitors such as mepacrine and dexametasone inhibited both the early α,β-meATP-induced release of AA and the long-term astrocytic elongation; (3) exogenously added AA mimicked the gliotic effects associated to P2Y receptor activation, resulting in a long-term elongation of astrocytic processes (Bolego et al., 1997).

A functional importance for the astroglial PLA$_2$-coupled P2Y receptor is suggested by previous results. Astrocytes are believed to represent an important source for eicosanoids in the CNS (Murphy et al., 1988; Bruner and Murphy, 1990). Various stimuli, such as ischemia, anoxia, convulsions and electroconvulsive shock, which all cause activation of astrocytes (Eddleston and Mucke, 1993) also cause increased release of arachidonates (Axelrod et al., 1988; Stella et al., 1994) which may be involved in the subsequent functional changes of these and neighbouring cells. Moreover, in our experimental system the protein kinase C (PKC) inhibitor H7 and the PKC activator phorbol-12-13-dibutyrate abolished and mimicked, respectively, α,β-meATP-induced reactive astrogliosis (Bolego et al., 1997), which may suggest the involvement of an arachidonate-sensitive protein kinase C (Tanaka and Nishizuka, 1994).

In conclusion, a P2Y receptor linked to early PLA$_2$ activation seems to mediate induction of reactive astrogliosis by ATP analogues. Of course, we cannot rule out that this novel atypical P2Y receptor may be one of the already cloned P2Y receptors that simply couples to a different transduction mechanism in our experimental system. Interestingly, the gliotic receptor shares some characteristics (e.g., sensitivity to α,β-meATP) with a guinea pig taenia coli P2Y receptor (Windscheif et al., 1995). Cloning studies are needed to establish the nature of the gliotic P2Y receptor and to verify its possible identity with the taenia coli receptor.

P2Y receptor signalling to the nucleus: towards the identification of primary and late response genes

The results summarized above suggest that release of AA may play a key role in P2Y receptor signalling to the nucleus. AA may serve as a substrate of either cyclooxygenases (COX-1 and COX-2; see Wu, 1996) or lypooxygenase (LO), leading to the formation of various metabolites that may participate to the induction of specific genes involved in expression of reactive gliosis (see also below). Alternatively, as suggested above, AA may activate an arachidonate-sensitive PKC, which is also consistent with data obtained by other authors. Neary and coworkers (1994) have shown that both ATP and bFGF are mitogenic for rat cortical astrocytes, and that ATP can activate astrocytic mitogen-activated protein (MAP) kinases via a PKC-dependent signalling pathway (Neary and Zhu, 1994; Neary, 1996). The MAP kinase cascade is a key element of signal transduction pathways involved in cellular proliferation and differentiation by both polypeptidic growth factors (for reviews, see Davis, 1993; Avruch et al., 1994) and G-protein-coupled receptors (Clapham and Neer, 1993). MAP kinases affect c-fos induction via phosphorylation of p62, a key transcription factor, that participates in the activation of the serum-response element of the c-fos promoter (Gille et al., 1992); MAP kinases also regulate Jun by post-translation phosphorylation (Pulverer et al., 1991). Interestingly, in rat cortical astrocytes, ATP and bFGF have been shown to induce the formation of AP-1 complexes, functional DNA binding protein heterodimers consisting of Fos and Jun families of transcription factors (Neary et al., 1996b). Consistent with this finding, we have demonstrated that, also in our experimental system, a brief challenge of cultures with α,β-meATP results in induction of the nuclear accumulation of the Fos and Jun proteins (Bolego et al., 1997). Figure 1 shows a typical experiment on Fos induction in rat striatal astrocytes. In control cultures (a) immunor-

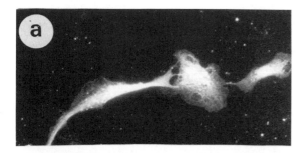
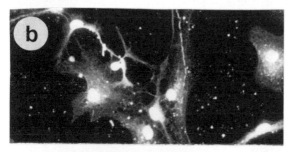
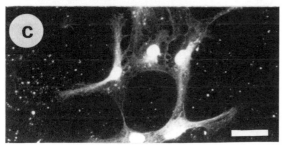

Fig. 1. Nuclear accumulation of the Fos protein induced by α,β-meATP and bFGF in rat brain astrocytes. At day 2 in culture, cells were exposed for 1 hour to either medium alone ((a) Control) or 10 μM α,β-meATP (b) or 1 ng/ml bFGF (c). At the end of the incubation period, cells were washed, fixed and immunostained with rabbit anti-Fos immunoglobulins, followed by biotinylated donkey anti-rabbit secondary antibody and by streptavidin-fluorescein, as previously described (Bolego et al., 1997). Cells were analysed under a fluorescent microscope equipped with a fluorescein filter, using a 20× objective. Note the intense accumulation of immunoreactivity in the nuclei of treated cells (b) and (c), compared to diffused immunoreactivity in control cultures (a). Similar results were obtained also for the Jun protein (see also Bolego et al., 1997). Scale Bar = 50 μm.

eactivity for the c-fos gene product was faint and non-specifically localized to the cytoplasm. A brief challenge of cultures with either α,β-meATP (b) or bFGF (c), utilized at the same concentrations that induced astrocytic elongation, produced a rapid and marked increase of Fos immunoreactivity, that in these treated cultures was exclusively associated to cell nuclei. Comparable data have been obtained in the same cells for Jun expression (Bolego et al., 1997). As expected for these primary response genes, induction of the nuclear accumulation of the Fos and Jun proteins by either α,β-meATP or bFGF was rapid and transient. A two-fold increase of the number of cells showing Fos- or Jun-positive nuclei was already evident after 1 h exposure to either α,β-meATP or bFGF, and peaked after additional 30–60 min (Bolego et al., 1997). Nuclear levels of both Fos and Jun started to decline after 120 min and were back to control values by 180–240 min (Bolego et al., 1997). Both c-fos and c-jun-induction by α,β-meATP could be completely prevented by pre-exposure of cultures to suramin, suggesting that these effects were also mediated by extracellular P2-purinoceptors. Instead, suramin did not affect induction of these genes by bFGF (Bolego et al., 1997).

Based on all these data, a picture emerges for astrocytic activation where both "classic" polypeptide growth factors (e.g., bFGF) and novel trophic agents (e.g., ATP), following interaction with their specific extracellular receptors, can activate distinct and independent transduction pathways that merge at the MAP kinase cascade (Abbracchio et al., 1996; Fig. 2). Tyrosine-kinase receptors for bFGF activate the *ras–raf* kinase pathway; conversely, P2Y purinoceptor signalling to MAPkinase, like other G-protein-coupled receptors (Lange-Carter et al., 1993), involves activation of PLA₂, early release of AA (Bolego et al., 1997), and subsequent activation of MEK kinase via PKC (Neary and Zhu, 1994; Neary, 1996; Bolego et al., 1997; Fig. 2). Hence, exposure to either agent results in activation of MAP kinase, leading to rapid and transient induction of primary response genes (e.g., c-fos and c-jun, Bolego et al., 1997; Fig. 1), which in turn may regulate late response genes mediating long-term phenotypic changes of astroglial cells (Abbracchio et al., 1996). In addition, P2Y recep-

338

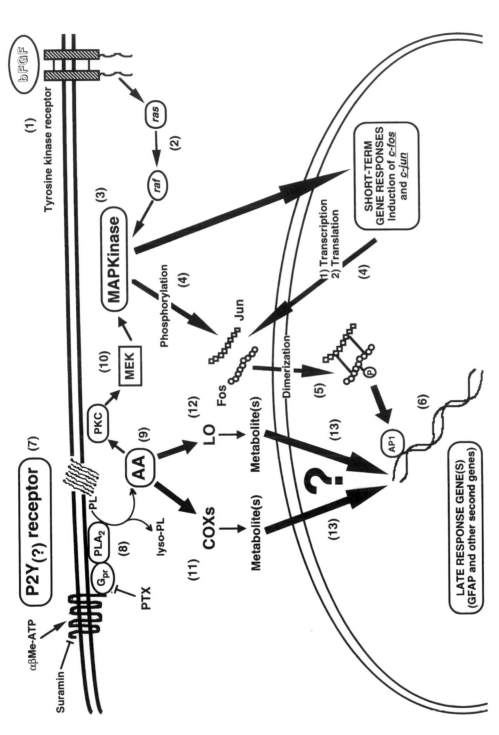

Fig. 2. Proposed pathways of astrocytic activation by α,β-meATP and bFGF. bFGF and other growth factors interact with membrane tyrosine kinase receptors (1), whose activation leads to the autophosphorylation of the two subunits and starts a series of kinase reactions (involving the *ras/raf* pathway, (2)). The final activation of the MAP kinase cascade (3) promotes the transcription and the post-translational phosphorylation of the so-called "immediate early genes" *c-fos* and *c-jun* (4). Their protein products Fos and Jun form heterodimers (5), which enter the nucleus, where they may bind to the AP-1 binding sites of late response genes (e.g., the GFAP gene (6)). All these events lead to the typical morphological changes observed in reactive astrocytes (see text for further details). α,β-meATP may act as a trophic agent via a yet-to-be identified G-protein coupled P2Y receptor, antagonized by suramin (7) and linked to phospholipase A_2 (PLA$_2$, (8)). As a consequence of PLA$_2$ activation, an early release of arachidonic acid (AA, (9)) from membrane phospholipids (PL) is detected. AA can stimulate MEK kinase via PKC (10) and consequently merge on the MAP kinase cascade (see above). Alternatively, AA can be metabolized by cyclooxygenase 1 or 2 (COXs, (11)) or by lypooxygenase (LO, (12)) to several different metabolites, which may also participate to the induction of specific genes via a yet-to-be identified pathway (13).

tor-induced AA accumulation may lead to the formation of various cyclooxygenase (COX) and LO (lypooxygenase) metabolites, that may also participate to induction of specific genes via a yet-to-be determined pathway (Fig. 2).

A final exciting issue is the identification of the late response genes eventually activated by the gliotic P2Y receptor. The GFAP gene contains in its promoter a binding site for AP-1 complexes (Masood et al., 1993), which would be highly consistent with the increased expression of this protein following exposure of astroglial cells to purine analogues and bFGF (Neary and Norenberg, 1992; Abbracchio et al., 1995b; Neary et al., 1996; Bolego et al., 1997; Fig. 2). We have recently raised the hypothesis that the cyclooxygenase-2 (COX-2) gene may also be specifically induced by α,β-meATP in our experimental model (Brambilla et al., 1998, 1999). This hypothesis is based on the fact that either AA or its metabolites may induce the COX-2 gene (Minghetti et al., 1997), that has been shown to contain an AP-1 binding site (Wu, 1996). To confirm that induction of COX-2 may contribute to α,β-meATP-induced reactive astrogliosis, a brief challenge of rat astrocytes with the purine analog resulted, 24 h later, in a significant increase of the COX-2 protein (Brambilla et al., 1998; 1999).

Conclusions and future perspectives

A better knowledge of the endogenous factors that regulate reactive gliosis may provide a means to manipulate the gliotic reaction in vivo, by either increasing the permissivity of reactive astrocytes towards regenerating axons, or by preventing or decreasing the detrimental effects of post-traumatic inhibitory gliosis. We and others have demonstrated that ATP may represent one of the endogenous factors involved in the formation of reactive astrocytes, in a similar way to classic polypeptidic growth factors, but via the activation of specific membrane P2Y receptors. We have shown that an atypical G-protein-coupled P2Y receptor activated by α,β-meATP and linked to PLA$_2$ stimulation is likely responsible for the formation of reactive astrocytes. Transductional studies have suggested that P2Y receptor signalling to the nucleus occurs

through an early release of AA and induction of the fos and jun gene products, likely via the PKC/MAPK pathway or through a yet-to-be determined intracellular pathway involving products of arachidonate metabolism (Fig. 2). Formation of Fos and Jun heterodimers upon challenge of astroglial cells with ATP has been suggested by Neary (1996); these complexes may hence affect the transcription of genes specifically involved in the gliotic reaction. An exciting and entirely new field concerns the identification of the molecules that specifically contribute to purine-induced reactive astrogliosis and it is anticipated that in the next few years several researchers will focus their attention on this intriguing topic.

The finding that ATP has a role in reactive astrogliosis may also disclose novel therapeutic approaches to the modulation of post-traumatic and post-ischemic brain repair. In particular, our recent data demonstrating induction of COX-2 in astrocytes upon stimulation of the gliotic P2Y receptor (Brambilla et al., 1998; 1999) may have intriguing implications in the therapy of neurological disorders characterized by excessive astroglial reaction and inflammation. Evidence has been accumulating in recent years that upregulation of COX-2 is a common feature of both chronic pain and inflammation (Dolan et al., 1998), brain ischemia (Planas et al., 1995; Ohtsuki et al., 1996), and chronic neurodegenerative diseases (e.g., Alzheimer's disease, Tocco et al., 1997). Enhanced COX-2 activity in neurological diseases may contribute to neuronal damage via production of free radicals during AA conversion (Katsuki and Okuda, 1995). Moreover, PGE$_2$, a main product of the COX-2 enzyme, largely potentiated release of the pro-inflammatory cytokine interleukin-6 from both astroglioma cell lines and post-mortem astrocytes (Blom et al., 1997), and has also been recently associated with release of reactive oxygen and other radicals in neurotoxicity induced by the prion protein (Brown et al., 1996). These data have hence raised the hypothesis that selective inhibitors of COX-2 may be beneficial for the therapy of central nervous system diseases characterized by neurodegenerative events (Ohtsuki et al., 1996; Miettinen et al., 1997; Blom et al., 1997), chronic inflammation and pain (Dolan et al., 1998), while avoiding the gastrointestinal, renal and haemato-poietic adverse effects typical of mixed COX-1/COX-2 inhibitors (Flower, 1996). Our data showing induction of COX-2 upon stimulation of astrocytes by purine analogues (Brambilla et al., 1998, 1999) suggests that down-regulation of COX-2 in inflammatory neurological diseases may be achieved by selectively blocking the gliotic P2Y receptor. Unfortunately, no selective antagonists for this receptor are currently available. Suramin, which concentration-dependently blocked the formation of reactive astrocytes by ATP analogues (Abbracchio et al., 1995b; Bolego et al., 1997), is a non-selective antagonist at both P2X and P2Y receptors and also has a variety of additional activities on membrane channels, G-proteins and enzymes (Voogdt et al., 1993), Hence, selective antagonists at the astroglial P2Y receptor devoid of effects on other biological targets are hence highly desirable; these may represent a novel class of anti-inflammatory agents of potential value in various nervous system pathologies.

Acknowledgements

The work described in this paper was partially supported by the European Union BIOMED2 Programme BMH4 CT96-0676. Authors are grateful to Professor G. Burnstock, Autonomic Neuroscience Institute, Royal Free Hospital, London, UK, for invaluable suggestion and support.

References

Abbracchio, M.P. (1997) ATP in brain function. In: K.A. Jacobson and M.F. Jarvis (Eds.), *Purinergic Approaches in Experimental Therapeutics.* Wiley-Liss, New York, pp. 383–404.

Abbracchio, M.P. and Burnstock, G. (1994) Purinoceptors: Are there families of P2X and P2Y purinoceptors? *Pharmac. Ther.,* 64: 445–475.

Abbracchio, M.P., Ceruti, S., Burnstock, G. and Cattabeni, F. (1995a) Purinoceptors on glial cells of the central nervous system: Functional and pathological implications. In: L. Belardinelli and A. Pelleg (Eds.), *Adenosine and Adenine Nucleotides: from Molecular Biology to Integrative Physiology.* Kluwer, Norwell, MA, pp. 271–280.

Abbracchio, M.P., Ceruti, S., Langfelder, R., Cattabeni, F., Saffrey, M.J. and Burnstock, G. (1995b) Effects of ATP analogues and basic fibroblast growth factor on astroglial cell differentiation in primary cultures of rat striatum. *Int. J. Devl. Neurosci.,* 13: 685–693.

Abbracchio, M.P., Ceruti S., Bolego, C., Puglisi, L., Burnstock, G. and Cattabeni, F. (1996) Trophic roles of P2 purinoceptors in central nervous system astroglial cells. In: *Proceedings of the Ciba Foundation Symposium 198 on P2 Purinoceptors: Localization, Function and Transduction Mechanisms.* Wiley, Chichester, pp. 142–148.

Alonso, G. and Privat, A. (1993) Reactive astrocytes involved in the formation of lesional scars differ in the mediobasal hypothalamus and in other forebrain regions. *J. Neurosci. Res.,* 34: 523–538.

Anderson, K., Launer, L.J., Ott, A., Hoes, A.W., Breteler, M.M. and Hofman, A. (1995) Do non-steroidal anti-inflammatory drugs decrease the risk for Alzheimer's disease? The Rotterdam study. *Neurology,* 45: 1441–1445.

Avruch, J., Zhang, X.-F. and Kyriakis, J.M. (1994) Raf meets Ras: Completing the framework of a signal transduction pathway.*Trends Biochem. Sci.,* 19: 279–283.

Axelrod, J., Burch, R.M. and Jelsema, C.L. (1988) Receptor-mediated activation of phospholipase A_2 via GTP-binding proteins: Arachidonic acid and its metabolites as second messengers. *Trends Neurosci.,* 11: 117–123.

Blom, M.A.A., van Twillert, M.G.H., de Vries, S.C., Engels, F., Finch, C.E., Veerhuis R. and Eikelenboom, P. (1997) NSAIDS inhibit the IL-1β-induced IL-6 release from human post-mortem astrocytes: The involvement of prostaglandin E2. *Brain Res.,* 777: 210–218.

Bolego, C., Ceruti, S., Brambilla, R., Puglisi, L., Cattabeni, F., Burnstock, G. and Abbracchio, M.P. (1997) Characterization of the signalling pathways involved in ATP- and basic fibroblast growth factor-induced astrogliosis. *Br. J. Pharmacol.,* 121: 1692–1699.

Boyer, J.L., Lazarowski, E.R., Chen, X.-H. and Harden, T.K. (1993) Identification of a P2y-purinergic receptor that inhibits adenylyl cyclase. *J. Pharmacol. Exp. Ther.,* 267: 1140–1146.

Brambilla, R., Cattabeni, F., Burnstock, G., Folco, G.C., Bonazzi, A., Hernandez, A. and Abbracchio, M.P. (1998) P2 receptors mediate cyclo-oxygenase 2 induction in rat astroglial cells. *Drug Dev. Res.,* 43: 14.

Brambilla, R., Burnstock, G., Bonazzi, A., Ceruti, S., Cattabeni, F. and Abbrachio, M.P. (1999) Cyclooxygenase-2 mediates P2Y receptor-induced reactive astrogliosis. *Br. J. Pharmacol.,* 126: 563–567.

Breitner, J.C.S., Gau, B.A., Welsh, K.A., Plassman, B.L., McDonald W.M., Helms, M.J. and Anthony, J.C. (1994) Inverse association of anti-inflammatory treatments and Alzheimer's disease: Initial results of a control co-twin study. *Neurology,* 44: 227–232.

Brown, D.R., Schmidt, B. and Kretzschmar, H.A. (1996) Role of microglia and host prion protein in neurotoxicity of a prion protein fragment. *Nature,* 380: 345–347.

Bruner, G. and Murphy, S. (1990) ATP-evoked arachidonic acid mobilization in astrocytes is via a P2Y-purinergic receptor. *J. Neurochem.,* 55: 1569–1575.

Cameron, R.S. and Rakic, P. (1991) Glial cell lineage in the cerebral cortex: A review and synthesis. *Glia*, 4: 124–137.

Centemeri, C., Bolego, C., Abbracchio, M.P., Cattabeni, F., Puglisi, L., Burnstock, G. and Nicosia, S. (1997) Characterization of the Ca^{2+} responses evoked by ATP and other nucleotides in mammalian brain astrocytes. *Br. J. Pharmacol.*, 121: 1700–1706.

Clapham, D.E. and Neer, E.J. (1993) New roles for G protein βγ dimers in transmembrane signalling. *Nature*, 365: 403–406.

Davis, R.J. (1993) The mitogen-activated protein kinase signal transduction pathway. *J. Biol. Chem.*, 268: 14553–14556.

Dolan, S., O'Shaughnessy, P.J. and Nolan, A.M. (1998) Up-regulation of COX-2 and prostaglandin EP3 receptor mRNA expression in spinal cord in a clinical model of chronic inflammation. *Eur. J. Neurosci.*, 10 (suppl.10): 77.

Eddleston, M. and Mucke, L. (1993) Molecular profile of reactive astrocytes—implication for their role in neurologic diseases. *Neuroscience*, 54: 15–36.

Evans, R.J., Surprenant, A. and North, R.A. (1997) P2X Receptors: Cloned and Expressed. In: J.T. Turner, G.A. Weisman and J.S. Fedan (Eds.), *The P2 Nucleotide Receptors*. Humana Press, Totowa, pp. 43–61.

Flower, R.J. (1996) New directions in cyclooxygenase research and their implications for NSAID-gastropathy. *Italian J. Gastroenterol.*, 28: 23027.

Fredholm, B.B., Abbracchio, M.P., Burnstock, G., Daly, J.W., Harden, T.K., Jacobson, K.A., Leff, P. and Williams, M. (1994) Nomenclature and classification of purinoceptors. *Pharm. Rev.*, 46: 143–156.

Fredholm, B.B., Abbracchio, M.P., Burnstock, G., Dubyak, G.R., Harden, T.K., Jacobson, K.A., Schwabe, U. and Williams, M. (1997) Towards a revised nomenclature for adenosine and P2 receptors. *Trends Pharmacol. Sci.*, 18: 79–82.

Gille, H., Sharrocks, A.D. and Shaw, P.E. (1992) Phosphorylation of transcription factor P62TCF by MAP kinase stimulates ternary complex formation at c-fos promoter. *Nature*, 358: 414–417.

Hatten, M.E., Liem, R.K.H., Shelanski, M.L. and Mason C.A. (1991) Astroglia in CNS injury. *Glia*, 4: 233–243.

Harden, T.K., Nicholas, R.A., Schachter, J.R., Lazarowski, E.R. and Boyer, J.L. (1997) Pharmacological selectivities of molecularly defined subtypes of P2Y receptors. In: J.T. Turner, G.A. Weisman and J.S. Fedan (Eds.), *The P2 Nucleotide Receptors*. Humana Press, Totowa, pp. 109–134.

Katsuki, H. and Okuda, S. (1995) Arachidonic acid as a neurotoxic and neurotrophic substance. *Prog. Neurobiol.*, 46: 607–636.

Lange-Carter, C.A., Pleiman, C.M., Gardner, A.M., Blumer, K.J. and Johnson, G.L. (1993) A divergence in the MAP kinase regulatory network defined by MEK kinase and Raf. *Science*, 260: 315–319.

Masood, K., Bresnard, F., Su, Y. and Brenner, M. (1993) Analysis of a segment of the human glial fibrillary acidic protein gene that directs astrocyte-specific transcription. *J. Neurochem.*, 61: 160–166.

Miettinen, S., Fusco, R.F., Yrjanheikki, J., Keinaken, R., Hirvonen, T., Roivainen, R., Narhi, M., Hokfelt, T. and Koistinaho, J. (1997) Spreading depression and focal brain ischemia induce cyclooxygenase-2 in cortical neurons through N-methyl-D-aspartatic acid-receptors and phospholipase A_2. *Proc. Natl. Acad. Sci. USA*, 94: 6500–6505.

Minghetti, L., Polazzi, E., Nicolini, A., Creminon, C. and Levi, G. (1997) Up-regulation of cyclooxygenase-2 expression in cultured microglia by prostaglandin E2, cyclic AMP and nonsteroidal anti-inflammatory drugs. *Eur. J. Neurosci.*, 9: 934–940.

Murphy, S., Pearce, B., Jeremy, J. and Dandona, P. (1988) Astrocytes as eicosanoids-producing cells. *Glia*, 1: 241–245.

Neary, J.T. (1996) Trophic actions of extracellular ATP on astrocytes, synergistic interactions with fibroblast growth factors and underlying signal transduction mechanisms. In: *Proceedings of the Ciba Foundation Symposium 198 on P2 Purinoceptors: Localization, Function and Transduction Mechanisms*. Wiley, Chichester, pp. 130–141.

Neary, J.T. and Norenberg, M.D. (1992) Signaling by extracellular ATP: Physiological and pathological considerations in neuronal-astrocytic interactions. *Prog. Brain Res.*, 94: 145–151.

Neary, J.T. and Zhu, Q. (1994) Signaling by ATP receptors in astrocytes. *NeuroReport*, 5: 1617–1620.

Neary, J.T., Rathbone, M.P., Cattabeni, F., Abbracchio, M.P. and Burnstock, G. (1996a) Trophic actions of extracellular nucleotides and nucleosides on glial and neuronal cells. *Trends Neurosci.*, 19: 13–18.

Neary, J.T., Zhu, Q., Kang, Y. and Dash, P.K. (1996b) Extracellular ATP induces formation of AP-1 complexes in astrocytes via P2 purinoceptors. *NeuroReport*, 7: 2893–2896.

Norton, W.T., Aquino, D.A., Hozumi, I., Chiu, F.C. and Brosnan, C.F. (1992) Quantitative aspects of reactive astrogliosis: a review. *Neurochem. Res.*, 17: 877–885.

O'Banion, M.K., Miller, J.C., Chang, W.J., Kaplan, M.D. and Coleman, P.D. (1996) Interleukin-1β induces prostaglandin G/H synthase-2 (cyclooxygenase-2) in primary murine astrocyte cultures. *J. Neurochem.*, 66: 2532–2540.

Ohtsuki, T., Kitagawa, K., Yamagata, K., Mandai, K., Mabuchi, T, Matsushita, K., Yanagihara, T. and Matsumoto, M. (1996) Induction of cyclooxygenase-2 mRNA in gerbil hippocampal neurons after transient forebrain ischemia. *Brain Res.*, 736: 353–356.

Pearce, B., Murphy, S., Jeremy, J., Morrow, C. and Dandona, P. (1989) ATP-evoked Ca^{2+} mobilization and prostanoid release from astrocytes: P2-purinergic receptors linked to phosphoinositide hydrolysis. *J. Neurochem.*, 52: 971–977.

Planas, A.M., Soriano, M.A., Rodriguez-Farre, E. and Ferrer, I. (1995) Induction of cyclooxygenase-2 mRNA and protein following transient focal ischemia in the rat brain. *Neurosci. Lett.*, 200: 187–190.

Pulverer, B.J., Kyriakis, J.M., Avruch, J., Nikolakaki, E. and Woodgett, J.R. (1991) Phosphorylation of c-Jun mediated by MAP kinases. *Nature*, 353: 670–674.

Ridet, J.L., Malhotra, S.K., Privat, A. and Gage, F.H. (1997) Reactive astrocytes: Cellular and molecular cues to biological function. *Trends Neurosci.*, 20: 570–577.

Rogers, J., Kirby, L.C., Hempelman, S.R., Berry, D.L., McGeer, P.L., Kaszniak, A.W., Zalinski, J., Cofield, M., Mansukhani, L., Wilson, P. et al. (1993) Clinical trial of indomethacin in Alzheimer's disease. *Neurology*, 34: 1609–1613.

Salter, M.W. and Hicks, J.L. (1995) ATP causes release of intracellular Ca^{2+} via the phospholipase $C\beta$/IP3 pathway in astrocytes from the dorsal spinal cord. *J. Neurosci.*, 15: 2961–2971.

Stella, N., Tence, M., Glowinski, J. and Premont, J. (1994) Glutamate-evoked release of arachidonic acid from mouse brain astrocytes. *J. Neurosci.*, 14: 568–575.

Tanaka, C. and Nishizuka, Y. (1994) The protein kinase C family for neuronal signalling. *Annu. Rev. Neurosci.*, 17: 551–567.

Tocco, G., Freire-Moar, J., Schreiber, S.S., Sakhi, S.H., Aisen, P.S. and Pasinetti, G.M. (1997) Maturational regulation and regional induction of cyclooxygenase-2 in rat brain: Implications for Alzheimer's disease. *Exp. Neurol.*, 144: 339–349.

Turner, J.T., Weisman, G.A. and Fedan, J.S. (Eds.), (1997) *The P2 Nucleotide Receptors*. Humana Press, Totowa, New Jersey.

Voogdt, T.E., Vansterkenburg, E.L.M., Wilting, J. and Janssen, L.H.M. (1993) Recent research on the biological activity of suramin. *Pharmacol. Rev.*, 45: 177–203.

Weinstein, D.E., Shelanski, M.L. and Liem, R. (1991) Suppression by antisense mRNA demonstrates a requirement for the glial fibrillary acidic protein in the formation of stable astrocytic processes in response to neurons. *J. Cell Biol.*, 112: 1205–1213.

Windscheif, U., Pfaff, G., Ziganshin, A.U., Hoyle, C.H., Baumert, H.G., Mutschler, E., Burnstock, G. and Lambrecht, G. (1995) Inhibitory action of PPADS on relaxant responses to adenine nucleotides or electrical field stimulation in guinea pig taenia coli and rat duodenum. *Br. J. Pharmacol.*, 115: 1509–1517.

Wu, K.H. (1996) Cyclooxygenase 2 induction: Molecular mechanisms and pathophysiological roles. *J. Lab. Clin. Med.*, 128: 242–245.

Immunomodulatory effects
of ATP

P. Illes and H. Zimmermann (Eds.)
Progress in Brain Research, Vol 120
© 1999 Elsevier Science BV. All rights reserved

CHAPTER 28

Purinoceptors in human B-lymphocytes

F. Markwardt[1,*], M. Klapperstück[1], M. Löhn[2], D. Riemann[2], C. Büttner[3] and
G. Schmalzing[3]

[1]*Julius-Bernstein-Institute for Physiology, Martin-Luther-University Halle-Wittenberg, D-06097 Halle/S., Germany*
[2]*Institute for Medical Immunology, Martin-Luther-University Halle-Wittenberg, D-06097 Halle/S., Germany*
[3]*Department of Pharmacology, Biocenter, Johann-Wolfgang-Goethe-University, D-60439 Frankfurt, Germany*

Introduction

ATP can be released under several circumstances by platelets, mast cells, cytolytic T-lymphocytes or erythrocytes. Another common source of the nucleotide is the intracellular ATP liberated from damaged cells. It has been known for many years that extracellular ATP is an extracellular messenger acting on P_2-purinoreceptors in different cell types, including blood cells. Although in leukocytes the precise in vivo function of extracellular ATP is still not known, ATP is believed to act as a modulator of the immune response at concentrations < 100 μmol/l. At higher concentrations extracellular ATP seems to be cytolytic (El-Moatassim et al., 1992, Murgia et al., 1992, Dubyak and El-Moatassim, 1993, Di Virgilio, 1995).

The P2Z-receptor, probably a subtype of the P2X-receptor family, is a purinoceptor found in cells of the immune and inflammatory system. It is characterized by binding of only free ATP^{4-} in the concentration range of about 100 μM and by the absence of desensitization. In fibroblasts, macrophages and mast cells, ATP^{4-} induces non-selective membrane pores, allowing the permeation of molecules up to 900 Da (DiVirgilio, 1995). In fibroblasts and mast cells the pore

diameter was even found to increase with agonist concentration and application duration (Tatham and Lindau, 1990; Saribas et al., 1993). The structure of these pores is not yet clear.

In human leukemic B-lymphocytes, effects of extracellular ATP have been investigated by flux studies and by measurement of intracellular Ca^{2+} and H^+ concentrations with fluorescent dyes. Within < 1 min, micromolar concentrations of ATP open pores in the plasmalemmal membrane mediating influx of cations with molecular weights < 320 Da. Under physiological conditions, this leads to an increase of the intracellular calcium ion concentration ($[Ca^{2+}]_i$) (Wiley et al., 1990, 1992, 1993; Chen et al., 1994). In human tonsillar B-lymphocytes it was shown that the increase of $[Ca^{2+}]_i$ after extracellular ATP-application was due to both influx of Ca^{2+} through the cell membrane and Ca^{2+}-release from intracellular stores triggered by the activation of phospholipase C (Padeh et al., 1991).

Electrophysiological measurements for further characterization of purinoceptors in B-lymphocytes are rare. In order to characterize the properties of purinoceptors in human B-lymphocytes, ATP-induced membrane currents and $[Ca^{2+}]_i$ were measured simultaneously in Epstein–Barr-virus-transformed and tonsillar human single B-lymphocytes by means of the tight-seal voltage-clamp- and Fluo-3/Fura-red-fluorescence-technique.

---

*Corresponding author. Tel.: +49-345-557-1390; fax: +49-345-55-27899; e-mail: fritz.markwardt@medizin. uni-halle.de

Materials and methods

Cell culture

Human B-lymphocytes, immortalized by Epstein–Barr-virus (MSAB, Reiner Laus, Stanford blood bank, USA) were kindly provided by Dr. J. Steinmann, Institute for Immunology, Kiel, Germany. The cells were grown in sterile filtered RPMI 16–40 medium (GIBCO BRL, Berlin, Germany) supplemented with 15 mmol/l HEPES, 2 mmol/l N-acetyl-L-alanyl-L-glutamine (Biochrom, Berlin, Germany), 33 μg/ml Gentamicinsulfate (Merck, Darmstadt, Germany), 0.25 μg/ml Amphotericin B (GIBCO) and 10% heat-inactivated (56°C, 30min) fetal calf serum (FCS) (CCPRO, Karlsruhe, Germany). Vitality was screened by methylene blue staining. Only cell cultures with >90% vital cells were chosen for experiments. For long-term storage, cells were frozen to −87°C in aliquots of 1ml (10^6 cells/ml) in a solution of DMSO (SIGMA, St. Louis, USA): FCS = 1 : 9.

Tonsillar B-lymphocytes were purified from surgically excised tonsils by centrifugation over a Ficoll-Hypaque (Pharmacia, Uppsala, Sweden) gradient and by rosetting T-lymphocytes with sheep erythrocytes. Depletion of monocytes was achieved by adhesion to plastic wells. After this procedure, the contamination of the B cell population with T-lymphocytes and monocytes was less than 5% as measured by FACS (flow assisted cell scan). Tonsillar B-cells were held in primary culture in a medium containing RPMI 1640, 10% fetal calf serum, 100 U/ml interleukin 2 (IL-2) and 100 U/ml IL-4 (Briere et al., 1994).

Electrophysiology

For electrophysiological measurements, the cells were fixed with poly-L-lysine onto the glass bottom (thickness 0.17 mm) of a 0.2 ml chamber. Voltage-clamp measurements were performed in the whole-cell and outside-out configuration of the tight-seal patch-clamp technique. The pipette resistance was between 4 to 7 MΩ. Rapid ligand applications (within 0.2 s) were performed by an U-tube technique (Bretschneider and Markwardt, 1999). Membrane currents and fluorescence signals were stored and analyzed on an IBM-compatible computer with software developed in our laboratory. All measurements were carried out at room temperature.

Nonlinear approximations and presentation of data were performed using the program Sigmaplot (Jandel, Corte Madera, USA). Statistical significance was verified by variance analysis and Scheffé's test ($P < 0.05$).

Fluorescence-analysis

The B-lymphocytes were loaded with Fluo-3 (40 μM) and Fura-red (100 μM) directly by the patch pipette, containing the fluorescent dye dissolved in the pipette solution. A rapid increase of the fluorescence level of Fluo-3 indicated successful loading of the B-lymphocyte with these dyes after establishing the whole cell configuration. A xenon arc light source was used for the excitation at 460–490 nm. Fluorescence signals were collected at wavelengths between 520–540 nm (Fluo-3) and >610 nm (Fura-red) by means of two photomultipliers.

Calibration of the fluorescence signals was performed by using eight different extracellular solutions with Ca^{2+}-concentrations between 0 (10 mmol/l EGTA in nominal Ca^{2+}-free solution) and 1 mmol/l. Approximation of the Ca^{2+}-dependence of the Fluo-3/Fura-red fluorescence ratio by a Hill equation yielded an apparent Ca^{2+}-binding constant for the Fluo-3/Fura-red-mixture of 687 nmol/l.

Solutions

The cells were superfused by the standard bath solution consisting of (mmol/l): 140 NaCl, 5.4 KCl, 1 CaCl$_2$, 10 HEPES, 10 glucose, pH 7.4 with NaOH. In some experiments, choline$^+$ was used as extracellular cation instead of Na$^+$. The patch pipettes were filled (if not otherwise stated) with the intracellular solution containing (mmol/l): 65 X-aspartate, 65 XCl, 0.5 MgCl$_2$, 5 Mg-ATP, 10 HEPES, 10 glucose, 3 EGTA, 3 BAPTA, pH 7.2 with XOH, where X = K$^+$ or Cs$^+$, respectively. In measurements with fluorimetric measurement of $[Ca^{2+}]_i$, EGTA and BAPTA were replaced by Fluo-3 (40 μmol/l) and Fura-red (100 μmol/l).

The free concentrations of Ca^{2+}, ATP (ATP^{4-}), and BzATP ($BzATP^{4-}$) of the U-tube solution, respectively, were calculated by means of a computer programme kindly provided by R. Schubert (1990). The U-tube solutions had the same concentration of free cations as the corresponding bathing solution, but additionally contained different concentrations of ATP^{4-} or $BzATP^{4-}$, respectively.

Results

Figure 1 demonstrates the effect of application of 1 mmol/l ATP^{4-} on a single MSAB lymphocyte

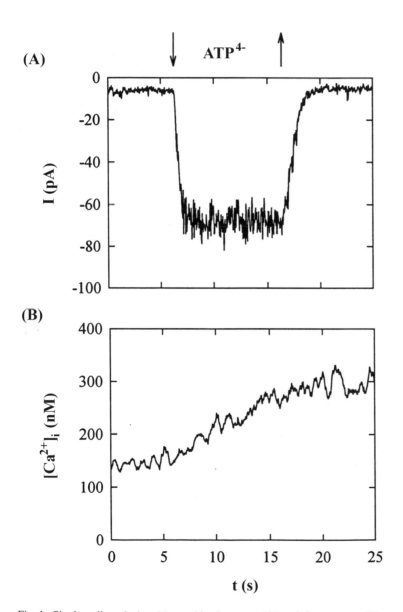

Fig. 1. Single cell analysis with combined current (A) and fluorescence (B) measurement. The cell (human Epstein–Barr-virus-transformed (MSAB) B-lymphocyte) was superfused with a Na^+-containing extracellular solution. The application of 1 mmol/l ATP^{4-} evokes an inward current and an enhancement of the intracellular free calcium concentration (holding potential -80 mV).

348

clamped at -80 mV. Within about 1 s, the whole cell current increased to a steady-state level accompanied by an enlargement of the current noise. During the 10 s of ATP^{4-}-application, $[Ca^{2+}]_i$ rises continuously. Within 1 s the removal of ATP^{4-} returns current noise and amplitude to the resting level and stops the $[Ca^{2+}]_i$-increase. This measurement indicates the existence of P2X-receptor-

MSAB, $V_h = -120$ mV

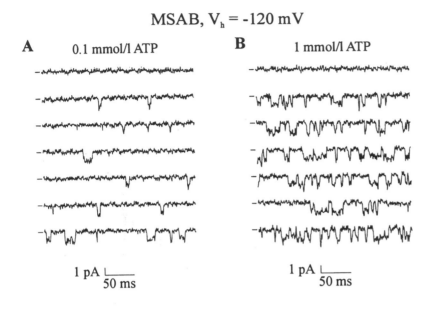

Tonsillar B-cell, $V_h = -80$ mV

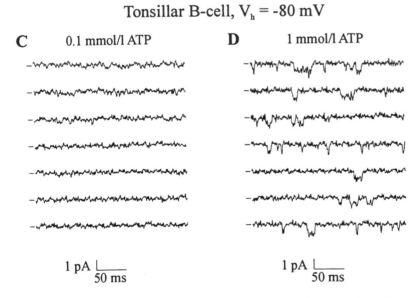

Fig. 2. Unitary currents evoked by ATP^{4-}. (A), (B) Response of an excised outside-out patch from a human B-cell (MSAB) bathed in an external Na^+-containing solution. The pipette solution contained Cs^+. The extracellular solution was replaced at the beginning of the second trace by a solution additionally containing 0.1 (A) or 1 mmol/l ATP^{4-} (B), respectively. (C), (D) Recordings from an outside-out patch of a human tonsillar B-cell before (C) and after (D) the addition of 1 mmol/l ATP^{4-} to the extracellular solution.

dependent Ca^{2+}-permeable ion channels in human B-lymphocytes.

To investigate lymphocytic purinoceptors on the single channel level, measurements were per-

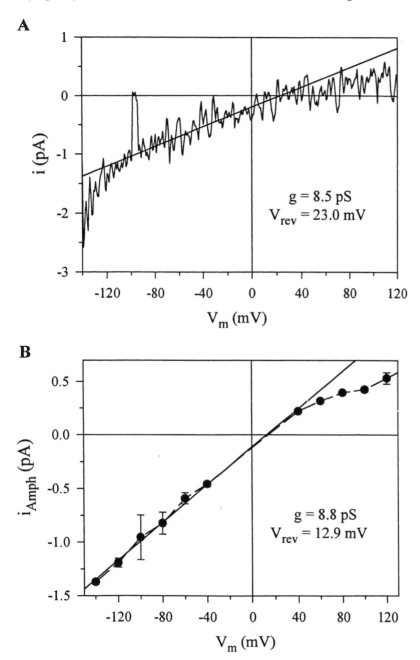

Fig. 3. Dependence of unitary purinoceptor-dependent single channel currents on membrane voltage. Same solutions as in Fig. 2. (A) ATP^{4-}-induced single-channel currents during voltage ramps. The straight line indicates the linear fit in the range between −90 and +40 mV. (B) Dependence of the mean single channel current i on membrane voltage in MSAB cells (means ± SD from 5 to 17 different cells). For the determination of the single channel conductance g and the reversal potential V$_{rev}$, the data in the voltage range between −140 and +40mV were approximated by a straight line with a correlation coefficient r of 0.999.

formed in the outside-out configuration of the patch clamp technique. As shown in Fig. 2, ATP^{4-} evokes single channel currents of about 1 pA in both EBV-transformed and tonsillar human B-lymphocytes. An increase of the ATP^{4-}-concentration (Fig. 2A vs. B) mainly decreases the duration of closures between channel openings.

The single channel conductance was determined in outside-out patches by fitting both currents evoked by voltage ramps going from -140 to $+120$ mV within 100 ms and single channel currents at different holding potentials by a straight line in the range of -90 to $+40$ mV (Fig. 3). Under both circumstances, a single-channel con-

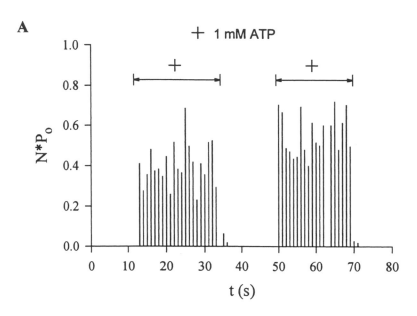

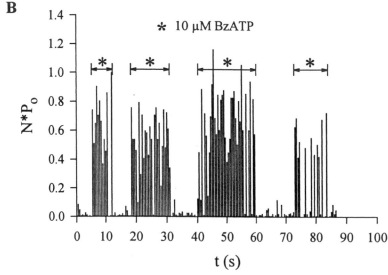

Fig. 4. Reversible activation and deactivation of single purinoceptor-gated ionic channels. Outside-out patches from two different MSAB-cells Na^+-containing solution. The pipette solution contained Cs^+. The holding potential was -80 mV. $N \times P_o$ was calculated by dividing the mean open channel current during the 0.5 s lasting sweep by the single channel current. The horizontal bars represent the time of ATP^{4-}-(A) or $BzATP^{4-}$-application (B), respectively.

(A)

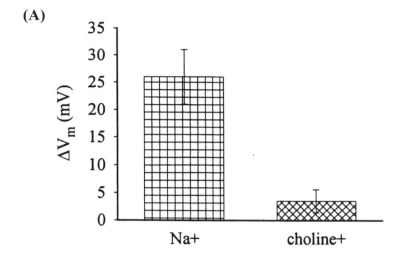

(B)

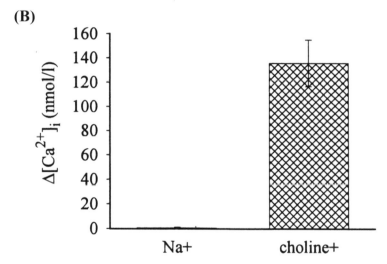

Fig. 5. Purinoceptor-dependent changes of membrane potential and intracellular free Ca^{2+}-concentration in single human B-lymphocytes. (A) MSAB cells are depolarized by ATP^{4-} in Na^+- but hyperpolarized in $choline^+$-containing extracellular solutions, respectively (seven cells). The change in membrane potential is significantly different in both conditions. (B) Under current clamp, only in extracellular solutions with Na^+ replaced by $choline^+$ $[Ca^{2+}]_i$ is significantly increased by ATP^{4-}. (10 s application of 1 mmol/l ATP^{4-}, means ± SEM from seven different cells).

ductance of about 9 pS was measured. This conductance was independent of the duration of ATP^{4-}-application. That means that no indication was found for induction of large increasing pores by ATP^{4-} in human B-cells. This finding was confirmed by the measurement of Fluo-3 and Fura-red fluorescence using the FACS where the cells were loaded with the fluorescent dyes by using their AM-esters. Even the application of ATP^{4-} for more than 10 min did not evoke an irreversible

decrease of the Fluo-3 or Fura-red fluorescence which would be indicative for an ATP^{4-}-induced leakage of the dyes (Löhn et al., 1998).

Like in whole-cell measurements, the activation and deactivation of purinoceptor-dependent channels by the agonists ATP^{4-} or $BzATP^{4-}$ takes place within 1 s or less (Fig. 4). This observation was independent of the Ca^{2+}-concentration in the extracellular solution, i.e. the fast deactivation persisted in low Ca^{2+} (0.2 mmol/l) and even in

352

Ca^{2+}-free U-tube solutions. Furthermore, the channels like the whole-cell currents did not desensitize in the range of minutes or less (Markwardt et al., 1997).

The ATP^{4-}-induced increase in [Ca^{2+}]$_i$ as shown in Fig. 1 implied a substantial Ca^{2+}-permeability of the purinoceptor-channels. This was confirmed by measurements of the Ca^{2+}-permeability of purinoceptor-dependent whole-cell currents (Bretschneider et al., 1995). Nevertheless, measurements using the FACS revealed almost no increase in [Ca^{2+}]$_i$ after ATP^{4-}-application for more than 5 min. One reason for this behaviour of human B-cells is the inhibition of Ca^{2+}-permeation by permeating Na$^+$-ions (Löhn et al., 1998). Another reason is the depolarizing current through the purinoceptor channels carried by Na$^+$ in normal extracellular solutions. This current decreases the driving force for the Ca^{2+}-ions to enter the cell. The much less permeant choline$^+$ if substituted for Na$^+$ is much less able to block the Ca^{2+}-permeation as well as to induce an ATP^{4-}-evoked membrane depolarization (Fig. 5).

Discussion

The results demonstrate a 9 pS ion channel which is at least partly involved in the ATP^{4-}-induced membrane currents and depolarization in human B-lymphocytes, respectively. In contrast to P2X$_1$ and P2X$_4$ (Surprenant et al., 1995) receptors, for which mRNA was found to be present in human B-lymphocytes (Klapperstück et al., 1998), the channels do not desensitize and bind free ATP^{4-} only (Bretschneider et al., 1995, Markwardt et al., 1997). With respect to the agonist binding characteristics, these channels are rather similar to the P2X$_7$ subtype for which the mRNA was also found in human B-cells (Klapperstück et al., 1998). But purinoceptors in human B-lymphocytes differ considerably from human P2X$_7$ receptors heterogeneously expressed in HEK cells (Rassendren et al., 1997). We found a rapid deactivation of ATP^{4-} and BzATP^{4-}-evoked currents. Furthermore, ATP^{4-} never induced leakage of fluorescent dyes with molecular weights of 831 (Fluo-3) or 725 (Fura-red), respectively. Therefore, in human B-lymphocytes the purinoceptor may be composed of additional components. Future experiments will show if these constituents are P2X$_1$ and/or P2X$_4$ subunits or other as yet undefined components.

Despite a substantial Ca^{2+}-permeability of the lymphocytic purinoceptors, the ATP^{4-}-induced [Ca^{2+}]$_i$-increase is small if not absent in unclamped cells investigated in our experiments. In these cells the role of the purinoceptors is not to increase the global free Ca^{2+}-concentration. Certainly, the ability of Na$^+$-ions to depolarize the B-lymphocyte may depend on the open probability of other ionic channels of the cell membrane. If the opening of such channels as well as of the purinoceptors depends on the state of differentiation, the effects of extracellular ATP on membrane potential and [Ca^{2+}]$_i$ may also differ.

Conclusions

In human B-lymphocytes the existence of an ATP-gated cation channel with a single-channel conductance of 9 pS was shown by patch clamp measurements. The channels are activated only by free ATP^{4-} or BzATP^{4-} within less than 1 s, do not desensitize during continuous application of the agonist for several seconds and are deactivated within 1 s after agonist removal. Human B-lymphocytes are not permeabilized to the fluorescent dye Fura-red (MW 725) by an application of 1 mmol/l ATP^{4-} for more than 5 min. The agonist binding characteristics of purinoceptors in human B-lymphocytes are similar to cloned P2X$_7$, purinoceptors. The permeation characteristics, on the other hand, are rather similar to the P2X subtypes 1 to 4.

Despite a high Ca^{2+}-permeability, ATP^{4-}-application to unclamped cells does not significantly increase the global intracellular Ca^{2+}-concentration. This is mainly due to a block of the Ca^{2+}-permeation by Na$^+$ and depolarization of the cells by the ATP^{4-}-induced inward current carried mainly by Na$^+$. It is concluded that under physiological conditions, this purinergic receptor on B-lymphocytes does not increase the global intracellular free Ca^{2+}-concentration.

Acknowledgement

Supported by DFG, Ma 1581/2-2.

References

Bretschneider, F., Klapperstück, M., Löhn, M. and Markwardt, F. (1995) Nonselective cationic currents elicited by extracellular ATP in human B-lymphocytes. *Pflügers Arch.*, 429: 691–698.

Bretschneider, F. and Markwardt, F. (1999) Methods in enzymology. Vol. 294, 180–189.

Briere, F., Servet-Delprat, C., Bridon, J.M., Saint-Remy, J. and Banchereau, J. (1994) Human interleukin 10 induces naive surface immunoglobulin D$^+$ (sIgD$^+$) B cells to secrete IgG$_1$ and IgG$_3$. *J. Exp. Med.*, 179: 757–762.

Chen, J.R., Jamieson, G.P. and Wiley, J.S. (1994) Extracellular ATP increases NH$_4^+$ permeability in human lymphocytes by opening a P$_{2Z}$ purinoceptor operated ion channel. *Biochem. Biophys. Res. Commun.*, 202: 1511–1516.

Di Virgilio, F. (1995) The P2Z purinoceptor: An intriguing role in immunity, inflammation and cell death. *Immunol. Today*, 16: 524–528.

Dubyak, G.R. and El-Moatassim, C. (1993) Signal transduction via P$_2$-purinergic receptors for extracellular ATP and other nucleotides. *Am. J. Physiol.*, 265: 577–606.

El-Moatassim, C., Dornand, J. and Mani, J.C. (1992) Extracellular ATP and cell signalling. *Biochim. Biophys. Acta*, 1134: 31–45.

Klapperstück, M., Büttner, C., Schmalzing, G. and Markwardt, F. (1998) P2X$_1$, P2X$_4$ and P2X$_7$ Purinoceptors are expressed in human B lymphocytes. *Pflügers Arch.*, 435 (S6): R106.

Löhn, M., Klapperstück, M. and Markwardt, F. (1998) Block by monovalents of Ca^{2+}-entry through P2X-Purinoceptors in human B-lymphocytes. *Pflügers Arch.*, 435 (S6): R107.

Markwardt, F., Löhn, M., Böhm, T. and Klapperstück, M. (1997) Purinoceptor-operated cationic channels in human B-lymphocytes. *J. Physiol. (Lond.)*, 498: 143–151.

Murgia, M., Pizzo, P., Steinberg, T.H. and Di Virgilio, F. (1992) Characterization of the cytotoxic effect of extracellular ATP in J774 mouse macrophages. *Biochem. J.*, 288: 897–901.

Padeh, S., Cohen, A. and Roifman, C.M. (1991) ATP-induced activation of human B lymphocytes via P$_2$-purinoceptors. *J. Immunol.*, 146: 1626–1632.

Rassendren, F., Buell, G.N., Virginio, C., Collo, G., North, R.A. and Surprenant, A. (1997) The permeabilizing ATP receptor, P2X$_7$—cloning and expression of a human cDNA. *J. Biol. Chem.*, 272: 5482–5486.

Saribas, A.S., Lustig, K.D., Zhang, X.K. and Weisman, G.A. (1993) Extracellular ATP reversibly increases the plasma membrane permeability of transformed mouse fibroblasts to large macromolecules. *Anal. Biochem.*, 209: 45–52.

Schubert, R. (1990) A program for calculating multiple metal-ligand solutions. *Comp. Meth. Progr. Biomed.*, 33: 93–94.

Surprenant, A., Buell, G. and North, R.A. (1995) P$_{2X}$ receptors bring new structure to ligand-gated ion channels. *Trends Neurosci.*, 18: 224–229.

Tatham, P.E.R. and Lindau, M. (1990) ATP-induced pore formation in the plasma membrane of rat peritoneal mast cells. *J. Gen. Physiol.*, 95: 459–476.

Wiley, J.S., Jamieson, G.P., Mayger, W., Cragoe, Jr., E.J. and Jopson, M. (1990) Extracellular ATP stimulates an amiloride-sensitive sodium influx in human lymphocytes. *Arch. Biochem. Biophys.*, 280: 263–268.

Wiley, J.S., Chen, R., Wiley, M.J. and Jamieson, G.P. (1992) The ATP^{4-} Receptor-operated ion channel of human lymphocytes: Inhibition of ion fluxes by amiloride analogs and by extracellular sodium ions. *Arch. Biochem. Biophys.*, 292: 411–418.

Wiley, J.S., Chen, R. and Jamieson, G.P. (1993) The ATP^{4-} receptor-operated channel (P$_2$Z class) of human lymphocytes allows Ba^{2+} and ethidium$^+$ uptake: Inhibition of fluxes by suramin. *Arch. Biochem. Biophys.*, 305: 54–60.

P. Illes and H. Zimmermann (Eds.)
Progress in Brain Research, Vol 120
© 1999 Elsevier Science BV. All rights reserved

CHAPTER 29

The P2Z/P2X$_7$ receptor of microglial cells: A novel immunomodulatory receptor

Francesco Di Virgilio*, Juana M. Sanz, Paola Chiozzi and Simonetta Falzoni

Department of Experimental and Diagnostic Medicine, Section of General Pathology, University of Ferrara, Italy

Introduction

Microglial cells are resident cells of the central nervous system (CNS) endowed with the ability to phagocytose, present antigen and secrete inflammatory cytokines (Perry and Gordon, 1988; Dickson et al., 1991). They are thought to derive from bone marrow precursors (monocyte-like cells) that enter the CNS after it has been vascularized, and thus can be considered the CNS equivalent of tissue macrophages, with which they share various functions and membrane markers. Morphologically, microglial cells are often subclassified into ameboid and ramified microglia. Ramified microglia are branched, small cells, while ameboid microglia are round cells with a smaller number of cell processes. Microglial cells, especially of the ramified subtype, are often closely associated with central neurons, thus suggesting a functional significance for this close proximity.

As microglial cells are equipped with a vast array of effector mechanisms (phagocytosis, NO production, free radical generation, release of pro-inflammatory cytokines), it is very likely that they take part to responses to damage or infection in the CNS, and can thus also play an important role in the mechanism of neuronal damage and/or degeneration (Kreutzberg, 1996). It is also likely that a complex bidirectional network of communication exists between neurons and the microglia. We believe that extracellular ATP and the purinergic receptors may be a crucial pathway for this communication, thus we have investigated the role of the P2X$_7$ receptor, formerly also known as P2Z, in microglia activation.

Microglial cells express the P2X$_7$ receptor

Effects of purinergic agonists on rat microglial cells have been extensively investigated by Peter Illes and co-workers, who identified by electrophysiological techniques the presence of both P2Y and P2X receptors in these cells, although the receptor subtype was not further characterized (Norenberg et al., 1994; Illes et al., 1996a, 1996b). These authors ascribed to stimulation of a P2X receptor an early rapidly desensitizing inward current activated by moderately high ATP (10–100 μM) concentrations. In these studies, however, it was unclear whether microglial cells also expressed the P2Z/P2X$_7$ receptor, although this was somehow expected by analogy with other immune cells. Responses of microglial cells to ATP have also been reported by Kettenmann and Cavorkey (Haas et al., 1996).

In 1996 our laboratory provided the first evidence in support of the expression by microglial cells of an ATP-gated plasma membrane pore closely resembling the macrophage P2Z receptor (Ferrari et al., 1996). This was inferred on the basis of the ability of ATP to cause: (a) fast plasma

*Corresponding author. Tel.: +39 532-291353; fax: +39 532-247278; e-mail: fdv@ifeuniv.unife.it

membrane depolarization, (b) uptake of ethidium bromide or lucifer yellow, and (c) release of cytoplasmic markers (e.g. lactate dehydrogenase), responses generally accepted as functional evidence for the expression of P2Z. This strong but indirect evidence was then confirmed (Ferrari et al., 1997a) by immunoblot analysis with an antiserum raised against the COOH tail of the $P2X_7$ receptor that was in the meantime cloned from a rat brain library by Gary Buell and colleagues (Surprenant et al., 1996). It is interesting that a detailed investigation in the CNS gave no evidence for the presence of $P2X_7$ in central neurons, thus suggesting that this receptor might in fact have been originally cloned from microglia. The finding of a $P2Z/P2X_7$ type receptor in microglia is consistent with the known peculiar expression of this molecule by immune cells, and in particular by mononuclear phagocytes (Di Virgilio et al., 1996).

Molecular structure of the $P2X_7$ receptor

We owe the near totality of our information on $P2X_7$ structure to the impressive effort of the Geneva group, once at the Glaxo-Wellcome Institute for Molecular Biology, led by Gary Buell, Annmarie Surprenant and Alan North. These investigators isolated a cDNA that when expressed in human embryonic kidney cells (HEK293) conferred some of the responses typically associated to P2Z activation in immune cells (Surprenant et al., 1996). The cDNA encoded a protein of 595 AA that in the first 395 AA was 35 to 50% identical to the other six P2X receptors so far cloned, while the COOH-terminal domain was much longer than in the other receptors and showed no sequence homology with known proteins. Within the first 395 AA, two hydrophobic domains were present that likely spanned the membrane, and within these two domains, a large extracellular loop rich in cysteine residues.

The $P2X_7$ receptor is not only the most distantly related member of the P2X subfamily, as it only shares limited homology with the other members, but also the one that shows the lowest interspecific relatedness, being the rat protein only 80% identical to the human (in comparison, rat/human identity of $P2X_1$, $P2X_3$ and $P2X_4$ is 91%, 94% and 88%, respectively) (Rassendren et al., 1997). Surprenant et al. (1996) showed that transfection of $P2X_7$ into HEK293 cells made it possible to load into these cells low MW extracellular solutes upon ATP stimulation, in other words they were able to reproduce the extensively described but little understood "ATP-dependent plasma membrane permeabilization", thus supporting the conclusion that $P2X_7$ is the permeabilizing P2Z receptor previously described in mast cells and macrophages (Cockcroft and Gomperts, 1979; Steinberg and Silverstein,1987). Ability to mediate the non-selective increase in plasma membrane permeability resides in the exclusive COOH tail of $P2X_7$, as transfection of a $P2X_7$ receptor truncated to 418 amino acids ($P2X_7\Delta C$) was unable to sustain plasma membrane permeabilization, although channel activity was preserved (Surprenant et al., 1996). This suggests that a single protein can function as ion channel and non-selective pore, depending on whether a brief pulse or a sustained stimulation with ATP is applied. No other plasma membrane intrinsic ion channel so far investigated shares this intriguing property.

It is reasonable to hypothesize that the functional ion channel is formed by the aggregation of more than one $P2X_7$ subunit, but the stoichiometry has not yet been determined. Among P2X receptors, the quaternary structure has been resolved only for $P2X_1$ and $P2X_3$, with the surprising discovery that these receptors form homo-trimers/hexamers, a novel structural motif for ligand-gated ion channels (Nicke et al., 1998). Thus, we may assume that $P2X_7$ may also undergo a similar process of homo-oligomerization to generate a functional ion channel, although this issue clearly requires an in-depth investigation. The channel/pore transition is likely triggered by the recruitment of further $P2X_7$ subunits mediated by the COOH tail, maybe via protein–protein interaction. The studies of the Geneva group suggest that no accessory proteins, besides the $P2X_7$ molecule itself, are necessary to generate the pore, and accordingly no evidence of interaction between this receptor and cytoplasmic proteins has been so far reported, but again further extensive investigation is needed before any conclusion can be drawn.

There are no obvious consensus sequences for ATP binding in the P2X$_7$ primary structure, thus raising the possibility that the ATP-binding site is formed by aminoacid residues lying far apart in the peptide chain, and brought in proximity only once the polypeptide is properly folded. The abundant cystein residues present in the extracellular domain may play a central role in the establishment of the correct intrachain bonds.

In mouse macrophages, a Hill plot analysis of ATP-dependent $[Ca^{2+}]_i$ rise and depolarization yielded coefficients of 2.00 and 2.23, respectively (Pizzo et al., 1991), suggesting that binding of at least two ATP molecules is required to activate P2X$_7$. No data are available for activation of the microglia P2X$_7$ receptor, but we think that a similar ligand/receptor stoichiometry is likely.

Identification of the ATP-gated pore of microglial cells as P2X$_7$

In 1996 our laboratory reported that two mouse microglial cell lines exhibited responses to extra-cellular ATP suggestive of the expression of a P2Z-like receptor (Ferrari et al., 1996). Our experiments showed that ATP and some ATP analogues, most notably benzoyl–benzoyl ATP (BzATP), caused a fast and long-lasting plasma membrane depolarization. Threshold ATP concentration in Ca^{2+} and Mg^{2+}-containing saline solution was about 100 μM, not far from the threshold for P2Z receptor activation in mouse macrophages. In one of these lines (N9 line) the ATP dose-dependence curve was biphasic (see Fig. 1), showing an early shoulder with a threshold just above 1 μM and a maximal response at 30 μM, followed by a second phase starting at 100 μM and levelling off at 300 μM. The first response could be due to activation of another high affinity non- (or slowly)-desensitizing P2X receptor (P2X$_2$? P2X$_4$?) co-expressed by N9 cells, or to the presence of two subpopulations of cells expressing either P2X$_7$ or the other putative high affinity P2X receptors. This issue is at the moment unresolved as microglial cells have not been thoroughly investigated for expression of P2X receptors. Other nucleotides were ineffective or rather caused a transient

hyperpolarization, suggestive of the presence of Ca^{2+}-activated K^+ channels, and thus of a P2Y receptor linked to Ca^{2+} mobilization.

Following an approach initially proposed by Steinberg and colleagues in mouse macrophages several years ago, we have selected a number of ATP-resistant microglial cell clones from the parental N9 and N13 cell lines by growing these lines in the presence of 5 mM ATP. This procedure lead to killing of the near totality of the microglial cells but for a few. The surviving cells can then be cloned and expanded to yield ATP-resistant cell clones. The ATP-resistant clones lack expression of P2X$_7$ (see Fig. 2) by Western blot analysis, and accordingly they also lack all P2X$_7$-dependent functional responses, i.e. sustained plasma membrane depolarization, uptake of low MW hydrophylic solutes, membrane blebbing and swelling, and obviously are refractory to ATP-dependent cytotoxicity. On the contrary, they retain those responses typically associated to the expression of P2Y receptors such as ATP-dependent Ca^{2+} mobilization from intracellular stores. Interestingly, some of these clones, e.g. the N13ATPR5 clone, can still be transiently depolarized by ATP. Although we have not further investigated this issue to ascertain, for example, whether the transient depolarization is due to expression of another P2X receptor subtype, this anecdotal observation is interesting because on the one hand it reinforces the hypothesis that microglia may express other P2X receptors besides P2X$_7$ (see above), and on the other clearly demonstrates that only a non-inactivating P2X receptor may sustain the cytotoxic response. The ATP-resistant clones also show some difference in the pattern of cell-to-cell interaction as they preferentially grow in clusters and are more firmly adherent. Besides the N9 and N13 lines, P2X$_7$ expression has also been demonstrated in the mouse microglia line NTW8 (Collo et al., 1997).

Most biochemical and functional studies have been performed in microglial cell lines; however we have also briefly checked P2Z/P2X$_7$ expression in primary microglial cell cultures from mouse brain. These cells are isolated from a mixed population of newborn mouse brain cells after 14 days of in vitro culture. The classical lucifer yellow uptake technique clearly suggests expression of a typical "permeabilizing ATP receptor", preferentially by ameboid rather than by ramified microglia. This observation, although preliminary, may be of interest in the light of the recent report by Collo and co-workers on the high reactivity to a $P2X_7$ riboprobe of ischemic areas from rat brain (Collo et al., 1997). Collo et al. (1997) also confirmed localization to this area of the $P2X_7$ by immunocytochemistry. Morphology of cells positive to

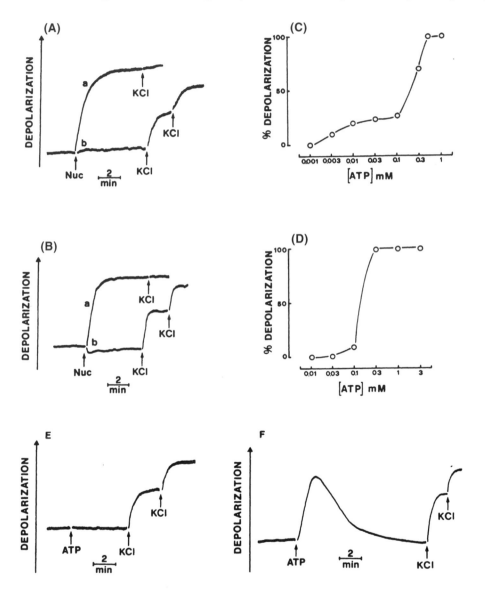

Fig. 1. Effect of exogenous ATP on plasma membrane potential in wild-type N9 and N13 microglial cell lines and in ATP-resistant cell clones. Microglial cells were incubated at 37°C in Ca^{2+}-containing medium at a concentration of 2.5×105/ml in the presence of 100 nM bisoxonol. (A) N9 cells: at the arrow 1 mM ATP (*trace a*) or UTP (*trace b*) was added. Each Kcl addition was 30 mM. (B) N13 cells: at the arrow, 1 mM ATP (*trace a*) or GTP (*trace b*) was added. (C) Dose dependency for ATP-triggered depolarization of N9 cells. (D) Dose-dependency for ATP-triggered depolarization of N13 cells. (E) Lack of response to ATP of a N9 ATP-resistant clone. (F) Transient response to ATP of a N13 ATP-resistant clone. (Reprinted with permission from Ferrari et al. (1996). Copyright, 1996. The American Association of Immunologists).

KD

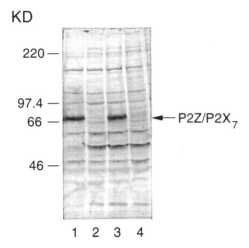

220 —

97.4 —
66 — ← P2Z/P2X₇

46 —

1 2 3 4

Fig. 2. P2X₇ receptor expression by N13 and N9 microglial cell lines. Lane 1, N9 cells; lane 2, ATP-resistant clone N9R17; lane 3, N13 cells; lane 4, ATP-resistant clone N13R5. (Reprinted with permission from Ferrari et al. (1997a)).

hybridization with the P2X₇ riboprobe and the anti-P2X₇ Ab suggests that they are activated microglial cells. These findings raise the appealing possibility that the P2X₇ receptor is preferentially, if not exclusively, expressed by activated microglia, for example during an inflammatory response or at sites of brain damage or neurodegeneration, in keeping with previous reports on the upregulation of macrophage P2Z/P2X₇ by the potent inflammatory cytokine IFN-γ (Blanchard et al., 1991; Falzoni et al., 1995).

Stimulation of microglial cells with ATP also triggers an increase in the cytoplasmic Ca^{2+} concentration ($[Ca^{2+}]_i$) with the typical shape suggestive of the presence of both P2Y and P2Z/P2X₇ receptors, i.e. an early spiking increase followed by a delayed long-lasting plateau. As previously shown for J774 mouse macrophages (Murgia et al., 1993), pre-treatment with the selective P2X antagonist oxidized ATP (oATP) obliterates the delayed plateau but not the early spiking rise. ATP dose-dependence for the $[Ca^{2+}]_i$ increase is also biphasic, while that obtained with other nucleotides is monophasic, again clearly indicating that ATP acivates two different receptor subtypes. Thus, all available data point to the presence of a P2X₇ receptor in microglia. The obvious question at this stage is why?

P2X₇-dependent cell responses in microglia

The first short term P2X₇-dependent response described in microglia was cytotoxicity (Ferrari et al., 1996). ATP-challenged cells rapidly round and swell and are irreversibly injured within 30–45 min of stimulation. The ATP threshold for release of lactic dehydrogenase from N9 and N13 cells is between 100 and 200 μM, while maximal effect is reached at about 500 μM. Only a fraction (about 40%) of microglial cells is killed by ATP. This fraction is increased to 70% and 85%, respectively, when BzATP or ATPγS are used. This may suggest that the vast majority of the cells express P2X₇, but that due to low affinity of the receptor for ATP and to the short life of the extracellular nucleotide, the threshold lethal activation is reached only in a relatively small number of cells. With BzATP as an agonist, the dose-dependence curve is also shifted to the left (see Fig. 3). Microglia is fully protected by the injurious effects of ATP by pre-treatment with oATP. Although not highly specific for the P2X₇ subtype (however, it discriminates pretty well between P2Y and P2X receptors) (Murgia et al., 1993), oATP is an interesting compound as it covalently labels, and thus irreversibly inhibits, P2X₇. This allows stable blocking of P2X₇-mediated responses in the absence of significant side effects. We have found that pyridoxalphosphate-6-azophenyl-2′,4′-disulphonic acid (PPADS) is also an inhibitor of microglial cell P2X₇ receptor, although not as good as oATP (J.M. Sanz and F. Di Virgilio, unpublished).

It is well known that in different cell types ATP can be an apoptotic agent acting at P2X₁ or P2X₇ receptors. This is also the case in microglia, as shown by the pattern of DNA degradation in ATP-pulsed N9 and N13 cells (Ferrari et al., 1997). ATP can also be a powerful necrotic agent but it is unclear which factors shift the balance from apoptosis to necrosis, even in a typical in vitro experiment. Likewise, the intracellular mechanisms involved in ATP-dependent apoptosis are completely unknown. We think that a key role must be played by the dramatic inbalance in the homeostasis of intracellular ions: activation of P2X₇ causes a fast and massive depletion of intracellular K^+ and a large increase in $[Ca^{2+}]_i$ followed by

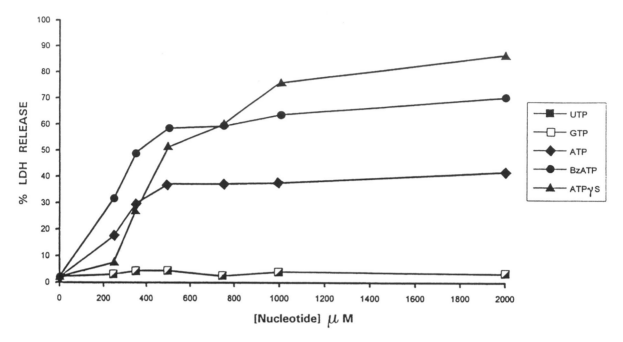

Fig. 3. Dependency of lactic dehydrogenase (LDH) release on the nucleotide concentration. (Reprinted with permission from Ferrari et al. (1997a)).

efflux of low MW cytoplasmic metabolites (nucleotides, cofactors, etc.). We speculate that among these possible injurious events, release of cytoplasmic K^+ might be an important factor in the activation of the apoptotic pathway.

Reasons to implicate K^+ as a trigger factor for ATP-dependent apoptosis stem from the known observation that in vitro activity of the prototypical cystein protease ICE (Interleukin-1-Converting Enzyme) is sensitive to high K^+, being optimal conditions for enzymatic assay at a K^+ concentration < 100 mM (Kostura et al., 1989). Although it is now believed that ICE is out of the main pathway of apoptosis (Nagata, 1997), it cannot be excluded that other proteases, e.g. caspase 8, can also be activated by a decrease in cytoplasmic K^+, and thus trigger the whole cystein protease cascade. Alternatively, it can be speculated that ICE, once stimulated by the fall in cytoplasmic K^+, can proteolytically activate other caspases that will then set in motion the whole apoptotic machinery. With regard to Ca^{2+}, we think it more unlikely that this cation plays a major role in ATP-dependent apoptosis as in regular physiological saline solution its presence is irrelevant; however, in certain experi-

mental conditions, for example in low salt buffers, Ca^{2+} greatly accelerates the rate of ATP-dependent cell death (Murgia et al., 1992). ATP stimulation of P2X_7-expressing cells also causes a dramatic rearrangement of the cytoskeletal network, another process intimately linked to apoptosis, and of course the typical fragmentation of genomic DNA into modules corresponding to nucleosomes or multiple integers of this structure. DNA fragmentation can be fully inhibited by oATP.

Although there is agreement that extracellular ATP can trigger apoptosis in phagocytes (Molloy et al., 1994; Lammas et al., 1997; Ferrari et al., 1997a), the experimental conditions to observe this phenomenon are not always straightforward. In our hands, microglia requires a stimulation of at least 4–6 h with medium-high (e.g. 3 mM) ATP concentrations in complete RPMI medium in order to undergo apoptosis, but we have often observed that other factors, such as cell density and phase of cell cycle, may greatly affect the outcome, either by rendering microglia temporary refractory to ATP-dependent cell death, or by preferentially inducing necrosis rather than apoptosis. It is of interest that microglial cells in Go phase (induced by 24 h serum starvation) are partially refractory to ATP-mediated cytotoxicity and accordingly down-regulate P2X$_7$ (Ferrari et al., 1996). Some investigators have pointed out that the ATP-permeabilizing receptor is up-regulated in transformed or tumoral cell lines, and this observation has also prompted attempts to develop a novel anti-tumor therapeutical approach based on the infusion of ATP. Our observation, albeit anecdotal, might suggest that rather than being associated to the neoplastic phenotype, P2X$_7$ up-regulation is linked to the growth phase. This interpretation is supported by reports from our laboratory indicating that, at least in T lymphocytes, the P2Z/P2X$_7$ receptor is involved in proliferation (Baricordi et al., 1996).

Although cytotoxicity has been the first and most striking response associated to P2X$_7$ activation, it is likely that this receptor, like many other plasma membrane receptors, can mediate more than one response, depending on the level of activation. In the original work reporting the first description of the P2X$_7$ receptor in mouse microglia (Ferrari et al., 1996), we showed that this receptor was also involved in the release of the proinflammatory cytokine IL-1β. The intracellular mechanism responsible for ATP-dependent IL-1β release has not been fully elucidated, but at least some steps have been clarified. In the first place, it is now well established that this is an event exclusively mediated by P2X$_7$, as microglial cell clones lacking P2X$_7$ are specifically incapable to produce IL-1β in response to ATP, yet they release this cytokine when challenged with stimulants unrelated to P2 receptors. Furthermore, in wild-type microglia oATP completely blocks IL-1β secretion. Secondly, P2X$_7$ activation by itself is inefficient unless it is preceded by a short (1–2 hours) priming with bacterial endotoxin (LPS). This suggests that P2X$_7$ acts on the pro-cytokine to promote its maturation and release, rather than trigger gene transcription. Gene transcription and pro-IL-1β accumulation are, on the contrary, promoted by LPS, that however appears to be very inefficient to turn on ICE. It is indeed well known that LPS is a weak stimulus for release of mature IL-1β release from resident mononuclear phagocytes (e.g. tissue macrophages and microglia), in contrast to circulating

monocytes (Hogquist et al., 1991; Perregaux and Gabel, 1994; Ferrari et al., 1996, 1997b).

We have followed the kinetics of cytoplasmic accumulation of pro-IL-1β (35 kDa) and release of the mature form (17 kDa) in microglia activated with LPS. As expected, there is a relatively fast synthesis of the 35 kDa procytokine that, however, appears to be very inefficiently cleaved and released over a time course of several hours (Sanz and Di Virgilio, 1999, in preparation). Addition of ATP (1–3 mM) triggers an almost immediate burst of mature IL-1β release that leads to a rapid mobilization of the accumulated pro-cytokine. The most straight-forward interpretation of these data is that P2X$_7$ activates ICE thus leading to pro-IL-1β processing and release. In support of this hypothesis, the specific tetrapeptide ICE-inhibitor YVAD (Tyrosine–Valine–Alanine–Aspartate) fully blocks ATP-dependent processing and release, but not LPS-mediated accumulation of the pro-cytokine. How is ICE activated via P2X$_7$?

The pathways involved in ICE activation are rather mysterious even with other less unusual stimuli (Cohen, 1997). In principle, ICE being one of those cystein proteases that are activated by proteolytic cleavage, it has to be expected that either another still unknown protease is first triggered to proteolytically activate ICE, or ICE is triggered in the first place and then autocatalytically cleaves itself. So far, no in-depth investigation on caspase activation via P2X$_7$ has been performed, but for a preliminary observation by Ferrari et al. (1997d), thus most of the evidence is indirect. However, some educated guesses on the possible pathways involved in ICE stimulation by extracellular ATP are possible. Scattered evidence suggests that K$^+$ movement is an important factor for ICE activity. Some cell types, e.g. B cells, stimulated with LPS rapidly synthesize and increase up to 1000-fold the number of voltage-gated K$^+$ channels (Amigorena et al., 1990); the K$^+$ channel blockers TEA and 4-AP inhibit IL-1β maturation (Walev et al., 1995); agents causing cytoplasmic K$^+$ depletion (nigericin, valinomycin, gramicidin) also trigger IL-1β maturation (Perregaux and Gabel, 1994). Even a brief hypotonic shock can promote IL-1β maturation (Walev et al., 1995). Last but not least, the in vitro activity of ICE is best measured at low K$^+$ (Kostura et al., 1989). In support of a major role of K$^+$ depletion for ICE activation, we have observed in human macrophages that incubation in high extracellular K$^+$ almost completely blocks ATP-dependent IL-1β maturation and release without blocking pore formation (Ferrari et al., 1997b). Vice versa, incubation in sucrose medium (low extracellular K$^+$) increases several-fold IL-1β release. Thus, a possible mechanism for ICE activation by ATP is cytoplasmic K$^+$ depletion through the large pore generated in the plasma membrane by the P2X$_7$ receptor. Experiments are in progress in our laboratory to further test this hypothesis.

Whether other cytokines are also released by ATP stimulation in microglia is not known. Preliminary results obtained in our laboratory show that in human fibroblasts IL-6 release can be stimulated via P2X$_7$ (Solini et al., 1999); microglia is also known to efficiently release this cytokine, thus it cannot be excluded that ATP may also release IL-6 from microglia. Nonetheless, at this stage we think that a major role of P2X$_7$ in supporting IL-6 release is unlikely. This bias comes from an investigation of the comparative effect of oATP on IL-1β and IL-6 release from N9 and N13 microglial cells stimulated with bacterial endotoxin. We noticed that pretreatment with oATP prevented almost completely IL-1β but not IL-6 release triggered by a 24 h incubation in the presence of LPS (Ferrari et al., 1997c). This would suggest that, although it cannot be excluded that P2X$_7$ might be involved, this receptor is not a major participant in IL-6 release. These experiments deserve a comment, irrespectively of their implications for IL-6 secretion. Release of IL-1β triggered by LPS (no extracellular ATP added) can be blocked not only by pre-incubation with oATP but also with an ATP-consuming enzyme such as apyrase (Ferrari et al., 1997c), thus suggesting that, even in the absence of added ATP, purinergic receptors take part in IL-1β release, maybe through an autocrine/paracrine loop initially set in motion by LPS-dependent ATP secretion.

Besides cytokine release, ATP appears to play a role also in secretion of other microglial factors. A recent evidence supports a role in plasminogen release, and also in this case the receptor implicated is P2X$_7$ probably via an increase in [Ca2 +]$_i$ (Inoue et al., 1998). There have been attempts to check whether NO is also released by microglia in response to ATP stimulation, but they have been so far unsuccesful. In view of previous positive results in macrophages (Tonetti et al., 1994; Tonetti et al., 1995), it is possible that this point requires further studies.

Most observations in this newborn field have been thus far rather phenomenological. An interesting exception is the recently published study by Schulze-Osthoff and colleagues (Ferrari et al., 1997d). These authors discovered that P2X$_7$ stimulation in microglia resulted in a potent activation of the transcription factor NF-κB. NF-κB activation required proteolytical ICE cleavage, and the time course of both NF-κB activation and ICE processing were closely matched. Furthermore, YVAD and the less specific inhibitor DEVD (Aspartate–Glutamate–Valine–Aspartate) also blocked NF-κB activation. Rather intriguingly, the subunit composition of the ATP-induced NF-κB-DNA complex was unusual in that the sole appearance of a p65 homodimer was detected, whereas LPS induced the formation of the more common p50 homodimer and p50/p65 heterodimer. However, the p65 homodimer was fully transcriptionally competent. This study describes a novel activation pathway for purinergic receptors in microglia that proceeds from the proteolytical activation of caspases up to activation of transcription factors and gene expression. It would be of great interest to test whether a similar pathway is also active in other cell types and which genes are turned on.

Other possible roles can be extrapolated from data obtained in other mononuclear phagocytes. There is evidence that in mouse and human macrophages P2X$_7$ may be involved in the process of cell fusion that leads to giant cell formation (Falzoni et al., 1995; Chiozzi et al., 1997), an event common at sites of chronic granulomatous inflammation. Microglia is also known to generate such multinucleated structures during chronic inflammatory diseases of the CNS, thus, although we have not directly investigated this issue, this might be another relevant function of P2X$_7$ in microglia physiology.

An integrated view of P2X₇ in the physiopathology of microglia

Microglia is thought to be the ontogenetic and functional equivalent of mononuclear phagocytes in CNS, and likely plays a central role in tissue repair and host defense. Since the first reports on the expression of P2 receptors on immune cells, macrophages and lymphocytes, we wondered what would be the best anatomical site for immune cells to fully exploit their ability to respond to extra-cellular nucleotides. ATP is now firmly established as a key neuromediator in the CNS (Burnstock, 1997; Burnstock, 1999), thus we thought it obvious to extend our investigation to microglial cells, on the assumption that no other mononuclear phago-cyte cells would enjoy a best environmental condition to fully exploit its equipment of pur-inergic receptors. On the basis of our previous studies in mouse and human macrophages we expected a certain pattern of responses to ATP stimulation in microglia, an expectation that was satisfactorily fulfilled. Thus, microglia express a fully functional P2X₇ receptor. The question arises as to how we can integrate this molecule in the complex network of cellular interactions in the CNS.

A major obstacle to the acceptance of a physio-logical role for P2 receptors in the immune system has been until recently the lack of convincing demonstration that ATP can accumulate to sig-nificant levels in the extracellular milieu. Nowadays this picture has drastically changed as there are few doubts that ATP (and very likely UTP) is released by many different cell types via non-lytic pathways. Not only by central and peripheral neurons (White, 1978; von Kugelgen and Starke, 1991), but also by non-neuronal cells such as platelets (Meyers et al., 1982; Colman, 1990), endothelial and epithelial cells (Pearson and Gordon, 1979; Hassessian et al., 1993; Reisin et al., 1994), T lymphocytes (Filippini et al., 1990), macrophages and microglial cells (Ferrari et al., 1997c). The pathway is not always identified (secretory exocytosis, transporters belonging to the ABC family?), but it is clear that ATP secretion is a phenomenon much more common that previously thought. Not to mention that, being the cytoplasmic

ATP concentration in the range of 5–10 mM, even a transient damage of the plasma membrane is expected to cause a burst of release of this nucleotide (but see Galietta et al. (1997), for the inability of hypotonic shock to cause ATP release from epithelial cells). Thus, all available evidence strongly suggests that presence of ATP in the pericellular space is the rule rather than the exception.

Over the last few years several investigators have elaborated on the hypothesis that ATP, besides being an obvious neurotransmitter, could also be an early inflammatory mediator and immunomodula-tor (Filippini et al., 1990; Di Virgilio et al., 1990; Di Virgilio, 1995; Dubyak et al., 1996; Fisette et al., 1996; Robson et al., 1997). The core of this hypothesis is that cell and tissue damage is the main trigger of inflammation, and at the same time inflammation is a cause of cell injury. This predicts that ATP will be present in the interstitium since the initial steps of an inflammatory response. Accumu-lation of extracellular ATP may result not only from release but also from inhibition of hydrolysis. At least two types of enzymes are involved in ATP hydrolysis, ecto-ATP diphosphohydrolase (apyr-ase) and ecto-ATPase, both of which are widely distributed in mammalian tissues and are also expressed in the brain (Kegel et al., 1997; Zimmer-mann and Braun, 1999). Other enzymes involved in the extracellular hydrolysis chain are ecto-ADPase and 5′-nucleotidase. The sequential activity of all these enzymes on released ATP leads to the generation af adenosine, a metabolite whose anti-inflammatory activity is increasingly recognized (Morabito et al., 1998). Given the presence of this hydrolytic machinery it has to be expected that ATP will be rapidly cleaved as soon as it is released into the extracellular environment, thus decreasing the chance that it might act as an intercellular mes-senger, unless during inflammation ecto-enzymes are down-modulated. A recent report by Robson et al., 1997) indicates that this might indeed be the case as these authors have observed that activation of endothelial cells by TNFα, or exposure to oxidative stress, leads to a rapid loss of ATP-diphosphohydrolase (ATPDase) activity. Furthermore, a marked reduction in ATPDase activity was also observed in rat kidney glomeruli subjected to riperfusion injury. If confirmed, these data will help solve the relevant paradox of how significant extracellular ATP concentrations can accumulate in the presence of such powerful hydrolytic pathways. If degrading enzymes are inhibited, it can be reasonably expected that in a protected environment ATP may reach concentrations high enough to activate even the low affinity $P2X_7$ receptor. By turning on $P2X_7$, ATP may act as an early amplification signal that triggers release of inflammatory cytokines or other pro-inflammatory factors from the microglia. If ATP release is massive, a cytotoxic effect may occur, thereby spreading the injury to nearby cells and tissues.

The cytotoxic activity of ATP on cells expressing $P2X_7$ deserves an additional comment as it might not only be an unwanted injurious side effect, but under some circumstances it might also be a favourable event aimed at enhancing defence mechanisms in the host. This interpretation is based on experiments published over the last few years by Kaplan and co-workers (Molloy et al., 1994) and Lammas et al. (1997) on the effect of $P2X_7$ stimulation on the viability of intracellular pathogens.

It is well known that some microorganisms, although normally ingested by phagocytes, survive and even proliferate within the phagosome, and are thus a cause of serious inflammatory diseases. Initially, Kaplan and co-workers showed that challenge with ATP of human monocyte-derived macrophages that had phagocytosed bacillus Calmette-Guerin (BCG, *Mycobacterium bovis*) led to killing of both the phagocyte and the phagocytosed microorganism (Molloy et al., 1994). Another cytotoxic stimulus such as hydrogen peroxide was ineffective to kill the microorganism although the phagocyte was easily killed. As ATP in human macrophages is a good apoptotic stimulus, these authors proposed that killing of the phagocytosed bacterium was due to a side effect of apoptosis itself. These findings were confirmed and further extended by the subsequent study by Lammas et al. (1997) who showed that the cytotoxic effect on intracellular BCG was specific to ATP, as other cytotoxic stimuli such as anti-CD95 or anti-CD69 antibodies, or anti-MCH class II antibodies plus complement were ineffective. Lammas and col-

leagues also made some attempts to identify the P2 receptor involved by using two known antagonists of $P2X_7$, oATP (Murgia et al., 1993) and KN-62 (Gargett and Wiley, 1997): not surprisingly these agents fully inhibited killing of the intracellular pathogen. Thus, an additional hypothesis on the physiological significance of $P2X_7$ is that the immune system may exploit this membrane receptor to get rid of phagocytic cells that are parasitized by microrganisms that survive phagocytosis, eliminating at the same time both the parasitized cell and the parasite ("Samson shall die with all the Philistines") (Di Virgilio et al., 1998). Whether this mechanism also applies to microglia we do not know as no experiments of this kind have yet been performed, but we would not be surprised if a similar mechanism of defence was also operating in this cell type.

Under less extreme conditions, $P2X_7$ stimulation will not cause cell death but rather cell activation. In synergy with bacterial products or other early inflammatory mediators, ATP will cause secretion of IL-1 and IL-6 from macrophages and microglial cells, and will trigger transcription of genes controlled by NF-κB and possibly other transcription factors. It is of interest that one of the genes controlled by NF-κB is the IL-8 gene; this means that under ATP stimulation microglia will cause synthesis and release this powerful chemotactic cytokine for polymorphonuclear leukocytes, thus leading to a further amplification of the inflammatory response. Figure 4 reports a schematic representation of the possible activatory mechanisms mediated by ATP.

A complex relationship exists between $P2X_7$ and inflammatory cytokines because this receptor is not only involved in the release of some of this pro-inflammatory molecules, but is also controlled by some of them. Various groups have reported that IFN-γ upregulates expression of $P2X_7$ and renders macrophages much more susceptible to all ATP-mediated effects (Blanchard et al., 1991; Falzoni et al., 1995; Humphreys and Dubyak, 1996). IFN-γ is secreted to high levels at sites of chronic inflammatory lesions and strongly potentiates ability of inflammatory cells to cope with pathogens. Membrane expression of molecules involved in antigen presentation is also up-modulated by IFN-γ. Thus

SOURCES OF ATP IN THE CNS AND POSSIBLE MECHANISM OF ACTIVATION OF MICROGLIA

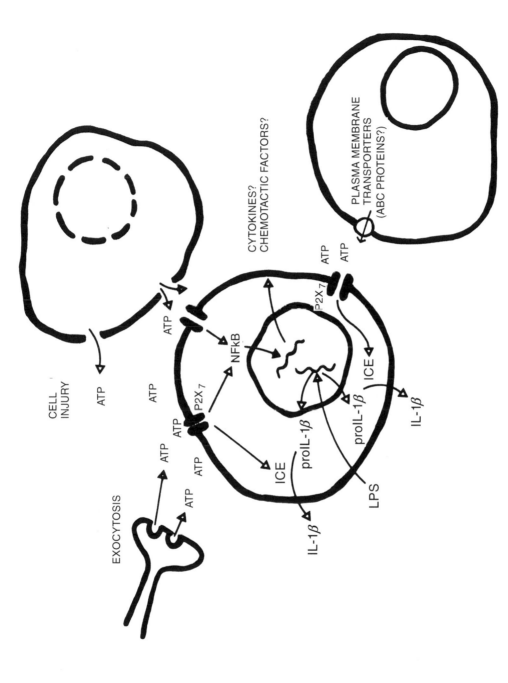

Fig. 4. Schematic representation of microglia activation by extracellular ATP. It is postulated that ATP in the CNS can be released by different processes (secretory exocytosis, cell damage, transmembrane transport). Once in the pericellular space, this nucleotide will stimulate the P2X$_7$ receptor of microglial cells (center cell) thus triggering NFκB activation and transcription of the IL-8 gene and possibly of other cytokines and chemotactic factors as well. Interleukin 1-β-converting enzyme will also be activated. In macrophages primed with bacterial endotoxin (LPS), under conditions therefore where a large amount of the ICE substrate pro-IL-1β is accumulated in the cytoplasm, activated ICE will also trigger formation and release of mature IL-1β.

the intriguing possibility arises that P2X$_7$ is more deeply implicated in immune modulation than previously anticipated on the basis of its currently known functions.

Conclusions and future directions

Only three years ago the P2Z receptor of immune cells was more a hypothesis than a real molecule. Now, not only has the molecular structure been identified, but the physiological role is being unveiled. Clearly, this is the beginning of the most interesting part of the story, a time in which the most crucial questions may be close to be answered: what is the quaternary structure of P2X$_7$? How is the channel/pore transition operated? How is P2X$_7$ expression regulated in vivo? Is there an association to major human diseases? Can P2X$_7$ being a target for pharmacological intervention? It is reasonable to believe that the generation of a P2X$_7$ knock-out mouse will help answer these questions.

Acknowledgements

This work was supported by grants from the National Research Council of Italy (target project on Biotechnology), the Ministry of Scientific Research (MURST), the Italian Association for Cancer Research (AIRC), the X AIDS Project, the II Tubercolosis Project and Telethon of Italy. JMS is supported by a fellowship by the Spanish Ministry of Science and Education.

References

Amigorena, S., Choquet, D., Teillaud, J.L., Korn, H. and Fridman, W.H. (1990) Ion channel blocker inhibit B cell activation at a precise stage of the G1 phase of the cell cycle. Possible involvement of K$^+$ channels. J. Immunol., 144: 2038–2045.

Baricordi, O.R., Ferrari, D., Melchiorri, L., Chiozzi, P., Hanau, S., Chiari, E., Rubini, M. and Di Virgilio, F. (1996) An ATP-activated channel is involved in mitogenic stimulation of human T lymphocytes. Blood, 87: 682–690.

Blanchard, D.K., McMillen, S. and Djeu, J.Y. (1991) IFN-γ enhances sensitivity of human macrophages to extracellular ATP-mediated lysis. J. Immunol., 147: 2579–2585.

Burnstock, G. (1997) The past, present and future of purine nucleotides as signalling molecules. Neuropharmacology, 36: 1127–1139.

Burnstock, G. (1999) Current status of purinergic signalling in the nervous system. Progress Brain Res., 120: 3–10.

Chiozzi, P., Sanz, J.M., Ferrari, D., Falzoni, S., Aleotti, A., Buell, G.N., Collo, G. and Di Virgilio, F. (1997) Spontaneous cell fusion in macrophage cultures expressing high levels of the P2Z/P2X$_7$ receptor. J. Cell Biol., 138: 697–706.

Cockcroft, S. and Gomperts, B.D. (1979) ATP induces nucleotide permeability in rat mast cells. Nature, 279: 451–452.

Cohen, G.M. (1997) Caspases: The excutioners of apoptosis. Biochem. J., 326: 1–16.

Collo, G., Neidhart, S., Kawashima, E., Kosco-Vilbois, M., North, R.A. and Buell, G. (1997) Tissue distribution of the P2X$_7$ receptor. Neuropharmacology, 36: 1277–1283.

Colman, R. (1990) Aggregin: A platelet ADP receptor that mediates activation. FASEB J., 4: 1425–1435.

Dickson, D.W., Mattiace, L.A., Kure, K., Hutchins, K., Lyman, W.D. and Brosnan, C.F. (1991) Biology of disease. Microglia in human disease, with an emphasis on acquired immune deficiency syndrome. Lab. Invest., 64: 135–156.

Di Virgilio, F. (1995) The P2Z purinoceptor: An intriguing role in immunity, inflammation and cell death. Immunol. Today, 16: 524–528.

Di Virgilio, F., Chiozzi, P., Falzoni, S., Ferrari, D., Sanz, J.M., Vishwanath, V. and Baricordi, O.R. (1998) Cytolytic P2X purinoceptors. Cell Death Diff., 5: 191–199.

Di Virgilio, F., Ferrari, D., Chiozzi, P., Falzoni, S., Sanz, J., Dal Susino, M., Mutini, C., Hanau, S. and Baricordi, O.R. (1996) Purinoceptor function in the immune system. Drug Dev. Res., 39: 319–329.

Di Virgilio, F., Pizzo, P., Zanovello, P., Bronte, V. and Collavo, D. (1990) Extracellular ATP as a possible mediator of cell-mediated cytotoxicity. Immunol. Today, 11: 274–277.

Dubyak, G.R., Clifford, E.E., Humphreys, B.D., Kertsey, S.B. and Martin, K.A. (1996) Expression of multiple ATP receptor subtypes during the differentiation and inflammatory activation of myeloid leukocytes. Drug Dev. Res., 39: 269–278.

Falzoni, S., Munerati, M., Ferrari, D., Spisani, S., Moretti, S. and Di Virgilio, F. (1995) The purinergic P2Z receptor of human macrophage cells. Characterization and possible physiological role. J. Clin. Invest., 95: 1207–1216.

Ferrari, D., Villalba, M., Chiozzi, P., Falzoni, S., Ricciardi-Castagnoli, P. and Di Virgilio, F. (1996) Mouse microglial cells express a plasma membrane pore gated by extracellular ATP. J. Immunol., 156: 1531–1539.

Ferrari, D., Chiozzi, P., Falzoni, S., Dal Susino, M., Collo, G., Buell, G.N. and Di Virgilio, F. (1997a) ATP-mediated cytotoxicity in microglial cells. Neuropharmacology, 36: 1295–1301.

Ferrari, D., Chiozzi, P., Falzoni, S., Dal Susino, M., Melchiorri, L., Baricordi, O.R. and Di Virgilio, F. (1997b) Extracellular ATP triggers IL-1κ release by activating the purinergic P2Z receptor of human macrophages. J. Immunol., 159: 1451–1458.

Ferrari, D., Chiozzi, P., Falzoni, S., Hanau, S. and Di Virgilio, F. (1997c) Purinergic modulation of interleukin-1κ release from microglial cells stimulated with bacterial endotoxin. J. Exp. Med., 185: 579–582.

Ferrari, D., Wesselborg, S., Bauer, M.K.A. and Schulze-Osthoff, K. (1997d) Extracellular ATP activates transcription factor NF-κB through the P2Z purinoceptor by selectively targeting NF-κB p65 (RelA). *J. Cell Biol.*, 139: 1635–1643.

Filippini, A., Taffs, R.E., Agui, T. and Sitkovsky, M.V. (1990) EctoATPase activity in cytolytic T lymphocytes: protection from the cytolytic effects of extracellular ATP. *J. Biol. Chem.*, 265: 334–340.

Fisette, P.L., Denlinger, L.C., Proctor, R.A. and Bertics, P.J. (1996) Modulation of macrophage function by P2Y-purinergic receptors. *Drug Dev. Res.*, 39: 377–387.

Galietta, J.V., Falzoni, S., Di Virgilio, F., Romeo, G. and Zegarra-Moran, O. (1997) Characterization of volume-sensitive taurine- and Cl⁻ permeable channels. *Am. J. Physiol.*, 273: C57–C66.

Gargett, C.E. and Wiley, J.S. (1997) The isoquinoline derivative KN-62: A potent antagonist of the P2Z receptor of human lymphocytes. *Br. J. Pharmacol.*, 120: 1483–1490.

Haas, S., Brockaus, J., Verkhratsky, A. and Kettenmann, H. (1996) ATP-induced membrane currents in ameboid microglia acutely isolated from mouse brain slices. *Neuroscience*, 75: 257–261.

Hasséssian, H., Bodin, P. and Burnstock, G. (1993) Blockade by glibenclamide of the flow-evoked endothelial release of ATP that contributes to vasodilatation in the pulmonary vascular bed of the rat. *Br. J. Pharmacol.*, 109: 466–472.

Hogquist, K.A., Nett, M.A., Unanue, E.R. and Chaplin, D.D. (1991) Interleukin-1 is processed and released during apoptosis. *Proc. Natl. Acad. USA*, 88: 8445–8450.

Humphreys, B.D. and Dubyak, G.R. (1996) Induction of the P2Z/P2X₇ nucleotide receptor and associated phospholipase D activity by lipopolysaccharide and IFN-γ in the human THP-1 monocytic cell line. *J. Immunol.*, 157: 5627–5637.

Illes, P., Nieber, K., Frohlich, R. and Norenberg, W. (1996a) P2 purinoceptors and pyrimidinoceptors of catecholamine-producing cells and immunocytes. *Ciba F. Symp.*, 198: 110–129.

Illes, P., Norenberg, W. and Gebicke-Haerter, P.J. (1996b) Molecular mechanisms of microglial activation. B. Voltage and purinoceptor-operated channels in microglia. *Neurochem. Int.*, 29: 13–24.

Inoue, K., Nakajima, K., Morimoto, T., Kikuchi, Y., Koizumi, S., Illes, P. and Kohsaka, S. (1998) ATP stimulation of Ca²⁺-dependent plasminogen release from cultured microglia. *Br. J. Pharmacol.*, 123: 1304–1310.

Kegel, B., Braun, N., Heine, P., Maliszewski, C.R. and Zimmermann, H. (1997) An ecto-ATPase and an ecto-ATP diphosphohydrolase are expressed in rat brain. *Neuropharmacology*, 36: 1189–1200.

Kostura, M.J., Tocci, M.J., Limjuco, G., Chin, J., Cameron, P., Hillman, A., Chartrain, N.A. and Schmidt, J.A. (1989) Identification of a monocyte specific pre-interleukin 1κ convertase activity. *Proc. Natl. Acad. Sci. USA*, 86: 5227–5231.

Kreutzberg, G.W. (1996) Microglia: A sensor for pathological events in the CNS. *Trends Neurosci.*, 19: 312–318.

Lammas, D.A., Stober, C., Harvey, C.J., Kendrick, N., Pan-chalingam, S. and Kumararatne, D.S. (1997) ATP-induced killing of mycobacteria by human macrophages is mediated by purinergic P2Z/P2X₇ receptors. *Immunity*, 7: 433–444.

Meyers, K.M., Holmsen, H. and Seachord, C.L. (1982) Comparative study of platelets dense granules constituents. *Am. J. Physiol.*, 243: R454–R461.

Molloy, A., Loachunmroonvorapong, P. and Kaplan, G. (1994) Apoptosis, but not necrosis, of infected monocytes is coupled with killing of intracellular bacillus Calmette-Guerin. *J. Exp. Med.*, 180: 1499–1509.

Morabito, L., Montesinos, M.C., Schreibman, D.M., Balter, L., Thompson, L., Resta, R., Carlin, G., Hule, A. and Cronstein, B.N. (1998) Methotrexate and sulfasalazine promote adenosine release by a mechanism that requires ecto-5′-nucleotidase-mediated conversion of adenine nucleotides. *J. Clin. Invest.*, 101: 295–300.

Murgia, M., Hanau, S., Pizzo, P., Rippa, M. and Di Virgilio, F. (1993) Oxidised ATP: An irreversible inhibitor of the macrophage P2Z receptor. *J. Biol. Chem.*, 268: 8199–8203.

Murgia, M., Pizzo, P., Steinberg, T.H. and Di Virgilio, F. (1992) Characterization of the cytotoxic effect of extracellular ATP in J774 macrophages. *Biochem. J.*, 288: 897–901.

Nagata, S. (1997) Apoptosis by death factor. *Cell*, 88: 355–365.

Nicke, A., Baumert, H.G., Rettinger, J., Eichele, A., Lambrecht, G., Mutschler, E. and Schmalzing, G. (1998) P2X₁ and P2X₃ receptors form stable trimers: A novel structural motif of ligand-gated ion channels. *EMBO J.*, 17: 3016–3028.

Norenberg, W., Langosch, J.M., Gebicke-Haerter, P.J. and Illes, P. (1994) Characterization and possible function of adenosine 5′-triphosphate receptors in activated rat microglia. *Br. J. Pharmacol.*, 111: 942–950.

Pearson, J.D. and Gordon, J.L. (1979) Vascular endothelium and smooth muscle cells in culture selectively release adenine nucleotides. *Nature*, 281: 384–386.

Perregaux, D. and Gabel. C.A. (1994) Interleukin-1κ maturation and release in response to ATP and nigericin. *J. Biol. Chem.*, 269: 15195–15203.

Perry, V.H. and Gordon, S. (1988) Macrophages and microglia in the nervous system. *Trends Neurosci.*, 6: 273–277.

Pizzo, P., Zanovello, P., Bronte, V. and Di Virgilio, F. (1991) Extracellular ATP causes lysis of mouse thymocytes and activates a plasma membrane ion channel. *Biochem. J.*, 274: 139–144.

Rassendren, F., Buell, G.N., Virginio, C., Collo, G., North, R.A. and Surprenant, A. (1997) The permeabilizing ATP receptor P2X₇: cloning of a human cDNA. *J. Biol. Chem.*, 272: 5482–5486.

Reisin, I.L., Prat, A.G., Abraham, E.H., Amara, J.F., Gregory, R.J., Ausiello, D.A. and Cantiello, H.F. (1994) The cystic fibrosis transmembrane conductance regulator is a dual ATP and chloride channel. *J. Biol. Chem.*, 269: 20584–20591.

Robson, S.C., Kaczmarek, E., Siegel, J.B., Candinas, D., Koziak, K., Millan, M., Hancock, W.W. and Bach, F.H. (1997) Loss of ATP diphosphohydrolase activity with endothelial cell activation. *J. Exp. Med.*, 185: 153–163.

Solini, A., Chiozzi, P., Fellin, R. and Di Virgilio, F. (1999) *J. Cell Sci.*, 112: 297–305.

Steinberg, T.H. and Silverstein, S.C. (1987) Extracellular ATP4- promotes cation fluxes in the J774 mouse macrophage cell line. *J. Biol. Chem.*, 262: 3118–3122.

Surprenant, A., Rassendren, F., Kawashima, E., North, R.A. and Buell, G. (1996) The cytolytic P2Z receptor for extracellular ATP identified as a P2X receptor (P2X$_7$). *Science*, 272: 735–738.

Tonetti, M., Sturla, L., Bistolfi, T., Benatti, U. and De Flora, A. (1994) Extracellular ATP potentiates nitric oxide synthase expression induced by lipopolysaccharide in RAW 264.7 murine macrophages. *Biochem. Biophys. Res. Commun.*, 203: 430–435.

Tonetti, M., Sturla, L., Giovine, M., Benatti, U. and De Flora, A. (1995) Extracellular ATP enhances mRNA levels of nitric oxide synthase and TNF-α in lipopolysaccharide-treated RAW 264.7 murine macrophages. *Biochem. Biophys. Res. Commun.*, 214: 125–130.

von Kügelgen, I. and Starke, K. (1991) Release of noradrenaline and ATP by electrical stimulation and nicotine in guinea-pig vas deferens. *Naunyn-Schmiedeberg's Arch. Pharmacol.*, 344: 419–429.

Walev, I., Reske, K., Palmer, M., Valeva, A. and Bhakdi, S. (1995) Potassium-inhibited processing of IL-1κ in human monocytes. *EMBO J.*, 14: 1607–1614.

White, T.D. (1978) Release of ATP from a synaptosomal preparation by elevated extracellular K$^+$ and by veratridine. *J. Neurochem.*, 30: 329–336.

Zimmermann, H. and Braun (1999) Ecto-nucleotidases in the nervous system. *Progress Brain Res.*, 120: 371–386.

Ecto-nucleotidases and ecto-protein kinase

P. Illes and H. Zimmermann (Eds.)
Progress in Brain Research, Vol 120
© 1999 Elsevier Science BV. All rights reserved

CHAPTER 30

Ecto-nucleotidases—molecular structures, catalytic properties, and functional roles in the nervous system

Herbert Zimmermann* and Norbert Braun

AK Neurochemie, Biozentrum der J.W. Goethe-Universität, Marie-Curie-Str. 9, D-60439 Frankfurt am Main, Germany

Introduction

ATP and other nucleotides serve as important extracellular signaling substances in the central and peripheral nervous system. They are involved in synaptic transmission and synaptic modulation, and equally act at neurons and glial cells. Presumably, they play an important role in the development of the nervous system (Kidd et al., 1998). The ionotropic P2X and the metabotropic P2Y receptors both have a broad distribution in brain (Buell et al., 1996; Dunwiddie et al., 1996). The action of nucleotides is terminated by a surface-located enzyme cascade that sequentially degrades nucleoside 5'-triphosphates to their respective nucleosides and free phosphate or pyrophosphate. In the case of ATP the extracellularly formed adenosine can itself modulate neural functions via specific adenosine receptors (Fredholm et al., 1996) or serve purine salvage after reuptake via plasma membrane-located adenosine transporters (Thorn and Jarvis, 1996). The extracellular ecto-nucleotidase chain of the nervous system has been best characterized in intact cellular systems. These include cultured neural cells, neuroblastoma cells, chromaffin cells and the related PC12 cells, astrocytes, oligodendrocytes, isolated synaptosomes from a variety of vertebrate sources, cultured endothelial cells or also

the neuromuscular junction and superfused peripheral organ preparations such as the vas deferens or the taenia coli (references in Zimmermann, 1996a, 1996b). A general feature of this hydrolysis cascade is that ATP is degraded with ADP, AMP and adenosine sequentially appearing in the assay medium. Furthermore, not only ATP but essentially all physiologically occurring purine and pyrimidine nucleoside 5'-tri, di- and mono-phosphates are hydrolyzed.

The time course of nucleotide degradation in these experiments depends on the amount of enzyme-containing cellular surface and the concentration of nucleotide applied. Generally, the half decay time observed in these experiments was in the order of minutes at the least. It can be expected that the time course of extracellular ATP degradation needs to be considerably faster in situ. More recently, monitoring the postsynaptic effects of adenosine apparently derived from released ATP at hippocampal synapses suggests that in the synaptic cleft complete hydrolysis can occur within less than 200 ms (Dunwiddie et al., 1997; Cunha et al., 1998). This comes close to the extracellular hydrolysis rate of acetylcholine by acetylcholinesterase at the neuromuscular junction (Miledi et al., 1984).

The presence of the ecto-nucleotidase chain in the nervous system is further supported by enzyme cytochemical analyses at the electron microscopical level. Ecto-ATPase reaction product is located at the outer surface of the plasma membrane

---

*Corresponding author. Tel.: +49 69 798 29602; fax: +49 69 798 29606; e-mail: H.Zimmermann@zoology.uni-frankfurt.de

surrounding neurosecretory nerve endings and pituicytes in the neurohypophysis (Thirion et al., 1996), at glial and neuronal plasma membranes including the synaptic cleft in the medial habenula (Sperlagh et al., 1995; Kittel et al., 1996), or at the neuronal plasma membrane/Schwann cell interface at the frog neuromuscular junction (Pappas and Kriho, 1988). Reaction product for the hydrolysis of AMP can be observed around glial cells but also inside the synaptic cleft (Bailly et al., 1995; Schoen and Kreutzberg, 1995). Reaction product for both, ATPase (Mata and Fink, 1989) and pyrophosphatase (László and Bodor, 1981) activity have been identified even within synaptic vesicles.

A major question is that of the molecular identity of the enzymes involved in the hydrolysis cascade as well as their exact location within the tissue. Several surface-located, enzymes capable of catalyzing the hydrolysis of extracellular nucleotides have been identified (Fig. 1). Recently, considerable progress has been made with the molecular identification of two novel enzyme families that catalyze the extracellular hydrolysis of nucleoside 5′-tri- and di-phosphates. At present a uniform nomenclature has not yet been developed. They are referred to as ecto-apyrase or also E-ATPase family (Plesner, 1995; Zimmermann, 1996a, 1996b) and ecto-phosphodiesterase/pyrophosphatase, PC-1, or

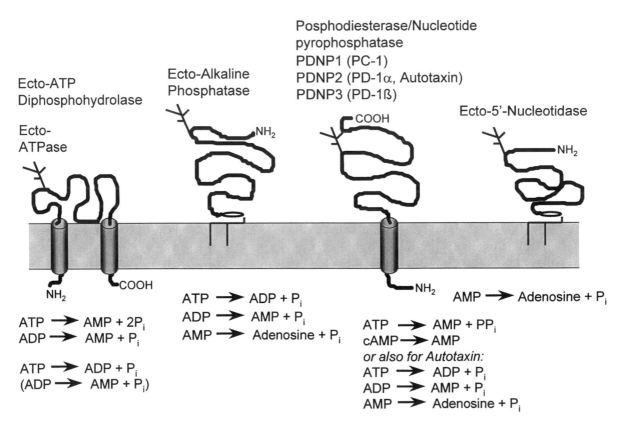

Fig. 1. Predicted membrane topography and demonstrated catalytic properties of key enzymes involved in the extracellular hydrolysis of nucleotides. Ecto-5′-nucleotidase and alkaline phosphatase are anchored to the plasma membrane via a glycosylphosphatidyl inositol anchor. Some of the enzymes occur as dimers or possibly also trimers. The enzymes hydrolyze not only adenine nucleotides (as shown) but all purine and pyrimidine nucleotides.

also phosphodiesterase/nucleotide pyrophosphatase (PDNP) family (Goding et al., 1998). Alkaline phosphatases represent an additional enzyme family and catalyze the entire hydrolysis chain from the nucleoside 5′-trisphosphate to the nucleoside. One of the major enzymes in terminating the hydrolysis chain by catalyzing the hydrolysis of nucleoside 5′-monophosphates is thought to be ecto-5′-nucleotidase.

The alkaline phosphatase family

Alkaline phosphatase represents an ecto-enzyme with a broad substrate specificity. It is capable of releasing inorganic phosphate from a variety of organic compounds including the degradation of ATP, ADP, AMP, and other nucleotides. One single enzyme could thus catalyze the entire hydrolysis chain from the nucleoside 5′-triphosphate to the respective nucleoside. Alkaline phosphatase comprises a group of glycosylphosphatidyl inositol (GPI)-anchored glycoproteins with molecular masses in the order of 70 kDa. At least four genes encode alkaline phosphatase isoenzymes in humans (Whyte et al., 1995). Very little is known concerning the distribution and functional role of alkaline phosphatase in the nervous system. As revealed by enzyme histochemistry, activity of alkaline phosphatase is specifically associated with endothelial cells of microvessels forming the blood-brain barrier and absent from other vessels. It is thus regarded as a marker for the blood-brain barrier. Conditioned media of glial cell lines can induce alkaline phosphatase activity in cultured artery endothelial cells (Takemoto et al., 1994). Alkaline phosphatase was also found in association with select neurons in the guinea-pig small intestine (Song et al., 1994).

Ecto-5′-nucleotidase

Catalytic properties and phylogenetic relationship

Ecto-5′-nucleotidase catalyzes the final hydrolysis step from the nucleoside 5′-monophosphate to the nucleoside. The molecular structure of this GPI-anchored plasma membrane-associated enzyme has been studied in detail (references in Zimmermann, 1992; 1996a, 1996b; Resta et al., 1998). The primary structures were determined for a number of vertebrate species including man. The protein occurs mainly as a dimer and the apparent molecular weight of the monomer ranges from 62 to 74 kDa. The K_m value is in the lower micromolar range. Ecto-5′-nucleotidase is a zinc binding metalloenzyme and metal ion chelators such as EDTA inhibit enzyme activity. It has a wide tissue distribution and may occur also in a soluble form originating from cleavage of the GPI-anchor by GPI-specific phospholipase C (Vogel et al., 1992).

At present only one ecto-5′-nucleotidase gene is known in vertebrates. Related enzymes are found in Arthropods and also in Archaea and Eubacteria (Fig. 2). Interestingly, the sequence of the putative 5′-nucleotidase from *Haemophilus influenzae* is more closely related to the arthropod and vertebrate sequences than to those of other bacteria. The substrate specificities of the distantly related bacterial enzymes are quite different from those of vertebrates. The periplasmatic 5′-nucleotidase encoded by the *ushA* gene of *E. coli* for example hydrolyzes nucleoside 5′-tri-, 5′-di-, and 5′-monophosphates as well as nucleotide diphosphate sugars (Glaser et al., 1967; Neu, 1968). Other bacterial enzymes hydrolyze 3′-nucleotides or 2′,3′ cyclic phosphodiesters (Zimmermann, 1992). 5′-Nucleotidase-related enzymes, found in the saliva of blood-sucking arthropods are phylogenetically more closely related to the mammalian enzymes. However, the catalytic properties of the 5′-nucleotidase-like enzyme from the tick *Boophilus microplus* are rather similar to those of the *E. coli* enzyme. This enzyme hydrolyzes nucleoside 5′-tri-, 5′-di-, and 5′-mono-phosphates as well as UDP glucose (Willadsen et al., 1993). In this context it is of interest that ATP and ADP still bind to the catalytic site of mammalian ecto-5′-nucleotidase but are no longer hydrolyzed. They act as competitive inhibitors with K_i-values in the low micromolar range or even below. But the phylogenetic relation reaches even further. There are three sequence motifs conserved between a variety of enzymes capable of phosphoester bond cleavage, the "phosphoesterase signature motif" (Koonin, 1994; Zhuo et al., 1994). The motive is not only present in vertebrate ecto-5′-nucleotidase, the ecto-

5'-nucleotidase-related enzymes, or bacterial diadenosine tetraphosphatase but also in a variety of phosphoesterases including Ser/Thr phosphoprotein phosphatases and sphingomyelin phosphomonoesterases (Blanchin-Roland et al., 1986; Zhuo et al., 1994).

Distribution in the nervous system

In the adult central nervous system the enzyme appears to be mainly associated with astrocytes or peripheral Schwann cells (Grondal et al., 1988)

whereas it is transiently associated with migrating neurons and developing synaptic contacts in the immature nervous system (Schoen et al., 1988; Fenoglio et al., 1995) (for a review see Zimmermann, 1996b). In the retina it is associated with Müller cells (Kreutzberg and Hussain, 1982; Braun et al., 1995) and in the olfactory system with the dark/horizontal basal cells at the basal side of the olfactory epithelium and with microvillar cells dispersed at the luminal side of the epithelium (Braun and Zimmermann, 1998).

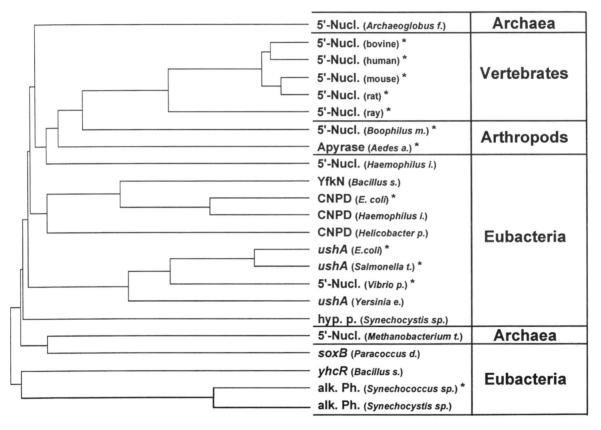

Fig. 2. Hypothetical phylogenetic tree revealing the relatedness between members of the ecto-5'-nucleotidase family. The dendrogram was constructed using the CLUSTAL program included in the PC/GENE software package (IntelliGenetics, Mountain View, USA). The Genbank™ accession numbers for the ecto-5'-nucleotidase family are as follows: putative 5'-nucleotidase (5'-Nucl.) of *Archaeoglobus fulgidus*, AE001044; bovine 5'-Nucl., D14541; human 5'-Nucl., X55740; mouse 5'-Nucl., L12059; rat 5'-Nucl., J05214; electric ray (*Discopyge ommata*) 5'-Nucl., X62278; 5'-Nucl. of *Boophilus microplus*, U80634; soluble apyrase of *Aedes aegypti*, L12389; putative 5'-Nucl. of *Haemophilus influenzae*, U32705; gene product YfkN *of Bacillus subtilis*, D83967; 2',3'-cyclic-nucleotide 2'-phosphodiesterase (CNPD) of *E.coli*, *Haemophilus influenzae*, and *Heliobacter pylori*, M13464, U32740, and AE000532, respectively; periplasmic UDP-sugar hydrolase gene (*ushA*) of *E. coli*, *Salmonella typhimurium*, and *Yersinia enterocolitica*, X03895, X04651, and U46859, respectively; 5'-Nucl. of *Vibrio parahaemolyticus*, X57711; hypothetical protein (hyp. p.) of *Synechocystis spec.*, D64005; 5'-Nucl. of *Methanobacterium thermoautotrophicum*, AE000909; *soxB* gene of *Paracoccus denitrificans*, X79242; hypothetical protein (hyp. p.) of *Bacillus subtilis*, X96983, atypical alkaline phosphatase of *Synechococcus sp.* (strain: PCC7942), M77507; alkaline phosphatase of *Synechocystis sp.* (strain:PCC6803), D90902. * Catalytic activity determined.

Functional roles in the nervous system

Ecto-5′-nucleotidase plays a critical role in the extracellular hydrolysis cascade of nucleotides and in neural development. The enzyme is transiently expressed at the surface of developing nerve cells and at synapses during synapse development and remodeling (Schoen and Kreutzberg, 1994, 1995). In the kitten occipital cortex ecto-5′-nucleotidase carries the cell adhesion epitope HNK-1 during a transient period of synapse formation by the thalamic afferents (Vogel et al., 1993). The enzyme plays a crucial role in the neuritic differentiation and survival of PC12 cells and rat cultured cerebellar granule cells. Coating coverslips with ecto-5′-nucleotidase leads to a significant increase in the length of NGF-induced neuritic processes. Conversely, application to the culture medium of antibodies that inhibit activity of ecto-5′-nucleotidase results in a time-dependent decrease in NGF-induced neurite formation by PC12 cells (Heilbronn and Zimmermann, 1995). Similarly, the selective inhibition of ecto-5′-nucleotidase synthesis by application of antisense-oligonucleotides causes a dramatic reduction of the ability of PC12 cells or cerebellar granule cells for neurite formation. The effect can be relieved by the addition of soluble 5′-nucleotidase to the medium (Heilbronn et al., 1995). Like other ecto-enzymes 5′-nucleotidase can display functional properties in addition to its catalytic activity. The protein binds to extracellular matrix proteins such as the laminin/nidogen complex and fibronectin (Stochaj et al., 1989, 1990; Stochaj and Mannherz, 1992) and may therefore be involved in cell/matrix interactions.

Recent evidence from our laboratory further supports the notion that ecto-5′-nucleotidase is involved in neural development (Kohring and Zimmermann, 1998). Ecto-5′-nucleotidase becomes upregulated when a neural phenotype is induced in the human neuroblastoma cell line SH-SY5Y by all *trans*-retinoic acid or phorbol-12-myristate-13-acetate. Both agents initiate a reduction in proliferation and an increase in differentiation and polarity of SH-SY5Y cells. Non-specific ecto-phosphatase activity which is also present at the surface of these cells remains unaltered. Northern hybridization experiments reveal that the increase in ecto-5′-nucleotidase activity is due to increased enzyme synthesis rather than due to enzyme activation.

A catalytically active and soluble form of ecto-5′-nucleotidase is now available for drug screening, or protein binding assays (Servos et al., 1998). The baculovirus system was used to express the glycosylated form of the protein in insect cells. Glutathione *S*-transferase (GST) as a fusion partner was included into the expressed protein. The GST can be used to immobilize the protein on agarose-coupled reduced glutathione for sensing molecular interactions. The catalytic properties of the recombinant protein correspond to those of the native protein isolated from animal tissues. A first series of screening experiments suggests that antagonists of P2 receptors as well as inhibitors of ecto-nucleotidases can inhibit also ecto-5′-nucleotidase activity.

The E-NTPase (E-ATPase) family

Catalytic properties and phylogenetic relationship

Members of this family hydrolyze either nucleoside-5′-triphosphates or both nucleoside-5′-triphosphates and nucleoside-5′-diphosphates. Where investigated these enzymes hydrolyze not only ATP or ADP but a large variety of purine and pyrimidine nucleoside 5′-tri- or diphosphates. The family would therefore more appropriately be addressed as E-NTPase rather than as E-ATPase family. Initially, primary structures were obtained for an ATP diphosphohydrolase (ecto-apyrase) (Maliszewski et al., 1994; Christoforidis et al., 1995; Kaczmarek et al., 1996; Wang and Guidotti, 1996; Marcus et al., 1997) that can hydrolyze either ATP or ADP and an ecto-ATPase (Kegel et al., 1997; Kirley, 1997) with a high preference for ATP hydrolysis. Both were shown to be cell surface-located enzymes. The primary structures of ecto-apyrase from mouse (Maliszewski et al., 1994) and rat (Wang et al., 1997) share 75% sequence identity with human ecto-apyrase (also referred to as CD39). The open reading frames of rat ecto-ATPase and rat ecto-apyrase encode 495 (54.4 kDa) and 511 (57.4 kDa) amino acid residues, respectively with an overall amino acid identity of 40% (Kegel et al., 1997; Wang et al., 1997). They represent integral membrane proteins with a single putative transmembrane domain at the N- and the C-terminus and with seven and six potential N-glycosylation sites, respectively. Ecto-apyrase and ecto-ATPase are members of a larger family of nucleotidases that occur also in plants and protozoans. Not all of them are surface-located, some of the enzymes are soluble. They share four major sequence motifs referred to as apyrase conserved regions (Handa and Guidotti, 1996; Kegel et al., 1997).

Apyrase-related proteins are widely distributed but have not been found in bacteria. They were identified in protozoans, yeast, plants and animals (Fig. 3). Multiple sequence alignment of 24 members of the family depicts four major groups that may be further subdivided. The central group (group 2) contains three closely related enzymes. These include ecto-apyrase (CD39-related) that revealed (where analyzed) ATPase as well ADPase activity (CD39 human and rat) (Kaczmarek et al., 1996; Wang and Guidotti, 1996; Wang et al., 1997), and ecto-ATPase (CD39L1) that showed a high

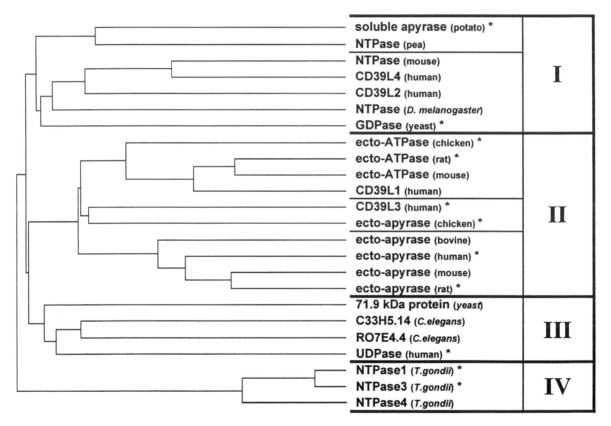

Fig. 3. Hypothetical phylogenetic tree revealing the relatedness between members of the ecto-apyrase or E-NTPase (E-ATPase) family. The dendrogram was constructed as indicated in Fig. 2. The Genbank™ accession numbers for the ecto-apyrase family are as follows: potato soluble apyrase, U58597; nucleoside triphosphatase (NTPase) of the garden pea, Z32743; putative nucleotidase (NTPase) of mouse, AF006482; human gene product CD39L4, AF039918; human gene product CD39L2, AF039916; putative nucleotidase (NTP) of *Drosophila melanogaster*, AF041048; yeast Golgi localized GDPase, L19560; ecto-ATPase of chicken, rat, mouse, and human (CD39L1), U74467, Y11835, AF042811, and U91510, respectively; human ecto-ATPase/apyrase CD39L3, AF039917 (similar sequence HB6, AF034840); chicken ecto-apyrase similar to CD39, AF041355; ecto-apyrase (CD39) of bovine, human, mouse, and rat, AF005940, S73813, AF037366, and U81295, respectively; yeast hypothetical 71.9 kDa protein (YER005W), U18778; gene products C33H5.14 and R07E4.4 of *Caenorhabditis elegans*, U41007 and U39652, respectively; human Golgi-localized UDPase, AF016032; nucleotidases NTPase1, NTPase3 and NTPase4 of *Toxoplasma gondii*, U14322, U14324, and U28353. * Catalytic activity determined.

preference for ATP hydrolysis in the examples investigated (chicken ecto-ATPase, rat ecto-ATPase) (Kegel et al., 1997; Kirley, 1997). Another closely related and recently sequenced human enzyme (CD39L3, HB6) (Chadwick and Frischauf, 1998; Smith and Kirley, 1998) reveals considerable sequence identity to both, ecto-apyrase and ecto-ATPase. When expressed in COS-1 cells (Smith and Kirley, 1988) it hydrolyzes ATP and ADP at a ratio of about 3 : 1 and thus appears to be also a functional intermediate between ecto-apyrase and ecto-ATPase. It is closely related to the recently cloned chicken oviduct ecto-ATP diphosphohydrolase (Nagy et al., 1998) suggesting that the diversity in molecular forms is conserved between vertebrate groups. All of these enzymes have a single predicted transmembrane domain at the N- and the C-terminus and are expected to be membrane anchored.

Group I contains as one of the subgroups plant apyrases such as the well characterized soluble potato apyrase and the nucleoside triphosphatase form the garden pea (Handa and Guidotti, 1996). The second subgroup consists a blend of interesting members. These include recently cloned and functionally as yet unidentified proteins from human (CD39L2, CD39L4), mouse, and *Drosphila* (Chadwick and Frischauf, 1997, 1998; Chadwick et al., 1998) as well as a more distantly related guanosine diphosphatase located in the Golgi apparatus of yeast (Abeijon et al., 1993). The yeast enzyme has a (predicted) uncleaved hydrophobic leader sequence. In contrast to the proteins referred to above the human proteins CD39L2 and CD39L4 have a single predicted transmembrane segment at the N-terminus and no hydrophobic domain at the C-terminus. Cleavage of the N-terminal hydrophobic domain appears possible and might result in the formation of a soluble and secreted protein.

A recently cloned member of group 3 is a human UDP preferring nucleotidase allocated to the Golgi apparatus (Wang and Guidotti, 1998). It has predicted transmembrane domains at the N- and C-terminus, respectively and is related to proteins of unknown functional properties in yeast and *C. elegans*.

The final group (4) comprises the distantly related but well characterized soluble nucleotide triphosphatases of the protozoan parasite *Toxoplama gondii* (Bermudes et al., 1994; Asai et al., 1995). Like some of the other members of the protein family they lack the hydrophobic C-terminus and possess a cleaved N-terminus. Further functional studies are required in particular for the mammalian members of the family to define their functional properties as well as their cellular location.

Distribution in the nervous system

As revealed by Northern-hybridization or RT-PCR and Southern analysis ecto-apyrase and ecto-ATPase occur in many tissues including the brain (Kaczmarek et al., 1996; Kegel et al., 1997; Wang et al., 1997). Ecto-apyrase can also be detected in immunoblots of primary cultures of neurons and astrocytes (Wang et al., 1997). Immunohistochemical data suggest that apyrase has a very broad distribution in rat brain including hippocampus and cerebral cortex. Pyramidal cells of the CA1 subfield and granule cells of the dentate gyrus were labeled. The antibody revealed no cross-reaction with heterologously expressed human ecto-ATPase (Wang et al., 1997). At present a potential cross-reactivity with other related rat sequences cannot be excluded. The expression in the human brain of interesting additional candidates has recently been demonstrated by Northern hybridization. These include CD39L3 which might act as an ecto-enzyme in the hydrolysis of both ATP and ADP as well as the functionally undefined CD39L2 (Chadwick and Frischauf, 1998). The latter is of particular interest since the mature protein may be soluble. The stimulation-induced release from nerve endings of soluble nucleotidases with undefined molecular structure has been reported (Todorov et al., 1997).

Functional roles

The molecular and functional properties of rat-derived ecto-ATPase and ecto-ATP diphosphohydrolase can be directly compared after heterologous expression in CHO cells (Heine et al., 1999). Sequence-specific polyclonal antibodies differentiate between the two proteins and reveal identical molecular weights of 70-80 kDa. The broad migration in polyacrylamide gels of the two proteins implicates a high yield of glycosylation. Both enzymes are stimulated by either Ca^{2+} or Mg^{2+} and reveal a broad substrate specificity towards purine- and pyrimidine nucleotides (including ATP, GTP, UTP, ITP, and CTP). They should therefore be addressed more accurately as ecto-nucleoside triphosphatase (E-NTPase) or ecto-nucleotide diphosphohydrolase (E-NTPDase). Both the dependence on Ca^{2+} and Mg^{2+} and the broad substrate specificity correspond to those of unidentified ecto-nucleotidases observed in many previous experiments with cultured cell of various origin (Zimmermann et al., 1996a, 1996b). Whereas ecto-apyrase hydrolyzes nucleoside 5'-diphosphates at a rate approximately 20–30% lower than nucleoside 5'-triphosphates, ecto-ATPase hydrolyzes nucleoside 5'-diphosphates only to a marginal extent. The sensitivity of the two enzymes to the inhibitors of P2 receptors suramin, PPADS and reactive blue differs. Evans blue has a strong inhibitory effect only on ecto-apyrase whereas suramin is a more potent inhibitor of ecto-ATPase. These results suggest that it might be possible to develop inhibitors with different potency on ecto-ATPase and ecto-apyrase.

Interestingly, ecto-apyrase expressed in CHO cells dephosphorylates ATP directly to AMP. These results are very similar to those obtained with an apyrase preparation previously purified to near homogeneity from porcine zymogen granule membranes (Laliberte and Beaudoin, 1983). In contrast, hydrolysis of ATP by ecto-ATPase leads to the accumulation of extracellular ADP as an intermediate product in the assay medium. Coexpression of ecto-ATPase and ecto-apyrase results in the sequential formation of ADP and AMP in the assay medium. These data suggest that on hydrolysis of ATP by ecto-apyrase the γ- and β-phosphates are cleaved at the identical site or that the intermediate ADP is transferred to a nearby second catalytic site without release from the protein. Our results are of relevance for interpreting data on extracellular ATP hydrolysis obtained from cultured cells or superfused organ preparations (Zimmermann, 1996a, 1996b). There, the degradation of ATP is followed by sequential appearance of

ADP, AMP, and finally adenosine in the medium. Since the accumulation of ADP cannot be accounted for by activity ecto-apyrase alone, another enzyme with properties corresponding to ecto-ATPase must be coexpressed in these cells. Coexpression of ecto-ATPase and ecto-apyrase in PC12 cells has been demonstrated by Northern-hybridization (Kegel et al., 1997). That ecto-apyrase and ecto-ATPase are coexpressed in situ is supported by a further piece of evidence. Both purified ecto-apyrase (Christoforidis et al., 1995; Picher et al., 1996) and ecto-apyrase expressed in CHO cells reveal very similar rates of ATP and ADP hydrolysis. But the rates of ATP hydrolysis in intact cellular systems is generally about twice as high as that of ADP hydrolysis (Zimmermann, 1996a, 1996b, Wang et al., 1997) This suggests that in situ an ecto-ATPase is coexpressed with ecto-apyrase increasing the hydrolysis rate of ATP.

Upregulation of the ecto-nucleotidase chain on ischemia

An increase in 5'-nucleotidase as revealed by enzyme histochemistry has previously been observed in the developing and damaged brain. Following postganglionic axotomy in the rat superior cervical ganglion 5'-nucleotidase is expressed on the plasma membrane of Schwann cells, proliferating satellite cells and fibroblasts (Nacimiento and Kreutzberg, 1990). Similarly, transection of the facial nerve causes a substantial proliferation of microglial cells in the facial nucleus with induction of 5'-nucleotidase. 5'-Nucleotidase on microglial cells can be demonstrated both by enzyme cytochemistry (Kreutzberg and Barron, 1978; Kreutzberg et al., 1986) and by immunohistochemistry (Schoen et al., 1992). On lesion-induced sprouting within the adult rat dentate gyrus 5'-nucleotidase is detected by enzyme cytochemistry not only on microglia and astrocytes. It is transiently present also at synapses that are related to the regenerative sprouting responses and synaptogenesis in this system (Schoen and Kreutzberg, 1994). It thus appears that 5'-nucleotidase is expressed on reactive glial as well as neural cells. We therefore analyzed the distribution and activity patterns of nucleotide-hydrolyzing ecto-enzymes following cerebral ischemia in the rat.

In a first series of experiments the effect of permanent middle cerebral artery occlusion (MCAO) on the reactive expression of the last enzyme of the hydrolysis cascade, of ecto-5′-nucleotidase, was investigated in rat brain (Braun et al., 1997). Activity of 5′-nucleotidase was analyzed by enzyme histochemistry and immunohistochemistry between 6 h and 7 days after vessel occlusion. The experiments reveal that ischemia following permanent MCAO results in an upregulation of the capacity for the extracellular hydrolysis of AMP and the extracellular formation of adenosine within the tissue adjacent to the infarcted volume. Increased 5′-nucleotidase activity was mainly associated with glial cells. At the rim of the necrotic tissue a strong colocalization was observed of ecto-5′-nucleotidase staining and accumulation of microglia (antibody OX42). First alterations in immunoreactive astrocytes were observed after 6 h but the further increase in 5′-nucleotidase activity occurred over a period of days. The results also imply that nucleotides may be released following ischemia for an extended period of time and that release may not be restricted to AMP but may include also other nucleotides such as ATP.

In a subsequent study we induced transient forebrain ischemia by bilateral common carotid artery occlusion and hypertension in rats, and probed for the expression of the entire ecto-nucleotidase chain from ATP to adenosine (Braun et al., 1998). Northern-hybridization using poly-adenylated RNA isolated from hippocampi 7 days after transient forebrain ischemia or sham operation revealed an increase in ecto-apyrase and ecto-5′-nucleotidase expression. This suggested that the entire ecto-nucleotidase chain for the hydrolysis of ATP to adenosine was upregulated. The tissue distribution of the nucleotidases was then analyzed in cryosections by enzyme cytochemistry 6 h, and 2, 3, 7, 14, and 28 days following ischemia. This was compared to the distribution of markers for astrocytes and microglia such as glial fibrillary acidic protein (GFAP), vimentin, and complement receptor type 3 (antibody OX42). For all substrates analyzed by enzyme cytochemistry (ATP, UTP, ADP, AMP) a parallel and strong postischemic increase in staining intensity could be demonstrated in the lesioned regions of the hippocampus. These included the pyramidal cell layer of CA1, the stratum lacunosum-moleculare, and the dentate gyrus. The development of staining paralleled that of neural cell death. It was significant after 2 days and persisted after 28 days. An analysis of the immunostaining for glial markers revealed a close similarity between the pattern of enzyme staining and that of markers for microglia. These data suggest that the increased expression of nucleotidases is mainly associated with activated glia, mainly microglia, in the regions of damaged nerve cells. Interestingly, $P2X_7$ receptor-immunoreactive microglia was found after MCAO in the penumbral region around an area of necrosis (Collo et al., 1997). Since the expression of the $P2X_7$ receptor by activated microglia can confer susceptibility to ATP-mediated cytotoxicity (Ferrari et al., 1997), the high levels of ecto-nucleotidase expression may represent a mechanism of self protection.

The upregulation of the ecto-nucleotidase chain implicates an ischemia-induced increased and sustained cellular release of nucleotides. The upregulation could have several functional implications. It could support the formation of extracellular adenosine as a neuroprotective agent and facilitate purine salvage. It could also relieve any potential cytotoxic, Ca^{2+}-mediated effects of ATP on neurons or glial cells. The data lend further support to a functional role of ATP and other nucleotides in cerebral physiology and pathology and a related regulatory function of ecto-nucleotidases.

The phosphodiesterase/pyrophosphatase or PDNP family

Catalytic properties and phylogenetic relationship

The members of the family of phosphodiesterase/pyrophosphatases have an even broader spectrum of substrate specificities (for a review see Goding et al., 1998). They possess phosphodiesterase as well as nucleotide pyrophosphatase activity which are properties of the same enzyme molecule. Members of the enzyme family are capable of hydrolyzing 3′,5′-cAMP to AMP, ATP to AMP and PP$_i$, ADP to AMP and P$_i$, or NAD$^+$ to AMP and nicotinamide mononucleotide. The first member of the enzyme family identified in molecular terms was the membrane glycoprotein PC-1 (5'-nucleotide phosphodiesterase) (Goding et al., 1998). Three members of the family have now been cloned from human sources and related sequences are present in mouse and rat (Stracke et al., 1992; Narita et al., 1994; Deissler et al., 1995; Jin-Hua et al., 1997; Scott et al., 1997). Individual members of the PDNP family have been given different names. For simplicity the gene symbol (*PDNP*) assigned to the human proteins (Jin-Hua et al., 1997) is commonly used here to denominate the various types of enzymes. They are referred to as PC-1 (PDNP1), PD-Iα/autotaxin (PDNP2) and PD-Iβ, B10, gp130[RB13-6] (PDNP3). Human PDNP-Iα and autotaxin presumably represent splice variants of the same gene (Murata et al., 1994; Kawagoe et al.,

1995). Additional, presumably alternatively spliced isoforms have been described (Fuss et al., 1997). All members of the PDNP family have apparent molecular masses in the order of 110–120 kDa and appear to occur in a membrane-bound as well as in a soluble, proteolytically cleaved form (Belli et al., 1993; Clair et al., 1997). Of the three presently cloned mammalian members of the PDNP family, PDNP1 and PDNP3 are more closely related to each other than to PDNP2 (PC-1) (Fig. 4).

In addition to their complex nucleotide hydrolyzing capacity the mammalian members of the PDNP family have a number of additional common features. They contain the somatomedin B-like domain of vitronectin. The consensus sequence of the EF-hand putative calcium-binding region which is essential for enzymatic activity (Belli et al., 1994; Kawagoe et al., 1995; Goding et al., 1998), is

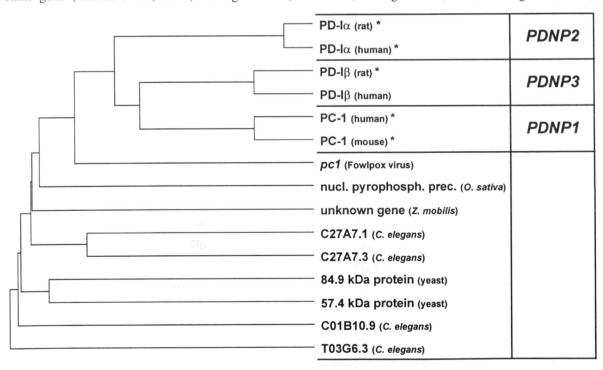

Fig. 4. Hypothetical phylogenetic tree revealing the relatedness between members of the ecto-phosphodiesterase/pyrophosphatase or PDNP family. The dendrogram was constructed as indicated in Fig. 2. The Genbank™ accession numbers for the PDNP family are as follows: rat PD-Iα, D28560; human PD-Iα, D45421 (similar related sequence of human autotaxin, L46720); rat PD-Iβ, D30649 (similar sequences U78787 and U78788); human PD-Iβ, AF005632; human PC-1, D12485; mouse PC-1 using the second methionin as initiator, J02700; Fowlpox virus gene *pc1*, AJ006408; nucleotide pyrophosphatase precuser of *Oryza sativa*, U25430; unknown gene of *Zymomonas mobilis*, AF060218; hypothetical proteins C27A7.1 and C27A7.3 of *Caenorhabditis elegans*, Z81041; hypothetical 84.9 kDa and 57.4 kDa yeast proteins, P25353 and P39997, respectively; and hypothetical proteins C01B10.9 and T03G6.3 of *Caenorhabditis elegans*, U58757 and U40940, respectively. * Catalytic activity determined.

present in its complete form in PC-1 (PDNP1) and with minor alterations also in the other mammalian PDNPs. The RGD-tripeptide which is recognized by several integrins is contained in PDNP2 and PDNP3 but not in PDNP1. The functional role of these motifs is not yet understood but they imply a functional Ca^{2+}-dependency and a potential interaction with cell surface receptors.

A search in Genbank™ revealed functionally undefined relatives of the mammalian members of the PDNP family in bacteria, yeast, *C. elegans* and rice (Fig. 4). The PC-1 gene is contained in the genome of the Fowlpox virus. All these sequences lack the somatomedin B-like domain of vitronectin, the consensus sequence of the EF-hand putative calcium-binding region, and the RGD-tripeptide. At the active site threonine (Goding et al., 1998) a consensus sequence ($TXPX_6TG$) is detected in all sequences analyzed. A second ubiquitously present consensus sequence which may also be related to catalytic activity is NX_5DHG (N^{365} in human PC-1).

Distribution and functional roles in the nervous system

The enzymes have a broad tissue distribution. PC-1 and PDNP2 (PD-Iα/autotaxin) occur also in the nervous system. Immunocytochemical data on mouse tissues using a monoclonal antibody against PC-1 suggest that the enzyme is associated with blood capillaries in the brain only (Harahap and Goding, 1988). Isoforms of PDNP2 are highly expressed in brain and spinal cord (Kawagoe et al., 1995; Fuss et al., 1997). The mRNA is located to oligodendrocytes and the choroid plexus. PDNP2 may play an important role in oligodendrocyte function. The expression during development reveals an intermediate peak around the time of active myelination and the dysmyelinating mouse mutant *jimpy* has decreased levels of mRNA. Furthermore, mRNA levels are reduced in the CNS at onset of clinical symptoms in experimental autoimmune encephalomyelitis (Fuss et al., 1997). PDNP2 (autotaxin) has surprising catalytic properties. In addition to its phosphodiesterase and pyrophosphatase activities it is capable of catalyzing the hydrolysis also of ATP or GTP to ADP or GDP, respectively. It also hydrolyzes AMP to adenosine (Belli et al., 1993; Clair et al., 1997). This suggests that not only ecto-5′-nucleotidase but also PDNP2 can catalyze the hydrolysis of nucleoside 5′-monophosphates in nervous tissue. By its ability to hydrolyze cAMP it could provide adenosine from exogenous cAMP, a pathway that exists in the brain (Brundege et al., 1997).

PDNP2 (autotaxin) becomes autophosphorylated but reveals no significant protein kinase activity towards a number of protein kinase substrates. GTP, NAD, AMP and PP_i all can serve as phosphate donors in the phosphorylation. Another intriguing feature of PDNP2 is its ability to promote tumor cell motility (Murata et al., 1994). Motility stimulation requires an intact 5′-nucleotide phosphodiesterase active site but the carbohydrate side chains are not involved. (Stracke et al., 1995; Lee et al., 1996). Both isoforms of PDNP2 (PD-Iα and autotaxin) are expressed in human neuroblastoma tumor tissues (Kawagoe et al., 1997).

Summary and outlook

It has become clear that several families of ecto-nucleotidases function in the nervous system. They differ in their catalytic and also other functional properties and consist of several members each. At present neither the functional diversity nor the apparent redundancy of the ecto-nucleotidase pathway is understood. But the notion is further supported that nucleotidergic signaling pathways are widely distributed in the nervous system and that the ecto-nucleotidases play a significant part in it. The functional properties of the presently known ecto-nucleotidases and also of the sequenced but as yet uncharacterized potential ecto-nucleotidases need to be further investigated. An evaluation of their role in defined signaling pathways requires that the cellular location of the enzymes is determined. Knock-out mice in which individual ecto-nucleotidases have been deleted from the germline promise to be important tools to unravel their functional roles in the nervous system and also in other tissues.

Acknowledgements

This study was supported by grants from the Deutsche Forschungsgemeinschaft (SFB 269, A4), by the European Community (BMH4-CT96-0676) and the Fonds der Chemischen Industrie.

References

Abeijon, C., Yanagisawa, K., Mandon, E.C., Häusler, A., Moremen, K., Hirschberg, C.B. and Robbins, P.W. (1993) Guanosine diphosphatase is required for protein and sphingolipid glycosylation of *Saccharomyces cerevisiae*. *J. Cell. Biol.*, 122: 307–323.

Asai, T., Miura, S., Sibley, L.D., Okabayashi, H. and Takeuchi, T. (1995) Biochemical and molecular characterization of nucleoside triphosphate hydrolase isozymes from the parasitic protozoan *Toxoplasma gondii*. *J. Biol. Chem.*, 270: 11391–11397.

Bailly, Y., Schoen, S.W., Delhaye-Bouchaud, N., Kreutzberg, G.W. and Mariani, J. (1995) 5'-Nucleotidase activity as a synaptic marker of parasagittal compartmentation in the mouse cerebellum. *J. Neurocytol.*, 24: 879–890.

Belli, S.J., van Driel, I.R. and Goding ,J.W. (1993) Identification and characterization of soluble form of the plasma cell membrane glycoprotein PC-1 (5'-nucleotide phosphodiesterase). *Eur. J. Biochem.*, 217: 421–428.

Belli, S.I., Sali, A. and Goding, J.W. (1994) Divalent cations stabilize the conformation of plasma cell membrane glycoprotein PC-1 (alkaline phosphodiesterase I). *Biochem. J.*, 304: 75–80.

Bermudes, D., Peck, K.R., Afifi, M.A., Beckers, C.J.M. and Joiner, K.A. (1994) Tandemly repeated genes encode nucleoside triphosphate hydrolase isoforms secreted into the parasitophorous vacuole of *Toxoplasma gondii*. *J. Biol. Chem.*, 269: 29252–29260.

Blanchin-Roland, S., Blanquet, S., Schitter, J.-M. and Fayat, G. (1986) The gene for *Echerichia coli* diadenosine tetraphosphatase is located immediately clockwise to *fol*A and forms an operon with *ksg*A. *Mol. and Gen. Genet.*, 205: 515–522.

Braun, N., Brendel, P. and Zimmermann, H. (1995) Distribution of 5'-nucleotidase in the developing mouse retina. *Brain. Res. Dev. Brain. Res.*, 88: 79–86.

Braun, N., Lenz, C., Gillardon, F., Zimmermann, M. and Zimmermann, H. (1997) Focal cerebral ischemia enhances glial expression of 5'-nucleotidase. *Brain. Res.*, 766: 213–226.

Braun, N., Zhu, Y., Krieglstein, J., Culmsee, C. and Zimmermann, H. (1998) Upregulation of the enzyme chain hydrolyzing extracellular ATP following transient forebrain ischemia in the rat. *J. Neurosci.*, 18: 4891–4900.

Braun, N. and Zimmermann, H. (1998) Association of ecto-5'-nucleotidase with specific cell types in the adult and developing rat olfactory organ. *J. Comp. Neurol.*, 393: 528–537.

Brundege, J.M., Diao, L.H., Proctor, W.R. and Dunwiddie, T.V. (1997) The role of cyclic AMP as a precursor of extracellular adenosine in the rat hippocampus. *Neuropharmacology*, 36: 1201–1210.

Buell, G., Collo, G. and Rassendren, F. (1996) P2X receptors: An emerging channel family. *Eur. J. Neurosci.*, 8: 2221–2228.

Chadwick, B.P. and Frischauf, A.M. (1997) Cloning and mapping of a human and mouse gene with homology to ecto-ATPase genes. *Mamm. Genome*, 8: 668–672.

Chadwick, B.P. and Frischauf, A.M. (1998) The CD39-like gene family: Identification of three new human members (CD39L2, CD39L3 and CD39L4), their murine homologues and a member of the gene family from *Drosophila melanogaster*. *Genomics*, 50: 357–367.

Chadwick, B.P., Williamson, J., Sheer, D. and Frischauf, A.M. (1998) cDNA cloning and chromosomal mapping of a mouse gene with homology to NTPases. *Mamm. Genome*, 9: 162–164.

Christoforidis, S., Papamarcaki, T., Galaris, D., Kellner, R. and Tsolas, O. (1995) Purification and properties of human placental ATP diphosphohydrolase. *Eur. J. Biochem.*, 234: 66–74.

Collo, G., Neidhart, S., Kawashima, E., Kosco-Vilbois, M., North, R.A. and Buell, G. (1997) Tissue distribution of the P2X$_7$ receptor. *Neuropharmacology*, 36: 1277–1283.

Cunha, R.A., Sebastiao, A.M. and Ribeiro, J.A. (1998) Inhibition by ATP of hippocampal synaptic transmission requires localized extracellular catabolism by ecto-nucleotidases into adenosine and channeling to adenosine A(1) receptors. *J.Neurosci.*, 18: 1987–1995.

Clair, T., Lee, H.Y., Liotta, L.A. and Stracke, M.L. (1997) Autotaxin is an exoenzyme possessing 5'-nucleotide phosphodiesterase/ATP pyrophosphatase and ATPase activities. *J. Biol. Chem.*, 272: 996–1001.

Deissler, H., Lottspeich, F. and Rajewsky, M.F. (1995) Affinity purification and cDNA cloning of rat neural differentiation and tumor cell surface antigen gp130[RB13-6] reveals relationship to human and murine PC-1. *J. Biol. Chem.*, 270: 9849–9855.

Dunwiddie, T.V., Abbracchio, M.P., Bischofberger, N., Brundege, J.M., Buell, G., Collo, G., Corsi, C., Diao, L., Kawashima, E., Jacobson, K.A., Latini, S., Lin, R.C.S., North, R.A., Pazzagli, M., Pedata, F., Pepeu, G.C., Proctor, W.R., Rassendren, F., Surprenant, A. and Cattabeni, F. (1996) Purinoceptors in the central nervous system. *Drug. Develop. Res.*, 39: 361–370.

Dunwiddie, T.V., Diao, L.H. and Proctor, W.R. (1997) Adenine nucleotides undergo rapid, quantitative conversion to adenosine in the extracellular space in rat hippocampus. *J. Neurosci.*, 17: 7673–7682.

Fenoglio, C., Scherini, E., Vaccarone, R. and Bernocchi, G. (1995) A re-evaluation of the ultrastructural localization of 5'-nucleotidase activity in the developing rat cerebellum, with a cerium-based method. *J. Neurosci. Methods*, 59: 253–263.

Ferrari, D., Chiozzi, P., Falzoni, S., Dal Susino, M., Collo, G., Buell, G. and Di Virgilio, F. (1997) ATP-mediated cytotoxicity in microglial cells. *Neuropharmacology*, 36: 1295–1301.

Fredholm, B.B., Arslan, C., Kull, B., Kontny, E. and Svennings-son, P. (1996) Adenosine (P1) receptor signalling. *Drug. Develop. Res.*, 39: 262–268.

Fuss, B., Baba, H., Phan, T., Tuohy, V.K. and Macklin, W.B. (1997) Phosphodiesterase I, a novel adhesion molecule and/ or cytokine involved in oligodendrocyte function. *J. Neurosci.*, 17: 9095–9103.

Glaser, L., Melo, A. and Paul, R. (1967) Uridine diphosphate sugar hydrolase. Purification of enzyme and protein inhibitor. *J. Biol. Chem.*, 242: 1944–1954.

Goding, J.W., Terkeltaub, R., Maurice, M., Deterre, P., Sali, A. and Belli, S.I. (1998) Ecto-phosphodiesterase/pyrophospha-tase of lymphocytes and non-lymphoid cells: Structure and function of the PC-1 family. *Immunol. Rev.*, 161: 11–26.

Grondal, E.J.M., Janetzko, A. and Zimmermann, H. (1988) Monospecific antiserum against 5'-nucleotidase from *Torpedo* electric organ: Immunocytochemical destribution of the enzyme and its association with Schwann cell membranes. *Neuroscience*, 24: 351–363.

Handa, M. and Guidotti, G. (1996) Purification and cloning of a soluble ATP-diphosphohydrolase (apyrase) from potato tubers (*Solanum tuberosum*). *Biochem. Biophys. Res. Commun.*, 218: 916–923.

Harahap, A.R. and Goding, J.W. (1988) Distribution of murine plasma cell antigen PC-1 in non-lymphoid tissues. *J. Immunol.*, 141: 2317–2320.

Heilbronn, A. and Zimmermann, H. (1995) 5'-Nucleotidase activates and an inhibitory antibody prevents neuritic differentiation of PC12 cells. *Eur. J. Neurosci.*, 7: 1172–1179.

Heilbronn, A., Maienschein, V., Carstensen, C., Gann, W. and Zimmermann, H. (1995) Crucial role of 5'-nucleotidase in differentiation and survival of developing neural cells. *Neuroreport*, 7: 257–261.

Heine, P., Kegel, B., Heilbronn, A. and Zimmermann, H. (1999) Functional characterization of rat ecto-ATPase and ecto-ATP diphosphohydrolase after heterologous expression in CHO cells. *Eur. J. Biochem.* in press.

Jin-Hua, P., Goding, J.W., Nakamura, H. and Sano, K. (1997) Molecular cloning and chromosomal localization of PD-I*b* (PDNP3), a new member of the human phosphodiesterase I genes. *Genomics*, 45: 412–415.

Kaczmarek, E., Koziak, K., Sévigny, J., Siegel, J.B., Anrather, J., Beaudoin, A.R., Bach, F.H. and Robson, S.C. (1996) Identification and characterization of CD39 vascular ATP diphosphohydrolase. *J. Biol. Chem.*, 271: 33116–33122.

Kawagoe, H., Soma, O., Goji, J., Nishimura, N., Narita, M., Inazawa, J., Nakamura, H. and Sano, K. (1995) Molecular cloning and chromosomal assignment of the humans brain-type phosphodiesterase I/nucleotide pyrophosphatase gene (*PDNP2*). *Genomics*, 30: 380–384.

Kawagoe, H., Stracke, M.L., Nakamura, H. and Sano, K. (1997) Expression and transcriptional regulation of the PD-Ia/autotaxin gene in neuroblastoma. *Cancer. Res.*, 57: 2516–2521.

Kegel, B., Braun, N., Heine, P., Maliszewski, C.R. and Zimmermann, H. (1997) An ecto-ATPase and an ecto-ATP

diphosphohydrolase are expressed in rat brain. *Neuropharmacology*, 36: 1189–1200.

Kidd, E.J., Miller, K.J., Sansum, A.J. and Humphrey, P.P.A. (1998) Evidence for P2X$_3$ receptors in the developing rat brain. *Neuroscience*, 87: 533–539.

Kirley, T.L. (1997) Complementary DNA cloning and sequencing of the chicken muscle Ecto-ATPase—Homology with the lymphoid cell activation antigen CD39. *J. Biol. Chem.*, 272: 1076–1081.

Kittel, A., Siklós, L., Thuróczy, G. and Somosy, Z. (1996) Qualitative enzyme histochemistry and microanalysis reveals changes in ultrastructural distribution of calcium and cal-cium-activated ATPases after microwave irradiation of the medial habenula. *Acta. Neuropath.*, 92: 362–368.

Koonin, E.V. (1994) Conserved sequence pattern in a wide variety of phosphoesterases. *Protein. Sci.*, 3: 356–358.

Kohring, K. and Zimmermann, H. (1998) Upregulation of ecto-5'-nucleotidase in human neuroblastoma SH-SY5Y cells on differentiation by retinoic acid or phorbolester. *Neurosci. Lett.*, 258: 127–130.

Kreutzberg, G.W. and Barron, K.D. (1978) 5'-Nucleotidase of microglial cells in the facial nucleus during axonal reaction. *J. Neurocytol.*, 7: 601–610.

Kreutzberg, G.W. and Hussain, S.T. (1982) Cytochemical heterogeneity of the glial plasma membrane: 5'-nucleotidase in retinal Müller cells. *J. Neurocytol.*, 11: 53–64.

Kreutzberg, G.W., Heymann, D. and Reddington, M. (1986) 5'-Nucleotidase in the nervous system. In: G.W. Kreutzberg, M. Reddington and H. Zimmermann (Eds.), *Cellular Biology of Ectoenzymes*, Springer, Berlin, pp. 147–164..

Laliberte, J.F. and Beaudoin, A.R. (1983) Sequential hydrolysis of the γ- and β-phosphate groups of ATP by the ATP diphosphohydrolase from pig pancreas. *Biochim. Biophys. Acta*, 742: 9–15.

László, I. and Bodor, T. (1981) Localization of thiamine pyrophosphatase activity in motor end plates. *J. Histochem. Cytochem.*, 29: 658–662.

Lee, H.Y., Clair, T., Mulvaney, P.T., Woodhouse, E.C., Azna-voorian, S., Liotta, L.A. and Stracke, M.L. (1996) Stimulation of tumor cell motility linked to phosphodiester-ase catalytic site of autotaxin. *J. Biol. Chem.*, 271: 24408–24412.

Maliszewski, C.R., DeLepesse, G.J.T., Schoenborn, M.A., Armitage, R.J., Fanslow, W.C., Nakajima, T., Baker, E., Sutherland, G.R., Poindexter, K., Birks, C., Alpert, A., Friend, D., Gimpel, S.D. and Gayle, R.B. (1994) The CD39 lymphoid cell activation antigen: Molecular cloning and structural characterization. *J. Immunol.*, 153: 3574–3583.

Marcus, A.J., Broekman, M.J., Drosopoulos, J.H.F., Islam, N., Alyonycheva, T.N., Safier, L.B., Hajjar, K.A., Posnett, D.N., Schoenborn, M.A., Schooley, K.A., Gayle, R.B. and Malis-zewski, C.R. (1997) The endothelial cell ecto-ADPase responsible for inhibition of platelet function is CD39. *J. Clin. Invest.*, 99: 1351–1360.

Mata, M. and Fink, D.J. (1989) Ca^{++}-ATPase in the central nervous system: An EM cytochemical study. *J. Histochem. Cytochem.*, 37: 971–980.

Miledi, R., Molenaar, P.C. and Polak, R.L. (1984) Acetylcholinesterase activity in intact and homogenized skeletal muscle of the frog. *J. Physiol. (Lond.)*, 349: 663–686.

Murata, J., Lee, H.J., Clair, T., Krutzsch, H.C., Arestad, A.A., Sobel, M.E., Liotta, L.A. and Stracke, M.L. (1994) cDNA cloning of the human motility-stimulating protein, autotaxin, reveals a homology with phosphodiesterases. *J. Biol. Chem.*, 269: 30479–30484.

Nacimiento, W. and Kreutzberg, G.W. (1990) Cytochemistry of 5′-nucleotidase in the superior cervical ganglion of the rat: Effects of pre- and postganglionic axotomy. *Exptl. Neurol.*, 109: 362–373.

Nagy, A.K., Knowles, A.F. and Nagami, G.T. (1998) Molecular cloning of the chicken oviduct ecto-ATP-diphosphohydrolase. *J. Biol. Chem.*, 273: 16043–16049.

Narita, M., Goji, J., Nakamura, H. and Sano, K. (1994) Molecular cloning, expression and localization of a brain-specific phosphodiesterase I/nucleotide pyrophosphatase (PD-Iα) from rat brain. *J. Biol. Chem.*, 269: 28235–28242.

Neu, H.C. (1968) The 5′-nucleotidases (uridine diphosphate sugar hydrolases) of the *Enterobacteriaceae*. *Biochemistry*, 7: 3766–3773.

Pappas, G.D. and Kriho, V. (1988) Fine structural localization of Ca^{2+}-ATPase activity at the frog neuromuscular junction. *J. Neurocytol.*, 17: 417–423.

Picher, M., Sévigny, J., D'Orléans-Juste, P. and Beaudoin, A.R. (1996) Hydrolysis of P$_2$-purinoceptor agonists by a purified ectonucleotidase from the bovine aorta, the ATP- diphosphohydrolase. *Biochem. Pharmacol.*, 51: 1453–1460.

Plesner, L. (1995) Ecto-ATPases: Identities and functions. *Int. Rev. Cytol.*, 158: 141–214.

Resta, R., Yamashita, Y. and Thompson, L.F. (1998) Ecto-enzyme and signaling functions of lymphocyte CD73. *Immunol. Rev.*, 161: 95–109.

Schoen, S.W. and Kreutzberg, G.W. (1994) Synaptic 5′-nucleotidase activity reflects lesion-induced sprouting within the adult dentate gyrus. *Exp. Neurol.*, 127: 106–118.

Schoen, S.W. and Kreutzberg, G.W. (1995) Evidence that 5′-nucleotidase is associated with malleable synapses—An enzyme cytochemical investigation of the olfactory bulb of adult rats. *Neuroscience*, 65: 37–50.

Schoen, S.W., Graeber, M.B., Tóth, L. and Kreutzberg, G.W. (1988) 5′-Nucleotidase in postnatal ontogeny of rat cerebellum: A marker for migrating nerve cells? *Develop. Brain. Res.*, 39: 125–136.

Schoen, S.W., Graeber, M.B. and Kreutzberg, G.W. (1992) 5′-Nucleotidase immunoreactivity of perineuronal microglia responding to rat facial nerve axotomy. *Glia*, 6: 314–317.

Scott, L.J., Delautier, D., Meerson, N.R., Trugnan, G., Goding, J.W. and Maurice, M. (1997) Biochemical and molecular identification of disctinct forms of alkaline phosphodiesteraseI expressed on the apical and basolateral plasma membrane surfaces of rat hepatocytes. *Hepatology*, 25: 995–1002.

Servos, J., Reiländer, H. and Zimmermann, H. (1998) Catalytically active soluble ecto-5′-nucleotidase purified after

heterologous expression as a tool for drug screening. *Drug. Develop. Res.*, 45: 269–276.

Smith, T.M. and Kirley, T.L. (1998) Cloning, sequencing and expression of a human brain ecto-apyrase related to both the ecto-ATPases and CD39 ecto-apyrases. *Biochim. Biophys. Acta*, 28: 65–78.

Song, Z.M., Brookes, S.J.H. and Costa, M. (1994) Characterization of alkaline phosphatase-reactive neurons in the guinea-pig small intestine. *Neuroscience*, 63: 1153–1167.

Sperlagh, B., Kittel, A., Lajtha, A. and Vizi, E.S. (1995) ATP acts as fast neurotransmitter in rat habenula: Neurochemical and enzymecytochemical evidence. *Neuroscience*, 66: 915–920.

Stochaj, U. and Mannherz, H.G. (1992) Chicken gizzard 5′-nucleotidase functions as a binding protein for the laminin/nidogen complex. *Eur. J. Cell. Biol.*, 59: 364–372.

Stochaj, U., Dieckhoff, J., Mollenhauer, J., Cramer, M. and Mannherz, H.G. (1989) Evidence for the direct interaction af chicken gizzard 5′-nucleotidase with laminin and fibronectin. *Biochim. Biophys. Acta*, 992: 385–392.

Stochaj, U., Richter, H. and Mannherz, H.G. (1990) Chicken gizzard 5′-nucleotidase is a receptor for the extracellular matrix component fibronectin. *Eur. J. Cell. Biol.*, 51: 335–338.

Stracke, M.L., Krutzsch, H.C., Unsworth, E.J., Arestad, A.A., Cioce, V., Schiffmann, E. and Liotta, L.A. (1992) Identification, purification and partial sequence analysis of autotaxin, a novel motility-stimulating protein. *J. Biol. Chem.*, 267: 2524–2529.

Stracke, M.L., Arestad, A., Levine, M., Krutzsch, H.C. and Liotta, L.A. (1995) Autotaxin is an *N*-linked glycoprotein but the sugar moieties are not needed for its stimulation of cellular motility. *Melanoma. Res.*, 5: 203–209.

Takemoto, H., Kaneda, K., Hosokawa, M., Ide, M. and Fukushima, H. (1994) Conditioned media of glial cell lines induce alkaline phosphatase activity in cultured artery endothelial cells—Identification of interleukin-6 as an induction factor. *FEBS. Lett.*, 350: 99–103.

Thirion, S., Troadec, J.D. and Nicaise, G. (1996) Cytochemical localization of ecto-ATPses in rat neurohypophysis. *J. Histochem. Cytochem.*, 44: 103–111.

Thorn, J.A. and Jarvis, S.M. (1996) Adenosine transporters. *Gen. Pharmacol.*, 27: 613–620.

Todorov, L.D., MihaylovaTodorova, S., Westfall, T.D., Sneddon, P., Kennedy, C., Bjur, R.A. and Westfall, D.P. (1997) Neuronal release of soluble nucleotidases and their role in neurotransmitter inactivation. *Nature*, 387: 76–79.

Vogel, M., Kowalewski, H., Zimmermann, H., Hooper, N.M. and Turner, A.J. (1992) Soluble low-Km 5′-nucleotidase from electric ray (*Torpedo marmorata*) electric organ and bovine cerebral cortex is derived from the glycosyl–phosphatidylinositol-anchored ectoenzyme by phospholipase C cleavage. *Biochem. J.*, 284: 621–624.

Vogel, M., Zimmermann, H. and Singer, W. (1993) Transient association of the HNK-1 epitope with 5′-nucleotidase during development of the cat visual cortex. *Eur. J. Neurosci.*, 5: 1423–1425.

Wang, T.F. and Guidotti, G. (1996) CD39 is an ecto-(Ca^{2+},Mg^{2+})-apyrase. *J. Biol. Chem.*, 271: 9898–9901.

Wang, T.F. and Guidotti, G. (1998) Golgi localization and functional expression of human uridine diphosphatase. *J. Biol. Chem.*, 273: 11392–11399.

Wang, T.F., Rosenberg, P.A. and Guidotti, G. (1997) Characterization of brain ecto-apyrase: Evidence for only one ecto-apyrase (CD39) gene. *Mol. Brain. Res.*, 47: 295–302.

Whyte, M.P., Landt, M., Ryan, L.M., Mulivor, R.A., Henthorn, P.S., Fedde, K.N., Mahuren, J.D. and Coburn, S.P. (1995) Alkaline phosphatase: Placental and tissue-nonspecific isoenzymes hydrolyze phosphoethanolamine, inorganic pyrophosphate and pyridoxal 5′-phosphate—substrate accumulation in carriers of hypophosphatasia corrects during pregnancy. *J. Clin. Invest.*, 95: 1440–1445.

Willadsen, P., Riding, G.A., Jarmey, J. and Atkins, A. (1993) The nucleotidase of *Boophilus microplus* and its relationship to enzymes from the rat and *Escherichia coli*. *Insect. Biochem. Mol. Biol.*, 23: 291–295.

Zhuo, S., Clemens, J.C., Stone, R.L. and Dixon, J.E. (1994) Mutational analysis of a Ser/Thr phosphatase—Identification of residues important in phosphoesterase substrate binding and catalysis. *J. Biol. Chem.*, 269: 26234–26238.

Zimmermann, H. (1992) 5′-Nucleotidase—molecular structure and functional aspects. *Biochem. J.*, 285: 345–365.

Zimmermann, H. (1996a) Extracellular purine metabolism. *Drug. Develop. Res.*, 39: 337–352.

Zimmermann, H. (1996b) Biochemistry, localization and functional roles of ecto-nucleotidases in the nervous system. *Prog. Neurobiol.*, 49: 589–618.

P. Illes and H. Zimmermann (Eds.)
Progress in Brain Research, Vol 120
© 1999 Elsevier Science BV. All rights reserved

CHAPTER 31

Immunolocalization of ATP diphosphohydrolase in pig and mouse brains, and sensory organs of the mouse

A.R. Beaudoin*, G. Grondin and F.-P. Gendron

Département de biologie, Faculté des Sciences, Université de Sherbrooke, Sherbrooke (Québec), Canada J1K 2R1

Introduction

Extracellular nucleotides are released from mammalian cells by both exocytotic, and as yet undefined, non-exocytotic mechanisms. These nucleotides interact with specific purine and pyrimidine receptors localized on cell surfaces. Duration and intensity of these interactions are in great part determined by ectonucleotidases which metabolize these nucleotides into nucleosides. Several ectoenzymes participate in these catalytic functions including: ecto-ATPase, ecto-ATPDase, ecto-protein kinase, nucleotide pyrophosphohydrolase, alkaline phosphatase and 5′-nucleotidase (Zimmermann, 1996). The end product of these dephosphorylations is a nucleoside which in turn can interact with another class of membrane receptors. Since the discovery of the first mammalian ATPDase by LeBel et al. (1980) our laboratory was involved in the biochemical characterization of this particular family of enzymes (see Laliberté et al., 1982, 1983). So far, different isoforms have been demonstrated (Sévigny et al., 1995, 1996, 1997a). The third isoform (Picher et al., 1993) is not yet clearly established. All these isoforms are recognized by a polyclonal antibody named "Ringo". More specifically, this polyclonal antibody reacts with a common sequence shared by all the mammalian ATPDases and is believed to be part

of the catalytic site. A second enzyme, the ecto-ATPase, also appears to play an important role as an ectonucleotidase (Kegel et al., 1997; Kirley, 1997; Chadwick and Frischauf, 1997). Analysis of its encoding cDNA revealed a high level of homology with ATPDase (Kegel et al., 1997). However, in contrast to the latter enzyme, which catalyses the sequential hydrolysis of the γ and β phosphate residues of triphospho- and diphospho-nucleosides, the ecto-ATPase is mainly involved in the hydrolysis of the γ-phosphate residues of nucleotides. ATPDase was reported for the first time in the mammalian brain by Schadeck et al. (1989) who demonstrated its existence in the rat by an enzymatic approach. Since then, its presence in the brain has been confirmed in different species by Western and Northern blots (Kegel et al., 1997; Sévigny, 1997b). The purpose of this study was to localize the ATPDase in mouse and pig brains, and mouse sensory organs.

Materials and methods

Experiments were carried out on the brain of a 17.5 day old mouse and a 5 week old piglet. Freshly dissected pieces of brain were fixed overnight in 4% paraformaldehyde, 4% sucrose, buffered at pH 7.4 with 0.1 M of sodium cacodylate. Tissues were dehydrated in graded ethanol solutions and embedded in paraffin. Sections were cut at 4 μm thickness and mounted on polyionic slides (Superfrost Plus Fisher, Montréal, Canada). Paraffin was

---

*Corresponding author. Tel.: + 1 819 821 8049; e-mail: abeaudoi@courrier.usherb.ca

removed with xylene. Sections were rehydrated through graded concentrations of ethanol to water and rinsed in 150 mM NaCl, 0.1 M Tris, pH 7.5 (TBS). Slides were incubated for 10 min in TBS containing 0.1 M glycine and subjected to pressure cooker "heat-induced epitope retrieval" procedure, by incubating in 1 mM EDTA, 10 mM Tris, pH 8.0 for 9min (Miller and Estran, 1995). After a wash of 10 min in TBS, at room temperature, non-specific binding sites were blocked with 1% BSA and 1% fat-free skimmed milk in TBS for 30 min, at room temperature. Sections were incubated overnight at 4°C with the ATPDase antiserum or the preimmune serum (1:100), washed several times with TBS, and incubated with mouse monoclonal anti-rabbit IgG (conjugated to alkaline phosphatase) at a dilution of 1:100 for 2 h, at room temperature. After several washings in TBS, visualization was obtained by the alkaline phosphatase reaction with NBT/BCIP. Sections were mounted in 5% gelatin, 27% glycerin, and 0.1% sodium azide, preheated at 45oC. Photographs were taken under bright field illumination, using a Zeiss photomicroscope on Kodak T Max 100 film.

Results

As shown in Fig. 1A, strong immunoreactivity of the ATPDase antibody was observed in brain and sensory organs of the mouse embryo. A parasagital section shows reactions at the level of brain cortex (1C), in the periphery of the hippocampus region (2A, 2B), in choroid plexus (2C) and ventricular layer of neuroepithelium (2D). Sensory organs also displayed strong reactions: the inner ear (3A, 3B), vibrissae (4A, 4B), the eye (5A, 5B). Figure 1B is a control section incubated in parallel with pre-immune serum. A more detailed analysis of the distribution of the immunoreactive material follows: Fig. 1C is a higher magnification of the mouse brain cortex showing that the pia mater, the granula layer and the different layers of pyramidal cells, are all highly reactive. In contrast, the molecular layer and the underlying white matter are much less reactive. As mentioned above, the enzyme appears to be concentrated in a zone localized in the periphery of the hippocampal region. All the sensory organs exhibited strong

immunoreactivity. Figure 2A shows the inner ear, more specifically the ductus cochlearis, where both cartilaginous capsule and sensory epithelium are positive. At the level of vibrissae, the stratum corneum, granulosum, and germinativum were all strongly positive (Fig. 2B). In the eye, the inner and outer epithelia of the lid, cornea, lens, and retina layers were also highly reactive (Fig. 3). A horizontal section through the lower part of the brain, illustrated in Fig. 4, shows a concentration of the enzyme in structures identified as pia mater, ependyma, and choroid plexus. High magnification confirms the concentration of the enzyme in pia mater and in a zone corresponding to the granular layer.

The pig brain cortex was examined in parallel with the mouse brain. Figure 5 shows hematoxylin-eosin staining together with its immunolabeling reactivity. The pia mater and the different layers of pyramidal cells were all reactive, most particularly the granular layer as in the mouse. The entire cell body and dendrites of pyramidal cells were, also, positive. Figure 5C is a control section incubated in parallel with preimmun serum.

Discussion

There are several lines of evidence for the co-release of ATP and acetylcholine (Richardson and Brown, 1987), ATP and noradrenaline (Burnstock, 1986) and ATP and serotonin (Potter and White, 1980) in the central nervous system. ATP diph-osphohydrolase is presumably one of the main ectonucleotidases involved in the hydrolysis of this extracellular ATP. Our immunolocalizations demonstrate the presence of this enzyme in mouse and pig brains. Immunoreactive sites are concentrated in mouse cerebrum, more specifically cortex and hippocampal, as well as choroid plexus, ependyma, and cerebellum. In both mouse and pig cortices, the molecular layer is much less reactive than the different layers of pyramidal cells, and granular layer. The enzyme is associated with neurones and other cell types as well. Indeed, primary cultures of mouse astrocytes showed a level of ATPDase activity in the order of 150 nmoles/min/mg protein, at 37°C, which is comparable to the highest activities found in other mammalian tissues. Our immunohistochemical localizations are in good agreement with those of Wang et al. (1997) who recently reported the presence of an ecto-apyrase in primary neurones and astrocytes in cell culture. They are also in good agreement with immunohistochemical studies showing that this enzyme is widely distributed in rat brain, more specifically neurons of the cerebral cortex, hippocampus and cerebellum (Wang and Guidotti, 1998). In addition, our study reveals that in these different cell types the enzyme is not restricted to a specific portion of the plasma membrane, but rather uniformly spread all over the cell body and processes.

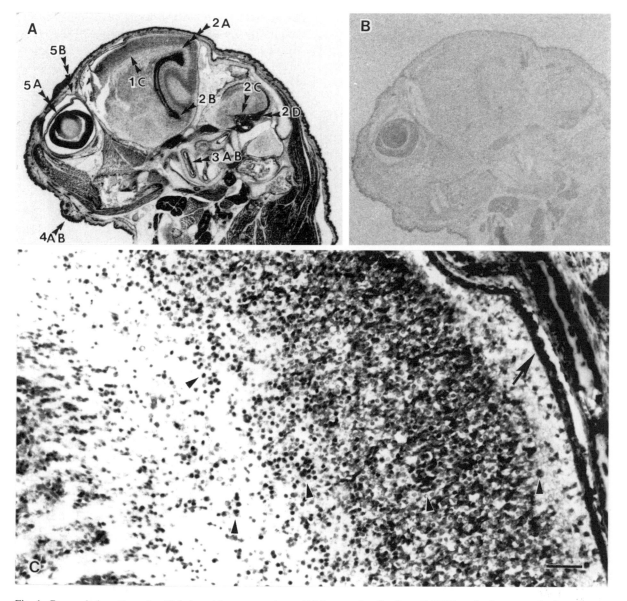

Fig. 1. Parasagital section of a 17.5 day old mouse embryo. (A) Immunolocalization of ATPDase in the mouse brain and sensory organs: (1C) Cortex; (2A,B) hippocampus; (2C) choroid plexus; (2D) ventricular layer of neuroepithelium; (3A,B) inner ear; (4A,B) vibrissae; (5A,B) Eye (11 ×). (B) Control section with preimmune serum (11 ×). (C) Cerebral cortex. ATPDase is localized in the pia mater (arrow), the granular layer and different layers of pyramidal cells (arrowheads). The molecular layer and white matter show little reactivity. Bar = 50 μm (200 ×).

390

We also report in this work particularly intense reactions in sensory organs. This observation, in addition to the detection of purinoceptors in these same organs, is indicative that ATP and perhaps other nucleotides are involved as neurotransmitters or neuromodulators of normal perception and nociception. As mentioned by Thorne et al. (1997) in the past few years considerable evidence has been accumulated for a complex signaling role for extracellular purines, particularly in the inner ear. In this respect, purinoceptors have been identified in neural, sensory and secretory structures of that organ. There is also evidence for endogenous release of ATP into the fluids and the presence of ectonucleotidases. These and other observations led these authors to speculate on a neuromodulatory and humoral role for extracellular ATP in cochlear function and hence on our sense of hearing.

One can wonder if variations in the expression of this enzyme are related to some pathologies in the brain. The co-localization of the ATPDase gene with the susceptibility gene involved in human partial epilepsy with audiogenic symptoms coincides with reports on the deficiency of ecto-apyrase

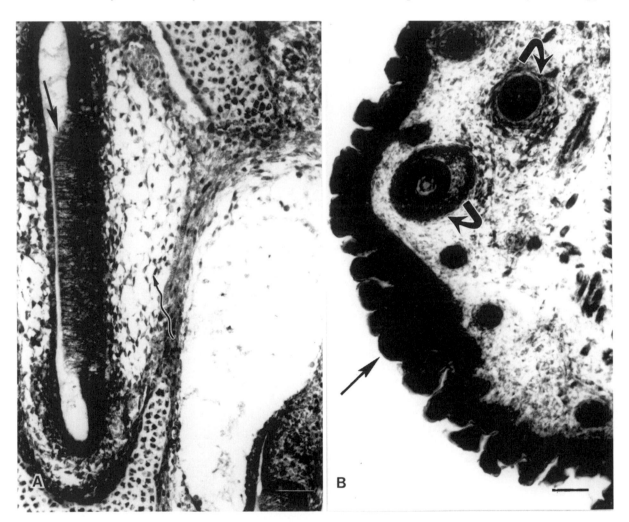

Fig. 2. Distribution of ectonucleotidases in mouse sensory organs. (A) Immunolocalization of ATPDase in the ductus cochlearis of the ear. Positive reaction in cartilaginous capsule (wavy arrow) and sensory epithelium (arrow) (200 ×). (B) Immunoreaction of ATPDase at the level of vibrissae (hair follicles). Stratum corneum, granulosum, germinativum (arrow) and vibrissae (curved arrow) are all strongly reactive. (200 ×). Bar = 50 μm.

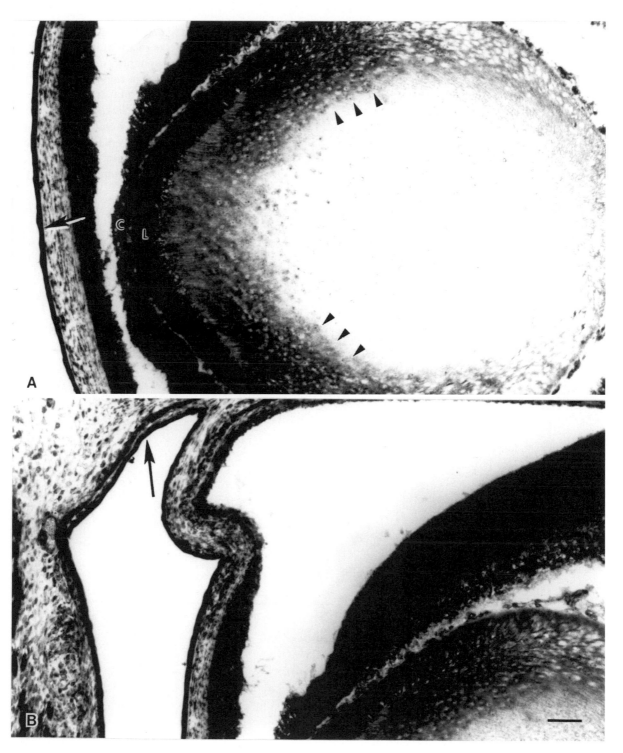

Fig. 3. Immunoreaction of ATPDase in the eye of mouse. (A) Corneal epithelium (C); lens epithelium (L), proliferative area of the lens (arrowheads) and skin (arrow) are positive (200×). (B) Upper conjunctiva is also positive (arrow) (200×). Notice the reaction at the level of skin. Bar = 50 μm.

activity in the brain of humans with temporal lobe epilepsy and in mice with audiogenic seizures (Wang et al., 1997). The high level of activity found in the hippocampus is of particular interest since it has been suggested that P_2-purinoceptors are involved in memory and learning (Inoue et al., 1996). Bonan et al. (1997) have found that aminotetrahydroacridine, a centrally active reversible acetylcholinesterase inhibitor which improves cognitive function in Alzheimer's patients, is also an ATPDase inhibitor. According to the latter authors, one of the effects of the drug would be to maintain ATP in the synaptic cleft. A direct implication of the ATPDase in memory and cognition is still, however, to be demonstrated.

From the physiological viewpoint, one can consider that ATPDase and other ectonucleotidases, like purinoceptors, would be operating for very short periods of time following the release of nucleotides. The most critical determinant of the

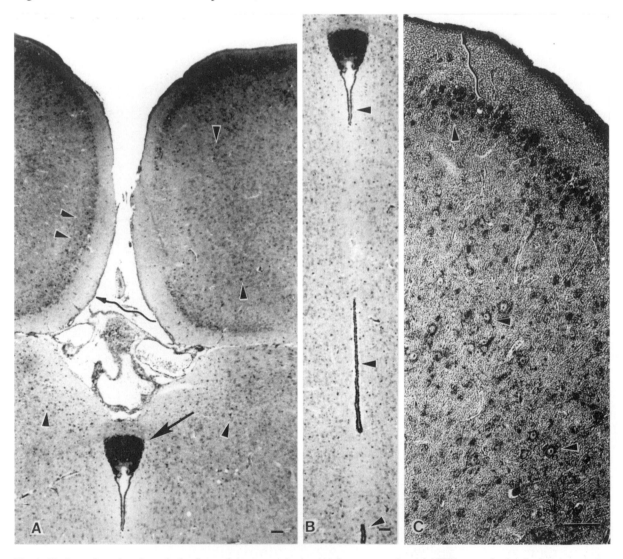

Fig. 4. Horizontal section through the floor of the mouse brain. (A) Immunoreaction of ATPDase at the level of pia mater (wavy arrow), cortex granula layer (arrowheads) and choroid plexus (arrow) (50×). (B) Same section showing immunoreaction at the level of ependyma and choroid plexus (arrowheads) (50×). (C) High magnification shows a detailed view at the different levels of cerebral cortex (arrowheads) pia mater (wavy arrow) (200×).

physiological action appears to be the nucleotide release processes. In addition to the well known process whereby nucleotides stored in synaptosomes are released by exocytosis, a release of ATP by cells devoid of secretory granules indicates the existence of an additional mechanism, yet to be defined. In this respect, it appears improbable that ATP is generated at the cell surface by an electrochemical gradient, since the extracellular concentration of ADP, required to prime such a reaction, is maintained at a very low level by the ectonucleotidases. It appears also improbable that ATP diffuses from an intracellular store to the extracellular medium via the cytosol since ATP would be in contact with enzymes with Km in the low micromolar range (protein kinases). This leads us to believe that there could be an organelle, or a protein, involved in the transport and release

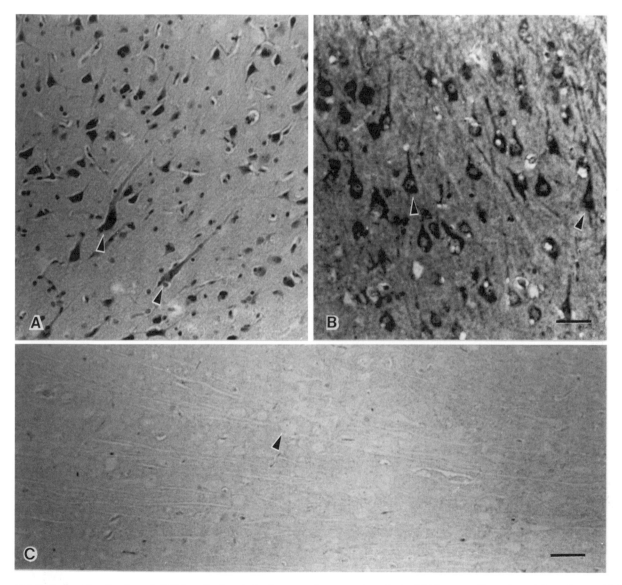

Fig. 5. Immunolocalization of ATPDase in the pig brain cortex (transversal section). (A) Hematoxylin-eosin staining of pyramidal cells (200×). (B) Immunoreaction of pyramidal cells (arrowheads). (C) Control: preimmune serum (200×). Bar = 50 μm.

process. In this respect, several years ago, we found a network of tubules in the pancreas acinar cell, identified as snake-like tubules. Ramifications of the network were found to stick to the mitochondrial surface, whereas other branches extented to the cell surface. This unusual configuration led us, at the time, to wonder if this network was involved in some form of energy transfer to the cell surface (Beaudoin et al., 1985). Morphologically similar tubules have been found in other cell types, including smooth muscle cells (Robinson et al., 1986) and hepatocytes. These tubules can accumulate quinacrine, a drug that binds to adenine nucleotides (see White et al., 1995, for an additionnal observation). This brings support to our speculation that snake-like tubules are involved in the transport and release of ATP.

In summary, ATPDase was observed in the cortex of mesencephalon and cerebellum. In the mouse, signals were particularly intense in anterior and posterior choroid plexus (ependyma) and in the periphery of hippocampus. In both pig and mouse cortices, ATPDase was concentrated in granular and pyramidal cell layers, whereas the molecular layer and white matter showed little reactivity. The enzyme was also found in high concentrations in mouse sensory organs: ear, vibrissae (hair follicles), nasal epithelium and eye.

Acknowledgements

This work was supported by grants from NSERC, of Canada, and le Fonds FCAR, Québec, Canada. We thank Francine Mongeau and Sheila McLean for their help in the preparation of the manuscript.

References

Beaudoin, A.R., Grondin, G., Lord, A. and Pelletier, M. (1985) β-NADPHase and TMPase positive "snake-like tubules" in the exocrine pancreas: Cytochemical and immunocytochemical studies. *J. Histochem. Cytochem.*, 33: 569–575.

Bonan, C.D., Battastini, A.M.O., Schetinger, M.R.C., Moreira, C.M., Frassetto, S.S., Dias, R.D. and Jarkis, J.J.F. (1997) Effects of 9-amino-1,2,3,4 tetrahydroacridine (THA) on ATPDiphosphohydrolase and 5′nucleotidase from rat brain synaptosomes. *Gen. Pharmacol.*, 28: 761–766.

Burnstock, G. (1986) The changing face of autonomic transmission. *Acta Physiol. Scand.*, 126: 76–91.

Chadwick, B.P. and Frischauf, A.M. (1997) Cloning and mapping of a human and mouse gene with homology to ecto-ATPase genes. *Mamm. Genome*, Sep., 8(9): 668–672.

Inoue, K., Koizumi, S. and Ueno, S. (1996) Implications of ATP receptors in brain functions. *Prog. Neurobiol.*, 50: 483–492.

Kegel, B., Braun, N., Heine, P., Maliszewski, C.R. and Zimmermann, H. (1997) An Ecto-ATPase and an Ecto-ATP diphosphohydrolase are expressed in rat brain. *Neuropharmacol.*, 36: 1189–1200.

Kirley, T.L. (1997) Complementary DNA cloning and sequencing of the chicken muscle exto-ATPase—homology with the lymphoid cell activation antigen CD39. *J. Biol. Chem.*, 272: 1076–1081.

Laliberté, J.-F., St-Jean, P. and Beaudoin, A.R. (1982) Kinetic effects of Ca^{2+} and Mg^{2+} on ATP hydrolysis by the purified ATP diphosphohydrolase. *J. Biol. Chem.*, 257: 3869–3874.

Laliberté, J.-F. and Beaudoin, A.R. (1983) Sequential hydrolysis of the γ- and β-phosphate groups of ATP by the ATP diphosphohydrolase from pig pancreas. *Biochim. Biophys. Acta*, 742: 9–15.

Lebel, D., Poirier, G.G., Phaneuf, S., St-Jean, P., Laliberté, J.-F. and Beaudoin, A.R. (1980) Characterization and purification of a calcium sensitive ATP diphospho-hydrolase from a pig pancreas. *J. Biol. Chem.*, 255: 1227–1233.

Miller, R.T. and Estran, C. (1995) Heat-induced epitope retrieval with a pressure cooker. *Appl. Immunohistochem.*, 3: 190–193.

Picher, M., Côté, Y.P., Béliveau, R., Potier, M. and Beaudoin, A.R. (1993) Demonstration of a novel type of ATP diphosphohydrolase in the bovine lung. *J. Biol. Chem.*, 7: 4699–4703.

Potter, P. and White, T.D. (1980) Release of adenosine and adenosine 5′ triphosphate from synaptosomes from different regions of rat brain. *Neuroscience*, 5: 1351–1356.

Richardson, P.J. and Brown, S.J. (1987) ATP release from affinity-purified cholinergic nerve terminals. *J. Neurochem.*, 48: 622–630.

Robinson, J.M., Okada, T., Castellot, J.J. and Karnosky, M.J. (1986) Unusual lysosomes in aortic smooth muscle cells. Presence in living and rapidly frozen cells. *J. Cell. Biol.*, 102: 1615–1622.

Schadeck, R.J.G., Sarkis, J.J.F., Dias, R.D., Araujo, H.M.M. and Souza, D.O.G. (1989) Synaptosomal apyrase in the hypothalamus of adult rats. *Brazilian J. Med. Biol. Res.*, 22: 303–314.

Sévigny, J., Côté, Y.P. and Beaudoin, A.R. (1995) Purification of pancreas Type I ATP diphosphohydrolase and identification by affinity labelling with 5′-p-fluorosulfonyl benzoyl adenosine ATP analogue. *Biochem. J.*, 312: 351–356.

Sévigny, J., Dumas, F. and Beaudoin, A.R. (1996) Purification identification by immunological techniques of different isoforms of mammalian ATP diphosphohydrolases. In: L. Plesner, T.L. Kirley and A.F. Knowles (Eds.), *Ecto-ATPases: Recent Progress on Structure and Function.* Plenum Publishing Corporation, New York. pp. 143–151.

Sévigny, J., Lévesque, F.P., Grondin, G. and Beaudoin, A.R. (1997a) Purification of the blood vessel ATP diphosphohydrolase. Identification and localization by immunological techniques. *Biochim. Biophys. Acta.*, 1334: 73–88.

Sévigny, J. (1997b) Purification, caractérisation et localisation des ATP diphosphohydrolases de mammifères et clonage de l'ADNc. Université de Sherbrooke. Ph.D. Thesis.

Thorne, P.R., Housley, G.D., Vlajkovic, S.M. and Munoz D.J.B. (1997) Ectonucleotidases and purinoceptors in the cochlea and their putative role in hearing. In: L. Plessner, T.L. Kirley and A.F. Knowles (Eds.), *Ecto-ATPases: Recent Progression Structure and Function.* Plenum Press. New York. pp. 239–246.

Wang, T.F. and Guidotti, G. (1998) Widespread expression of ecto-apyrase (CD39) in the central nervous system. *Brain Res.*, 790: 318–322.

Wang, T.F., Rosenberg, P.A. and Guidotti, G. (1997) Characterization of brain ecto-apyrase: Evidence for only one ecto-apyrase (CD39) gene. *Brain Res. Mol.*, 47: 295–302.

White, P.N., Thorne, P.R., Housley, G.D., Mockett, B.M., Billett, T.E. and Burnstock, G. (1995) Quinacrine staining of marginal cells in the stria vascularis of the guinea pig cochlea: A possible source of extracellular ATP? *Hear Res.*, 90: 97–105.

Zimmermann, H. (1996) Biochemistry, localization and functional roles of ecto-nucleotidases in the nervous system. *Prog. Neurobiol.*, 49: 589–618.

P. Illes and H. Zimmermann (Eds.)
Progress in Brain Research, Vol 120

CHAPTER 32

Diadenosine polyphosphates, extracellular function and catabolism

M. Teresa Miras-Portugal*, Javier Gualix, Jesús Mateo, Miguel Díaz-Hernández, Rosa Gómez-Villafuertes, Enrique Castro and Jesús Pintor

Departamento de Bioquímica, Facultad de Veterinaria, UCM, 28040 Madrid, Spain

Introduction

Diadenosine polyphosphates (Ap_nA $n=2–6$), are natural compounds that play an important role inside the cell in both the nucleus and at the cytosol (for review see McLennan, 1992.). Their levels are increased in cellular proliferation and stress situations resulting in stimulation of DNA duplication and DNA repair (Rapaport and Zamecnick, 1976; Baker and Jacobson, 1986; Baxi and Vishwanatha, 1995). Besides, Ap_nA can be strong inhibitors of the enzymes that balance the energetic phosphate groups among the cytosolic nucleotides, such as adenylate kinase and adenosine kinase (Leinhardt and Secemsky, 1973; Rotllan and Miras-Portugal, 1985). Recent findings indicate that the cytosolic face of some plasma membrane proteins exhibits regulatory sites modulated by various types of Ap_nA. Among the most important of these proteins are the ATP regulated K^+ channels, and the nucleoside transporters (Delicado et al., 1994; Jovanovic and Terzic, 1996; Ripoll et al., 1996).

The transport of Ap_nA to cytoplasmatic storage granules provides both the end of their intracellular actions, and at the same time the way to be released by controlled exocytosis (Gualix et al., 1996, 1997). The storage organella so far found to contain

Ap_nA, include the serotoninergic dense granules of platelets, the noradrenergic and adrenergic granules from adrenal medulla, and the acethylcholinergic vesicles from the *Torpedo* electric organ (Lüthje and Ogilvie, 1983; Rodriguez del Castillo et al., 1988; Pintor et al., 1991a, 1992a, 1992b; Schlüter et al., 1994).

For further interpretations it is relevant to consider the concentration levels expected once released to the extracellular media. These are in the low μM range in the surrounding area of platelets aggregation and at the vicinity of the exocytotic event in the cultured neurochromaffin cell model. More diluted concentrations of Ap_nA could be expected after it has diffused or undergone local enzymatic hydrolysis. Values could then be expected in the low nM range (Pintor et al., 1991b). Values in the nM range have also been detected in brain perfusion after amphetamine stimulation (Pintor et al., 1995). Binding studies and autoradiography indicate the existence of receptors with very high and high affinity (K_d values for Ap_4A in rat brain synaptic terminals of 0.1 nM and 0.57 μM respectively), which is concordant with this concentration range. The values in neurochromaffin cells from adrenal medulla and plasma membranes from the *Torpedo* are very similar (Pintor et al., 1991b, 1993; Walker et al., 1993; Rodríguez-Pascual et al., 1997).

From a functional point of view (and given to its chemical structure) Ap_nA could act through the

---

*Corresponding author. Tel.: +34-91-394-3894; fax: +34-91-394-3909; e-mail: mtmiras@eucmax.sim.ucm.es

ATP receptors, which are well characterised, and have been cloned and classified. Two main families of nucleotide receptors exist: $P2X_{1-7}$ and the $P2Y_{1-11}$ both of them with a still increasing number of components (Abbracchio and Burnstock, 1994; Boarder et al., 1995; Fredholm et al., 1997). Dinucleoside polyphosphates have proved to be agonists for some of these receptors, but considering the concentrations required and their physiological concentrations, these actions could perhaps be considered to be "pharmacological". This is the case for the action on the $P2X_1$ receptors of the vas deferens (Hoyle et al., 1995). Another possibility for the diadenosine polyphosphates is to behave as allosteric effectors. This is the case for the homomeric $P2X_2$ expressed in oocytes, where the Ap_5A activates the calcium entrance induced by ATP at low nM range (Pintor et al., 1996). The dinucleotide effect on P2Y metabotropic receptors has been documented in endothelial cells on a $P2Y_1$ and cloned $P2Y_2$ (Lazarowski et al., 1995; Mateo et al., 1996). The existence of dinucleotide specific metabotropic and ionotropic receptors can be deduced from the data reported below, although they have not yet been cloned. Recently, potentiation effects of Ap_5A on the Ca^{2+} signals mediated by P2Y receptors, stimulated by 2-MeS-ATP or UTP in cerebellar astrocytes have been reported, and it is worth noting that the concentrations required were in the nM concentration range (1–10 nM), and that potentiation effect lasted for almost 6 h. In this case the structure and second messenger pathway of this receptor is still unknown (Jimenez et al., 1998).

The ionotropic presynaptic dinucleotide receptor from rat midbrain synaptic terminals, also called the P4 receptor, is the best characterised of the ionotropic dinucleotide receptors, and most of the data presented in this report refers to it (Pintor and Miras-Portugal, 1995). In well characterised neural preparations such as the neurochromaffin cell cultures, only the cells containing noradrenaline exhibit ionotropic receptors for ATP, these correspond to the $P2X_2$ subtype, and the Ap_nA has neither an agonistic, nor an antagonistic effect (Castro et al., 1995). A similar situation occurs in the cultured Purkinje cells from immature rat brains, where the ionotropic receptors for ATP at the cell soma, that are mainly $P2X_2$, do not respond to Ap_nA (Mateo et al., 1998).

Dinucleotide receptors from rat midbrain synaptic terminals

Brain synaptic terminals, known as synaptosomes, are among the most frequently employed neural preparations when studying the functioning of the nervous system. Nevertheless, they are extremely heterogeneous, both in their neurotransmitter molecular species and the presence of presynaptic modulatory receptors. However, in spite of these drawbacks, and in the absence of any specific model, this preparation provides a first approach to the study of a substance expected to be a neurotransmitter or a neuromodulator.

In our studies, rat midbrain synaptic terminals were used because this area is very rich in aminergic and cholinergic terminals, and by analogy with the peripheral models, Ap_nA should be co-stored with them and ATP. In these synaptic terminals, Ap_5A, which is the best effector (Fig. 1A), was able to induce Ca^{2+} transients, which were not cross desensitised by ATP or its non-hydrolysable analogues. On the other hand, they were not blocked by any of the toxins available for the voltage dependent calcium channels (VDCC) and only modulated after the first initial calcium spike by the ω-conotoxin G-VI-A, which is an N-type VDCC inhibitor (Pintor and Miras-Portugal, 1995), similar ionotropic responses were found in guinea pig and deer mouse terminals (Pintor et al., 1995, 1997c; Pivorum and Nordone, 1996).

The diadenosine polyphosphates are not the only agonists for the presynaptic receptor; the guanine dinucleotides and the ethenoderivatives exhibit similar agonistic effects. Figure 2 shows the effect of various agonists, Ap_5A being the best natural agonist with an EC_{50} value of 55 μM. It is relevant to point out that both parameters, the affinity and the maximal calcium response should be considered. The dose response studies gave the following potency and effectiveness ranking: for the EC_{50} values $Gp_5G > Ap_5A = Ap_4A > Gp_4G = \varepsilon\text{-}Ap_4A = \varepsilon\text{-}Ap_5A > Gp_3G > Ap_3A = \varepsilon\text{-}Ap_3A$ and for the maximal values in calcium response $Ap_5A = Gp_5G = \varepsilon\text{-}Ap_5A > Ap_4A = Gp_4G = \varepsilon\text{-}Ap_4A >$

(A)

Ap₅A

(B)

Ip₅I

(C)

ε-Ap₅A

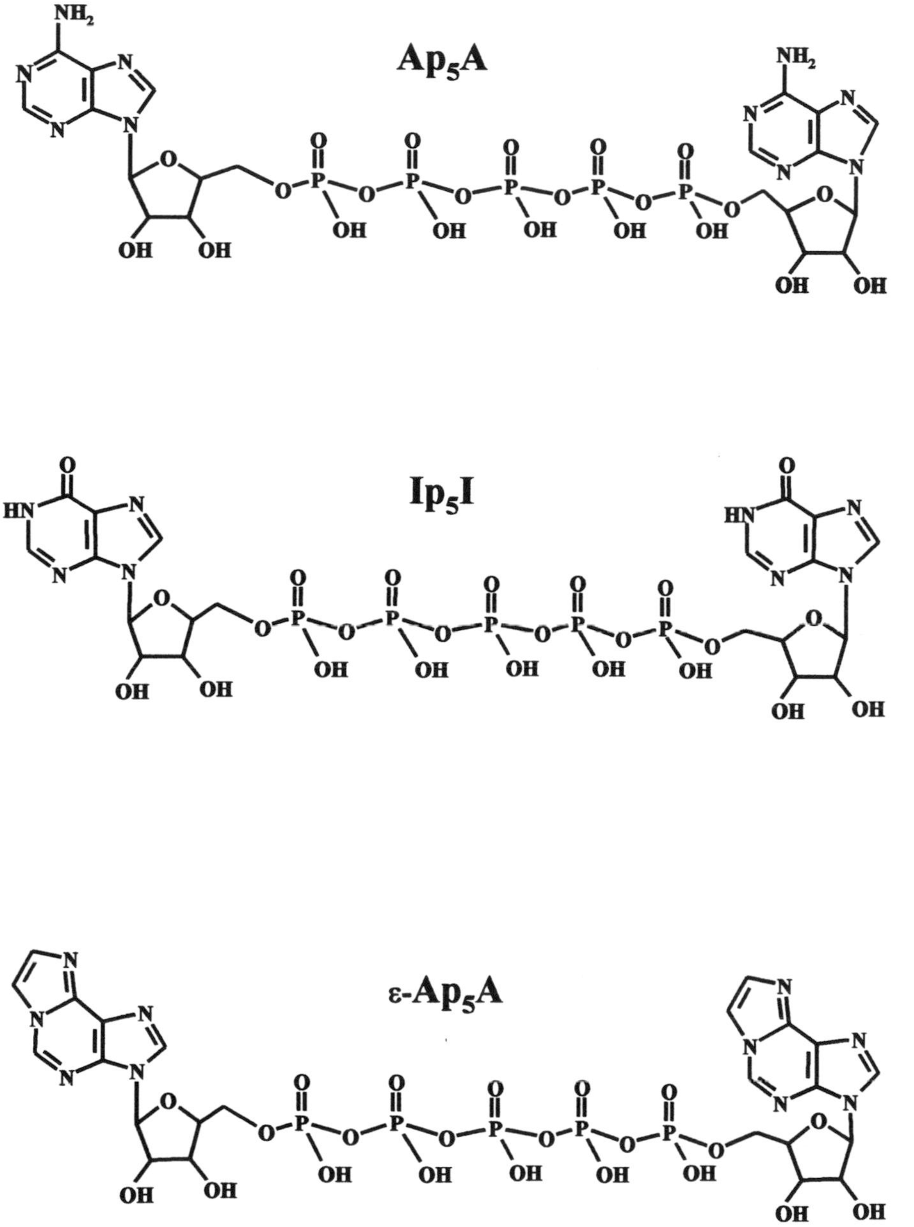

Fig. 1. Structure of the diadenosine pentaphosphate and main derivatives. (A) Diadenosine pentaphosphate. Natural agonist of dinucleotide-P4-receptors. Its presence has been described in the dense granules of platelets, chromaffin granules from adrenal medulla, and synaptic vesicles from rat brain and *Torpedo* electric organ. (B) Diinosine pentaphosphate. Antagonist of dinucleotide-P4-receptors. It is synthesised by the enzymatic action of adenylic acid deaminase from *Aspergillus sp.* (Pintor et al., 1997a). It has not yet been found as a natural compound, although it could be supposed. (C) Ethenoadenosine pentaphosphate. Fluorescent derivative employed in continuous fluorescent measurements of enzymatic hydrolysis. The etheno-derivatives are synthesised by condensation reaction with 2-Cl-acetaldehide (Rotllán et al., 1991).

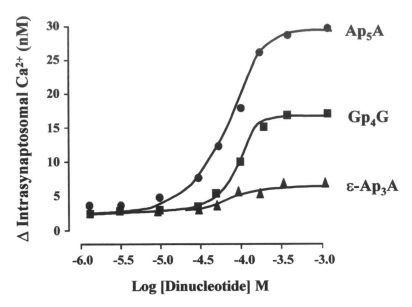

Fig. 2. Dose response curve of Ap$_5$A, and structural analogues. Concentration–response relationship for Ap$_5$A, Gp$_4$G and ε-Ap$_3$A on intracellular calcium increase in rat brain synaptosomes. Synaptosomes were prepared from midbrains of male Wistar rats that had been cervically dislocated and decapitated (Pintor and Miras-Portugal, 1995). Synaptosomal pellets containing 1 mg of protein were used. The cytosolic free calcium concentration was determined with fura-2 as described by Grynkiewicz et al. (1985).

Gp$_3$G = Ap$_3$A = ε-Ap$_3$A. Three main families appear to exist when ranked for maximal effect: Np$_5$N > Np$_4$N > Np$_3$N.

In this experimental model diinosine pentaphosphate (Fig. 1B), (which is enzymatically synthesised from Ap$_5$A by the action of the enzyme adenylic acid deaminase from *Aspergillus sp.*) proved to be an excellent inhibitor of this receptor (IC$_{50}$ = 4.2 nM), having an affinity 6,000 times higher than that exhibited with the ionotropic presynaptic ATP receptor, also described in this preparation (IC$_{50}$ = 27.7 μM) (Pintor et al., 1997a). In the vas deferens, where the ATP receptors are supposed to be mostly homomeric P2X$_1$, the Ip$_5$I shows a pA$_2$ value of 6,4 (which is roughly 0.2 μM) with respect to ATP (Hoyle et al., 1997). This differential inhibitory effect clearly confirms the existence of separate receptors for ATP and Ap$_n$A on the presynaptic terminals.

Regulation of dinucleotide receptors by protein kinases and phosphatases

A large number of metabotropic receptors have been found to be involved in the modulation of presynaptic terminals. This raises the question of how the dinucleotide ionotropic receptor (which is also presynaptic) could be influenced by this coexistence (Herrero et al., 1992, 1996). It should be borne in mind that Ap$_n$A are co-stored with most of the aminergic neurotransmitters, and also with acetylcholine, in all models studied so far (Richardson and Brown, 1987). Moreover, these compounds have well characterised presynaptic receptors. As there is a wide range of possible membrane receptors, the first experimental approaches were undertaken by directly acting on protein kinases and protein phosphatases, or intracellular second messengers (Pintor et al., 1997b).

As is shown in Fig. 3A, direct activation of PKC by phorbol esters inhibits the maximal response of the Ap$_5$A induced calcium signal. Conversely, inhibition of PKC by staurosporin or the inhibitory peptide results in an increase in the calcium signal, with respect to the control. The activation of PKA by dibutyril cAMP and activation of adenylylcyclase by forskolin both result in a reduction of the calcium signal. The inhibition of the enzyme with the PKA-inhibitory peptide results in a clear increase in the effect through the dinucleotide receptor. Further studies have shown that PKA activation causes a drastic decrease in the receptor affinity for Ap$_5$A, and PKA inhibition produced receptor affinity states with EC$_{50}$ values close to the nM range (Miras-Portugal et al., 1998).

Protein phosphatase inhibition by okadaic acid (which at the concentrations used is non-specific),

microcystin (a better inhibitor of protein phosphatase 2A), and cyclosporin-A (a substance which is specific for protein phosphatase 2B, also known as calcineurin) all brought about a drastic reduction in the calcium transients elicited by Ap$_5$A (Fig. 3B). Both protein phosphatases 2A and 2B appear to

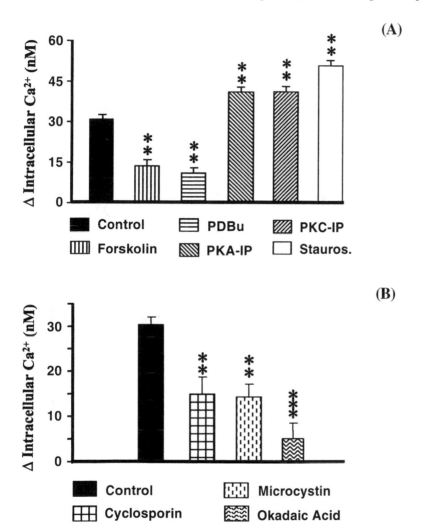

Fig. 3. Effect of protein kinase and protein phosphatase modulators on the calcium responses elicited by Ap$_5$A. Rat midbrain synaptic terminals were pre-treated for 2 min with the different effectors and the calcium response to 50 μM Ap$_5$A measured. When non-permeable substances were assayed, such as the inhibitory peptides IP-PKA, IP-PKC, or the protein phosphatase inhibitor microcystin, the brain homogeneization was carried out in the presence of these compounds to allow their entrance before the resealing of synaptic terminals. (A) Effect of Protein Kinases modulation. Forskolin, an adenylate cyclase activator, was used at 100 μM concentration. PDBu, the allosteric activator of PKC was used at a 1 μM concentration. The inhibitory peptides PKA-IP and PKC-IP were added at the first homogeneization step at concentrations of 25 μM and 5 μM, respectively. Staurosporine, the PKC inhibitor, was used at a concentration of 100 nM. (B) Effect of protein phosphatase inhibitors. Cyclosporin–A, the specific inhibitor of PPase 2B was used at 1 μg/ml. Microcystin an inhibitor of PPase 2A, was added to the first homogenization step at a concentration of 15 μM. Okadaic acid was added at a concentration of 100 nM. Under these conditions the both previously cited PPases were inhibited.

Fig. 4. $[Ca^{2+}]_i$ responses in isolated single nerve terminals. *Panel A*: viewfield image of isolated nerve terminals (synaptosomes) glued to a coverslip with poly-L-lysine and loaded with fura-2. Arrows indicate terminals analysed in other panels. *Panel B*: representative images of analysed terminals (1 and 2) before, during and after the stimulation. Pseudocolour scale represent F_{380} intensity (lower at higher $[Ca^{2+}]_i$). The consecutive images correspond to time points (a)-(h) identified with lines in the time-course traces. *Panel C*: time course of fluorescence changes (as F/Fo after correcting photo-bleaching) recorded for terminals 1 and 2 in panel A. The stimulation periods are indicated by the solid bars and images in B by a-h lines. *Methods*: synaptosomes were diluted (0.5 mg/ml) and spreaded onto coverslips coated with poly-L-lysine for attachment and then loaded with fura-2/AM (1 h, 37°C). The coverslips were washed with saline solution and mounted in a small superfusion chamber in the stage of a Nikon TE-200 microscope. Fura-2 fluorescence was excited at 380 nm and collected at 510 nm (Omega Optical bandpass filters, 430 nm dichroic mirror) through a fluor × 100, 1.3 NA, objective (Nikon). The 12-bit images were collected every 1.023 s with a Hamamatsu C4880-80 CCD camera controlled by Kinetic (UK) software. Time course data represent the average light intensity in a small elliptical region inside each terminal. Continuous fading due to photo-bleaching was corrected by local fitting of an exponential function to the data (Raw(t) = F(t)·exp(-k·t)). The data is represented as the normalized ratio F/F_0, that decreases with $[Ca^{2+}]_i$ increases (Lev-Ram et al., 1992).

play a significant role in the signal recovery after phosphorylation and the inhibition of both is required for the maximal inhibitory effect.

The logical question that arises from these results is how to know which are the presynaptic receptors that (by means of protein kinases and phosphatases) are relevant in the synaptic terminal containing the dinucleotide receptor. As has been mentioned, it is a complex picture, and one that needs to be studied one step at a time. As ATP is always co-stored with dinucleotides, and could also be at the extracellular level after controlled release or due to the fragility of synaptosomal preparations. Thus its effects were the first to be analysed. Activation of metabotropic P2Y receptors resulted in a decrease in the calcium signal induced by Ap_5A. Furthermore, as ATP breaks down to adenosine at the extracellular level via the ecto-nucleotidases cascade, this nucleoside was studied and found to exert a powerful modulatory effect through the A1 presynaptic receptor. The main effect of this modulatory action is to increase the affinity of the dinucleotide receptors to Ap_5A. Changes in the EC_{50} values from 55 μM to 10 nM, for Ap_5A, in the absence or presence respectively of an A1 receptor agonist is currently obtained (Miras-Portugal et al., 1998). A scheme on the complexity of presynaptic dinucleotide receptors regulation is represented in Fig. 5.

Thus, in spite of the complexity that may be envisaged at the presynaptic terminals, the regulation of dinucleotide receptors by the action of protein kinases and phosphatases could be a subtle way of adapting the receptors to the physiological levels of their natural agonists, or to prepare the terminal to be more effective in the exocytotic event that is a Ca^{2+} dependent process.

Studies of calcium responses in isolated single synaptic terminals by microfluorimetry

Until now the Ca^{2+} responses induced by Ap_nA have been studied in synaptosomal preparations using aproximately 1 mg of protein to obtain an optimal fluorimetric response. However, in our laboratory single cell microfluorimetry techniques have now being used for some time to study the presence of receptors, and their pharmacology, in cultured or recently isolated single cells (Castro et al., 1996; Mateo et al., 1998). Today, this technique has been improved, and it is possible to study the calcium responses induced by Ap_nA in perfused single synaptic terminals by using the calcium dye Fura–2. Figure 4 shows the response elicited by Ap_5A in a synaptic terminal compared with the Ca^{2+} transients induced by K^+ depolarisation. Most of the synaptic terminals exhibit a clear response to K^+ depolarisation, but only 20% of them showed a clear dinucleotide response in midbrain synaptosomal preparations.

This technique opens new possible experimental approaches to the study of ionotropic presynaptic receptors, not only for Ap_nA, but also for more classical transmitters such as acetylcholine (nicotinic) receptors, glutamate ionotropic receptors, and ATP receptors (Wonnacott, 1997; Clarke et al., 1997; Gu and Macdermott, 1997; Kaiser et al., 1998). The coexistence of these ionotropic presynaptic receptors and their mutual interactions through ionotropic or metabotropic receptors is

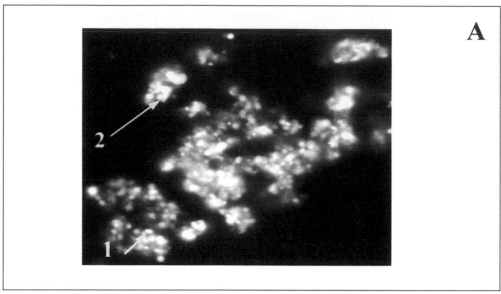

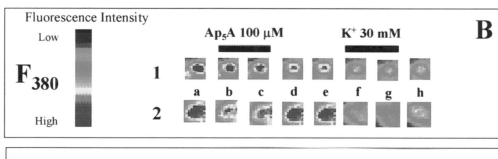

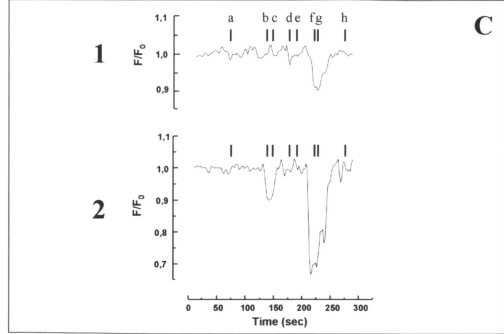

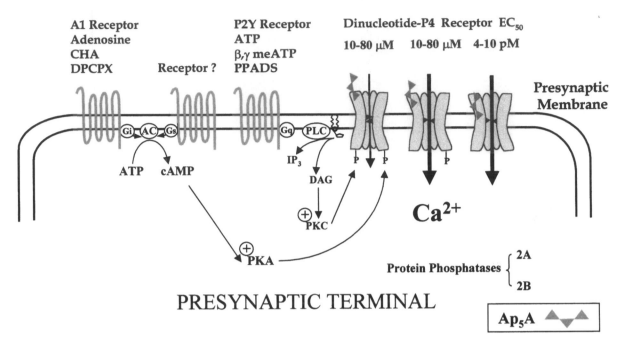

Fig. 5. Regulation of presynaptic dinucleotide -P4- receptors by metabotropic A1 and P2Y receptors and their coupled second messengers. Activation of protein kinases A and/or C results in a clear inhibition in the calcium response elicited by Ap_5A. Inhibition of protein phosphatases 2A and/or 2B maintains the phosphorylated form of this receptor and a reduced calcium response. When the ATP metabotropic receptors are stimulated, the PLC is activated producing diacylglycerol (DG) and IP3, resulting in PKC activation. This enzyme is able to phosphorylate many proteins from the presynaptic terminals, and phosphorylation of the P4 receptor by PKC results in a considerable reduction of calcium influx, without a significant change on the P4-receptor affinity for diadenosine polyphosphates. Adenosine receptors appear to play a leading role in P4-receptor modulation, through the A1 receptor, whose signalling pathway via G_i protein avoids adenylate cyclase activation, further increase in cAMP, and the subsequent activation of PKA. It should be noted that if the PKA phosphorylation pathway was blocked at whatever level, the affinity of the P4-receptor was able to reach EC_{50} values in the low pM range. Inhibition of protein-phosphatases results in a significant decrease of the Ca^{2+} influx to synaptic terminals induced by diadenosine polyphosphates.

expected to be one of the most productive and challenging areas on the road to understanding presynaptic terminal functioning (Fig. 5). Thus perhaps the question about the physiological reasons for the existence of a presynaptic ionotropic receptor, could have a suitable answer.

Ecto-diadenosine polyphosphate hydrolase from neural origin

Ecto-enzymes able to hydrolyse diadenosine polyphosphates are required to terminate the extracellular activity of Ap_nA, and at the same time

to lead the resultant mononucleotides to further degradation via the ecto-nucleotidases cascade (Zimmermann, 1996). The presence of Ap_nA in the dense granules of platelets, make the vascular endothelial cells a suitable place to search for ecto-diadenosine polyphosphate hydrolase activity, and it was the first tissue in which this it was reported (Goldman et al., 1986; Ogilvie et al., 1989). Besides, the presence of these compounds in neurosecretory granules has provided a stimulus to the search for ecto-diadenosine polyphosphate hydrolase activities in neural tissues (Rodríguez-Pascual

Fig. 6. Hydrolysis of ethenodiadenosine polyphosphates. The ε-Ap_5A structure and the hydrolysis products, ε-Ap_4 and ε-AMP, after the ecto-diadenosine polyphosphate hydrolase action is represented. The enzyme always acts in an asymmetrical way, AMP being always one of the reaction products. The HPLC profiles illustrating the time dependent hydrolysis of ε-Ap_5A by *Torpedo* plasma membranes (100 μg/ml), incubated at 37°C in the presence of 1 μM ε-Ap_5A and represented as consecutive chromatograms.

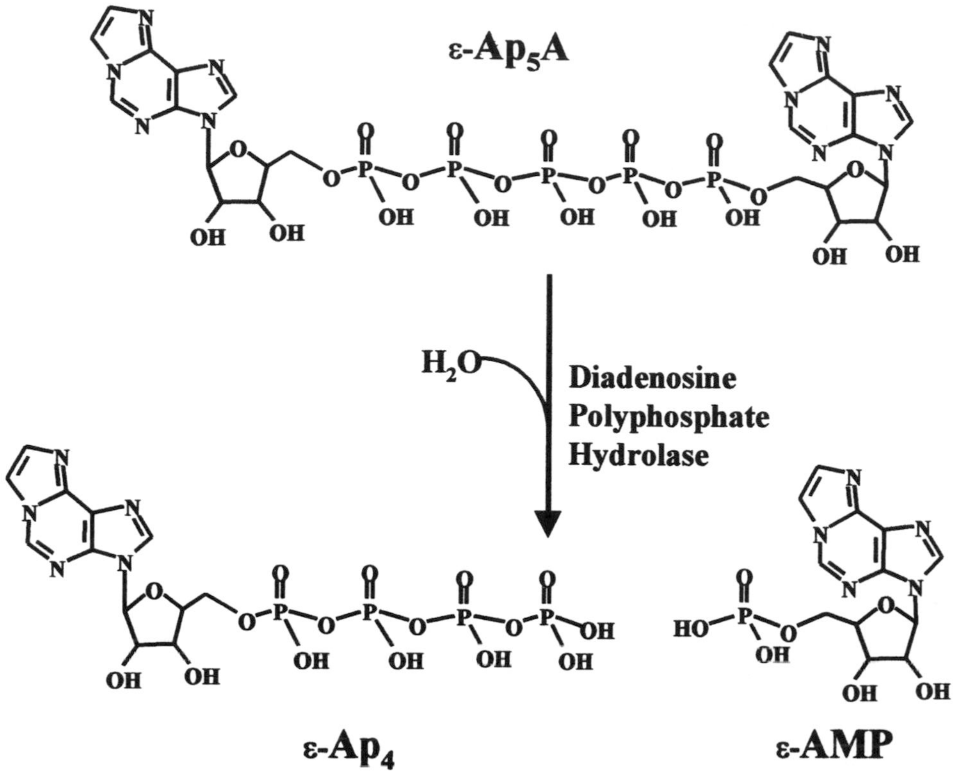
ε-Ap5A
H2O
Diadenosine
Polyphosphate
Hydrolase
ε-Ap4
ε-AMP

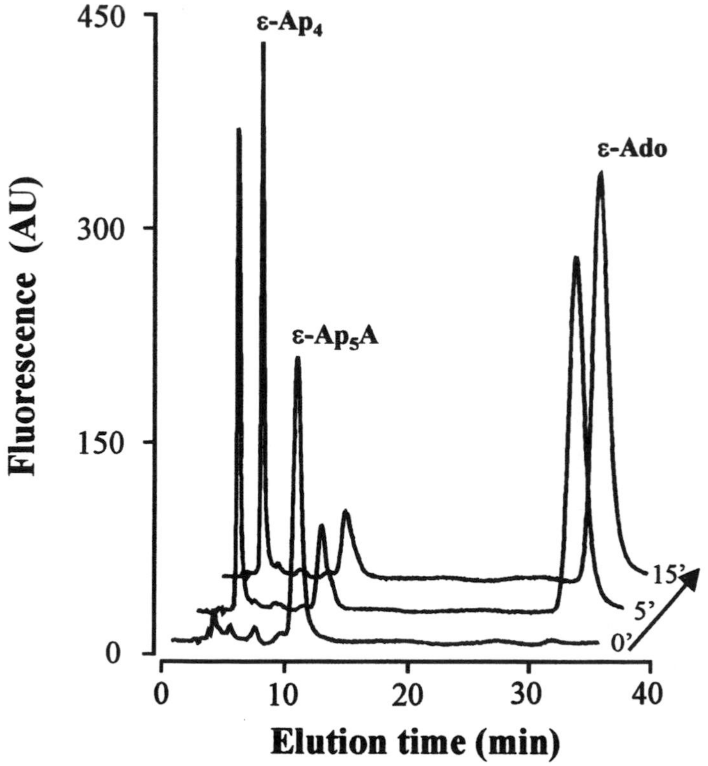
ε-Ap4
ε-Ado
ε-Ap5A
Fluorescence (AU)
450
300
150
0
15'
5'
0'
0
10
20
30
40
Elution time (min)

et al., 1992a, 1992b). From an experimental point of view, the first enzymatic assays required the processing of samples by HPLC, to make an accurate quantification of substrate hydrolysis, or to employ radiometric techniques. Thus, the continuous fluorimetric technique developed by Rotllán et al. (1991), using the etheno-Ap_nA derivatives as substrate analogues, allowed an accurate and rapid approach to the enzymes present in both the isolated cells and the plasma membrane preparations.

In Fig. 6 the etheno-Ap_5A is represented together with the hydrolysis products, ε-adenosine tetraphosphate (ε-Ap_4) and ε-AMP. Independent of their origin, the ecto-enzymes studied so far always cut the phosphate bridge asymmetrically, i.e. producing AMP and the mononucleotide with $n-1$ phosphates (Fig. 6). This is the case for the endothelial enzymes from bovine adrenal capillaries and aortic vessels (Goldman et al., 1986; Ogilvie et al., 1989; Mateo et al., 1997b) and all types of neural enzymes studied so far (Ramos et al., 1995; Mateo et al., 1997a).

Comparative studies of enzymes from different sources have shown there to be variation in ionic requirements, and at least two groups can be distinguished. First, the enzymes of vascular endothelium origin, which are inhibited by Ca^{2+}, and second, neural enzymes in which Ca^{2+} is an activator (Mateo et al., 1997a, 1997b). Studies carried out with isolated synaptic terminals showed a hydrolysis rate 20–50 times slower for these compounds than that of ATP. It is also noteworthy that in cultured chromaffin cells the V_{max} for the ecto-ATPase activity is between 500–1,000 times higher than that for the ecto-diadenosine polyphosphate hydrolase (Rodríguez-Pascual et al., 1992a). This finding explains the longer half-life of Ap_nA in biological samples. In the case of perfussed brain from conscious rats, after amphetamine stimulation, ATP and ADP are present in such small quantities that they are almost undetectable, but Ap_4A and Ap_5A are present at nM concentration levels (Pintor et al., 1995).

The presynaptic neural enzyme was approached in the colinergic model of *Torpedo*. This enzyme exhibited the same ionic requirements as the neurochromaffin one, and the affinities for the etheno-derivatives showed K_m values of 0.39 μM, 0.42 μM and 0.37 μM for ε-Ap_3A, ε-Ap_4A and ε-Ap_5A respectively. This submicromolar affinity is perhaps related to the levels reached by the natural compounds after exocytosis and strengths their role in cellular and neural signalling.

Figure 6 shows the hydrolysis of ε-Ap_5A by the *Torpedo* electric organ membranes as a function of the time. It is worthy to notice that the nucleotide adenosine tetraphosphate –Ap_4– accumulates with the experimental time, exhibiting a very low rate of hydrolysis. The enzymatic production of Ap_4, at the synaptic terminal, made it worth studying the effect of this compound on the nucleotide receptors. Ap_4 behaved as an agonist on the ATP ionotropic presynaptic receptor, but not on the dinucleotide receptor, and its affinity was similar to that of ATP, but with much better stability, and it should be considered a natural agonist of ATP receptors (Gómez-Villafuertes, 1998).

Conclusions

Diadenosine polyphosphates Ap_nA are co-stored and co-released from a large variety of synaptic terminals. Their physiological extracellular concentration range can be expected to be between the nM and low μM. Binding studies have detected the presence of high affinity sites, correlating with the physiological concentrations reported. From a functional point of view, specific receptors, called dinucleotide receptors, or P4 receptors, have been described in presynaptic terminals, and elicit calcium entrance through ionotropic receptors, which are not susceptible to inhibition by voltage dependent calcium channel toxins. Diinosine polyphosphates are good inhibitors of the P4 receptor, Ip_5I exhibiting an IC_{50} value of 4 nM.

The presynaptic dinucleotide receptors are modulated by the action of protein kinases and protein phosphatases. An increase in the levels of phosphorylation results in a decrease in both receptor affinity to its agonists and maximal calcium entrance through the receptor. Adenosine, via its A1 receptor, coupled to a G_i protein and inhibition of the adenylyl cyclase, therefore prevents phosphorylation through PKA, and is able to increase the affinity of this receptor to the pM range of Ap_5A. The effects of adenosine are mimicked by CHA (cyclohexyladenosine) and reversed by adenosine deaminase, or the addition of the antagonist DPCPX (8-cyclopentyl-1,3-dipropylxanthine). Activation of the P2Y receptors by ATP, or its non-hydrolyzable analogues, results in PKC activation and a significant reduction on the maximal calcium response, with minor effects on receptor affinity. The effect of ATP can be reversed by hydrolysis of the extracellular nucleotides by alkaline phosphatase, or by using the antagonist PPADS.

The extracellular actions of diadenosine polyphosphates are terminated by the enzymatic hydrolysis by an ecto-diadenosine polyphosphate hydrolase. The neural type, from *Torpedo* synaptic terminals, exhibit K_m values close to 0.4 μM, for all the dinucleotides tested and present similar ionic requirements than other neural enzymes.

Acknowledgements

This work was supported by grants from DGICYT (PM 95-0072), EU. Biomed-2 (PL 95-0676), Multidisciplinar U.C.M. (1995) and the Areces Foundation Neurosciences programme. J. G. is a fellowship from the Spanish Ministry of Education and Culture, M. D-H from the U.C.M. and R. G-V from the C.A.M. We thank Duncan Gilson for his help in preparing the manuscript.

References

Abbracchio, M.P. and Burnstock, G. (1994) Purinoceptors: are there families of P2X and P2Y purinoceptors?. *Pharmac. Ther.*, 64: 445–475.

Baker, J.C. and Jacobson, M. (1986) Alteration of adenyl dinucleotide metabolism by enviromental stress. *Proc. Natl. Acad. Sci. USA*, 83: 2350–2352.

Baxi, M.D. and Vishwanatha, J.K. (1995) Uracil DNA-glycosylase/glyceraldehyde-3-phosphate dehydrogenase is an Ap₄A binding protein. *Biochemistry*, 34: 9700–9707.

Boarder, M.R., Weisman, G.A., Turner, J.T. and Wilkinson, G.F. (1995) G protein-coupled P2 purinoceptors: from molecular biology to functional responses. *Trends. Pharmacol. Sci.*, 16: 133–139.

Castro, E., Mateo, J., Tomé, A.R., Barbosa, M., Miras-Portugal, M.T. and Rosario, L.M. (1995) Cell-specific purinergic receptors coupled to Ca^{2+} entry and Ca^{2+} release from internal stores in adrenal chromaffin cells: differential sensitivity to UTP and suramin. *J. Biol. Chem.*, 270: 5098–5106.

Clarke, V.R.J. et al. (1997) A hippocampal GluR5 kainate receptor regulating inhibitory synaptic transmission. *Nature*, 389: 599–602.

Delicado, E.G., Casillas, T., Sen, R.P. and Miras-Portugal, M.T. (1994) Evidence that adenine nucleotides modulate the functionality of nucleoside transporter. Characterization of uridine transport in chromaffin cells and plasma membrane vesicles. *Eur. J. Biochem.*, 225: 355–362.

Fredholm, B.B., Abbracchio, M.P., Burnstock, G., Dubyak, G.R., Harden, T.K., Jacobson, K.A., Schwabe, U. and Williams, M. (1997) Towards a revised nomenclature for P1 and P2 receptors. *TIPS*, 18: 79–82.

Goldman, S.J., Gordon, E.L. and Slakey, L.L. (1986) Hydrolisis of diadenosine 5′,5″-P′,P″-triphosphate (Ap₃A) by porcine aortic endothelial cells. *Circ. Res.*, 59: 362–366.

Gómez-Villafuertes, R. (1998) Acción del adenosina tetrafosfato sobre receptores nucleotídicos de cerebro de rata: Caracterización del receptor. *M.D. Thesis*. Biological Sciences. University Complutense de Madrid.

Grynkiewicz, G., Ponie, M. and Tsien, R.Y. (1985). A new generation of Ca^{2+} indicators with greatly improved fluorescence properties. *J. Biol. Chem.*, 260: 3440–3450.

Gu, J.G. and Macdermott, A.B. (1997) Activation of ATP P2X receptors elicits glutamate release from sensory neuron synapses. *Nature*, 389: 749–753.

Gualix, J., Abal, M., Pintor, J., García-Carmona, F. and Miras-Portugal, M.T. (1996) Nucleotide vesicular transporter of bovine chromaffin granules. Evidence for mnemonic regulation. *J.Biol. Chem.*, 271: 1957–1965.

Gualix, J., Fideu, M.D., Pintor, J., Rotllán, P., García-Carmona, F. and Miras-Portugal, M.T. (1997) Characterization of diadenosine polyphosphate transport into chromaffin granules from adrenal medulla. *FASEB*, 11: 981–990.

Herrero, I., Miras-Portugal, M.T. and Sánchez-Prieto, J. (1992) Positive feedback of glutamate exocytosis by metabotropic glutamate receptor stimulation. *Nature*, 360: 163–165.

Herrero, I., Castro, E., Miras-Portugal, M.T. and Sánchez-Prieto, J. (1996) Two components of glutamate exocytosis differentially affected by presynaptic modulation. *J. Neurochem.*, 67: 2346–2354.

Hoyle, C.H.V., Postorino, A. and Burnstock, G. (1995) Pre- and postjunctional effects of diadenosine polyphosphates in guinea-pig vas deferens. *J. Pharm. Pharmacol.*, 47: 928–933.

Hoyle, C.H.V., Pintor, J., Gualix, J. and Miras-Portugal, M.T. (1997) Antagonism of P2X receptors in guinea-pig vas deferens by diinosine pentaphospahte. *Eur. J. Pharmacol.*, 333: R1–R2.

Jimenez, A.I., Castro, E., Delicado, E.G. and Miras-Portugal, M.T. (1998) Potentiation of ATP calcium responses by diadenosine pentaphosphate in individual cerebellar astrocyte. *Neurosci. Lett.*, 246: 1–3.

Jovanovic, A. and Terzic, A. (1996) Diadenosine tetraphosphate-induced inhibition of ATP-sensitive K^+ channels in patches excised from ventricular myocites. *Br. J. Pharmacol.*, 117: 233–235.

Kaiser, S.A., Soliakov, L., Harvey, S.C., Luetje, C.W. and Wonnacott, S. (1998) Differential inhibition by alfa-conotoxin-MII of the nicotinic stimulation of dopamine release from rat striatal synaptosomes and slices. *J. Neurochem.*, 70: 1069–1076.

Lazarowski, E., Watt, W., Stutts, M.J., Boucher, R. and Harden, T.K. (1995) Pharmacological selectivity of the cloned human P2U-purinoceptor: Potent activation by diadenosine tetraphosphate. *Br. J. Pharmacol.*, 116: 1619–1627.

Leinhardt, G.E. and Secemsky, I.I. (1973) P1,P5-Di(adenosine-5′)-pentaphosphate, a potent multisubstrate inhibitor of adenylate kinase. *J. Biol. Chem.*, 248: 1121–1123.

Lev-Ram, V., Miyakawa, H., Lasser-Ross, N. and Ross, W.N. (1992) Calcium transients in cerebellar Purkinje neurons evoked by intracellular stimulation. *J. Neurophysiol.*, 68: 1167–1177

Lüthje, J. and Ogilvie, A. (1983) The presence of diadenosine 5′,5‴-P1,P3-triphosphate (Ap₃A) in human platelets. *Biochem. Biophys. Res. Commun.*, 115: 253–260.

Mateo, J., Miras-Portugal, M.T. and Castro, E. (1996) Coexistence of P2Y- and PPADS-insensitive P2U-purinoceptors in endothelial cells from adrenal medulla. *Br. J. Pharmacol.*, 119: 1223–1232.

Mateo, J., Miras-Portugal, M.T. and Rotllán, P. (1997b) Ectoenzymatic hydrolisis of diadenosine polyphosphates by cultured adrenomedullary vascular endothelial cells. *Am. J. Physiol.*, 273: C918–C927.

Mateo, J., Rotllán, P., Marti, E., Gómez de Aranda, I., Solsona, C. and Miras-Portugal, M.T. (1997a) Diadenosine polyphosphate hydrolase from presynaptic plasma membranes of *Torpedo* electric organ. *Biochem. J.*, 323: 677–684.

Mateo, J., Garcia-Lecea, M., Miras-Portugal, M.T. and Castro, E. (1998) Ca²⁺ signals mediated by P2X-type purinoceptors in cultured cerebelar Purkinje cells. *J. Neurosci.*, 18: 1704–1712.

McLennan, A.G. (1992) *Ap₄A and Other Dinucleoside Polyphosphates*. CRC Press, Boca Raton, Florida.

Miras-Portugal, M.T., Gualix, J. and Pintor, J. (1998) The neurotransmitter role of diadenosine polyphosphates. *FEBS Lett.*, 430: 78–82.

Ogilvie, A., Lüthje, J., Pohl, U. and Busse, R. (1989) Identification and partial characterization of an adenosine-(5′)tetraphospho(5′)-adenosine hydrolase on intact bovine aortic endothelial cells. *Biochem. J.*, 259: 97–103.

Pintor, J. and Miras-Portugal, M.T. (1995) A novel receptor for diadenosine polyphosphates coupled to calcium increase in rat midbrain synaptosomes. *Br. J. Pharmacol.*, 115: 895–902.

Pintor, J., Torres, M. and Miras-Portugal, M.T. (1991a) Carbachol induced release of diadenosine polyphosphates -Ap₄A and Ap₅A- from perfused bovine adrenal medulla and isolated chromaffin cells. *Life Sci.*, 48: 2317–2324.

Pintor, J., Torres, M., Castro, E. And Miras-Portugal, M.T. (1991b) Characterization of diadenosine tetraphosphate Ap₄A binding sites in cultured chromaffin cells: Evidence for a P2Y site. *Br. J. Pharmacol.*, 103: 1980–1984.

Pintor, J., Diaz-Rey, M.A., Torres M. and Miras-Portugal, M.T. (1992a) Presence of diadenosine polyphosphates -Ap₄A and Ap₅A- in rat brain synaptic terminals. Ca²⁺ dependent release evoked by 4-aminopyridine and veratridine. *Neurosci.Lett.*, 136: 114–144.

Pintor, J., Kowalewsky, H.J., Torres, M., Miras-Portugal, M.T. and Zimmermann, H. (1992b) Synaptic vesicle storage of diadenosine polyphosphates in the *Torpedo* electric organ. *Neurosci. Res. Commun.*, 10: 9–15.

Pintor, J., Díaz-Rey, M.A. and Miras-Portugal, M.T. (1993) Ap₄A and ADP-β-S binding to P2-purinergic receptors present on rat brain synaptic terminals. *Br. J .Pharmacol.*, 108: 1094–1099.

Pintor, J., Porras, A., Mora, F. and Miras-Portugal, M.T. (1995) Dopamine receptor blockade inhibits the amphetamine-induced release of diadenosine polyphosphates –Ap₄A and Ap₅A- from neostriatum of the conscious rat. *J. Neurochem.*, 64: 670–676.

Pintor, J., King, B.F., Miras-Portugal, M.T. and Burnstock, G. (1996) Selectivity and activity of adenine dinucleotides at recombinant P2X2 and P2Y1 purinoceptors. *Br. J. Pharmacol.*, 119: 1–7.

Pintor, J., Gualix, J. and Miras-Portugal, M.T. (1997a) Diinosine polyphosphates, a group of dinucleotides with antagonistic effects on diadenosine polyphosphate receptor. *Mol. Pharmacol.*, 51: 277–284.

Pintor, J., Gualix, J. and Miras-Portugal, M.T. (1997b) Dinucleotide receptor modulation by protein kinases (protein kinases A and C) and protein phosphatases in rat brain synaptic terminals. *J. Neurochem.*, 68: 2552–2557.

Pintor, J., Puche, J.A., Gualix, J. Hoyle, C.H.V. and Miras-Portugal, M.T. (1997c) Diadenosine polyphosphates evoke Ca²⁺ transients in guinea-pig brain via receptors distinct from those for ATP. *J. Physiol.*, 504: 327–335.

Pivorum, E.B. and Nordone, A. (1996) Brain synaptosomes display a diadenosine tetraphosphate (Ap₄A)-mediated Ca²⁺ influx distinct from ATP-mediated influx. *J. Neurosci. Res.*, 44: 478–489.

Ramos, A., Pintor, J., Miras-Portugal, M.T. and Rotllan, P. (1995) Use of fluorogenic substrates for detection and investigation of ectoenzymatic hydrolysis of diadenosine polyphosphates: A fluorometric study on chromaffin cells. *Anal. Biochem.*, 228: 74–82.

Rapaport, E. and Zamecnik, P.C. (1976) Presence of diadenosine 5′,5‴-P1,P4-tetraphosphate (Ap₄A) in mammalian cells in levels varying widely with proliferative activity of the tissue: A possible positive "pleiotypic activator". *Proc. Natl. Acad. Sci. USA*, 73: 3984–3988.

Richardson, P.J. and Brown, S.J. (1987) ATP release from affinity-purified cholinergic nerve terminals. *J. Neurochem.*, 48: 622–630.

Ripoll, C., Martín, F., Rovira, J.M., Pintor, J., Miras-Portugal, M.T. and Soria, B. (1996) Diadenosine polyphosphates: A novel class of glucose-induced intracellular messengers in the pancreatic β-cell. *Diabetes*, 45: 1431–1434.

Rodríguez del Castillo, A., Torres, M., Delicado, E.G. and Miras-Portugal, M.T. (1998) Subcellular distribution studies of diadenosine polyphosphates -Ap₄A and Ap₅A- in bovine adrenal medulla: Presence in chromaffin granules. *J. Neurochem.*, 51: 1699–1703.

Rodríguez-Pascual, F., Torres, M. and Miras-Portugal, M.T. (1992a) Studies on the turnover of ecto-nucleotidases and ecto-dinucleoside polyphosphate hydrolase in cultured chromaffin cells. *Neurosci. Res. Commun.*, 11: 101–107.

Rodríguez-Pascual, F., Torres, M., Rotllán, P. and Miras-Portugal, M.T. (1992b) Extracellular hydrolysis of diadenosine polyphosphates, Ap_nA, by bovine chromaffin cells in culture. *Arch. Biochem. Biophys.*, 297: 176–183.

Rodríguez-Pascual, F., Cortés, R., Torres, M., Palacios, J.M. and Miras-Portugal, M.T. (1997) Distribution of $[^3H]$-Ap_4A binding sites in rat brain. *Neurosci.*, 77: 247–255.

Rotllán, P. and Miras-Portugal, M.T. (1985) Adenosine kinase from bovine adrenal medulla. *Eur. J. Biochem.* 151: 365–371.

Rotllán, P., Ramos, A., Pintor, J., Torres, M. and Miras-Portugal, M.T. (1991) Di(1, N^6-ethenoadenosine)5′,5‴-P1, P4-tetraphosphate, a fluorescent enzymatically active derivative of Ap_4A. *FEBS Lett.*, 280: 371–374.

Schlüter, H., Offers, E., Brüggemann, G., van der Giet, M., Tepel, M., Nordhoff, E., Karas, M., Spieker, C., Witzel, H. and Zidek, W. (1994) Diadenosine polyphosphates and the biological control of blood pressure. *Nature*, 367: 186–188.

Walker, J., Bossman, P., Lackey, B.R., Zimmerman, J.K., Dimmick, M.A. and Hilderman, R.H. (1993) P1,P4-tetraphosphate receptor is at the cell surface of heart cells. *Biochemistry*, 32: 14009–14014.

Wonnacott, S. (1997) Presynaptic nicotinic ACh receptors. *TINS*, 20: 92–98.

Zimmermann, H. (1996) Extracellular purine metabolism. *Drug Develop. Res.*, 39: 337–352.

P. Illes and H. Zimmermann (Eds.)
Progress in Brain Research, Vol 120
© 1999 Elsevier Science BV. All rights reserved

CHAPTER 33

Ecto-protein kinases as mediators for the action of secreted ATP in the brain

Yigal H. Ehrlich[1,*] and Elizabeth Kornecki[2]

[1]*The Program in Neuroscience, Department of Biology and the CSI/IBR Center for Developmental Neuroscience, The College of Staten Island of The City University of New York, Staten Island, NY 10314 USA*
[2]*The Department of Anatomy and Cell Biology, The State University of New York, Downstate Medical Center, Brooklyn, NY 11203, USA*

Introduction

Protein phosphorylation is a ubiquitous step in intracellular pathways that produce rapid and transient changes in neuronal activity, and is also known to operate as a key mechanism of molecular adaptations which produce long lasting alterations in neuronal function, such as those operating in memory formation (reviewed by Ehrlich, 1984; 1987; and Ehrlich et al., 1987). The discovery of neuronal ecto-protein kinase activity (Ehrlich et al., 1986) has revealed that the powerful regulatory-mechanism of protein phosphorylation operates not only within nerve cells, but also in the *extra*cellular environment of the nervous system. This phosphorylative activity may thus regulate neuronal responsiveness and neuronal adaptation by using as phosphoryl-donor molecules of ATP that are secreted to the extracellular space by stimulated neurons.

The storage of ATP in synaptic vesicles and its secretion from stimulated neurons, described first in the 1950s in peripheral neurons and later in the CNS, is now extensively documented (reviewed by Gordon, 1986; see also Zimmermann and Braun

*Corresponding author. Tel.: +1 718 982-3932; fax: +1 718 982-3944; e-mail: Ehrlich@Postbox.csi.cuny.edu

(1999), in this volume). ATP-release from neurons has recently become the subject of intensive investigation with increasing evidence of a role for extracellular ATP in synaptic transmission (Edwards et al., 1992; Evans et al., 1992; Silinsky et al., 1992; Shen and North, 1993) and in neuronal events associated with synaptic plasticity and implicated in memory formation (Wieraszko and Seyfried, 1989; Fujii et al., 1995a, 1995b). The mechanisms by which secreted ATP can exert effects on neuronal function include interaction with purinergic receptors (reviewed by Burnstock, 1981, 1990; see also Burnstock, 1999, in this volume), hydrolysis by ecto-ATPase (Nagy et al., 1986, Nagy and Shuster 1995; Nagy, 1997; see also Zimmermann and Braun, 1999, this volume), and the extracellular phosphorylation of surface and matrix proteins by ecto-protein kinases (reviewed by Ehrlich, 1987; Ehrlich et al., 1990). Studies of cloned neural cell lines of peripheral origin have implicated ecto-protein kinase (ecto-PK) activity in the regulation of events that are initiated at the cell surface, including Ca^{2+}-influx (Ehrlich et al., 1988) and norepinephrine uptake (Hendley et al., 1988; Hardwick et al., 1989). Ecto-PK was also implicated in the action of extracellular regulators of neuronal development: nerve growth factor (NGF) in PC12 cells (Pawlowska et al., 1993), and gangliosides in GOTO neuroblastoma cells (Tsuji et al., 1988). Here we review recent studies on the identification, activity and function of ecto-PKs in the central nervous system.

Since an earlier review of studies on ecto-PK activities in various cell types by Ehrlich et al., (1990), there have appeared additional reports on the presence of ecto-PK and endogenous substrates for its activity in cells that release ATP, including smooth muscle cells (Fedan and Lamport, 1990), platelets (Naik et al., 1991; Babinska et al., 1996), L6 myoblasts (Chen and Lo, 1991a), neutrophils (Skubitz asnd Ehresmann, 1992; Skubitz and Goueli, 1994), cloned hybrid hippocampal cells (Pawlowska et al., 1994) endothelial cells on blood vessel walls (Pirrotton et al., 1992; Hartmann and Schrader, 1992), as well as granule neurons (Merlo et al., 1997). The phosphorylation of surface proteins appears to influence cell communication as well as cell–environment interactions (Pfeifle et al., 1981). The inhibition by N-CAM antibodies of the phosphorylation of specific surface proteins suggested a role for ecto-PK in neuronal development (Ehrlich et al., 1986). This finding also demonstrated that antibodies directed against substrates of ecto-protein kinase can inhibit their phosphorylation without penetrating cells, and hence provide important tools for the investigation of the precise role of specific surface protein phosphorylation systems in the regulation of biological functions.

The functions assigned to several ecto-PKs to date stem from the identification of their naturally occuring substrates. Ecto-PKs have been shown to phosphorylate two types of substrates: (1) cell surface proteins (Ehrlich et al., 1986; Chen and Lo, 1991a; Vilgrain and Baird, 1991; Volonte et al., 1994), and (2) protein components that reside in the extracellular environment (Kalafatis et al., 1993; Kübler et al., 1992; Vilgrain and Baird, 1991). Chen and Lo (1991b) have shown the involvement of an ecto-PK and its 112TK cell surface protein-substrate in the process of myogenesis. Two independent studies (Skubitz and Goueli, 1991; Vilgrain and Baird, 1991) have found that basic fibroblast growth factor serves as a substrate for ecto-PK associated with the outer surface of the target cells. Studies which detail the characterization of a cAMP-dependent ecto-PK capable of phosphorylating atrialnatriuretic peptide (Kübler et

al., 1992) and the plasma protein fibrinogen (Sonka et al., 1989; Dorner et al., 1989), have been reported. The major plasma membrane differentiation antigen called PC-1 (Rebbe et al., 1991) was found to have a threonine-specific ecto-protein kinase activity stimulated by acidic fibroblast growth factor (Oda et al., 1991, 1993), similar to the NGF stimulated ecto-phosphorylation of a 53 K protein we have described in PC-12 cells (Pawlowska et al, 1993). Phosphorylation and dephosphorylation events involving theextracellularly-acting blood protein, fibrinogen, have been shown to affect thrombin-induced gelation and plasmin degradation of fibrinogen (Forsberg and Martin, 1991; Martin et al., 1991). A releasable surface phosphoprotein substrate of ecto-PK has been implicated in the regulation of cell growth and malignant transformation (Friedberg et al., 1995). Herbert et al. (1991) have characterized extracellular binding sites for the PK inhibitor staurosporine on the surface of capillary endothelial cells, and we have found interactions of cyclosporine, a phosphoprotein phosphatase inhibitor, with the surface of platelets (Fernandez et al., 1993). A membrane impermeable protein kinase inhibitor, K-252b, was found to inhibit NGF-induced neuritogenesis in PC12 cells (Nagashima et al., 1991; Pawlowska et al., 1993), to inhibit synaptogenesis in cortical neurons (Muramoto et al., 1988, 1992; Kuroda et al., 1992) and to inhibit the maintenance phase of long-term potentiation (Chen et al., 1994; Fujii et al., 1995a and 1995b). Finally, a casein kinase-like ecto-PK has been purified (Walter et al., 1996), and the use of antibodies directed to the catalytic domain of PKC have identified novel and atypical isozymes of protein kinase C as ecto-enzymes in brain neurons (Hogan et al., 1995; see below). This activity could be inhibited by Alzheimer's β-amyloid peptides (Hogan et al., 1995; see below). Furthermore, phosphorylation sites in the ecto-domain of the β-Amyloid Precursor Protein (β-APP) itself were directly demonstrated (Walter et al., 1997), indicating that an ecto-PK responsible for the regulation of βAPP is located on the surface of brain neurons.

Stimulus-induced exocytosis of vesicular ATP occurs not only in the nervous system, from neurons, but also in the cardiovascular system, from human platelets, where we have described an ecto-protein kinase/ecto-phosphoprotein phosphatase system (Ehrlich, 1987; Ehrlich and Kornecki, 1987; Naik et al., 1991) that plays a role in the cascade of cell surface interactions with extracellular factors that control blood coagulation (Ehrlich, 1987; Ehrlich and Kornecki, 1987; Kalafatis et al., 1993). Dephosphorylation of a specific ecto-domain in the platelet collagen-receptor (a protein called CD36) was found to control its ligand specificity by shifting its affinity from collagen binding, in the phosphorylated state, to high-affinity for the extracellular adhesion protein thrombospondin, that CD36 acquires in the dephosphorylated state (Asch et al., 1993). In an additional extensive study of ecto-PK in circulating cells, the phosphorylation of ecto-domains in the α, β T-cell receptor (TCR) on the surface of T-lymphocytes, was shown to affect TCR-cognate interactions and TCR-multimolecular complex formation (Apasov et al., 1996).

In summary, extracellular phosphorylative events can regulate the function of receptors (Ehrlich et al., 1990; Asch et al., 1993; Apasov et al., 1996), ion channels (Ehrlich et al., 1988), uptake carriers (Hendley et al., 1988; Hardwick et al., 1989), integrins (Ehrlich, 1987; House and Kemp, 1987; Asch et al., 1993), adhesion proteins (Ehrlich et al., 1986, 1990), growth factors (Ehrlich et al., 1990; Nagashima et al., 1991; Pawlowska et al., 1993) and coagulation factors (Kalafatis et al., 1993) in the nervous and cardiovascular-systems, and thus may play critical roles in homeostasis, hemostasis, thrombosis, neuritogenesis, neuronal adhesion, and synaptogenesis.

Criteria fulfilled as evidence for the presence of ecto-protein kinase and phosphoprotein substrates on the surface of brain neurons

Several criteria that were set as a requirement for evidence of the presence of ecto-enzymes on the surface of cultured cells by Kreutzberg et al., (1986) have been supplemented with specific requirements for proving the presence of ecto-protein kinases (Ehrlich et al., 1990). These criteria demonstrate directly the presence of phosphorylated sites in ecto-domains of cell-surface proteins,

and exclude the contribution by intracellular substrates of protein kinases to the reaction products detected in ecto-PK assays, as follows: (1) Rigorous measurements of the intact state and viability of cells are carried-out, and intact cells are assayed while still attached to the tissue-culture plates in which they are grown. (2) The substrates of ecto-PK phosphorylation by extracellular [γ^{32}P] ATP, are carefully compared with those phosphorylated substrates labeled by intracellular ATP following preincubation of cells with inorganic phosphate [^{32}Pi]. (3) Excess concentrations (100–1000-fold) of nonlabeled inorganic phosphate are used during ecto-PK phosphorylation reactions in order to compete with the uptake of ^{32}P-generated molecules formed as a result of the hydrolysis of ATP. The excess of inorganic phosphate thus eliminates the possibility of radiolabeling of intracellular proteins. (4) The ATP-consuming enzyme, apyrase, is added to the medium to intentionally hydrolyze the radiolabeled ATP, which serves as a control in these experiments. (5) Low concentrations of trypsin are added to surface-phosphorylated cells to demonstrate removal of phosphorylated products from the intact cell surface. Likewise, preincubation of intact cells with trypsin, prior to [γ^{32}P]ATP labeling of the cell surface, is used to inactivate the cell surface ecto-protein kinase and/or its endogenous substrates, and hence to prevent the phosphorylation of surface proteins. (6) Partial permealization of the plasma membrane with low concentrations of nonionic detergents is used to label intracellular proteins. Similarly, addition of lysed cells (and their released contents) to reactions of intact cells carried out with radiolabeled ATP is used to exclude the possibility that the substrates observed in ecto-PK reactions emerged from intracellular compartments. (7) The [γ^{32}P]ATP-phosphorylation of exogenously-added substrates, such as casein, myelin basic protein and protamine, by intact cells is another evidence for the presence of ecto-protein kinase activity on the cell surface. (8) The availability of specific membrane-impermeable inhibitors to various protein kinases, such as specific antibodies directed to the catalytic domain of the kinase, and prototype pseudosubstrate inhibitory-peptides that mimick specific sequences in unique substrates, serve to both demonstrate the surface location of an ecto-protein kinase and to determine the potential similarity of its catalytic properties to those of known intracellular protein kinases.

All of the criteria required as direct evidence for the presence of ecto-protein kinase (ecto-PK) in cultured cells have been fulfilled in experiments conducted with brain neurons that were obtained from embryonic chick telencephalon or mouse hippocampus and maintained in culture in a chemically-defined, serum-free medium. As shown in Fig. 1, the results reported by Hogan et al. (1995) provided evidence that seven surface phosphoproteinsof brain neuron with apparent MW of 116,

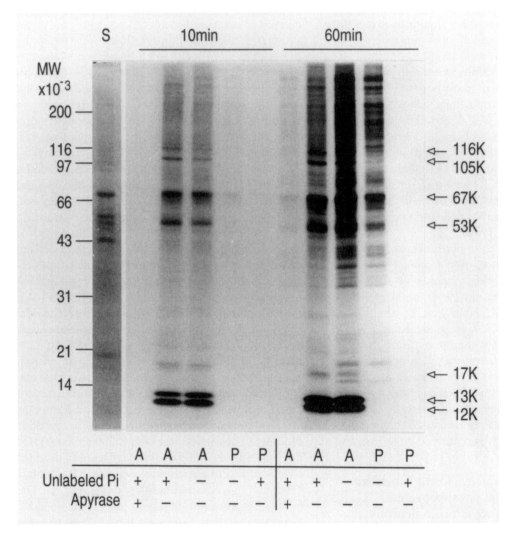

Fig. 1. Protein substrates of ecto-PK on the surface of brain neurons. Primary neurons from the telencephalon of chick embryos were maintained for 3 days in culture and then assayed for ecto-PK activity. In each reaction, neurons attached to wells of 48-well plates were incubated for 10 min or 60 min with 0.1 μM [γ^{32}P]ATP (10 μCi/well; wells marked A) or 10 μCi/well of inorganic ^{32}Pi (wells marked P). The radiolabeled proteins were resolved by SDS-PAGE followed by autoradiography. Reactions were carried out with (+) or without (−) addition of 1 mM Na_2HPO_4 (unlabled Pi) OR 5 U/ml apyrase, as marked at the bottom. S = protein staining pattern. The arrows on the right side of the figure identify specific substrates of ecto-PK by the criteria that their labeling by extracellular [γ^{32}P]ATP is seen already after 10 min incubation, it is sensitive to addition of apyrase, and that the addition of unlabled Pi completely prevents the labeling of intracellular phosphoproteins by ^{32}Pi (right-most lane), but has no effect on substrates of ecto-PK. Data from Hogan et al. (1995).

105, 67, 53, 17, 13, and 12 K are phosphorylated on ecto-domains. Figure 1 also depicts the testing of criteria no. 3 and 4 (see above): adding an excess of nonlabeled inorganic phosphate (Pi) and examining the effects of apyrase revealed that the four surface proteins of MW of 116, 105, 67, and 53 K are phosphorylated by *intra*cellular PK as well as by ecto-PK activity. In contrast, the surface phospho-proteins of 17, 13, and 12 K are not labeled by inorganic ^{32}Pi (even after 60 min) and thus appear to be phosphorylated *exclusively* by ecto-PK. The neuronal specific ecto-PK substrate of 105 K discussed above was previously detected in cloned neural cells of peripheral origin (Ehrlich et al., 1986; Pawlowska et al., 1993). However, the surface phosphoproteins of 13 and 12 K appear to be ecto-PK substrates unique to brain neurons.

Novel and atypical isozymes of protein kinase C as ecto-enzymes

Hogan et al. (1995) have reported that a specific inhibitor of PKC, the pseudosubstrate peptide PKC 19-36 (added at 10 μM), completely inhibited the phosphorylation of the 67 K substrate of ecto-PK by extracellular [γ^{32}P] ATP, and this PKC peptide caused a $74.3 \pm 2.4\%$ and $84.2 \pm 0.9\%$ inhibition of the phosphorylation of the 13 and 12 K proteins, respectively (mean $\pm$ SEM, $n = 7$; $p < 0.05$). Under the same reaction conditions, the active phorbol ester PMA, an activator of the regulatory domain of PKC, did *not* stimulate the phosphorylation of surface proteins by ecto-PK indicating that ecto-PK is PMA-insensitive. We employed a monoclonal antibody (termed M.Ab. 1.9), developed by Mochly-Rosen and Koshland (1987, 1988) against the catalytic domain of purified rat brain PKC, to shed more light on the similarity to PKC of the neuronal ecto-PK activity. M.Ab. 1.9 inhibits the activity of all PKC isozymes and of the phorbol-*in*sensitive proteolytic product of PKC, PKM, but does not inhibit cyclic-AMP or calmodulin-dependent protein kinases (Mochly-Rosen and Koshland, 1987). The addition of 0.1 μg/well of M.Ab. 1.9 to our standard assays with embryonic chick neurons produced a $51.4 \pm 7.4\%$ inhibition in the phosphorylation of the 67 K protein, and $62.2 \pm 4.2\%$ and $44.0 \pm 5.6\%$ inhibition of the phosphorylation of the 13 and 12 K substrates,

respectively (means $\pm$ SEM, $n = 5$; $p < 0.05$). Poly-clonal antibodies that recognize sequences in the catalytic regions of eight different PKC isozymes (α, β, γ, δ, ε, θ, η, and ζ) were used to characterize the ecto-PKC activity of brain neurons. Two of these PKC antibodies (anti-δ and anti-ζ) produced significant and selective inhibitory effects. An antibody prepared by Dr. P. J. Parker against the 11 amino-acid sequence at the C-terminus in the catalytic region of the isozyme PKC-δ (used at a dilution of 1:300) selectively inhibited the endogenous phosphorylation of the 12 and 13 K proteins by ecto-PK in intact brain neurons. Quantitation of these results determined that anti-PKC-δ caused a $60.4 \pm 2.2\%$ and $55.9 \pm 5.1\%$ inhibition (means $\pm$ SEM, $n = 3$; $p < 0.05$) in the phosphorylation of the 13 and 12 K protein substrates, respectively. This anti-PKC δ antibody, however, did not inhibit the endogenous phosphorylation by ecto-PKC of the 67 K protein substrate (Hogan et al., 1995). As a control for the specificity of these results, we found that an antibody directed against the V3 region in the regulatory domain of PKC-δ had *no* effect on ecto-PK activity. In contrast to the effects of anti-PKC-δ, another antibody directed to a specific sequence in the C-terminus of PKC ζ (zeta; from Gibco, used at 1:300) caused a $36.8 \pm 6.3\%$ inhibition (mean $\pm$ SEM, $n = 3$, $p < 0.05$) in the surface phosphorylation of the 67 K substrate of ecto-PKC in brain neurons, without significant effects on the phosphorylation of the 12 and 13 K surface proteins. We conclude that neuronal ecto-protein kinase activity is exerted by several different protein kinases, each with a different catalytic specificity. Each ecto-protein kinase phosphorylates different endogenous substrates on the surface of the neuronal plasma membrane, and therefore may be expected to regulate different neuronal functions.

Ecto-PKC activity of brain neurons is regulated by development and by Alzheimer's β-amyloid peptides

The phosphorylation of ecto-domains of membrane proteins by ecto-PK on the surface of immature brain neurons (obtained from embryonic chick-telencephalon) was followed for 1-7 days in culture. The results of these experiments are shown in Fig. 2. Rapid neurite outgrowth begins in these cells within 1 DIV, and an extensive network of neurites can be seen at 7 DIV. It can be clearly seen that the peak in surface phosphorylation of the 12 and 13 K proteins by ecto-PK in these cultured neurons occurs between 3–4 DIV, which coincides with the maximal rate of new neurite outgrowth (Hogan et al., 1995). At 7 DIV the surface phosphorylation of the 12 and 13 K protein decreased by over 75%. These results suggest that the ecto-PK which phosphorylates these two surface proteins may contribute to molecular mechanisms that regulate neuritogenesis. The preparation of specific antibodies that inhibit selectively the phosphorylation of sites in the ecto-domain of these surface proteins will enable us to determine directly the role of this developmentally regulated process in neuritogenesis.

The 12 and 13 K protein-substrates of neuronal ecto-PKC whose maximal surface phoshorylation occurs at the onset of neuritogenesis are likely candidates for regulation by factors that stimulate or inhibit neurite outgrowth, such as β-amyloid peptides. Using ecto-PKC assays conducted for 10 min with immature brain neurons (3 DIV), Hogan et al. (1995) found that the peptide β-amyloid 25–35 applied at the neurotrophic concentration of 0.1 nM (Yankner et al., 1990), significantly stimulated the $[\gamma^{32}P]ATP$ phosphorylation of the 12 K ($125 \pm 13\%$ of control; mean $\pm$ SEM, $n = 6$; $p < 0.05$) and the 13 K ($130 \pm 11\%$, $n = 6$, $p < 0.05$) neuronal surface proteins. The β-amyloid peptide

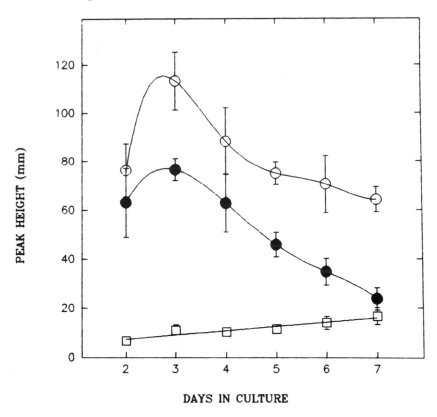

Fig. 2. Surface phosphorylation of the 12 and 13 K proteins by ecto-PK is developmentally regulated. Primary neurons were cultured from the telencephalon of chick embryos, maintained up to 7 days in vitro (DIV) and assayed for ecto-PK activity as described in Hogan et al. (1995). Every day, neurons attached to wells of 48-well plates were incubated for 10 min with $[\gamma^{32}P]ATP$. The radiolabeled proteins were resolved by SDS-PAGE and the autoradiographic results were quantitated by densitomertry ($n = 3$, $\pm$ SD) as described by Hogan et al.(1995). Note that while the surface phosphorylation of a 105 K protein (open squares) did not change much with development, surface phosphorylation of the 12 K (open circles) and the 13 K (filled circles) protein components peaked at 3 days in culture.

1–40, when added at 0.1 nM, stimulated the [γ^{32}P]ATP phosphorylation of the 13 K (134±20%, $n=3$; $p<0.05$) surface protein. In contrast, when the β-amyloid peptides 1-40 or 25-35 were added to neurons at 7 DIV, there was a very specific inhibitory effect. At 0.1 μM, β1–40 inhibited selectively the [γ^{32}P] ATP phosphorylation of the 67, 13, and 12 K substrates of ecto-PKC in brain neurons. The main effect was over 80% inhibition in surface phosphorylation of the 13 K protein. Examination of the structure/function relationships of β-amyloid peptides also revealed specificity. At 0.1 μM, β1–28, a fragment that has no neurotoxic effects, but has been reported to be neurotrophic (Whitson et al., 1989), did not inhibit the extracellular [γ^{32}P] ATP phosphorylation of the 13 and 12 K surface proteins but stimulated the phosphorylation of the ecto-PK substrate proteins of 116 and 105 K. The β25-35 peptide, on the other hand, has potent neurotoxic effects (Koh et al., 1990; Yankner et al., 1990; Pike et al., 1993) and we anticipated that it may inhibit the phosphorylation of specific proteins involved in neuritogenesis. The β25-35 peptide indeed inhibited the surface phosphorylation of all three specific substrates of brain ecto-PKC, with the rank order of potency being 13 > 67 > 12 K. Interestingly, the inhibition of surface protein phosphorylation in cultured brain neurons by β25-35 which was observed in our laboratory, showed the *same* concentration-dependence as that observed for the neurodegenerative action of this peptide (Koh et al., 1990; Yankner et al., 1990). The extracellular location of ecto-PKC in brain neurons, its selectivity towards developmentally-regulated surface protein substrates, and the specificity of its dual responses (stimulation and inhibition) to different regions within the sequence of β-amyloid protein (1–28 vs 25–35), render this enzyme most suitable as a candidate biochemical mediator for the neurotrophic and neurotoxic actions of Alzheimer's amyloid peptides.

The hydrolysis of extracellular ATP plays a role in the maintenance of long term potentiation (LTP)

The accumulation of increasing concentrations of ATP in the synaptic cleft that is induced by a train of repetitive stimuli delivered at high frequency (Silinsky, 1975), together with the utilization of this ATP by an ecto-PK with a relatively high K_m for ATP (Ehrlich et al., 1988) were proposed (Ehrlich, 1987) as biochemical events that are *unique* to the process of long-term potentiation (LTP) of synaptic strength. Indeed, exocytosis of vesicular ATP induced by high frequency stimulation (HFS) that causes LTP has been demonstrated directly in hippocampal slices (Wieraszko et al., 1989). Furthermore, low-frequency presynaptic stimulation delivered in hippocampal slices *together* with exogenously added ATP has been reported to induce LTP (Wieraszko and Seyfried, 1989). The continuation of these studies (Wieraszko and Ehrlich, 1994) revealed that application of extracellular ATP (100–500 nM) or its slowly hydrolyzable analogue adenosine 5'-0-(3-thiotriphosphate) (ATP-γ-S; 2.5 μM), to hippocampal slices together with *low*-frequency stimulation, amplified permanently the magnitude of the population spike. This effect was antagonized by adenylimidodiphosphate (AMPPNP), a nonhydrolyzable analogue of ATP. AMPPNP, other ATP analogues [2-methylthioadenosine triphosphate (2-MeSATP) and α, β-methyleneadenosine 5'-triphosphate (α, β-ATP)], or a purinergic receptor antagonist (Cibacron Blue 3G) tested in the concentration range of 3–40 μM did not exert agonistic activity similar to that of ATP or ATP-γ-S, suggesting that ATP hydrolysis is required to exert this effect. All the nonhydrolyzable analogues tested in these experiments are known agonists of ATP receptors. However, they did not stimulate, but rather reduced or prevented the establishment of stable, nondecremental LTP induced by high frequency electrical stimulation, without blocking the short-lasting increase in the magnitude of the population spike seen immediately after electrical stimulation (short-term potentiation). These results indicate that ATP released by high-frequency stimulation contributes to the maintenance of stable LTP. The underlying mechanism operating in this process may involve a new type of ATP receptor or hydrolysis by ecto-ATPase. However, the finding that ATP-γ-S is less potent than ATP and that other ATP analogues known to act as agonists of purinergic receptors did not induce LTP, but rather inhibited its maintenance, is more consistent with the involvement of an ecto-protein kinase, using extracellular ATP as a co-substrate, in mechanisms of LTP underlying synaptic plasticity. Experimental testing of this hypothesis required first direct demonstration of the presence of ecto-protein kinase on the surface of hippocampal pyramidal neurons. In these studies, pyramidal neurons were cultured from the hippocampus of 16-day-old mice embryos, maintained for 3 weeks in serum-free Neurobasal medium with supplement B27 (Gibco), until achieving maturation and synaptogenesis (Chen et al., 1994, 1996). All the criteria detailed above as required evidence for the presence of ecto-PK in intact neurons were fulfilled in experiments carried out with these hippocampal pyramidal neurons. The surface phosphoprotein substrates of ecto-PK in these hippocampal cells were identified as components migrating with apparent molecular mass of 105 K; a 66/68 K duplex; a 48/50 K duplex; 43, 32; 17 and a 12/13 K duplex. Phosphorylation by ecto-PK of the 12/13 K duplex on the surface of pyramidal neurons was detected only in cells younger than 8 DIV, and that of the 48/50 K began to appear at 8 DIV and reached maximum at 14-21 DIV (Chen et al., in preparation). The surface phosphorylation of the 48/50 K duplex in 18-21 DIV pyramidal neurons was extensively characterized in our laboratory. We found that the phosphorylation of this duplex increased with reaction time and reached a maximum after 2 h. However, even after 2 h labeling with inorganic ^{32}Pi this 48/50 K duplex was not labeled, indicating that in pyramidal neurons at 18–21 DIV these proteins are phosphorylated *only* on ecto-domains, exclusively by ecto-PK.

Ecto-protein kinase activity is required for the maintenance of stable LTP

As reported by Chen et al. (1996), the phosphorylation of a 48/50 K surface protein duplex by extracellular ATP identified in cultured pyramidal neurons and in hippocampal slices could be blocked by adding to the medium a monoclonal antibody termed M.Ab.1.9, directed against the catalytic domain of PKC. To study the involvement of this surface phosphorylation system in LTP,

M.Ab.1.9 was added to hippocampal slices 2 h prior to the induction of LTP by high frequency stimulation (HFS). The experiment has been conducted with slices placed at the interface of a recording chamber and maintained at 33°C. A bipolar stimulating-electrode was placed on Schaffer collateral–commissural fibers and recording electrodes were guided into pyramidal or radiatum layer of CA1 to record population spike or excitatory-post-synaptic potentials (EPSP).

The influence of M.Ab.1.9 on the slope of EPSP and on the population spike was found to be as follows: compared to the slope of EPSP in control slices which was permanently amplified by the HFS, in M.Ab.1.9-treated slices only a brief period of amplification was observed immediately after HFS . This was followed by a sharp decrease of EPSP and return to the control, prestimulus value within 30 to 40 min after HFS application. Parallel changes were recorded in the magnitude of the population spike: the increase in population spike measured immediately after HFS applied in the presence of M.Ab.1.9 was identical to that of control slices, but the potential recorded 30 min later was 40% lower than that of control slices. In slices incubated with a control antibody that binds to PKC but that cannot inhibit its activity (called M.Ab.1.12), LTP was the *same* as in slices tested without any antibody (control slices). Furthermore, both the slope of the EPSP and the magnitude of the population spike measured just before the application of HFS were not altered during very long preincubations with M.Ab.1.9, and remained the same as those recorded before adding the antibody, which did not differ from those seen in the control slices. Thus, M.Ab.1.9 does *not* influence basal synaptic transmission. Finally, the size and shape of the potentials recorded 30 min after HFS was applied in the presence of the antibody returned to the very same values observed before-eHFS, again confirming that M.Ab.1.9 does not interfere with basal activity. The selective inhibitory effects of the PKC antibody M.Ab.1.9 on the maintenance phase of LTP (seen 30 min after HFS) but not on its induction phase (1 min after HFS) are summerized in the histograms depicted in Fig. 3. These results have provided the first direct evidence of a causal role for ecto-protein kinase in the maintenance of stable LTP (Chen et al., 1996). We anticipate that our ongoing studies, focusing on the purification and cloning of the ecto-PK inhibited by M.Ab.1.9, and on the selective 48 and 50 K surface phosphoproteins substrates of ecto-PK of brain neurons, will reveal novel biochemical mechanisms operating in theregulation of processes underlying synaptic plasticity.

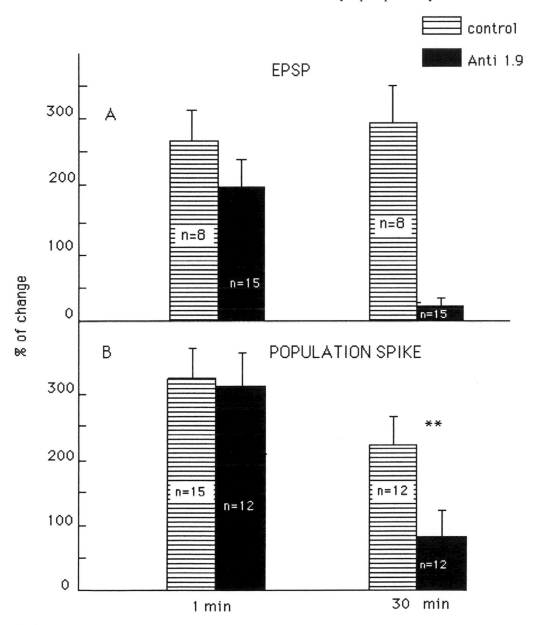

Fig. 3. Inhibition of ecto-PK by an anti-PKC antibody blocks the maintenance of LTP. The increase in excitatory post-synaptic potentials (EPSP) and the amplification of the population spike induced by high frequency electrical stimulation (HFS) were recorded from hippocampal slices as described in Wieraszko and Ehrlich (1994) under control conditions, and following preincubation with the anti-PKC monoclonal antibody-M.Ab.1.9 (Anti 1.9). The values for the mean ± SEM for the indicated number (*n*) of the percent of change from prestimulation values as recorded 1 min (left side) and 30 min (right side) after the application of HFS, are depicted in the histograms above. Data adapted from Chen et al. (1996).

Localization of the neuronal ecto-PKC within the synaptic cleft

Monoclonal antibody M.Ab.1.9 recognizes an epitope in the catalytic domain of cytoplasmic PKC (Mochly-Rosen and Koshland, 1988), and inhibits ecto-PKC activity when applied to the extracellular medium of cultured brain neurons (Hogan et al., 1995; Chen et al.,1996). Therefore, it could be expected that when this monoclonal antibody is applied to permealized cells or fixed synaptosomes, it should recognize epitopes of *intra*cellular PKC as well as sites on the cell surface. However, when applied to intact cells or to unfixed synaptosomes, M.Ab.1.9 binding should be restricted to the catalytic domain of ecto-PKC, and observed as binding only to externally-oriented epitope(s) localized on the outer surface of the plasma membrane. Studies designed to examine these events were carried out by Dr. Robert Lasher at the University of Colorado Medical Center. As reported by Ehrlich et al. (1997), immunoperoxidase staining with M.Ab.1.9 localized a labeled epitope on the external surface of the junctional plasma membrane of unlysed synaptosomes, and this staining was restricted to locations within the synaptic cleft. When M.Ab.1.9 was added to permealized synaptosomes, after fixation, this antibody recognized intracellular epitopes associated with synaptic vesicles and mitochondria, in addition to the staining of junctional plasma membranes. Thus, immunoperoxidase localization of ecto-PKC revealed that the externally-oriented epitope recognized by M.Ab.1.9 is present primarily on the surface of pre- and post-synaptic membranes within the synaptic cleft .

The binding of extracellular ATP to a 43 k protein on the neuronal cell-surface is blocked by M.Ab.1.9: Identification of ecto-PKC

Ecto-PKC is a protein kinase that operates on the surface of intact hippocampal pyramidal neurons and catalyzes reactions carried out with low concentrations (50–100 nM) of extracellular ATP. Preincubation with M.Ab.1.9, that binds to the catalytic domain of this kinase and inhibits its activity as well as blocks the stabilization of LTP,

should inhibit the binding of ATP to this ecto-PK. We have successfully accomplished the identification of an ecto-PK with these properties by using the procedures for photoaffinity labeling of ATP binding proteins with [α^{32}P]8-azido ATP as described by Nagy and Shuster (1995). Photoaffinity labeling was conducted with chick telencephalon neurons and mature hippocampal pyramidal neurons. The neurons were labeled as intact cells, while still attached to plates under conditions identical to those in which we assay ecto-PK activity in these cells (Pawlowska et al., 1993; Hogan et al., 1995; Chen et al., 1996). Using these procedures we have determined that a protein migrating with apparent molecular mass of 43 Kda is labeled when this membrane-impermeable photoaffinity probe is added to the extracellular medium at a concentration of 100 nM (Ehrlich et al., 1997 and 1999); identical to the concentration of ATP used in our routine ecto-PK assays. In one of the experiments conducted to prove the extracellular orientation of this ATP-binding protein, we incubated intact, attached neurons pre-labeled with [α^{32}P]8-azido ATP, with very low concentrations of trypsin (as described in Ehrlich et al., 1986). As expected from a surface protein, this mild proteolysis degraded the 43 K binding-protein without harming cell viability. As expected from the ATP binding site of a protein kinase, an excess of nonradioactive ATP fully competed with the labeling of this 43 K surface protein with 100 nM [α^{32}P]8-azidoATP.

We have examined the relationships between the high-affinity binding of ATP to the 43 K surface protein, and the activity of the pyramidal ecto-PK which is involved in the maintenance of LTP, by using several selective monoclonal antibodies to PKC. Preincubation with 1 μg/ml of M.Ab.1.9 (the concentration used to inhibit the phosphorylation of the 48 and 50 K substrates of ecto-PK in pyramidal neurons, Chen et al., 1996), inhibited the binding of 100 nM [α^{32}P]8-azido ATP to the 43 K protein on the neuronal cell surface. M.Ab.M6, however, that binds to PKC alpha but not to ecto-PKC, did not inhibit this binding. The correspondence in the ability of several specific PKC antibodies to inhibit the phosphorylation of the 48 and 50 K proteins on the surface of pyramidal neurons by ecto-PKC, *and* to block the binding of $[\alpha^{32}P]$8-azido ATP to the surface 43 K protein of these neurons, *and* to stain intrajunctional surface epitopes within the synaptic cleft, *and* to prevent the stabilization of LTP in the hippocampus, collectively identify the ecto-protein kinase molecule (ecto-PKC) and the specific surface protein substrates that play a role in synaptic plasticity.

Activity of ecto-PKC on the surface of human platelets: cardiovascular implications

Stimulus-induced exocytosis of vesicular ATP occurs not only from neurons, but also in the cardiovascular system, from platelets. We have described in human platelets an ecto-protein kinase/ectophosphoprotein phosphatase system which can utilize extracellular ATP as a co-substrate (Ehrlich, 1987; Ehrlich and Kornecki, 1987; Naik et al., 1991). The ecto-PK activity of platelets could be inhibited by the PKC antibody M.Ab.1.9. The effect was dose-dependent with 50% inhibition exerted by about 5 μg of M.Ab.1.9 per 5×10^8 platelets (Babinska et al., 1996). These results indicated that, like in neurons, one of the ecto-PKs acting on the surface of platelets has the catalytic specificity of PKC. Most interestingly, as depicted here in Fig. 4, incubation of washed, intact human platelets with antibody M.Ab-1.9 resulted in direct activation of the platelets, inducing aggregation followed by granule-secretion (Babinska et al., 1996). Two control antibodies (M.Ab.1.12 and 1.3) directed to the regulatory domain of PKC, did *not* activate platelets. Increasing the concentration of M.Ab.1.9 had no effect on the extent of aggregation, but in a dose-dependent manner shortened the latency of antibody-induced activation, similar to the manner induced by the platelet-stimulating antibodyM.Ab.F11 (Kornecki et al., 1990). Using procedures of flow cytometry and immunofluorescence (shown here in Fig. 5), we have demonstrated that the ecto-PKC whose inhibition by M.Ab.1.9 causes platelet aggregation is localized on the external surface of the platelet plasma membrane (Babinska et al., 1996). Normal hemostasis maintains platelets in a discoid shape and prevents cells from aggregating spontaneously in the circulation. Constitutive phosphorylation of surface proteins by

a platelet ecto-PKC with a low K_m for ATP (ATP concentration in circulating plasma is <1 μM) may play a role in hemostasis by maintaining a steady-state of phosphorylation of these platelets surface proteins. We tested this hypothesis by using a membrane-impermeable inhibitor of phosphatase activity, microcystin-LR (Honkanen et al.,1990). Microcystin-LR is a cyclic heptapeptide which inhibits type 1 and type 2A protein phosphatases. The addition of microcystin to the extracellular medium prevented the induction of platelet aggregation caused by M.Ab.1.9. This result provided evidence that the inhibition of surface ecto-protein kinase *together* with continuing ecto-protein phosphatase activity causes an agonist-independent aggregation of human platelets by M.Ab.1.9. Thus,

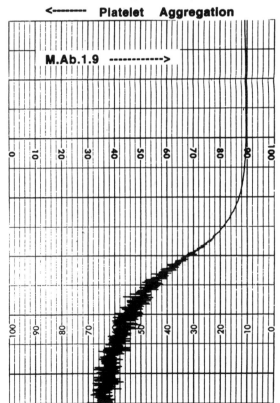

Fig. 4. Inhibition of ecto-PK by an anti-PKC antibody induces aggregation of human platelets. Aggregation of washed human platelets (4×10^8 ml) is shown here as a deflection to the left in the graph of light-transmission which begins approximately 2 min (a latency period) after the addition (marked by the arrow) of M.Ab.1.9 (20 μg/ml) to the suspension of stirred platelets. Data from Babinska et al. (1996).

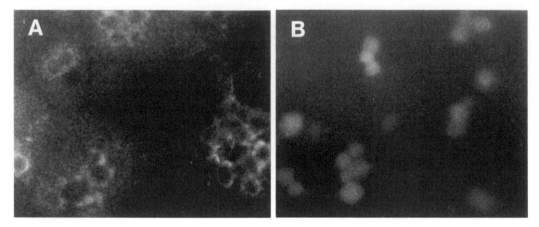

Fig. 5. Localization of the ecto-PK recognized by M.Ab.1.9 on the surface of intact platelets. Human platelets incubated with M.Ab.1.9 were subjected to immunofluorescent staining using fluorescein-conjugated second antibody, then photographed at a total magnification of 660×. Note that a rim-staining pattern characteristic of surface-localization can be seen with intact platelets preincubated with M.Ab.1.9 (panel A), but not with control samples preincubated with non-immune immunoglobulin instead of M.Ab.1.9 (panel B). Data from Babinska et al. (1996).

the platelet activation process which is triggered by M.Ab.1.9 begins with changes in the phosphorylation state of surface proteins. In the same study we have determined that the next step in this process involves membrane rearrangements that bring about the exposure of fibrinogen-binding sites on the platelet surface (Babinska et al., 1996). This represents a novel homeostatic mechanism in which the maintenance of a proper steady-state of phosphorylation of proteins on the cell surface controls the state of cell activity. In the circulation, this mechanism protects platelets from spontaneous aggregation. We propose that in the nervous system, rearrangements of plasma-membrane proteins that are controlled by surface phosphorylation events may serve as signals, or regulators, of neuritogenesis and synaptogenesis.

Conclusions and future directions

Activity-dependent neuronal events in which vesicular-exocytosis from presynaptic nerve terminals influences the connectivity between cells was shown in recent years to play critical roles in neuronal development and in processes underlying synaptic plasticity. These processes depend on neuronal stimulation and neurotransmitter release from nerve terminals. Neuronal exocytosis releases not only neurotransmitters, but also the nucleotide ATP is secreted to the synaptic cleft during activity-dependent events. Nucleotides represent a ubiquitous class of important signalling molecules in all tissues, including the nervous system. In particular a final breakdown product of ATP, adenosine, is a well known neuroregulator. However, the purpose for the secretion of the high energy nucleotide, ATP, to the extracellular space was not clear. The discovery of neuronal ecto-protein kinase revealed that in molecular mechanisms operating in activity-dependent neuronal events, the ATP co-secreted together with the neurotransmitter molecules into the synaptic cleft can be utilized by surface protein phosphorylation systems for the regulation of cellular interactions involved in neuronal development and synaptic plasticity.

Our studies of the extracellular phosphorylation of unique proteins on the surface of brain neurons have implicated a specific, developmentally-regulated, ecto-PK activity in neuritogenesis and synaptogenesis. This ecto-PK activity is inhibited selectively by extracellular application of monoclonal antibodies directed to the catalytic domain of protein kinase C (PKC) and by PKC pseudosubstrate peptides. Therefore, we termed this enzyme ecto-PKC. The onset of neuritogenesis correlated temporally with phosphorylation by ecto-PKC of 12 and 13 K surface proteins, whereas the onset of synaptogenesis corresponded with phosphorylation of 48 and 50 K surface proteins by ecto-PKC. In electrophysiological studies of the induction of LTP of synaptic strength in hippocampal slices, a role for extracellular ATP has been established. The critical involvement of ecto-PK in molecular mechanisms operating in LTP was revealed by use of the PKC inhibitory antibody M.Ab.1.9. Application of M.Ab.1.9 inhibited the phosphorylation of the 48 and 50 K proteins on the surface of pyramidal neurons and blocked the stabilization of LTP induced by high-frequency stimulation in hippocampal slices. Photoaffinity labeling by $[\alpha^{32}P]$8-azido-ATP identified ecto-PKC as a 43 K cell-surface protein whose binding of ATP is blocked by M.Ab.1.9. The externally-oriented epitope of the ecto-PKC recognized by M.Ab.1.9 was localized by immuno-electron microscopy of synaptosomes in intrajunctional locations; demonstrating that ecto-PKC operates within the synaptic cleft - where ATP is secreted upon neuronal stimulation. Studies of the ecto-PKC of human platelets revealed that changes in the phosphorylation state of surface proteins can trigger the membrane rearrangements associated with cellular activation.

The hypothesis we have formulated on the basis of all these findings states that alterations in the phosphorylation state of ecto-domains in surface proteins of developing neurons induce conformational changes leading to rearrangements of membrane components which trigger the onset of neuritogenesis and synaptogenesis. We propose that a similar process operates in the mature brain in LTP: the phosphorylation of specific surface proteins by ecto-PKC, which occurs uniquely after the enhanced release of presynaptic ATP by high-frequency stimulation, signals for the initiation of cellular events involved specifically in the formation of long-lasting alterations in neuronal structure and function, including the dendritic-sprouting and de novo synaptogenesis involved in memory formation.

Future studies based on the synthesis of pseudo-substrate-peptides that inhibit specifically the phosphorylation of ecto-domains in the 12, 13, 48 and 50 K surfaceprotein-substrates of ecto-PKC, on the preparation of new antibodies specific to different ecto-PKs, and studies of the expression of genes to be cloned for all these proteins, will undoubtedly reveal novel mechanisms with critical roles in the regulation of neuronal development and synaptic plasticity, and could provide a new generation of therapeutic agents for the treatment of developmental disabilities, dementia, and certain neurodegenerative disorders.

Acknowledgments

The research in the authors' laboratories described in this chapter has been supported by grants from the National Institute of Health, the US Air Force Office for Scientific Research and the American Heart Association, and by research awards from the PSC-CUNY and the HEAT-Program of the state of New York.

References

Asch, A.S., Liu, I., Briccetti, F.M., Barnwell, J.W., Kwakye-Berko, F., Dokun, A., Goldberger, J. and Pernambuco, M. (1993) Analysis of CD36 binding domains: Ligand specificity controlled by dephosphorylation of an ectodomain. *Science*, 262: 1436–1440.

Apasov, S.G., Smith, P.T., Jelonek, M.T., Margulies, D.H. and Sitkovsky, M.V. (1996) Phosphorylation of extracellular domains of T-lymphocyte surface proteins. *J. Biol. Chem.* 271: 25677–25683.

Babinska, A., Ehrlich, Y.H. and Kornecki, E. (1996) Activation of human platelets by protein kinase C antibody: Role of ecto-protein kinase in platelet homeostasis. *Am. J. Physiol.*, 27: H2134–2144.

Burnstock, G.J. (1981) Neurotransmitters and trophic factors in the autonomic nervous system. *J.Physiol.*, 313: 1–35.

Burnstock, G. (1990) Purinergic mechanisms. *Ann. NY Acad. Sci.*, 603: 1–18.

Burnstock, G. (1999) Current status of purinergic signalling in the nervous system. Chapter 1 of this volume, pp. 3–10.

Chen, W., Hogan, N.V., Wieraszko, A., Pawlowska, Z., Soifer, D. and Ehrlich, Y.H. (1994) NMDA-regulated ecto-protein kinase in hippocampal neurons: Role in LTP. *Society for Neuroscience Abstracts,* 20: 263, no. 119.6.

Chen, W., Wieraszko, A., Hogan, M.V., Yang, W., Kornecki, E. and Ehrlich, Y.H. (1996) Surface protein phosphorylation by ecto-protein kinase is required for the maintenance of hippocampal long-term potentiation. *Proc. Natl. Acad. Sci., USA*, 93: 8688–93.

Chen, X.Y. and Lo, T.C. (1991a) Phosphorylation of a cell surface 112 Kd protein by an ecto-protein kinase in rat L6 myoblasts. *Biochem J.*, 279: 467–474.

Chen, X.Y. and Lo, T.C. (1991b) Involvement of a cell surface protein and an ecto-protein kinase in myogenesis. *Biochem. J.*, 279: 475–482.

424

Chen et al. (in preparation).

Dorner, T., Gagelmann, M., Geller, S., Herbst, F. and Forssmann, W.G. (1989) Phosphorylation and dephosphorylation of the natriuretic peptide urodilatin (CDD-ANP-95-126) and the effect on biological activity. *Biochem. Biophys. Res. Commun.*, 163: 830–835.

Edwards, F.A., Gibb, A.J. and Colquhoun, D. (1992) ATP receptor-mediated synaptic currents in the central nervous system. *Nature,* 359: 144–147.

Ehrlich, Y.H. (1984) Protein phosphorylation. Role in the function, regulation, and adaptation of neural receptors. In: *Handbook of Neurochemistry.* 2nd edition, Vol. 6. Plenum Press, New York, pp. 541–574.

Ehrlich, Y.H., Davis, T.B., Bock, E., Kornecki, E. and Lenox, R.H. (1986) Ecto protein kinase activity on the external surface of intact neural cells. *Nature*, 320: 67–69.

Ehrlich, Y.H., Lenox, R.H., Kornecki, E. and Berry, W. (eds.) (1987) Molecular mechanisms of neuronal responsiveness. *Adv. Exper. Med. Biol.*, Vol. 221. Plenum Press, New York.

Ehrlich, Y.H. (1987) Extracellular protein phosphorylation in neuronal responsiveness and adaptation. *Adv. Exp. Med. Biol.*, 221: 187–199.

Ehrlich, Y.H. and Kornecki, E. (1987). In: M.C. Cabot and W.L. McKeehan (Eds.) *Mechanisms of Signal Transduction by Hormones and Growth Factors.* A. Liss, New York, pp. 193–205.

Ehrlich, Y.H., Snider, R.M., Kornecki, E., Garfield, M.K. and Lenox, R.H. (1988) Modulation of neuronal signal transduction system by extracellular ATP. *J. Neurochem.* 50: 295–301.

Ehrlich, Y.H., Hogan, M.V., Pawlowska, Z., Naik, U. and Kornecki, E. (1990) Ectoprotein kinase in the regulation of cellular responsiveness to extracellular ATP. *Ann. NY Acad. Sci.* 603: 401–416.

Ehrlich, Y.H., Lasher, R., Hogan, M.V., Babinska, A., Wieraszko, A. and Kornecki, E. (1997) Molecular identification and synaptic localization of the ecto-protein kinase required for the stabilization of LTP. *Society for Neuroscience Abstracts.*, 23: 1393, no. 549.1.

Ehrlich, Y.H., Hogan, M.V., Babinska, A., Sobocki, T. and Kornecki, E. (1999) Identification of ecto-protein kinase C as a 43K ATP-binding protein on the surface of neurons and human platelets. *J. Neurochem.*, 72: S36C.

Evans, R., Derkach, V. and Surprenant, A. (1992) ATP mediates fast synaptic transmission in mammalian neurons. *Nature*, 357: 503–505.

Fedan, J.S. and Lamport, S.J. (1990) P2-purinoceptor antagonists. *Ann. New York Acad. Sci.*, 603: 182–97.

Fernandez, J., Naik, U., Ehrlich, Y.H. and Kornecki, E. (1993) Comparative investigation of the effects of the immunosuppressants cyclosporine A, cyclosporine G and FK-506 on platelet activation. *Cell. Mol. Biol. Res.*, 39: 257–264.

Forsberg, P.O. and Martin, S.C. (1991) Phosphorylation/dephosphorylation and the regulation of fibrinogen and complement factor C3. *Upsala J. Med. Sci.* 96: 75–93.

Friedberg, H., Belzer, I., Ogad-Plesz, O. and Kuebler, D. (1995) Activation of cell growth inhibitor by ectoprotein kinase-mediated phosphorylation in transformed mouse fibroblasts. *J. Biol. Chem.* 270: 20560–20567.

Fujii, S., Kato, H., Furuse, H, Osada, H. and Kuroda, Y. (1995a) The mechanism of ATP-induced long-term potentiation involves extracellular phosphorylation of membrane proteins in guinea-pig hippocampal CA1 neurons. *Neurosci. Lett.*, 187: 130–132.

Fujii, S., Osada, H., Hamaguchi, T., Kuroda, Y. and Kato, H. (1995b) Extracellular phosphorylation of membrane protein in hippocampal slices. *Neurosci. Lett.*, 187: 133–136.

Gordon, J.L. (1986) Extracellular ATP: Effects, sources and fate. *Biochem. J.* 233: 309–319.

Hardwick, J.C., Ehrlich, Y.H. and Hendley, E.D. (1989). Extracellular ATP stimulates norepinephrine uptake in PC12 cells. *J. Neurochem.*, 53: 1512–1518.

Hartmann, M. and Shrader, J. (1992) Exo-protein kinase release from intact cultured aortic endothelial cells. *Biochim. Biophys. Acta.*, 1136(2): 189–195.

Hendley, E.D., Whittemore, S.R., Chaffee, J.E. and Ehrlich, Y.H. (1988). Regulation of norepinephrine uptake by adenine nucleotides and divalent cations: Role for extracellular protein phosphorylation. *J. Neurochem.*, 50: 263–273.

Herbert, J.M., Daviet, I. and Maffrand, J.P. (1991) Characterization of extracellular binding for [^{3}H]-staurosporine on capillary endothelial cells. *Cell Bio. Int. Rep.*, 15: 883–890.

Hogan, M.V., Pawlowska, Z., Yang, H.A., Kornecki, E. and Ehrlich, Y.H. (1995) Surface phosphorylation by ecto-protein kinase C in brain neurons: a target for Alzheimer's β-amyloid peptides. *J. Neurochem.*, 65: 2022–2030.

Honkanen, R.E., Zwiller, J. and Boynton, A.L. (1990) Characterization of microcystin-LR, a potent inhibitor of type 1 and type 2A protein phosphatases. *J. Biol. Chem.*, 265: 19401–19404.

House, C. and Kemp, B.E. (1987) Protein kinase C contains a pseudosubstrate prototope in its regulatory domain. *Science*, 238: 1726–1728.

Kalafatis, M., Rand, M.D., Jenny, R.J., Ehrlich, Y.H. and Mann, K.G. (1993). Phosphorylation of Factor Va and Factor VIIIa by activated platelets. *Blood*, 81: 704–719.

Kornecki, E., Walkowiak, B., Naik, U. and Ehrlich, Y. H. (1990) Activation of human platelets by a stimulatory monoclonal antibody. *J. Biol. Chem.*, 265: 10042–10048.

Koh, J., Yang, L.L. and Cotman, C.W. (1990) β-Amyloid protein increases the vulnerability of cultured cortical neurons to excitotoxic damage. *Brain Res.*, 533: 315–320.

Kreutzberg, G.W., Reddington, M. and Zimmermann, H. (1986) In: *Cellular Biology of Ectoenzymes: Proceedings of the International Erwin-Riesch-Symposium on Ectoenzymes.* Springer-Verlag, Berlin.

Kübler, D., Reinhardt, D., Reed, J., Pyerin, W. and Kinzel, V. (1992) Atrial natriuretic peptide is phosphorylated by intact cells through cAMP-dependent ecto-protein kinase. *Eur. J. Biochem.*, 206: 10638–10645.

Kuroda, Y., Ichikawa, M., Muramoto, K., Kobayashi, K., Matsuda, Y., Ogura, A. and Kudo, Y. (1992) Block of synapse formation by a protein kinase inhibitor. *Neurosci. Lett.*, 135: 255–258.

Martin, S.C., Forsberg, P.O. and Eriksson, S.D. (1991) The effects of in vitro phosphorylation and dephosphorylation on the thrombin-induced gelation and plasmic degradation of fibrinogen. *Thromb. Res.*, 61: 243–252.

Merlo, D., Anelli, R., Calissano, P., Ciotti, M.T. and Volonte, C. (1997) Characterization of an ecto-phosphorylated protein of cultured cerebellar granule neurons. *J. Neurosci. Res.*, 47: 500–508.

Mochly-Rosen, D. and Koshland, D.E. (1987) Domain structure and phosphorylation of protein kinase C. *J. Biol. Chem.*, 262: 2291–2297.

Mochly-Rosen, D. and Koshland, D.E. (1988) A general procedure for screening inhibitory antibodies: Application for identifying anti-protein kinase C antibodies. *Anal. Biochem.*, 170: 31–37.

Muramoto, K., Kobayashi, K., Nakanishi, S. and Kuroda, Y. (1988) Functional synapse formation between cultured neurons of rat cerebral cortex. *Proc. Jpn. Acad. Sci.* [Ser. B] 64: 319–322.

Muramoto, K., Kawahara, M., Kobayashi, K., Taniguchi, H. and Kuroda, Y. (1992) Phosphorylation of extracellular domains of microtubule-associated protein 1B may be involved in synapse formation between cortical neurons. *Neurosci.Lett.*, 135, 255.

Nagashima, K., Nakanishi, S. and Matsuda, Y. (1991) Inhibition of nerve growth factor-induced neurite outgrowth of PC12 cells by a protein kinase inhibitor which does not permeate the cell membrane. *FEBS Lett.*, 293: 119–123.

Nagy, A.K., Shuster, T.A. and Delgado-Escueta, A.V. (1986) Ecto-ATPase of mammalian synaptosomes. Identification and enzymic characterization. *J. Neurochem.* 47: 976–86.

Nagy, A.K. and Shuster, T.A. (1995) ATP-binding proteins on the external surface of synaptic membranes: Identification by photoaffinity labeling. *J. Neurochem.*, 65: 1849–1858.

Nagy, A.K. (1997) Ecto-ATPases of the nervous system. In: Plesner, L., Kirley, T.L. and Knowles, A.F. (Eds.) *Ecto-ATPases. Recent Progress on Structure and Function*. Plenum Press, New York.

Naik, U.P., Kornecki, E. and Ehrlich, Y.H. (1991) Phosphorylation and dephosphorylation of human platelet surface proteins by an ecto-protein kinase/phosphatase system. *Biochem. Biophys. Acta*, 1092: 256–264.

Oda, Y., Kuo, M., Huang, S.S. and Huang, J.S. (1993) Acidic fibroblast growth factor receptor purfied from bovine liver plasma membranes has aFGF-stimulated kinase, autoadenylation, and alkaline nucleotide phosphodiesterase activities. *J. Biol. Chem.*, 268: 27318–27326.

Oda, Y., Kuo, M., Huang, S.S. and Huang, J.S. (1991) The plasma cell membrane glycoprotein, PC-1, is a threonine-specific protein kinase stimulated by acidic fibroblast growth factor. *J. Biol. Chem.*, 266: 16791–16795.

Pawlowska, Z., Hogan, M.V., Kornecki, E. and Ehrlich, Y.H. (1993). Ecto-protein kinase and surface protein phosphorylation in PC12 cells: Interactions with nerve growth factor. *J. Neurochem.*, 60: 678–686.

Pawlowska, Z., Chen, W., Hogan, M.V., Yang, H-A. and Ehrlich, Y.H. (1994) Neuronal surface phosphoproteins associated with neuritogenesis and synaptogenesis. *Society for Neuroscience Abstracts*, 20: 1298, no. 534.5.

Pfeifle, J., Hagmann, W. and Anderer, A. (1981) Cell adhesion-dependent differences in endogenous protein phosphorylation on the surface of various cell lines. *Biochim.Biophys.Acta*, 670: 274.

Pike, C.J., Burdick, D. and Walencewicz, A.J. (1993) Neurodegeneration induced by b-amyloid peptides in-vitro: The role of peptide assembly state. *J. Neurosci.*, 13(4): 1676–1687.

Pirotton, S., Boutherin-Falson, O., Robaye, B. and Boeynaems, J.M. (1992) Ecto-phosphorylation on aortic endothelial cells. Exquisite sensitivity to staurosporine. *J. Biochem.*, 285: 585–591.

Rebbe, N.F., Tong, B.D., Finley, E.M. and Hickman, S. (1991) Identification of nucleotide pyrophosphatase/alkaline phosphodiesterase I activity associated with the mouse plasma cell differentiation antigen PC-1. *Proc. Natl. Acad. Sci. USA*, 88: 5192–5196.

Shen, K.Z. and North, R.A. (1993) Excitation of rat locus coeruleus neurons by adenosine 5′-triphosphate: Ionic mechanism and receptor characterization. *J. Neurosci.*, 13: 894–899.

Silinsky, E.M. (1975) On the association between transmitter secretion and the release of adenine nucleotides from mammalian motor nerve terminals. *J. Physiol.*, 247: 145–162.

Silinsky, E.M., Gerzanichne, V. and Vanner, S.M. (1992). ATP mediates excitatory synaptic transmission in mammalian neurons. *Br. J. Pharm.*, 106: 762–763.

Skubitz, K.M. and Ehresmann, D.D. (1992) The angiogenesis inhibitor beta-cyclodextrin tetradecasulfate inhibits ecto-protein kinase activity. *Cell Mol. Biol.*, 38: 543–560.

Skubitz, K.M. and Goueli, S.A. (1994) CD31 (PECAM-1), CDw32 (Fc gamma RII), and anti-HLA class I monoclonal antibodies recognize phosphotyrosine-containing proteins on the surface of human neutrophils. *J. Immunol.*, 152: 5902–5911.

Sonka, J., Kubler, D. and Kinzel, V. (1989) Phosphorylation by cell surface protein kinase of bovine and human fibrinogen and fibrin. *Biochim. Biophys. Acta*, 997: 268–277.

Tsuji, S., Yamashita, T. and Nagai, Y. (1988) A novel, carbohydrate signal-mediated cell surface protein phosphorylation: Ganglioside GQ1b stimulates ecto-protein kinase activity on the cell surface of a human neuroblastoma cell line, GOTO. *J. Biochem. (Japan)*, 104: 498–503.

Vilgrain, I. and Baird, A. (1991) Phosphorylation of basic fibroblast growth factor by a protein kinase associated with the outer surface of a target cell. *Mol. Endocrinol.*, 5: 1003–1012.

Volonte, C., Merlo, D., Ciotti, M.T. and Calissano, P. (1994) Identification of an ectokinase activity in cerebellar granule primary neuronal cultures. *J. Neurochem.*, 63: 2028–2037.

Walter, J., Schnolzer, M., Pyerin, W., Kinzel, V. and Kubler, W. (1996) Induced release of cell surface protein kinase yields

426

CK1- and CK2-like enzymes in tandem. *J. Biol. Chem.*, 271: 111–119.

Walter, J., Capell, A., Hung, A.Y., Langen, H., Schnolzer, M., Thinakaran, G., Sisodia, S. S., Selkoe, D.J. and Haass, C. (1997) Ectodomain phosphorylation of β-amyloid precursor protein at two distinct cellular locations. *J. Biol. Chem.* 272: 1896–1903.

Whitson, J.S., Selkoe, D.J. and Cotman, C.W. (1989) Amyloid β-protein enhances the survival of hippocampal neurons in vitro. *Science*, 243: 1488–1490.

Wieraszko, A., Goldsmith, G. and Seyfried, T.N. (1989) Stimulation-dependent release of adenosine triphosphate from hippocampal slices. *Brain Res.* 485: 244–250.

Wieraszko, A. and Seyfried, T. (1989) ATP-induced synaptic potentiation in hippocampal slices. *Brain Res.*, 491: 356–359.

Wieraszko, A. and Ehrlich, Y.H. (1994) On the role of extracellular ATP in the induction of Long-Term Potentiation (LTP) in the hippocampus. *J. Neurochem.*, 63: 1731–1738.

Yankner, B.A., Duffy, L.K. and Kirschner, D.A. (1990). Neurotrophic and neurotoxic effects of amyloid β-protein: Reversal by tachykinin neuropeptides. *Science*, 250: 279–282.

Zimmermann, H. and Braun, M. (1999) Ecto-nucleotidase–molecular structures, catalytic properties and functional roles in the nervous system. pp. 371–386.

Subject Index

acetylcholine release   96, 111, 148–149, 153, 161–164, 185, 189
activated microglia   203
adenosine deaminase   146, 184, 189, 215–252, 260
ϵ-adenosine 5'-diphosphate (ϵ-ADP)   13
ϵ-adenosine 5'-monophosphate (ϵ-AMP)   13
adenosine receptor, genomic variations   304
adenosine receptor, phenptypic variations   305
A_1 adenosine receptor   113, 146, 176, 173, 185, 189
A_2 adenosine receptor   187, 265–272, 304–305
A_{2B} adenosine receptor   266–267
A_3 adenosine receptor   265, 277–284
adenylate cyclase   127, 135
5'-adenylic acid deaminase   261
ADPβS   175
adrenal chromaffin cells   34
α_1 adrenoceptor   5, 160
β_2 adrenoceptor   13, 14, 37
afterhyperpolarization (AHP) response   278–279
airways, Cl$^-$ secretion   45
ATI-082   100
alkaline phosphatase   373
Alzheimer disease   100, 101, 334
amiloride   75
D-2-amino-phonovalerate (APV)   194, 198, 199
aminophylline   195, 196
4-aminopyridine (4AP)   260
AMP deaminase   252
AMPA receptor   67
angiotensin receptor   14
anxiety   94
AP1 transcription complex   328, 336
A3P5P   95
Ap$_4$A   256–260
[^{3}H] Ap$_4$A, binding site   48, 49
Ap$_5$A   184, 256–260
ϵ-Ap$_5$A   405–406
Ap$_n$A, calcium response   406
apoptosis   4, 98–99, 319, 359–360
apyrase treatment   135, 138
arachidonic acid (AA)   6, 34, 194

arachidonic acid release   336
ARL67156 (N-N,N-diethyl-D-β-γ-dibromomethylene ATP)   16, 95
arterial natriuretic receptor   14
astrocytes   33, 323–331
   neuron interaction   318
   modulation by nucleotides, nucleosides   334
   spinal cord P2Y   312–313, 317
astrocytoma cells, 1321N1 human   101, 135–142
astrogliosis   100, 333–340
arterial natriuretic peptide receptor   13
ATP
   CA1 region of hippocampus   242–244
   central dopamiergic neurons   224
   cotransmitter   11, 13
   FGF2   327–329
   immunomodulator   363
   inflammatory mediator   363
   inhibition of hydrolysis   363–364
   neurotransmitter   16–18
   noradrengic neurons   224
   p38, MAPK   328
   plasminogen release   362
   prejunctional modulation of cotransmitter   13
   storage   11–12
   vascular smooth muscle   14
ATP binding cassette protein   7, 64, 94
ATP binding domain (P2X)   70
ATP receptors, postjunctional   13
ATP release   12–13, 145–157, 159–167
   carbamylcholine   164
   inhibition by A_2 adenosine receptor   267–270
   from astrocytoma cells   135–142
   from motor nerve endings   145–157
   postsynaptic   159–167
   synchronous quantal   148–149
   with acetylcholine   161–162
   with noradrenaline   160
ϵ-ATP   3
[^{35}S] ATPαS binding site   47–48, 50–51
[^{35}S] ATPβS   259
[^{35}S[ATPγS   120
ATPγS   137, 175, 195, 242

ATP-modulated potassium channels   94
atropine   15, 113, 162, 164

barbiturates   67
benzodiazepines   67, 96
Bergmann glia   34
bFGF   100
bicuculline   197–199
blood-brain-barrier   28
B-lymphocytes   342–352
   $[Ca^{2+}]$; ATP^{4-} induced   348, 352
   ATP-gated cation channel   351
   ATP^{4-} 348
   BzATP   351
   FURA red   351
brain ischemia   334
brain trauma   100
BzATP   96

C1-IB-MECA   279–282
C6 glioma cells   34, 127, 136, 141
CA1 area of hippocampus   188, 202, 275–284
CA1 hippocampal neurons
   adenosine A_3 receptor   277–284
   CNQK   239, 240
   D-APV   239
   EPSC   239
Ca^{2+} imaging   193
Ca^{2+} oscillation   194
Ca^{2+} response   138, 305
Ca^{2+} signals in astrocytes   317
Ca^{2+} waves in astrocytes   318
CA3 area of hippocampus   188
Caenorhabditis elegans   75
calcium channel inhibition   266
cAMP, increase by A_2 adenosine receptor   266
cAMP-regulated Cl^- channels   45
cancer   99
capsacin receptor   63
carbamylcholine = carbachol   161, 162, 164, 165, 278–279
CD39   95, 376–377
CGS-22494 (A_{2A} adenosine receptor agonist)   194
CHA (A_1 adenosine receptor agonist)   194
charybodotoxin   63
CHO-K1 cells   25, 54
chP2X$_4$ receptor
   development   87–88
   N-glycosylation sites   86
   recombinant expression   86–88
   topology   86
cibacron blue 3GA   173

coformycin   184
concanavalin A   186–187
ω-conotoxin   178
ω-conotoxin GVIA   260
COX-2 induction   329
CNQX   194, 198–200, 224, 228
CP66713   194
CPX   151, 185
cyclooxygenase   336
cyclopentyltheophylline   146, 194, 195, 256, 260
cysteine protease, caspase   99
cystic fibrosis (CF)   99–100
cystic fibrosis transmembrane conductance regulator (CFTR)   7, 45, 141

4-DAMP / M_3-receptor antagonist   162
DARPP-32   266
degenerins   75
diabetes   100
diadenosine polyphosphates (Ap_nA)
   catabolism   397–407
   extracellular function   397–407
   intracellular calcium increase   398, 400
diazepam   96
DIDS   95, 107
dinucleotide receptors (P4 receptors)
   rat midbrain synaptic terminals   398–400
   regulation protein kinases   400–402
   regulation protein phosphatases   400–402
dipyridamole   185, 189
dopamine D2 receptor   13, 224, 270
dopamine release   198, 227–228
dopaminergic neuron   224–232

ecto-5'-nucleotidase   95, 145, 166, 184, 188, 373–75
   catalytic properties   373
   distribution in nervous system   374
   functional role in nervous system   374
   phylogenetic relationship   373
   recombinant   375
ecto-apyrase (ATPDase)   16, 94, 375–76, 404–406
   immunolocalization   387–394
ecto-ATPase, inhibitor ARL67156   16, 95
ecto-nucleotidases, ischemia   378
ecto-nucleotidases   114, 162, 183, 371–381
ecto-protein kinase C (ecto-PKC)   420–421
ecto-protein kinases (PK)   411–423
   α, β T-cell receptor (TCR)   413
   β-APP (β-Amyloid Precursor Protein)   412, 416–417
   casein-kinase like ecto-PK   412
   cyclosporine   412

ecto-phosphoprotein phosphatase   412–413

  inhibition by anti-PKC-antibodies   415

  K-252b   412

  long term potentation (LTP)   417

  regulation   415–416

  stauroporine   412

  substrates on the surface of brain neurons   413

EEG (electroencephalogram)   230

end plate potentials (EPPs)   145

endothelin ET3 receptor   13

E-NTPase (E-ATPase) family

  apyrase related proteins   376

  catalytic properies   375–377

  CD 39, L1,–2,–3,–4   376–377

  distribution in nervous system   377

  ecto-apyrase   16, 94, 375–376

  ecto-ATPase   16, 94, 159, 166, 184, 375–376

  functional role   377

  phylogenetic relationship   375–377

epilepsy   93, 97

epileptiform activity   260, 261

epithelial Na^+ channel, degenerin gene superfamily   74

ERK (extracellular signal-regulated kinase) cascade   324–328

Evans blue   2, 95, 107

excitatory ATP receptors   198

excitatory junction potentials (EJPs)   14

excitatory muscarine receptor   218

excitatory P2X-receptors   177–178

excitatory polysynaptic potentials (EPSP)   6, 151

extracellular nucleotides, calcium response   314–317

extracellular nucleotides, sources in nervous system   311–312

FGF (fibroblast growth factor) receptor   324–328

forskolin   28

forskolin-stimulated cAMP accumulation   27

Fos/Jun heterodimer   336–337

GABA   96, 189

  receptor   65

  release, inhibition by A_{2A} receptor   266

  release, stimulation by ATP   203

$GABA_A$ receptor   67, 96, 224

$GABA_C$ receptor   67–68

GFAP (glial fibrillary acidic protein)   100

glutamate   96, 189

glutamate receptor   202

glutamate release   194, 197, 199, 202, 239

glycine   202

glycine receptor   65

glycine release   270

guanethidine   226

guanosine-5′-O (2-thiodiphosphate), GDPβS   212

[^{35}S] GTPγS binding site   51, 52

HEK 293 cells   96, 266

hexamethonium   149, 162

2-(hexylthio) adenosine 5′-monophosphate   119

hippocampus   188–190, 193–204

hippocampus

  ATP receptors in synaptic transmission   193–204

  Ca^{2+} oscillation   194–198

  excitatory synaptic transmission   237–247

  glutamate release   194–195

  paroxysmal depolarising bursts   251–261

HMA (5-(N-N hexamethylene) amiloride)   96

5-HT receptor   65

5-HT$_3$R-AS   66

α7/5-HT$_3$ receptor, chimeric   66

6-hydroxydopamine   225

hypertension   93

hypoxia   81, 287–294

hypoxia (brain)

  adenosine   291

  inactivation of Na^+, K^+-ATPase   288

  intracellular ATP   288

  K_{ATP}-channels   290

  membrane potential   288

idazoxan   225

IL-1β 98

IL-1β release, ATP dependent   100, 200, 361

IL-6 release   339

IL-6 release, P2X$_7$ 362, 100

ileum (guinea-pig), P2Y   111, 113

inositol 1,4,5-trisphosphate (InsP$_3$) receptor   63

INS 316   100

insulin   100

integrins   38–39

interleukin-1β convertase (ICE)   99

ionotropic glutamate receptor   67–69

ionotropic glycine receptor (GlyR)   66

ionotropic serotonin receptor (5-HT$_3$R)   66

ischemia   100, 161, 202, 292

isolectin B$_4$ 7

isoproterenol   278–279

430

K+ channel (inward rectifying)   63–65, 74
K+ channel (outward rectifying)   28
K+ channel, Shaker   63
K562 cells   37
K_{ATP} channels   64, 100, 290–291
K_{ir} 64
KN-04   95
KN-62   96
kynurenic acid   225

La^{3+} 149
lactic dehydrogenase (LDH) release,
   microglia   359–360
leukotriene B_4 receptor   24
ligand-gated ion channels   61–63
locus coeruleus (LC)   224–225, 288
long term potentiation (LTP)   194
long-term (trophic) actions   6–7
LPS (lipopolysaccaride)   201
lypoxygenase (LO)   336

MAPK cascades   324, 328–329
mechanically induced ATP release   142
mechanosensitive channel of E. coli   74
α,β-MeATP   3, 107, 111, 145, 176, 184,
   195, 212–213, 227, 242, 256, 260
[^{3}H] α,β-MeATP   71
2MeS-ATP   4, 126, 228, 315
microglia /
   ATP-resistant   357
   membrane potential   357
   $P2X_7$ dependent cell response   359–363
mitogen / MAPK kinase   324
mitogen / MAPK kinase kinase   324
mitogenesis / FGF2   327–329
motor neurotransmission   6
MRS 2142   123
MRS 2159   122
MRS 2160   122
MRS 2166   122
MRS 2179   95, 107, 127–129
MRS 2191   123
MRS 2192   123, 129
MRS 2206   122
MRS 2219   123, 125
MRS 2220   95, 107, 125
MRS 2559   95
MRS 2560   95
M-type K+ currents   27, 28, 34
multinucleated giant cells (MGCs)   98
Munc-13   155
muscarinic M_1 receptor   13
muscarinic M_3 receptor   111, 113, 161, 164
muscarinic receptor   6, 99, 135, 162
myasthenia gravis   145

N1E-115 neuroblastoma cells   33
N^6-cyclopentyl adenosine (CPA)   151, 173
NANC inhibitory nerves   6
NCB-20   66
necrosis, microglia   359–360
N-ethylmaleimide   175
neurogenic inflammation   98
neuromuscular depression   145–157
neuronal nACHR   66
neuropeptide Y (NPY)   5, 11, 96
neuropeptide Y receptor   13
neurotransmitter release, controlling by P2-
   receptors   173–180
neurotransmitter release, single type of
   synapse   270–272
neuro-urology   99
NF023   95, 108–115
NF031   108–115
NF279   108–115
NG108-15   26, 27, 33–40, 66
NGF (nerve growth factor)   100
NGF treatment (PC12 cells)   305–307
NHO1   95
nicotinic acetylcholine receptor   6, 65–66,
   162
N-methyl-D-aspartic acid (NMDA)
   receptor   67, 201
NO (nitric oxide)   99, 194
   release   201–202
   synthase   201
nociception   7
non-adrenergic, non-cholinergic enteric
   nerves   6
non-NMDA receptor   67, 224
noradrenaline - norepinephrine   96
noradrenaline release   175–177, 225
noradrenergic neurons   223–224
NR1,2 (NMDA receptor subunit)   68–69
NT3 (neurotrophin-3)   100
5'-nucleotidase   256, see also ecto-
   5'-nucleotidase
nucleus accumbens (NAc)   228–232

oligodendrocytes   34
opoid receptor   13
oxo-M   216–218

P1 receptor   3, 6, 183, 256
P2 receptor   4, 13–14, 305
P2 receptor, therapeutics   93–101
P2D receptor   5, 26, 260
P2T receptor   3, 25
P2U receptor   3–4

P2X receptor   4, 13–14, 69–70, 212
  architecture   69–70
  ATP binding site   70–71
  Ca^{2+}-permeable ion channel   349
  homo/heterooligomers   70
  K^+ channel like H5-domain   61
  N-glycosylation sites   70
  novel antagonists   122–126
  quarternary structure   72–74
P2X$_1$ receptor   4, 70, 71, 73, 112, 123, 237, 305
P2X$_2$ receptor   4, 70, 71, 97–98, 112, 237, 245–46, 305
P2X$_2$ / P2X$_5$-like receptor   216
P2X$_3$ receptor   4, 7, 70, 96–97, 101, 112, 120, 245
P2X$_3$ receptor / DGR   97
P2X$_4$ receptor   4, 71, 81–88, 95, 97, 112, 237, 305
P2X$_6$ receptor   4, 97, 238
P2X$_7$ receptor   4, 95–96, 99, 202
  activation of NF-$_\kappa$B   362
  ICE (Interleukin-1-Converting-Enzyme) activation   360–361
  microglia   355–366
  molecular structure   356–357
  physiopathology of microglia   362
P2Y receptors   4, 23–28, 112–114, 312–314
  AP1 transcription complex   328
  ERK cascade   334–338
  mediated reactive astrogliosis   335–335
  mitogenic signaling   326–327
  neonatal hippocampus   325
  novel antagonists   126–128
  physiological role   28
  reactive astrocytes   335
  transduction mechanism   25
P2Y$_1$ receptor   4, 25–27, 119–122, 317, 319
P2Y$_2$ receptor   4, 25–27, 33–40, 48, 100, 305, 319
  calcium channels   28
  integrin receptor   39
  vinculin   39
P2Y$_3$ receptor   25
P2Y$_4$ receptor   4, 25–26, 138, 135–142
P2Y$_6$ receptor   25–27, 135–142
P2Y$_7$ (leukotriene B$_4$ receptor)   4
P2Y$_{11}$ receptor   4, 25, 138, 141
P2Z receptor   4
P3 receptor   14, 266
P4 receptor   398–402
pain transmission   93, 97
parasympathetic neurotransmission   6
Parkinson's disease   101

PC12 cells   218, 266, 301–310, 375
PDNP (phosphodiesterase / pyrophosphatase) family   379–381
PDNP1 (PC-1)   379–381
PDNP2 (PD1α/autotaxin)   380–381
PDNP2 (PD1β, B10, gp130^{RB13-6})   380–381
pertussin toxin   175
phospholipase A$_2$ 34
phospholipase C (PLC)   34, 37, 120, 126, 135, 153, 266
picrotoxin   225
PKC / MAPK pathway   336
PLA$_2$ stimulation   336
plasminogen release   193, 200–203, 362
postganglionic sympathetic neurons   173–180
PPADS binding domain (P2X)   70
PPADS   4, 70, 71, 95, 107, 129, 214, 227, 230, 232, 239, 242, 246, 256
prazosin   16, 160
prostaglandin   34, 97, 175
protaglandin E2 receptor   13
protein kinase A   266, 283–284
protein kinase activation   283–284
protein kinase C (PKC)   27, 153, 325–326
protein kinase, calcium/calmodulin dependent (CaM-K)   95, 201, 283
protein phosphatase-I   266
purinergic receptors, mitogenic signaling pathway   328

quinacrine   3

rabphilin, rab3a complex   155
Reactive blue   2, 4, 95, 107, 173
rEPSC (residual excitatory post-synaptic current)   241–246
ryanodine receptor   63

Schaffer collaterals   241
schizophrenia   224
Schwann cells   27
serotonin receptor   189
Sf9 cells   63, 70
SH58261 (A$_{2A}$ receptor antagonist)   306
spermine   64
spontaneously hypertensive rats (SHRs)   7
Sr^{2+} 149, 153
sulfonylurea receptor   64–65, 100
superventricular tachycardia   94
suramin   6, 70, 71, 97, 107, 149, 175, 198, 214, 225, 267
synapsin   307–309

syntaxin   157

taenia coli (guinea-pig)   111
tetrodotoxin   113, 178, 194, 197
theophylline   186, 187, 260
thoracolumbal sympathetic neurons   212–218
TNP-adenosine   95
TNP-ATP   95–96, 107
TNP-GTP   95
ToK   64
trophic signaling pathways   323–331
trypan blue   95
TwiK (K$^+$ channel)   64

tyrosinehydroxylase-like immunreactivity   210

UDP-glucose pyrophosphorylase   139
[^{3}H] UTP binding site   48

vanilloid VR-1 receptor   63, 97
vasoactive intestinal polypeptide (VIP)   6
ventral tegmental area (VTA) neurons   226–227
vinculin   39
voltage-gated ion channels   61–65

XAMR0271   95